T0251372

Decentralized Control and Filtering in Interconnected Dynamical Systems

Decentralized Control and Filtering in Interconnected Dynamical Systems

Magdi S. Mahmoud

CRC Press
Taylor & Francis Group
Boca Raton London New York

CRC Press is an imprint of the
Taylor & Francis Group, an **informa** business
AN AUERBACH BOOK

CRC Press
Taylor & Francis Group
6000 Broken Sound Parkway NW, Suite 300
Boca Raton, FL 33487-2742

Printed in the United States of America on acid-free paper
10 9 8 7 6 5 4 3 2 1

International Standard Book Number: 978-1-4398-3814-3 (Hardback)

Library of Congress Cataloging-in-Publication Data

Mahmoud, Magdi S.
Decentralized control and filtering in interconnected dynamical systems / Magdi S. Mahmoud.
p. cm.
Includes bibliographical references and index.
ISBN 978-1-4398-3814-3 (hardback)
1. Decomposition method. 2. Large scale systems--Mathematical models. 3. Mechanics, Analytic--Mathematical models. 4. Intelligent control systems. I. Title.

QA402.2.M34 2010
003'.71015118--dc22 2010035473

Visit the Taylor & Francis Web site at
http://www.taylorandfrancis.com

and the CRC Press Web site at
http://www.crcpress.com

In the Name of

GOD

the All-Compassionate, the All-Merciful.
"And of knowledge, you (mankind) have been given only a little."

Dedicated

To The Memory of My Parents
(Source of My Big Dreams)
To My Family: Salwa, Medhat, Monda, Mohamed
(For Their Love and Gratitude)
To My Grandchildern: Malak, Mostafa
(Representative of New Generations)

MsM
Dhahran, Saudi Arabia, 2010

Contents

Preface

Large-scale systems (LSS) is a diverse field that can be presented in different ways. The past several decades have witnessed an accelerating attention paid to this field. The title *Decentralized Control and Filtering in Interconnected Dynamical Systems* is selected to reflect the attitude of the present volume where the role of interconnections is emphasized in wide classes of dynamical systems. The accelerating attention on LSS comes quite naturally from the relatively rapid growth of our societal needs, which often result in multidimensional, highly interacting, complex systems that are frequently stochastic in nature. Though the existence of LSS as objects for understanding and management is repeatedly affirmed, there has yet been proposed no precise definition for largeness nor generally acceptable quantitative measures of scale. From the standpoint of developing analytical models, a system is large when its input-output behavior cannot be understood without curtailing it, partitioning it into modules, and/or aggregating its modularized subsystems. On the other hand, from a systems viewpoint, a system is large if it exceeds the capacity of a single control structure. Based thereon, one can enumerate several viewpoints regarding scale. The definition of an LSS we will adopt in this volume corresponds to *a dynamical system which contains a number of interdependent constituents which serve particular functions, share resources, and are governed by a set of interrelated goals and constraints.* Therefore the pattern and consequences of coupling pattern play a crucial role in the analysis and design. Alternatively, the phrase *interconnected systems (IS)* is frequently employed to imply the same definition as LSS and hence avoids the debate about largeness and scale.

From this perspective, the notion of LSS has been introduced in the context of control engineering problems when it became clear that there are real world control problems that cannot be solved by using one-shot approaches. Such typical motivating problems arise in the control of interconnected power systems with strong interactions, water systems that are widely distributed in space, traffic systems with many external signals, or large-space flexible structures. These problems necessitate new ideas for dividing the analysis and synthesis of the overall system into independent or almost independent subproblems, for dealing with the incomplete information about the system, for coping with the uncertainties, and for dealing with time-delays.

It is manifested that "complexity" is an essential and dominating problem in systems theory and practice. It leads to severe difficulties that are encountered in the tasks of analyzing, designing, and implementing appropriate control strategies and algorithms. These difficulties arise mainly from the following well-known reasons:

1. *Dimensionality*,
2. *Information structure constraints*,
3. *Uncertainty*,
4. *Gain perturbations*, and
5. *Delays*.

Several approaches to deal with problems of LSS have been developed based on key ideas from economics, management sciences, and operations research. Over the years, such approaches have been dynamically evolved into a body of "large-scale systems (LSS) theories."

This book is written about large-scale systems theories. It aims at providing a rigorous framework for studying analysis, stability, and control problems of LSS while addressing the dominating sources of difficulties due to dimensionality, information structure constraints, parametric uncertainty, and time-delays. The primary objective is threefold: to review past methods and results from a contemporary perspective, to examine present trends and approaches, and to provide future possibilities, focusing on robust, reliable, and/or resilient decentralized design methods based on linear matrix inequalities framework.

For the purpose of simplicity in exposition, the approach followed in this book is to trace the research progress along three eras:

1. *Fundamental era* in which the basic conceptual frameworks, major ideas, and operational methodologies are laid down,

2. *Contemporary era* in which several of the workable methods and techniques are established and applied to many application areas,

3. *Advanced era* in which different high-level schemes and configurations are being developed to meet accelerated technological advancements.

There are no sharp edges for each era (period); rather they are overlapped. As illustrated by the outline of the book contents, we will do our best to follow a chronological order and attempt to interrelate the scattered results although the literature is extremely vast.

Throughout this book, the following terminologies, conventions, and notations have been adopted. All of them are quite standard in the scientific media and only vary in form or character. Matrices, if their dimensions are not explicitly stated, are assumed to be compatible for algebraic operations. In symmetric block matrices or complex matrix expressions, we use the symbol $\bullet$ to represent a term that is induced by symmetry.

Acknowledgments

The subject matter of decentralized control and filtering is fascinating and is predominantly concerned with interaction among subsystems. In retrospect, the “interaction” with people in general and colleagues, students, and friends in particular manifests our lives. This is certainly true in my academic career where I benefited from listening, discussing, and collaborating with several colleagues across the globe. The various topics discussed in this book have constituted an integral part of my academic research investigations over the past several years. The idea of writing the book was aroused and developed during fall 2005 and was revived after joining KFUPM where I gratefully took full advantage of the supportive scientific environment.

In writing this volume, I took the approach of referring within the text to papers and/or books that I believe taught me some concepts, ideas, and methods. I further complemented this by adding some remarks, observations, and notes within and at the end of each chapter to shed some light on other related results. I apologize in advance in case I committed an injustice and assure all of the colleagues that any mistake was definitely unintentional.

I am indebted to the people who taught me the subject of large-scale systems and to the people who made the writing possible. At KFUPM, I owe a measure of gratitude to *Dr. Ahmed Al-Homoud*, Dean of Scientific Research, for providing a superb competitive environment of research activities through internal funding grants. This book is written based on the research project IN 100015 and in this regard, the unfailing assistance of the DSR team is gratefully acknowledged. The continuous encouragements of *Dr Omar Al-Turki*, Dean of Computer Science and Engineering, and *Professor Fouad AL-Sunni*, Chairman of the Systems Engineering Department, are wholeheartedly acknowledged. Special thanks must go to my colleagues at the Systems Engineering Department—*Professors Moustafa El-Shafie, Hosam E. Emara-Shabaik*, and *Drs. Abdul-Wahed Saif, Smair Al-Amer*, and *Sami ElFreik* for continuous interactions, helpful comments, and assistance throughout my stay at KFUPM.

This book is intended for graduate students and researchers in control systems design including robust, reliable, and resilient methods with the hope to provide a guided tour through large-scale systems (LSS) results and a source of new research problems. I had the privilege of teaching graduate courses on decentralized control, large-scale systems, and hierarchical systems at Cairo University (Egypt), University of Manchester Institute of Science and Technology (UK), Kuwait University (Kuwait), and KFUPM (Saudi Arabia). The course notes, updated and expanded over years, were instrumental in generating different chapters of this book and valuable comments and suggestions as well as several detailed critiques made by graduate students were greatly helpful. This is particularly true for the graduate students of SE 509 course offered during the fall semester (2009-2010) at the Systems Engineering Department. Several of my colleagues have been very generous with suggestions for improving the accuracy and clarity of the exposition.

For the development of the book, I am immensely pleased for many stimulating and fruitful discussions with and helpful suggestions from colleagues, students, and friends worldwide throughout my academic career who have definitely enriched my knowledge and experience. I have had the good fortune to interact with and be inspired by conversations with international experts in systems and control theory, including *Profs. T. Basar (University of Illinois), H. K. Khalil (Michigan State University), M. H. Mickle* and *W. G. Vogt (University of Pittsburgh), P. Shi (University of Glamorgan), E. K. Boukas (Polytechnique of Montreal), M. Toma (University of Hanover), A. Title (LAAS), M. G. Singh (University of Manchester Institute of Science and Technology)*, and *Yuanqing Xie (Beijing Institute of Technology)*. I deeply appreciate the efforts of *Mohamed Tawfik, Ali Al-Rayyah, Rohmat Widodo*, and *Muhammad Sabih* for their help in preparing portions of the manuscript. I would like to thank *Publisher Richard O'Hanley* and *Project Editor Michele Dimont* of CRC Press at Taylor & Francis Group for making this project viable.

Most of all, however, I would like to express my deepest gratitude to my parents who taught me the value of the written word and to all the members of my family and especially my wife *Salwa* for her elegant style and for proofreading parts of the manuscript. Without their constant love, incredible amount of patience, and (mostly) enthusiastic support this volume would not have been finished.

I would appreciate any comments, questions, criticisms, or corrections that readers may take the trouble of communicating to me at msmahmoud@kfupm.edu.sa or magdim@yahoo.com.

Magdi S. Mahmoud
Dhahran, Saudi Arabia

Notations and Symbols

I^+	:=	*the set of positive integers*
$\Re$	:=	*the set of real numbers*
$\Re_+$	:=	*the set of nonnegative real numbers*
$\Re^n$	:=	*the set of all n-dimensional real vectors*
$\Re^{n\times m}$	:=	*the set of* $n \times m$*-dimensional real matrices*
$\mathcal{C}^-$	:=	*the open right-half complex plane*
$\mathcal{C}^+$	:=	*the closed right-half complex plane*
$\in$	:=	*belong to or element of*
$\subset$	:=	*subset of*
$\bigcup$	:=	*union*
$\bigcap$	:=	*intersection*
$>>$	:=	*much greater than*
$<<$	:=	*much less than*
A^t	:=	*the transpose of matrix A*
A^{-1}	:=	*the inverse of matrix A*
I	:=	*an identity matrix of arbitrary order*
I_s	:=	*the identity matrix of dimension* $s \times s$
e_j	:=	*the jth column of matrix I*
x^t *or* A^t	:=	*the transpose of vector x or matrix A*
$\lambda(A)$	:=	*an eigenvalue of matrix A*
$\varrho(A)$	:=	*the spectral radius of matrix A*
$\lambda_j(A)$	:=	*the* jth *eigenvalue of matrix A*
$\lambda_m(A)$	:=	*the minimum eigenvalue of matrix A where* $\lambda(A)$ *are real*
$\lambda_M(A)$	:=	*the maximum eigenvalue of matrix A where* $\lambda(A)$ *are real*
A^{-1}	:=	*the inverse of matrix A*

$A^\dagger$	:=	*the Moore-Penrose-inverse of matrix* A
$P > 0$	:=	*matrix P is real symmetric and positive-definite*
$P \geq 0$	:=	*matrix P is real symmetric and positive semidefinite*
$P < 0$	:=	*matrix P is real symmetric and negative-definite*
$P \leq 0$	:=	*matrix P is real symmetric and negative semidefinite*
$A(i,j), A_{ij}$	:=	*the* ij*-th element of matrix* A
$\mathbf{det}(A)$	:=	*the determinant of matrix* A
$\mathbf{trace}(A)$	:=	*the trace of matrix* A
$\mathbf{rank}(A)$	:=	*the rank of matrix* A
$\mathcal{L}_2(-\infty,\infty)$	:=	*space of time domain square integrable functions*
$\mathcal{L}_2[0,\infty)$	:=	*subspace of* $\mathcal{L}_2(-\infty,\infty)$ *with functions zero for* $t < 0$
$\mathcal{L}_2(-\infty,0]$	:=	*subspace of* $\mathcal{L}_2(-\infty,\infty)$ *with functions zero for* $t > 0$
$\mathcal{L}_2(j\Re)$	:=	*square integrable functions on* $\mathcal{C}_0$ *including at* ∞
$\mathcal{H}_2$	:=	*subspace of* $\mathcal{L}_2(j\Re)$ *with functions*
	:=	*analytic in* $Re(s) > 0$
$\mathcal{L}_\infty(j\Re)$	:=	*subspace of functions bounded*
	:=	*on* $Re(s) = 0$ *including at* ∞
$\mathcal{H}_\infty$	:=	*the set of* $\mathcal{L}_\infty(j\Re)$ *functions analytic in* $Re(s) > 0$
$\|a\|$	:=	*the absolute value of scalar* a
$\|\|x\|\|$	:=	*the Euclidean norm of vector* x
$\|\|A\|\|$	:=	*the induced Euclidean norm of matrix* A
$\|\|x\|\|_p$	:=	*the* ℓ_p *norm of vector* x
$\|\|A\|\|_p$	:=	*the induced* ℓ_p *norm of matrix* A
$Im(A)$	:=	*the image of operator/matrix* A
$Ker(A)$	:=	*the kernel of operator/matrix* A
$max\ \mathbf{D}$	:=	*the maximum element of set* **D**
$min\ \mathbf{D}$	:=	*the minimum element of set* **D**
$sup\ \mathbf{D}$	:=	*the smallest number that is larger than or equal to each element of set* **D**
$inf\ \mathbf{D}$	:=	*the largest number that is smaller than or equal to each element of set* **D**
$arg\ max\mathbf{D}$	:=	*the index of maximum element of ordered set* **S**
$arg\ min\mathbf{D}$	:=	*the index of minimum element of ordered set* **S**
$\mathbf{B}_r$	:=	*the ball centered at the origin with radius* r
$\mathbf{R}_r$	:=	*the sphere centered at the origin with radius* r
$\mathcal{N}$	:=	*the fixed index set* $\{1, 2, ..., N\}$
$[a, b)$	:=	*the real number set* $\{t \in \Re : a \leq t < b\}$
$[a, b]$	:=	*the real number set* $\{t \in \Re : a \leq t \leq b\}$
$\mathbf{S}$	:=	*the set of modes* $\{1, 2, \ ..., \ s\}$
iff	:=	*if and only if*
$\otimes$	:=	*the Kronecker product*

$\mathbf{O}(.)$	:=	*order of (.)*
$diag(...)A$	:=	*diagonal matrix with given diagonal elements*
$\mathbf{spec}(A)$	:=	*the set of eigenvalues of matrix A (spectrum)*
$\mathbf{min-poly}(A)(s)$	:=	*the minimal polynomial of matrix A.*

List of Acronyms

AGC	*automatic-generation control*
ARE	*algebraic Riccati equation*
BTD	*block tridiagonal form*
CIP	*centralized information processing*
DC	*decentralized control*
DFC	*decentralized feedback control*
DHC	*decentralized* $\mathcal{H}_\infty$ *control*
DIP	*distributed information processing*
DNS	*decentralized nonlinear systems*
$DSMP$	*decentralized servomechanism problem*
DTS	*discrete-time systems*
EVP	*eigenvalue problem*
$GEVP$	*generalized eigenvalue problem*
HC	*hierarchical control*
LBD	*Lyapunov-based design*
LKF	*Lyapunov-Krasovskii functional*
$LMCR$	*liquid-metal cooled reactor*
LMI	*linear matrix inequality*
$LMIP$	*linear matrix inequality problem*
LQC	*linear quadratic control*
LSS	*large-scale systems*
LTI	*linear time-invariant*
$MIMO$	*multi-input multi-output*
MIQ	*machine intelligence quotient*
OLC	*overlapping control*
OLD	*overlapping decomposition*
QMI	*quadratic matrix inequality*

$SISO$	*single-input single-output*
SMC	*sliding-mode control*
SVD	*singular value decomposition*
TDS	*time-delay system*
$TDUS$	*time-delay uncertain system*
UTD	*uncertain time-delay*

Author

Magdi S. Mahmoud earned his Ph.D. degree in systems engineering from Cairo University in 1974. He has been a professor of systems engineering since 1984. He is now on the faculty of KFUPM, Saudi Arabia. He worked at different universities worldwide in Egypt, Kuwait, UAE, UK, USA, Singapore, Saudi Arabia, and Australia. He has given invited lectures in Venezuela, Germany, China, UK, and USA. He has been actively engaged in teaching and research in the development of modern methodologies of computer control, systems engineering, and information technology. He is the principal author of 11 books and 9 book chapters and the author/co-author of more than 450 peer-reviewed papers. He is the recipient of two national, one regional, and a university prize for outstanding research in engineering. He is a fellow of the IEE, a senior member of the IEEE and the CEI (UK), and a registered consultant engineer of information engineering and systems (Egypt).

Chapter 1

Introduction

The chapter overviews the past and present results pertaining to the area of decentralized control of large-scale systems. An emphasis is laid on decentralization, decomposition, and robustness. These methodologies serve as effective tools to overcome specific difficulties arising in large-scale complex systems such as high dimensionality, information structure constraints, uncertainty, and delays. The overview is focused on recent decomposition approaches in interconnected dynamic systems due to their potential in providing the extension of decentralized control into networked control systems, switching systems, and dynamic systems with abrupt-change parameters.

1.1 Overview

Any scholarly account of the history of the general subject area of large-scale (complex, interconnected) systems would have to span the past several decades, as we have witnessed, during this span of time, an increasing amount of attention paid to problems of modeling, characterization, structural properties, control analysis, optimization, and feedback design strategies [136, 164, 303, 306, 323, 324, 335, 337]. This steady growing interest comes quite naturally from the relatively rapid expansion of our societal needs which often result in multidimensional, highly interacting, complex systems which are frequently stochastic in nature. Though the existence of large-scale systems as objects for understanding and management is repeatedly affirmed, there has yet been proposed no precise definition for largeness nor generally acceptable quantitative measures of scale.

From the viewpoint of developing analytical models, a system is large when its input-output behavior cannot be understood without curtailing it, partitioning it into modules, and/or aggregating its modularized subsystems. On the other hand, from a systems viewpoint, a system is large if it exceeds the capacity of a single control structure. Thus one can enumerate several viewpoints regarding scale. The class of large-scale systems we will deal with in the subsequent chapters is *the class of systems that contains a number of interdependent constituents that serve particular functions, share resources, and are governed by a set of interrelated goals and constraints* [206, 224].

The notion of large-scale systems is therefore adopted when it becomes clear that one-shot approaches to the associated control problems would not be appropriate. Therefore, the approach adopted in this book is that a system is considered large scale if it can be decoupled or partitioned into a number of interconnected subsystems or "small-scale" systems for either computational or practical reasons [106, 136, 206]. Figure 1.1 displays this approach. Such typical motivating problems arise in the control of interconnected power systems with strong interactions, water systems that are widely distributed in space, traffic systems with many external signals, or large-space flexible structures. It is quite evident in these cases that the systems to be controlled are too large and the problems to be solved are computationally demanding or complicated. New ideas are thought for dividing the analysis and synthesis of the overall system into almost independent subproblems, for handling the incomplete information about the system, for dealing treating with the uncertainties, and for accommodating delays. One prevailing approach has been that the complexity is an essential and dominating problem in systems theory and practice [106].

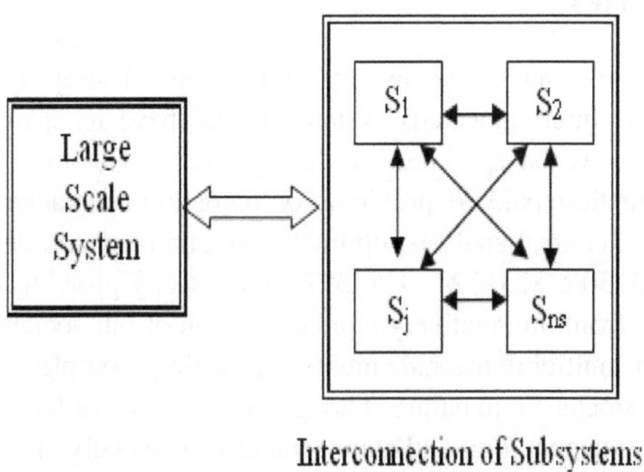

Figure 1.1: Large-scale system structure

Among the major difficulties that are encountered in the tasks of analyzing, designing, and implementing appropriate control strategies and algorithms are:

1. *Dimensionality*,
2. *Information structure constraints*,
3. *Parameter uncertainty*,
4. *Delays*.

It turns out that the theoretical framework of large-scale systems provides answers to the fundamental question of *how to decompose a given control problem into manageable subproblems that are only weakly related to each other and can be solved independently*. It is observed that the overall system is no longer controlled by a single controller but by several independent controllers that all together represent a hierarchical or decentralized controller. This is one of the fundamental differences between feedback control of small and large systems. To shed more light on this, we recall the basic centralized system depicted in Figure 1.2.

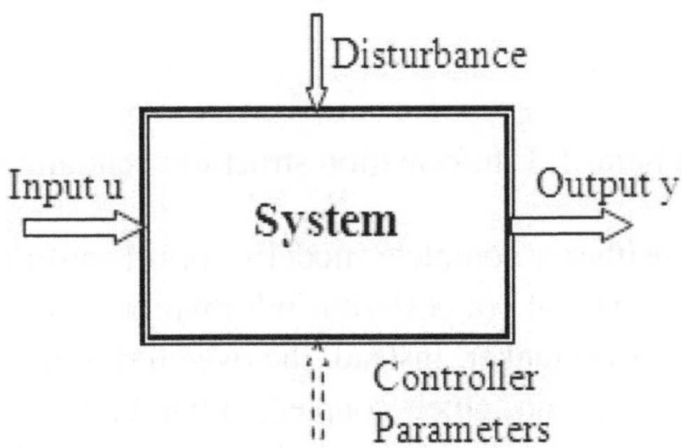

Figure 1.2: Centralized system

1.1.1 Information structure

One of the major issues that manifests large-scale systems is the role governed by the idea of information structure. Initially, in the case of centralized systems (refer to Figure 1.2), the basic feedback problem is to find control input vector $u(.)$ on the basis of the a priori knowledge of the plant S described by its *design model* in the presence of a class of disturbances $v(.)$ and the control goal given in the form of the design requirements $\{C\}$ and the a posteriori information about the outputs $y(.)$ and the command signals $r(.)$. Classical information structure corresponds to centralized control as illustrated by Figure 1.3. It is important to note that the controller receives all sensor data available and determines all input signals of the plant. In other words, all information is assumed to be available for a single unit that designs and applies the controller to the plant. In present-day technologies where several different units coexist

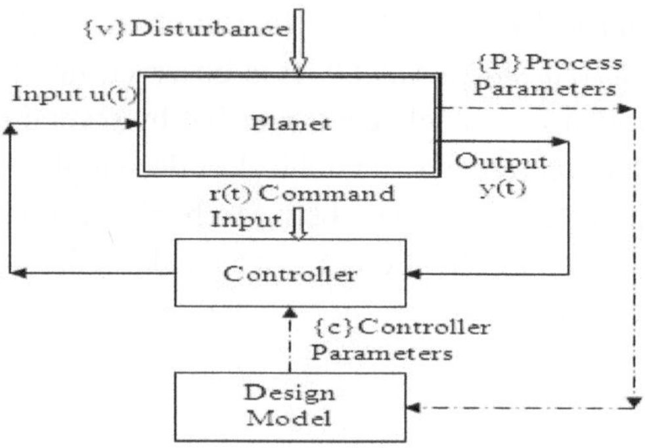

Figure 1.3: Information structure: centralized control

side by side, neither a complete model (a priori information) nor a complete set of measurement data (a posteriori information) can be made available for a centralized decision maker. Instead, the overall design problem has to be broken down into different, albeit coupled, subproblems. As a result, the overall plant is no longer controlled by a single controller but by several independent controllers constituting a decentralized controller structure. Moreover, these controllers are no longer designed simultaneously on the basis of a complete knowledge of the plant, but in different design steps by means of models that describe only the relevant parts of the plant. This amounts to a nonclassical information structure that arises in decentralized design schemes as shown in Figure 1.4.

1.1.2 System representation

There are available two main structures of the models of large-scale systems distinguished by the degree to which they reflect the internal structure of the

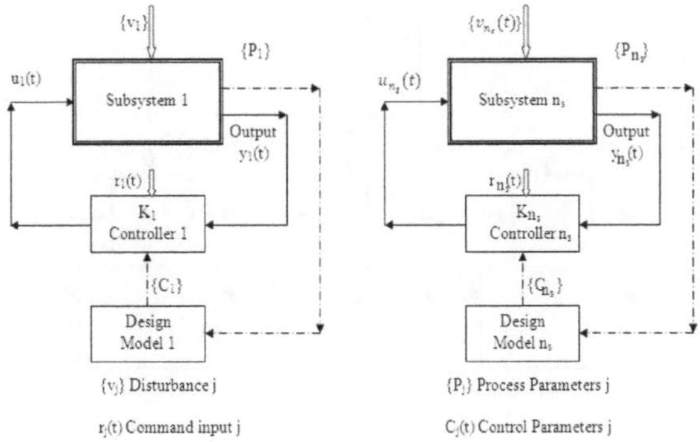

Figure 1.4: Information structure: decentralized control

overall dynamic system. These structures are called multichannel systems entailing the presence of multicontrollers and interconnected systems incorporating coordinated controllers as illustrated in Figures 1.5 and 1.6.

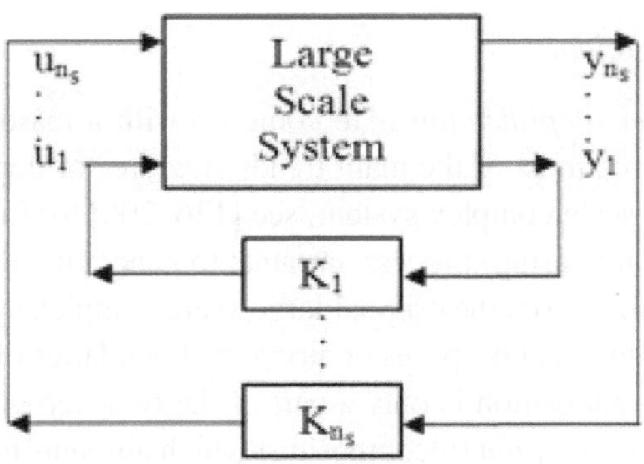

Figure 1.5: Multicontroller structure

In multichannel systems, the associated input and output vectors are decomposed into subvectors constituting n_s channels, while the system is considered as one whole. More on these types of systems will be mentioned in later chapters.

Interconnected systems operate with interactions among subsystems. They are represented by signals through which subsystems interact among themselves. These signals are internal signals of the overall system.

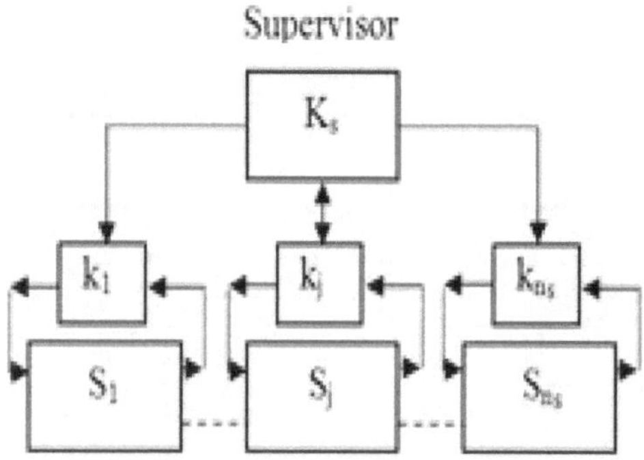

Figure 1.6: Coordinated control structure

To cope with the aforementioned appearance of the complexity issues, several general methodologies have been and are being elaborated. Most of them belong to one of the following three groups [206, 323, 335, 337]:

1. *Model simplification*,
2. *Decomposition*,
3. *Decentralization*.

The idea of *model simplification* is to come up with a reasonable model that preserves or inherits most of the main trends (features or dominant modes) of the original large-scale/complex system; see [136, 206] for further elaboration. The *decomposition* (tearing) process amounts to generating a group of subsystems (smaller in size) from the original large-scale/complex system. This could be achieved for numerical purposes or along the boundaries of coupled units. It turns out that decomposition is only a part of the two-step procedure, the second of which is *coordination* (recomposing) which amounts to synthesizing the overall solution from the generated solutions of the subsystems (subsolutions). There are two aspects of *decentralization*: the first issue is concerned with the information structure inherent in the solution of the given control problem and refers to the subdivision of the process in terms of the model and the design goals. The other issue is associated with on-line information about the state and the command to generate the decentralized control law. The net result is that a completely independent implementation of the controllers is made viable. There are a variety of different motivating reasons for the decentralization of the design process such as weak coupling of subsystems, subsystems have contradictory goals, subsystems are assigned to different authorities, or the high dimensionality of the overall system. Following [192], the principal ways of decentralizing the design tasks belong to two groups: *decentralized design for strongly coupled subsystems* and *decentralized design for weakly coupled subsystems.*

The *decentralized design for strongly coupled subsystems* means that at least an approximate model of all other subsystems must be considered for the design of any subsystem under the current design, while the coupling can be neglected during the design of individual control stations when considering the decentralized design for weakly coupled subsystems.

1.1.3 Hierarchical systems

One of the fundamental approaches in dealing with large-scale static systems was the idea of *decomposition* treated theoretically in mathematical programming by treating large linear programming problems possessing special structures. The objective was to gain computational efficiency and design simplification. There are two basic approaches for dealing with such problems:

1. The *coupled* approach where the problem's structure kept intact while taking advantage of the structure to perform efficient computations [106], and

2. The *decoupled* approach which divides the original system into a number of subsystems involving certain values of parameters. Each subsystem is solved independently for a fixed value of the so-called "decoupling" parameter, whose value is subsequently adjusted by a coordinator in an appropriate fashion so that the subsystems resolve their problems and the solution to the original system is obtained.

Four decades ago, Mesarovic and his coworkers [271] introduced a formalism of hierarchically controlled systems. Out of the many concepts found in their work, three sets of ideas have subsequently proved useful. The first idea centers on the description of a large system as an interconnected system as a collection of subsystems coupled by interaction variables in such a way that if these are considered to be completely fixed or completely free, then the system decomposes into independent subsystems. By the second idea, this system decomposition suggests a corresponding division of control into separate *lower level* or infimal controllers, one for each subsystem. Through the third idea, the interaction variables actually present (and ignored in the decomposition) are taken into account by an *upper level* or supremal controller whose task is to coordinate the infimal units. Inspired by results in large-scale mathematical programming. several coordination principles were suggested in [271], notably the so-called interaction prediction and interaction balance principles. Several decomposition-coordination approaches have been subsequently elaborated to simplify the analysis and synthesis tasks for large-scale systems, mainly in the context of optimal control. The underlying multilevel computation structures have the following features:

1. *The decomposition of systems with fixed designs at one level followed by coordination at another level is often the only alternative available.*
2. *In situations where available decision units have limited capabilities, the associated problem is formulated in a multilayer hierarchy of subproblems.*
3. *Generally, there will be an increase in system reliability and flexibility.*
4. *System resources will be better utilized.*
5. *Systems are commonly described only on a stratified basis.*

A comprehensive application of the hierarchical multilevel approach to water resources systems has been documented in [97]. Applications to closed-loop optimal control problems have been reported in [224, 225, 226, 335].

1.1.4 Structure of interconnections

With focus on interconnections, approaches of large-scale systems can be broadly categorized as follows:

1. *Approaches for subsystems with strong interconnections,*
2. *Approaches for subsystems with weak interconnections.*

The first category embraces the following classes:

1. *Decoupled subsystems,*
2. *Overlapping subsystems,*
3. *Symmetric composite systems.*

The second category includes the following classes:

1. *Multitime scale systems,*
2. *Hierarchically structured systems.*

More on these methods will be discussed in the subsequent chapters. A common feature among these approaches is best illustrated by looking at the role of the interconnections, as depicted in Figures 1.7 through 1.9.

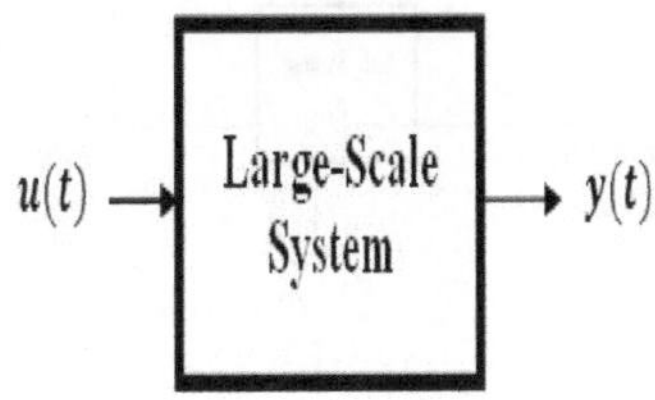

Figure 1.7: Unstructured model

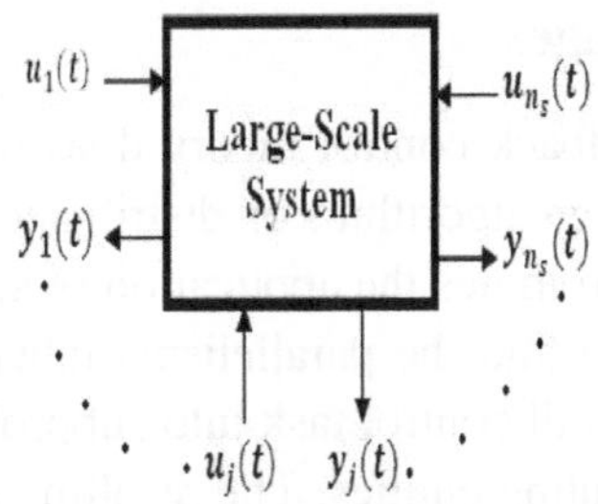

Figure 1.8: Input-output oriented model

Throughout this book, we will take into account the possibility of parametric uncertainties and perform robustness analysis to exploit the character of uncertainties mainly on the bases of the stability analysis of coupled systems.

Another important issue that we will seriously tackle is that of *delays* since the incorporation of modern devices and computing facilities eventually incurs delays in one form or another. This will be a major milestone in the subsequent analysis. In fact, in some respects we start with large-scale time-delay systems and after deriving the desired results, we extract the corresponding results for none-delay systems.

It must be stressed that the origin and the rapid development of decentralized control design methods began more than four decades ago. Various decentralized control design structures and algorithms have been developed to

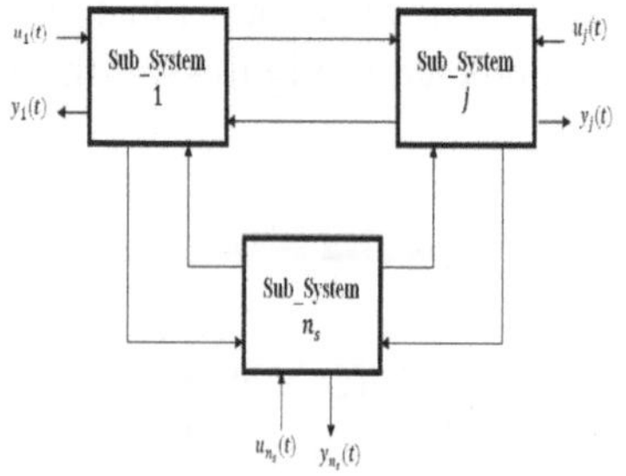

Figure 1.9: Interaction-oriented model

present the flexibility and superiority of this approach for different classes of interconnected systems.

1.1.5 Revived challenges

It is well known that feedback control theory does not tackle the problem of implementing control design algorithms on distributed microprocessor systems or Transputer. Rather, it promotes the application of such modern hardware facilities. In fact, it helps to find the parallelisms inherent in control problems and to decompose the overall control task into subproblems that can be implemented on separate computing entities. The resulting control design algorithm is fault tolerant if the subproblems, which are allocated to the individual processors, are sufficiently autonomous. We expect that communication delays, noise, synchronization difficulties, or other failures to occur in the processors or processor links merely degrade the overall system performance but operation can continue.

In this regard, large-scale systems present feedback control theory with revived challenges in the following manner:

1. *The large dimension of the plant and, thus, the complexity of the plant model have to be confronted. As a result of restrictions on computer time and memory space the system cannot be analyzed and controlled as a whole.*

2. *Practically speaking, every model of a large-scale system has severe uncertainties that have to be considered explicitly in all analysis and design steps. The possible effects of modeling errors cannot be evaluated*

heuristically but necessitate a systematic procedure.

3. *The decentralization of both the design and the controller imposes restrictions on the model and on the on-line information links introduced by the controller. This decentralization is crucial because the subsystems are under the control of different agents, which make their own models and design their own control stations, or because the design problem has to be split into independent subproblems to become manageable.*

4. *The design aims include not only stability or optimality, but also a variety of properties such as reliability, flexibility, robustness against structural perturbations of the plant, or restrictions on the interactions between the subsystems.*

Admittedly, the prime objective of decentralized control and filtering is to get a reasonable solution with reasonable effort in modeling, designing, and implementing the controller/filter. To realize this objective, methods for dealing with large-scale systems have to be based on a new methodological background. The structural properties of the plant have to be exploited in order to derive suitable design and control structures. A fundamental question concerns the conditions on the interconnection structure, under which the subsystems are "weakly coupled" and, thus, can be analyzed separately while ignoring their interactions. Decomposition and decentralization methods help to answer this question and lead to new means for testing the stability of interconnected systems, for analyzing the input-output behavior of a certain subsystem under the influence of other subsystems without using a complete model of the overall plant, and for designing decentralized controllers.

Concerning the interlinks among the theories of large-scale systems and of *small* multivariable systems should emphasize this point. Initially, many problems that can be solved by the methods of multivariable control theory turn out to be subproblems encountered in large-scale systems theory. This is not surprising, for here it is a principal aim to break down complex problems into a combination of easier ones. It is actually the way to reduce the complexity that needs novel ideas, which then have to be elaborated. These aspects will be clarified in the subsequent chapters.

Alternatively, large-scale systems give rise to problems that are known in multivariable systems theory but have not yet been solved satisfactorily. Rigorous methods are missing for some multivariable control problems, because solutions to these *small* problems could have been found by engineering common sense rather than by universal algorithms [136, 164, 206, 224, 303, 306, 323, 324, 335, 337].

1.2 Outline of the Book

During the past several decades, there have been real world system applications for which the associated control design problems cannot be solved by using one-shot approaches. Typical applications arise in the areas of interconnected power systems with strong coupling ties among network elements, water systems that are widely distributed in space, traffic systems with many external signals, or large-space flexible structures with interacting modes. Models of such systems are frequently complex in nature, multidimensional and/or composed of highly interacting subsystems. Several approaches to deal with these systems have been developed based on key ideas from economics, management sciences, and operations research. Over the years, such approaches have been dynamically evolved into a body of "large-scale systems (LSS) theories." This book is written about the wide spectrum of large-scale system theories.

1.2.1 Methodology

Throughout the book, our methodology in each chapter/section is composed of five steps:

- **Mathematical Modeling**
  Here we discuss the main ingredients of the state-space model under consideration.

- **Definitions and/or Assumptions**
  Here we state the definitions and/or constraints on the model variables to pave the way for subsequent analysis.

- **Analysis and Examples**
  This signifies the core of the respective sections and subsections which contains some solved examples for illustration.

- **Results**
  These are provided most of the time in the form of theorems, lemmas, and corollaries.

- **Remarks**
  These are given to shed some light of the relevance of the developed results vis-à-vis published work.

In the sequel, theorems (lemmas, corollaries) are keyed to chapters and stated with **bold titles**, for example, **Theorem 3.4** means Theorem 4 in Chapter 3 and so on. For convenience, we have grouped the references in one major bibliography cited toward the end of the book. Relevant notes and research issues are offered at the end of each chapter for the purpose of stimulating the reader. We hope that this way of articulating the information will attract the attention of a wide spectrum of readership.

This book aims at providing a rigorous framework for studying analysis, stability, and control problems of LSS while addressing the dominating sources of difficulties due to dimensionality, information structure constraints, parametric uncertainty, and time-delays. The primary objective is threefold: to review past methods and results from a contemporary perspective; to examine present trends and approaches; and to provide future possibilities, focusing on robust, reliable, and/or resilient decentralized design methods based on linear matrix inequalities framework.

In brief, the main features of the book are:

1. *It provide an overall assessment of the large-scale systems theories over the past several decades.*

2. *It addresses several issues like model-order reduction, parametric uncertainties, time-delays, control/estimator gain perturbations.*

3. *It presents key concepts with their proofs followed by efficient computational methods.*

4. *It establishes decentralized control and filtering techniques.*

5. *It treats equally both continuous-time and discrete-time representations.*

6. *It gives some representative applications and end-of-chapter problems.*

1.2.2 Chapter organization

Large-scale systems have been investigated for a long time in the control literature and have attracted increasingly more attention for more that two decades. The literature grew progressively and quite a number of fundamental concepts and powerful tools have been developed from various disciplines. Despite the rapid progress made so far, many fundamental problems are still either unexplored or less well understood. In particular, there still lacks a unified framework that can cope with the core issues in a systematic way. This motivated us to write the current book. The book presents theoretical explorations on several fundamental problems for large-scale systems.

The book is primarily intended for researchers and engineers in the system and control community. It can also serve as complementary reading for linear/nonlinear system theory at the postgraduate level.

After the first introductory chapter, the core material of the book is divided into five parts.

Part 1 is comprised of two chapters: Chapter 2 reviews some basic elements of mathematical analysis, calculus, and algebra of matrices to build up the foundations for the remaining topics on stability, stabilization, control, and filtering of LSS. Chapter 3 introduces different forms of description and motivation of the study of LSS and presents several analytical tools and theories that were developed three decades ago.

Part 2 establishes a detailed characterization of the class of LSS with strongly coupled subsystems and consists of two chapters: Chapter 4 presents the decentralized stabilization and the robust decentralized servomechanism problems in order to develop conditions and algorithms for obtaining the decentralized controllers. Chapter 5 focuses on LSS where the model structure admits an overlapping decomposition and addresses the main concepts and ideas pertaining to system analysis and control synthesis.

Part 3 is devoted to recent developments in decentralized stabilization and feedback control methods and is subsumed of four chapters: Chapter 6 puts emphasis on time-delay LSS and discusses interaction effects, whereas Chapter 7 attends to the actuator/sensor failure issues seeking the design of decentralized reliable control. Of prime interest is the explanation of new phenomena that are encountered in interconnected systems and the new ideas that enable the control engineer to cope with the various problems raised by the need and the aims of decentralization. In Chapter 8, we examine the implementation issue in LSS with particular consideration to gain perturbation in feedback gains. Then in Chapter 9, we deal with LSS under sliding-mode control for both delay-free and time-delay systems. A complete analysis of static output-feedback sliding-mode control is provided,

Part 4 provides results on decentralized filtering methods and is summarized in two chapters: Chapter 10 concentrates on decentralized resilient $\mathcal{H}_\infty$ filters and Chapter 11 is left to decentralized Kalman filtering and decentralized fault detection.

In the last chapter "Epilogue," a concise summary is given to overview the results on decentralized systems and to furnish a "recipe" to future research. An appendix containing some relevant mathematical lemmas and basic algebraic inequalities is provided at the end of the book. Several illustrative numerical examples are provided throughout the book and relevant questions as well as problem sets are situated at the end of each chapter.

Throughout the book and seeking computational convenience, all the developed results are cast in the format of a family of linear matrix inequalities (LMIs). In writing up the different topics, emphasis is primarily placed on major developments attained thus far and then reference is made to other related work.

1.3 Some Notes

This book covers a wide spectrum of design tools for decentralized control and filtering in dynamical interconnected systems. For the purpose of unifying the results, we provide theoretical results of large-scale systems that are developed for linear, stationary, continuous-time deterministic systems. Afterward, whenever deemed convenient, we establish the discrete-time counterpart. Some other extensions to classes of nonlinear systems are also presented.

In most of the recent developments, we start the model description and analysis for large-scale systems with time-delays within the subsystems and along the coupling links. We feel that this set-up provides a general framework from which we can extract several important LSS descriptions of interest. For example, suppressing the delayed-coupling terms yields a class of LSS with time-delay subsystems having instantaneous interactions. On the other hand, deleting the "delay-" terms within the subsystems results in a class of LSS where the subsystems have delayed communications due to finite processing and/or use of networks.

The analytical tools developed in this volume are supplemented with rigorous proofs of closed-loop stability properties and simulation studies. The material contained in this book is not only organized to focus on the new developments in the control methodologies for continuous-time and discrete-time LSS systems, but it also integrates the impact of the delay factor, complexity, and uncertainties on important issues like delay-dependent stability and control design. For the purpose of clarity, it is intended to split the book into self-contained chapters with each chapter being equipped with illustrative examples, problems, and questions. The book will be supplemented by an extended bibliography, appropriate appendices, and indexes. At the end of each chapter, a summary of the results with the relevant literature is given and pertinent discussions of future advances and trends are outlined. Some selected problems are organized for the reader to solve. The bibliographical notes cannot give a complete review of the literature but indicate the most influential papers/books in which the results presented here have been described and provide interested readers with adequate entry points to the still-growing literature in this field.

The symbols and conventions used throughout the book are summarized in the List of Symbols.

It was planned while organizing the material that this book would be appropriate for use as a graduate-level textbook in applied mathematics as well as different engineering disciplines (electrical, mechanical, civil, chemical, systems), a good volume for independent study, or a suitable reference for graduate students, practicing engineers, interested readers, and researchers from a wide spectrum of engineering disciplines, science, and mathematics.

Chapter 2

Mathematical Foundations

This chapter contains a collection of useful mathematical concepts and tools, which are useful directly or indirectly, for the subsequent development to be covered in the main portion of the book. While much of the material is standard and can be found in classical textbooks, we also present a number of useful items that are not commonly found elsewhere. Essentially, this chapter serves as a brief overview and as a convenient reference when necessary.

2.1 Basic Mathematical Concepts

Let $x_j,\ y_j,\quad, j = 1, 2, ..., n \in \Re(or\ \mathbf{C})$. Then the n-dimensional vectors $x,\ y$ are defined by $x = [x_1\ x_2\ ...\ x_n]^t,\ y = [y_1\ y_2\ ...\ y_n]^t\ \in \Re^n$, respectively.

A nonempty set $\mathcal{X}$ of elements $x,\ y, ...$ is called the *real (or complex) vector space (or real (complex) linear space)* by defining two algebraic operations, *vector additions* and *scalar multiplication*, in $x = [x_1\ x_2\ ...\ x_n]^t$.

2.1.1 Euclidean space

The n-dimensional Euclidean space, denoted in the sequel by $\Re^n$, is the linear vector space $\Re^n$ equipped by the inner product

$$\langle x, y\rangle = x^t\, y = \sum_{j=1}^{n} x_j y_j$$

Let $\mathcal{X}$ be a linear space over the *field* **F** (typically **F** is the field of real numbers $\Re$ or complex numbers C). Then a function

$$||.|| : \mathcal{X} \rightarrow \Re$$

that maps $\mathcal{X}$ into the real numbers $\Re$ is a norm on $\mathcal{X}$ iff
1. $||x|| \geq 0, \; \forall x \in \mathcal{X}$ (nonnegativity)
2. $||x|| = 0, \iff x = 0$ (positive definiteness)
3. $||\alpha\, x|| = |\alpha|||x||\forall x \in \mathcal{X}$ (homogeneity with respect to $|\alpha|$)
4. $||x + y|| \leq ||x|| + ||y||, \;\; \forall x, y \in \mathcal{X}$ (triangle inequality)
Given a linear space $\mathcal{X}$, there are many possible norms on it. For a given norm $||.|| \; on \; \mathcal{X}$, the pair $(\mathcal{X}, \; ||.||)$ is used to indicate $\mathcal{X}$ endowed with the norm $||.||$.

2.1.2 Norms of vectors

The class of L_p-norms is defined by

$$\begin{aligned} ||x||_p &= \left(\sum_{j=1}^{n} |x_j|^p\right)^{1/p}, \quad for 1 \leq p < \infty \\ ||x||_\infty &= \max 1 \leq j \leq n \; |x_j| \end{aligned}$$

The three most commonly used norms are $||x||_1, \; ||x||_2$, and $||x||_\infty$. All p-norms are equivalent in the sense that if $||x||_{p1}$ and $||x||_{p2}$ are two different p-norms, then there exist positive constants c_1 and c_s such that

$$c_1 \; ||x||_{p1} \; \leq \; ||x||_{p2} \; c_2 \; ||x||_{p1}, \qquad \forall x \in \Re^n$$

induced norms of matrices

For a matrix $A \in \Re^{n \times n}$, the *induced* $p-norm$ of A is defined by

$$||A||_p := \sup_{x \neq 0} \frac{||Ax||_p}{||x||_p} \; = \; \sup_{||x||_p = 1} \; ||Ax||_p$$

Obviously, for matrices $A \in \Re^{m \times n}$ and $A \in \Re^{n \times r}$, we have *the triangle inequality*

$$||A + B||_p \; \leq \; ||A|||_p \; + \; ||B||_p$$

It is easy to show that the *induced norms* are also equivalent in the same sense as for the vector norms, and satisfying

$$||AB||_p \; \leq \; ||Ax||_p \; ||B||_p, \;\; \forall \; A \in \Re^{n \times m}, \; B \in \Re^{m \times r}$$

which is known as the *submultiplicative property*. For $p = 1, 2, \infty,$ we have the corresponding induced norms as follows:

$$\begin{aligned} ||A||_1 &= \max_j \sum_{s=1}^{n} |a_{sj}|, \quad (column\ sum) \\ ||A||_2 &= \max_j \sqrt{\lambda_j(A^t A)} \\ ||A||_\infty &= \max_s \sum_{j=1}^{n} |a_{sj}|, \quad (row\ sum) \end{aligned}$$

2.1.3 Convex sets

A set $\mathbf{S} \subset \Re^n$ is said to be *open* if every vector $x \in \mathbf{S}$, there is an $\epsilon-$neighborhood of x

$$\mathcal{N}(x, \epsilon) = \{z \in \Re^n | ||z - x|| < \epsilon\}$$

such that $\mathcal{N}(x, \epsilon) \subset \mathbf{S}$.

A set is *closed* iff its complement in $\Re^n$ is open; *bounded* if there $r > 0$ such that $||x|| < r, \ \forall \ x \in \mathbf{S}$; and *compact* if it is closed and bounded; *convex* if for every $x, y \in \mathbf{S}$, and every real number α, $0 < \alpha < 1$, the point $\alpha\, x + (1 - \alpha)\, x \in \mathbf{S}$.

A set $\mathbf{K} \subset \Re^n$ is said to be *convex* if for any two vectors x and y in $\mathbf{K}$ any vector of the form $(1 - \lambda)x + \lambda y$ is also in $\mathbf{K}$ where $0 \leq \lambda \leq 1$. This simply means that given two points in a convex set, the line segment between them is also in the set. Note in particular that subspaces and linear varieties (a linear variety is a translation of linear subspaces) are convex. Also the empty set is considered convex. The following facts provide important properties for convex sets.

1. Let $\mathcal{C}_j$, $j = 1, ..., m$ be a family of m convex sets in $\Re^n$. Then the intersection $\mathcal{C}_1 \cap \mathcal{C}_2 \cap .... \cap \mathcal{C}_m$.

2. Let $\mathcal{C}$ be a convex set in $\Re^n$ and $x_o \in \Re^n$. Then the set $\{x_o + x : x \in \mathcal{C}\}$ is convex.

3. A set $\mathbf{K} \subset \Re^n$ is said to be *convex cone* with vertex x_o if $\mathbf{K}$ is convex and $x \in \mathbf{K}$ implies that $x_o + \lambda x \in \mathbf{K}$ for any $\lambda \geq 0$.

An important class of convex cones is the one defined by the positive semidefinite ordering of matrices, that is, $A_1 \geq A_2 \geq A_3$. Let $P \in \Re^{n \times n}$ be a positive semidefinite matrix. The set of matrices $X \in \Re^{n \times n}$ such that $X \geq P$ is a convex cone in $\Re^{n \times n}$.

2.1.4 Continuous functions

A function $f : \Re^n \longrightarrow \Re^m$ is said to be *continuous* at a point x if $f(x + \delta x) \longrightarrow f(x)$ whenever $\delta x \longrightarrow 0$. Equivalently, f is continuous at x if, given $\epsilon > 0$, there is $\delta > 0$ such that

$$||x - y|| < \epsilon \Longrightarrow ||f() - f(y)|| < \epsilon$$

A function f is continuous on a set of **S** if it is continuous at every point of **S**, and it is uniformly continuous on **S** if given $\epsilon > 0$, there is $\delta(\epsilon) > 0$ (dependent only on ϵ), such that the inequality holds for all $x, y \in \mathbf{S}$.

A function $f : \Re \longrightarrow \Re$ is said to be *differentiable* at a point x if the limit

$$\dot{f}(x) = \lim_{\delta x \rightarrow 0} \frac{f(x + \delta x) - f(x)}{\delta x}$$

exists. A function $f : \Re^n \longrightarrow \Re^m$ is *continuously differentiable* at a point x (a set **S**) if the partial derivatives $\partial f_j / \partial x_s$ exist and continuous at x (at every point of **S**) for $1 \leq j \leq m$, $1 \leq s \leq n$ and the *Jacobian matrix* is defined as

$$\mathbf{J} = \left[\frac{\partial f}{\partial x}\right] = \begin{bmatrix} \partial f_1/\partial x_1 & \cdots & \partial f_1/\partial x_n \\ \vdots & \ddots & \vdots \\ \partial f_m/\partial x_1 & \cdots & \partial f_m/\partial x_n \end{bmatrix} \in \Re^{m \times n}$$

2.1.5 Function norms

Let $f(t) : \Re_+ \longrightarrow \Re$ be a continuous function or piecewise continuous function. The $p-$norm of f is defined by

$$\begin{aligned} ||f||_p &= \left(\int_0^\infty |f(t)|^p \, dt\right)^{1/p}, \quad for\ p \in [1, \infty) \\ ||f||_\infty &= \sup t \in [0, \infty) |f(t)|, \quad for\ p = \infty \end{aligned}$$

By letting $p = 1, 2, \infty$, the corresponding normed spaces are called $\mathcal{L}_\infty$, $\mathcal{L}_\in$, $\mathcal{L}_\infty$, respectively. More precisely, let $f(t)$ be a function on $[0, \infty)$

of the signal spaces; they are defined as

$$\mathcal{L}_{\infty} := \Big\{ f(t) : \Re_{+} \longrightarrow \Re | ||f||_1 = \int_0^{\infty} |f(t)| \, dt < \infty, \text{ convolution kernel} \Big\}$$

$$\mathcal{L}_{\in} := \Big\{ f(t) : \Re_{+} \longrightarrow \Re | ||f||_2 = \int_0^{\infty} |f(t)|^2 \, dt < \infty, \text{ finite energy} \Big\}$$

$$\mathcal{L}_{\infty} := \Big\{ f(t) : \Re_{+} \longrightarrow \Re | ||f||_{\infty} = \sup_{t \in [0,\infty)} |f(t)| < \infty, \text{ bounded signal} \Big\}$$

From a signal point of view, the *1-norm*, $||x||_1$ of the signal $x(t)$ is the integral of its absolute value, the square $||x||_2^2$ of the *2-norm* is often called the energy of the signal $x(t)$, and the ∞-*norm* is its absolute maximum amplitude or peak value. It must be emphasized that the definitions of the norms for vector functions are not unique.

In the case of $f(t) : \Re_{+} \longrightarrow \Re^n$, $f(t) = [f_1(t) \;\; f_2(t)...f_n(t)]^t$ which denote a continuous function or piecewise continuous vector function, the corresponding $p-$norm spaces are defined as

$$\mathbf{L_p^n} := \Big\{ f(t) : \Re_{+} \longrightarrow \Re^n | ||f||_p = \int_0^{\infty} ||f(t)||^p \, dt < \infty, \text{ for } p \in [1, \infty) \Big\}$$

$$\mathbf{L_{\infty}^n} := \Big\{ f(t) : \Re_{+} \longrightarrow \Re^n | ||f||_{\infty} = \sup_{t \in [0,\infty)} ||f(t)|| < \infty \Big\}$$

2.2 Calculus and Algebra of Matrices

In this section, we solicit some basic facts and useful relations from linear algebra and calculus of matrices. The materials are stated along with some hints whenever needed but without proofs unless we see the benefit of providing a proof. Reference is made to matrix M or matrix function $M(t)$ in the form:

$$M = \begin{bmatrix} M_{11} & \cdots & M_{1n} \\ \vdots & \ddots & \cdots \\ M_{m1} & \cdots & M_{mn} \end{bmatrix}, \quad or$$

$$M(t) = \begin{bmatrix} M_{11}(t) & \cdots & M_{1n}(t) \\ \vdots & \ddots & \cdots \\ M_{m1}(t) & \cdots & M_{mn}(t) \end{bmatrix}$$

2.2.1 Fundamental subspaces

A nonempty subset $\mathcal{G} \subset \Re^n$ is called a *linear subspace* of $\Re^n$ if $x + y$ and αx are in $\mathcal{G}$ whenever x and y are in $\mathcal{G}$ for any scalar α. A set of elements $X = \{x_1,\ x_2,\ ...,\ x_n\}$ is said to be a *spanning set* for a linear subspace $\mathcal{G}$ of $\Re^n$ if every element $g \in \mathcal{G}$ can be written as a linear combination of the $\{x_j\}$. That is, we have

$$\mathcal{G} = \{g \in \Re :\ g = \alpha_1 x_1 + \alpha_2 x_2 +\ ...\ \alpha_n x_n$$

for some scalars $\alpha_1,\ \alpha_2, \ldots, \alpha_n$. A spanning set X is said to be a *basis* for $\mathcal{G}$ if no element x_j of the spanning set X of $\mathcal{G}$ can be written as a linear combination of the remaining elements $x_1,\ x_2,\ ...,\ x_{j-1},\ x_{j+1},\ ...,\ x_n,$ that is, $x_j,\ 1 \leq i \leq n$ form a linearly independent set. Frequently $x_j = [0\ 0\ ...\ 0\ 1\ 0\ ...\ 0]^t$ is used as the kth unit vector.

The geometric ideas of linear vector spaces had led to the concepts of *"spanning a space"* and a "*basis for a space*." The idea now is to introduce four important subspaces which are useful. The entire linear vector space of a specific problem can be decomposed into the sum of these subspaces.

The *column space* of a matrix $A \in Re^{n \times m}$ is the space spanned by the columns of A, also called the *range space* of A, denoted by $\mathcal{R}[A]$. Similarly, the *row space* of A is the space spanned by the rows of A. Since the column rank of a matrix is the dimension of the space spanned by the columns and the row rank is the dimension of the space spanned by the rows, it is clear that the spaces $\mathcal{R}[A]$ and $\mathcal{R}[A^t]$ have the same dimension $r = rank(A)$.

The *right null space* of $A \in Re^{n \times m}$ is the space spanned by all vectors x that satisfy $A\ x\ =\ 0$, and is denoted $\mathcal{N}[A]$. The right null space of A is also called the *kernel* of A. The *left null space* of A is the space spanned by all vectors y that satisfy $y^t\ A\ =\ 0$. This space is denoted $\mathcal{N}[A^t]$, since it is also characterized by all vectors y such that $A^t\ y\ =\ 0$.

$$
\begin{array}{rcl}
r & := & rank(A) = \textit{dimension of column space } \mathcal{R}[A] \\
dim\ \mathcal{N}[A] & := & \textit{dimension of right null space } \mathcal{N}[A] \\
n & := & \textit{total number of columns of } A \\
r & := & rank(A^t) = \textit{dimension of row space } \mathcal{R}[A^t] \\
dim\ \mathcal{N}[A^t] & := & \textit{dimension of left null space } \mathcal{N}[A^t] \\
m & := & \textit{total number of rows of } A \\
\mathcal{R}[A^t] & := & \textit{row space of } A : \textit{dimension } r \\
\mathcal{N}[A] & := & \textit{right null space of } A : \textit{dimension } n - r \\
\mathcal{R}[A] & := & \textit{column space of } A : \textit{dimension } r \\
\mathcal{N}[A^t] & := & \textit{left null space of } A : \textit{dimension } n - r
\end{array}
$$

The dimensions of the four spaces $\mathcal{R}[A]$, $\mathcal{R}[A^t]$, $\mathcal{N}[A]$, and $\mathcal{N}[A^t]$ are to be determined in the sequel. Since $A \in \Re^{n \times m}$ and the fact that $rank(A) = rank(A^t)$, we have the dimension of the null space $dim\ \mathcal{N}[A] = n - r$ and the dimension of the null space $dim\ \mathcal{N}[A^t] = m - r$. The following is a summary of these results: Note from these facts that the entire $n-$dimensional space can be decomposed into the sum of the two subspaces $\mathcal{R}[A^t]$ and $\mathcal{N}[A]$. Alternatively, the entire $m-$dimensional space can be decomposed into the sum of the two subspaces $\mathcal{R}[A]$ and $\mathcal{N}[A^t]$.

An important property is that $\mathcal{N}[A]$ and $\mathcal{R}[A^t]$ are *orthogonal subspaces*, that is, $\mathcal{R}[A^t]^{\perp} = \mathcal{N}[A]$. This has the meaning that every vector in $\mathcal{N}[A]$ is orthogonal to every vector in $\mathcal{R}[A^t]$. In the same manner, $\mathcal{R}[A]$ and $\mathcal{N}[A^t]$ are *orthogonal subspaces*, that is, $\mathcal{R}[A]^{\perp} = \mathcal{N}[A^t]$. The construction of the fundamental subspaces is appropriately attained by the singular value decomposition.

2.2.2 Calculus of vector-matrix functions of a scalar

The differentiation and integration of time functions involving vectors and matrices arises in solving state equations, optimal control, and so on. This section summarizes the basic definitions of differentiation and integration on vectors and matrices. A number of formulas for the derivative of vector-matrix products are also included.

The derivative of a matrix function $M(t)$ of a scalar is the matrix of the derivatives of each element in the matrix

$$
\frac{dM(t)}{dt} = \begin{bmatrix} \frac{dM_{11}(t)}{dt} & \cdots & \frac{dM_{1n}(t)}{dt} \\ \vdots & \ddots & \cdots \\ \frac{dM_{m1}(t)}{dt} & \cdots & \frac{dM_{mn}(t)}{dt} \end{bmatrix}
$$

The integral of a matrix function $M(t)$ of a scalar is the matrix of the integral of each element in the matrix

$$\int_a^b M(t)dt = \begin{bmatrix} \int_a^b M_{11}(t)dt & \cdots & \int_a^b M_{1n}(t)dt \\ \vdots & \ddots & \cdots \\ \int_a^b M_{m1}(t)dt & \cdots & \int_a^b M_{mn}(t)dt \end{bmatrix}$$

The Laplace transform of a matrix function $M(t)$ of a scalar is the matrix of the Laplace transform of each element in the matrix

$$\int_a^b M(t)e^{-st}dt = \begin{bmatrix} \int_a^b M_{11}(t)e^{-st}dt & \cdots & \int_a^b M_{1n}(t)e^{-st}dt \\ \vdots & \ddots & \cdots \\ \int_a^b M_{m1}(t)e^{-st}dt & \cdots & \int_a^b M_{mn}(t)e^{-st}dt \end{bmatrix}$$

The scalar derivative of the product of two matrix time-functions is

$$\frac{d(A(t)B(t))}{dt} = \frac{A(t)}{dt}B(t) + A(t)\frac{B(t)}{dt}$$

This result is analogous to the derivative of a product of two scalar functions of a scalar, except caution must be used in reserving the order of the product. An important special case follows.

The scalar derivative of the inverse of a matrix time-function is

$$\frac{dA^{-1}(t)}{dt} = -A^{-1}\frac{A(t)}{dt}A(t)$$

2.2.3 Derivatives of vector-matrix products

The derivative of a real scalar-valued function $f(x)$ of a real vector $x = [x_1, \ldots, x_n]^t \in Re^n$ is defined by

$$\frac{\partial f(x)}{\partial x} = \begin{bmatrix} \frac{\partial f(x)}{\partial x_1} \\ \frac{\partial f(x)}{\partial x_2} \\ \vdots \\ \frac{\partial f(x)}{\partial x_n} \end{bmatrix}$$

where the partial derivative is defined by

$$\frac{\partial f(x)}{\partial x_j} := \lim_{\Delta x_j \to 0} \frac{f(x + \Delta x) - f(x)}{\Delta x_j}, \quad \Delta x = [0 \ldots \Delta x_j \ldots 0]^t$$

An important application arises in the Taylor's series expansion of $f(x)$ about x_o in terms of $\delta x := x - x_o$. The first three terms are

$$f(x) = f(x_o) + (\frac{\partial f(x)}{\partial x})^t\ \delta x + \frac{1}{2}\delta x^t \left[\frac{\partial^2 f(x)}{\partial x^2}\right] \delta x$$

where

$$\frac{\partial f(x)}{\partial x} = \begin{bmatrix} \frac{\partial f(x)}{\partial x_1} \\ \vdots \\ \frac{\partial f(x)}{\partial x_n} \end{bmatrix},$$

$$\frac{\partial^2 f(x)}{\partial x^2} = \frac{\partial}{\partial x}\left(\frac{\partial f(x)}{\partial x}\right)^t = \begin{bmatrix} \frac{\partial^2 f(x)}{\partial x_1^2} & \cdots & \frac{\partial^2 f(x)}{\partial x_1 \partial x_n} \\ \vdots & \ddots & \cdots \\ \frac{\partial^2 f(x)}{\partial x_n \partial x_1} & \cdots & \frac{\partial^2 f(x)}{\partial x_n^2} \end{bmatrix}$$

The derivative of a real scalar-valued function $f(A)$ with respect to a matrix

$$A = \begin{bmatrix} A_{11} & \cdots & A_{1n} \\ \vdots & \ddots & \cdots \\ A_{n1} & \cdots & A_{nn} \end{bmatrix} \in Re^{n\times n}$$

is given by

$$\frac{\partial f(A)}{\partial A} = \begin{bmatrix} \frac{\partial f(A)}{\partial A_{11}} & \cdots & \frac{\partial f(A)}{\partial A_{1n}} \\ \vdots & \ddots & \cdots \\ \frac{\partial f(A)}{\partial A_{n1}} & \cdots & \frac{\partial f(A)}{\partial A_{nn}} \end{bmatrix}$$

A vector function of a vector is given by

$$v(u) = \begin{bmatrix} v_1(u) \\ \vdots \\ \vdots \\ v_n(u) \end{bmatrix}$$

where $v_j(u)$ is a function of the vector u. The derivative of a vector function of a vector (the *Jacobian*) is defined as follows:

$$\frac{\partial v(u)}{\partial u} = \begin{bmatrix} \frac{\partial v_1(u)}{\partial u_1} & \cdots & \frac{\partial v_1(u)}{\partial u_m} \\ \vdots & \ddots & \cdots \\ \frac{\partial v_n(u)}{\partial u_1} & \cdots & \frac{\partial v_n(u)}{\partial u_m} \end{bmatrix}$$

Note that the *Jacobian* is sometimes defined as the transpose of the foregoing matrix. A special case is given by

$$\frac{\partial(S\ u)}{\partial u} = S, \quad \frac{\partial(u^t Ru)}{\partial u} = 2\ u^t R$$

for arbitrary matrix S and symmetric matrix R.

The following section includes useful relations and results from linear algebra.

2.2.4 Positive definite and positive semidefinite matrices

A matrix P is positive definite if P is real, symmetric, and $x^t Px > 0, \ \ \forall x \neq 0$; equivalently, if all the eigenvalues of P have positive real parts. A matrix S is positive semidefinite if S is real, symmetric, and $x^t Px \geq 0, \ \ \forall x \neq 0$.

Since the definiteness of the scalar $x^t Px$ is a property only of the matrix P, we need a test for determining definiteness of a constant matrix P. Define a *principal submatrix* of a square matrix P as any square submatrix sharing some diagonal elements pf P. Thus the constant, real, symmetric matrix $(P \in \Re^{n \times n}$ is positive definite $(P > 0)$ if either of these equivalent conditions holds:
1. All eigenvalues of P are positive.
2. The determinant of P is positive.
3. All successive principal submatrices of P (minors of successively increasing size) have positive determinants.

2.2.5 Trace properties

The trace of a square matrix P, trace (P), equals the sum of its diagonal elements or equivalently the sum of its eigenvalues. A basic property of the trace is invariant under cyclic perturbations, that is,

$$trace(AB) = trace(BA)$$

where AB is square. Successive applications of the above results yield

$$trace(ABC) = trace(BCA) = trace(CAB)$$

where ABC is square. In general, $trace(AB) = trace(B^t A^t)$. Another result is that

$$trace(A^t BA) = \sum_{k=1}^{p} a_k^t B a_k$$

where $A \in \Re^{n\times p},\ B \in \Re^{n\times n}$, and $\{a_k\}$ are the columns of A. A list of some identities of trace derivatives and related results is given below:

$$
\begin{aligned}
\frac{\partial(trace(AB))}{\partial A} &= \frac{\partial(trace(A^tB^t))}{\partial A} = \frac{\partial(trace(B^tA^t))}{\partial A},\\
&= \frac{\partial(trace(BA))}{\partial A} = B^t,\\
\frac{\partial(trace(AB))}{\partial B} &= \frac{\partial(trace(A^tB^t))}{\partial B} = \frac{\partial(trace(B^tA^t))}{\partial B},\\
&= \frac{\partial(trace(BA))}{\partial B} = A^t,\\
\frac{\partial(trace(BAC))}{\partial A} &= \frac{\partial(trace(B^tC^tA^t))}{\partial A} = \frac{\partial(trace(C^tA^tB^t))}{\partial A},\\
&= \frac{\partial(trace(ACB))}{\partial A} = \frac{\partial(trace(CBA))}{\partial A},\\
&= \frac{\partial(trace(A^tB^tC^t))}{\partial A} = B^t\,C^t,\\
\frac{\partial(trace(A^tBA))}{\partial A} &= \frac{\partial(trace(BAA^t))}{\partial A} = \frac{\partial(trace(AA^tB))}{\partial A},\\
&= (B\,+\,B^t)A,\\
\frac{\partial(trace(AX^t))}{\partial X} &= A,\quad \frac{\partial(trace(AXB))}{\partial X} = A^t\,B^t,\\
\frac{\partial(trace(AX^tB))}{\partial X} &= B\,A,\quad \frac{\partial(trace(AX))}{\partial X^t} = A,\\
\frac{\partial(trace(AX^t))}{\partial X^t} &= A^t,\quad \frac{\partial(trace(AXB))}{\partial X^t} = B\,A,\\
\frac{\partial(trace(AX^tB))}{\partial X^t} &= A^t\,B^t,\quad \frac{\partial(trace(XX))}{\partial X} = 2\,X^t,\\
\frac{\partial(trace(XX^t))}{\partial X} &= 2\,X,\\
\frac{\partial(trace(AX^n))}{\partial X} &= \left(\sum_{j=0}^{n-1} X^j\,A\,X^{n-j-1}\right)^t,\\
\frac{\partial(trace(AXBX))}{\partial X} &= A^tX^tB^t\,+\,B^tX^tA^t,\\
\frac{\partial(trace(AXBX^t))}{\partial X} &= A^tXB^t\,+\,AXB,\\
\frac{\partial(trace(X^{-1}))}{\partial X} &= -(X^{-2})^t,
\end{aligned}
$$

$$\frac{\partial(trace(AX^{-1}B))}{\partial X} = -\left(X^{-1}BAX^{-1}\right)^t,$$
$$\frac{\partial(trace(AB))}{\partial A} = B^t + B - diag(B)$$

2.2.6 Partitioned matrices

Given a partitioned matrix (matrix of matrices) of the form

$$M = \begin{bmatrix} A & B \\ C & D \end{bmatrix}$$

where $A,\ B,\ C$, and D are of compatible dimensions, then
(1) if A^{-1} exists, a Schur complement of M is defined as $D - CA^{-1}B$, and
(2) if D^{-1} exists, a Schur complement of M is defined as $A - BD^{-1}C$.
When $A,\ B,\ C$, and D are all $n \times n$ matrices, then:

$$a) \qquad det \begin{bmatrix} A & B \\ C & D \end{bmatrix} = det(A)det(D - CA^{-1}B),\ det(A) \neq 0$$

$$b) \qquad det \begin{bmatrix} A & B \\ C & D \end{bmatrix} = det(D)det(A - BD^{-1}C),\ det(D) \neq 0$$

In the special case, we have

$$det \begin{bmatrix} A & B \\ C & 0 \end{bmatrix} = det(A)det(C)$$

where A and C are square. Since the determinant is invariant under row, it follows

$$\begin{aligned} det \begin{bmatrix} A & B \\ C & D \end{bmatrix} &= det \begin{bmatrix} A & B \\ C - CA^{-1}A & D - CA^{-1}B \end{bmatrix} \\ &= det \begin{bmatrix} A & B \\ 0 & D - CA^{-1}B \end{bmatrix} \\ &= det(A)det(D - CA^{-1}B) \end{aligned}$$

which justifies the foregoing result.

Given matrices $A \in \Re^{m\times n}$ and $B \in \Re^{n\times m}$, then

$$det(I_m - AB) = det(I_n - BA)$$

In case that A is invertible, then $det(A^{-1}) = det(A)^{-1}$.

2.2.7 Matrix inversion lemma

Suppose that $A \in \Re^{n\times n}$, $B \in \Re^{n\times p}$, $C \in \Re^{p\times p}$, and $D \in \Re^{p\times n}$. Assume that A^{-1} and C^{-1} both exist. Then

$$(A+BCD)^{-1} = A^{-1} - A^{-1}B(DA^{-1}B+C^{-1})^{-1}DA^{-1}$$

In the case of partitioned matrices, we have the following result:

$$\begin{bmatrix} A & B \\ C & D \end{bmatrix}^{-1} = \begin{bmatrix} A^{-1}+A^{-1}B\Xi^{-1}CA^{-1} & -A^{-1}B\Xi^{-1} \\ -\Xi^{-1}CA^{-1} & \Xi^{-1} \end{bmatrix}$$
$$\Xi = (D - CA^{-1}B)$$

provided that A^{-1} exists. Alternatively,

$$\begin{bmatrix} A & B \\ C & D \end{bmatrix}^{-1} = \begin{bmatrix} \Xi^{-1} & -\Xi^{-1}BD^{-1} \\ -D^{-1}C\Xi^{-1} & D^{-1}+D^{-1}C\Xi^{-1}BD^{-1} \end{bmatrix}$$
$$\Xi = (D - CA^{-1}B)$$

provided that D^{-1} exists.

For a square matrix Y, the matrices Y and $(I+Y)^{-1}$ commute, that is, given that the inverse exists

$$Y\,(I+Y)^{-1} = (I+Y)^{-1}\,Y$$

Two additional inversion formulas are given below:

$$\begin{aligned} Y\,(I+XY)^{-1} &= (I+YX)^{-1}\,Y, \\ (I+YX)^{-1} &= I - YX\,(I+YX)^{-1} \end{aligned}$$

The following result provides conditions for the positive definiteness of a partitioned matrix in terms of its submatrices. The following three statements are equivalent:

$$
\begin{aligned}
&1) \quad \begin{bmatrix} A_o & A_a \\ A_a^t & A_c \end{bmatrix} > 0, \\
&2) \quad A_c > 0, \quad A_o - A_a A_c^{-1} A_a^t > 0, \\
&3) \quad A_a > 0, \quad A_c - A_a^t A_o^{-1} A_a > 0
\end{aligned}
$$

2.2.8 Singular value decomposition

The singular value decomposition (SVD) is a matrix factorization that has found a number of applications to engineering problems. The SVD of a matrix $M \in Re^{n \times m}$ is

$$M = U\, S\, V^{\dagger} = \sum_{j=1}^{p} \sigma_j\, U_j\, V_j^{\dagger}$$

where $U \in Re^{\alpha \times \alpha}$ and $V \in Re^{\beta \times \beta}$ are unitary matrices ($U^{\dagger}\, U = U\, U^{\dagger} = I$ and $V^{\dagger}\, V = V\, V^{\dagger} I$); $S \in Re^{\alpha \times \beta}$ is a real, diagonal (but not necessarily square); and $pmin(\alpha, \beta)$. *The singular values* $\{\sigma_1, \sigma_2, ..., \sigma_\beta\}$ *of* M *are defined as the positive square roots of the diagonal elements of* $S^t S$, and are ordered from largest to smallest.

To proceed further, we recall a result on unitary matrices. If U is a unitary matrix ($U^{\dagger}\, U = I$), then the transformation U preserves length, that is:

$$
\begin{aligned}
||U\, x|| &= \sqrt{(Ux)^{\dagger}(Ux)} = \sqrt{x^{\dagger}\, U^{\dagger}\, U\, x}, \\
&= \sqrt{x^{\dagger}\, x} = ||x||
\end{aligned}
$$

As a consequence, we have

$$
\begin{aligned}
||M\, x|| &= \sqrt{x^{\dagger}\, M^{\dagger}\, M\, x} = \sqrt{x^{\dagger}\, V S^t U^{\dagger}\, U S V^{\dagger}\, x}, \\
&= \sqrt{x^{\dagger}\, V S^t S V^{\dagger}\, x}
\end{aligned}
$$

To evaluate the maximum gain of matrix M, we calculate the maximum norm of the above equation to yield

$$\max_{||x||=1} ||M\, x|| = \max_{||x||=1} \sqrt{x^{\dagger}\, V S^t S V^{\dagger}\, x} = \max_{||\tilde{x}||=1} \sqrt{\tilde{x}^{\dagger}\, V S^t S\, \tilde{x}}$$

Note that maximization over $\tilde{x} = Vx$ is equivalent to maximizing over x since V is invertible and preserves the norm (equals 1 in this case). Expanding the norm yields

$$\begin{aligned}\max_{||x||=1} ||M\,x|| &= \max_{||\tilde{x}||=1} \sqrt{\tilde{x}^{\dagger}\,V S^t S\,\tilde{x}},\\ &= \max_{||\tilde{x}||=1} \sqrt{\sigma_1^2|\tilde{x}_1|^2 + \sigma_2^2|\tilde{x}_2|^2 + ... + \sigma_\beta^2|\tilde{x}_\beta|^2}\end{aligned}$$

The foregoing expression is maximized, given the constraint $||\tilde{x}|| = 1$, when $\tilde{x}$ is concentrated at the largest singular value; that is, $|\tilde{x}| = [1\ 0\ ...\ 0]^t$. The maximum gain is then

$$\max_{||x||=1} ||M\,x|| = \sqrt{\sigma_1^2|1|^2 + \sigma_2^2|0|^2 + ... + \sigma_\beta^2|0|^2} = \sigma_1 = \sigma_M$$

In words, this reads *The maximum gain of a matrix is given to be the maximum singular value* σ_M. Following similar lines of development, it is easy to show that

$$\begin{aligned}\min_{||x||=1} ||M\,x|| &= \sigma_\beta = \sigma_m,\\ &= \begin{cases} \sigma_p & \alpha \geq \beta \\ 0 & \alpha < \beta \end{cases}\end{aligned}$$

A property of the singular values is expressed by

$$\sigma_M(M^{-1}) = \frac{1}{\sigma_m(M)}$$

2.3 Directed Graphs

A graph $\mathsf{G}(\mathcal{V},\ \mathcal{E})$ is described by a set $\mathcal{V} = [v_1,\ v_2, ...]$ of vertices and a set $\mathcal{E} = [e_1,\ e_2, ...]$ of edges. The edges can be represented by their end points as $e_j = (v_k,\ v_\ell)$, which means that the edge e_j connects the vertices v_k and v_ℓ and is directed from v_k to v_ℓ; see Figure 2.1. Considerations concerning graphs in which for any pair v_k and v_ℓ there is at most one edge $(v_k \quad v_\ell)$ and one edge $(v_\ell \quad v_k)$ are presented in the sequel. Associated with graph there is an $(n,\ n)$ *adjacency matrix* $\mathbf{A} = (a_{jk})$ with n being the number of vertices of the graph to signify which vertices of the graph are connected by an edge

$$a_{jk} = \begin{cases} \bullet & if\ there\ exists\ an\ edge\ (v_k \quad v_j) \\ 0 & otherwise \end{cases}$$

A graph $\mathsf{G}(\mathcal{V},\ \mathcal{E})$ is completely described by the matrix $\mathbf{A}$.

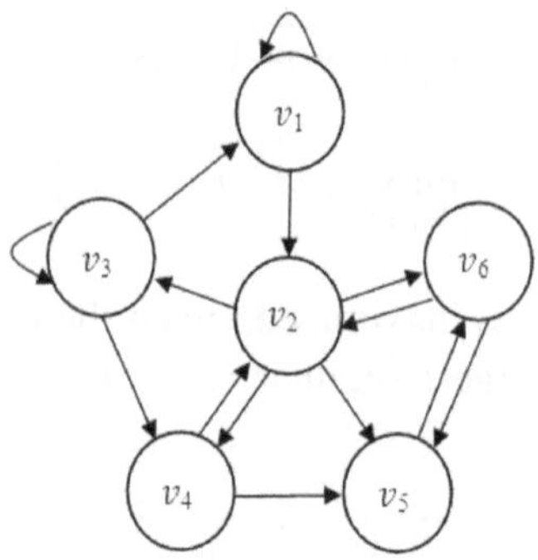

Figure 2.1: Directed graph

A *path* is a sequence of edges $[(v_{j1} \;\; v_{j2})(v_{j2} \;\; v_{j3}) \cdots (v_k \;\; v_\ell)]$ such that the final vertex and the initial vertex of succeeding edges are the same. For example, $[(v_3 \;\; v_1)(v_1 \;\; v_2)(v_2 \;\; v_4)]$ is a path from v_3 to v_4 in the graph of Figure 2.1, but there is no path from v_5 to v_1.

Definition 2.3.1 *Two vertices v_k and v_ℓ are said to be strongly connected if there is a path from v_k to v_ℓ as well as a path from v_ℓ to v_k. The graph is called strongly connected if every pair of vertices v_k and v_ℓ is strongly connected.*

In the graph $\mathsf{G}(\mathcal{V}, \;\mathcal{E})$ the subset of vertices that are strongly connected to a given vertex v_j forms an equivalence class $\mathcal{K}(v_j)$ within the set $\mathcal{V}$. For example, $\mathcal{K}(v_1) = [v_1, \; v_2, \; v_3, \; v_4]$; see Figure 2.1. The *adjacency matrix* $\mathbf{A} = (a_{jk})$ is given by

$$\mathbf{A} = \begin{bmatrix} \bullet & 0 & \bullet & 0 & 0 & 0 \\ \bullet & 0 & 0 & \bullet & 0 & \bullet \\ 0 & \bullet & \bullet & 0 & 0 & 0 \\ 0 & \bullet & \bullet & 0 & 0 & 0 \\ 0 & \bullet & 0 & \bullet & 0 & \bullet \\ 0 & \bullet & 0 & 0 & \bullet & 0 \end{bmatrix}$$

The *reachability matrix* $\mathbf{R} = (r_{jk})$ describes which pairs of vertices are connected by a path

$$r_{jk} = \begin{cases} \bullet & if \; there \; exists \; a \; path \; from \; v_k \; to \; v_j \\ 0 & otherwise \end{cases}$$

The *reachability matrix* $\mathbf{R}$ and the *adjacency matrix* $\mathbf{A}$ are related by

$$\mathbf{R} = \sum_{j=1}^{n-1} \mathbf{A}^j$$

where the multiplication and addition of the elements of $\mathbf{A}$ are carried out according to the following

$$a_{jk}a_{km} = \begin{cases} \bullet & if\ a_{jk} = \bullet\ and\ a_{km} = \bullet \\ 0 & otherwise \end{cases}$$

$$a_{jk} + a_{m\ell} = \begin{cases} 0 & if\ a_{jk} = 0\ and\ a_{m\ell} = 0 \\ \bullet & otherwise \end{cases}$$

A *cycle* is a path with identical initial and final vertices. For example in Figure 2.1, $[(v_3\ \ v_1)(v_1\ \ v_2)(v_2\ \ v_3)]$ form a *cycle*. A set of vertex-disjoint cycles is said to be a cycle family. The cycle mentioned above represents a cycle family with only a single cycle. There is also another cycle family consisting of the cycles $[(v_3\ \ v_4)(v_4\ \ v_2)(v_2\ \ v_3)]$ and indeed the self-cycle $[(v_1\ \ v_1)]$.

2.4 Notes and References

The topics covered in this chapter are meant to provide the reader with a general platform containing the basic mathematical information needed for further examination of switched time-delay systems. These topics are properly selected from standard books and monographs on mathematical analysis. For further details, the reader is referred to the standard texts [147, 152, 404] where fundamentals of linear algebra and related math facts are provided. Graph search algorithms for determining paths, cycles, reachability matrices, etc. can be found in [69] .

Chapter 3

Historical Perspective and Scope

Many real problems are considered to be "large scale" by nature and not by choice. Some important attributes of large-scale systems are

1. *They often represent complex, real-life systems*,
2. *Their structures follow hierarchical (multilevel) order*,
3. *Their dynamics involve time-delays and are subject to uncertainties*,
4. *Decentralized information structures are the rule and not the exception.*

These attributes depict systems dealing with societal, business, and management organizations, the economy, the environment, data networks, electric power, transportation, information systems, aerospace (including space structures), water resources, and, last but not least, energy. Such systems that are eventually used in support of human life are complex in nature. As a result of these important properties and potential applications, several researchers have paid a great deal of attention to various facets of large-scale systems such as modeling, model reduction, control, stability, controllability, observability, optimization, and feedback stabilization. These concepts have been applied to various problems and have helped with the development of different notions of systems analysis, design, control, and optimization.

In this chapter, we follow a chronological order in providing a unified discussion about the concepts and techniques of large-scale systems with particular emphasis on fundamental issues, basic definitions, and main directions of development. Whenever deemed convenient, we compare among their merits, features, and computational requirements.

3.1 Overview

An integral assumption in almost all of the conventional control theories and methods, for quite some time, has been "centralization" [307], meaning that all the computations and measurements based upon system information are localized at a given center, very often a geographical position. A prime example of a centralized system is a computer-controlled experimental testbed physically located in a laboratory setting. This concept becomes unreliable when the system is composed of several subsystems and/or is geographically distributed over wide areas. Consider, for example, water systems.

It has been recognized [22, 303, 323] that a notable characteristic of most large-scale systems is that centrality fails to hold due to either the lack of centralized computing capability or centralized information. The important points regarding large-scale systems are that some of these systems are often separated geographically, and their treatment requires consideration of not only economic costs, as is common in centralized systems, but also such important issues as reliability of communication links, value of information, environmental consciousness, and machine intelligence quotient (MIQ), to name a few. It is for the decentralized and hierarchical control properties and potential applications of such exciting areas as intelligent large-scale systems in which fuzzy logic and neural networks are incorporated within the control architecture that many researchers have devoted a great deal of effort to large-scale intelligent systems in recent years.

3.2 Decomposition-Coordination Methods

A fairly general dynamic optimization problem could be described as follows:

$$\begin{aligned} x_j(k+1) &= \mathbf{f}_j[x_j(k), u_j(k), k] \\ x_j(0) &= x_{j0},\ j=1,...,Nn_s,\ k=0,...,P-1 \end{aligned} \tag{3.1}$$

where the interconnections and terminal constraints are specified by

$$for \ \ j=1,...,n_s,$$

$$\sum_{j=1}^{n_s} \mathbf{h}_j[x_j(k), u_j(k), k] \ \le \ 0, \ \ k=0,...,P-1 \tag{3.2}$$

$$\mathbf{R}_j[x_j(k), u_j(k), k] \ \le \ 0, \ , \ k=0,...,P-1 \tag{3.3}$$

$$\mathbf{Q}_j[x_j(P)] \ \le \ 0 \tag{3.4}$$

The performance measure is

$$J = \sum_{j=1}^{n_s} \{\mathbf{S}_j[x_j(P)] + \sum_{k=0}^{P-1} \mathbf{F}_j[x_j(k), u_j(k), k]\} \tag{3.5}$$

Equations (3.1)-(3.5) represent n_s-coupled subsystems in discrete-time format where $x_j(k)$ and $u_j(k)$ are the respective state and control vectors. The objective is to determine the optimal sequences $[\{x_j * (k), u_j * (k)\},\ j = l, , n_s]$ which minimize the performance measure (3.5) subject to the constraints (3.1)-(3.4). This is a fairly general class of constrained optimization problems. It has been shown in [137, 138, 164] that the introduction of the dual problem results in a natural two-level formulation of the optimization problem with the first level solving the primal problem and the second level solving the dual problem. Thus the procedure involves the formulation and solution of the Lagrangian dual of the above problem. The dual problem is

$$\max_{\lambda}\ \Phi(\lambda), \quad \lambda \geq 0 \tag{3.6}$$

where

$$\Phi(\lambda) := \min_{x,u}\ \mathcal{L}(x, u, \lambda) \tag{3.7}$$

and the Lagrangian functional is

$$\begin{aligned} \mathcal{L}(x, u, \lambda) &= \sum_{j=1}^{n_s} \{\mathbf{S}_j[x_j(P)] + \sum_{k=0}^{P-1} \mathbf{F}_j[x_j(k), u_j(k), k] \\ &+ \lambda_j^t \mathbf{h}_j[x_j(k), u_j(k), k]\} \end{aligned} \tag{3.8}$$

Due to the fact that the interconnection constraints (3.2) provide the only coupling between the subsystems, (3.8) could be separated into additive sub-Lagrangians, and consequently n_s independent minimization problems are defined. These constitute the first level, while solution of (3.6) and (3.7) specifies the second level.

Observe from (3.6)-(3.8), the gradient of the dual problem is exactly the value of the coupling constraint. While this observation enhances the dual optimization approach, it implies that the overall optimum is reached only when the second-level solution converges. On the other hand, since the dual function is concave without any convexity requirements on the original system, the global optimum is assured. In addition, the computational efforts at the first level are still burdensome, although the dimensions are reduced. However, they can be implemented by making use of parallel processing capabilities since each subsystem Lagrangian can be operated from the remainder. Generally, it is impractical for the first level to minimize the Lagrangian as an explicit function of the multipliers, and therefore it is always desirable to change the problem into a sequential search for the optimum. The first-level minimization problems can be further decomposed in time, thereby leading to the development of a three-level optimization algorithm. This algorithm utilizes the dual concept to parameterize the local problems by the index k, the discrete instant, and the potential of this algorithm is explained in [331]. Also, it was pointed out in [331, 332] the fact that most of the available real-time multilevel techniques are best suited for slow dynamic systems due to the requirement for repeated iterations.

3.2.1 Hierarchical structures

Extending on the ideas of the foregoing section, we must mention that one of the earlier attempts in dealing with large-scale systems was to "decompose" a given system into a number of subsystems for computational efficiency and design simplification. The "decomposition" approach divides the original system into a number of subsystems involving certain values of parameters. Each subsystem is solved independently for a fixed value of the so-called "decoupling" parameter, whose value is subsequently adjusted by a coordinator in an appropriate fashion so that the subsystems resolve their problems and the solution to the original system is obtained. The goal of the coordinator is to arrange the activities of the subsystems to provide a feasible solution to the overall system. This exchange of solution (by the subsystems) and coordination (interaction) vector (by the coordinator) will continue until convergence has been achieved.

A unified approach to multilevel control and optimization of dynamic systems has been developed in [196]-[199] [201] using the generalized gradients technique. The basic motivations behind this approach were the following:

1. *In the context of two-level structures, the ordering of an optimizing level (control) and a coordination level (supervisory) is immaterial,*
2. *There is no restriction on the variables to be handled (optimized or manipulated) on each level,*
3. *Problems of physical realizability of the generated control actions, singular arcs, and interaction feasibility are critical in any two-level control structure,*
4. *Within the existing capabilities of digital computers, a single processor*

is always used for problem solution, and hence, the computational effort should be adequately shared between levels.

Several candidate computation structures have been developed; for a detailed analysis see [194]-[199].

3.2.2 Decentralized control

Most practical large-scale systems are characterized by a great multiplicity of measured outputs and inputs. For example, an electric power system has several control substations, each being responsible for the operation of a portion of the overall system. This situation arising in a control system design is often referred to as decentralization. The designer for such systems determines a structure for control that assigns system inputs to a given set of local controllers (stations), which observe only local system outputs. In other words, this approach, from now onward called *decentralized control*, attempts to avoid difficulties in data gathering, storage requirements, computer program debugging, and geographical separation of system components. A preliminary comparison between decentralized and hierarchical control can be given here. In hierarchical control, a decomposition in system structure will lead to computational efficiency. In decentralized control, on the other hand, a decomposition takes place with the system's output information leading to simpler controller structures and computational efficiency.

Experience research results have concluded that the application of the hierarchical multilevel structures to systems engineering problems is guaranteed to yield a "coordinated control" which, iteratively, approach a "decentralized control."

A decentralized approach to hierarchical optimization can be made effective if the large system is composed of weakly coupled subsystems, or a decomposition and rearrangement of variables are used to achieve weak coupling as is the case of sparse systems. In situations when the subsystems are strongly connected and cannot be simply reconstituted to reduce the strength of the couplings, the assumption of weak coupling may produce gross inaccuracies of the obtained results. Even in cases when the assumption is appropriate, it is often not clear what roles should be played by the coordinator and the information structure to achieve the benefits of the decentralized optimization strategies.

3.3 Multilevel Optimization of Nonlinear Systems

In this section, we consider the optimal control of a class of large-scale nonlinear dynamical systems described by

$$\dot{x}(t) = \mathbf{f}(t, x, u) \tag{3.9}$$

where $x(t) \in \Re^n$ is the state vector and $u(t) \in \Re^{m_j}$ is the control input at time $t \in \Re$. The function $\mathbf{f} : \Re \times \Re^n \times \Re^m$ is continuous on a bounded region $\mathcal{D} \in \Re^{n+1}$ and is locally Lipschitzian with respect to $x \in \mathcal{D}$, so that for every fixed control function $u(t)$, a unique solution $x(t; t_o, x_o)$ exists for all initial conditions $(t_o, x_o) \in \mathcal{D}$ and all $t \in [t_o, t_1]$ and $\mathcal{D} = \{(t, x) : t_o \leq t \leq t_1,\ ||x|| \leq \varrho < +\infty\}$.

Now we view (3.9) as a large-scale or interconnected system. In either case, dynamic optimization by standard methods becomes impractical due the excessive requirements of computer storage or processing time. Following [136, 324, 335], the problem becomes tractable by an appropriate decomposition of the system and optimization of the subsystems. At start, we re-express (3.9) as a collection of n_s coupled subsystems described by

$$\dot{x}_j(t) = g_j(t, x_j, u_{jc}) + h_j(t, x), \quad j = 1, 2, ..., n_s \tag{3.10}$$

where $x_j(t) \in \Re^{n_j}$ is the state vector of the jth subsystem and $u_{jc}(t) \in \Re^{n_j}$ is the local control input of the jth subsystem so that

$$\begin{aligned} \Re^n &= \Re^{n_1} \times \Re^{n_2} \times \cdots \times \Re^{n_{n_s}} \\ \Re^m &= \Re^{m_1} \times \Re^{m_2} \times \cdots \times \Re^{m_{n_s}} \\ n &= \sum_{j=1}^{n_s} n_j, \quad m = \sum_{j=1}^{n_s} m_j \end{aligned} \tag{3.11}$$

In addition, $g_j : \Re \times \Re^{n_j} \times \Re^{m_j} \rightarrow \Re^{n_j}$ represents the dynamics of the decoupled subsystems

$$\dot{x}_j(t) = g_j(t, x_j, u_{jc}), \quad j = 1, 2, ..., n_s \tag{3.12}$$

which are all completely locally controllable about any admissible trajectory $x_j(t; t_o, x_{jo})$ and $h : \Re \times \Re^n \rightarrow \Re^{n_j}$ is the function that represents the coupling pattern of the jth subsystem within the overall system (3.9).

3.3.1 Local-level optimization

We now seek local dynamic optimization using the control law

$$u_{jc}(t) = K_j(t, x_j), \quad j = 1, 2, ..., n_s \tag{3.13}$$

For each isolated subsystem (3.12) is optimized with respect to the local performance index

$$J_j(t_o, x_{jo}, u_{jc}) = p_j[t_o, x_j(t_1)] + \int_{t_o}^{t_1} M_j[t, x_j(t), u_{jc}(t)]dt \qquad (3.14)$$

where $p_j : \Re \times \Re^{n_j} \rightarrow \Re_+$, $M_j : \Re \times \Re^{n_j} \times \Re^{m_j} \rightarrow \Re_+$ are functions of the class C^2 in all arguments [168], and $x_j(t)$ denotes the solution $x_j(t; t_o, x_{jo})$ of (3.12) for the local control function $u_{jc}(t)$.

It must be born in mind that the decomposition of (3.9) into (3.12) entails that the jth subsystem has low order and/or has a simple structure so that it is a straightforward task to determine the optimal control law

$$u^*_{jc}(t) = K_{jc}(t, x_j), \quad j = 1, 2, ..., n_s \qquad (3.15)$$

which yields the optimal cost

$$J^*_j(t, x_{jo}) = J_j(t, x_{jo}, K^*_j(t, x_{jo})) \qquad (3.16)$$

Remark 3.3.1 *One should realize that there are two sources of performance degradation: one is due to the interconnection pattern which imposes computational constraints and the other arises from certain structural considerations that should be taken into account when designing controllers for interconnected systems. Generally speaking, a system that is composed of coupled subsystems may undergo structural perturbations whereby subsystems are disconnected (and again connected) in various ways during the operation. Typical examples are found in power systems, chemical plants, and traffic networks.*

Therefore, in order to cope with the participation nature of the subsystems while guaranteeing a satisfactory performance of the system, we should preserve as much as possible the autonomy of each isolated subsystem (3.12). This leads to expressing the performance index for the overall system as:

$$\begin{aligned} J(t, x_o, u_c(t)) &= \sum_{j=1}^{n_s} J_j(t, x_o, u_{jc}(t)) \\ u_c(t) &= [u^t_{1c}, \ u^t_{2c}, \ \cdots, \ u^t_{n_sc}]^t, \in \Re^m \end{aligned} \qquad (3.17)$$

This implies that each isolated subsystem (3.12) is optimized with respect to its own performance index (3.16) regardless of the behavior of the other subsystems. In this regard, the couplings among the subsystems are regarded as perturbation terms.

A crucial point to observe is that maintaining the subsystem autonomy means ignoring possible beneficial effects of the interconnection patterns. More about this point will be mentioned in the subsequent sections. For the time being, we take the view that any use of the beneficial effects of couplings will naturally increase the dependence among the subsystems and, hence, increase the liability to malfunctioning under structural perturbations.

In brief, the optimal performance index

$$J^*(t, x_o) \;=\; \sum_{j=1}^{n_s} J_j^*(t, x_{jo}) \tag{3.18}$$

cannot be achieved by using only the local control $u_c(t)$, unless all the subsystems are decoupled ($h_j \equiv 0,\ j = 1, 2, .*, s$). This simply means that $u_c(t)$ is near-optimal. Let $\tilde{x}_j = \tilde{x}_j(t; t_o, x_{jo})$ denote the solution of the coupled system (3.10), then the value of the associated performance index becomes

$$\tilde{J}(t, x_o) = \sum_{j=1}^{n_s} \left(p_j[t_o, \tilde{x}_j(t_1)] + \int_{t_o}^{t_1} M_j[t, \tilde{x}_j(t), K_{jc}(t, \tilde{x}_j)]dt \right) \tag{3.19}$$

Define the degree of suboptimality as ε. Hence it follows that [322]

$$\tilde{J}(t, x_o) \;\leq\; (1 \,+\, \varepsilon)\; J^*(t, x_o), \quad \forall (t, x_o) \in \mathcal{D} \tag{3.20}$$

It is readily evident that the suboptimality index ε resulting from the optimal local control

$$\dot{x}_j(t) = g_j(t, x_j, K_{jc}(t, x_j)) + h_j(t, x), \quad j = 1, 2, ..., n_s \tag{3.21}$$

depends on the size of the couplings $h_j(t, x)$; then it could furnish a measure of the deterioration of the performance.

In the sequel, we seek to derive conditions on $h_j(t, x)$ to guarantee a prescribed value of the suboptimality ε.

Remark 3.3.2 *It is significant to observe that any procedure to derive the suboptimality index ε would involve only bounds on the norms of the coupling functions $h_j(t, x)$. The obtained results are naturally valid for a class of $h_j(t, x)$ and, thus, do not depend on the actual form of these nonlinear functions. This particular aspect is of major importance in the context of modeling uncertainties and possible variations in the shape of nonlinear couplings during operation. An improvement in the system performance is possible if $||h_j(t, x)||$ can be reduced. It was proposed in [322] to accomplish this improvement by using additional control functions that neutralize the effect of interconnections. These functions are generated by a global controller on a higher level using the states of the subsystems.*

In this regard, the interconnected subsystems are described by

$$\dot{x}_j(t) = g_j(t, x_j, u_{jc}) + h_j(t, x, u_{jg}), \quad j = 1, 2, ..., n_s \tag{3.22}$$

where

$$\begin{aligned} h &= \left[\; h_1^t, \; h_2^t, \cdots, h_{n_s}^t \; \right] \in \Re \times \Re^n \times \Re^m \to \Re^n \\ u_g &= \left[\; u_{1g}^t, \; u_{2g}^t, \cdots, u_{n_s g}^t \; \right]^t \in \Re^m \to \Re^n \end{aligned} \tag{3.23}$$

Define the functions

$$\mathsf{V} : \Re \times \Re^n \to \Re_+, \; p : \Re \times \Re^n \to \Re_+, \; M : \Re \times \Re^n \times \Re^m \to \Re_+$$

such that

$$\begin{aligned} \mathsf{V}(t, x) &= \sum_{j=1}^{n_s} \mathsf{V}_j(t, x_j), \; p(t, x) = \sum_{j=1}^{n_s} p_j(t, x_j), \\ M(t, x, u_c) &= \sum_{j=1}^{n_s} M_j(t, x_j, u_{jc}) \\ K_c &= \left[\; K_{1c}^t, \; K_{2c}^t, \cdots, K_{n_s c}^t \; \right]^t \in \Re \times \Re^n \to \Re^m \end{aligned} \tag{3.24}$$

where $\mathsf{V}_j(t, x_j)$ belongs to the class C^2 in both arguments and satisfies the Hamilton-Jacobi equation

$$\begin{aligned} &\frac{\partial \mathsf{V}_j(t, x_j)}{\partial t} + \nabla \mathsf{V}_j^t(t, x_j) g_j[t, x_j, K_{jc}(t, x_j)] \\ &M_j[t, x_j, K_{jc}(t, x_j)] \;=\; 0, \qquad j = 1, 2, ..., n_s, \;\; \forall (t, x) \in \mathcal{D} \end{aligned} \tag{3.25}$$

The following result is established in [322, 324].

Theorem 3.3.3 *Consider the subsystem (3.21) and let the couplings $h_j(t, x)$ satisfy the constraint*

$$\nabla \mathsf{V}^t(t, x) h(t, x) \;\leq\; \frac{\varepsilon}{1 + \varepsilon} M[t, x, K_c(t, x)], \quad \forall (t, x) \in \mathcal{D} \tag{3.26}$$

Then the overall system (3.21) is suboptimal with index ε.

Remark 3.3.4 *It must be observed that we can view inequality (3.26) as imposing constraints on the couplings $h_j(t, x)$ so as to guarantee a prescribed suboptimality index ε. Another view is that, given a particular pattern of coupling $h_j(t, x)$, inequality (3.26) provides a way to estimate the suboptimality index ε.*

Building on Remark 3.3.4 and given the global control function $u_g(t)$, the following constraint on couplings $h_j(t, x, u_{jg})$ was presented in [322]:

$$||\nabla \mathsf{V}(t, x)|| \, ||h(t, x, u_g)|| \;\leq\; \frac{\varepsilon}{1+\varepsilon} \, M[t, x, K_c(t, x)], \quad \forall (t, x) \in \mathcal{D} \tag{3.27}$$

It is obvious that inequality (3.27) implies inequality (3.26). Interestingly enough, provided that expressions for the function V and M are available, inequality (3.27) can then be used to get an explicit relationship between the index ε and the norm $||h(t, x, u_g)||$. In turn, this can be used in the course of designing $u_g(t)$. One strong candidate for this is the case of local linear quadratic regulators (where local free subsystems are linear and the optimization cost is quadratic). We will elaborate more on this case in later sections.

3.4 Decentralized Nonlinear Systems

In this section, we take another direction and look at the control of multi-input multi-output (MIMO) systems where constraints are placed on the information flow. In particular we consider the control of decentralized nonlinear systems (DNS) where there are constraints on information exchange among subsystems. In this context, decentralized control systems often arise from either the physical inability of subsystem information exchange or the lack of computing capabilities required by a single central controller. In addition, it is proved more convenient to design a controller with decentralized structure than a controller for a composite MIMO system.

Within a decentralized framework, the overall system is broken into n_s subsystems each with its own inputs and outputs. A decentralized control law is then defined using local subsystem signals. Thus subsystem j does not have access to the signals associated with subsystem m, $m \neq j$. Figure 3.1 shows a decentralized system with four subsystems, each influenced by couplings to one or more of the other subsystems, where the notation S_j represents the jth subsystem and c_{jm} is a coupling that defines how the jth subsystem influences the mth subsystem. In this section, we consider a standard MIMO system described by

$$\begin{aligned} \dot{\eta}(t) &= \mathbf{f}_\eta(\eta) + g_\eta u \\ y &= h_\eta(\eta) \end{aligned} \tag{3.28}$$

where $\eta(t) \in \Re^n$ is the state vector, $u(t) \in \Re^m$ is the control input, and $y(t) \in \Re^m$ is the output at time $t \in \Re$. For the purpose of simplicity in exposition, we let $\eta := [\eta_1^t, \ \eta_2^t, \ \cdots, \ \eta_{n_s}^t]^t$ where n_s signifies the number of

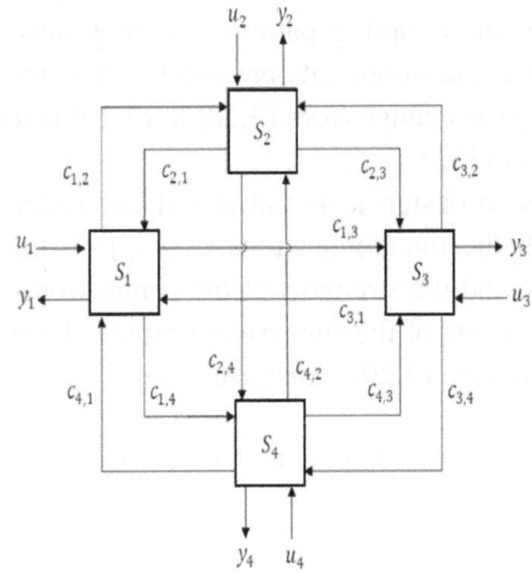

Figure 3.1: A decentralized system structure

subsystems. We will assume that the decomposition of (3.28) into subsystems is achieved by some means. When communication constraints are incorporated into the design of a control system, the subsystems may be easily defined. Alternatively, choosing a decentralized design approach for computational or other design considerations, the definition of a subsystem may often be determined due to physical grounds such as strong input-output pairing relationships. To illustrate this, consider two bodies are being connected by a spring. Removing the spring, the subsystems would then be independent from one another. Thus, the spring acts as an interconnection between the subsystems. Hereafter, we will assume that the definition of the subsystems is obvious from the problem statement.

To identify the subsystems, we let $T_j : \Re^{n_j} \rightarrow \Re^{n_j},\ x_j = T_j(\eta_j)$ be an appropriate transformation such that the subsystem $\mathbf{S}_j$ has dynamics defined by:

$$\mathbf{S}_j : \quad \begin{aligned} \dot{x}_j(t) &= f_j(x_j) + g_j u_j + \mathbf{c}_j(t, x) \\ y_j &= h_j(x_j) \end{aligned} \tag{3.29}$$

where for $j = 1,\ ...,\ n_s,\ x_j(t)\ \in \Re^n$ is the subsystem state vector, $u_j(t)\ \in$

$\Re^m$ is the subsystem control input, and $y(t) \in \Re^m$ is the subsystem output at time $t \in \Re$. Also, we let $x := [x_1^t,\ x_2^t,\ \cdots,\ x_{n_s}^t]^t$. The vector function $\mathbf{c}_j(t,x)$ accounts for the coupling pattern affecting subsystem T_j. For well-posedeness, only local measurements are used in the transformation T_j. The procedure underlying the change from (3.28) to (3.29) belongs to the family of feedback linearization [152].

The goal now is to design a decentralized controller for systems of the type (3.29) in which the interconnections $\mathbf{c}_j(t,x)$ do not necessarily satisfy matching conditions, and the strengths of the connections may be bounded by arbitrary smooth functions of the subsystem outputs. To simplify the notation, we consider the subsystem (3.29) transformed into

$$\mathbf{S}_j : \quad \begin{aligned} \dot{x}_{j,1}(t) &= f_{j,1}(\tilde{x}_{j,1}) + \tilde{x}_{j,1} + c_{j,1}(t,x) \\ &\vdots \\ \dot{x}_{j,n_s-1}(t) &= f_{j,n_s-1}(\tilde{x}_{j,n_s-1}) + \tilde{x}_{j,n_s} + c_{j,n_s-1}(t,x) \\ \dot{x}_{j,n_s}(t) &= f_{j,n_s-1}(\tilde{x}_{j,n_s-1}) + u_j + c_{j,n_s-1}(t,x) \\ y_j &= x_{j,1} \end{aligned} \tag{3.30}$$

with $\tilde{x}_{j,k} = [x_{j,1},\ ...,\ x_{j,k}]^t$ and $c_{j,m}$ represents the coupling among subsystems. Our immediate task is to design a controller which forces $y_j \to 0,\ j = 1,\ ...,\ n_s$ when the couplings $c_{j,m}$ satisfy certain bounding conditions which will be defined shortly. We observe here that each $c_{j,m}$ is influenced by subsystems $\mathbf{S}_1, ..., \mathbf{S}_{n_s}$. In line of the development of the previous section, we will consider that the couplings are modeled by

$$\mathsf{C}_j : \quad \begin{aligned} \dot{\zeta}_j(t) &= q_j(t,x) \\ \mathbf{c}_j(t,x) &= s_j(t,\zeta_j,x) \end{aligned} \tag{3.31}$$

where $\zeta_j(t) \in \Re^{m_j}$ and $\mathbf{c}_j(t,x) \in \Re^{n_j}$. Notice that C_j combines the effects of the individual couplings $\mathsf{C}_{j,1},\ \mathsf{C}_{j,n_s}$. It is obvious that model (3.31) includes the case of static couplings ($\mathbf{c}_j(t,x) = s_j(t,x)$) and complete decoupling ($\mathbf{c}_j(t,x) \equiv 0$) corresponding to isolated subsystems.

3.4.1 Diagonal dominance condition

In this section, the immediate task is to develop a static controller for decentralized systems of the type (3.30). Our tool is Lyapunov-based design (LBD) to help in placing bounds on states which may be used as inputs to a finite approximator used within the control law. An integral portion of this design is finding robust controllers when the interconnections are ignored (case of isolated subsystems). By making each controller "very robust," the effects of the couplings are dominated by the inherently robust local controllers. In this regard, a powerful tool often used in the design of decentralized controllers is the concept of diagonal dominance for which the following theorem provides one possible application.

Theorem 3.4.1 *Given matrices* $W \in \Re^{m\times n}$, $S \in \Re^{n\times n}$ *where* $W = [w_{j,k}]$ *and* $S = diag(s_1, \ ..., \ s_n)$. *The inequality*

$$x^t(W + S)x \ \geq 0$$

holds for all x if

$$s_j \geq n(1 + \sum_{k}^{n} \ w_{k,j}^t)$$

defined along the columns of W *or*

$$s_k \geq n(1 + \sum_{k}^{n} \ w_{k,j}^t)$$

defined along the rows of W.

The proof of **Theorem 3.4.1** can be found in [344].

Remark 3.4.2 *The relevance of* ***Theorem 3.4.1*** *stems from its basic role to determine how to accommodate the couplings while stabilizing each subsystem individually. That is, the couplings do not cause system instability. It should be noted that* ***Theorem 3.4.1*** *only provides sufficient, and not necessary, conditions on the choice of the diagonal terms. This is in fact the case of almost all decentralized techniques. It simply implies that the results of* ***Theorem 3.4.1*** *may be rather conservative for certain applications.*

3.4.2 Static controller design

A decentralized control design is now developed for the family of subsystems (3.30) under the coupling dynamics governed by

$$\mathsf{C}_j : \quad \begin{aligned} \dot{\zeta}_j(t) \ &= \ q_j(t, \zeta_j, x) \\ \mathbf{c}_{j,k} \ &= \ s_{j,k}(t, \zeta_j \tilde{x}_{j,k}, y) \end{aligned} \qquad (3.32)$$

where the coupling is bounded by

$$|\mathbf{c}_{j,k}| = \varrho + \xi_{j,k}(\tilde{x}_{j,k}) \sum_{m=1}^{n_s} \psi_{j,m}(|y_m|) \tag{3.33}$$

where $\xi_{j,k} : \Re^k \rightarrow \Re^+$ and $\psi_{j,m} : \Re \rightarrow \Re^+$ are smooth nonnegative functions which are assumed bounded for all bounded inputs and $\varrho \in \Re$. It must be asserted that the bound on $\mathbf{c}_{j,k}$ is defined in terms of other subsystem outputs through the elements $\psi_{j,m}$. Additionally, it is required that the C_j-dynamics are input-to-state stable [152] so that there exists some $V_j(t, \zeta_j)$ such that

$$\begin{aligned} \kappa_{1j}(\zeta_j) \ \leq \ V_j(t, \zeta_j) \ \leq \ \kappa_{2j}(\zeta_j) \\ \dot{V}_j(t, \zeta_j) \ \leq \ -\kappa_{3j}(\zeta_j) + \theta(x) \end{aligned} \tag{3.34}$$

where $\kappa_{1j}(\zeta_j),\ \kappa_{2j}(\zeta_j)$, and $\kappa_{3j}(\zeta_j)$ are class-$\mathcal{K}_\infty$. Observe that if the controlled state x is bounded so is the coupling vector ζ_j. It follows from [152] that given a nonnegative continuous function $\psi_{j,m} : \Re \rightarrow \Re^+$ there exists a nonnegative continuous function $\pi_{j,m} : \Re \rightarrow \Re^+$ such that

$$\psi(|x|) \ \leq \ |x|^n \ \pi(|x|) \ + \ e \tag{3.35}$$

with $n > 0,\ \ e \geq 0$ are finite constants. Proceeding further, we assume that there exist known constants $e_{j,k}$ and smooth functions $\pi_{j,k}(|y_k|)$ such that

$$\psi_{j,k}(|y_k|) \ \leq \ \pi_{j,k}(|y_k|)\sqrt{|y_k|} \ + \ e_{j,k}, \quad \forall \ y_k \in \Re \tag{3.36}$$

Our objective now is to design a decentralized controller to render $y_j \rightarrow 0,\ j = 1,\ ...,\ n_s$. At this stage, we introduce the subsystem error

$$\varepsilon_{j,k} \ = \ x_{j,k} \ - \ \sigma_{j,k-1}(\tilde{x}_{j,k-1}), \ \ j = 1,\ ...,\ n_s,\ k = 1,\ ...,\ n_j \tag{3.37}$$

where the term $\sigma_{j,k}$ will be defined shortly with $\sigma_{j,0} = 0$. On taking the derivative of (3.37), we get:

$$\begin{aligned} \dot{\varepsilon} \ &= \ f_{j,k} \ + \ x_{j,k+1} \ + \ c_{j,k} \\ &- \ \sum_{m=1}^{k-1} \frac{\partial \sigma_{j,k-1}}{\partial x_{j,m}} [f_{j,m} + x_{j,m+1} + c_{j,m}] \\ &= \ f_{j,k} \ + \ \varepsilon_{j,k+1} \ + \ \sigma_{j,k} \ - \ \sum_{m=1}^{k-1} \frac{\partial \sigma_{j,k-1}}{\partial x_{j,m}} [f_{j,m} + x_{j,m+1}] \\ &+ \ \ c_{j,m} \ - \ \sum_{m=1}^{k-1} \frac{\partial \sigma_{j,k-1}}{\partial x_{j,m}} c_{j,m}, \ \ j = 1,\ ...,\ n_s,\ k = 1,\ ...,\ n_j \end{aligned} \tag{3.38}$$

Now letting

$$\begin{aligned}\sigma_{j,k} &= -(\delta_j + \nu_{j,k}(\tilde{x}_{j,k}))\varepsilon_{j,k} - \varepsilon_{j,k-1} \\ &- f_{j,k} + \sum_{m=1}^{k-1} \frac{\partial \sigma_{j,k-1}}{\partial x_{j,m}}[f_{j,m} + x_{j,m+1}]\end{aligned} \tag{3.39}$$

we get

$$\begin{aligned}\dot{\varepsilon}_{j,k} &= -(\delta_j + \nu_{j,k})\varepsilon_{j,k} - \varepsilon_{j,k-1} + \varepsilon_{j,k+1} + c_{j,m} \\ &- \sum_{m=1}^{k-1} \frac{\partial \sigma_{j,k-1}}{\partial x_{j,m}} c_{j,m}\end{aligned} \tag{3.40}$$

for $k = 1, \ldots, n_j - 1$ where $\varepsilon_{j,0} = 0$ and the terms $\nu_{j,k}$ will be decided upon to accommodate the couplings. On defining the feedback control law as

$$\begin{aligned}u_j &= -(\delta_j + \nu_{j,k}(\tilde{x}_{j,k}))\varepsilon_{j,k} - \varepsilon_{j,k-1} \\ &- f_{j,n_j} + \sum_{m=1}^{n_j-1} \frac{\partial \sigma_{j,n_s-1}}{\partial x_{j,m}}[f_{j,m} + x_{j,m+1}]\end{aligned} \tag{3.41}$$

we obtain

$$\begin{aligned}\dot{\varepsilon}_{j,n_s} &= -(\delta_j + \nu_{j,k})\varepsilon_{j,n_s} - \varepsilon_{j,n_s-1} + c_{j,n_s} \\ &- \sum_{m=1}^{n_s-1} \frac{\partial \sigma_{j,n_s-1}}{\partial x_{j,m}} c_{j,m}\end{aligned} \tag{3.42}$$

In terms of $\varepsilon = \varepsilon_{j,1}, \ldots, \varepsilon_{j,n_s}]^t$, we express the subsystem error dynamics into the form

$$\begin{aligned}\dot{\varepsilon} &= \mathcal{A}_j\varepsilon_j + \mathcal{D}_j \\ \mathcal{A}_j &= \begin{bmatrix} -\delta_j & 1 & 0 & \cdots & 0 \\ -1 & -\delta_j & 1 & & \vdots \\ 0 & -1 & -\delta_j & & \\ \vdots & & & \ddots & 1 \\ 0 & \cdots & & -1 & -\delta_j \end{bmatrix}, \\ \mathcal{D}_j &= \begin{bmatrix} -\nu_{j,k}\varepsilon_{j,1} + c_{j,1} \\ \vdots \\ -\nu_{j,n_s}\varepsilon_{j,n_s} + c_{j,n_s} - \sum_{m=1}^{n_s-1} \frac{\partial \sigma_{j,n_s-1}}{\partial x_{j,m}} c_{j,m} \end{bmatrix}\end{aligned} \tag{3.43}$$

The following theorem summarizes the resulting closed-loop stability properties of the proposed static controller.

Theorem 3.4.3 *Given the subsystem (3.30) with interconnections bounded by (3.33), the decentralized control law (3.41) with*

$$\begin{aligned}
\nu_{j,k} &= \mu_j z_{j,k}^2 + \zeta_j(y_j), \ \mu_j > 0 \\
z_{j,k} &= \varrho\Big(1 + \sum_{m=1}^{k-1} |\frac{\partial \sigma_{j,k-1}}{\partial x_{j,m}}|\Big) + \Big(\xi_{j,k} + \sum_{m=1}^{k-1} \xi_{j,m} |\frac{\partial \sigma_{j,k-1}}{\partial x_{j,m}}| \sum_{r=1}^{n_s} e_{j,r}\Big) \\
&+ \frac{n_s}{4\sqrt{c}}\Big(\xi_{j,k} + \sum_{m=1}^{k-1} \xi_{j,m} |\frac{\partial \sigma_{j,k-1}}{\partial x_{j,m}}|\Big)^2, \quad c > 0
\end{aligned}$$

will ensure that the subsystem error ε_j (and thus each subsystem output y_j) is uniformly ultimately bounded.

The proof of **Theorem 3.4.3** can be found in [344] based on some algebraic bounding inequalities.

While the results attained guarantee decentralized controlled subsystems, the price is a residual error of the subsystem output trajectories in the form of a ball around the origin.

Remark 3.4.4 *It is crucial at this stage to mark our position. In the previous two sections, we focused on decentralized control of nonlinear systems and showed how difficult the design problem is. We have seen that the chief effort has to be concentrated on solving a nonlinear Hamilton-Jacobi equation or estimating bounds on the nonlinearities. To reach at practical results, we will relax the situation a little bit and start by dealing with a class of dynamical interconnected systems, which we call "nominally linear uncertain systems." Then we end by "linear uncertain systems." We added uncertainties to obtain robust design results for the continuous-time and discrete-time cases.*

3.4.3 Illustrative example 3.1

Consider the following large-scale system:

$$A = \begin{bmatrix} 0.6 & 0 & 0 & 1 & 0.2 & 0 \\ 0.2 & 2.3 & 0.1 & 0 & 0 & 0.3 \\ 0 & 0 & -1.5 & 0 & 0 & 0 \\ 0 & 0 & 0.7 & 1.1 & 0 & 0 \\ 0 & 0 & 0 & 0 & -0.8 & 0 \\ 0 & 0 & 0 & 0 & 0.5 & 0.8 \end{bmatrix}$$

$$
\begin{aligned}
G_o &= [0.1\ 0.2\ 0.4\ 0.3],\ G_{do} = [0.01\ 0\ 0.01\ 0] \\
B_o &= \begin{bmatrix} 0 \\ 0.3 \end{bmatrix},\ C_o = \begin{bmatrix} 1 & 10 & 0 & 0 \\ 0 & 0 & 10 & 10 \end{bmatrix},\ C_d = \begin{bmatrix} 2 & 1 & 0 & 0 \\ 0 & 0 & 1 & 3 \end{bmatrix}, \\
D_o &= \begin{bmatrix} 0.4 & 0.2 \end{bmatrix},\ \Psi_o = \begin{bmatrix} 0.01 & 0.01 \end{bmatrix},\ \Phi = 0.1
\end{aligned}
$$

Using the MATLAB®-LMI solver, the feasible solutions of the respective design methods yield the feedback gain matrices:

$$
\begin{aligned}
K_s &= \begin{bmatrix} -34.5094 & -4.9061 & -52.4436 & -130.5408 \\ 17.5392 & -12.7566 & 79.5070 & 157.7008 \end{bmatrix} \\
K_d &= \begin{bmatrix} -34.5094 & -4.9061 & -52.4436 & -130.5408 \\ 17.5392 & -12.7566 & 79.5070 & 157.7008 \end{bmatrix} \\
K_o &= \begin{bmatrix} -55.2134 & 15.8031 \\ 69.2668 & -18.6633 \\ -18.4083 & 5.0529 \\ 5.2346 & -0.5445 \end{bmatrix}
\end{aligned}
$$

3.5 Nominally Linear Systems—Continuous Case

Consider a class of continuous nominally linear uncertain systems **S** composed of n_s subsystems where the jth subsystem is depicted in Figure 3.2 and described by

$$
\begin{aligned}
\dot{x}_j(t) &= [A_j + \Delta A_j(r_j(t))]x_j(t) + [B_j + \Delta B_j(s_j(t))]u_j(t) + C_j v_j(t) \\
&+ Z_j + h_j(t),\ \ j = 1,\ 2,\ ...,\ n_s,\ x_j(0) = x_{jo} \qquad (3.44) \\
Z_j(t) &= \sum_{k \neq j}^{n_s} A_{jk} x_k(t),\ \ h_j(t) = \sum_{k \neq j}^{n_s} H_{jk}(t, r_j(t), x_k(t)) \\
y_j(t) &= x_j(t) + w_j(t) \qquad (3.45)
\end{aligned}
$$

where for $j \in \{1, ..., n_s\}$, $x_j(t) \in \Re^{n_j}$ is the state vector, $u_j(t) \in \Re^{m_j}$ is the control input, $y_j(t) \in \Re^{n_j}$ is the measured output, $v_j(t) \in \Re^{s_j}$ is the bounded disturbance, $w_j(t) \in \Re^{n_j}$ is the measurement error, and $Z_j(t) \in \Re^{n_j}$ is the interconnection vector with other subsystems. The matrices $A_j \in \Re^{n_j \times n_j}$, $B_j \in \Re^{n_j \times m_j}$, $C_j \in \Re^{n_j \times s_j}$, $A_{jk} \in \Re^{n_j \times n_k}$ are real and constants. The uncertain matrices $\Delta A_j(r_j(t))$ and $\Delta B_j(s_j(t))$ depend continuously on the Lebesgue-measurable parameters $r_j \in \Im_j \subset \Re^{p_j}$ and $s_j \in \wp_j \subset \Re^{q_j}$, respectively, where $\Im_j$ and $\wp_j$ are appropriate compact bounding sets. Also, the term $C_j v_j(t)$ accounts for external input uncertainty,

$Z_j(t) \in \Re^{n_j}$ is the coupling vector with other subsystems, and $h_j(t) \in \Re^{n_j}$ contains the uncertainties in the interconnections between the jth subsystem and the other subsystems. This system is depicted in Figure 3.2.

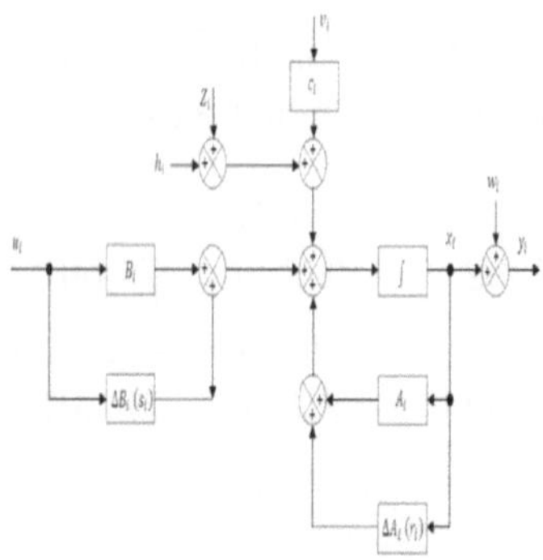

Figure 3.2: Additive-type uncertain subsystem

Assumption 3.1: *The pair* (A_j, B_j) *is stabilizable for all* $j \in \{1, ..., n_s\}$
Assumption 3.2:

1. *There exist matrix functions* $D_j(.)$ *and* $E_j(.)$ *of appropriate dimensions whose entries are continuous on* $\Re^{p_j}$ *and* $\Re^{q_j}$*, respectively, such that*

$$\Delta A_j(r_j) = B_j D_j(r_j), \quad \Delta B_j(s_j) = B_j E_j(s_j) \tag{3.46}$$

2. *There exists a constant matrix function* F_j *such that*

$$C_j = B_j F_j \tag{3.47}$$

3. $\varrho_{sj} = \max_{s_j \in \wp_j} ||E_j(s_j)|| < 1$

Assumption 3.3: *The interactions* $H_{jk}(t, r_j(t), x_k(t))$ *are assumed to be carathedory functions (see the Appendix). Furthermore, they are bounded in the form*

$$\begin{aligned} ||H_{jk}(t, r_j(t), x_k(t))|| &\leq \sum_{k \neq j}^{n_s} ||\Delta A_{jk}(t, r_j)|| ||x_k|| \\ &\leq \sum_{k \neq j}^{n_s} \gamma_{jk} ||x_k||, \quad j = 1, ..., n_s \\ &\forall \quad (t, r_j(t), x_k(t)) \in \Re \times \Re^{p_j} \times \Re^{n_j} \end{aligned} \tag{3.48}$$

where $\gamma_{jk} \geq 0,\ j = 1, ..., n_s,\ k = 1, ..., n_s$ are n_s^2 scalars representing upper bounds for the uncertainties among the subsystems. The overall system under consideration can be written as

$$\begin{aligned}
\dot{x}(t) &= [A + \Delta A(r) + M]x(t) + [B + \Delta B(s)]u(t) \\
&+ Cv + H(t, x, r), \quad x(0) = x_o \qquad (3.49) \\
y(t) &= x(t) + w(t) \qquad (3.50)
\end{aligned}$$

where

$$x = [x_1^t,\ x_2^t,\ ...,\ x_{n_s}^t]^t \in \Re^n, \quad u = [u_1^t,\ u_2^t,\ ...,\ u_{n_s}^t]^t \in \Re^m$$

$$y = [y_1^t,\ y_2^t,\ ...,\ y_{n_s}^t]^t \in \Re^n, \quad v = [v_1^t,\ v_2^t,\ ...,\ v_{n_s}^t]^t \in \Re^s$$

$$w = [w_1^t,\ w_2^t,\ ...,\ w_{n_s}^t]^t \in \Re^n$$

are the state, control, measured state, disturbance, and measurement error of the overall system, respectively, with

$$n = \sum_{k=1}^{n_s} n_k, \quad m = \sum_{k=1}^{n_s} m_k, \quad s = \sum_{k=1}^{n_s} s_k$$

The vectors M and H denote the couplings and the associated uncertainties within the overall system. Furthermore, we have

$$\begin{aligned}
A &= diag(A_j), \quad B = diag(B_j), \quad C = diag(C_j), \\
\Delta A(.) &= diag(\Delta A_j(.)), \quad \Delta B(.) = diag(\Delta B_j(.))
\end{aligned}$$

In view of (3.44) and (3.49), it should be observed that both M and H have zero diagonal elements, that is, ($M_{jj} = 0,\ H_{jj} = 0,\ j = 1, ..., n_s$).

Remark 3.5.1 *In view of the results of Chapter 2, it is emphasized that if the family of subsystems (3.44)-(3.45) has unstable fixed modes, we resort to the technique of [370] to ensure that the subsystems (3.44)-(3.45) are amenable for decentralized control.*

In the absence of unstable fixed modes, our objective is to design either a decentralized or two-level control strategies to stabilize the class of nominally linear uncertain systems **S** where the perturbations are subject to **Assumption 3.1** and **Assumption 3.2**.

3.5.1 Decentralized control

It follows from **Assumption 3.1** that there exists a constant matrix $G_j \in \Re^{m_j \times n_j}$, such that the eigenvalues of $(A_j + B_j G_j)$ or $\lambda_j(A_j + B_j G_j)$ are strictly in the complex left half plane. Consider the decentralized feedback control

$$u_j(t) = G_j y_j(t) + g_j(t, y_j(t)) \tag{3.51}$$

$$g_j(t, y_j(t)) = \begin{cases} \frac{-B_j^t \mathcal{P}_j y_j}{\|B_j^t \mathcal{P}_j y_j\|} \sigma_j(y_j) & for\ \|B_j^t \mathcal{P}_j y_j\| > \epsilon_j \\ \\ \frac{-B_j^t \mathcal{P}_j y_j}{\epsilon_j} \sigma_j(y_j) & for\ \|B_j^t \mathcal{P}_j y_j\| \le \epsilon_j \end{cases} \tag{3.52}$$

where $\epsilon_j > 0$ is a prespecified parameter and $\mathcal{P}_j$ is a solution of the Lyapunov equation

$$\mathcal{P}_j(A_j + B_j G_j) + (A_j + B_j G_j)^t \mathcal{P}_j = -\mathcal{Q}_j, \ \ 0 < \mathcal{Q}_j = \mathcal{Q}_j^t \tag{3.53}$$

and $\sigma_j(y_j) : \Re^{n_j} \to \Re^+$ is a nonnegative function chosen to satisfy

$$\begin{aligned} \sigma_j(y_j) &= \Big[1 - \varrho_{sj}\Big]^{-1} \Big\{ \max_{r_j \in \Im_j} \|D_j(r_j) y_j\| + \max_{v_j(t) \in \Re^{s_j}} \|F_j v_j\| \\ &+ \max_{w_j(t) \in \Re^{n_j}} \|G_j w_j\| + \max_{s_j \in \wp_j} \|E_j(s_j) G_j y_j\| \\ &+ \max_{r_j \in \mathbf{Im}_j,\, w_j(t) \in \Re^{n_j}} \|D_j(r_j) w_j\| \Big\} \end{aligned} \tag{3.54}$$

where the indicated inverse exists in view of **Assumption 3.2**. For simplicity in exposition, we introduce the following norm quantities:

$$\begin{aligned} \varrho_{rj} &= \|D_j(r_j)\|, \ \ \varrho_{rj} = \max_{v_j(t) \in \Re^{s_j}} \|F_j v_j\|, \ \ \varrho_{wj} = \|w_j\| \\ \varrho_{gj} &= \|G_j\|, \ \ \varrho_{sgj} = \max_{s_j \in \wp_j} \|E_j(s_j) G_j\| \end{aligned} \tag{3.55}$$

It is evident by simple algebraic manipulations from (3.54) and (3.55) that the functions $\sigma_j(y_j)$, $j = 1, ..., n_s$ are cone bounded in the sense that

$$\|\sigma_j(y_j)\| \ \le \ \alpha_j + \beta_j \|y_j\| \tag{3.56}$$

Remark 3.5.2 *It is important to observe that the controller (3.52) is of switching type, dependent on the local measured output* y_j *and is independent of the interactions. In view of (3.54),* $\sigma_j(y_j)$ *is formulated based only on the possible bound of local uncertainties. This paves the way to designing controllers based on completely decentralized information.*

Remark 3.5.3 *It should be emphasized that the cone-boundedness form (3.56) is always satisfied because in view of (3.45), (3.55), and (3.56), we can deduce that*

$$\begin{aligned} ||\sigma_j(y_j)|| &\leq ||\sigma_j(x_j)|| := \hat{\alpha}_j + \hat{\beta}_j \, ||x_j|| \\ \hat{\alpha}_j &:= [(2\varrho_{rj} + \varrho_{gj} + \varrho_{sgj})\varrho_{wj} + \varrho_{vj}]/(1 - \varrho_{sj}) \\ \hat{\beta}_j &:= (\varrho_{rj} + \varrho_{sgj})/(1 - \varrho_{sj}) \end{aligned} \tag{3.57}$$

Based on the foregoing remarks and recalling the notion of *practically stabilizability* from the Appendix, the following decentralized stabilization results can now be developed.

Theorem 3.5.4 *The overall system (3.50) satisfying* **Assumption 3.1**-**Assumption 3.3** *is practically stabilizable by the decentralized control laws (3.51)-(3.52) if the test matrix $L = [\ell_{jk}]$ given by*

$$\ell_{jk} = \begin{cases} \lambda_m(\mathcal{Q}_j) & for\ j = k \\ -\lambda_M(\mathcal{P}_j)[||A_{jk}|| + ||A_{kj}|| + 2\sum_{j=1}^{n_s}\sum_{k=1}^{n_s}\gamma_{jk}] & for\ j \neq k \end{cases} \tag{3.58}$$

is positive-definite.

The proof of **Theorem 3.5.4** can be found in [10] based on constructive use of Lyapunov analysis.

3.5.2 Two-level control

We have seen so far that the overall uncertain system can be stabilized using decentralized controllers which do not need any exchange of state information among the subsystems. This, in turn, is economical from the point of view of the control realization. A possible side effect is that a completely decentralized control methodology implies that the dynamic behavior of interconnections are completely neglected thereby yielding a loss of information. In the sequel, we investigate the stabilization problem via a multilevel (hierarchical) technique [136, 236]. For a comprehensive overview about the hierarchical control theory and applications, the readers are referred to [192, 194]. The principal idea of a two-level control structure is to define a set of subproblems that can be considered independent at a certain level (subsystem level). Through the manipulation of the interplaying effect at a higher level (coordinator), we obtain the global solution. Accordingly, the decentralized control (3.51) is modified to be in the hierarchical form

$$u(t) = G_d y(t) + g(t, y(t)) \tag{3.59}$$

where $G_d = Blockdiagonal(G_j)$, and G_j is a stabilizing gain matrix such that the eigenvalues of $(A_j + B_j G_j)$ or $\lambda_j(A_j + B_j G_j)$ are strictly in the complex left half plane and $g(.,.) : \Re^n \rightarrow \Re^m$ given by

$$g(t, y(t)) = \begin{cases} \sigma(y)\frac{-B^t \mathcal{P} y}{||B^t \mathcal{P} y||} & for \ ||B^t \mathcal{P} y|| > \epsilon \\ \\ -\sigma(y)\frac{B^t \mathcal{P} y}{\epsilon} & for \ ||B^t \mathcal{P} y|| \leq \epsilon \end{cases} \quad (3.60)$$

where $\epsilon > 0$ is a prespecified parameter, $\mathcal{P} = diag(\mathcal{P})_j$ is the positive-definite solution of (3.53), and $\sigma(y) : \Re^n \rightarrow \Re^+$ is a nonnegative function chosen to satisfy

$$\begin{aligned} \sigma(y) &= \max_j \left\{ \left[1 - \varrho_{sj} \right]^{-1} \left[\max_{r_j \in \Im_j} ||D_j(r_j) y_j|| + \max_{v_j(t) \in \Re^{s_j}} ||F_j v_j|| \right. \right. \\ &+ \max_{w_j(t) \in \Re^{n_j}} ||G_j w_j|| + \max_{s_j \in \wp_j} ||E_j(s_j) G_j y_j|| \\ &+ \left. \left. \max_{r_j \in \mathbf{Im}_j,\, w_j(t) \in \Re^{n_j}} ||D_j(r_j) w_j|| \right] \right\},\ j = 1, .., n_s \end{aligned} \quad (3.61)$$

where the indicated inverse exists in view of **Assumption 3.2**. In the same manner, it is easy to see that

$$\begin{aligned} ||\sigma(y)|| &\leq ||\sigma(x)|| := \tilde{\alpha} + \tilde{\beta}\, ||x|| \\ \tilde{\alpha} &:= \max_j [(2\varrho_{rj} + \varrho_{gj} + \varrho_{sgj})\varrho_{wj} + \varrho_{vj}]/(1 - \varrho_{sj}) \\ \tilde{\beta} &:= \max_j (\varrho_{rj} + \varrho_{sgj})/(1 - \varrho_{sj}),\ j = 1, 2, ..., n_s \end{aligned} \quad (3.62)$$

Following the development of **Theorem 3.5.4**, we are ready to provide the following result.

Theorem 3.5.5 *The overall system (3.50) satisfying* **Assumption 3.1-Assumption 3.3** *is practically stabilizable by the decentralized control laws (3.59)-(3.60) if the test matrix $L = [\ell_{jk}]$ given by (3.58) is positive-definite. Furthermore, the resulting closed-loop state trajectories are bounded in a set $\Omega^c(\tilde{\eta})$, where $\Omega^c(\tilde{\eta})$ is the complement of set $\Omega(\tilde{\eta})$, having radius $\tilde{\eta}$ given by*

$$\begin{aligned} \tilde{\eta} &= \left[\tilde{\mu}_a + \sqrt{(\tilde{\mu}_a^2 + 4\tilde{\mu}_c \lambda_m(L))} \right] \\ \tilde{\mu}_a &= \tilde{\alpha}(\frac{\epsilon}{2} + 4||B^t \mathcal{P}||\varrho_w) \\ \tilde{\mu}_c &= \tilde{\beta}(\frac{\epsilon}{2} + 4||B^t \mathcal{P}||\varrho_w) \\ \varrho_w &= \max_j \{\varrho_{wj}\},\quad j = 1, .., n_s \end{aligned} \quad (3.63)$$

The proof of **Theorem 3.5.5** can be found in [10] based on constructive use of Lyapunov analysis.

Remark 3.5.6 *We emphasize that the hierarchical control structure outlined in **Theorem 3.5.5** has the following features:*

1. *The second term of the present control (the coordinator gain) is nonlinear and depends on the initial conditions.*

2. *Except for the term $\sigma(y)$, all the computations are on the subsystem level. In light of (3.62), however, the computation of $\sigma(y)$ is performed separately for each subsystem.*

3. *The exchange of information, in the present control implementation, between the coordinator and the subsystems is minimal, and hence the control is easy to realize, especially for large-scale geographically distributed systems.*

Remark 3.5.7 *In the absence of state uncertainty, that is, $w_j \equiv 0$, and if very large feedback gains are permitted ($\epsilon_j \rightarrow 0$), then the nonlinear (switching) term of the controller (3.51) will render the subsystem asymptotically stable. Same result holds when using the hierarchical controller.*

Remark 3.5.8 *The implementation of hierarchical control necessitates the use of many communication links, to facilitate transmitting pertinent information and corrective signals. However, in many situations, physical perturbations took place which tended to modify the data transmission network and caused information losses. This may affect the stability of the global system and frequently undergoes instability. Extensions of **Theorems 3.5.4** and **3.5.5** to deal with the effects of structural perturbations have been developed in [10] and will be demonstrated in the numerical example to follow.*

3.5.3 Illustrative example 3.2

We consider a gas-absorber system as detailed in [193]. For numerical simulation we take the case of six plates that we consider to be split into two subsystems; each is composed of three plates (each subsystem is a 3-plate gas absorber). The definitions of the various elements of the model are presented in Figure 3.3, where the variables x_3 and x_4 represent the interconnection between the two subsystems. Using typical data values, the matrices of the model (3.44)-(3.45) are given by

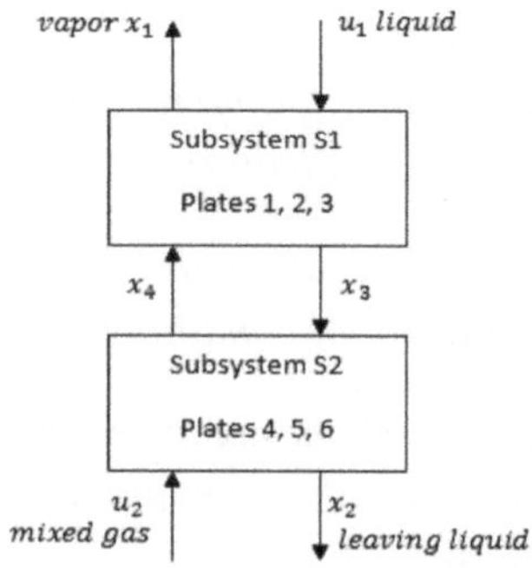

Figure 3.3: Six-plate gas absorber representation

$$A = \begin{bmatrix} -1.173 & 0.634 & 0 & 0 & 0 & 0 \\ 0.538 & -1.173 & 0.634 & 0 & 0 & 0.3 \\ 0 & 0.538 & -1.173 & 0.634 & 0 & 0 \\ 0 & 0 & 0.538 & -1.173 & 0.634 & 0 \\ 0 & 0 & 0 & 0.538 & -1.173 & 0.634 \\ 0 & 0 & 0 & 0 & 0.538 & -1.173 \end{bmatrix}$$

$$\begin{aligned} B &= \begin{bmatrix} 0.538 & 0 \\ 0 & 0 \\ 0 & 0 \\ 0 & 0 \\ 0 & 0 \\ 0 & 0.88 \end{bmatrix}, \ \Delta A(r) = \begin{bmatrix} \Delta A_1(r) & 0 \\ 0 & \Delta A_2(r) \end{bmatrix} \\ \Delta B(s) &= \begin{bmatrix} \Delta B_1(s_1) & 0 \\ 0 & \Delta B_2(s_2) \end{bmatrix}, \\ H(t,r,x) &= \begin{bmatrix} 0 & \Delta A_{12}(r) \\ \Delta A_{12}(r) & 0 \end{bmatrix} \\ \Delta A_1(r) &= \Delta A_2(r) = \begin{bmatrix} r_1 & r_2 & 0 \\ r_3 & r_1 & r_2 \\ 0 & r_3 & r_1 \end{bmatrix} \\ \Delta B_1(s_1) &= \begin{bmatrix} s_1 \\ 0 \\ 0 \end{bmatrix}, \ \Delta B_2(s_2) = \begin{bmatrix} 0 \\ 0 \\ s_2 \end{bmatrix} \end{aligned}$$

$$\Delta A_{12}(r) = \begin{bmatrix} 0 & 0 & 0 \\ 0 & 0 & 0 \\ r_2 & 0 & 0 \end{bmatrix}, \ \Delta A_{21}(r) = \begin{bmatrix} 0 & 0 & r_3 \\ 0 & 0 & 0 \\ 0 & 0 & 0 \end{bmatrix}$$

where the compact bounding sets $\Im_j,\ \wp_j,\ j = 1, 2$ are given by

$$\begin{aligned} \Im_1 &= \Im_2 = \{r \in \Re^3 : -0.46 \leq r_1 \leq 0.39, \\ & -0.16 \leq r_1 \leq 0.16,\ -0.24 \leq r_1 \leq 0.30\} \\ \wp_1 &= \{s_1 : -0.24 \leq s_1 \leq 0.30\},\ \wp_2 = \{s_2 : -0.22 \leq s_1 \leq 0.22\} \end{aligned}$$

Checking on **Assumption 3.2**-1 and -2, it follows that

$$\begin{aligned} D_1(r) &= D_2(r) = 1.859 \begin{bmatrix} r_1 & r_2 & 0 \end{bmatrix} \\ E_1(s_1) &= 1.859 s_1,\ E_2(s_2) = 1.136 s_2 \end{aligned}$$

Simple calculations give

$$\varrho_{r_1} = \varrho_{r_2} = 0.904,\ \varrho_{s_1} = 0.563,\ \varrho_{s_2} = 0.251$$

and hence **Assumption 3.2**-3 is satisfied.

Invoking the linear quadratic regulator theory [3] with unity weighting matrices and degree of stability $\alpha = 1.6$, computation of the decentralized gains yields

$$\begin{aligned} G_1 &= \begin{bmatrix} -6.6712 & -16.3014 & -17.1622 \end{bmatrix} \\ G_2 &= \begin{bmatrix} -9.2401 & -9.9440 & -4.4264 \end{bmatrix} \end{aligned}$$

Proceeding further to solve (3.53) with $\mathcal{Q}_1 = \mathcal{Q}_2 = I$ and then computing the nonlinear controller (3.52) with $\epsilon_1 = \epsilon_2 = 0.5$, we reach

$$\begin{aligned} \sigma_1(\mathbf{x}_1) &= 33.750\ ||\mathbf{x}_1||,\ \mathbf{x}_1 = \begin{bmatrix} x_1 & x_2 & x_3 \end{bmatrix} \\ \sigma_2(\mathbf{x}_2) &= 5.991\ ||\mathbf{x}_2||,\ \mathbf{x}_2 = \begin{bmatrix} x_4 & x_5 & x_6 \end{bmatrix} \\ \mathbf{g} &= [g_1(\mathbf{x}_1) \quad g_2(\mathbf{x}_2)]^t \\ g_j(\mathbf{x}_j) &= \begin{cases} -\sigma(\mathbf{x}) \dfrac{B_j^t \mathcal{P}_j \mathbf{x}_j}{||B_j^t \mathcal{P}_j \mathbf{x}_j||} & for\ ||B_j^t \mathcal{P}_j \mathbf{x}_j|| > 0.5 \\ -\sigma(\mathbf{x}) \dfrac{B_j^t \mathcal{P}_j \mathbf{x}_j}{0.5} & for\ ||B_j^t \mathcal{P}_j \mathbf{x}_j|| \leq 0.5 \end{cases}, \quad j = 1, 2 \end{aligned}$$

Simulation of the closed-loop system using a decentralized control structure is shown in Figure 3.4, where the controller's components were given before for the following computer runs:

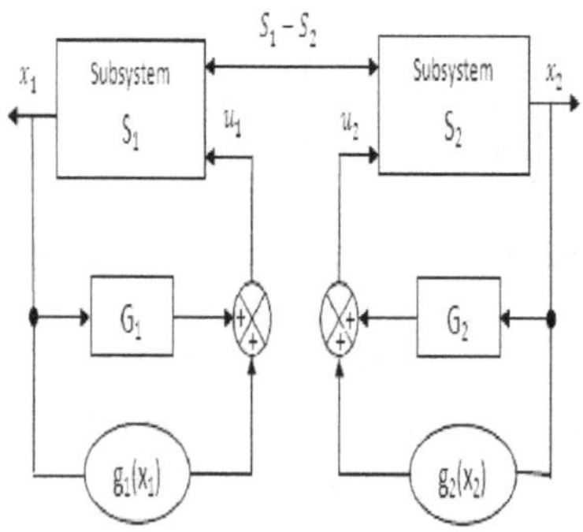

Figure 3.4: Decentralized control (DC) structure

1. $r_1 = 0.2,\ r_2 = 0.1,\ r_3 = 0.25,\ s_1 = 0.28,\ s_2 = 0.2.$
2. $r_1 = -0.46,\ r_2 = -0.16,\ r_3 = -0.24,\ s_1 = -0.24,\ s_2 = -0.22.$
3. $r_1 = 0.39,\ r_2 = 0.16,\ r_3 = 0.30,\ s_1 = 0.30,\ s_2 = 0.22.$

Representative plots are presented in Figures 3.5-3.7 for the three different runs and markings. On the other hand, simulation of the closed-loop system

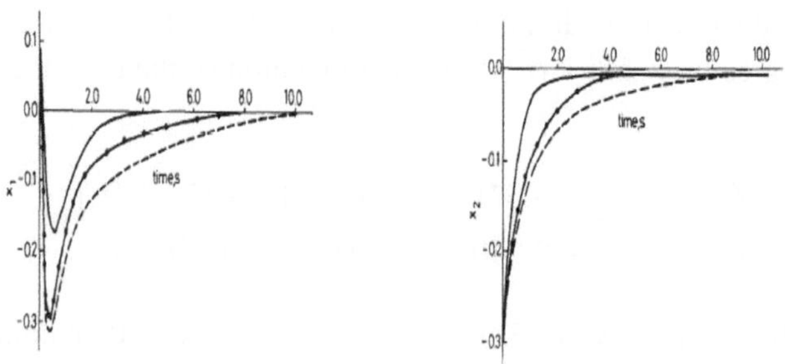

Figure 3.5: Trajectories of first (left) and second (right) states under DC

using hierarchical control structure, shown in Figure 3.8, was performed. The ensuing results are depicted in Figures 3.9-3.11 for the three different runs with the same markings.

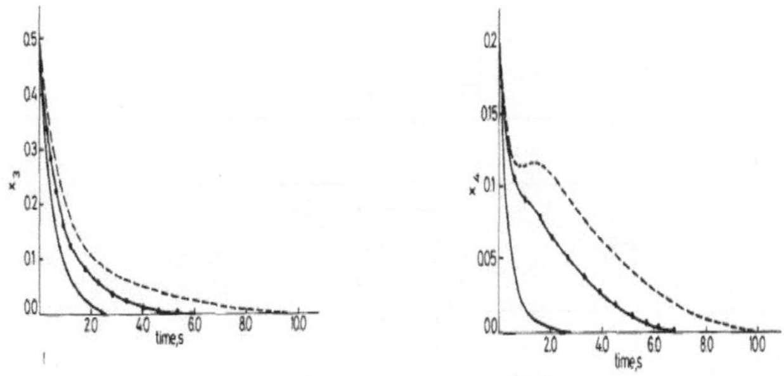

Figure 3.6: Trajectories of third (left) and fourth (right) states under DC

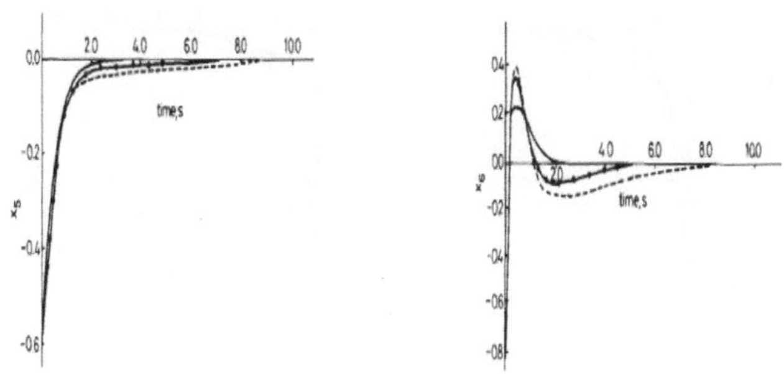

Figure 3.7: Trajectories of fifth (left) and sixth (right) states under DC

3.6 Linear Systems

In this section, we focus on the class of systems composed of n_s coupled linear subsystems with nonlinear interconnections described by

$$\dot{x}_j(t) = A_j x_j + B_j u_{jc} + h_j(t, x), \quad j = 1, 2, ..., n_s \tag{3.64}$$

where $x_j(t) \in \Re^{n_j}$ is the state vector, $u_{jc}(t) \in \Re^{n_j}$ is the local control input of the jth subsystem, and the matrices $A_j \in \Re^{n_j \times n_j}$, $B_j \in \Re^{n_j \times m_j}$ are constants. Following the development of Section 3.3, each of the free subsystems

$$\dot{x}_j(t) = A_j x_j + B_j u_{jc}, \quad j = 1, 2, ..., n_s \tag{3.65}$$

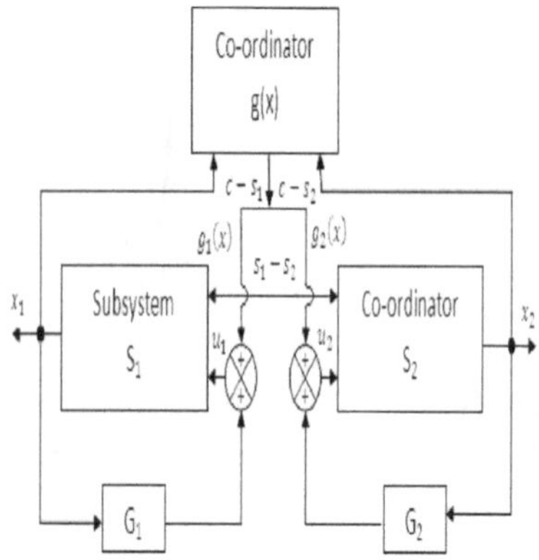

Figure 3.8: Hierarchical control (HC) structure

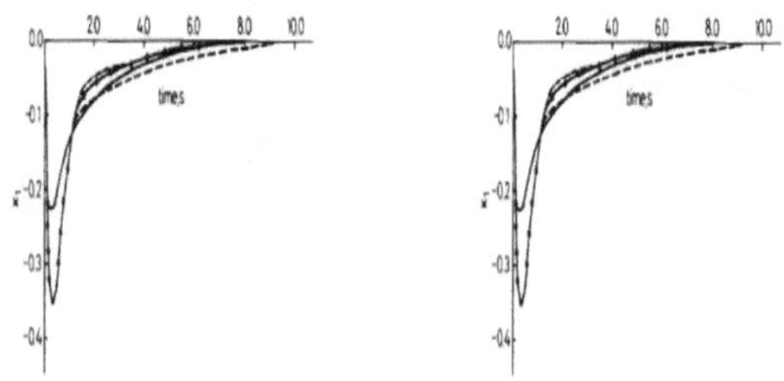

Figure 3.9: Trajectories of first (left) and second (right) states under HC

are optimized with respect to the quadratic performance index

$$J_j(t_o, x_{jo}, u_{jc}) = \int_{t_o}^{\infty} e^{\pi t} \left[x_j^t(t) \mathcal{Q}_j x_j(t) + u_{jc}^t(t) \mathcal{R}_j u_{jc}(t) \right] dt \quad (3.66)$$

where $0 < \mathcal{Q}_j \in \Re^{n_j \times n_j}$, $0 < \mathcal{R}_j \in \Re^{m_j \times m_j}$ are weighting matrices and π is a nonnegative number. Under the standard assumption of complete controllability of the subsystem pair $(A_j,\ B_j)$, there exists a unique linear optimal control law [3]

$$u_{jc}^*(t) \ = \ -K_{jc}\, x_j(t), \quad K_{jc} = \mathcal{R}_j^{-1} B_j^t \mathcal{P}_j \quad (3.67)$$

where $0 < \mathcal{P}_j \in \Re^{n_j \times n_j}$ is the solution of the algebraic Riccati equation

$$\mathcal{P}_j(A_j + \pi I_j) + (A_j + \pi I_j)^t \mathcal{P}_j - \mathcal{P}_j B_j \mathcal{R}_j^{-1} B_j^t \mathcal{P}_j + \mathcal{Q}_j = 0 \quad (3.68)$$

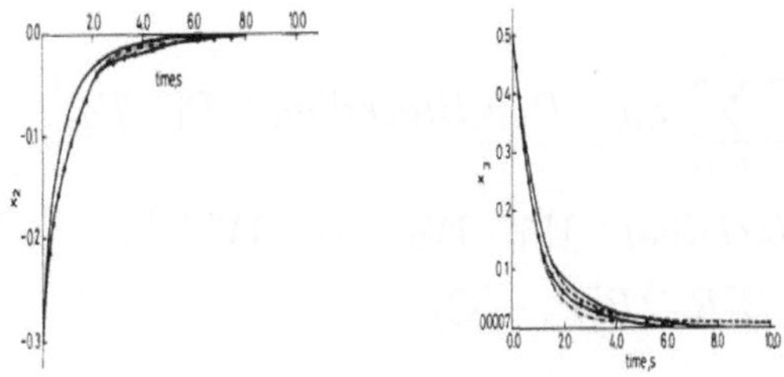

Figure 3.10: Trajectories of third (left) and fourth (right) states under HC

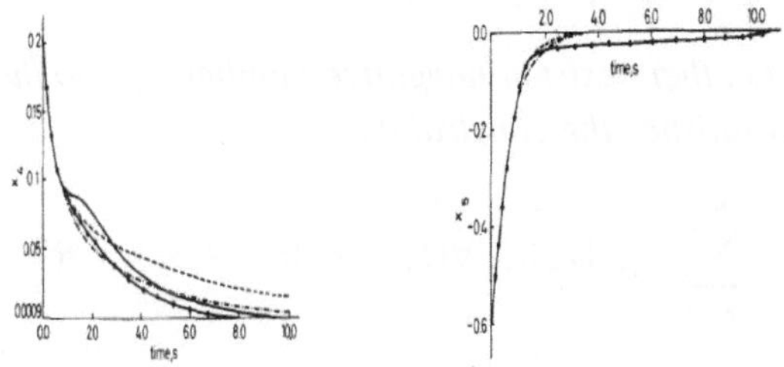

Figure 3.11: Trajectories of fifth (left) and sixth (right) states under HC

such that $u_{jc}^*(t)$ yields the minimum of $J_j(t_o, x_{jo}, u_{jc})$ in (3.66) where the optimal performance index is

$$J_j^*(t_o, x_{jo}) = e^{2\pi t_o}\; x_{jo}^t \mathcal{P}_j x_{jo} \qquad (3.69)$$

When the matrix $\mathcal{Q}_j$ can be factored as $\mathcal{Q}_j = C_j C_j^t$, where $C_j \in \Re^{n_j \times n_j}$ is a constant matrix, so that the subsystem pair $(A_j,\ C_j)$, is completely observable, it follows [3] that each closed-loop subsystem

$$\dot{x}_j(t) = (A_j - B_j \mathcal{R}_j^{-1} B_j^t \mathcal{P}_j) x_j, \quad j = 1, 2, ..., n_s \qquad (3.70)$$

is globally exponentially stable with the degree π. That means that the solution $x_j(t; t_o, x_{jo})$ of (3.70) approaches the equilibrium at the origin at least as fast as $e^{-\pi t}$ for all $(t_o, x_{jo}) \in \Re^{n_j+1}$, $t \in [t_o,\ +\infty)$. Introduce the following quantities:

$$
\begin{aligned}
\xi &= \sum_{j=1}^{n_s}\sum_{k=1}^{n_s} \xi_{jk}, \;\; \mathcal{P} = Blockdiag\left[\; \mathcal{P}_1 \;\; \mathcal{P}_2 \;\; \cdots \;\; \mathcal{P}_{n_s} \;\right], \\
\mathcal{W} &= Blockdiag\left[\; \mathcal{W}_1 \;\; \mathcal{W}_2 \;\; \cdots \;\; \mathcal{W}_{n_s} \;\right], \\
\mathcal{W}_j &= \mathcal{P}_j B_j \mathcal{R}_j^{-1} B_j^t \mathcal{P}_j + \mathcal{Q}_j
\end{aligned} \tag{3.71}
$$

The following theorem presented in [322] establishes condition on the interaction $h_j(t,x)$ to guarantee a prescribed degree of suboptimality of the interconnected system

$$
\dot{x}_j(t) = (A_j - B_j \mathcal{R}_j^{-1} B_j^t \mathcal{P}_j)x_j + h_j(t,x), \quad j = 1,2,...,n_s \tag{3.72}
$$

Theorem 3.6.1 *Let there exist nonnegative numbers ξ_{jk} so that the interaction $h_j(t,x)$ in (3.72) satisfies the constraints*

$$
\begin{aligned}
||h_j(t,x)|| &\leq \sum_{k=1}^{n_s} \xi_{jk}||x_k||, \;\; \forall (t,x) \in [t_o, \; +\infty) \times \Re^n, \; \forall j = 1,2,...,n_s \\
\xi &\leq \frac{\epsilon}{1+\epsilon}\, \frac{\lambda_m(\mathcal{W})}{2\lambda_M(\mathcal{P})}
\end{aligned} \tag{3.73}
$$

Then the overall system (3.72) is suboptimal with index ϵ and globally exponentially stable with degree π.

Observe that $\lambda_M(\mathcal{P})$ and $\lambda_m(\mathcal{W})$ are the maximum and minimum eigenvalues of matrices $\mathcal{P}$ and $\mathcal{W}$, respectively, with $\lambda_M(\mathcal{P}) = \max_j\{\lambda_M(\mathcal{P}_j)\}$ and $\lambda_m(\mathcal{W}) = \min_j\{\lambda_m(\mathcal{W}_j)\}$.

In the case of a linear interaction pattern, the overall system becomes

$$
\begin{aligned}
\dot{x}(t) &= Ax + Bu + Hx \\
A &= Blockdiag\left[\; A_1 \;\; A_2 \;\; \cdots \;\; A_{n_s} \;\right], \\
B &= Blockdiag\left[\; B_1 \;\; B_2 \;\; \cdots \;\; B_{n_s} \;\right]
\end{aligned} \tag{3.74}
$$

for which we introduce the two-level control

$$
u(t) = u_c(t) + u_g(t) \tag{3.75}
$$

which consists of two components: $u_c(t)$ is the subsystem (lower-level) control as given by (3.67) and $u_g(t)$ is the global (higher-level) control chosen linearly of the form

$$
u_g(t) = -K_g x(t) \tag{3.76}
$$

where $K_g \in \Re^{m\times n}$ is a constant matrix to be selected such that [322]

$$\inf_{K_g} ||(H - BK_g)x||$$

is achieved for all $x \in \Re^n$. Taking into consideration the fact that $||(H - BK_g)x|| \leq ||(H - BK_g)|| ||x||$ holds for all $x \in \Re^n$, then the selection of K_g reduces to finding $\min_{K_g} ||H - BK_g||$ for which the solution is known to be

$$K_g = B^\dagger \, H$$

where $B^\dagger$ is the Moore-Penrose generalized inverse of B [162]. When $rank(B) = m$, then $B^\dagger = (B^tB)^{-1}B^t$ and thus

$$K_g = (B^tB)^{-1}B^tH \tag{3.77}$$

implying that the closed-loop system (3.74) assumes the form

$$\dot{x}(t) = \Big(A - B\mathcal{R}^{-1}B^t\mathcal{P} + \Big[I - B(B^tB)^{-1}B^t\Big]H\Big)x(t) \tag{3.78}$$

which would be exponentially stable with a prescribed degree π.

Remark 3.6.2 *It is interesting to observe that when B is a square and nonsingular matrix, the choice (3.77) reduces to $Kg = B^{-1}H$, which eventually leads to a perfect neutralization of the interaction effects and hence $\epsilon = 0$.*

An important structural property for the two-level control systems is that suboptimality index is invariant under structural perturbations whereby subsystems are disconnected from each other and again connected together in various ways during operation. The notion of connective suboptimality introduced here is, therefore, an extension of the connective concept originated in stability studies [321] of large-scale systems.

3.6.1 Illustrative example 3.3

Consider the following seventh-order system [179], which we treat in the sequel as composed of three subsystems:

$$A = \begin{bmatrix} A_1 & 0 & A_2 \\ 0 & A_3 & A_4 \\ A_5 & A_6 & A_7 \end{bmatrix}, \; B = \begin{bmatrix} B_1 & 0 & 0 \\ 0 & B_2 & 0 \\ 0 & 0 & B_3 \end{bmatrix},$$

$$A_1 = \begin{bmatrix} 6 & -10 \\ 10 & 20 \end{bmatrix}, \; A_2 = \begin{bmatrix} 2.5 & -10 \\ 1 & 7.5 \end{bmatrix}, \; A_5 = \begin{bmatrix} -1 & -0.5 \\ 0.1 & 0.3 \end{bmatrix},$$

$$A_3 = \begin{bmatrix} 10 & 5 & 8 \\ 2 & -10 & 2 \\ 10 & 0 & -2.5 \end{bmatrix}, \; A_4 = \begin{bmatrix} -20 & -10 \\ 5 & 4 \\ 1 & -1 \end{bmatrix},$$

$$A_6 = \begin{bmatrix} -0.3 & 0.4 & 0.2 \\ -0.1 & 0.1 & 0.1 \end{bmatrix}, \; A_7 = \begin{bmatrix} -0.8 & 0.3 \\ -1 & 0.2 \end{bmatrix}, \; B_2 = \begin{bmatrix} 2.5 \\ 5 \\ 10 \end{bmatrix},$$

$$B_1 = \begin{bmatrix} 2 \\ 10 \end{bmatrix}, \; B_3 = \begin{bmatrix} 0.1 & 0.25 \\ -1 & 1 \end{bmatrix}$$

Applying the linear quadratic regulator theory [3] with weighting matrices $\mathcal{R} = Blockdiag[I_2 \;\; I_3 \;\; I_2]$, $\mathcal{Q} = Blockdiag[I_2 \;\; I_3 \;\; 2I_2]$ where I_j is the $j \times j$ unit matrix for the full system, the two-level system, and the block arrow structure (BAS) system [179], the ensuing results of the performance indices with $x_o = [I_2 \;\; I_3 \;\; I_2]^t$ are given by

1. *The performance index for the full (centralized) system* $J^* = 5.4231$.
2. *The performance index for the two-level system* $J^+ = 5.7485$.
3. *The performance index for the BAS system* $J^- = 6.1285$.

We record that the performance degradation afforded by the two-level system is 6%, while that of the BAS system is 13%, an observation that confirms the superiority of the two-level with interaction neutralization. Representative plots are presented in Figures 3.12-3.15. In the following section, we address a perturbational approach to large-scale optimization and control of linear systems.

3.7 A Perturbational Approach

Since a large-scale system will invariably be an interconnection of several subunits, it has been noted in the foregoing section that one of the important phenomena that must be accounted for in the design controllers is the occurrence of

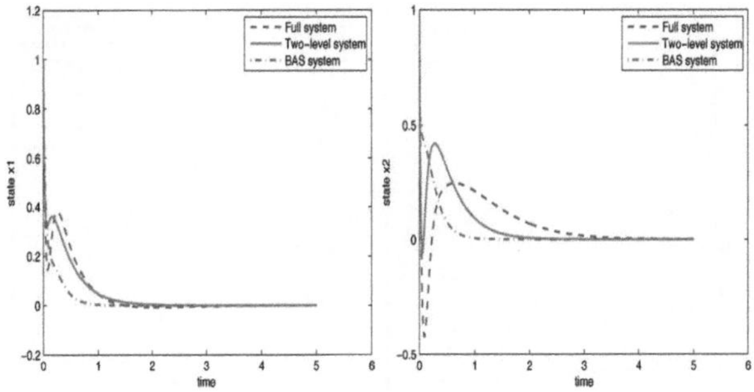

Figure 3.12: Trajectories of first (left) and second (right) states under different controls

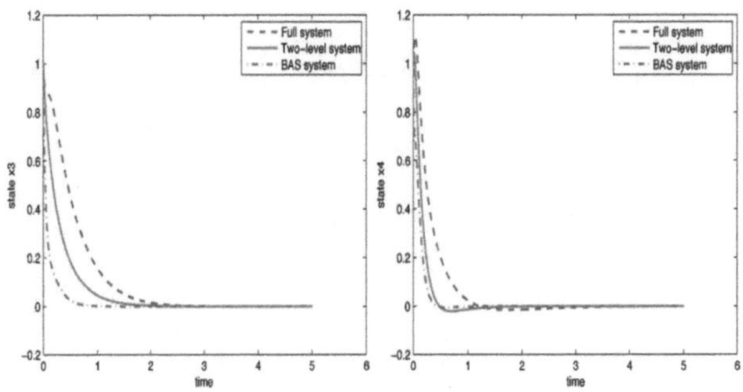

Figure 3.13: Trajectories of third (left) and fourth (right) states under different controls

structural perturbations, that is, changes in the interconnection pattern within the system during operation. It has been recognized [340] that reliability of system performance in the face of such structural changes is intimately related to the allowance for the subsystems to perform with as much autonomy as possible; a typical example occurs in interconnected power systems where the reliability of operation is of significant importance [334].

From the preceding sections, we learned that an interconnected system composed of n_s coupled linear subsystems with linear interconnections can be described by

$$\dot{x}_j(t) = A_j x_j(t) + B_j u_j(t) + \sum_{k=1}^{n_s} H_{jk} x_k(t) \tag{3.79}$$

where for $j \in \{1, \ldots, n_s\}$, $x_j(t) \in \Re^{n_j}$ is the state vector and $u_j(t) \in \Re^{m_j}$ is the control input. The constant matrices $A_j \in \Re^{n_j \times n_j}$ and $B_j \in \Re^{n_j \times m_j}$

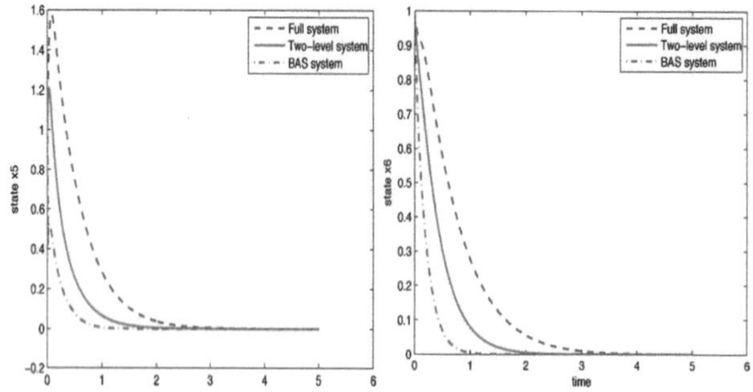

Figure 3.14: Trajectories of fifth (left) and sixth (right) states under different controls

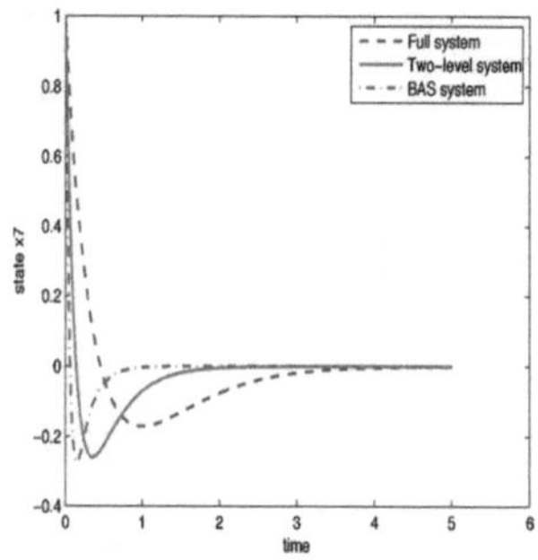

Figure 3.15: Trajectories of seventh state under different controls

describe the dynamics of the isolated subsystem $\mathbf{S}_j$ and $H_{jk} \in \Re^{n_j \times n_k}$ specifies the interaction from $\mathbf{S}_k$ to $\mathbf{S}_j$. The complete controllability of the pairs $(A_j,\ B_j)$ is assumed in the sequel. For the purpose of regulation, the performance of each subsystem is measured by an associated quadratic cost

$$\begin{aligned} J_j(t_o, x_j(t_o), u_j(t_o)) &= \int_{t_o}^{\infty} \left[x_j^t(t) \mathcal{Q}_j x_j(t) + u_j^t(t) \mathcal{R}_j u_j(t) \right] dt, \\ & j \in \{1, ..., n_s\} \end{aligned} \tag{3.80}$$

where $0 \leq \mathcal{Q}_j^t = \mathcal{Q}_j \in \Re^{n_j \times n_j}$, $0 < \mathcal{R}_j^t = \mathcal{R}_j \in \Re^{m_j \times m_j}$ are weighting matrices. To guarantee a goal harmony among the subsystems, the overall quadratic cost

$$J(t_o, x(t_o), u(t_o)) = \sum_{j=1}^{n_s} J_j(t_o, x_j(t_o), u_j(t_o)) \tag{3.81}$$

is sought to be optimized where

$$x(t) = [\ x_1^t(t)\ \ x_2^t(t)\ \ \cdots\ \ x_{n_s}^t(t)\ ]^t,$$
$$u(t) = [\ u_1^t(t)\ \ u_2^t(t)\ \ \cdots\ \ u_{n_s}^t(t)\ ]^t$$

are the overall state and control input vectors.

Our objective in this section is to design a decentralized controller structure where the local (subsystem) controller $u_j(t)$ will be generated from the local information only $\{A_j(t),\ B_j(t),\ \mathcal{Q}_j,\ \mathcal{R}_j\}$ without information exchange among the local controllers. One must bear in mind that such decentralized structure with information constraints may lead to performance loss in comparison to the centralized optimal controllers, a fact that poses a compromise between performance versus implementation.

When the subsystems are decoupled, that is, when $H_{jk} \equiv 0,\ \ j,k \in \{1,...,n_s\}$, the decentralized optimal controls $u_j^*(t)$ that minimize J_j in (3.80) subject to the dynamic constraints

$$\dot{x}_j(t) = A_j x_j(t) + B_j u_j(t),\ \ j \in \{1,...,n_s\} \tag{3.82}$$

are given by

$$u_j^*(t) = -K_j^*\, x_j(t),\ \ K_j^* = \mathcal{R}_j^{-1} B_j^t \mathcal{P}_j \tag{3.83}$$

where $0 < \mathcal{P}_j^t = \mathcal{P}_j \in \Re^{n_j \times n_j}$ is the solution of the algebraic Riccati equation

$$\mathcal{P}_j(A_j + \pi I_j) + (A_j + \pi I_j)^t \mathcal{P}_j - \mathcal{P}_j B_j \mathcal{R}_j^{-1} B_j^t \mathcal{P}_j + \mathcal{Q}_j = 0 \tag{3.84}$$

with the associated optimal costs

$$J_j^*(t_o, x_j(t_o)) = x_j^t(t_o) \mathcal{P}_j x_j(t_o),\ \ j \in \{1,...,n_s\} \tag{3.85}$$

It is obvious that the decoupled optimal cost

$$J^*(t_o, x_j(t_o)) = \sum_{j=1}^{n_s} J_j^*(t_o, x_j(t_o)) \tag{3.86}$$

would be markedly different from $J^o(t_o, x_j(t_o))$ that would result from the minimization of (3.81) subject to the full dynamics (3.79) depending on the interaction pattern. The real challenge is to seek methods that result in improved performance whenever possible for the overall system. It must be emphasized that $J*$ may in general be less or greater than J^o and the presence of interconnections may generally be termed "nonbeneficial" or "beneficial" to the optimization [205, 354]. The following theorem provides a possible characterization of an interconnection pattern [354].

Theorem 3.7.1 *Let*

$$\mathcal{P} = Blockdiag \left[\begin{array}{cccc} \mathcal{P}_1 & \mathcal{P}_2 & \cdots & \mathcal{P}_{n_s} \end{array} \right]$$

where $0 < \mathcal{P}_j^t = \mathcal{P}_j \in \Re^{n_j \times n_j}$ *is the solution of (3.84). Then the optimum cost condition*

$$J^* = J^o$$

is attained if and only if the interaction matrix $H = [H_{jk}] \equiv 0, \; j,k \in \{1, ..., n_s\},$ *can be factorized,*

$$H \; = \; \mathcal{S} \, \mathcal{P} \tag{3.87}$$

where $\mathcal{S} \in \Re^{n \times n}$ *is an arbitrary skew-symmetric matrix.*

Remark 3.7.2 *We note that* ***Theorem 3.7.1*** *identifies a class of interconnection patterns H that ensure that the overall optimum* J^o *is attained by the decentralized local controls* $u_j^*(t)$. *Consequently, the stability of the closed-loop overall system*

$$\dot{x}_j(t) = [A_j - B_j K_j^*] x_j(t) + \sum_{k=1}^{n_s} H_{jk} x_k(t), \;\; j \in \{1, ..., n_s\} \tag{3.88}$$

is also guaranteed since when (3.87) holds, $\mathcal{P} \in \Re^{n \times n}$ *is the positive definite solution of the Riccati equation associated with the overall system (3.79) together with the cost (3.85). It is quite evident that a system equipped with these controls would perform reliably (without loss of stability and optimality of performance) when the structural perturbations can be limited to result in interconnection patterns belonging to the above class.*

3.7.1 Multilevel state regulation

In the case of the interconnection pattern H when it does not satisfy the equality constraint (3.87), the application of the decentralized local controls $u_j^*(t)$ will result in a performance that deviates from J^o. In the preceding section, bounds on the resulting suboptimality of performance and the norm of the interconnections have been developed. There multilevel control schemes to improve the performance by neutralizing the interconnection effects have been designed.

In line of [205, 353, 354], using different ideas to develop an alternative multilevel scheme does not ignore the possible beneficial aspects of the interconnections, but exploits them to permit resulting costs less than J^o to be attained. We will achieve this within a two-level hierarchical structure in which an additional control is to be generated by a supervisory coordinator at a higher level, to whom the subsystems agree to submit their states and, in turn, is capable of assessing the extent of interactions at any instant. The information structure of this coordinator consists of $\{x_j(t), t \leq t_o, j = 1, 2, ..., n_s\}$, H_{jk}, $j, k = 1, 2, ..., n_s\}$ together with the *a priori* information of A_j, B_j, $\mathcal{P}_j$, $j = 1, 2, ..., n_s$.

Remark 3.7.3 *Had the subsystem cost matrices* $\{\mathcal{Q}_j, \mathcal{R}_j, j = 1, 2, \ldots, n_s\}$, *been made available to the supervisory coordinator, he would then have all the information necessary to compute the centralized optimal control* $u^o(t)$ *and substitute the local controls* $u_j^*(t)$ *by the components* $u_j^c(t)$. *Generally, this is not encouraged for the following reasons:*

1. *The associated data processing difficulties in computing this control,*

2. *The fact that such a strategy would oppose our desire for keeping the autonomy of the subsystems. This comes in full support with the results reported in [322] as a centralized optimal control is often unreliable when the system is expected to undergo structural perturbations.*

As a practical rule (the reader is referred to the foregoing chapter for issues on decomposition-coordination structures) the supervisory coordinator must deploy a minimum intervention providing only corrective controls that deal with the interactions existing at any given period.

Now let the input $u_j(t)$ to each subsystem consist of two components

$$u_j(t) = u_j^*(t) + u_j^c(t), \quad j \in \{1, ..., n_s\} \tag{3.89}$$

where $u_j^*(t) = -K_j^* \; x_j(t)$ is the subsystem (lower-level) control with K_j^* being given by (3.83) and $u_j^c(t)$ is the component of the coordinator control chosen linearly of the form [353, 354]

$$u_j^c(t) = -\sum_{k=1}^{n_s} N_{jk} x_k(t) \tag{3.90}$$

where $N_{jk} \in \Re^{m_j \times n_j}$ is the component of the coordinator control gain matrix. Using (3.89), the closed-loop subsystem becomes:

$$\dot{x}_j(t) = [A_j - B_j K_j^*] x_j(t) + \sum_{k=1}^{n_s} [H_{jk} - B_j N_{jk}] x_k(t) \tag{3.91}$$

By selecting the gains N_{jk} through the equality relation

$$\begin{aligned} B\,N &= B^c\,K^* \\ K^* &= Blockdiag\left[\begin{array}{cccc} K_1^* & K_2^* & \cdots & K_{n_s}^* \end{array}\right], \\ N &= [N_{jk}],\ j,k \in \{1,...,n_s\} \end{aligned} \tag{3.92}$$

and introducing

$$\begin{aligned} A &= Blockdiag\left[\begin{array}{cccc} A_1 & A_2 & \cdots & A_{n_s} \end{array}\right], \\ B &= Blockdiag\left[\begin{array}{cccc} B_1 & B_2 & \cdots & B_{n_s} \end{array}\right] \end{aligned} \tag{3.93}$$

we express the closed-loop overall system

$$\dot{x}(t) = (A+H)x(t) - (B+B^c)K^*x(t) \tag{3.94}$$

where $B^c \in \Re^{n\times m}$ is the coordinator perturbation in matrix B.

Remark 3.7.4 *It is significant to note that B as in (3.93) is block-diagonal and hence the components of $-BK^*x(t)$ applied to each subsystem involve their own state only, which determines the local controls. On the other hand, the matrix B^c is not block-diagonal and this implies that the components of $-B^cK^*x(t)$ involve the states of the other subsystems thereby characterizing the global control.*

It remains to seek ways to the determination of the global gains B^c that yield a bounded performance deviation from $J*$ without causing any change in the feedback gain matrix K^*. One possible way to solve this problem is attained by the following theorem [353, 354].

Theorem 3.7.5 *Let $\mathcal{S} \in \Re^{n\times n}$ be a skew-symmetric solution of the matrix equation*

$$\mathcal{S}(A+H) + (A+H)^t\mathcal{S} + H^t\mathcal{P}A - A^t\mathcal{P}H = 0 \tag{3.95}$$

and such that

$$\widehat{\mathcal{P}} = \mathcal{P} + (\mathcal{S} - \mathcal{P}H)(A+H)^{-1} > 0 \tag{3.96}$$

The perturbation matrix B^c given by

$$B^c = -\widehat{\mathcal{P}}^{-1}(\mathcal{S} - \mathcal{P}H)(A+H)^{-1}B \tag{3.97}$$

yields a cost

$$\widehat{J} = x^t(t_o)\widehat{\mathcal{P}}x(t_o)$$

for the overall system (3.94). A bound on the suboptimality index is given by

$$\hat{\epsilon} := \frac{\widehat{J} - J^*}{J^*} \leq \frac{\lambda_M((\mathcal{S} - \mathcal{P}H)(A + H)^{-1})}{\lambda_m(\mathcal{P})}$$

Moreover, the control

$$u(t) = -(K^* + N)x(t)$$

is a stabilizing controller for the overall system.

Remark 3.7.6 *We observe that the perturbation matrix B^c helps retain the unchanged feedback gain matrix K^* while solving the Lyapunov matrix (3.95) as compared with two-level coordination algorithms [196]-[203]. Observe also that the suboptimality is assessed with respect to the decoupled optimum J^*, instead of the overall optimum J^o by virtue of considerations of reliability and subsystem autonomy. This performance indicator is essentially used to measure the deviation caused by the beneficial or nonbeneficial effects of interactions.*

3.7.2 Generalized scheme

In the sequel, we provide a multicontroller structure to handle the information flow within hierarchical control systems. Extending on the foregoing section, we consider here n_s linear output regulators of the form

$$\dot{x}_j(t) = A_j x_j(t) + B_j u_j(t) + g_j[x(t), u(t)] \quad (3.98)$$

$$g_j[x(t), u(t)] = \sum_{k=1}^{n_s} D_{jk} x_k(t) + G_{jk} u_k(t) \quad (3.99)$$

$$y_j(t) = C_j x(t), \quad j \in \{1, ..., n_s\} \quad (3.100)$$

For the decoupled case $g_j[x(t), u(t)] \equiv 0,$ the optimal policy $u^*(t)$ of minimizing the performance index

$$J_j(t_o, y_j(t_o), u_j(t_o)) = \int_{t_o}^{\infty} \left[y_j^t(t) \mathcal{Q}_j y_j(t) + u_j^t(t) \mathcal{R}_j u_j(t) \right] dt, \quad j \in \{1, ..., n_s\} \quad (3.101)$$

and satisfying the isolated dynamics is expressed as [3]:

$$u_j^*(t) = -K_j^+ x_j(t), \quad K_j^+ = \mathcal{R}_j^{-1} B_j^t \mathcal{P}_j \quad (3.102)$$

where $0 < \mathcal{P}_j^t = \mathcal{P}_j \in \Re^{n_j \times n_j}$ is the solution of the algebraic Riccati equation

$$\mathcal{P}_j A_j + A_j^t \mathcal{P}_j - \mathcal{P}_j B_j \mathcal{R}_j^{-1} B_j^t \mathcal{P}_j + C_j^t \mathcal{Q}_j C_j = 0 \quad (3.103)$$

In the subsequent analysis, we seek an approach of handling a general form of interactions considering both undesirable and favorable effects. From an engineering design point of view, these effects are deemed noninteracting although they might exist simultaneously. Therefore, adding up the extent of the coupling terms eventually implies placing an additional task on the control policies in order to master the information flow. To preserve the autonomy of the individual problems, this task is reflected in suitably superimposing further components, directly or indirectly, to the control effort. Specifically, the jth control policy is first written of the form (3.89) where $u^c(t)$ is treated hereafter as the corrective control component responsible for constraining the information flow against the existing coupling modes. Substituting (3.89), (3.99), and (3.102) into (3.98) and considering the term $u_j(t)$ in (3.99) as an externally applied signal $u_j^e(t)$, thus,

$$\dot{x}_j(t) = [A_j - B_j K_j^+]x_j(t) + B_j u_j^c(t) + \sum_{k=1}^{n_s} D_{jk}x_k(t) + G_{jk}u_k^e(t) \quad (3.104)$$

Building on the results of the previous section, we select the corrective control as

$$u_j^c(t) = -\sum_{k=1}^{n_s} M_{jk}x_k(t) \quad (3.105)$$

which is in agreement with (3.90) with M_{jk} being the feedback coupling gains. It is useful at this stage to recognize the role of the various control policies (controllers) within the context of hierarchically structured multilevel systems. The group of controllers $\{u_j^*(t)\},\ j = 1, ..., n_s$ is performing the basic regulation task, thus it has a local attribute and it is quite natural to attach these controllers to the lower hierarchical control level. Alternatively, the function of $\{u_j^c(t)\},\ \{u_j^e(t)\},\ j = 1, \ldots, n_s$ being associated with the intercouplings among the subsystems has global attributes. This motivates assigning them to higher hierarchical control levels. Our task now is to derive the global control policies. The integrated system is considered by summing up equation (3.104) over the n_s subsystems

$$\dot{x}(t) = [A - BK^+]x(t) + [D - BM]x(t) + Gu^e(t) \quad (3.106)$$

where

$$\begin{aligned} A &= Blockdiag[A_j],\ B = Blockdiag[B_j],\ K^+ = Blockdiag[K_j^+], \\ u^e(t) &= \begin{bmatrix} u_1^e(t) & u_2^e(t) & \cdots & u_{n_s}^e(t) \end{bmatrix} \end{aligned} \quad (3.107)$$

Some relevant observations that deserve emphasis are:

1. *The term* $[D - BM]$ *quantifies an effective interconnection matrix whose structural form and strength, identified in some sense, depend entirely on the particular choice of P. Evidently, in order to absorb the undesirable information mode without affecting the autonomous behavior of the subsystems, this term should be kept at minimal strength. This is classically equivalent to designing the controllers* $\{u_j^c(t)\}$ *to have compensatory features. Our approach to achieving this situation is to take the norm of the effective matrix as an appropriate quantity for the strength of information. Then, obviously, the minimum of the norm corresponds to the least strength. It turns out that the problem at hand reduces to that of an algebraic minimization problem, whose solution is only unique if* $rank[B \;\; D] = rank[B] = m$ *and it therefore has the form* [87]:

$$M = (B^t B)^{-1} B^t D \tag{3.108}$$

If the rank condition is met, the strength would be equal to zero, which ultimately entails that the interconnections among the state variables are completely absorbed.

2. *Some crucial factors such as the inaccessibility of system states and the difficulty involved in data processing may pose real design problems that may render the rank constraint far reaching. In such case, it is worthwhile to define an equivalent interconnection matrix E to designate the capacity of information flow:*

$$E = M - (B^t B)^{-1} B^t M \tag{3.109}$$

The effect of (3.109) converts (3.106) to

$$\dot{x}(t) = [A + E]x(t) + Gu^e(t) - BK^+ x(t) \tag{3.110}$$

3. *The global control policy* $u^e(t)$ *is now designed based on a different philosophy. The procedure underlying this philosophy stems from utilizing the favorable attitude of the external signals to produce potentially useful regulatory effects for the overall system. One intuitive way is through representing the control policy in global feedback form. That is,*

$$Gu^e(t) = -ZK^+ x(t) \tag{3.111}$$

where $Z \in \Re^{n \times m}$ *is an arbitrary gain matrix. Rewriting (3.110) using (3.111) is the form*

$$\dot{x}(t) = [A + E]x(t) - [B + Z]K^+ x(t) \tag{3.112}$$

Determination of the global gain matrix Z is established by the following theorem.

Theorem 3.7.7 *If the effective system matrix $(A + E)$ is nonsingular, then the gain matrix Z satisfying (3.112) while minimizing the performance index is given by*

$$Z = -(\mathcal{W} + \mathcal{P})^{-1}\mathcal{W}B \tag{3.113}$$

with an improvement in performance index given by

$$\tilde{J} = x_o^t \mathcal{W} x_o \tag{3.114}$$

Proof: Rewriting system (3.112) in the form

$$\begin{aligned} \dot{x}(t) &= [A + E]x(t) - [B + Z]u(t), \\ y(t) &= Cx(t) \end{aligned} \tag{3.115}$$

for which the associated the performance index (3.101) becomes

$$\begin{aligned} J(t_o, y(t_o), u(t_o)) &= \int_{t_o}^{\infty} \Big[y^t(t)\mathcal{Q}y(t) + u^t(t)\mathcal{R}u(t) \Big] dt, \\ C &= Blockdiag[C_j], \;\; \mathcal{Q} = Blockdiag[\mathcal{Q}_j], \\ \mathcal{R} &= Blockdiag[\mathcal{R}_j], \\ y(t) &= \left[\; y_1(t) \;\; y_2(t) \;\; \cdots \;\; y_{n_s}(t) \; \right] \end{aligned} \tag{3.116}$$

The minimizing feedback control is given by

$$u(t) = -\mathcal{R}^{-1}(B + Z)^t \mathcal{G} x(t) \tag{3.117}$$

where $0 < \mathcal{G}^t = \mathcal{G} \in \Re^{n \times n}$ is the solution of the algebraic Riccati equation

$$\mathcal{G}(A + E) + (A + E)^t\mathcal{G} - \mathcal{G}(B + Z)\mathcal{R}^{-1}(B + Z)^t\mathcal{G} + C^t\mathcal{Q}C = 0 \tag{3.118}$$

Imposing the relation

$$\mathcal{R}^{-1}((B + Z))^t\mathcal{G} = \mathcal{R}^{-1}B^t\mathcal{P} = K^+ \tag{3.119}$$

where $0 < \mathcal{P}^t = \mathcal{P} \in \Re^{n \times n}$ is the solution of

$$\mathcal{P}A + A^t\mathcal{P} - \mathcal{P}B\mathcal{R}^{-1}B^t\mathcal{P} + C^t\mathcal{Q}C = 0 \tag{3.120}$$

It follows from (3.118)-(3.120) with $\mathcal{W} = \mathcal{G} - \mathcal{P}$ that

$$\mathcal{W}(A + E) + (A + E)^t\mathcal{W} + \mathcal{P}E + E^t\mathcal{P} = 0 \tag{3.121}$$

which yields Z in (3.113) and consequently the improvement in the overall performance is $\tilde{J} = x_o^t\mathcal{G}x_o - x_o^t\mathcal{P}x_o$ from which (3.114) is obtained.

Remark 3.7.8 *We note that the interpretation of (3.121) is simple and interesting. It implies that if the effective interconnection matrix (D - BP) is constrained to possess a skew-symmetry profile, then the feedback matrix remains unchanged and hence the local autonomy is preserved. Under this condition, the gains P are installed within the local boundaries and this, in turn, simplifies the design considerations.*

3.7.3 Illustrative example 3.4

To demonstrate the application of the above theoretical analysis, a typical system problem frequently encountered in power engineering is considered. The problem represents the optimal design of network frequency and tie-line power controllers of two coupled steam electric networks sharing the supply of a group of loads. In modeling the first electric network, five state variables are considered. They represent the deviations in the network frequency, power generation, steam valve position, servomotor angular displacement, and angular velocity. The second network state vector comprises six variables; the first five of these are similar to those of the first network and the sixth component denotes the change in the tie-line power level. For both networks, the control vector signifies the signal command to reset the control unit and the telemetry signal expresses the change in the load. Reducing these deviations to minimal is the desired objective of the power optimization problem for balanced and continuous operation.

Using reasonable data values in per-unit representation, the associated quantities of the dynamic model (3.98)-(3.100) are defined by

$$A_1 = \begin{bmatrix} -0.284 & 0.227 & 0 & 0 & 0 \\ 0 & -3.846 & 3.846 & 0 & 0 \\ -19.23 & 0 & -3.846 & 3.846 & 0 \\ 0 & 0 & 0 & 0 & -3.846 \end{bmatrix},$$

$$A_2 = \begin{bmatrix} -1.284 & 0.227 & 0 & 0 & 0 & 0.227 \\ 0 & -3.846 & 3.846 & 0 & 0 & 0 \\ -19.23 & 0 & -3.846 & 3.846 & 0 & 0 \\ 0 & 0 & 0 & 0 & 3.846 & 0 \\ -0.1 & 0 & 0 & 0 & 0 & 0 \end{bmatrix},$$

$$B_1 = \begin{bmatrix} 0 & -0.227 \\ 0 & 0 \\ 0 & 0 \\ 0 & 0 \\ 1 & 0 \end{bmatrix}, \; B_2 = \begin{bmatrix} 0 & 0.227 \\ 0 & 0 \\ 0 & 0 \\ 0 & 0 \\ 1 & 0 \\ 0 & 0 \end{bmatrix},$$

$$\begin{aligned} D_{11} &= [0],\ D_{22} = [0],\ D_{12} = [0]\ except\ d_{16} = -0.227, \\ D_{21} &= [0]\ except\ d_{61} = 0.1, \\ G_{11} &= [0],\ G_{22} = [0],\ G_{12} = [0]\ except\ g_{22} = 0.15, \\ G_{21} &= [0]\ except\ g_{22} = 0.12, \\ Q_1 &= Blockdiag[10,\ 0,\ 0,\ 1,\ 1], \\ Q_2 &= Blockdiag[10,\ 0,\ 0,\ 1,\ 1,\ 1], \\ R_1 &= Blockdiag[0.1],\ R_2 = Blockdiag[0.1] \end{aligned}$$

Applying the linear quadratic regulator theory [3] yields the local decoupled gain matrices

$$K_1 = \begin{bmatrix} 9.826 & 0.311 & -0.051 & -1.512 & -0.812 \\ 0.311 & -0.295 & 0.029 & 0.850 & 0.444 \\ -0.051 & 0.029 & 0.347 & -0.763 & 1.045 \\ -1.512 & 0.850 & -0.763 & 0.428 & -0.75 \\ -0.812 & 0.444 & 1.045 & -0.75 & 0.49 \end{bmatrix},$$

$$K_2 = \begin{bmatrix} -9.826 & -0.205 & 0.025 & 0.805 & 0.544 & -0.061 \\ -0.205 & 0.349 & 0.423 & 0.189 & -2.105 & 1.764 \\ 0.025 & 0.423 & 1.608 & 0.242 & 0.195 & -0.126 \\ 0.805 & 0.189 & 0.242 & 0.794 & -0.324 & -0.447 \\ 0.544 & -2.105 & 0.195 & -0.324 & 2.788 & 5.243 \\ -0.061 & 1.764 & -0.126 & -0.447 & 5.243 & 0.425 \end{bmatrix}$$

It is easy to check that rank condition is not satisfied thereby yielding an equivalent matrix $E \in \Re^{11 \times 11}$ given by

$$E = [0]\ except\ \ e_{1,11} = -0.227,\ e_{11,1} = 0.1$$

Algebraic computation of **Theorem 3.7.7** yields an overall improvement in the performance index of 6%.

Remark 3.7.9 *The role of interaction patterns in large multilevel control systems is viewed as representing undesirable (acting against the direct control policies) and/or favorable (supporting the basic regulation) information* *modes. The rationale adopted in the foregoing analysis has been based on building up hierarchical control structure with the objective of treating a combined form of both modes. Three noninteracting control policies are employed to implement this structure. The first policy provides for a basic regulation task directly influencing the decoupled subsystems. The function of the second policy is to neutralize (completely or partially) the undesirable modes of information flow. Assigned to the third policy is the task of exploiting the beneficial features of information and therefore to establish an improved overall feedback strategy. It is worth mentioning that the basic regulation job is performed on a decentralized basis. Although the global policies are implemented in a centralized manner, the computational load is not a burden since the calculation is off-line. Also, it has the advantage of adequately making use of the characteristic features of the information pattern.*

3.8 Observation and Estimation

In this section, we focus on the dual problem to the foregoing control problem, that is, we look at the problem of observation and/or estimation in large-scale systems. An efficient approach to the development of multilevel control and estimation schemes for large-scale systems with a major emphasis on the reliability of performance under structural perturbations is described in the sequel. As we shall see, the study is conducted within a decomposition-decentralization framework and leads to simple and noniterative control and estimation schemes. From the previous sections, it turns out that the solution to the control problem involves the design of a set of locally optimal controllers for the individual subsystems in a completely decentralized environment and a global controller on a higher hierarchical level that provides corrective signals to account for the interconnection effects. Similar principles are employed hereafter to develop an estimation scheme, which consists of a set of decentralized optimal estimators for the subsystems, together with certain compensating signals for measurements.

3.8.1 Model set-up

We consider the class of interconnected systems described by

$$\begin{aligned} \dot{x}(t) &= Ax + Bu + h(t,x) \\ y &= Cx \end{aligned} \qquad (3.122)$$

$$\begin{aligned} A &= Blockdiag\left[\begin{array}{cccc} A_1 & A_2 & \cdots & A_{n_s} \end{array}\right], \\ B &= Blockdiag\left[\begin{array}{cccc} B_1 & B_2 & \cdots & B_{n_s} \end{array}\right] \\ C &= Blockdiag\left[\begin{array}{cccc} C_1 & C_2 & \cdots & C_{n_s} \end{array}\right] \end{aligned} \qquad (3.123)$$

where $x(t) \in \Re^n$, $u(t) \in \Re^m$, $y(t) \in \Re^p$ are the composite state, input and output vectors at time $t \in \Re$ given by

$$\begin{aligned} x &= Blockdiag\left[\begin{array}{cccc} x_1^t & x_2^t & \cdots & x_{n_s}^t \end{array}\right]^t, \\ u &= Blockdiag\left[\begin{array}{cccc} u_1^t & u_2^t & \cdots & u_{n_s}^t \end{array}\right]^t, \\ y &= Blockdiag\left[\begin{array}{cccc} y_1^t & y_2^t & \cdots & y_{n_s}^t \end{array}\right]^t \end{aligned} \qquad (3.124)$$

and $h(t,x)$ is the overall interaction pattern. In terms of the subsystems, we express system (3.122) in the form

$$\begin{aligned} \dot{x}_j(t) &= A_j x_j + B_j u_j + h_j(t,x), \quad j = 1, 2, ..., n_s \\ y_j &= C_j x_j(t) \end{aligned} \qquad (3.125)$$

where $x_j(t) \in \Re^{n_j}$ is the state vector , $u_j(t) \in \Re^{n_j}$ is the local control input of the jth subsystem, and the matrices $A_j \in \Re^{n_j \times n_j}$, $B_j \in \Re^{n_j \times m_j}$ are constants. We shall assume that the n_s-pairs $\{A_j,\ C_j\}$ are completely observable.

3.8.2 Observation scheme I

When the subsystems (3.125) are decoupled, that is, $h_j(t,x) \equiv 0,\ \forall j = 1, ..., n_s$, it is a simple design task to construct s independent observers (of full order for simplicity in the exposition) based only on the dynamics of the subsystems. Such observers are linear and are of relatively small dimensions. More precisely, since $\{A_j,\ C_j\}$ are completely observable, it is possible to construct the local observers

$$\dot{\hat{x}}_j(t) = (A_j - K_j C_j)\hat{x}_j + K_j y_j + B_j u_j, \quad j = 1, 2, ..., n_s \qquad (3.126)$$

where $\hat{x}_j(t) \in \Re^{n_j}$ is the state estimate of $x_j(t)$ for the decoupled subsystems

$$\dot{x}_j(t) = A_j x_j + B_j u_j, \quad j = 1, 2, ..., n_s$$

$$y_j = C_j x_j(t) \qquad (3.127)$$

and $K_j \in \Re^{n_j \times p_j}$ are the unknown gains to be determined to ensure any desired degree of convergence of the observation scheme. It is well known [158] that a convergence at least as fast as an exponential decay $e^{-\pi t}$, $\pi > 0$ can be achieved, that is, $l\{x_j(t) - \hat{x}(t)\}exp(\pi t) \rightarrow 0 \; as \; t \rightarrow \infty$ by the selection

$$K_j = \mathcal{G}_j =_j C_j^t, \quad j = 1, 2, ..., n_s \tag{3.128}$$

where $\mathcal{G}_j \in \Re^{n_j \times n_j}$ is the symmetric positive definite solution of the Riccati equation

$$\mathcal{G}_j(A_j^t + \pi I_j) + (A_j + \pi I_j)\mathcal{G}_j - \mathcal{G}_j C_j^t C_j \mathcal{G}_j + \mathcal{Q}_j = 0 \tag{3.129}$$

where $\mathcal{Q}_j \in \Re^{n_j \times n_j}$ is an arbitrary symmetric and nonnegative definite matrix.

Remark 3.8.1 *It is readily evident that the observation scheme (3.126) developed under the decoupling assumption $h_j(t, x) \equiv 0$, $\forall j = 1, ..., n_s$, will eventually be far from satisfactory when used with the composite system (3.122). This is expected, since no knowledge of the interconnection functions or of the outputs of the other subsystems is used in constructing the observers for the subsystems.*

Hence, the problem of interest is to seek modifications of the decentralized observation scheme (3.126), which would result in the same degree of convergence π when used with the composite system (3.126). This problem of tailoring the observation scheme for the overall system to attain a specified degree of convergence is generally more appealing than the intuitively simple and rather brute force method of choosing arbitrary large gains K_j to locate the local observer poles lying quite deep in the left-half complex plane, hoping that the interconnection effects are suitably swamped. Note that this latter strategy would make the observer very sensitive to any noise that may be present.

A simple result established in [355] for the case of linear interconnections

$$h_j(t, x) = \sum_{k=1}^{n_s} H_{jk} x_k(t), \quad H_{jk} \in Re^{n_j \times n_j}$$

is summarized by the following theorem.

Theorem 3.8.2 *Let the overall interconnection matrix $H = [H_{jk}]$, $j, k = 1, 2, ..n_s$ satisfy the following rank condition*

$$rank \begin{bmatrix} C \\ H \end{bmatrix} = rank[C] = p = \sum_{j=1}^{n_s} p_j \tag{3.130}$$

Then the modified observation scheme

$$\begin{aligned} \dot{\hat{x}}_j(t) &= (A_j - K_jC_j)\hat{x}_j + K_jy_j + B_ju_j \\ &\quad + \sum_{k=1}^{n_s} \mathcal{K}_{jk}y_k(t), \quad j = 1, 2, \ldots, n_s \end{aligned} \tag{3.131}$$

where $\mathcal{K} = [\mathcal{K}_{jk}],\ \mathcal{K}_{jk} \in Re^{n_j \times p_j}$ *given by*

$$\mathcal{K} = HC^t(CC^t)^{-1} \tag{3.132}$$

ensures the degree of convergence π *for the observation of the states of the overall system.*

Remark 3.8.3 *We notice from* ***Theorem 3.8.2*** *that the developed decentralized observation scheme incorporates only the outputs of the other subsystems as additional input signals to each observer with the gains* $\mathcal{K}_{jk}$ *being determined by the matrices* $H,\ C$. *The degree of convergence* π *is dependent only on spectrum* $(A - KC)$ *and it is not necessary to design each decentralized local observer to achieve the same degree of convergence. Rather, each* K_j *may be computed corresponding with a desired* π_j*, and the convergence of the overall scheme is determined by the number* $\pi = \min_j \pi_j$.

3.8.3 Observation scheme II

The following theorem established in [355] based on Lyapunov theory provides an alternative decentralized observation scheme which relies on using the estimated states of the other subsystems in generating the additional inputs to be applied to each local observer.

Theorem 3.8.4 *Consider the linear interconnection pattern*

$$h_j(t, x) = \sum_{k=1}^{n_s} H_{jk}(t)x_k(t), \quad H_{jk}(t) \in Re^{n_j \times n_j}$$

and let the overall interconnection matrix $H(t) = [H_{jk}(t)],\ j, k = 1, 2, ..n_s$ *satisfy the following rank condition:*

$$H(t) \ = \ G[U(t) \ - \ S(t)] \tag{3.133}$$

where $U(.) : \Re \to \Re^{n \times n}$ *is an arbitrary skew-symmetric matrix, and* $S(.) : \Re \to \Re^{n \times n}$ *is an arbitrary symmetric matrix that satisfies the condition*

$$\begin{aligned} \Pi(t) &= \mathcal{Q}_j + \mathcal{G}C^tC\mathcal{G} + 2\mathcal{G}S(t)\mathcal{G} \geq 0, \\ \Pi(t) &: \Re \to \Re^{n \times n} \end{aligned} \tag{3.134}$$

with

$$\mathcal{G} = Blockdiag \left[\begin{array}{cccc} \mathcal{G}_1 & \mathcal{G}_2 & \cdots & \mathcal{G}_{n_s} \end{array} \right]$$

and $\mathcal{G}_j > 0$ *is the solution of (3.129). Then the modified observation scheme*

$$\begin{aligned} \dot{\hat{x}}_j(t) &= (A_j - K_j C_j)\hat{x}_j + K_j y_j + B_j u_j \\ &+ \sum_{k=1}^{n_s} \mathcal{K}_{jk}\hat{x}_k(t), \quad j = 1, 2, ..., n_s \end{aligned} \tag{3.135}$$

ensures the degree of convergence π *for the observation of the states of the overall system.*

Remark 3.8.5 *In it readily evident that* **Theorems 3.8.2** *and* **3.8.4** *identify conditions under which decentralized observation scheme with desired properties may be designed for large-scale systems with linear interconnection patterns. A noteworthy feature of the underlying result is that every one of these conditions as well as the solution of the associated Riccati equations and the calculation of the required eigenvalues are performed on the subsystem level, which is of relatively small dimensions thereby rendering minimal computational requirements. In addition, only the observability of the decoupled subsystems is assumed, and any test for the observability of the overall system is avoided. The characterization of the interconnections among subsystems as perturbation terms affecting the satisfactory behavior of the decoupled subsystems helps in reaching the desired design goal. An alternative read-out of the results is that if the strength of the interconnections is suitably limited, or the interconnection pattern satisfies suitable symmetry conditions, simple modifications of the decentralized observation scheme may be developed for the state observation, with desired convergence properties, of the composite system.*

3.8.4 Estimation scheme

The ideas used for the generation of multilevel control schemes may be extended by duality to the linear least squares estimation problem. The objective is the determination of conditions for obtaining optimal or near-optimal estimates of the state of the system by designing locally optimal estimates for the individual subsystems.

Consider the large-scale system formed by an interconnection of subsystems

$$\begin{aligned} \dot{x}_j(t) &= A_j x_j(t) + B_j w_j(t) + \sum_{k=1}^{n_s} H_{jk} x_k(t) \\ y_j(t) &= C_j x_j(t) + v_j(t) \end{aligned} \tag{3.136}$$

where for $j \in \{1, ..., n_s\}$, $x_j(t) \in \Re^{n_j}$ is the state vector, $w_j(t) \in \Re^{m_j}$ is the input noise, $y_j(t) \in \Re^{p_j}$ is the output, and $v_j(t) \in \Re^{p_j}$ is the measurement noise. In the sequel, it will be assumed that the stochastic processes $\{w_j(t);\ t \geq t_o,\ j \in \{1, ..., n_s\}\}$ and $\{v_j(t);\ t \geq t_o,\ j \in \{1, ..., n_s\}\}$ are zero mean Gaussian stationary white noise processes of known covariances $0 \leq \mathcal{Q}_j \in \Re^{m_j \times m_j}$ and $0 < \mathcal{R}_j \in \Re^{p_j \times p_j}$, respectively. The processes $u_j(t)$ and $v_j(t)$ are assumed to be independent of each other, that is, $\mathbf{E}[u_j(t)v_j(\tau)] = 0,\quad \forall \tau,\ t \geq t_o,$ for all $j \in \{1, ..., n_s\}$. The initial states $x_j(to)$ are Gaussian random vectors of known mean and are independent of $u_j(t),\ t \geq to$ and $v_j^t(t),\ t \geq to$. Moreover, we assume that the s-pairs $(A_j,\ C_j)$ are completely observable.

By similarity to the state-regulation case, when the subsystems are decoupled (that is, when $H_{jk} \equiv 0$), independent decentralized optimal estimators may be designed for the individual subsystems and are given by [305]:

$$\begin{aligned} \dot{\hat{x}}_j(t) &= (A_j - \mathsf{K}_j C_j)\hat{x}_j(t) + \mathsf{K}_j y_j(t), \quad j = 1, 2, ..., n_s \\ \hat{x}_j(t_o) &= \mathbf{E}[x_j(t_o)] \end{aligned} \tag{3.137}$$

where $\mathsf{K}_j \in \Re^{n_j \times p_j}$ is given by

$$\mathsf{K}_j = \mathcal{W}_j C_j^t \mathcal{R}_j^{-1} \tag{3.138}$$

and $\mathcal{W}_j \in \Re^{n_j \times n_j}$ is the covariance of the error process $e_j(t) = x_j(t) - \hat{x}_j(t)$, satisfying the Riccati equation

$$\mathcal{W}_j A_j^t + A_j \mathcal{W}_j - \mathcal{W}_j C_j^t \mathcal{W}_j^{-1} C_j \mathcal{G}_j + B_j \mathcal{Q}_j B_j^t = 0 \tag{3.139}$$

We recall that $Tr(\mathcal{W}_j)$ serves as a measure of the performance of the estimation scheme (3.139). In line of the regulation case, it is of interest to determine conditions on the interconnection matrix H such that the decentralized estimators (3.137), or equivalently

$$\dot{\hat{x}}(t) = (A - \mathsf{K}C)\hat{x}(t) + \mathsf{K}y(t) \tag{3.140}$$

remain optimal for the overall system, that is,

$$\begin{aligned} \dot{x}(t) &= (A + H)x(t) + Bw(t) \\ y(t) &= Cx(t) + v(t) \end{aligned} \tag{3.141}$$

where $x,\ \hat{x},\ y,\ w$, and v are composite vectors of $x_j,\ \hat{x}_j,\ y_j,\ w_j$, and v_j, respectively. Such conditions are given in the following theorem [354].

Theorem 3.8.6 *Let*

$$\mathcal{W} = Blockdiag \left[\mathcal{W}_1 \quad \mathcal{W}_2 \quad \cdots \quad \mathcal{W}_{n_s} \right]$$

where $0 < \mathcal{W}_j$ *is the solution of (3.139). The decentralized estimation scheme (3.140) gives optimal state estimates of the overall system (3.141) if and only if the the linear interconnection pattern* H *admits the factorization*

$$H = \mathcal{W}\mathcal{S} \tag{3.142}$$

where $\mathcal{S} \in \Re^{n \times n}$ *is an arbitrary skew-symmetric matrix.*

Proof: The proof runs parallel to that of **Theorem 3.7.1**.

Remark 3.8.7 *In some relevant classes of the interconnection pattern, condition (3.142) fails to be satisfied. This leads to the conclusion that the decentralized estimators (3.140) no longer generate optimal estimates of the states of (3.141). This is quite expected since the estimator for any subsystem does not incorporate the estimates of the states of the rest of the subsystems. This defect may be remedied by suitably modifying the dynamics of the estimators and consequently obtaining improved estimates.*

Following the approach of decentralized state regulation, our strategy hereafter involves retaining the gain matrix K unchanged. For this purpose, consider the new estimators described by

$$\begin{aligned} \dot{\hat{x}}_j(t) &= (A_j - \mathsf{K}_j C_j)_j(t) + \mathsf{K}_j y_j(t) \\ &- \sum_{k=1}^{n_s} \mathsf{K}_j C^*_{jk} \hat{x}_k(t), \quad j = 1, 2, ..., n_s \\ \hat{x}_j(t_o) &= \mathbf{E}[x_j(t_o)] \end{aligned} \tag{3.143}$$

where $C^*_{jk} \in \Re^{p_j \times n_j}$ are the perturbation matrices to be determined. Equivalently (3.143) can be represented by

$$\dot{\hat{x}}(t) = [A - \mathsf{K}(C + C^*)]\hat{x}(t) + \mathsf{K}y(t) \tag{3.144}$$

where $C^* = [C^*_{jk}] \in \Re^{p_j \times n_j}, \ j, k = 1, 2, .., s$. Assuming that the $n \times n$ matrix $(A + H)$ is nonsingular, a relation among the perturbation matrix $C*$, the interconnection matrix H, and the performance of the estimation scheme (3.144) compared with that of the decentralized scheme (3.137) is now given by the following theorem.

Theorem 3.8.8 *Let $\mathcal{S} \in \Re^{n \times n}$ be an arbitrary skew-symmetric matrix of the matrix equation*

$$\mathcal{S}(A + H)^t + (A + H)\mathcal{S} + (A\mathcal{W}H^t \; - \; H\mathcal{W}A) = 0 \tag{3.145}$$

and the matrix $\widehat{\mathcal{W}} \in \Re^{n \times n}$ given by

$$\widehat{\mathcal{W}} = (A + H)^{-1}(\mathcal{S} \; - \; H\mathcal{W}) \tag{3.146}$$

be such that $(\mathcal{W} + \widehat{\mathcal{W}})$ is positive definite. Then the estimation scheme (3.144) with

$$C^* = -C\widehat{\mathcal{W}}(\mathcal{W} + \widehat{\mathcal{W}})^{-1}$$

generates suboptimal estimates of the states of the composite system (3.141) with an index of suboptimality given by

$$\epsilon = \frac{Tr(\widehat{\mathcal{W}})}{Tr(\mathcal{W})}$$

Remark 3.8.9 *It must be emphasized that* ***Theorem 3.8.8*** *may be regarded as the dual of* ***Theorem 3.7.5*** *and hence the proof proceeds along almost identical steps to that of* ***Theorem 3.7.5****. In addition, all features and remarks made earlier for the regulation hold with appropriate dualizations to the present result. In essence, subsystem autonomy is a factor of major importance in the design of controllers and estimators for large-scale systems from considerations of reliability of performance under structural perturbations.*

Remark 3.8.10 *It is worthy to note that efficient two-level algorithms based on the multiple projection approach were developed for the global Kalman filter in [99] and for parameter estimation in [101]. In both cases, the optimal minimum variance estimate is achieved using a fixed number of iterations. This equally applies to both the recursive and nonrecursive versions of the algorithms.*

In the remaining part of this chapter, we direst attention to the class of large-scale discrete-time systems. Given the rationale that continuous-time and discrete-time systems are equally important, the main reason behind this direction is the fact that there is a little exposition in the literature about this topic. Virtually almost all of the texts on large-scale systems have focused on continuous-time systems. Therefore, addressing basic results related to discrete-time systems deserves special consideration.

3.9 Interconnected Discrete-Time Systems

A wide class of interconnected discrete-time systems (DTS) Σ composed of n_s coupled subsystems $\Sigma_j,\ j = 1, 2, ..., n_s$ is represented by:

$$\begin{aligned} \Sigma: x(k+1) &= Ax(k) + Bu(k) + g(k,x) \\ y(k) &= Cx(k) \end{aligned} \tag{3.147}$$

where for $k \in Z_+ := \{0, 1, ...\}$ and $x = (x_1^t, ..., x_{n_s}^t)^t \in \Re^n$, $n = \sum_{j=1}^{n_s} n_j$, $u = (u_1^t, ..., u_{n_s}^t)^t \in \Re^p$, $p = \sum_{j=1}^{n_s} p_j$, $y = (y_1^t, ..., y_{n_s}^t)^t \in \Re^q$, $q = \sum_{j=1}^{n_s} q_j$ being the state, control, and output vectors of system Σ. The function $g : Z_+ \times \Re^n \to \Re^n$, $g(k, x(k)) = (g_1^t(k, x(k)), .., g_{n_s}^t(k, x(k)))^t$ is a piecewise-continuous vector function in its arguments. The associated matrices are real constants and modeled as $A = diag\{A_1, .., A_N\}$, $A_j \in \Re^{n_j \times n_j}$, $B = diag\{B_1, .., B_N\}$, $B_j \in \Re^{n_j \times p_j}$, and $C = diag\{C_1, .., C_N\}$, $C_j \in \Re^{q_j \times n_j}$. Exploiting the structural form of system (3.147), a model of the jth subsystem Σ_j can be described by

$$\begin{aligned} \Sigma_j: \ x_j(k+1) &= A_j x_j(k) + B_j u_j(k) + g_j(k, x(k)) \\ y_j(k) &= C_j x_j(k) \end{aligned} \tag{3.148}$$

where $x_j(k) \in \Re^{n_j}$, $u_j(k) \in \Re^{p_j}$, and $y_j(k) \in \Re^{q_j}$ are the subsystem state, control input, and output, respectively.

By similarity to the case of continuous-time systems [155], there are discrete-time control systems that exhibit singular perturbation or mode-separation phenomena. Basically, there are two sources of modeling the discrete-time systems [279, 280]: from pure difference equations or from discrete-time modeling of continuous-time systems. Regarding the first source, we look at a general linear, time-invariant discrete-time system:

$$\begin{aligned} x(k+1) &= A_{11} x(k) + \epsilon^{1-j} A_{12} z(k) + B_1 u(k), \\ \epsilon^{2i} z(k+1) &= \epsilon^i A_{21} x(k) + \epsilon A_{22} z(k) + \epsilon^j B_2 u(k) \end{aligned} \tag{3.149}$$

where $i \in \{0,\ 1\}$, $j \in \{0,\ 1\}$, $x(k) \in \Re^n$, $z(k) \in \Re^m$ are state-vectors and $u(k) \in \Re^r$ is the control vector. Depending on the values for i and j, there are three limiting cases of (3.149) described by [280]:

1. *C-model* $i = 0,\ j = 0,$ *where the small parameter* ϵ *appears in the column of the system matrix,*

2. *R-model* $i = 0,\ j = 1,$ *where we see the small parameter* ϵ *in the row of the system matrix*, and

3. *D-model* $i = 1,\ j = 1,$ *where* ϵ *is positioned in an identical fashion to that of the continuous-time system described by differential equations.*

Turning the second source, either numerical solution or sampling of singularly perturbed continuous-time systems normally results in discrete-time models. It is reported in [279] that by applying a block diagonalization transformation to a continuous-time singularly perturbed system, the original state variables $x(t)$ and $z(t)$ can be expressed in terms of the decoupled system consisting of slow and fast variables $x_s(t)$ and $z_f(t)$, respectively. Using a sampling device with the decoupled continuous-time system, we get a discrete-time model that critically depends on the sampling interval T.

Depending on the sampling interval, we get a fast(subscripted by f) or slow (subscripted by s) sampling model. In a particular case, when $T_f = \epsilon$, we get the *fast sampling model* as

$$\begin{aligned} x(n+1) &= (I_s + \epsilon D_{11})x(n) + \epsilon D_{12}z(n) + \epsilon E_1 u(n), \\ z(n+1) &= D_{21}x(n) + D_{22}z(n) + E_2 u(n) \end{aligned} \tag{3.150}$$

where n denotes the fast sampling instant (not to be confused with the system order described previously). Similarly, if $T_s = 1$, we obtain the slow sampling model as

$$\begin{aligned} x(k+1) &= E_{11}x(k) + \epsilon E_{12}z(k) + E_1 u(k), \\ z(k+1) &= E_{21}x(k) + \epsilon E_{22}z(k) + E_2 u(k) \end{aligned} \tag{3.151}$$

where k represents the slow sampling instant, and $n = k[1/\epsilon]$. Also, the D and E matrices in (3.150) and (3.151) are related to the matrices A, B, and transformation matrices; see [188]. Note that the fast sampling model (3.150) can be viewed as the discrete analog (either by exact calculation using the exponential matrix or by using the Euler approximation) of the continuous-time system

$$\begin{aligned} \frac{x}{\tau} &= \epsilon A_{11}x(\tau) + \epsilon A_{12}z(\tau) + \epsilon B_1 u(\tau), \\ \frac{z}{\tau} &= A_{21}x(\tau) + A_{22}z(\tau) + E_2 u(\tau) \end{aligned} \tag{3.152}$$

which itself is obtained from its continuous-time system counterpart using the stretching transformation $\tau = t/\epsilon$.

In [187], composite control technique and related asymptotic results were extended from systems described by singularly perturbed differential equations to systems described by singularly perturbed difference equations. It laid down a theoretical framework for studying multirate sampling control for systems having slow and fast modes.

An alternative modeling formalism is to express system (3.147) into the form

$$\begin{aligned} \Sigma_s: \; x_a(k+1) &= A_a x_a(k) + A_e x_m(k) + B_a u(k), \; x_a(0) = x_{ao} \\ x_m(k+1) &= A_c x_a(k) + A_m x_m(k) + B_m u(k), \; x_m(0) = x_{mo} \end{aligned} \tag{3.153}$$

$$y(k) = C_a x_a(k) + C_m x_m(k) \tag{3.154}$$

where $x_a(k) \in \Re^r$, $x_m(k) \in \Re^{n-r}$, $u(k) \in \Re^m$, and $y(k) \in \Re^p$. The vectors $x(k) = [x_a^t(k) \;\; x_m^t(k)]^t$, $u(k)$, and $y(k)$ are the overall state, control, and output vectors at the discrete instant k, respectively. The overall systems matrices are expressed by

$$A = \begin{bmatrix} A_a & A_e \\ A_c & A_m \end{bmatrix}, \; B = \begin{bmatrix} B_a \\ B_m \end{bmatrix}, \; C^t = \begin{bmatrix} C_a^t \\ C_m^t \end{bmatrix} \tag{3.155}$$

Remark 3.9.1 *The material covered in this section constitutes, to a large extent, a distinct discrete version of the two-time-scale model of continuous-time systems [155, 280], a simple comparison with models (3.149)-(3.152). In the subsequent sections, we treat models (3.148) and (3.153) in relation to the original model (3.147). We will reveal relevant structural properties and attend to both regulation and observation (or estimation) problems. As we shall see later, the dimension r plays a crucial role in the analysis.*

3.9.1 Mode separation in discrete systems

Starting from model (3.153), we introduce the following notations,

$$\begin{aligned} \lambda(A) &= \begin{bmatrix} \lambda_1 & \lambda_2, \cdots & \lambda_n \end{bmatrix}, \\ \mathsf{V} &= \begin{bmatrix} \mathsf{v}_1 & \mathsf{v}_2, \cdots & \mathsf{v}_n \end{bmatrix}, \\ \mathsf{W} &= \begin{bmatrix} \mathsf{w}_1 & \mathsf{w}_2, \cdots & \mathsf{w}_n \end{bmatrix} \end{aligned} \tag{3.156}$$

to denote the eigen-spectrum of matrix A, the modal matrix and the matrix of reciprocal basis vectors, respectively. For simplicity in exposition, we assume that system (3.153) is a well-damped asymptotically stable system and is put forward such that the n eigenvalues are arranged in the following manner:

$$1 \, |\lambda_1| \geq \cdots\cdots \geq |\lambda_r| \geq |\lambda_{r+1}| \cdots\cdots \geq |\lambda_n| \tag{3.157}$$

The columns of V and $\mathsf{W} = \mathsf{V}^{-1}$ are arranged in the same manner. It is should be noted the ordering process can be conveniently achieved by standard permutation and/or scaling procedures in numerical analysis [87].

In the literature, system (3.153) is called two-time-scale [206]-[210] if $\lambda(A)$ can be separated by absolute values into nonempty sets S and F so that

$$|s_j| >> |f_k| \;\forall\, s_j \in \mathsf{S},\; \forall\, f_k \in \mathsf{F} \tag{3.158}$$

Let r be the number of eigenvalues in S, then in view of (3.157), we get

$$\begin{aligned} |s_{j+1}| &\leq |s_j|, \;\; j = 1, 2, \cdots\cdots, r-1, \\ |f_{k+1}| &\leq |f_k|, \;\; k = 1, 2, \cdots\cdots, n-r-1 \end{aligned} \tag{3.159}$$

It is shown in [209, 210] that for any two-time-scale discrete system, the ratio

$$\begin{aligned} \mu &= \frac{|\lambda_M(\mathsf{F})|}{|\lambda_m(\mathsf{S})|} = \frac{|f_1}{|s_r|}, \\ &= \frac{|\lambda_{r+1}|}{|\lambda_r|} << 1 \end{aligned} \tag{3.160}$$

defines a measure of the time-separation in discrete systems. It should be emphasized that the parameter μ signifies an intrinsic property of system (3.153) and it does not appear explicitly in the model under consideration.

When discrete system (3.153) possesses the mode-separation property (3.160), it follows that the eigen-spectrum $\lambda(A)$ is composed of r large eigenvalues of order 1 and $n-r$ small eigenvalues of order μ within the unit circle. The small eigenvalues represent a *fast* subsystem whereas the large eigenvalues designate a *slow* subsystem. In an asymptotically stable system, the fast modes are important only during a short transient period. After that period they are negligible and the behavior of the system can be described by its slow modes.

It has been reported that the slow and fast subsystems of system (3.153) are expressed as

Slow subsystem:

$$\begin{aligned} x_s(k+1) &= A_s x_s(k) + B_s u_s(k), \; x_s(0) = x_{ao} \\ y_s(k) &= C_s x_s(k) + D_s u_s(k) \end{aligned} \tag{3.161}$$

Fast subsystem:

$$\begin{aligned} x_f(k+1) &= A_m x_f(k) + B_m u_f(k), \; x_f(0) = x_{mo} - \bar{x_m}(0) \\ y_f(k) &= C_m x_f(k) \end{aligned} \tag{3.162}$$

where

$$\begin{aligned} A_s &= A_a + A_e(I - A_m)^{-1} A_c, \;\; B_s = B_a + A_e(I - A_m)^{-1} B_m \\ C_s &= C_a + C_m(I - A_m)^{-1} A_c, \;\; D_s = C_m(I - A_m)^{-1} B_m \end{aligned} \tag{3.163}$$

$$\bar{x}_m(k) = (I - A_m)^{-1}[A_c x_s(k) + B_m u_s(k)] \tag{3.164}$$

and the matrix $(I - A_m)^{-1} \in \Re^{n-r \times n-r}$ is assumed to exist and the quantity $\bar{x}_m(k)$ represents the quasi-steady state. The following definition is needed.

Definition 3.9.2 *A vector or matrix function* $\Psi(\mu)$ *of a positive scalar* μ *is said to be* $\mathbf{O}(\mu^m)$ *if there exist positive constants* $\mathbf{c}$ *and* μ^* *such that*

$$|\Psi(\mu)| \leq \mathbf{c}\mu^*, \quad \forall \mu \leq \mu^*$$

In view of **Definition 3.9.2**, the slow (reduced) subsystem (3.161) together with the fast (boundary layer) subsystem (3.162) provides $\mathbf{O}(\mu)$ approximation to the dynamic behavior of the original system (3.153) and both subsystems are of low order. This implies that

$$\begin{aligned} x_a(k) &= x_s(k) + \mathbf{O}(\mu) \\ x_m(k) &= x_f(k) + \mathbf{O}(\mu) \end{aligned} \tag{3.165}$$

The reader is referred to [209, 238]-[241] where alternative procedures and further details are given.

Our next objective is to improve the accuracy of the derived lower-order methods. We recall from [280] that the underlying assumption of multitime-scale theory is that during short-term studies the slow variables remain constant and that by the time their changes become noticeable, the fast transients have already reached their quasi-steady state. As pointed out in [155], the slow variables during transients are time-varying quantities, a fact that leads us to conclude that the true states $x_a(k)$ and $x_m(k)$ will differ from $x_s(k)$ and $\bar{x}_m(k)$ mainly by their fast parts. A constructive procedure for iterative separation of time-scales of the class of discrete-time systems (3.153) was reported in [212]. The result after $t-$L and $i-$K iterations is summarized by the model

$$\begin{aligned} \alpha_i(k+1) &= F_{ti}\alpha_i(k) + G_{ti}\beta_t(k) + R_{ti}u(k), \\ \beta_t(k+1) &= H_t\alpha_i(k) + P_{ti}\beta_t(k) + E_t u(k) \end{aligned} \tag{3.166}$$

where

$$\begin{aligned} F_{ti} &= F_t - K_{ti}H_t, \;\; G_{ti} = A_e - K_{ti}P_t + F_{ti}K_{ti}, \\ P_{ti} &= P_t - H_tK_{ti}, \;\; R_{ti} = (I - K_{ti}L_t)B_a - K_{ti}B_m, \\ F_t &= A_a - A_eL_t, \;\; H_t = A_c - A_mL_t + L_tF_t, \\ P_t &= A_m + L_tA_e, \;\; E_t = B_m + L_tB_a \end{aligned} \tag{3.167}$$

and

$$\begin{aligned} L_{t+1} &= L_t + (A_mL_t - L_tA_a + L_tA_e - A_c)(A_a - A_eL_j)^{-1}, \\ L_1 &= -(I - A_m)^{-1}A_c, \\ K_{ti+1} &= -F_t^{-1}K_{ti}(P_tH_tK_{ti}), \;\; K_{t1} = -F_t^{-1}A_e \end{aligned} \tag{3.168}$$

It should be noted that the slow and fast subsystems (3.166) are now very weakly coupled since G_{ti} is $\mathbf{O}(\mu^{i+t})$ and H_t is $\mathbf{O}(\mu^t)$. It is also evident that F_{ti} and P_{ti} are $\mathbf{O}(\mu^{i+t})$ approximations of the slow and fast modes, respectively. Obviously, setting $t = 1$ and $i = 0$ in (3.166)-(3.168) results in the lower-order models (3.161)-(3.162).

3.10 Feedback Control Design

In view of the mode-separation property, we now proceed to consider the feedback control design problem with the objective of developing two-stage schemes based on the dynamics of the slow and the fast subsystems.

3.10.1 State feedback design

A linear state feedback control is sought of the form

$$u_s(k) = K_s x_s(k), \quad u_f(k) = K_m x_f(k) \tag{3.169}$$

where the gain matrices $K_s \in \Re^{m\times r}$ and $K_m \in \Re^{m\times n-r}$ are to be designed based on the independent specifications of the slow and fast dynamics. A composite state feedback control is given by

$$\begin{aligned} u(k) &= ([I - K_f(I - A_m)^{-1}B_m]K_s - K_f(I - A_m)^{-1}A_c)x_a(k) \\ &+ K_m x_m(k) \end{aligned} \tag{3.170}$$

The following theorem in [209] establishes the properties of the feedback system (3.153).

Theorem 3.10.1 *Let the independent controls (3.169) and the composite control (3.170) be applied to the systems (3.161), (3.162), and (3.153), respectively. If matrix $A_m + B_m K_m$ is asymptotically stable, then*

$$\begin{aligned} x_a(k) &= x_s(k) + \mathbf{O}(\mu) \\ x_m(k) &= x_f(k) + (I - A_m)^{-1}(A_c + B_m K_s)x_s(k) + \mathbf{O}(\mu) \end{aligned} \tag{3.171}$$

hold for all finite $k \geq 0$. In addition, if matrix $A_s + B_s K_s$ is asymptotically stable then (3.171) holds for all $k \in [0, \infty)$.

3.10.2 Design algorithm I

A two-stage eigenvalue assignment algorithm for state feedback design consists of the following steps:

1. *Compute* K_f *to place the eigenvalues of* $A_m + B_m K_m$ *at* $n - r$ *desired locations.*

2. *Compute* K_s *to place the eigenvalues of* $A_s + B_s K_s$ *at* r *desired locations.*

3. *Compute* K_a *using the formula*

$$K_a = [I - K_f(I - A_m)^{-1} B_m] K_s - K_f (I - A_m)^{-1} A_c$$

4. *Implement the composite control*

$$u(k) = K_a x_a(k) + K_m x_m(k)$$

Theorem 3.10.2 *If* A_m *has a spectral radius less than 1, then the low-order control* $u(k) = K_s x_a(k)$ *can be used to stabilize system (3.153).*

3.10.3 Illustrative example 3.5

To demonstrate the application of the above theoretical analysis, we consider a fifth-order discrete model of a steam plant described by

$$A = \begin{bmatrix} 0.915 & 0.051 & 0.038 & 0.015 & 0.038 \\ -0.031 & 0.889 & -.001 & 0.046 & 0.111 \\ -0.006 & 0.468 & 0.247 & 0.014 & 0.048 \\ -0.715 & -0.022 & -0.021 & 0.240 & -0.024 \\ -0.148 & -0.003 & -0.004 & 0.090 & 0.026 \end{bmatrix},$$
$$B^t = \begin{bmatrix} 0.0098 & 0.122 & 0.036 & 0.562 & 0.115 \end{bmatrix}$$

The open-loop eigenvalues are $\{0.8828 \pm j0.0937,\ 0.2506 \pm j0.0252,\ 0.295\}$ which shows internal oscillations. It can be easily checked that this model has the mode-separation property (3.160) with $r = 2,\ n - r = 3$, and $\mu = 0.2646$.

The slow and fast subsystems (3.161) and (3.162) are given by

$$\begin{aligned}
A_s &= \begin{bmatrix} 0.8901 & 0.0727 \\ -0.099 & 0.8858 \end{bmatrix}, \ \lambda(A_o) = \{0.8879 \pm j0.0848\} \\
B_s^t &= \begin{bmatrix} 0.0306 & 0.1761 \end{bmatrix}, \\
A_m &= \begin{bmatrix} 0.247 & 0.014 & 0.048 \\ -0.021 & 0.240 & -0.024 \\ -0.004 & 0.090 & 0.026 \end{bmatrix}, \\
\lambda(A_m) &= \{0.2506 \pm j0.0252, \ 0.0295\}, \\
B_m^t &= \begin{bmatrix} 0.036 & 0.562 & 0.115 \end{bmatrix}
\end{aligned}$$

The pair $(A_s, \ B_s)$ is controllable and the desired slow eigenvalues are placed at $\{0.893, \ 0.825\}$ using the gain

$$K_s = [-0.5465 \quad 0.0402]$$

For the fast subsystem, the desired fast eigenvalues are placed at

$$\{0.251, \ 0.250, \ 0.0295\}$$

using the gain

$$K_m = [-0.0365 \quad -0.0195 \quad -0.0509]$$

This yields

$$K_a = [-0.5679 \quad 0.0432]$$

The closed-loop eigenvalues of the original fifth-order model using the composite control $u(k) = K_a x_a(k) + K_m x_m(k)$ are found to be

$$\{0.8891, \ 0.8890, \ 0.243, \ 0.242, \ 0.0342\}$$

and the original internal oscillations have been eliminated thereby leading to smooth behavior.

The foregoing design scheme depends on the availability of the state variables. In what follows, we look at the problem of designing observers in order to estimate the slow and fast states.

3.10.4 Observer design

By employing the available input and output measurements and assuming one-step delay between measuring and processing the information records, a full-order observer for system (3.153) is given by:

$$\begin{aligned}\hat{x}_a(k+1) &= A_a\hat{x}_a(k) + A_e\hat{x}_m(k) + B_a u(k) \\ &+ K_{oa}[y(k) - C_a\hat{x}_a(k) + C_m\hat{x}_m(k)] \quad (3.172) \\ \hat{x}_m(k+1) &= A_c\hat{x}_a(k) + A_m\hat{x}_m(k) + B_m u(k) \\ &+ K_{om}[y(k) - C_a\hat{x}_a(k) + C_m\hat{x}_m(k)] \quad (3.173)\end{aligned}$$

where $\hat{x}_a(k) \in \Re^r$, $\hat{x}_m(k)\Re^{n-r}$ are the estimates of $x_a(k)$, $x_m(k)$, and $K_{oa} \in \Re^{r\times p}$ and $K_{om} \in \Re^{n-r\times p}$ are the design parameters that may be properly selected to ensure any desired degree of convergence of the observation scheme. Indeed, the design will exploit the mode-separation property.

In terms of the observation error vectors $\tilde{x}_a(k) = x_a(k) - \hat{x}_a(k)$ and $\tilde{x}_m(k) = x_m(k) - \hat{x}_m(k)$, it follows from (3.153), (3.154), (3.172), and (3.173) that

$$\begin{aligned}\tilde{x}_a(k+1) &= (A_a - K_{oa}C_a)\tilde{x}_a(k) + (A_e - K_{oa}C_m)\tilde{x}_m(k) \quad (3.174) \\ \tilde{x}_m(k+1) &= (A_c - K_{om}C_a)\tilde{x}_a(k) + (A_m - K_{om}C_m)\tilde{x}_m(k) \quad (3.175)\end{aligned}$$

Indeed, system (3.174)-(3.175) will function as an observer for system (3.153) and (3.154) if the matrices can be selected such that system (3.172) and (3.173) is asymptotically stable. The following theorem in [208] provides the main results.

Theorem 3.10.3 *Suppose that $(I - A_m)^{-1}$ exists. If the (A_s, C_s) and (A_m, C_m) are observable pairs, then system (3.174)-(3.175) is asymptotically stable. The gain matrix K_{oa} is given by*

$$K_{oa} = K_{os}[I + C_f(I - A_m)^{-1}K_{om}] - A_e(I - A_m)^{-1}K_{om} \quad (3.176)$$

where K_{os} and K_{om} are any matrices for which $A_s - K_{os}C_s$ and $A_m - K_{om}C_m$, respectively, have spectral norms less than one.

3.10.5 Design algorithm II

A two-stage eigenvalue assignment algorithm for observer design consists of the following steps:

1. *Compute K_{om} to place the eigenvalues of $A_m - K_{om}C_m$ at $n-r$ desired locations.*

2. *Compute K_{os} to place the eigenvalues of $A_s - K_{os}C_s$ at n desired locations.*

3. *Compute K_{oa} using the formula (3.176).*

3.10.6 Observer-based control design

For simplicity, it is assumed that the fast subsystem is asymptotically stable. This allows us to construct a reduced-order observer of the slow subsystem (3.161) and (3.163) of the form:

$$\begin{aligned} \hat{x}_a(k+1) &= A_s\hat{x}_a(k) + B_s u(k) \\ &+ K_{ob}[y(k) - C_s\hat{x}_a(k) - D_s u(k)] \end{aligned} \tag{3.177}$$

An observer-based controller is then described by

$$u(k) = G_{ob}\hat{x}_a(k) \tag{3.178}$$

where $K_{ob} \in \Re^{r\times p}$ and $G_{ob} \in \Re^{m\times r}$ are the design parameters. The closed-loop system can be expressed by

$$\begin{aligned} \begin{bmatrix} x_a(k+1) \\ \hat{x}_a(k+1) \end{bmatrix} &= \begin{bmatrix} A_a & B_a G_{ob} \\ K_{ob}C_a & A_{mm} \end{bmatrix} \begin{bmatrix} x_a(k) \\ \hat{x}_a(k) \end{bmatrix} \\ &+ \begin{bmatrix} A_e \\ K_{ob}C_m \end{bmatrix} x_m(k) \end{aligned} \tag{3.179}$$

$$x_m(k+1) = [A_c - B_m G_{ob}] \begin{bmatrix} x_a(k) \\ \hat{x}_a(k) \end{bmatrix} + A_m x_m(k) \tag{3.180}$$

where

$$A_{mm} = A_s + B_s G_{ob} - K_{ob}C_s - K_{ob}D_s G_{ob}$$

System (3.179)-(3.180) exhibits the time-separation property with *fast subsystem* described by

$$x_m(k+1) = A_m x_m(k) \tag{3.181}$$

and *slow subsystem* given by

$$\begin{aligned} \widetilde{x}_s(k+1) &= \mathsf{F}\widetilde{x}_a(k) \\ \mathsf{F} &= \begin{bmatrix} A_a & B_a G_{ob} \\ K_{ob}C_a & A_{mm} \end{bmatrix} \end{aligned} \tag{3.182}$$

The following theorem in [208] presents the main result.

Theorem 3.10.4 *If*

1. *A_m is a stable matrix,*

2. *(A_s, B_s) is a controllable pair,*

3. *(A_s, C_s) is an observable pair*

then the observer-based control (3.178) is a stabilizing controller. The gains K_{ob} and G_{ob} are any matrices for which the matrices $(A_s + B_s G_{ob})$ and $(A_s - K_{ob} C_s)$, respectively, are stable matrices.

3.10.7 Design algorithm III

A two-stage eigenvalue assignment algorithm for observer-based feedback design consists of the following steps:

1. *Compute K_{ob} to place the observer eigenvalues of $A_s - K_{ob} C_s$ at r desired locations.*

2. *Compute G_{ob} to place the controller eigenvalues of $A_s + B_s G_{ob}$ at r desired locations.*

Next, we demonstrate the application of the foregoing feedback design.

3.10.8 Illustrative example 3.6

A variety of investigations on internal combustion engines have been carried out using engine-dynamometer test rigs [276, 277]. Recent concern with pollution has caused further studies on petrol and diesel engines to become necessary with a view to reducing their contribution to pollution of the atmosphere. For this purpose, it is desirable to be able to control the torque and speed developed by such engines independently so that the engines can be subjected to a test cycle that will simulate typical road driving conditions. The purpose of this simulation example is to study the design of control schemes for such a system to see how useful two-stage design methods are in providing independent control of torque and speed. An existing analogue model of an engine-dynamometer test rig [276] is used throughout this study, and two different models of the system are considered [277] having order 3 and S. The engine considered is a petrol engine representative of the type used to power medium-sized passenger cars. The energy developed by the engine is controlled by the input voltage to the throttle servo system used on the test rig. The dynamometer, which acts as a load for the engine, consumes energy at a rate determined by the input voltage to the dynamometer field-current controller. The engine-dynamometer configuration is shown schematically in Figures 3.16 and 3.17.

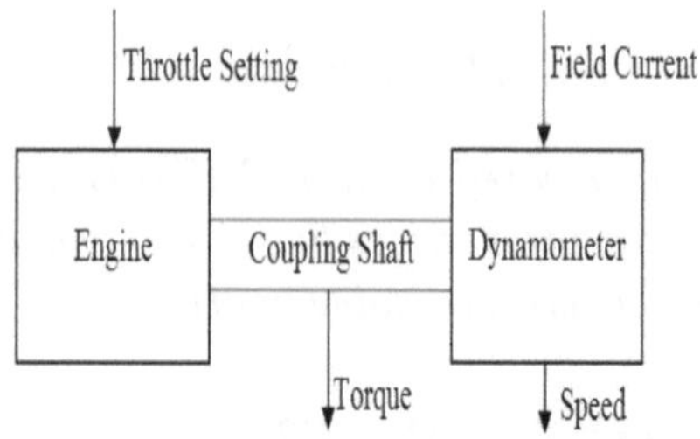

Figure 3.16: Engine-dynamometer schematic

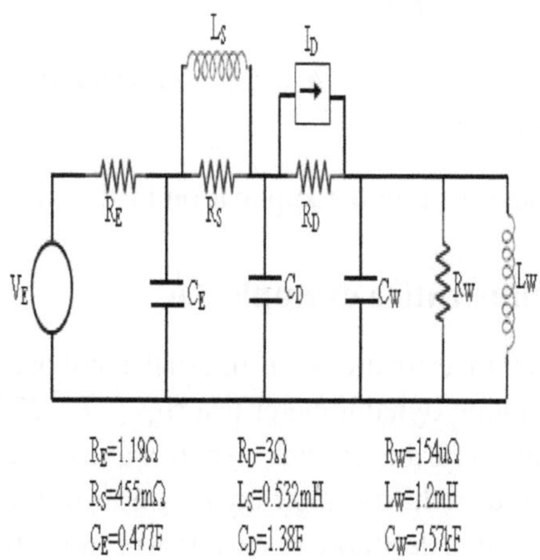

Figure 3.17: Electrical analogue model

By linearizing the whole system model about a typical operating point, an electrical analogue model [276] is shown in Figure 3.17 with experimental values for the various elements. The engine is represented as a constant velocity source V_E with internal resistance R_E. The inertia of the engine is represented by the capacitance C_E. The dynamometer is represented by the constant current source I_D with internal resistance R_D. The shaft connecting the engine to the dynamometer is modeled by the combination of the dissipater R_s and the inductance L_s, where the shaft is essentially considered as a rotational spring.

The dynamics of the dynamometer clamping gear are represented by the dissipater R_w, the capacitance C_w, and the inductance L_w. The control studies to be described here were carried out on two versions of this basic model by using a pertinent sampler plus hold units.

1. Third-order model

One model of the test rig considered is for the case where the dynamometer casing was firmly clamped to the reference frame. This mode of operation allows the design control systems with a marginally greater bandwidth [276] and results in a simple third-order model, where the parallel combination of R_w, C_w, and L_w are replaced by a direct connection. Using a sampling period of $0.18s$ with zero-order hold, the discrete model of the type (3.147) with $g(k,x) \equiv 0$ is given by the following matrices:

$$A = \begin{bmatrix} 0.7189 & 0.0866 & 0.0733 \\ 0.4312 & 0.4704 & -0.4206 \\ -0.3262 & 0.1731 & 0.2027 \end{bmatrix}$$

$$B = \begin{bmatrix} 0.1637 & -0.2056 \\ 0.2010 & -0.2155 \\ 0.0169 & 0.0152 \end{bmatrix}, \; C^t = \begin{bmatrix} 1 & 0 & 0 \\ 0 & 0 & 1 \end{bmatrix}$$

where the state vector consists of the dynamometer rotor speed, engine speed, and shaft torque while the control vector is composed of the throttle servo voltage and dynamometer source current. This model has eigenvalues $\{0.8275, \; 0.2830 \pm j0.3030\}$ which show that by tacking $r = 1, \; n - r = 2$ we have the mode separation ratio $\mu = 0.5$. The slow and fast subsystems (3.161) and (3.162) are given by

$$\begin{aligned} A_s &= 0.7885, \; B_s = \begin{bmatrix} 0.1970 & -0.2411 \end{bmatrix}, \\ C_s &= \begin{bmatrix} 1 \\ 1.1982 \end{bmatrix}, \; D_s = \begin{bmatrix} 0 & 0 \\ 0.0884 & -0.0591 \end{bmatrix}, \\ A_m &= \begin{bmatrix} 0.4704 & -0.4206 \\ 0.1731 & 0.2027 \end{bmatrix}, \; B_m = \begin{bmatrix} 0.2010 & -0.2155 \\ 0.0169 & 0.0152 \end{bmatrix}, \\ C_m &= \begin{bmatrix} 0 & 1 \end{bmatrix} \end{aligned}$$

The pair $(A_s, \; B_s)$ is controllable and let the desired slow eigenvalues be placed at $\{0.88\}$ using the gain

$$K_s = [-2.0726 \quad -2.0726]^t$$

For the fast subsystem, the desired fast eigenvalues are placed at $\{0.2 \pm j0.3\}$ using the gain

$$K_m = [-9.9615 \quad -6.8717]^t$$

This yields the composite control

$$u(k) = \begin{bmatrix} -4.652 & 0.0756 & -9.961 \\ -3.852 & 0.0128 & -6.872 \end{bmatrix} x(k)$$

2. Fifth-order model

This model is an extension of the third-order model by including the dynamics of an existing dynamometer field-current amplifier. The resulting model is now of order 5 and has the state variables as the dynamometer rotor speed, shaft torque, engine speed, and current amplifier states. Using a sampling period of $0.1928s$ and zero-order hold, which is selected to ensure matching the response of the continuous model, the discrete model of the type (3.147) with $g(k, x) \equiv 0$ is given by the following matrices:

$$A = \begin{bmatrix} 0.8070 & 0 & 0 & 0.0092 & 0 \\ -0.0267 & 0.5527 & 0.0171 & -0.0002 & 0.0012 \\ -0.1998 & 5.9560 & 0.1599 & -0.0018 & -0.2576 \\ -5.0795 & 0 & 0 & -0.0381 & 0 \\ 0.0243 & -7.8493 & 0.2311 & 0.0003 & -0.3805 \end{bmatrix},$$

$$B = \begin{bmatrix} 0 & 0.8511 \\ 0.0766 & -0.0106 \\ 0.7019 & -0.0832 \\ 0 & 22.3995 \\ 0.1418 & 0.0257 \end{bmatrix}, \; C^t = \begin{bmatrix} 0 & 1 & 0 & 0 & 0 \\ 0 & 0 & 0 & 0 & 1 \end{bmatrix}$$

This model has open-loop eigenvalues

$$\{0.7487, \; 0.7476, \; -0.2083 \pm j0.2274, \; 0.0213\}$$

which show that the static separation ratio has the value of 0.4125. It is readily seen that this model has two slow states $r = 2$ and three fast states $n - r = 3$. Direct calculation gives the slow subsystem matrices as

$$A_s = \begin{bmatrix} 0.7621 & 0 \\ -0.0294 & 0.6885 \end{bmatrix}, \; B_s = \begin{bmatrix} 0 & 1.0492 \\ 0.0899 & -0.0179 \end{bmatrix},$$

$$C_s = \begin{bmatrix} 0 & 1 \\ -0.2213 & 8.191 \end{bmatrix}, \; D_s = \begin{bmatrix} 0 & 0 \\ 0.7684 & -0.1439 \end{bmatrix}$$

and the fast subsystem matrices as

$$A_m = \begin{bmatrix} 0.1599 & -0.0018 & -0.2576 \\ 0 & -0.0381 & 0 \\ 0.2311 & 0.0003 & -0.3805 \end{bmatrix}, \; B_m = \begin{bmatrix} 0.7019 & -0.0832 \\ 0 & 22.3995 \\ 0.1418 & 0.0257 \end{bmatrix},$$
$$C_m = \begin{bmatrix} 0 & 0 & 1 \end{bmatrix}$$

Through the use of auxiliary devices, we can consider that all the state variables become available for generating feedback signals, and proceed to apply the state feedback design scheme. The desired eigenvalues are to be placed at

$$\{0.8,\ 0.7,\ 0.0999,\ -0.2026,\ -0.2173\}$$

to eliminate system oscillation. The slow gain matrix is

$$K_s = \begin{bmatrix} 0.0076 & -0.0913 \\ 0.0076 & -0.0913 \end{bmatrix}$$

The fast gain matrix is

$$K_m = \begin{bmatrix} -0.2861 & 0.0011 & -0.0787 \\ -0.2861 & 0.0011 & -0.0787 \end{bmatrix}$$

This yields the composite control

$$u(k) = \begin{bmatrix} 0.0541 & 0.0301 & -0.2877 & 0.0012 & -0.0784 \\ 0.0541 & 0.0301 & -0.2877 & 0.0012 & -0.0784 \end{bmatrix} x(k)$$

This two-stage feedback control yields the closed-loop eigenvalues as

$$\{0.8,\ 0.7001,\ 0.0998,\ -0.2179,\ -0.2021\}$$

which are very close to the desired ones.

From the foregoing results, we can draw the following conclusions:

1. *The two-stage feedback design schemes, based on the multiple-time-scale analysis, provide effective methods for separate eigenvalue assignment.*

2. *Since the output variables of the engine system model have different time responses, independent control actions can easily be obtained that enable the design engineer to improve the performance of the engine system.*

3. *In general, state feedback control provides robust results with respect to parameter variations on the expense of using more measuring devices.*

3.10.9 Decentralized state reconstruction

In this section, we direct attention to large-scale systems with subsystem model given by (3.148) and consider the problem of decentralized state reconstruction. Unlike the continuous-time case, we wish to show that depending on the available observation records for reconstructing the states, two different problems arise:

1. *The observation records consist of the measurements*

$$\{y_j(k-1),\ y_j(k-2),\ \cdots,\ y_j(0)\},\ j=1,2,...,n_s$$

2. *The observation records consist of the measurements*

$$\{y_j(k),\ y_j(k-1),\ \cdots,\ y_j(0)\},\ j=1,2,...,n_s$$

Whereas *Problem 1* entails a one-step delay in the measurement pattern on the subsystem level, *Problem 2* considers the situation whereby the subsystem is performing instantaneous measurements. In both problems, the n_s pairs $(A_j,\ C_j)$ are completely observable.

Considering *Problem 1* and letting $g_j(k,x(k)) \equiv 0$, it is fairly straightforward to design a state reconstructor for the jth subsystem of the form

$$\begin{aligned} \hat{x}_j(k+1) &= A_j\hat{x}_j(k) + B_ju_j(k) + K_j[y_j(k) - C_j\hat{x}_j(k)], \\ j &= 1,2,...,n_s \end{aligned} \tag{3.183}$$

where $\hat{x}_j(k) \in \Re^{n_j}$ is the estimate of $x_j(k)$ and $K_j \in \Re^{n_j \times p_j}$ is the design parameters that may be properly selected to ensure any desired degree of convergence of the observation scheme. In terms of the estimation error vector $\tilde{x}_j(k) = x_j(k) - \hat{x}_j(k)$, it follows from (3.153), (3.183) that

$$\tilde{x}_j(k+1) = (A_j - K_jC_j)\tilde{x}_j(k) \tag{3.184}$$

It is well known that $\tilde{x}_j(k) \to 0$ as $k \to \infty$ provided that the eigenvalues of the matrix $(A_j - K_jC_j)$ are located within the unit circle of the complex plane.

Theorem 3.10.5 *If the interaction pattern is a linear function of the state variables,* $g_j(k,x(k)) = \sum_{k=1}^{n_s} G_{jk}x_k(k)$, *with the matrix* $G = \{G_{jk}\}$ *satisfying the rank qualification*

$$rank\begin{bmatrix} C \\ G \end{bmatrix} = rank[C] = p = \sum_{j=1}^{n_s} p_j \tag{3.185}$$

then the addition of the quantity $\sum_{k=1}^{n_s} Mjky_k(k)$ *to each local state reconstructor (3.183) with*

$$M = \{M_{jk}\} = GC^t[CC^t]^{-1}$$

ensures the asymptotic convergence of the integrated state reconstruction scheme.

The proof of **Theorem 3.10.5** can be found in [207].

Remark 3.10.6 *It is needless to emphasize that the developed reconstruction scheme enjoys the merits of decentralized schemes. Only the local observability condition is required for the present analysis. As mentioned before, the additional constraint arises from the linear interconnection pattern that identifies several useful situations in practice.*

Next we consider *Problem 2*. An appropriate state reconstructor for the jth subsystem would be

$$\begin{aligned}\widehat{x}_j(k+1) &= A_j\widehat{x}_j(k) + B_ju_j(k) \\ &+ L_j[y_j(k+1) - A_jC_j\widehat{x}_j(k) - C_jB_ju_j(k)], \quad j = 1,2,...,n_s \\ &= [A_j - L_jC_jA_j]\widehat{x}_j(k) + [B_j - L_jC_jB_j]u_j(k) \\ &+ L_jy_j(k+1), \quad j = 1,2,...,n_s \end{aligned} \tag{3.186}$$

where $\widehat{x}_j(k) \in \Re^{n_j}$ is the estimate of $x_j(k)$ and $L_j \in \Re^{n_j \times p_j}$ is the design parameters that may be properly selected to ensure any desired degree of convergence of the observation scheme. Similarly, in terms of the estimation error vector $\widetilde{x}_j(k) = x_j(k) - \widehat{x}_j(k),$ it follows from (3.153), (3.186) that

$$\widetilde{x}_j(k+1) = [A_j - L_jC_jA_j]\widetilde{x}_j(k) \tag{3.187}$$

It is well known that $\widetilde{x}_j(k) \rightarrow 0$ as $k \rightarrow \infty$ provided that the eigenvalues of the matrix $[A_j - L_jC_jA_j]$ are located within the unit circle of the complex plane. It should be noted that the possibility of arbitrarily locating these eigenvalues depends on the observability of the pair $(A_j,\ C_jA_j)$. If the matrix A_j is nonsingular, then the observability of the pair $(A_j,\ C_j)$ is equivalent to the observability of the pair $(A_j,\ C_jA_j)$ [35]. On the other hand, if the matrix A_j is singular and the pair $(A_j,\ C_j)$ is observable, then it can be shown that the pair $(A_j,\ C_jA_j)$ is both detectable and reconstructible [158].

The following results can be found in [207].

Theorem 3.10.7 *The state reconstruction schemes (3.183) and (3.186) are identical under the condition*

$$K_j \;=\; A_j\, L_j \tag{3.188}$$

Theorem 3.10.8 *The modified state-reconstructor scheme*

$$\begin{aligned}\widehat{x}_j(k+1) &= A_j\widehat{x}_j(k) + B_j u_j(k)\\ &+ L_j[y_j(k+1) - A_jC_j\widehat{x}_j(k) - C_jB_ju_j(k)],\;\; j=1,2,...,n_s\\ &= [A_j - L_jC_jA_j]\widehat{x}_j(k) + [B_j - L_jC_jB_j]u_j(k)\\ &+ L_jy_j(k+1) + \sum_{k=1}^{n_s} S_{jk}y_k(k)\end{aligned} \tag{3.189}$$

ensures the asymptotic convergence of the integrated state vector if A_j is non-singular and the condition

$$rank\begin{bmatrix} C \\ G \end{bmatrix} = rank[C]$$

is satisfied where

$$S \;=\; \{S_{jk}\} \;=\; GC^t[CC^t]^{-1},\;\; L = Blockdiag[L_1,\; L_2,\cdots, L_{n_s}]$$

3.10.10 Illustrative example 3.7

To demonstrate the application of the above theoretical analysis, we consider three interconnected subsystems described by

$$\begin{aligned}A_1 &= \begin{bmatrix} 0.7 & 0.2 \\ 0.1 & 0.8 \end{bmatrix},\; B_1 = \begin{bmatrix} 0 \\ 1 \end{bmatrix},\; C_1^t = \begin{bmatrix} 1 \\ 1 \end{bmatrix},\\ G_{12} &= \begin{bmatrix} -6 & 3 \\ 0 & 0 \end{bmatrix},\; G_{13} = \begin{bmatrix} 0 & 0 \\ -3 & 3 \end{bmatrix},\\ A_2 &= \begin{bmatrix} 0.5 & 0.25 \\ 0 & 0.4 \end{bmatrix},\; B_2 = \begin{bmatrix} 1 \\ 0 \end{bmatrix},\; C_2^t = \begin{bmatrix} 2 \\ -1 \end{bmatrix},\\ G_{21} &= \begin{bmatrix} 0 & 0 \\ -2 & -2 \end{bmatrix},\; G_{23} = \begin{bmatrix} 2 & -2 \\ 0 & 0 \end{bmatrix},\\ A_3 &= \begin{bmatrix} 0.6 & 0 \\ 0.3 & 0.1 \end{bmatrix},\; B_3 = \begin{bmatrix} 1 \\ 1 \end{bmatrix},\; C_3^t = \begin{bmatrix} -1 \\ 1 \end{bmatrix},\\ G_{31} &= \begin{bmatrix} 1 & 1 \\ 0 & 0 \end{bmatrix},\; G_{32} = \begin{bmatrix} 0 & 0 \\ 4 & -2 \end{bmatrix}\end{aligned}$$

Observe that matrices $A_1,\ A_2,\ A_3$ are nonsingular, $rank[C_j] = 1,\ j = 1, 2, 3,$ and the pairs $(A_1,\ C_1),\ (A_2,\ C_2),\ (A_3,\ C_3)$ are all observable. In case of one-step delay in measurement, three eigenvalues $z_1 = 0.25,\ z_2 = 0.45,\ z_3 = 0.65$ could be assigned. The result of computations give

$$K_1 = \begin{bmatrix} 0.506 \\ 0 \end{bmatrix},\ K_2 = \begin{bmatrix} 0.025 \\ 0 \end{bmatrix},\ K_3 = \begin{bmatrix} 0.11 \\ 0 \end{bmatrix}$$

Because the rank qualification is met, the additional term for the local state reconstructors is

$$M = \begin{bmatrix} 0 & -3 & 0 \\ 0 & 0 & 3 \\ 0 & 0 & -2 \\ -2 & 0 & 0 \\ 1 & 0 & 0 \\ 0 & 2 & 0 \end{bmatrix}$$

On the other hand, if the measurement records were instantaneous and because of the nonsingularity of subsystem matrices, *Problem 2* was solved to yield

$$L_1 = \begin{bmatrix} 0.749 \\ -0.094 \end{bmatrix},\ L_2 = \begin{bmatrix} 0.05 \\ 0 \end{bmatrix},\ L_3 = \begin{bmatrix} 0.183 \\ -0.55 \end{bmatrix}$$

3.10.11 Linear quadratic control design

It was shown in the foregoing sections that a systematic procedure for complete separation of slow and fast regulator designs can be developed by extending the idea of two-stage eigenvalue assignment [208, 209]. Extending on this, we show in this section that by recomposing the fast and slow optimal controls based on the linear quadratic control (LQC) theory [3], an approximate feedback control can be readily obtained and subsequently formulated in a standard form. When the asymptotically stable fast modes are neglected, a reduced-order control can be derived and again put in a standard form.

Our starting point is the linear discrete system (3.153)-(3.154) and the corresponding slow-fast subsystems (3.161) and (3.162). We note from (3.161) and (3.162) that the slow control $u_s(k)$ and the fast control $u_m(k)$ produce the composite control $u_c(k)$ according to $u_c(k) = u_s(k) + u_m(k)$. In view of the mode-separation, it has been shown [210] that a linear feedback scheme of the type

$$u_s(k) = G_s x_s(k), \quad u_m(k) = G_m x_m(k) \tag{3.190}$$

can be designed using independent gains G_s and G_m.

Hereafter, the optimal state regulator problem is considered which consists of the linear dynamical system (3.153) and the associated quadratic performance measure

$$J = \frac{1}{2}\sum_{m=0}^{\infty} \{y^t(m)y(m) \; + \; u^t(m)Ru(m)\}, \quad R > 0 \tag{3.191}$$

Instead of tackling the regulator problem (3.153) and (3.191) directly, we decompose it appropriately into two discrete regulators. The first (slow) regulator consists of the slow subsystem (3.161) and a quadratic performance measure J_s. The second (fast) regulator consists of the fast subsystem (3.162) and a quadratic performance measure J_f. It must be emphasized that the construction of the subsystem measures is done such that $J = J_s + J_m$. By solving the fast and slow regulator problems independently, we obtain the slow control $u_s(k)$ and the fast control $u_m(k)$. Later we recompose these controls, which are subsystem optimal, to form the control $u_c(k)$ to be implemented on the system (3.153).

Slow regulator

The problem is to find $u_s(k)$ to minimize

$$J_s = \frac{1}{2}\sum_{m=0}^{\infty} \{y_s^t(m)y_s(m) \; + \; u_s^t(m)Ru_s(m)\}, \quad R > 0 \tag{3.192}$$

for the slow subsystem (3.161). Manipulating (3.192), we reach

$$\begin{aligned} J_s &= \frac{1}{2}\sum_{m=0}^{\infty} \{x_s^t(m)C_s^tC_sx_s(m) \; + \; 2u_s^t(m)D_s^tC_sx_s(m) \\ &+ \; u_s^t(m)R_su_s(m)\}, \quad R > 0 & (3.193) \\ R_s &= R + D_s^tD_s & (3.194) \end{aligned}$$

Recall from [158] that if the triplet $(A_s,\; B_s,\; C_s)$ is stabilizable-detectable, then there exists a stabilizing solution $K_s \geq 0$ for the algebraic Riccati equation

$$\begin{aligned} K_s &= (A_s^tK_sA_s + C_s^tC_s) \\ &- (B_s^tK_sA_s + D_s^tC_s)^t(R_s + B_s^tK_sB_s)^{-1} \\ &\times (B_s^tK_sA_s + D_s^tC_s) \end{aligned} \tag{3.195}$$

The optimal slow control law is given by

$$u_s(k) = -(R_s + B_s^tK_sB_s)^{-1}(B_s^tK_sA_s + D_s^tC_s)x_s(k) \tag{3.196}$$

Fast regulator

The problem is to find $u_m(k)$ to minimize

$$J_m = \frac{1}{2}\sum_{r=0}^{\infty} \{y_m^t(r)y_m(r) \; + \; u_m^t(r)Ru_m(r)\}, \quad R > 0 \tag{3.197}$$

for the slow subsystem (3.162). Manipulating (3.197), we reach

$$J_m \;\; = \;\; \frac{1}{2}\sum_{r=0}^{\infty} \{x_m^t(r)C_m^tC_mx_m(r) \; + \; u_m^t(r)Ru_m(r)\} \tag{3.198}$$

Recall from [158] that if the triplet $(A_m,\ B_m,\ C_m)$ is stabilizable-detectable, then there exists a stabilizing solution $K_m \geq 0$ for the algebraic Riccati equation

$$\begin{aligned} K_m &= (A_m^tK_mA_m + C_m^tC_m) \\ &- A_m^tK_mB_m(R + B_m^tK_mB_m)^{-1}B_m^tK_mA_m \end{aligned} \tag{3.199}$$

The optimal fast control law is given by

$$u_m(k) \;\; = \;\; -(R + B_m^tK_mB_m)^{-1}B_m^tK_mA_mx_m(k) \tag{3.200}$$

Remark 3.10.9 *The stabilizability-detectability conditions of the triples*

$$(A_s,\ B_s,\ C_s),\ \ (A_m,\ B_m,\ C_m)$$

are eventually independent. More importantly, it has been established [208, 209, 210] that they are equivalent to the stabilizability-detectability of the triple $(A,\ B,\ C)$ of the original system (3.153), where $B^t = [B_a^t \;\; B_m^t]$. The control laws $u_s(k)$ and $u_m(k)$ given by (3.196) and (3.200) are only subsystem optimal, that is, with respect to the slow and fast subsystem variables. However, it is much easier and computationally simpler to determine them than the optimal control for the overall system (3.153).

By comparing (3.196) and (3.200) to (3.190) we can identify the gains

$$\begin{aligned} G_s &= -(R_s + B_s^tK_sB_s)^{-1}(B_s^tK_sA_s + D_s^tC_s) \\ G_m &= -(R + B_m^tK_mB_m)^{-1}B_m^tK_mA_mx_m(k) \end{aligned} \tag{3.201}$$

A standard form of the composite control $u_c(k)$ is established by the following theorem which can be found in [223].

Theorem 3.10.10 *The composite control $u_c(k)$ can be put into the form*

$$\begin{aligned} u_c(k) &= -(R + B^t K_c B)^{-1} B^t K_c A_c x(k) \;=\; Lx(k) && (3.202)\\ B &= \begin{bmatrix} B_a \\ B_m \end{bmatrix},\; A_c = \begin{bmatrix} A_s & A_b \\ A_d & A_m \end{bmatrix},\; K_c = \begin{bmatrix} K_s & 0 \\ K_n & K_b \end{bmatrix} \\ A_b &= -B_a G_m,\; K_n^t = K_s A_1 (I - A_m)^{-1} \\ S &= (B_s - B_a) G_s + B_a G_m (I - A_m)^{-1} (B_m G_s + A_c) \\ A_d &= A_c - (I - A_m)^{-1}(B_m G_s + A_c) + B_m (I - A_m)^{-1} B_a^t K_s S \\ &+ K_m^{-1}[K_n S + (I - A_m)^{-1} C_m^t (C_s + D_s C_s)] && (3.203) \end{aligned}$$

We must emphasize that the approximate control (3.202) is in the standard form of optimal discrete regulator theory [158]. Recall that the exact optimal control for the complete dynamic problem (3.153) and (3.191) is given by

$$u_c(k) \;=\; -(R + B^t P_o B)^{-1} B^t P_o A x(k) \;=\; -G^o x(k) \qquad (3.204)$$

where P_o is the stabilizing solution of the algebraic Riccati equation

$$P_o = A^t P_o A - A^t P_o B (R + B^t P_o B)^{-1} B^t P_o A + C^t C \qquad (3.205)$$

and the associated optimal performance measure is $J_o = (1/2) x_o^t P_o x_o$ with x_o being the initial state. By virtue of (3.202), the approximate closed-loop system takes the form

$$x(k+1) \;=\; (A \;-\; BL)x(k) \qquad (3.206)$$

We know that minimizing (3.191) subject to (3.206) results in the suboptimal performance measure $J_c = (1/2) x_o^t P_c x_o$, where P_c is the positive definite solution of the discrete Lyapunov equation

$$P_c = (A - BL)^t P_c (A - BL) + C^t C + L^t R L \qquad (3.207)$$

Let the suboptimality index be defined by

$$P_c - P_o \;=\; \varepsilon P_c \qquad (3.208)$$

which, in the light of J_o and J_c, expresses the relative performance degradation, that is, $\varepsilon = (J_c - J_o)/J_o$. The following theorem provides a useful expression [223].

Theorem 3.10.11 *An upper bound on the suboptimality index ε is given by*

$$\varepsilon \;\leq\; \frac{||A^t P_o A|| + ||P_c||}{||C^t C||} \;-\; 1 \qquad (3.209)$$

It is important to observe that the relative performance degradation ε depends on the subsystem information. When the fast modes are asymptotically stable, we set $G_m = 0$ and obtain the reduced control

$$\begin{aligned} u_r(k) &= -(R_s + B_s^t K_s B_s)^{-1}(B_s^t K_s A_s + D_s^t C_s)x_1(k) \\ &= -L_r x(k) \end{aligned} \tag{3.210}$$

The next theorem gives an important special case.

Theorem 3.10.12 *The reduced-order control (3.210) can be expressed as*

$$\begin{aligned} u_r(k) &= -(R + B^t K_r B)^{-1} B^t K_r A_r x(k), \\ K_r &= = \begin{bmatrix} K_s & K_n^t \\ K_n & 0 \end{bmatrix}, \; A_r = \begin{bmatrix} A_s & 0 \\ A_w & 0 \end{bmatrix}, \\ A_w &= K_n B_a (B_a^t K_n^t K_n B_a)^{-1} D_s^t C_s - WV(B_s^t K_s A_s + D_s^t C_s), \\ W &= D_s^t D_s + B_m^t (I - A_m)^{-1} A_e^t K_n^t B_m, \\ V &= (R_s + B_s^t K_s B_s) \end{aligned} \tag{3.211}$$

which produces performance degradation bounded by

$$\varepsilon_r \le \frac{||A^t P_o A|| + ||P_c||}{||C^t C||} - 1 \tag{3.212}$$

where $P_r = P_c + P_d = (\varepsilon_r - \varepsilon)P_o$ *and*

$$\begin{aligned} P_r &= (A - BL_r)^t P_r (A - BL_r) + C^t C + L_r^t R L_r \\ L_r &= LT = L \begin{bmatrix} I & 0 \\ (I - A_m)^{-1}(B_m G_o + A_c) & 0 \end{bmatrix} \\ P_d &= (A - BL_r)^t P_d (A - BL_r) + A^t P_c BL(I - T) \\ &+ (I - T)^t L^t B^t P_c A - L^t B^t P_c BRL \\ &- L^t RL + T^t L^t B^t P_c BRLT + T^t L^t RLT \end{aligned} \tag{3.213}$$

A detailed treatment of **Theorem 3.10.12** can be found in [223].

3.10.12 Illustrative example 3.8

A power system model of the type (3.153) is described by [206].

$$A_a = \begin{bmatrix} 0.928 & -0.029 & 0.028 & 0.0318 \\ -0.253 & 0.882 & -0.09 & -0.0091 \\ 0 & 0 & 0.861 & 0.218 \\ 0 & 0 & 0 & 0.835 \end{bmatrix},$$

$$A_e = \begin{bmatrix} 0.06 & 1.073 & 0 & 0.04 \\ -0.03 & -0.455 & 0.5467 & -0.02 \\ 0 & 0 & 0 & 0.29 \\ 0 & 0 & 0 & 0 \end{bmatrix},$$

$$A_c = \begin{bmatrix} 0 & 0 & 0.1516 & 0.0218 \\ -0.077 & 0.0665 & -0.003- & 0.003 \\ -0.1727 & 0.152 & 0.152 & -0.0102 \\ 0 & 0 & 0 & 0.185 \end{bmatrix},$$

$$A_m = \begin{bmatrix} 0.165 & 0 & 0 & 0.046 \\ -0.007 & 0.156 & 0.118 & 0.007 \\ -0.016 & -0.31 & 0.008 & -0.012 \\ 0 & 0 & 0 & 0.011 \end{bmatrix},$$

$$B_a^t = \begin{bmatrix} -0.038 & 0.294 & 0 & 0 \\ 0.01 & -0.003 & 0.076 & 0.725 \end{bmatrix},$$

$$B_m^t = \begin{bmatrix} 0 & 0.081 & 0.207 & 0 \\ 0.006 & -0.0008 & -0.004 & 0.124 \end{bmatrix},$$

$$C_1 = \begin{bmatrix} 0 & 0 & 0 & 4.545 \\ 0 & 1 & 0 & 0 \\ 0 & 0 & 0 & 0 \\ 0 & 0 & 0 & 0 \end{bmatrix},$$

$$C_2 = \begin{bmatrix} 0 & 0 & 0 & 0 \\ 0 & 0 & 0 & 0 \\ 0 & 0 & 13.333 & 0 \\ 0 & 18.182 & 0 & 0 \end{bmatrix}$$

With $R = I_2$, direct computation yields the gain matrices

$$\begin{aligned}
G^o &= \left[\begin{array}{cc} G_1^o & G_2^o \end{array}\right], \\
G_1^o &= \left[\begin{array}{cccc} 0.818 & -0.744 & 0.038 & 0.026 \\ -0.019 & 0.005 & -0.006 & -1.061 \end{array}\right], \\
G_2^o &= \left[\begin{array}{cccc} 0.075 & 0.685 & -0.635 & 0.025 \\ -0.002 & -0.037 & -0.0014 & -0.0031 \end{array}\right], \\
L &= \left[\begin{array}{cc} G_1^o & G_2^o \end{array}\right], \\
L_1 &= \left[\begin{array}{cccc} 0.813 & -0.743 & 0.039 & 0.0259 \\ -0.021 & 0.0061 & -0.005 & -1.061 \end{array}\right], \\
L_2 &= \left[\begin{array}{cccc} 0.072 & 0.615 & -0.608 & -0.023 \\ -0.001 & -0.075 & -0.011 & -0.003 \end{array}\right], \\
L_r &= \left[\begin{array}{cc} L_{r1} & 0 \end{array}\right], \\
L_{r1} &= \left[\begin{array}{cccc} 0.806 & -0.736 & 0.523 & 0.03 \\ -0.02 & 0.006 & -0.006 & -1.061 \end{array}\right]
\end{aligned}$$

together with

$$\begin{aligned}
J_o &= \textbf{trace}(P_o) = 567.424, \quad J_c = \textbf{trace}(P_c) = 567.5176, \\
J_r &= \textbf{trace}(P_r) = 578.861
\end{aligned}$$

The index ε in (3.208) has the values 0.0168, whereas its upper bound in (3.209) is 0.19845. Also, $\varepsilon_r = 0.0202$ whereas its upper bound in (3.212) is 0.22232. It is clearly evident that the developed composite control produces good performance results.

The net result of this section illuminated the fact that procedures for decomposition, feedback design, and optimal control of linear discrete regulators with time-separation property are systematically disclosed and shown to be effective. The computations are carried out on the subsystem matrices. Bounds on suboptimality are given to clarify the degree of suboptimality.

3.11 Applications

In this section, we solve some examples that arise in different system applications.

3.11.1 Serially connected subsystems

A state-space model of four serially connected subsystems considered in [292] has the following data:

$$
\begin{aligned}
A &= \begin{bmatrix} A_1 & 0 & 0 & A_2 \\ A_2 & A_3 & A_4 & 0 \\ 0 & A_4 & A_3 & 0 \\ A_4 & 0 & A_2 & A_1 \end{bmatrix}, \ B = \begin{bmatrix} B_1 & 0 \\ 0 & 0 \\ 0 & B_1 \\ 0 & 0 \end{bmatrix}, \\
A_1 &= \begin{bmatrix} 0 & 1 & 0 \\ 0 & 0 & 1 \\ -1 & -3 & -2 \end{bmatrix}, \ A_2 = \begin{bmatrix} 0 & 0 & 0 \\ 0 & 0 & 0 \\ 1 & 0 & 0 \end{bmatrix}, \\
A_3 &= \begin{bmatrix} 0 & 1 & 0 \\ 0 & 0 & 1 \\ -1 & -2 & -3 \end{bmatrix}, \ A_4 = \begin{bmatrix} 0 & 0 & 0 \\ 0 & 0 & 0 \\ 0 & 1 & 0 \end{bmatrix}, \\
B_1 &= \begin{bmatrix} 0 \\ 0 \\ 1 \end{bmatrix}, \ A_2 = \begin{bmatrix} 0 & 0 & 0 \\ 0 & 0 & 0 \\ 1 & 0 & 0 \end{bmatrix}, \\
C &= Blockdiag \begin{bmatrix} C_1 & C_1 & C_1 & C_1 \end{bmatrix}, \ C_1 = \begin{bmatrix} 1 & 0 & 0 \\ 0 & 1 & 0 \end{bmatrix}
\end{aligned}
$$

Simulation results are shown in Figure 3.18.

3.11.2 Liquid-metal cooled reactor

In this section, we introduce the model and some numerical simulation results of a liquid-metal cooled reactor (LMCR) of a nuclear power system based on the material reported in [23]. Consider the LMCR module of a nuclear reactor shown in Figure 3.19. Each reactor module consists of an LMCR reactor with the corresponding primary and intermediate heat transport loops, a recirculating steam generator, and steam drum. All the modules are connected to a common steam heater that feeds the turbine. The equations describing the dynamic behavior of a reduced model can be cast into the nonlinear form:

$$
\begin{aligned}
\dot{x}_1 &= (a_1 + a_2 x_3 + a_3 x_4) a_4 x_1 + a_4 x_2 + (a_4 a_5) u_1, \\
\dot{x}_2 &= \lambda(x_1 - x_2), \\
\dot{x}_3 &= b_1 x_1 + b_2 (x_4 - x_3), \\
\dot{x}_4 &= -b_3 (x_4 - x_3) + b_4 (x_7 - x_4) u_2,
\end{aligned}
$$

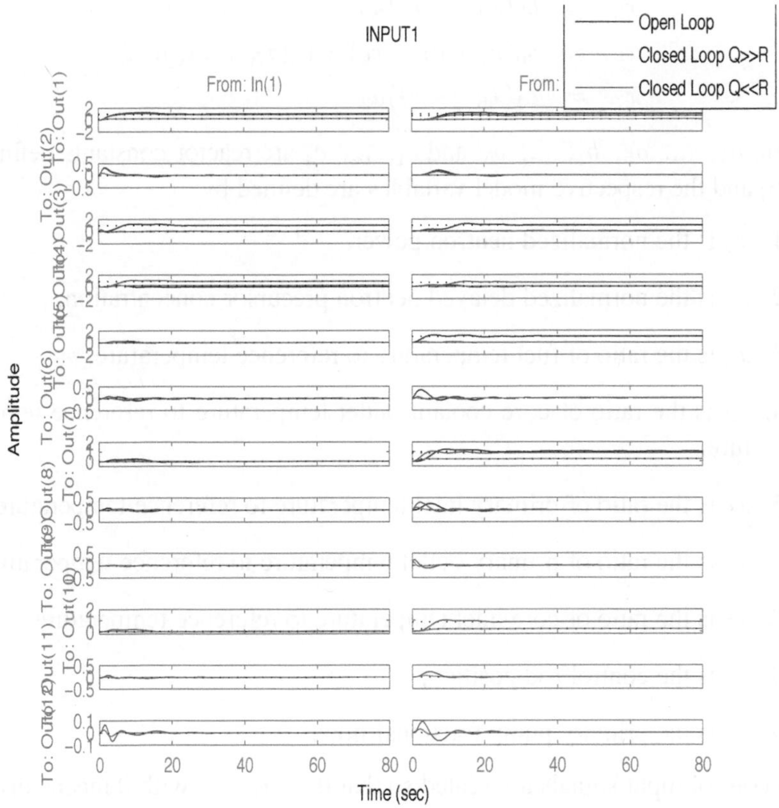

Figure 3.18: State trajectories

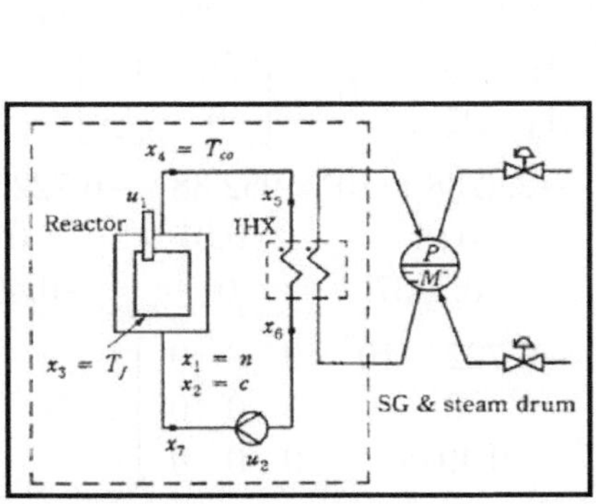

Figure 3.19: A liquid-metal cooled reactor

$$\begin{aligned}\dot{x}_5 &= b_5(x_4 - x_3)u_2,\\ \dot{x}_6 &= b_6(c_1 + c_2 - x_6) + c_3(x_5 - x_6)u_2,\\ \dot{x}_7 &= c_4(x_6 - x_7)u_2\end{aligned}$$

where $a_1, \ldots, a_5, b_1, \ldots, b_5$, and $c_1, \ldots, c_4$ are reactor constants defined in [136] and the respective model variables are defined by

1. x_1 is the normalized neutron power,
2. x_2 is the normalized delayed neutron precursor concentration,
3. x_3 is the ratio of fuel temperature to reference temperature,
4. x_4 is the ratio of core coolant outlet temperature to reference temperature,
5. x_5 is the ratio of primary inlet temperature to reference temperature,
6. x_6 is the ratio of primary outlet temperature to reference temperature,
7. x_7 is the ratio of core inlet temperature to reference temperature,
8. u_1 is the control rod position,
9. u_2 is the primary pump fractional flow.

The control input signals are scaled so that $0 \leq u_1 \leq 1$ with 0 means inserted (minimum or closed) and 1 corresponds to withdrawn (maximum or open). Evaluating the equilibrium operating values x_e, u_e from the data in [136] and performing standard linearization, we obtain the linearized model

$$\begin{aligned}\dot{z}_1 &= Az + Bv, \; z = x - x_e, \; v = u - u_e,\\ A &= \begin{bmatrix} A_1 & A_2 \\ A_3 & A_4 \end{bmatrix}, \; B = \begin{bmatrix} B_1 & 0 \\ 0 & B_2 \end{bmatrix},\\ A_1 &= \begin{bmatrix} -2.4238 \times 10^8 & 952.38 & -6.722 \times 10^7 \\ 0.01 & -0.01 & 0 \\ -0.1437 & 0 & -0.4948 \end{bmatrix},\\ A_2 &= \begin{bmatrix} -6.722 \times 10^7 & 0 & 0 & 0 \\ 0 & 0 & 0 & 0 \\ 0.4948 & 0 & 0 & 0 \end{bmatrix},\\ A_3 &= \begin{bmatrix} 0 & 0 & 0.3877 \\ 0 & 0 & 0 \\ 0 & 0 & 0 \\ 0 & 0 & 0 \end{bmatrix},\end{aligned}$$

$$A_4 = \begin{bmatrix} -0.4101 & 0 & 0 & 0.0224 \\ 0.06772 & -0.0677 & 0 & 0 \\ 0 & 0.0677 & -0.1754 & 0 \\ 0 & 0 & 0.0224 & \end{bmatrix},$$

$$B_1 = \begin{bmatrix} 3333.3 \\ 0 \\ 0 \end{bmatrix}, \; B_2 = \begin{bmatrix} -0.0101 \\ 0 \\ 0.0304 \\ 0 \end{bmatrix}$$

We solve the problem as a full-order linear-quadratic regulator and the ensuing step response is presented in Figures 3.20 and 3.21 for the respective control inputs. Alternatively, we apply the decentralized linear-quadratic regulator

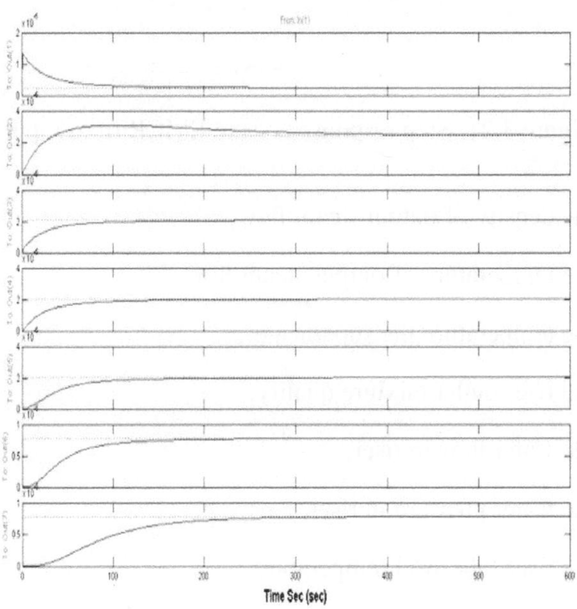

Figure 3.20: Step responses of LMCR from input 1

[200] by treating $A_2, \; A_3$ as interaction matrices. The corresponding system responses are provided in Figures 3.22-3.25, which show the effectiveness of the decentralized control approach.

3.11.3 Boiler system

The state-, input-, and output-variables of a boiler system comprising a superheater and riser in series with each other [284] are

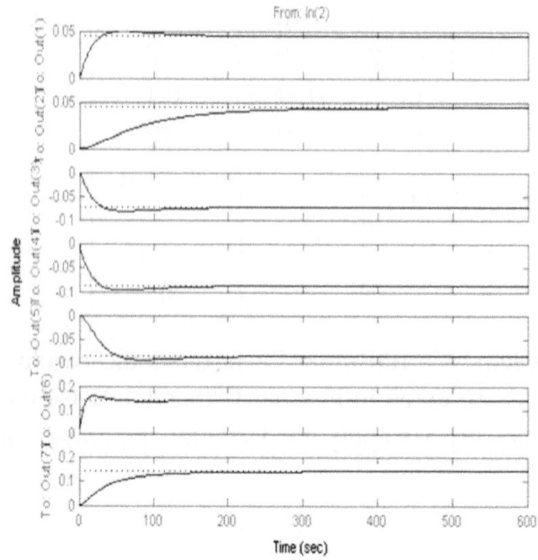

Figure 3.21: Step responses of the LMCR from input 2

- x_1 is the density of output steam flow,
- x_2 is the temperature of output steam flow,
- x_3 is the temperature the superheater,
- x_4 is the riser outlet mixture quality,
- x_5 is the water flow in riser,
- x_6 is the water pressure in riser,
- x_7 is the riser tube-wall temperature,
- x_8 is the temperature of water in boiler,
- x_9 is the level of water in boiler,
- u_1 is the input fuel,
- u_2 is the input water flow,
- y_1 is the temperature of output steam flow,
- y_2 is the riser outlet mixture quality,
- y_3 is the water flow in riser,

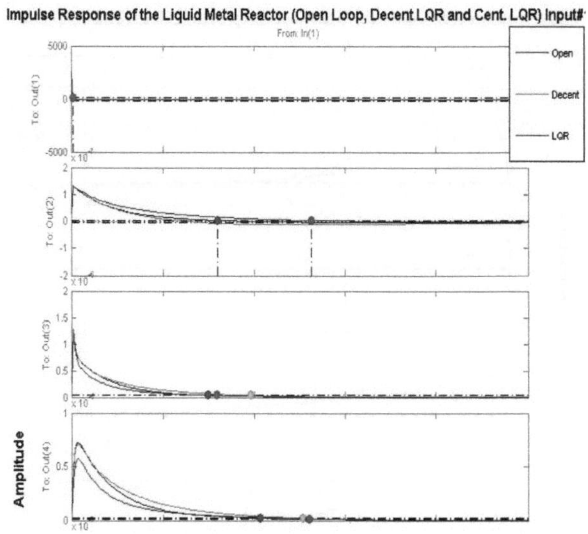

Figure 3.22: Different responses of the LMCR from input 1-part A

- y_4 is the water pressure in riser.

By simulating the ninth-order linear continuous model and its discretized version using first-order approximation, it is found that a sampling period of $0.5s$ yields a discrete model whose response matches very closely that of the continuous model. The data values use aE^x to mean $a \times 10^x$.

Using the permutation matrix

$$\mathsf{P} = \{e_9,\ e_8,\ e_2,\ e_5,\ e_7,\ e_6,\ e_4,\ e_3,\ e_1\}$$

and the scaling matrix

$$\mathsf{S} = Blockdiag[0.015,\ 0.015,\ 0.05,\ 0.1,\ 2,\ 0.5E^{-4},\ 0.15,\ 5,\ 0.2E^{4}\}$$

where e_j is the elementary column vector whose jth entry is 1, the transformed discrete system has the eigenvalues

$$\{1.0,\ 0.1452 \pm j0.0726,\ 0.2298,\ 0.89,\ 0.996,\ 0.9741 \pm j0.0905,\ 0.8461\}$$

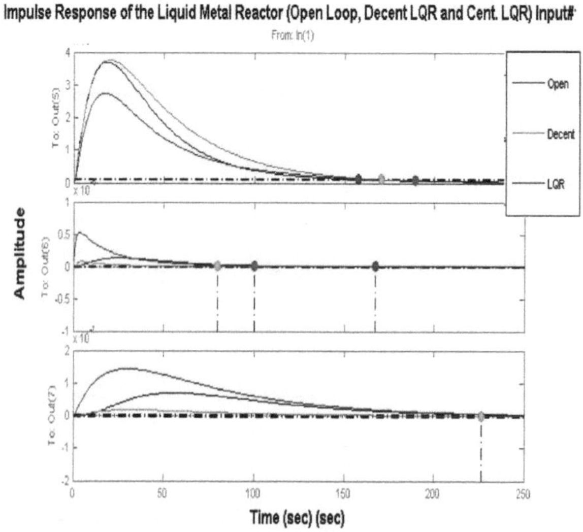

Figure 3.23: Different responses of the LMCR from input 1-part B

It is estimated that the model has six slow and three fast variables. In terms of system (3.153) and (3.154), the data values are

$$A_a = \begin{bmatrix} A_{a1} & A_{a2} \\ A_{a3} & A_{a4} \end{bmatrix},$$

$$A_{a1} = \begin{bmatrix} 1 & -0.1489E^{-3} & 0.1050E^{-3} \\ 0 & 0.9866 & -0.3550E^{-3} \\ 0 & -0.1389E^{-3} & 0.9866 \end{bmatrix},$$

$$A_{a2} = \begin{bmatrix} 0.1051E^{-3} & 0.0289 & 0.3127E^{-3} \\ -0.2745E^{-3} & 0.9544E^{-5} & -0.0195 \\ 0.3156E^{-3} & -0.0391 & 0.0257 \end{bmatrix},$$

$$A_{a3} = \begin{bmatrix} 0 & 0.8048E^{-2} & 0.2856E^{-2} \\ 0 & -0.2065E^{-2} & 0.3328E^{-2} \\ 0 & 0.7152E^{-2} & 0.2589E^{-1} \end{bmatrix},$$

$$A_{a4} = \begin{bmatrix} 0.9057 & -0.7275E^{-4} & 0.1951 \\ 0.7091E^{-3} & 0.8829 & 0.1479E^{-1} \\ 0.1980E^{-1} & -0.8358E^{-3} & 0.8705 \end{bmatrix},$$

$$A_e = \begin{bmatrix} -0.2667E^{-5} & -0.5914E^{-6} & -0.3823E^{-5} \\ -0.1585E^{-7} & 0.4712E^{-2} & 0.5030E^{-4} \\ 0.8717E^{-4} & 0.9676E^{-5} & -0.1144E^{-5} \\ 0.1169E^{-6} & 0.3265E^{-5} & 0.1673E^{-4} \\ -0.1071E^{-4} & -0.9028E^{-5} & 0.1334E^{-4} \\ 0.1445E^{-5} & 0.1345E^{-4} & 0.1143E^{-3} \end{bmatrix},$$

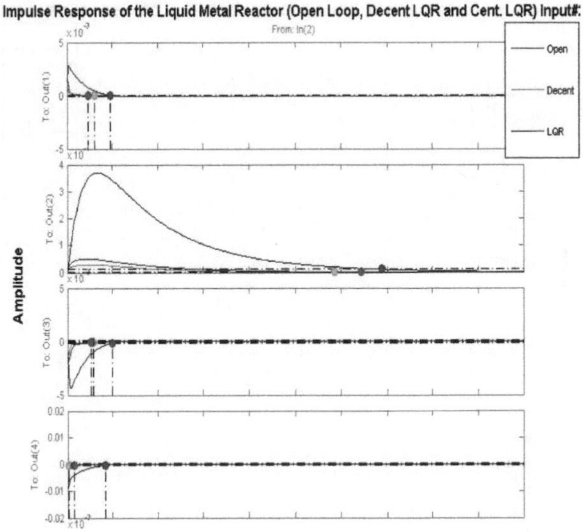

Figure 3.24: Different responses of the LMCR from input 2-part A

$$
\begin{aligned}
A_c &= \begin{bmatrix} A_{c1} & A_{c2} \end{bmatrix}, \\
A_{c1} &= \begin{bmatrix} 0 & -6.01465 & 0.3120E^2 \\ 0 & 0.2490E^3 & -0.8749 \\ 0 & -0.5153E^2 & 6.2408 \end{bmatrix}, \\
A_{c2} &= \begin{bmatrix} -0.1336E^1 & -0.2310E^3 & -0.1006E^3 \\ -0.6724 & 0.2564E^{-1} & -0.3105E^2 \\ 0.4815E^1 & -0.1692 & 0.3291E^3 \end{bmatrix}, \\
A_m &= \begin{bmatrix} 0.2375 & 0.0670 & -0.2622E^{-1} \\ -0.4447E^{-4} & 0.1998 & 0.8275E^{-1} \\ 0.2825E^{-3} & -0.1018 & 0.1490 \end{bmatrix}, \\
A_e &= \begin{bmatrix} -0.2667E^{-5} & -0.5914E^{-6} & -0.3823E^{-5} \\ -0.1585E^{-7} & 0.4712E^{-2} & 0.5030E^{-4} \\ 0.8717E^{-4} & 0.9676E^{-5} & -0.1144E^{-5} \\ 0.1169E^{-6} & 0.3265E^{-5} & 0.1673E^{-4} \\ -0.1071E^{-4} & -0.9028E^{-5} & 0.1334E^{-4} \\ 0.1445E^{-5} & 0.1345E^{-4} & 0.1143E^{-3} \end{bmatrix},
\end{aligned}
$$

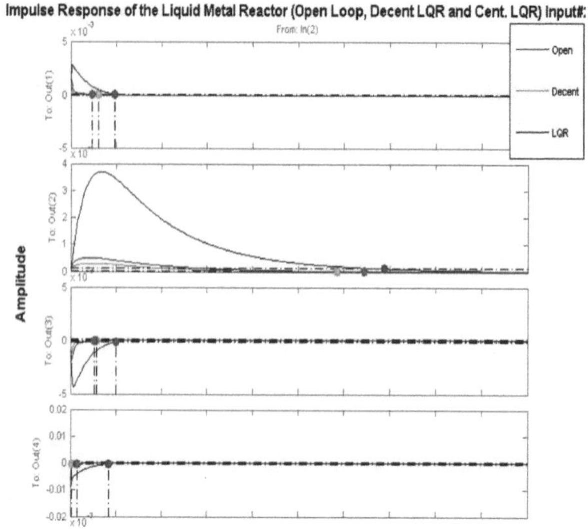

Figure 3.25: Different responses of the LMCR from input 2-part B

$$B_a = \begin{bmatrix} 0.1777E^{-4} & 0.4490E^{-5} \\ -0.3191E^{-3} & 0.1159E^{-1} \\ 0.2177E^{-3} & 0.3889E^{-4} \\ -0.6494E^{-4} & 0.1109 \\ -0.1159E^{-3} & -0.2689E^{-4} \\ -0.7698E^{-3} & 0.1239E^{-2} \end{bmatrix},$$

$$B_m = \begin{bmatrix} 2.3080 & 0.1651 \\ -0.7292 & 1.8098 \\ -0.4393 & -0.5085E^{-1} \end{bmatrix}, \; C_m = \begin{bmatrix} 0 & 0 & 0 \\ 0 & 0 & 0 \\ 0 & 0 & 0 \\ 0 & 1 & 0 \end{bmatrix},$$

$$C_a = \begin{bmatrix} 1 & 0 & 0 & 0 & 0 & 0 \\ 0 & 0 & 0 & 0 & 1 & 0 \\ 0 & 0 & 0 & 0 & 0 & 1 \\ 0 & 0 & 0 & 0 & 0 & 0 \end{bmatrix}$$

The slow subsystem is described by the level of water in boiler, temperature of the superheater, riser tube-wall temperature, riser outlet mixture quality, and water pressure in riser, in order of dominance. On the other hand, the fast subsystem is represented by water flow in riser, temperature of output steam flow, and density of output steam flow.

The numerical solution based on iterative separation of time-scales took *seven*-L and *one*-K iterations to converge to mode-separation ratio $\mu = 0.2069$.

This yields the fast subsystem as described by the triplet $A_m,\ B_m,\ C_m$, while the slow subsystem is expressed by the matrices

$$A_s = \begin{bmatrix} A_{s1} & A_{s2} \\ A_{s3} & A_{s4} \end{bmatrix},$$

$$A_{s1} = \begin{bmatrix} 1 & -0.1334E^{-3} & 0.5951E^{-3} \\ 0 & 0.9959 & -0.2997E^{-6} \\ 0 & 0.3450E^{-2} & 0.9721 \end{bmatrix},$$

$$A_{s2} = \begin{bmatrix} 0.8872E^{-4} & -0.2813E^{-1} & -0.7785E^{-3} \\ -0.2274E^{-6} & 0.1616E^{-7} & 0.6003E^{-5} \\ 0.1350E^{-3} & -0.6551E^{-1} & 0.1263E^{-1} \end{bmatrix}$$

$$A_{s3} = \begin{bmatrix} 0 & 0.7419E^{-2} & 0.2983E^{-2} \\ 0 & -0.6295E^{-2} & 0.2994E^{-2} \\ 0 & 0.1967E^{-2} & 0.2697E^{-1} \end{bmatrix},$$

$$A_{s4} = \begin{bmatrix} 0.9058 & -0.1132E^{-3} & 0.2015 \\ 0.8083E^{-3} & 0.8862 & 0.2149E^{-1} \\ 0.2045E^{-1} & -0.1308E^{-2} & 0.9145 \end{bmatrix}$$

$$B_s = \begin{bmatrix} 0.2953E^{-3} & 0.3310E^{-5} \\ -0.6407E^{-3} & 0.1168E^{-1} \\ 0.4760E^{-3} & 0.9704E^{-4} \\ -0.1528E^{-3} & 0.1109 \\ -0.2016E^{-3} & -0.2016E^{-5} \\ -0.1355E^{-2} & 0.1232E^{-2} \end{bmatrix},$$

$$C_s = \begin{bmatrix} 1 & 0 & 0 & 0 & 0 & 0 \\ 0 & 0 & 0 & 0 & 1 & 0 \\ 0 & 0 & 0 & 0 & 0 & 1 \\ 0 & 0.3012E^{3} & -0.3318 & -0.2520 & 0.1771E^{-1} & 1.1668 \end{bmatrix}$$

It is easy to check that both slow and fast subsystems are completely controllable and observable. A full-order observer can be designed to reconstruct the slow and fast states. Assigning three eigenvalues at $\{0.15,\ 0.13,\ 0.11\}$ yields

$$K_{om} = \begin{bmatrix} 0 & 0 & 0 & -61.6373 \\ 0 & 0 & 0 & -0.1962 \\ 0 & 0 & 0 & -0.9500 \end{bmatrix}$$

Next, assigning six eigenvalues at $\{0.99,\ 0.97,\ 0.95,\ 0.93,\ 0.91,\ 0.98\}$ gives

$$K_{os} = \begin{bmatrix} -0.3992E^{-1} & 0.8283E^{-2} & -0.7771E^{-2} & 0.1253E^{-4} \\ -0.1301E^{-2} & -0.2002E^{-3} & 0.3398E^{-3} & -0.2193E^{-3} \\ 0.4738E^{-1} & -0.1646E^{-1} & -0.3697E^{-4} & 0.7003E^{-2} \\ 0.6404 & 0.1308 & -0.2290 & 0.1758E^{-3} \\ 0.9804E^{-1} & 0.1859E^{-1} & -0.2819E^{-1} & 0.3910E^{-4} \\ -0.3736 & -0.4547E^{-1} & 0.5312E^{-1} & -0.1472E^{-3} \end{bmatrix}$$

From (3.176), we get the slow gain as

$$K_{oa} = \begin{bmatrix} -0.3992E^{-1} & 0.8283E^{-2} & -0.7771E^{-2} & 0.1996E^{-3} \\ -0.1301E^{-2} & -0.2002E^{-3} & 0.3398E^{-3} & -0.2585E^{-3} \\ 0.4738E^{-1} & -0.1646E^{-1} & -0.3697E^{-4} & 0.7004E^{-2} \\ 0.6404 & 0.1308 & -0.2290 & 0.2321E^{-3} \\ 0.9804E^{-1} & 0.1859E^{-1} & -0.2819E^{-1} & -0.8178E^{-3} \\ -0.3736 & -0.4547E^{-1} & 0.5312E^{-1} & 0.5173E^{-4} \end{bmatrix}$$

Proceeding further and since A_m is a stable matrix, $A_s,\ B_s$ is a controllable pair, and $A_s,\ C_s$ is an observable pair, then we employ **Theorem 3.10.4** to design a lower-order observer-based controller. Placing the observer eigenvalues at

$$\{0.83,\ 0.82,\ 0.81,\ 0.80,\ 0.79,\ 0.78\}$$

gives

$$K_{ob} = \begin{bmatrix} -0.5879 & 0.2398 & -0.2538E^{-1} & 0.2100E^{-4} \\ -0.9063E^{-1} & 0.5282E^{-1} & -0.6228E^{-2} & -0.6453E^{-3} \\ -0.1249E^{3} & 0.6905E^{2} & -0.8445E^{1} & 0.6829E^{-2} \\ 0.3669E^{2} & -0.1807E^{2} & 0.1703E^{1} & -0.18123E^{-2} \\ 0.19324E^{1} & 0.19968 & -0.1554 & 0.1285E^{-3} \\ -0.1232E^{2} & 0.7272E^{1} & -0.1058E^{1} & 0.7215E^{-3} \end{bmatrix}$$

and selecting the desired closed-loop eigenvalues to be

$$\{0.99,\ 0.897,\ 0.95,\ 0.93,\ 0.91,\ 0.89\}$$

results in

$$\begin{aligned} G_{ob} &= \begin{bmatrix} G_{ob1} & G_{ob2} \end{bmatrix}, \\ G_{ob1} &= \begin{bmatrix} -0.2035E^{3} & -0.7295E^{2} & 0.6145E^{2} \\ -0.3760E^{2} & -0.8199E^{1} & 0.5847E^{1} \end{bmatrix}, \\ G_{ob2} &= \begin{bmatrix} 0.6797E^{1} & 0.7904E^{1} & 0.3152E^{2} \\ 0.8749 & 0.5298E^{1} & 0.6536E^{1} \end{bmatrix} \end{aligned}$$

The matrices K_{ob} and G_{ob} are then the required gains to implement the observer-based feedback controller (3.177)-(3.178) to the discrete model of the boiler system.

3.11.4 Hydraulic system with electronic test gear

In process industry, electronic test gears are usually mounted near large tanks of liquid gases to help in evaluating the temperature profile. A state model of this type of hydraulic system equipped with insulation materials has the following data values:

$$A = \begin{bmatrix} -5 & 0 & 0 & 1 & 0.1 & -0.5 & -0.009 & 3 \\ 0 & -2 & 0 & 1 & -0.29 & 0 & -0.3 & 0.48 \\ -0.08 & -0.11 & -3.99 & -0.93 & 0 & 0.1 & 0 & 0 \\ 0 & 0 & 1.32 & -1.39 & -1 & -0.4 & 0 & 0 \\ 0 & 0 & -0.1 & -0.4 & -0.2 & 0 & 0 & 0 \\ 0 & 0 & 0 & 0 & 0 & -0.17 & 0 & 0 \\ 0 & 0 & 0 & 0 & 0 & 0 & -0.5 & 0 \\ 0 & 0 & 0 & 0 & 0 & 0.01 & 0 & -0.11 \end{bmatrix},$$

$$B^t = \begin{bmatrix} 4 & 0 & 0 & 0 & 0 & 0 & 0 & 0 \\ 0 & 0 & 0 & 0 & 0 & 0 & 10 & 0 \end{bmatrix}, \; C = \begin{bmatrix} 1 & 0 & 0 & 0 & 0 & 0 & 0 & 0 \\ 0 & 0 & 0 & 0 & 0 & 1 & 0 & 0 \end{bmatrix}$$

where the states are the internal pressures and heat flow rates and the control variables are the pressures of the liquid gas and surroundings. We treat the system under consideration as composed of two-coupled systems: each has four states, single input and single output. Our purpose is to compare three coordinated control methods as reported in [200, 322, 354]. Our scheme of comparison relies on the results of two fourth-order subsystems vis-à-vis the eighth-order system. The control gains are given by

Method of [322]

$$\mathsf{K}_1 = \begin{bmatrix} \mathsf{K}_{11} & \mathsf{K}_{12} \end{bmatrix},$$

$$\mathsf{K}_{11} = \begin{bmatrix} -0.0098 & -0.0685 & -0.8308 & -0.3792 \\ -0.0037 & -0.0261 & -0.3159 & -0.1442 \end{bmatrix},$$

$$\mathsf{K}_{12} = \begin{bmatrix} 0.1283 & 0.0001 & -0.0001 & 0.0001 \\ -0.1043 & -0.0001 & 0.0001 & -0.0001 \end{bmatrix}$$

$$
\begin{aligned}
&Method\ of\ [354] \\
\mathsf{K}_2 &= \begin{bmatrix} \mathsf{K}_{21} & \mathsf{K}_{22} \end{bmatrix}, \\
\mathsf{K}_{21} &= \begin{bmatrix} -0.0141 & -0.2290 & -0.7302 & -0.6726 \\ 0.0170 & 1.4904 & 0.0355 & 0.3876 \end{bmatrix}, \\
\mathsf{K}_{22} &= \begin{bmatrix} 0.2358 & 0.1156 & 0.0393 & -0.1314 \\ -0.5895 & 0.0536 & 0.2157 & -0.3400 \end{bmatrix} \\
&Method\ of\ [200] \\
\mathsf{K}_3 &= \begin{bmatrix} \mathsf{K}_{31} & \mathsf{K}_{32} \end{bmatrix}, \\
\mathsf{K}_{31} &= \begin{bmatrix} -0.0084 & -0.0592 & -0.7171 & -0.3273 \\ -0.0002 & 0.0001 & -0.025 & -0.1001 \end{bmatrix}, \\
\mathsf{K}_{32} &= \begin{bmatrix} -0.0020 & 0.0100 & -0.0002 & 0.0001 \\ -0.9512 & 0.0100 & -0.0002 & 0.0001 \end{bmatrix}, \\
||\mathsf{K}_1|| &= 0.9835, \quad ||\mathsf{K}_2|| = 0.9613, \quad ||\mathsf{K}_3|| = 1.7706
\end{aligned}
$$

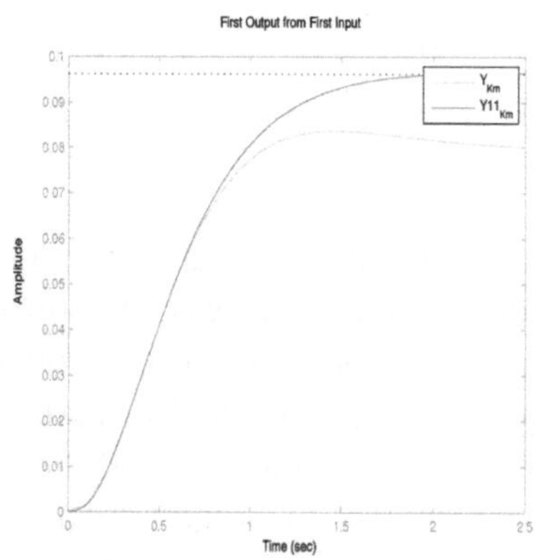

Figure 3.26: Responses of the main pressure using centralized gain and coordinated gain K_1

Simulation results are presented in Figures 3.26-3.28. It is noted that the difference between the centralized method based on the linear-quadratic theory and the coordinated control methods lies toward the end of the time-span and there is consistency during most of the trajectories.

3.12 Problem Set I

Problem I.1: Consider a large-scale system of the type (3.125) with linear

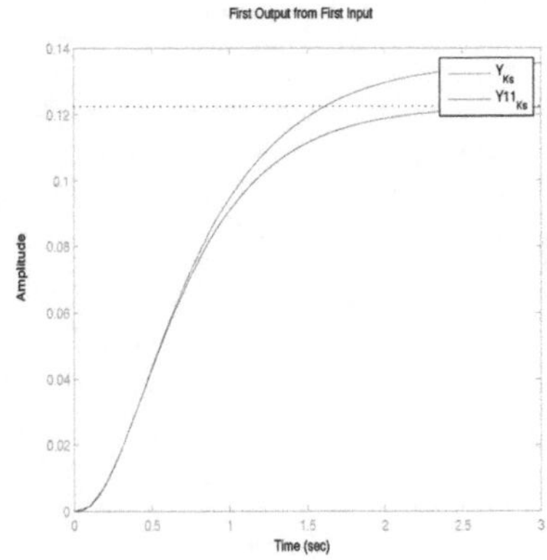

Figure 3.27: Responses of the main pressure using centralized gain and coordinated gain K_2

interconnection pattern. Let $0 < P_j = P_j^t$ satisfy the following ARE:

$$P_j A_j + A_j^t P_j + Q_j - P_j B_j R_j^{-1} B_j^t P_j = 0, \quad j = 1, 2, ...n_s$$

where $0 < Q_j = Q_j^t, \; 0 < R_j = R_j^t$. Use Lyapunov theory to show that there exists a stabilizing decentralized output feedback controller of the form $u(t) = -FCx(t)$ where

$$F = Blockdiag[F_1, ..., F_{n_s}], \; C = Blockdiag[C_1, ..., C_{n_s}]$$

if the following inequalities

$$\begin{aligned} &P_j^{-1} Q_j P_j^{-1} + 2 B_j P_j^{-1} B_j^t - 4 A_{jk} Q_k^{-1} A_{jk}^t + \\ &2(A_j P_j^{-1} + P_j^{-1} A_j^t) \; > \; 0, \quad j = 1, 2, ...n_s, \; k = 1, 2, ...n_s \end{aligned}$$

are satisfied.

Problem I.2: Consider the system described by

$$\begin{aligned} \dot{x}_1 &= x_1^2 \, x_2 \, + \, u_1, \quad \dot{x}_2 = x_2^2 \, x_3 \, + \, u_2, \\ \dot{x}_3 &= x_3^2 \, x_1 \, + \, u_3 \end{aligned}$$

where it is desired to develop static controllers in the spirit of Section 3.4.2 such that $x_j \;\rightarrow\; 0, \; j = 1, 2, 3$. Identify the bounding relations on the interconnections. By selecting appropriate controller parameters, simulate the resulting closed-loop system starting from $x_1(0) = 1, \; x_2(0) = -2, \; x_3(0) = 2$.

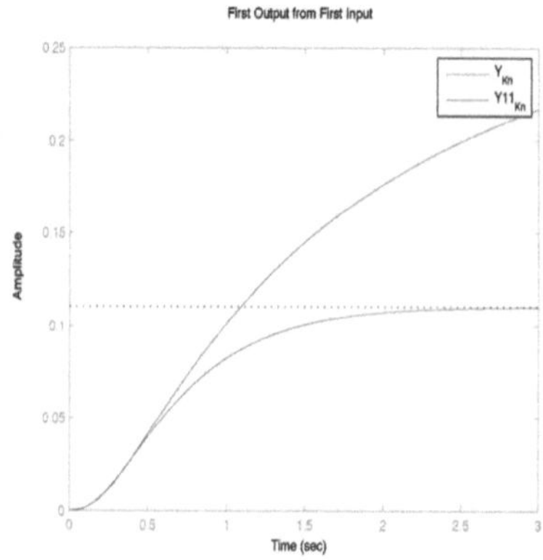

Figure 3.28: Responses of the main pressure using centralized gain and coordinated gain K3

Problem I.3: Consider a system composed of two pendulums connected by a spring. The dynamics for this system are given by

$$\begin{aligned}\dot{\theta}_1 &= \omega_1 \,, \\ J_1\dot{\omega}_1 &= \kappa\,(\theta_2 - \theta_1) + 1; u_1, \\ \dot{\theta}_2 &= \omega_2 \,, \\ J_1\dot{\omega}_2 &= \kappa\,(\theta_1 - \theta_2) + 1; u_2\end{aligned}$$

where θ_j and ω_j are the position and angular velocity for the jth pendulum, respectively. Here κ is an interconnection constant based on the spring stiffness and lever arm associated with the connection point. Derive decentralized controllers for each torque input u_j so that $\theta_j \rightarrow 0, \; j = 1, 2$.

Problem I.4: Consider a large-scale discrete-time system of order 6 with subsystem model given by (3.148) and explicitly consists of three subsystems with the following data:

$$\begin{aligned}A_1 &= \begin{bmatrix} 0.7 & 0.2 \\ 0.1 & 0.8 \end{bmatrix}, \; B_1 = \begin{bmatrix} 0 \\ 1 \end{bmatrix}, \; C_1 = \begin{bmatrix} 1 & 0 \\ 0 & 1 \end{bmatrix}, \\ A_2 &= \begin{bmatrix} 0.5 & 0.15 \\ 0 & 0.4 \end{bmatrix}, \; B_2 = \begin{bmatrix} 1 \\ 0 \end{bmatrix}, \; C_1 = \begin{bmatrix} 2 & 0 \\ 0 & -1 \end{bmatrix}, \\ A_3 &= \begin{bmatrix} 0.6 & 0 \\ 0.3 & 0.1 \end{bmatrix}, \; B_3 = \begin{bmatrix} 1 \\ 1 \end{bmatrix}, \; C_1 = \begin{bmatrix} -1 & 0 \\ 0 & 1 \end{bmatrix}\end{aligned}$$

along with the interconnection matrix

$$H = \begin{bmatrix} 0 & 0 & -6 & 0 & 0 & 0 \\ 0 & 0 & 0 & 0 & 0 & 3 \\ 0 & 0 & 0 & 0 & 0 & 0 \\ 0 & -2 & 0 & 0 & 2 & 0 \\ 1 & 0 & 0 & 0 & 0 & 0 \\ 0 & 0 & 0 & -2 & 0 & 0 \end{bmatrix}, \; B_1 = \begin{bmatrix} 0 \\ 1 \end{bmatrix}, \; C_1 = \begin{bmatrix} 1 & 0 \\ 0 & 1 \end{bmatrix}$$

It is required to design observers such that the desired eigenvalues for the respective subsystems are $\{0.1, \; -0.7\}$, $\{0.3, \; 0.4\}$, $\{-0.6, \; -0.2\}$ using instantaneous and one-step delay measurements.

Problem I.5: A fourth-order discrete system of the type (3.153)-(3.154) is described by:

$$\begin{aligned} A_a &= \begin{bmatrix} 0.9 & 0 \\ 0.1 & 0.8 \end{bmatrix}, \; A_e = \begin{bmatrix} 0 & 0.1 \\ 0.05 & -0.1 \end{bmatrix}, \; A_c = \begin{bmatrix} -0.1 & 0 \\ 0.12 & 0.03 \end{bmatrix}, \\ A_m &= \begin{bmatrix} 0.15 & 0 \\ 0 & 0.1 \end{bmatrix}, \; B_a = \begin{bmatrix} 1 & 0 \\ 0 & 1 \end{bmatrix}, \; B_m = \begin{bmatrix} 1 & 0.5 \\ 0.5 & 0 \end{bmatrix}, \\ C_a &= \begin{bmatrix} 0.1 & 0 \\ 0 & 0.1 \end{bmatrix}, \; C_m = \begin{bmatrix} 0 & 0.1 \\ 0.2 & 0 \end{bmatrix} \end{aligned}$$

Develop a reduced-order model that inherits the slow modes of the system. Then design a stabilizing dynamic output-feedback controller.

Problem I.6: A large-scale discrete-time system of the form

$$\dot{x}(t) = Ax(t) \; + \; Bu(t)$$

has the following data:

$$A = \begin{bmatrix} 0 & 1 & 0.5 & 1 & 0.6 & 0 \\ -2 & -3 & 1 & 0 & 0 & 1 \\ 0.5 & 1 & 0 & 2 & 1 & 0.5 \\ 0 & 0.5 & 1 & 3 & 0 & -0.5 \\ 1 & 0 & 0 & 1 & 0 & 1 \\ 0 & 0.5 & 0.5 & 0 & -3 & -4 \end{bmatrix},$$

$$B^t = \begin{bmatrix} 1 & 1 & 0 & 0 & 0 & 0 \\ 0 & 0 & 3 & 0 & 0 & 0 \\ 0 & 0 & 0 & 4 & 0 & 0 \\ 0 & 0 & 0 & 0 & 2 & 3 \end{bmatrix}$$

Derive the optimal controller based for the full-order system with weighting matrices $Q = I_6,\ R = I_4$. Then compare the results with the decentralized controllers based on three coupled subsystems.

Problem I.7: A linearized model of a two-link manipulator can be put into the form

$$\dot{x}(t) = Ax(t) \; + \; Bu(t)$$

with the following data:

$$A = \begin{bmatrix} -a_1 & 0 & 0 & 0 & 0 & 0 \\ 0 & 0 & 1 & 0 & 0 & 0 \\ a_2 & -a_2 & a_3 & 0 & 0 & a_4 \\ 0 & 0 & 0 & -a_5 & 0 & 0 \\ 0 & 0 & 0 & 0 & 0 & 1 \\ 0 & 0 & a_6 & a_7 & -a_7 & a_8 \end{bmatrix},$$

$$B^t = \begin{bmatrix} a_1 & 0 & 0 & 0 & 0 & 0 \\ 0 & 0 & 0 & a_5 & 0 & 0 \end{bmatrix}$$

Use the following parameter values:

$$a_1 = a_5 = 0.1,\ a_2 = a_7 = 10,\ a_3 = -1.8,\ a_4 = 0.1,\ a_6 = 0.1,\ a_8 = -2$$

Develop decentralized and hierarchical controllers with unity weighting matrices and compare between the closed-loop system performance. Then discretize the model using $\Delta t = 0.05$ and generate decentralized and hierarchical controllers with unity weighting matrices. Compare between the closed-loop system performance in each case. Would the results be preserved if the parameters $a_3,\ a_4,\ a_6$ are perturbed by 100%?

Problem I.8: A typical four-stage mixer-settler model [2] can be cast into the type (3.148) with $j = 1, 2$ and having the following values:

$$A_1 = \begin{bmatrix} -0.5672 & 0.4699 \\ 0.1022 & -0.5672 \end{bmatrix},\ B_1 = \begin{bmatrix} 0.0176 & 0 \\ 0.0039 & 0.1003 \end{bmatrix},$$

$$A_2 = \begin{bmatrix} -0.5672 & 0 \\ 0 & -0.5672 \end{bmatrix},\ B_2 = \begin{bmatrix} -0.1108 & 0 \\ -0.1607 & 0.1022 \end{bmatrix},$$

$$C_1 = \begin{bmatrix} 0.0230 & 0 \\ 0.0130 & 0 \end{bmatrix},\ C_2 = \begin{bmatrix} 0 & 0.0130 \\ 0 & 0.0230 \end{bmatrix},$$

$$g_1 = \begin{bmatrix} 0 & 0.1022 \\ 0.4649 & 0 \end{bmatrix} x_2,\ g_2 = \begin{bmatrix} 0 & 0.1022 \\ 0.4649 & 0 \end{bmatrix} x_1$$

Using unity weighting factors, it is desired to derive a decentralized controller

$$u(t) = -\begin{bmatrix} -G_1 x_1 & 0 \\ 0 & -G_2 x_2 \end{bmatrix}$$

to achieve exponential decay of $\alpha \geq 0.4$. To cope with a constant disturbance d applied to both subsystems, extend the developed controller to the form

$$u(t) = -\begin{bmatrix} -G_1 x_1 - \int_0^t L_1 x_1 dt & 0 \\ 0 & -G_2 x_2 - \int_0^t L_2 x_2 dt \end{bmatrix}$$

and compute the integral gains L_1 and L_2.

3.13 Notes and References

This section provided an overview to some of the fundamental approaches to large-scale systems using multilevel optimization (coordinated control) and decentralized control. The material covered on multilevel optimization offers advantages in the following situations:

1. *When a large-scale system is subject to structural perturbations [303, 304, 306, 321, 323], whereby subsystems are disconnected and again connected in various ways during operation, a trade-off between reliability and optimality can be established. A certain level of suboptimality and exponential stability is assured under arbitrary structural perturbations.*

2. *In cases when the individual subsystems have no information about the actual shape of interactions save that they are bounded, suboptimality and stability of the overall system can be accomplished using local controllers only.*

3. *When a system is so large that a straightforward optimization is either uneconomical because of an excessive computer time required, or impossible due to excessive computer storage needed to complete the optimization.*

4. *If the state of the overall system is not accessible for direct measurement, and a single observer is not feasible. The proposed multilevel optimization scheme can easily accommodate the use of observers on the subsystem level.*

5. *The above situations can be resolved successfully by the proposed multilevel optimization method at some cost involving the system performance.*

Therefore, the method should be judged satisfactory to the extent that the benefits of the advantages outlined above outweigh the sacrifice in the suboptimality of the system. Some relevant practical applications of hierarchical control can be found in [241] for stream water quality, in [218] for synchronous machine connected to an infinite bus, in [216] for an industrial fermentation process, and in [239] for freeway traffic control problems.

Decentralized control has long been used in the control of power systems and spacecraft since the problem may often be viewed as a number of interconnected systems. See [273] for general stability theory related to decentralized systems. An early form of diagonal dominance led to the concept of an M-matrix [132, 152, 386]. We have focused on dealing with linear subsystems with known or uncertain parameters. This has shown to facilitate the design task. In this chapter, we have allowed interconnections to be linear or bounded. In some sections, we focused on nonlinear subsystems to identify some structural features. A distinct feature of this chapter is the presentation of large-scale discrete-time systems with either bounded interconnections or operating on time-scale separation.

Chapter 4

Decentralized Servomechanism System

We initiate our guided tour through the development of control problems of large-scale systems, where we focus in Chapter 4 on conditions and algorithms for obtaining the dynamic controllers to solve the decentralized stabilization and the robust decentralized servomechanism problems (DSMP). The associated attributes including decentralized fixed modes of interconnected dynamical systems are carefully examined. We recall that in Chapter 3, the fundamental notion of decentralized control and its associated structures was briefly addressed. As it was introduced there and will be emphasized again here, the main motivation behind decentralized control is the limitations and/or failure of conventional methods of centralized control theory. Some basic techniques such as eigenvalue assignment, state feedback, optimal control, state estimation, and that similar to the latter (centralized control) theory demand complete information flow from all system sensors for the sake of feedback control. Clearly, these schemes are totally inadequate for feedback control of large-scale systems. Due to the physical configuration, high dimensionality, and/or interconnection patterns of such systems, a centralized control would be truly complex, which is neither economically feasible nor even necessary. Therefore, in many applications of feedback control theory to linear large-scale systems some degree of restriction is assumed to prevail on the information processing. In some cases a total decentralization is assumed; that is, every local control u_j is obtained from the local output y_j and possible external input w_i [307, 337]. In others, an intermediate restriction on the information is possible. Some related issues and discussions were reported in [5, 43, 45, 106, 303, 335].

4.1 Introduction

When control theory is applied to solve problems associated with large-scale systems (LSS), an important feature called *decentralization* often arises. In this chapter, three major problems related to decentralized structure of large-scale systems are addressed. The first problem is decentralized stabilization. In centralized systems, the problem of finding a state or an output feedback gain whereby the closed-loop system has all its poles on the left half-plane is commonly known as feedback "stabilization." Alternatively, the closed-loop poles of a controllable system may be preassigned through the state or output feedback. Clearly, the applications of these concepts in a decentralized fashion require certain extensions which will be discussed in Section 4.3.

The second problem addressed in this chapter is the decentralized "robust" control of large-scale linear systems with or without a known plant. This problem, first introduced by [45]-[52] and known as "general servomechanism," takes advantage of the tuning regulators and dynamic compensators to design a feedback that both stabilizes and regulates the system in a decentralized mode. The notion of "robust" feedback control will be defined and discussed in detail later; however, for the time being a control is said to be robust if it continues to provide asymptotic stability and regulation under perturbation of plant parameters and matrices.

The third problem is stochastic decentralized control of continuous and discrete-time systems. The scheme discussed here is based on the assumption of one sensor for each controller (channel or node) whose information, processed with a local Kalman estimator, is shared with all other controllers.

In Section 4.2, the problem of decentralized stabilization is mathematically formulated. The section reviews some of the appropriate schemes for decentralized feedback stabilization, including the notions of "fixed modes," "fixed polynomials" [45], and their role in dynamic compensation. Also discussed are the dynamic stabilization of large-scale systems and the notion of "exponential stability" applied to decentralized systems. The stabilization of linear time-varying systems [125, 126, 127] and the special case of time-invariant system stabilization will follow.

4.2 A Class of Large-Scale Systems

In this section we look at a basic mathematical model of large-scale linear time-invariant (LTI) time-varying systems using multicontroller structure. Then we move to study the problem of stabilizing this class of systems and present

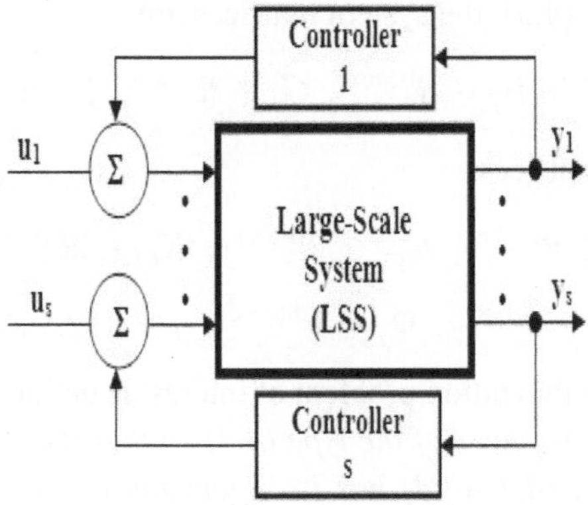

Figure 4.1: Large-scale system with multicontroller structure

conditions under which the overall system with decentralized control can be stabilized.

A class of large-scale LTI systems with n_s local control stations (channels) can be represented by:

$$\begin{aligned} \dot{x}(t) &= Ax(t) + \sum_{j=1}^{n_s} B_j u_j(t) \\ y_j(t) &= C_j x(t), \quad j = 1, 2, ..., n_s \end{aligned} \tag{4.1}$$

where $x \in \Re^n$ is the state, $u_j \in \Re^{m_j}$ are the local control inputs, and $y_j \in \Re^{r_j}$ are the local measurable outputs associated with control channel j. The overall system control and output orders m and r are given by

$$m = \sum_{j=1}^{n_s} m_j, \quad r = \sum_{j=1}^{n_s} r_j \tag{4.2}$$

This system is depicted in Figure 4.1. Consider now a p_j-th order *decentralized dynamic controller* of the form

$$\begin{aligned} \dot{\xi}_j(t) &= \Lambda_j \xi_j(t) + \Phi_j y_j(t) + \Gamma_j \omega_j(t) \\ u_j(t) &= K_{cj} \xi_j(t) + K_{oj} y_j(t) + \Psi_j \omega_j(t), \quad j = 1, 2, ..., n_s \end{aligned} \tag{4.3}$$

where $\xi \in \Re^{p_j}$ is the controller state vector and $\omega \in \Re^{q_j}$ is an external input vector. In the sequel, system (4.1) is termed *decentralized ν control-channel system*. In (4.1) and (4.3), the system matrices are

$$A \in \Re^{n \times n}, \ B_j \in \Re^{n \times m_j}, \ C_j \in \Re^{r_j \times n}, \ j = 1, 2, ..., n_s$$

and the controller matrices

$$\Lambda_j \in \Re^{p_j \times p_j}, \ \Phi_j \in \Re^{p_j \times r_j}, \ K_{cj} \in \Re^{m_j \times r_j}, \ K_{cj} \in \Re^{m_j \times p_j}, \ j = 1, 2, ..., n_s$$

$$\Gamma_j \in \Re^{p_j \times q_j}, \ \Psi_j \in \Re^{m_j \times q_j}, \ j = 1, 2, ..., n_s$$

The decentralized stabilization problem of interest is defined as follows: *Obtain n_s local output control laws of the type (4.3) so that the resulting closed-loop system described by (4.1)-(4.3) has its eigenvalues in a set $\aleph$*, where $\aleph$ is a nonempty symmetric open subset of complex s-plane [48]. It is quite evident that the membership of a closed-loop system eigenvalue $\lambda \in \aleph$ implies its complex conjugate $\lambda^* \in \aleph$ in a prescribed manner.

4.3 Decentralized Stabilization

In this section the problem of stabilizing large-scale LTI and time-varying systems is presented. Conditions under which the overall system with decentralized control can be stabilized will be given. Decentralized stabilization has been an active field of research for large-scale systems. The discussions on this topic are restricted to LTI systems for the most part and are based on the works of [45]-[52], [5, 43, 115, 304], and [323, 324]. Additional works on such subjects as large-scale linear systems with nonlinear interconnections [353] and the results of [125, 126, 127] for the time-varying systems and the case of time-invariant systems [4] will also be considered.

4.3.1 Fixed modes

The notions of fixed modes and the associated fixed polynomials are generalizations of the "centralized" systems pole placement problem, in which any uncontrollable and unobservable mode of the system must be stable [30], to the decentralized case. The idea of fixed modes for decentralized control was introduced by [45] which leads to necessary and sufficient conditions for stabilizability of decentralized systems (4.1)-(4.3). A more refined characterization of fixed modes and the associated decentralized control was presented later on by [5, 6, 43, 55, 359, 368]. Some of these approaches have led to conditions which involve difficult rank and eigenvalue computations [324].

At this point, the aggregate dynamic controllers can be written as

$$\begin{aligned} \dot{\xi}(t) &= \Lambda\xi(t) + \Phi y(t) + \Gamma\omega(t) \\ u(t) &= K_c\xi(t) + K_o y(t) + \Psi\omega(t) \end{aligned} \tag{4.4}$$

where

$$\begin{aligned} \Lambda &= Block - diag\{\Lambda_1,\ \Lambda_2,\ ...,\ \Lambda_{n_s}\}, \\ \Phi &= Block - diag\{\Phi_1,\ \Phi_2,\ ...,\ \Phi_{n_s}\} \\ \Gamma &= Block - diag\{\Gamma_1,\ \Gamma_2,\ ...,\ \Gamma_{n_s}\}, \\ \Psi &= Block - diag\{\Psi_1,\ \Psi_2,\ ...,\ \Psi_{n_s}\}, \\ K_c &= Block - diag\{K_{c1},\ K_{c2},\ ...,\ K_{cn_s}\}, \\ K_o &= Block - diag\{K_{o1},\ K_{o2},\ ...,\ K_{on_s}\} \end{aligned} \tag{4.5}$$

and

$$\begin{aligned} u(t) &= \mathbf{col}[u_1(t),\ ...,\ u_{n_s}(t)],\ \xi(t) = \mathbf{col}[\xi_1(t),\ ...,\ \xi_{n_s}(t)], \\ y(t) &= \mathbf{col}[y_1(t),\ ...,\ y_{n_s}(t)],\ \omega(t) = \mathbf{col}[\omega_1(t),\ ...,\ \omega_{n_s}(t)] \end{aligned} \tag{4.6}$$

By applying the aggregate controller (4.4) to system (4.1), we obtain the aggregate closed-loop system

$$\begin{aligned} \dot{\zeta}(t) &= \Pi\zeta(t) + \Omega\omega(t),\ \ \zeta(t) = \mathbf{col}[x(t)\ \ \xi(t)] \\ \Pi &= \begin{bmatrix} A + BK_oC & BK_c \\ \Phi C & \Lambda \end{bmatrix},\ \Omega = \begin{bmatrix} B\Gamma \\ \Psi \end{bmatrix} \end{aligned} \tag{4.7}$$

and

$$B := [B_1,\ B_2,\ ...,B_{n_s}],\ \ C := \mathbf{col}\begin{bmatrix} C_1 & C_2 & \cdots & C_{n_s} \end{bmatrix} \tag{4.8}$$

As mentioned earlier, the decentralized stabilization problem is to find the control laws (4.4) so that the aggregate closed-loop system (4.7) is asymptotically stable. Equivalently stated, by means of local output feedback, the closed-loop eigenvalues of the large-scale system are required to lie on the left half of the complex s-plane C.

An important structural property of the decentralized control system (4.1)-(4.3) using static output-feedback is disclosed by the following definition.

Definition 4.3.1 *Given system (4.1), we let*

$$\begin{aligned} \mathsf{K} &= \{\mathcal{K}|\ \mathcal{K} = block\ diag[\mathcal{K}_1,\ \mathcal{K}_2,\ ...,\mathcal{K}_s]; \\ &\quad \mathcal{K}_j\Re^{m_j\times r_j},\ j = 1, 2, ..., n_s\} \end{aligned} \tag{4.9}$$

then the **decentralized fixed modes** *of system (4.1) with respect $\mathcal{K}$ are given by*

$$\lambda(C, A, B, \mathsf{K}) = \bigcap_{\mathcal{K} \in \mathsf{K}} \mathbf{spec}(A + B\mathcal{K}C) \tag{4.10}$$

where $\mathbf{spec}(.)$ *denotes the set of eigenvalues of (.).*

Note that the associated "fixed polynomial" of (C, A, B) with respect K is the greatest common divisor (**gcd**) of the set of polynomials $|\lambda I - A - B\mathsf{K}C|$ for all $\mathcal{K} \in \mathsf{K}$ and is denoted by

$$\phi(\lambda, C, A, B, \mathcal{K}) = \mathbf{gcd}\{|\lambda I - A - B\mathcal{K}C|\} \tag{4.11}$$

Since $\mathcal{K}$ can take on the null matrix, therefore the set of "fixed modes" is contained in $\mathbf{spec}(A)$. In view of **Definition 4.3.1**, the fixed modes are the roots of the fixed polynomials in (4.11), that is,

$$\mathbf{spec}(A, B, \mathcal{K}, C) = \{\lambda | \lambda \in \mathbf{C} \subset \mathcal{C} \ and \ \phi(\lambda, C, A, B, \mathcal{K}) = 0\}$$

where $\mathcal{C}$ represents a set of values on the entire complex s-plane $\mathcal{C}$.

The following algorithm reported in [49] shows, in a simple way, how to calculate the decentralized fixed modes:

1. *Compute all the eigenvalues of matrix A, that is,* $\mathbf{spec}(A)$.

2. *Select "arbitrary matrices"* $\mathcal{K}_j,\ j = 1, 2, ..., n_s$

3. *Compute the eigenvalues of* $A + \sum_{j=1}^{n_s} B_j\mathcal{K}_jC_j$

Then the decentralized fixed modes are contained in those eigenvalues of $A + \sum_{j=1}^{s} B_j\mathcal{K}_jC_j$ *which are common with the eigenvalues of A.*

Simply stated, the decentralized fixed modes are the eigenvalues of A that cannot be moved, relocated, or assigned to other places, by decentralized control actions. There have have been numerous efforts to compute the decentralized feedback gains; the reader is referred to [136]. It has been reported in [45] that the fixed modes of a centralized system $(A, B, \mathbf{K}, C),\ \mathbf{K} \in \Re^{m \times r}$ eventually correspond to the uncontrollable and unobservable modes of the system. For the case of decentralized closed-loop system, the following theorem establishes the necessary and sufficient conditions for stabilizability.

Theorem 4.3.2 *Consider system (4.1) and the class of block-diagonal gain matrices in (4.9). Then the local feedback controllers () would asymptotically stabilize the system if and only if the set of fixed modes of* $(A, B, \mathbf{K}, C)$ *is contained in the open left-half s-plane, that is,* $\mathbf{spec}(A, B, \mathbf{K}, C) \in \mathcal{C}^{-}$.

Proof: The details can be found in [45] which is based on Kalman's canonical structure [147].

An alternative algebraic characterization of the decentralized fixed modes is summarized by the following theorem.

Theorem 4.3.3 *Consider system (4.1). Then necessary and sufficient condition for $\lambda \in \mathbf{spec}(A)$ to be a decentralized fixed mode of (4.1) is that for some partition of the set $\{1, 2, ..., s\}$ into disjoint sets $\{j_1,\ j_2,\ ...,\ j_k\}$ and $\{j_{k+1},\ j_{k+2},\ ...,\ j_s\}$ there exists*

$$rank \begin{bmatrix} A - \lambda I & B_{j_1} & B_{j_2} & \cdots & B_{j_2 k} \\ C_{j_{k+1}} & 0 & 0 & \cdots & 0 \\ C_{j_{k+2}} & 0 & 0 & \cdots & 0 \\ \vdots & \vdots & \vdots & & \vdots \\ C_{j_s} & 0 & 0 & \cdots & 0 \end{bmatrix} < 0 \tag{4.12}$$

Proof: The details can be found in [5].

More work has shown that the existence of decentralized fixed modes in a s control channel system always reduces to the existence of fixed modes in an $s-1$ control channel system. Effectively then, when studying the characterization of decentralized fixed modes in a s control channel system, it is only necessary to eventually examine the case of a two-channel system. The following theorem summarizes a pertinent result.

Theorem 4.3.4 *Consider system (4.1) with $s \geq 3$. Then $\lambda \in \mathbf{spec}(A)$ is not a decentralized fixed mode of (4.1) and only if λ is not a decentralized fixed mode of (4.1) of any of the following $s-1$ control channel systems for (4.1):*

$$1) \quad \left\{ \begin{bmatrix} \begin{bmatrix} C_1 \\ C_2 \end{bmatrix} \\ C_3 \\ \vdots \\ C_s \end{bmatrix} \quad A \quad \begin{bmatrix} (B_1, B_2) & B_3 & \cdots & B_s \end{bmatrix} \right\} < 0 \tag{4.13}$$

$$2) \quad \left\{ \begin{bmatrix} C_1 \\ \begin{bmatrix} C_2 \\ C_3 \end{bmatrix} \\ \vdots \\ C_s \end{bmatrix} \quad A \quad \begin{bmatrix} B_1 & (B_2, B_3) & \cdots & B_s \end{bmatrix} \right\} < 0 \tag{4.14}$$

$$\vdots$$

$$\text{3)} \quad \left\{ \begin{bmatrix} C_1 \\ \vdots \\ C_{s-2} \\ \begin{bmatrix} C_{s-1} \\ C_2 \end{bmatrix} \\ C_3 \\ \vdots \\ C_s \end{bmatrix} A \begin{bmatrix} B_1 & \cdots & B_{s-2} & (B_{s-1}, B_s) \end{bmatrix} \right\}$$
$$< 0 \tag{4.15}$$

$$\text{4)} \quad \left\{ \begin{bmatrix} C_1 \\ \vdots \\ C_{s-3} \\ \begin{bmatrix} C_{s-2} \\ C_s \end{bmatrix} \\ C_{s-1} \\ \vdots \\ C_s \end{bmatrix} A \begin{bmatrix} B_1 & \cdots & B_{s-3} & (B_{s-2}, B_s) & B_{s-1} \end{bmatrix} \right\}$$
$$< 0 \tag{4.16}$$

Proof: The details can be found in [55].

Let us shed some light on the foregoing result. From the conventional theory of centralized systems [147], it is well known that when the system matrix A is diagonal with distinct eigenvalues then system (4.1) is controllable and observable if and only if no row of B and no column of C are identically equal to zero. We can generalize this result to the case of decentralized systems for the type of (4.1) when $n_s = 2$. Thus, we let

$$\dot{x} = \begin{bmatrix} \lambda_1 & \vdots & & & \\ \cdots & \cdots & \cdots & \cdots & \cdots \\ & \vdots & \lambda_2 & & \\ & \vdots & & \ddots & \\ & \vdots & & & \lambda_n \end{bmatrix} x + \begin{bmatrix} b_1^t \\ \cdots \\ B_1 \end{bmatrix} u_1 + \begin{bmatrix} b_2^t \\ \cdots \\ B_2 \end{bmatrix} u_2 \tag{4.17}$$

$$y_1 = (c_1 \quad C_1)x, \quad y_2 = (c_2 \quad C_2)x \tag{4.18}$$

where $b_1 \in \Re^{m_1}$, $b_2 \in \Re^{m_2}$, $c_1 \in \Re^{r_1}$, $c_2 \in \Re^{r_2}$ are prescribed vectors, respectively. We assume that $\lambda_1, \lambda_2, ..., \lambda_n$ are all distinct and occur in complex conjugate pairs. For simplicity in exposition, we define

$$\begin{aligned} B_1 &:= \left[\begin{array}{cccc} b_{11} & b_{12} & \cdots & b_{1m_1} \end{array} \right], \\ B_2 &:= \left[\begin{array}{cccc} b_{21} & b_{22} & \cdots & b_{2m} \end{array} \right] \\ C_1 &:= \textbf{col}\left[\begin{array}{cccc} c_{11} & c_{12} & \cdots & c_{1r_1} \end{array} \right], \\ C_2 &:= \textbf{col}\left[\begin{array}{cccc} c_{21} & c_{22} & \cdots & c_{2r_2} \end{array} \right] \end{aligned} \tag{4.19}$$

The following result is established.

Theorem 4.3.5 *Consider the two-channel decentralized system (4.17)-(4.19). Then λ_1 is not a decentralized fixed mode if and only if the following conditions hold:*

$$1) \quad \left[\begin{array}{cc} B_1 & B_2 \end{array} \right] \neq 0 \; and \; \left[\begin{array}{cc} C_1^t & C_2^t \end{array} \right] \neq 0 \tag{4.20}$$

$$\begin{aligned} 2) \quad & The\ condition: \\ & B_1 = 0,\ and\ C_2 = 0,\ and\ \lambda_1\ is\ a\ transmission\ zero\ [47]\ of \\ & \left\{ \begin{array}{ccc} c_{2j} & diag\left[\begin{array}{ccc} \lambda_1 & \cdots & \lambda_s \end{array} \right] & b_{1k} \end{array} \right\} \\ & \forall\, j \in [1, 2, ..., r_2],\ \forall\, k \in [1, 2, ..., m_1]\ does\ not\ hold \end{aligned} \tag{4.21}$$

$$\begin{aligned} 3) \quad & The\ condition: \\ & B_2 = 0,\ and\ C_1 = 0,\ and\ \lambda_1\ is\ a\ transmission\ zero\ of \\ & \left\{ \begin{array}{ccc} c_{1j} & diag\left[\begin{array}{ccc} \lambda_1 & \cdots & \lambda_s \end{array} \right] & b_{2k} \end{array} \right\} \\ & \forall\, j \in [1, 2, ..., r_1],\ \forall\, k \in [1, 2, ..., m_2]\ does\ not\ hold \end{aligned} \tag{4.22}$$

Proof: The details can be found in [55].

Remark 4.3.6 *It is significant to note that in centralized systems [147], the controllability-observability of mode λ_1 depends only on $b_1,\ b_2,\ c_1,\ c_2$ as given by condition 1) and is independent of the values of $B_1,\ B_2,\ C_1,\ C2,\ \lambda_1,\ ...,\ \lambda_s$. Note that this is no longer the case with decentralized systems.*

We are now in a position to extend on the foregoing results and provide conditions for an interconnected system to have no decentralized fixed modes.

4.3.2 Decentralized fixed modes

In this section, we examine the existence of fixed modes in interconnected systems described by

$$\begin{aligned}
\dot{x}(t) &= \begin{bmatrix} A_{11} & A_{12} & \cdots & A_{1n_s} \\ A_{21} & A_{22} & \cdots & A_{2n_s} \\ \vdots & \vdots & & \vdots \\ A_{n_s1} & A_{n_s2} & \cdots & A_{n_sn_s} \end{bmatrix} x(t) + \begin{bmatrix} B_1 \\ 0 \\ \vdots \\ 0 \end{bmatrix} u_1(t) \\
&+ \begin{bmatrix} 0 \\ B_2 \\ \vdots \\ 0 \end{bmatrix} u_2(t) + ... + \begin{bmatrix} 0 \\ 0 \\ \vdots \\ B_{n_s} \end{bmatrix} u_{n_s}(t) \\
y_1(t) &= \begin{bmatrix} C_1 & 0 & \cdots & 0 \end{bmatrix} x(t) + \begin{bmatrix} 0 & C_2 & \cdots & 0 \end{bmatrix} x(t) \\
&+ \cdots\cdots \begin{bmatrix} 0 & 0 & \cdots & C_{n_s} \end{bmatrix} x(t) \qquad (4.23)
\end{aligned}$$

where

$$A_{jk} = T_{jk} K_{jk} R_{jk}, \; j = 1, 2, ..., n_s, \; k = 1, 2, ..., n_s, \; j \neq k$$

and K_{jk} is the interconnection gain matrix and $T_{jk}, \; R_{jk}$ are arbitrary matrices. The following theorem furnishes the desired results, the proof of which follows from [48, 52, 304]:

Theorem 4.3.7 *Consider the decentralized system (4.23) and assume that the triplets $(C_j, \; A_{jj}, \; B_j), \; j = 1, 2, ..., n_s$ are all controllable and observable. Then this implies that system (4.23) has no decentralized fixed modes for almost all interconnection gains*

$$K_{jk}, \; j = 1, 2, ..., n_s, \; k = 1, 2, ..., n_s, \; j \neq k$$

Remark 4.3.8 *Consider the decentralized system (4.23) with interconnection structure*

$$A_{jk} = B_j K_{jk} C_k, \; j = 1, 2, ..., n_s, \; k = 1, 2, ..., n_s, \; j \neq k$$

Then necessary and sufficient conditions for system (4.23) to have no decentralized fixed modes are that the triplets $(C_j, \; A_{jj}, \; B_j), \; j = 1, 2, ..., n_s$ are all controllable and observable.

Some pertinent remarks stand out:

Remark 4.3.9 *The approach that led to the establishment of* **Theorems 4.3.2-4.3.5** *can be labeled, from a system theoretic standpoint, as macroscopic view as they treat the large-scale system as one black box without emphasis on the internal coupling pattern. Alternatively, the approach that led to the establishment of* **Theorem 4.3.7** *can be labeled as microscopic view as they treat the large-scale system as interconnections of subsystems in black box without emphasis on the internal coupling pattern. This is crucial to understand since wider applicability and greater benefits could be gained from the implementation of the latter approach. We will follow such a trend throughout the book.*

Remark 4.3.10 *It is worth noting extensions of the definition decentralized fixed modes to other information structure constraints can be readily made by modifying the set of controllers (4.9); the reader is referred to [43, 45] for further accounts.*

4.4 Illustrative Examples

Some examples are presented to illustrate the foregoing results.

4.4.1 Illustrative example 4.1

Consider the following large-scale system of the type (4.1) with

$$A = \begin{bmatrix} A_1 & A_2 & 0 \\ A_3 & A_4 & A_5 \\ A_6 & A_7 & A_8 \end{bmatrix}, \; A_1 = \begin{bmatrix} -3.9 & -0.3 & 0 \\ 0.06 & -3 & 3 \\ 27 & 0.8 & -0.9 \end{bmatrix},$$

$$A_2 = \begin{bmatrix} 0 & 0 & 4 \\ 0 & 0 & -0.3 \\ 0 & 0 & -0.8 \end{bmatrix}, \; A_3 = \begin{bmatrix} -0.4 & -5.2 & 0 \\ -38 & 17 & -12 \\ 22 & 18 & 0 \end{bmatrix}$$

$$A_4 = \begin{bmatrix} -0.25 & -3.35 & 3.6 \\ -12 & -2.9 & -0.1 \\ -35 & -0.4 & -0.4 \end{bmatrix}, \; A_5 = \begin{bmatrix} 6.3 & 0.1 & 0 \\ 12 & 43 & 0 \\ 90 & 56 & 0 \end{bmatrix},$$

$$A_6 = \begin{bmatrix} 0 & 0 & 0.4 \\ 0 & 0 & 0 \\ -2.2 & -0.7 & 0 \end{bmatrix}, \; A_7 = \begin{bmatrix} 0 & 0 & 10 \\ -1.2 & 0 & 7 \\ -8 & 0 & 1.3 \end{bmatrix},$$

$$A_8 = \begin{bmatrix} -20 & 0 & 0 \\ 0 & -7 & 0 \\ 0.1 & 6 & -1 \end{bmatrix}$$

$$
\begin{aligned}
B &= \begin{bmatrix} B_1 & B_2 & B_3 \\ B_4 & B_5 & B_6 \\ B_7 & B_8 & B_9 \end{bmatrix}, \; B_1 = \begin{bmatrix} -0.1 & 0 & 0 \\ 0 & 1 & 0 \\ 0 & 15.6 & 0 \end{bmatrix}, \\
B_2 &= \begin{bmatrix} 0 & 0 & 4 \\ 0 & 0 & 0 \\ 0 & 0 & 0 \end{bmatrix}, \; B_3 = \begin{bmatrix} 0 & 0 & 0 \\ 1 & 0 & 0 \\ 0 & 0 & 1 \end{bmatrix}, \\
B_4 &= \begin{bmatrix} 0 & 0 & -5.6 \\ 52 & 8.2 & -1.5 \\ 0 & 0 & 1.7 \end{bmatrix}, \; B_5 = \begin{bmatrix} 0 & 0 & 0 \\ -0.4 & 0.9 & 0 \\ 0 & 0 & 0 \end{bmatrix}, \\
B_6 &= \begin{bmatrix} 0 & 0 & 1 \\ 0 & 1 & 0 \\ 0 & 0 & 0 \end{bmatrix}, \; B_7 = \begin{bmatrix} 0 & 2.9 & 0 \\ 0 & 0 & -0.4 \\ 0 & 0 & 2.9 \end{bmatrix}, \\
B_8 &= \begin{bmatrix} 0 & 0 & 0 \\ 0 & 0 & 0 \\ 0 & 0 & 3 \end{bmatrix}, \; B_9 = \begin{bmatrix} 0 & 0 & 10 \\ -1.2 & 0 & 7 \\ -8 & 0 & 1.3 \end{bmatrix}
\end{aligned}
$$

$$
\begin{aligned}
C &= \begin{bmatrix} I_3 & 0 & C_1 \\ C_2 & C_3 & C_4 \\ 0 & C_5 & C_6 \end{bmatrix}, \; C_1 = \begin{bmatrix} 0 & 1 & 0 \\ 1 & 0 & 0 \\ 0 & 0 & 1 \end{bmatrix}, \\
AC_2 &= \begin{bmatrix} 0 & 0 & 0 \\ 0 & 1 & 0 \\ 0 & 0 & 0 \end{bmatrix}, \; C_3 = \begin{bmatrix} 1 & 0 & 0 \\ 0 & 1 & 0 \\ 0 & 1 & 0 \end{bmatrix}, \\
C_4 &= \begin{bmatrix} 1 & 0 & 0 \\ 0 & 0 & 0 \\ 0 & 0 & 0 \end{bmatrix}, \; C_5 = \begin{bmatrix} 0 & 0 & 0 \\ 0 & 0 & 1 \\ 0 & & 0 \end{bmatrix}, \\
C_6 &= \begin{bmatrix} 0 & 1 & 0 \\ 0 & 0 & 0 \\ 0 & & 1 \end{bmatrix}
\end{aligned}
$$

Using MATLAB® to perform simulation, it is found that the open-loop eigenvalues are given by

$$
\begin{aligned}
\lambda(A) &= \{-1.0, \;\; -46.37, \;\; 26.96, \;\; 6.77, \;\; -11.55, \;\; 0.11, \\
&\quad -3.92 \pm 5.64j, \;\; -6.54\}
\end{aligned}
$$

By repeating generating a gain matrix $K \in \Re^{7\times 7}$ of arbitrary form like $K = diag[K_1, ..., K_7]$ and evaluating the eigenvalues of $A + BKC$, it turns out that all the open-loop eigenvalues can be relocated. Hence, there are no fixed modes and, therefore, it is concluded that the system under consideration can be stabilized by a decentralized control with dynamic compensator.

4.4.2 Illustrative example 4.2

Consider the following large-scale system of the type (4.1) with

$$A_o = \begin{bmatrix} 0 & 1 & 0 & 0 & 0 & 0 & 0 \\ 0 & 0 & 1 & 0 & 0 & 0 & 0 \\ -1 & -2 & -3 & 1 & 0 & 0 & 0 \\ 1 & 0 & 0 & -1 & -1 & 0 & 0 \\ 0 & 0 & 0 & 0 & 0 & 1 & 0 \\ 0 & 0 & 0 & 0 & 0 & 0 & 1 \end{bmatrix}$$

$$B_o = \begin{bmatrix} 0 & 0 & 0 \\ 0 & 0 & 0 \\ 1 & 0 & 0 \\ 0 & 0 & 0 \\ 0 & 1 & 0 \\ 0 & 0 & 0 \\ 0 & 0 & 1 \end{bmatrix}, \; C_o = \begin{bmatrix} 1 & 0 & 0 & 0 & 0 & 0 & 0 \\ 0 & 1 & 0 & 0 & 0 & 0 & 0 \\ 0 & 0 & 1 & 0 & 0 & 0 & 0 \\ 0 & 0 & 0 & 1 & 0 & 0 & 0 \\ 0 & 0 & 0 & 0 & 1 & 0 & 0 \end{bmatrix}$$

The open-loop eigenvalues are given by

$$\{-2.472, \quad -0.322 \pm 1.43j, \quad -1.0, \quad 0.24, \quad -0.562 \pm 0.68j\}$$

Consider two cases of the gain matrices

$$K_d = \begin{bmatrix} K_{d1} & 0 & 0 & 0 & 0 \\ 0 & K_{d2} & 0 & 0 & 0 \\ 0 & 0 & K_{d3} & K_{d4} & K_{d5} \end{bmatrix}, \; K_f = \begin{bmatrix} K_{f1} & K_{f2} & K_{f3} \\ K_{f4} & K_{f5} & K_{f6} \\ K_{f7} & K_{f8} & K_{f9} \end{bmatrix}$$

By randomly selecting several entries, it is found [136] that a fixed mode $\lambda = 1.0$ took place. On attempting other structures of K_d, it is found that a similar conclusion is reached.

On the other hand, consider a similar system with

$$A_c = \begin{bmatrix} 0 & 1 & 0 & 0 & 0 & 0 & 0 \\ 0 & 0 & 1 & 0 & 0 & 0 & 0 \\ -1 & -2 & -3 & 1 & 0 & 0 & 0 \\ 1 & 0 & 0 & -1 & -1 & 0 & 0 \\ 0 & 0 & 0 & 0 & 0 & 1 & 0 \\ 0 & 0 & 0 & 0 & 0 & 0 & 1 \end{bmatrix},$$

$$B_c = \begin{bmatrix} 0 & 0 & 0 & 0 & 0 & 0 & 0 \\ 0 & 0 & 0 & 0 & 0 & 0 & 0 \\ 1 & 0 & 0 & 1 & 0 & 0 & 1 \\ 0 & 0 & 0 & 0 & 0 & 0 & 0 \\ 0 & 1 & 0 & 0 & 1 & 0 & 0 \\ 0 & 0 & 0 & 0 & 0 & 0 & 0 \\ 0 & 0 & 1 & 0 & 0 & 1 & 0 \end{bmatrix},$$

$$C_c = \begin{bmatrix} 1 & 0 & 0 & 0 & 0 & 0 & 0 \\ 0 & 1 & 0 & 0 & 0 & 0 & 0 \\ 0 & 0 & 1 & 0 & 0 & 0 & 0 \\ 0 & 0 & 0 & 1 & 0 & 0 & 0 \\ 0 & 0 & 0 & 0 & 1 & 0 & 0 \\ 0 & 0 & 0 & 0 & 0 & 1 & 0 \\ 0 & 0 & 0 & 0 & 0 & 0 & 1 \end{bmatrix}$$

It is desired of find the fixed modes, if any, of this model having open-loop eigenvalues

$$\lambda(A) = \{-3.71, -2.45, 0.11, -0.76 + -0.90i, -0.21 + -0.83i\}$$

The result of simulation is that there is no eigenvalue that can be a fixed mode even though different controller structures are used.

4.4.3 Illustrative example 4.3

Consider a fourth-order, three-input and four-output system described by

$$A_o = \begin{bmatrix} 7 & 0 & 0 & 1 \\ 3 & 6 & 0 & 3 \\ 0 & 4 & 2 & 0 \\ 0 & 0 & 1 & 1 \end{bmatrix}, \quad B_o = \begin{bmatrix} 1 & 0 & 0 \\ 0 & 1 & 0 \\ 0 & 0 & 0 \\ 0 & 0 & 1 \end{bmatrix}, \quad C_o = \begin{bmatrix} 0 & 0 & 0 & 0 \\ 1 & 0 & 0 & 1 \\ 0 & 1 & 1 & 0 \\ 0 & 0 & 0 & 1 \end{bmatrix}$$

The design objective is to investigate the existence of fixed modes and structurally fixed modes. First, we note that the open-loop eigenvalues are

$$\lambda(A) = \{7.35, 6.0, 1.32 + -1.34j\}$$

Now, let the controller structure be

$$K = \begin{bmatrix} K_1 & 0 & 0 & 0 \\ 0 & K_2 & 0 & 0 \\ 0 & 0 & K_{d3} & K_{d4} \end{bmatrix}$$

As a result of extensive MATLAB simulations, it is found that $\lambda = 6$ is independent of the different of the controller. Therefore it represents a fixed mode and the system cannot be stabilized using decentralized feedback gain. It is important to note that on changing the element a_{22} by a small amount, the new system will not have a fixed mode with respect to decentralized feedback gain.

Next, to find the difference between the fixed mode and structurally fixed mode of the system, we set $a_{11} \equiv 0, \ a_{22} \equiv= 0$. In this new system, a fixed mode $\lambda = 0$ will remain regardless of the values of all nonzero elements of the new system. Thus we conclude that $\lambda = 0$ is a structurally fixed mode.

4.5 Decentralized Regulators

Thus far, we were concerned with the existence of fixed modes and decentralized fixed modes. In the control literature, an interconnected system is often referred to as a system with a set of interacting subsystems [342]. Systems with different types of interaction topologies have been investigated in the literature, among which the class of hierarchical interconnected systems has drawn special attention in recent publications due to its broad applications, that is, in formation flying, underwater vehicles, automated highway, robotics, satellite constellation, etc., which have leader-follower structures or structures with virtual leaders [170, 350]. To stabilize such a large-scale system with hierarchical fashion, one can design a set of stabilizing local controllers for the individual subsystems. In some cases, it is permissible that these local controllers partially exchange information [165, 352]. In general, the need for this type of structurally constrained controller can be originated from some practical limitations concerning, for instance, the geographical distribution of the subsystems or the computational complexity associated with the centralized controller design for large-scale systems [169]. The case when these local controllers operate independently, that is, they do not interact with each other, is referred to as decentralized feedback control [56, 170, 324]. Various aspects of the decentralized control theory have been extensively investigated in the past few decades. In [56, 89, 165, 166] the decentralized stabilizability of a system is studied by using the notions of decentralized fixed modes and quotient fixed modes.

To elaborate on the different methods, we start by looking at models that represent large-scale systems with structural information constraints.

4.5.1 The robust decentralized servomechanism problem

In the following, we present a mathematical model of a large-scale system which has n_s local control stations, each of which has outputs to be regulated, and in addition has m local control stations in which there are no outputs to be regulated but in which there do exist outputs that can be measured and utilized for control purposes. Such a framework is called the decentralized servomechanism problem, in which the m control stations can be used to assist in the control of the outputs of the n_s control stations that are to be regulated.

Thus the plant to be controlled, consisting of $n_s + m$ local control stations, is assumed to be described by the following LTI system:

$$\begin{aligned} \dot{x}_j(t) &= Ax(t) + \sum_{k=1}^{n_s+m} B_k u_k(t) + \Gamma w(t) \\ y_j(t) &= C_j x(t) + D_j u_j(t) + \Phi_j w(t), \;\; j = 1, 2, ..., n_s \\ z_j(t) &= G_j x(t) + H_j u_j(t) + \Psi_j w(t), \;\; j = 1, 2, ..., n_s + m \\ e_j(t) &= y_j(t) \; - \; y_{rj}(t), \;\; j = 1, 2, ..., n_s \end{aligned} \tag{4.24}$$

where for $x(t) \in \Re^n$ is the state vector, $u_j(t) \in \Re^{m_j}$ is the control input, $z_j(t) \in \Re^{r_j}$ is the measurable outputs of the jth control station $(j = 1, 2, ..., n_s + m)$, $w(t) \in \Re^k$ is a disturbance vector, and $e_j(t) \in \Re^{r_j}$, $j = 1, 2, ..., n_s$ is the output error, with the difference between the output $y_j(t)$ to be regulated and the specified reference input $y_{rj}(t)$, $j = 1, 2, ..., n_s$.

The following matrix notations are introduced:

$$\begin{aligned} B &:= \begin{bmatrix} B_1 & B_2 & \cdots & B_{n_s} \end{bmatrix}, \; C := \mathbf{col}\begin{bmatrix} C_1 & C_2 & \vdots & C_{n_s} \end{bmatrix} \\ \Phi &:= \mathbf{col}\begin{bmatrix} \Phi_1 & \Phi_2 & \vdots & \Phi_{n_s} \end{bmatrix}, \; \hat{B} := \begin{bmatrix} B_{n_s+1} & B_{n_s+2} & \cdots & B_{n_s+m} \end{bmatrix} \\ y_r &:= \mathbf{col}\begin{bmatrix} y_{r1} & y_{r2} & \vdots & y_{rn_s} \end{bmatrix}, \; e := \mathbf{col}\begin{bmatrix} e_1 & e_2 & \cdots & e_{n_s} \end{bmatrix} \\ D &:= Blockdiag\begin{bmatrix} D_1, & D_2, & \cdots, & D_{n_s} \end{bmatrix} \\ G &:= \mathbf{col}\begin{bmatrix} G_1 & G_2 & \vdots & G_{n_s} \end{bmatrix} \\ \hat{G} &:= \mathbf{col}\begin{bmatrix} G_{n_s+1} & G_{n_s+2} & \vdots & G_{n_s+m} \end{bmatrix} \end{aligned} \tag{4.25}$$

In the sequel, we consider that the disturbance inputs $w(t)$ arise from the following class of systems:

$$\dot{\zeta}_j(t) = S\,\zeta(t), \;\; w(t) = W_\zeta(t), \;\; \zeta(t) \in \Re^{g_w} \tag{4.26}$$

Additionally, for regulation purposes, we introduce the reference output $y_{rj}(t)$ which is specified by the model

$$\dot{\eta}(t) = R\,\eta(t), \;\; \xi(t) = X\eta(t), \;\; y_r = Y\xi, \;\; zeta(t) \in \Re^{g_s} \qquad (4.27)$$

Following [51, 119], it is assumed for nontriviality that the spectrum $\mathbf{spec}(S) \subset \mathcal{C}^+$, $\mathbf{spec}(R) \subset \mathcal{C}^+$ where $\mathcal{C}^+$ denotes the closed right-half complex plane. Without loss of generality, it is also assumed that the pairs (W, S) and (X, R) are observable and

$$rank(\Gamma^t \;\; \Phi^t)^t = rank(W) = dim(w), \;\; rank(Y) = rank(X) = dim(\xi)$$

The underlying objective for system (4.24)-(4.27) amounts to *finding a decentralized LTI controller so that*

1. *All the resultant closed-loop subsystems are asymptotically stable.*

2. *Asymptotic tracking is achieved, that is,*

$$\lim_{t\to\infty} e(t) = 0, \; \forall\, x(0) \in \Re^n, \; \forall\zeta(0) \in \Re^{g_w}, \; \forall\xi(0) \in \Re^{g_s}$$

   and for all controller initial conditions. This also holds for any arbitrary perturbations in the plant model.

The above problem is frequently termed *the robust decentralized servomechanism problem.*

The following theorem establishes conditions for the existence of solution to the robust decentralized servomechanism problem.

Theorem 4.5.1 *There exists a solution to the robust decentralized servomechanism problem (4.23) if and only if all of the following conditions hold:*

1) $\left\{ \begin{bmatrix} G \\ \hat{G} \end{bmatrix}, \; A, \; \begin{bmatrix} B & \hat{B} \end{bmatrix} \right\}$ $has\ no\ unstable\ decentralized\ fixed\ modes$

2) $The\ decentralized\ fixed\ modes\ of\ the\ q\ systems$

$$\left\{ \begin{bmatrix} G^* \\ \hat{G}^* \end{bmatrix}, \; \begin{bmatrix} A & 0 \\ C & \lambda_j I \end{bmatrix}, \; \begin{bmatrix} B & \hat{B} \\ H & 0 \end{bmatrix} \right\} \; j = 1, 2, ..., q$$

$do\ not\ contain\ \lambda_j\ j = 1, 2, ..., q,\ respectively.$

3) $The\ outputs\ y_j(t),\ j = 1, 2, ..., n_s\ are\ physically\ measurable,\ that\ is,$ $y_j(t) \subset z_j(t),\ j = 1, 2, ..., n_s$

where

$$
\begin{aligned}
G^* &:= \mathbf{col}\left[\begin{array}{cccc} G_1^* & G_2^* & \vdots & G_{n_s}^* \end{array}\right], \\
\hat{G}^* &:= \mathbf{col}\left[\begin{array}{cccc} G_{n_s+1}^* & G_{n_s+2}^* & \vdots & G_{n_s+m}^* \end{array}\right] \qquad (4.28)\\
G_1^* &:= \left[\begin{array}{ccccc} G_1 & 0 & 0 & \cdots & 0 \\ 0 & I_{r_j} & 0 & \cdots & 0 \end{array}\right], \\
G_{n_s+1}^* &:= \mathbf{col}\left[\begin{array}{ccccc} G_{n_s+1} & 0 & 0 & \cdots & 0 \end{array}\right] \\
G_2^* &:= \left[\begin{array}{ccccc} G_1 & 0 & 0 & \cdots & 0 \\ 0 & 0 & I_{r_j} & \cdots & 0 \end{array}\right], \\
G_{n_s+2}^* &:= \mathbf{col}\left[\begin{array}{ccccc} G_{n_s+1} & 0 & 0 & \cdots & 0 \end{array}\right] \\
\vdots &:= \qquad \vdots \\
G_{n_s}^* &:= \left[\begin{array}{ccccc} G_{n_s} & 0 & 0 & \cdots & 0 \\ 0 & 0 & 0 & \cdots & I_{r_j} \end{array}\right], \\
G_{n_s+1}^* &:= \mathbf{col}\left[\begin{array}{ccccc} G_{n_s+m} & 0 & 0 & \cdots & 0 \end{array}\right] \qquad (4.29)
\end{aligned}
$$

Proof: The details can be found in [48, 50].

4.5.2 The robust decentralized controller

It is natural now to seek a design procedure for system (4.24). To facilitate such a procedure, we introduce the following notations. Let the minimal polynomial of matrices S and R be denoted by $\mathbf{min-poly(S)(s)}$ and $\mathbf{min-poly(R)(s)}$, respectively, and we let the zeros of the least common multiple of $\mathbf{min-poly(S)(s)}$ and $\mathbf{min-poly(R)(s)}$ including multiplicities be given by

$$\lambda_1,\ \lambda_2,\ ...,\ \lambda_q \qquad (4.30)$$

Let the coefficients $\sigma_j,\ j = 1, 2, ..., q$ correspond to the coefficients of the polynomial $\prod_{k=1}^{q}(\lambda - \lambda_k)$ where λ_k are given by (4.30), that is,

$$\lambda^q + \sigma_q\,\lambda^{q-1} + \sigma_{q-1}\,\lambda^{q-2} + \cdots + \sigma_2\,\lambda + \sigma_1 = \prod_{k=1}^{q}(\lambda - \lambda_k) \qquad (4.31)$$

Finally, let

$$\Upsilon := \begin{bmatrix} 0 & 1 & 0 & \cdots & 0 \\ 0 & 0 & 1 & \cdots & 0 \\ \vdots & \vdots & \vdots & & \vdots \\ -\sigma_1 & -\sigma_2 & -\sigma_3 & \cdots & \sigma_q \end{bmatrix}, \in \Re^{q \times q}, \quad \delta := \begin{bmatrix} 0 \\ 0 \\ \vdots \\ 0 \\ 1 \end{bmatrix} \tag{4.32}$$

Now consider the large-scale system (4.24) with (4.26) and (4.27) and introduce the decentralized compensator with input $e_j(t) \in \Re^{r_j}, \ j = 1, 2, ..., n_s$ and output $f_j(t) \in \Re^{r_{jq}}, \ j = 1, 2, ..., n_s$:

$$\dot{f}_j(t) = \Upsilon_j^* \, f_j(t) + \Delta_j^* \, e_j(t), \ j = 1, 2, ..., n_s \tag{4.33}$$

$$\Upsilon_j^* := Blockdiag \left[\Upsilon, \ \Upsilon, \ \cdots, \ \Upsilon \right],$$

$$\Delta_j^* := Blockdiag \left[\delta, \ \delta, \ \cdots, \ \delta \right] \tag{4.34}$$

From the linear system theory [147] it follows that the decentralized compensator (4.33) is unique within the class of coordinate transformations and nonsingular input transformations. Next, when **Theorem 4.5.1** holds, the decentralized controller consists of the following structure [48, 50]:

$$u_j(t) = K_j \, f_j(t) + \hat{x}_j(t), \ j = 1, 2, ..., n_s$$

$$u_j(t) = \hat{x}_j(t), \ j = n_s + 1, n_s + 2, ..., n_s + m \tag{4.35}$$

where $f_j(t) \in \Re^{r_{jq}}$ is the output of *the decentralized compensator (4.33)* and $\hat{x}_j(t), \ j = 1, 2, ..., n_s + m$ is the output of *a decentralized stabilizing compensator* $\Im_j^*$ (with inputs $y_{rj}, \ z_j, \ f_j, \ u_j, \ j = 1, 2, ..., n_s$) and $\Im_j^*$ (with inputs $z_j, \ u_j, \ j = n + s + 1, n_s + 2, ..., n_s + m$). In addition, $\Im_j^*, \ j = 1, 2, ..., n_s + m$ and the gain matrices $K_j, \ j = 1, 2, ..., n_s$ are designed to guarantee that the following augmented system

$$\left[\begin{array}{cccc} \dot{x}(t) & \dot{f}_1(t) & \cdots & \dot{f}_{n_s}(t) \end{array}\right] :=$$
$$\left[\begin{array}{cccc} A & 0 & \cdots & 0 \\ \Delta_1^* C_1 & \Upsilon_1^* & \cdots & 0 \\ \vdots & \vdots & & \vdots \\ \Delta_{n_s}^* C_{n_s} & 0 & \cdots & \Upsilon_{n_s}^* \end{array}\right] \left[\begin{array}{cccc} x(t) & f_1(t) & \vdots & f_{n_s}(t) \end{array}\right] +$$
$$\left[\begin{array}{cc} B & \hat{B} \\ Blockdiag\left[\begin{array}{ccc} \Delta_1^* D_1, & \cdots, & \Delta_{n_s}^* D_{n_s} \end{array}\right] & 0 \end{array}\right] \left[\begin{array}{c} u_1 \\ \vdots \\ u_{n_s} \\ u_{n_s+1} \\ \vdots \\ u_{n_s+m} \end{array}\right]$$
$$\left[\begin{array}{c} G_j x(t) \\ f_j(t) \end{array}\right] := \left[\begin{array}{c} \hat{z}_j(t) \\ f_j(t) \end{array}\right], \; j = 1, 2, ..., n_s$$
$$\hat{z}_j(t) := G_j x(t), \; j = n_s + 1, n_s + 2, ..., n_s + m \tag{4.36}$$

is stabilizable with desired transient behavior, where $\hat{z}_j(t) := z_j(t) - H_j u_j(t), \; j = 1, 2, ..., n_s + m$ and system (4.36) has decentralized fixed modes corresponding to the decentralized fixed modes (if any) of

$$\left\{ \left[\begin{array}{c} G \\ \hat{G} \end{array}\right], \; A, \; \left[\begin{array}{cc} B & \hat{B} \end{array}\right] \right\}$$

4.5.3 Decentralized controller for interconnected systems

We proceed further and extend the results of the decentralized servomechanism results to the class of interconnected systems

$$\begin{aligned} \dot{x}_j(t) &= A_j x_j(t) + B_j u_j(t) + \Gamma_j w(t) + \sum_{k=1, k \neq j}^{n_s} A_{jk} x_k(t) \\ y_j(t) &= C_j x_j(t) + D_j u_j(t) + \Phi_j w(t) \\ z_j(t) &= G_j x_j(t) + H_j u_j(t) + \Psi_j w(t) \end{aligned} \tag{4.37}$$

where for $j \in \{1, ..., n_s\}$, $x_j(t) \in \Re^{n_j}$ is the state vector, $u_j(t) \in \Re^{m_j}$ is the control input, $y_j(t) \in \Re^{p_j}$ is the output to be regulated, $z_j(t) \in \Re^{q_j}$ is the measurable output. The matrices $A_j \in \Re^{n_j \times n_j}$, $B_j \in \Re^{n_j \times m_j}$, $D_j \in \Re^{q_j \times m_j}$, $A_{dj} \in \Re^{n_j \times n_j}$, $\Phi_j \in \Re^{q_j \times q_j}$, $\Omega_j \in \Re^{n_j \times q_j}$, $C_j \in \Re^{p_j \times n_j}$, $C_{dj} \in$

$\Re^{p_j \times n_j}$, $G_j \in \Re^{q_j \times n_j}$, $G_{dj} \in \Re^{q_j \times n_j}$, $A_{jk} \in \Re^{n_j \times n_k}$ are real and constants. In the sequel, we consider that the disturbance input $w_j(t) \in \Re^{q_j}$ arise from the following class of systems:

$$\dot{\zeta}_j(t) = S_j\, \zeta_j(t), \;\; w_j(t) = W_j \zeta_j(t), \;\; \zeta_j(t) \in \Re^{g_{wj}} \tag{4.38}$$

Additionally, for regulation purposes, we introduce the reference output $y_{rj}(t)$ which is specified by the model

$$\dot{\eta}_j(t) = R_j\, \eta_j(t), \;\; \xi_j(t) = X_j \eta_j(t), \;\; y_{rj} = Y_j \xi_j, \;\; zeta_j(t) \in \Re^{g_{sj}} \tag{4.39}$$

and define the tracking error

$$e_j(t) = y_j(t) - y_{rj}(t) \tag{4.40}$$

At this stage, we let the interconnections among subsystems be characterized by

$$A_{jk} = M_{jk} K_{jk} N_{jk}, \;\; j \in 1, ..., n_s, \;\; k \in 1, ..., n_s, \;\; j \neq k \tag{4.41}$$

where K_{jk} denotes the interconnection gain connecting subsystem j and subsystem k. The following theorem provides results reported in [48, 52] for the case when subsystem j has $A_{jk} = 0, \;\; j \in \{1, ..., n_s, \;\; k \in 1, ..., n_s, \;\; j \neq k$.

Theorem 4.5.2 *Consider the collection of subsystems (4.37) with (4.38)-(4.41), $j \in \{1, ..., n_s, \;\; k \in 1, ..., n_s, \;\; j \neq k$, then the following statements hold:*

1. *Assume that there exists a solution to the centralized, robust servomechanism problem for each subsystem of the type (4.37), then there exists a solution to the decentralized robust servomechanism problem for the interconnected system (4.37) provided the norms of the interconnection gains K_{jk} are "small enough."*

2. *Assume that there exists a solution to the centralized, robust servomechanism problem for each subsystem of the type (4.37) and in addition assume that the triplets $G_j, \; A, \; B_j, \; j \in 1, ..., n_s$ are controllable and observable, then there exists a solution to the decentralized robust servomechanism problem for the interconnected system (4.37) for almost all interconnection gains K_{jk}.*

3. *Assume that the interconnections among subsystems are characterized by (4.41) with $M_{jk} = B_j, \; N_{jk} = C_k$ and that $D_j = 0, \; j \in 1, ..., n_s,$*

then there exists a solution to the decentralized robust servomechanism problem for the interconnected system (4.37) if and only if there exists a solution to the centralized robust servomechanism problem for each subsystem of the type (4.37).

Remark 4.5.3 *It is significant to observe that the salient feature of the works reported in [45]-[56] is that it parameterizes all the decentralized controllers which reject the unmeasurable exogenous disturbances with known dynamics. In this section we observe the disturbance input generated by model (4.38) represents one possible approach of dealing with unknown input disturbances. In several portions throughout the book, there is an alternative approach to treat the disturbance input as a member of the space $\mathcal{L}_2[0,\infty)$. The difference between both representations is significant and the subsequent impact on control design and system performance brings about numerous interesting features. We will put this into further consideration to disclose the merits and demerits.*

4.5.4 Illustrative example 4.4

Consider the following large-scale system of the type (4.1) with

$$A_o = \begin{bmatrix} 3 & 2 & 1 \\ 0 & 2 & 1 \\ 0 & 0 & 1 \end{bmatrix},$$

$$B_o = \begin{bmatrix} 2 & 0 & 0 \\ 0 & 3 & 0 \\ 0 & 0 & 1 \end{bmatrix}, \; C_o = \begin{bmatrix} 1 & 0 & 0 \\ 0 & 1 & 0 \\ 0 & 0 & 1 \end{bmatrix}$$

For this system, we wish to design a decentralized output feedback controller of the form

$$\begin{aligned} \dot{z}(t) &= Fz(t) + Sy(t) \\ u(t) &= Hz(t) + Ky(t) \end{aligned}$$

such that a prescribed set of eigenvalues is achieved.

Simple calculations show that $\lambda(A) = \{3,\ 2,\ 1\}$ and by attempting several controller structures, no fixed mode is found. Now, let $n_c,\ n_o$ be the smallest integers such that

$$rank[B_o,\ A_oB_o, ..., A_o^{n_c}B_o] = n,\ rank[C_o^t,\ A_o^tC_o^t, ..., A_o^{t\,n_c}C_o] = n,$$
$$\eta = \min(n_c,\ n_o) = 1$$

to get

$$A_\eta = \begin{bmatrix} 3 & 2 & 1 & 0 \\ 0 & 2 & 1 & 0 \\ 0 & 0 & 1 & 0 \\ 0 & 0 & 0 & 0 \end{bmatrix},$$

$$B_\eta = \begin{bmatrix} 2 & 0 & 0 & 0 \\ 0 & 3 & 0 & 0 \\ 0 & 0 & 1 & 0 \\ 0 & 0 & 0 & 1 \end{bmatrix}, \; C_\eta = \begin{bmatrix} 1 & 0 & 0 & 0 \\ 0 & 1 & 0 & 0 \\ 0 & 0 & 1 & 0 \\ 0 & 0 & 0 & 1 \end{bmatrix}$$

Let

$$K_\eta = \begin{bmatrix} k_1 & 0 & 0 & h_1 \\ 0 & k_2 & 0 & 0 \\ 0 & 0 & 1k_3 & 0 \\ s_1 & 0 & 0 & f_1 \end{bmatrix}$$

which yields

$$A_\eta + B_\eta K_\eta C_\eta = \begin{bmatrix} 2k_1+3 & 2 & 1 & 2h_1 \\ 0 & 3k_2+2 & 1 & 0 \\ 0 & 0 & k_3+1 & 0 \\ s_1 & 0 & 0 & f_1 \end{bmatrix}$$

where the unknown scalars $k_1,\ k_2,\ k_3,\ s_1,\ h_1,\ f_1$ are to be determined such that the closed-loop system has the prescribed set of eigenvalues $\{-3,\ -4,\ -1,\ -2\}$. The solution procedure employs the determinant identity

$$det \begin{bmatrix} M_1 & M_2 \\ M_3 & M_4 \end{bmatrix} \; det[M_1] \; det[M_4 - M_3 M_1^{-1} M_2]$$

such that the greatest common divisor of

$$det[\lambda I - A_o - B_o K_o C_o] = det[\lambda I - A_\eta - B_\eta K_\eta C_\eta]$$

The result is given by

$$k_1 = -3,\ k_2 = -2,\ k_3 = -2,\ s_1 = 1,\ h_1 = 1,\ f_1 = -4$$

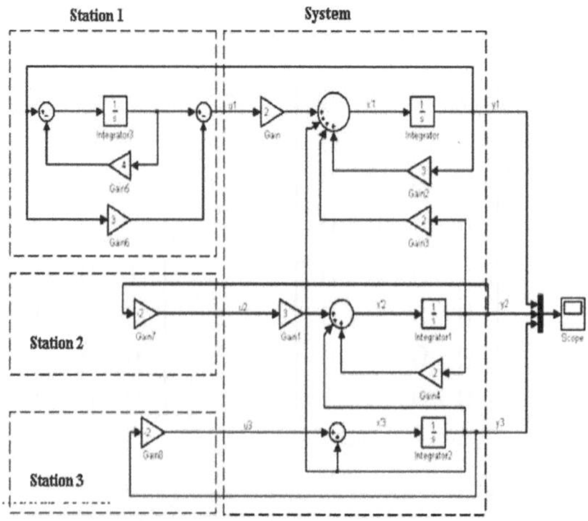

Figure 4.2: Simulink set-up

The foregoing data yield the dynamic compensator

$$
\begin{aligned}
\dot{z}_1(t) &= -4z_1(t) + y_1(t) \\
u_1(t) &= z_1(t) - 3y_1(t) \\
\dot{z}_2(t) &= 0 \\
u_2(t) &= -2y_2(t) \\
\dot{z}_3(t) &= 0 \\
u_3(t) &= -2y_3(t)
\end{aligned}
$$

A complete Simulink simulation diagram is depicted in Figure 4.2 for generating the output response as displayed in Figure 4.3.

4.6 Hierarchically Structured Servomechanism

High-performance decentralized control design techniques have been recently investigated for classes of interconnected systems with hierarchical structures. Several real-world systems are frequently subject to external disturbances; the controller for a hierarchical interconnected system is therefore desired to satisfy the following properties [168, 169, 171, 301]:

1. *The disturbances must be rejected in the steady state.*

2. *A predefined $\mathcal{H}_2$ performance index should be minimized to achieve a fast transient response with a satisfactorily small control energy.*

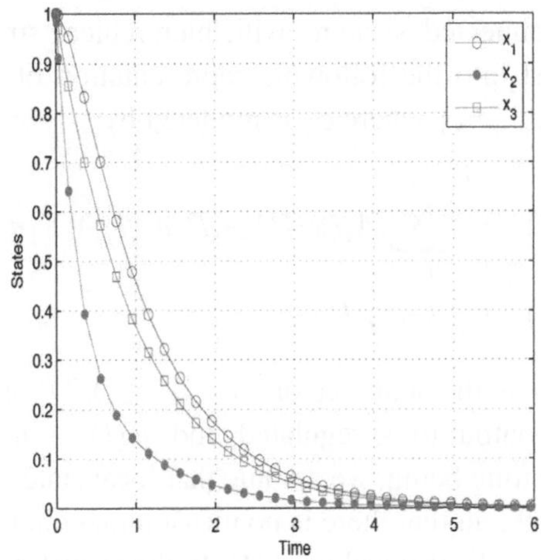

Figure 4.3: Output trajectories

3. *The structure of the controller to be designed should be decentralized.*

We have seen in the preceding sections the problem of designing a controller satisfying the properties (1) and (3) given above, and the controller obtained is regarded as decentralized servomechanism controller [48, 52, 54]. The design of a controller that meets the criteria (2) and (3), however, has been studied in [309, 343]. For structural convenience, the available techniques often seek a near-optimal solution, rather than a globally optimal one. The main shortcoming of this approach is that the controller obtained may destabilize the system, as a result of neglecting the system's interconnection parameters in the design procedure. An alternative approach for handling the underlying problem is to consider only static decentralized controllers [267]. This imposes a stringent constraint on the controller, which can lead to a poor closed-loop performance. It is to be noted that once a centralized controller is designed to achieve the properties (1) and (2), its decentralized version does not necessarily maintain the same properties. A technique yielding $\mathcal{H}_2$ near-optimal is developed in [169] for decentralization of any given centralized controller of desired performance. The only requirement of this technique is that the nominal model of the system is known by all control agents, that is, every local controller must have a belief about the model of the entire system.

In this section, we direct attention to the formalism of decentralized servomechanism problems of interconnected systems with hierarchical structures. A centralized controller is designed first, which should satisfy certain constraints [170]. This control design is then formulated in terms of LMI [171].

The class of interconnected systems with hierarchical structures considered hereafter can be cast into the following representation of the family of subsystems $\mathcal{S}_j,\ j \in \{1, ..., n_s\}$ where $\mathcal{S}_j$ is modeled by:

$$\begin{aligned} \dot{x}_j(t) &= \sum_{k=1}^{n_s} A_{jk}x_k(t) + B_j u_j(t) + \Gamma_j w(t) \\ y_j(t) &= C_j x_j(t) \end{aligned} \tag{4.42}$$

where $x_j(t) \in \Re^{n_j}$ is the state vector, $u_j(t) \in \Re^{m_j}$ is the control input, $y_j(t) \in \Re^{p_j}$ is the output to be regulated, and $w(t) \in \Re^q$ is the disturbance input vector. For the time being, we assume that local state $x_j(t)$ is measurable at the subsystem level and that there is no measurement noise. This means that the measured output of $\mathcal{S}_j$ is equal to $x_j(t)$. In the sequel, we consider that the disturbance input $w(t) \in \Re^q$ is generated by the following class of systems:

$$\dot{\zeta}(t) = S\,\zeta(t),\ \ w(t) = W\,\zeta(t),\ \ \zeta(t) \in \Re^g \tag{4.43}$$

where $S \in \Re^{g \times g}$, the pair $W,\ S$ is observable with the initial condition $\zeta(0)$ in (4.43), and the matrix Γ_j in () is arbitrary, nevertheless unknown.

For the purpose of future comparison, we represent the state-space model of the overall system S as follows:

$$\begin{aligned} \dot{x}(t) &= Ax(t) + Bu(t) + \Gamma w(t) \\ y(t) &= Cx(t) \end{aligned} \tag{4.44}$$

where

$$\begin{aligned} x(t) &= \left[\ x_1^t(t)\ \ x_2^t(t)\ \ \cdots\ \ x_{n_s}^t(t)\ \right]^t \in \Re^n, \\ u(t) &= \left[\ u_1^t(t)\ \ u_2^t(t)\ \ \cdots\ \ u_{n_s}^t(t)\ \right]^t \in \Re^m, \\ y(t) &= \left[\ y_1^t(t)\ \ y_2^t(t)\ \ \cdots\ \ y_{n_s}^t(t)\ \right]^t \in \Re^p, \end{aligned}$$

$$\begin{aligned} B &= Blockdiag\left[\ B_1\ \ B_2\ \ \cdots\ \ B_{n_s}\ \right]^t, \\ \Gamma &= \left[\ \Gamma_1^t\ \ \Gamma_2^t\ \ \cdots\ \ \Gamma_{n_s}^t\ \right]^t, \\ C &= Blockdiag\left[\ C_1\ \ C_2\ \ \cdots\ \ C_{n_s}\ \right], \\ n &= \sum_{j=1}^{n_s} n_j,\ \ m = \sum_{j=1}^{n_s} m_j,\ \ p = \sum_{j=1}^{n_s} p_j \end{aligned} \tag{4.45}$$

and

$$A = \begin{bmatrix} A_{11} & 0 & 0 & \cdots & 0 \\ A_{21} & A_{22} & 0 & \cdots & 0 \\ \vdots & \vdots & \vdots & & \vdots \\ A_{1n_s} & A_{2n_s} & A_{3n_s} & \cdots & A_{n_s n_s} \end{bmatrix} \tag{4.46}$$

having a lower block triangular form reflecting the desired hierarchical structure in (4.44). Moreover, we consider the initial state $x(0)$ as a random variable with mean x_m and covariance X_m which leads to

$$X_0 = \mathbf{E}[x(0)x^t(0)] = X_m + x_m x_m^t \tag{4.47}$$

with $\mathbf{E}[.]$ being the expectation operator. The following assumptions are needed before tackling the desired decentralized control problem.

Assumption 4.1: *The matrices B_j, C_j, and W satisfy the following rank conditions:*

$$rank(W) = q, \;\; rank(B_j) = m_j, \;\; rank(C_j) = p_j, \;\; j \in \{1, ..., n_s\} \tag{4.48}$$

Assumption 4.2: *The matrices*

$$\begin{bmatrix} A_{jj} - s_j I_j & B_k \\ & \\ C_k & 0 \end{bmatrix}, \;\; j \in \{1, ..., n_s\}, \;\; k \in 1, ..., g \tag{4.49}$$

are full-rank where $s_1, \; s_2, ..., s_g$ represents the eigenvalues of S and the inequality $m_j \geq p_j$ holds for all $j \in \{1, ..., n_s\}$.

Assumption 4.3: *The pair (A_{jj}, B_j) is stabilizable for all $j \in \{1, ..., n_s\}$*

Given the foregoing assumptions and model set-up, the decentralized control problem of interest is phrased below.

Design a decentralized LTI controller K_d with block diagonal information flow structure such that the following conditions hold:

1. *The state x(t) goes to zero as t $\rightarrow$ 0, provided $z(0) = 0$;*

2. *The output y(t) approaches zero as $t \rightarrow 0$ regardless of the initial state $z(0)$;*

3. *When $z(0)$ is a zero vector, the following closed-loop performance index*

$$J := \mathbf{E}\Big[\int_0^\infty \; [x^t(s)Qx(s) + u^t(s)Ru(s)]ds\Big] \tag{4.50}$$

   is satisfactorily small, where $0 < Q \in \Re^{n\times n}, \; 0 \leq R \in \Re^{m\times m}$.

4.6.1 Decentralized controller structure

In preparation of the main results on decentralized gain computation [169], we consider the case of free disturbance $\zeta(0) = 0$ and define the following quantities for $j \in \{1, ..., n_s\}$:

$$\begin{aligned} x^k(t) &= \left[\begin{array}{cccccc} x_1^t(t) & \cdots & x_{k-1}^t(t) & x_{k+1}^t(t) & \cdots & x_{n_s}^t(t) \end{array} \right]^t, \\ u^k(t) &= \left[\begin{array}{cccccc} u_1^t(t) & \cdots & u_{k-1}^t(t) & u_{k+1}^t(t) & \cdots & u_{n_s}^t(t)^t \end{array} \right]^t \end{aligned} \tag{4.51}$$

Then we introduce the following centralized dynamic controller K_c:

$$\begin{aligned} \dot{\pi}_c(t) &= A_c \pi_c(t) + E_c x(t), \quad \pi_c(t) \in \Re^{n_c} \\ u(t) &= C_c \pi_c(t) + D_c x(t) \end{aligned} \tag{4.52}$$

It was shown in [169] that there exist constant matrices $E_c,\ F_c,\ G_c,\ H_c,\ I_c,\ J_c,\ M_c,\ N_c$ such that controller (4.52) can be expressed by the following block-structure representation:

$$\begin{aligned} \dot{\pi}_c(t) &= A_c \pi_c(t) + E_c x^k(t) + F_c x_k(t) \\ u^k(t) &= G_c \pi_c(t) + I_c x^k(t) + J_c x_k(t) \\ u_k(t) &= H_c \pi_c(t) + M_c x^k(t) + N_c x_k(t) \end{aligned} \tag{4.53}$$

for any $k \in \{1, ..., n_s\}$. By the same token, there exist matrices $A^k,\ A^m,\ A_k,\ B^k$ derived from A and B in the manner of (4.51) such that the aggregated system (4.42) admits the following structure for any $k \in \{1, ..., n_s\}$:

$$\begin{aligned} \dot{x}^k(t) &= A^k x^k(t) + A_k x_k(t) + B^k u^k(t) \\ \dot{x}_k(t) &= A^m x^k(t) + A_{kk} x_k(t) + B_k u_k(t) \end{aligned} \tag{4.54}$$

Now for subsystem S_j, we introduce the following local dynamic controller K_{dj}:

$$\begin{aligned} \dot{\pi}_{dk}(t) &= \left[\begin{array}{cc} A^k + B^k I_c & B^k G_c \\ E_c & A_c \end{array} \right] \pi_{dk}(t) + \left[\begin{array}{c} A_k + B^k J_c \\ F_c \end{array} \right] x_k(t), \\ u_k(t) &= \left[\begin{array}{cc} M_c & H_c \end{array} \right] \pi_{dk}(t) + N_c x_k(t) \end{aligned} \tag{4.55}$$

Let K_d be a decentralized controller consisting of the local controllers

$$K_{d1},\ K_{d2},\ ...,\ K_{dn_s}$$

In view of the foregoing dynamical representations, the following result was reported in [169].

Theorem 4.6.1 *Assuming that $x(0)$ is a known vector, then the state and input patterns of the centralized system $\mathcal{S}$ under the centralized controller K_c are the same as those of system $\mathcal{S}$ under the decentralized controller K_d, if the initial state of the local controller K_{dk} as selected as*

$$\pi_{dk}(0) = \begin{bmatrix} x^k(0) \\ 0 \end{bmatrix} \tag{4.56}$$

This result implies that a centralized controller can be transformed to an equivalent decentralized controller under appropriate choice of the initial state and the exchange of all local controller initial states amongst dynamic controllers, which is a quite demanding constraint as it destroyed the desired decentralization. One way to remedy this is to recall upon a statistical nature of $x(0)$, which leads to modifying (4.56) into

$$\pi_{dk}(0) = \begin{bmatrix} x_m^k \\ 0 \end{bmatrix} \tag{4.57}$$

Next, to examine the internal stability of the system $\mathcal{S}$ under the decentralized controller $\{K_{dj}\},\ j \in \{1, ..., n_s\}$, we look at a modified system S^k generated from (4.44) with the following state-space representation:

$$\begin{aligned} \dot{x}(t) &= \tilde{A}^k x(t) + Bu(t) + \Gamma w(t) \\ y(t) &= Cx(t) \end{aligned} \tag{4.58}$$

where $\tilde{A}^k$ is derived from A by replacing the first $k-1$ block entries of its kth block row with zeros. Intuitively, we note that $\mathrm{S}^1 = \mathcal{S}$. Define the decoupled subsystem $\mathcal{S}_j,\ j \in \{1, ..., n_s\}$ by eliminating all of its incoming interconnections. This leads to the following result [170, 171].

Theorem 4.6.2 *The centralized system $\mathcal{S}$ is internally stable under the decentralized controller $\{K_{dj}\},\ j \in \{1, ..., n_s\}$, if and only if the system S^k is stable under the centralized controller K_d, for all $k \in \{1, ..., n_s\}$.*

Considering the centralized dynamic controller K_c (4.52) where the system matrices have the following properties:

1. $A_c := Blockdiag \begin{bmatrix} \Lambda & \Lambda & ,\cdots, & \Lambda \end{bmatrix}$;

2. $E_c := B_c C := Blockdiag \begin{bmatrix} \Upsilon & \Upsilon & ,\cdots, & \Upsilon \end{bmatrix}$;

3. The pair (A_c, B_c) is controllable

where the matrices A_c, E_c, Λ, and Υ are of appropriate dimensions. The control design problem is summarized by the following:

Determine the matrices E_c, C_c, D_c so that the centralized dynamic controller K_c (4.52) with the foregoing properties guarantees that the following conditions hold:

1. *The state x(t) goes to zero as t $\to 0$, provided $z(0) = 0$;*
2. *The output y(t) approaches zero as t $\to 0$ regardless of the initial state $z(0)$;*
3. *It satisfies all of the systems $\mathsf{S}^2, \cdots, \mathsf{S}^{n_s}$.*

It has been shown in [170] that there exists a solution to the above problem if and only if the matrices E_c, C_c, D_c satisfy a particular matrix inequality. The result stems from a generalization of [46] as applied to the augmented system of (4.52) and (4.58). Some appropriate computational procedures are presented, which will be illustrated by the following examples.

4.6.2 Illustrative example 4.5

Consider the following large-scale system of the type (4.1) with

$$
\begin{aligned}
A &= \begin{bmatrix} 0.6 & 0 & 0 & 1 & 0.2 & 0 \\ 0.2 & 2.3 & 0.1 & 0 & 0 & 0.3 \\ 0 & 0 & -1.5 & 0 & 0 & 0 \\ 0 & 0 & 0.7 & 1.1 & 0 & 0 \\ 0 & 0 & 0 & 0 & -0.8 & 0 \\ 0 & 0 & 0 & 0 & 0.5 & 0.8 \end{bmatrix} \\
G_o &= [0.1 \; 0.2 \; 0.4 \; 0.3], \; G_{do} = [0.01 \; 0 \; 0.01 \; 0] \\
B_o &= \begin{bmatrix} 0 \\ 0.3 \end{bmatrix}, \; C_o = \begin{bmatrix} 1 & 10 & 0 & 0 \\ 0 & 0 & 10 & 10 \end{bmatrix}, \; C_d = \begin{bmatrix} 2 & 1 & 0 & 0 \\ 0 & 0 & 1 & 3 \end{bmatrix}, \\
D_o &= \begin{bmatrix} 0.4 & 0.2 \end{bmatrix}, \; \Psi_o = \begin{bmatrix} 0.01 & 0.01 \end{bmatrix}, \; \Phi = 0.1
\end{aligned}
$$

Using the MATLAB-LMI solver, the feasible solutions of the respective design methods yield the feedback gain matrices:

$$
K_s = \begin{bmatrix} -34.5094 & -4.9061 & -52.4436 & -130.5408 \\ 17.5392 & -12.7566 & 79.5070 & 157.7008 \end{bmatrix}
$$

$$K_d = \begin{bmatrix} -34.5094 & -4.9061 & -52.4436 & -130.5408 \\ 17.5392 & -12.7566 & 79.5070 & 157.7008 \end{bmatrix}$$

$$K_o = \begin{bmatrix} -55.2134 & 15.8031 \\ 69.2668 & -18.6633 \\ -18.4083 & 5.0529 \\ 5.2346 & -0.5445 \end{bmatrix}$$

4.6.3 Illustrative example 4.6

Consider the following large-scale system of the type (4.1) with

$$A = \begin{bmatrix} 0.6 & 0 & 0 & 1 & 0.2 & 0 \\ 0.2 & 2.3 & 0.1 & 0 & 0 & 0.3 \\ 0 & 0 & -1.5 & 0 & 0 & 0 \\ 0 & 0 & 0.7 & 1.1 & 0 & 0 \\ 0 & 0 & 0 & 0 & -0.8 & 0 \\ 0 & 0 & 0 & 0 & 0.5 & 0.8 \end{bmatrix}$$

$$\begin{aligned} G_o &= [0.1\ 0.2\ 0.4\ 0.3],\ G_{do} = [0.01\ 0\ 0.01\ 0] \\ B_o &= \begin{bmatrix} 0 \\ 0.3 \end{bmatrix},\ C_o = \begin{bmatrix} 1 & 10 & 0 & 0 \\ 0 & 0 & 10 & 10 \end{bmatrix},\ C_d = \begin{bmatrix} 2 & 1 & 0 & 0 \\ 0 & 0 & 1 & 3 \end{bmatrix}, \\ D_o &= \begin{bmatrix} 0.4 & 0.2 \end{bmatrix},\ \Psi_o = \begin{bmatrix} 0.01 & 0.01 \end{bmatrix},\ \Phi = 0.1 \end{aligned}$$

Using the MATLAB-LMI solver, the feasible solutions of the respective design methods yield the feedback gain matrices:

$$K_s = \begin{bmatrix} -34.5094 & -4.9061 & -52.4436 & -130.5408 \\ 17.5392 & -12.7566 & 79.5070 & 157.7008 \end{bmatrix}$$

$$K_d = \begin{bmatrix} -34.5094 & -4.9061 & -52.4436 & -130.5408 \\ 17.5392 & -12.7566 & 79.5070 & 157.7008 \end{bmatrix}$$

$$K_o = \begin{bmatrix} -55.2134 & 15.8031 \\ 69.2668 & -18.6633 \\ -18.4083 & 5.0529 \\ 5.2346 & -0.5445 \end{bmatrix}$$

4.7 Problem Set II

Problem II.1: Consider a large-scale system described by:

$$\begin{bmatrix} \dot{x}_1(t) \\ \dot{x}_2(t) \\ \dot{x}_3(t) \\ \dot{x}_4(t) \end{bmatrix} = \begin{bmatrix} -1 & 2 & 3 & 1 \\ \alpha & -2 & 4 & 1 \\ 0 & 0 & -3 & 0 \\ 0 & 0 & 0 & -4 \end{bmatrix} \begin{bmatrix} x_1 \\ x_2 \\ x_3 \\ x_4 \end{bmatrix} + \begin{bmatrix} 1 & 0 \\ 0 & 0 \\ 0 & 1 \\ 0 & 0 \end{bmatrix} \begin{bmatrix} u_1 \\ u_2 \end{bmatrix},$$

$$\begin{bmatrix} y_1 \\ y_2 \end{bmatrix} = \begin{bmatrix} 1 & 0 & 0 & 0 \\ 0 & 0 & 1 & 0 \end{bmatrix} \begin{bmatrix} x_1 \\ x_2 \\ x_3 \\ x_4 \end{bmatrix}$$

Check the controllability and observability of the system and show that it is independent of the parameter α. Examine the decentralized modes of the system with respect to $\alpha = 0,\ \alpha \neq 0$. Identify the decentralized fixed modes.

Problem II.2: Consider a parametrized large-scale system described by:

$$\begin{bmatrix} \dot{x}_1(t) \\ \dot{x}_2(t) \\ \dot{x}_3(t) \end{bmatrix} = \begin{bmatrix} 0 & \alpha & 3 \\ 0 & \beta & 4 \\ 0 & 0 & 3 \end{bmatrix} \begin{bmatrix} x_1 \\ x_2 \\ x_3 \end{bmatrix} + \begin{bmatrix} 1 \\ 1 \\ 0 \end{bmatrix} u_1 + \begin{bmatrix} 0 \\ 0 \\ 1 \end{bmatrix} u_2,$$

$$\begin{bmatrix} y_1 \\ y_2 \end{bmatrix} = \begin{bmatrix} 1 & 2 & 0 \\ 0 & 0 & 1 \end{bmatrix} \begin{bmatrix} x_1 \\ x_2 \\ x_3 \end{bmatrix}$$

Analyze the complete controllability and observability of the system with respect to the parameters $\alpha,\ \beta$. Which set of values of $\alpha,\ \beta$ yields decentralized fixed modes?

Problem II.3: A class of models for pollution control studies can be cast into the form:

$$\begin{aligned} \dot{\mathbf{x}}(t) &= \begin{bmatrix} A_1 & 0 & 0 \\ A_{21} & A_2 & 0 \\ 0 & A_{32} & A_3 \end{bmatrix} \mathbf{x} + \begin{bmatrix} B_1 & 0 \\ 0 & 0 \\ 0 & B_3 \end{bmatrix} \mathbf{u}(t), \\ \mathbf{y}(t) &= \begin{bmatrix} C_1 & 0 & 0 \\ 0 & 0 & C_3 \end{bmatrix} \mathbf{x} \end{aligned}$$

Naturally, this model has fixed modes. Use the static decentralized controller

$$\mathbf{u}(t) = \begin{bmatrix} K_1 & 0 \\ 0 & K_2 \end{bmatrix} \mathbf{y}(t)$$

to identify the decentralized fixed modes. Interpret your results.

Problem II.4: Consider the following system:

$$\begin{aligned}\dot{\mathbf{x}}(t) &= \begin{bmatrix} 0 & 0 & 1 \\ 1 & -2 & 1 \\ 0 & 0 & -4 \end{bmatrix} \mathbf{x} + \begin{bmatrix} 0.2 & 0 \\ 0 & 0.5 \\ 0 & 0 \end{bmatrix} \mathbf{u}(t), \\ \mathbf{y}(t) &= \begin{bmatrix} 0 & 5 & 0 \\ 2 & 0 & 0 \end{bmatrix} \mathbf{x}\end{aligned}$$

It is required to examine the stabilizability of this system under decentralized control by randomly selecting the gain matrix $K = diag[k_1 \quad k_2$. Then design a suitable decentralized dynamic compensator.

Problem II.5: Consider the following system:

$$\begin{aligned}\dot{x}_1(t) &= \begin{bmatrix} 0 & 0 \\ 1 & 0.7 \end{bmatrix} x_1(t) + \begin{bmatrix} 1 \\ 0 \end{bmatrix} u_1(t), \quad y_1(t) = \begin{bmatrix} 1 & 0 \end{bmatrix} x_1(t), \\ \dot{x}_2(t) &= -0.6x_2(t) + u_2(t), \quad y_2(t) = x_2(t), \\ \dot{x}_3(t) &= \begin{bmatrix} 1 & -0.2 \\ 1 & -2 \end{bmatrix} x_3(t) + \begin{bmatrix} 1 \\ 1 \end{bmatrix} u_2(t), \quad y_3(t) = \begin{bmatrix} 1 & 1 \end{bmatrix} x_3(t)\end{aligned}$$

Develop a decentralized stabilizing control for this system.

Problem II.6: Illustrate the application of **Theorem 4.5.1** on the large-scale system described by:

$$\begin{aligned}\begin{bmatrix} \dot{x}_1(t) \\ \dot{x}_2(t) \\ \dot{x}_3(t) \end{bmatrix} &= \begin{bmatrix} -1 & 2 & 3 \\ 1 & -2 & 4 \\ 0 & 0 & -3 \end{bmatrix} \begin{bmatrix} x_1 \\ x_2 \\ x_3 \end{bmatrix} + \begin{bmatrix} 1 & 0 \\ 0 & 0 \\ 0 & 1 \end{bmatrix} \begin{bmatrix} u_1 \\ u_2 \end{bmatrix}, \\ \begin{bmatrix} y_1 \\ y_2 \end{bmatrix} &= \begin{bmatrix} 1 & 0 & 0 \\ 0 & 0 & 1 \end{bmatrix} \begin{bmatrix} x_1 \\ x_2 \\ x_3 \end{bmatrix}\end{aligned}$$

Give a detailed analysis of the results.

Problem II.7: Consider the large-scale system described by:

$$\begin{bmatrix} \dot{x}_1(t) \\ \dot{x}_2(t) \\ \dot{x}_3(t) \end{bmatrix} = \begin{bmatrix} 0 & 2 & 3 \\ 0 & -2 & 4 \\ 0 & 1 & -3 \end{bmatrix} \begin{bmatrix} x_1 \\ x_2 \\ x_3 \end{bmatrix} + \begin{bmatrix} 1 & 0 \\ 0 & 0 \\ 0 & 1 \end{bmatrix} \begin{bmatrix} u_1 \\ u_2 \end{bmatrix},$$

$$\begin{bmatrix} y_1 \\ y_2 \end{bmatrix} = \begin{bmatrix} \delta_1 & \delta_2 & 0 \\ \delta_3 & 0 & \delta_4 \end{bmatrix} \begin{bmatrix} x_1 \\ x_2 \\ x_3 \end{bmatrix}$$

For the two distinct cases

1. $\delta_1 = 0,\ \delta_2 = 1,\ \delta_3 = 1,\ \delta_4 = 0$

2. $\delta_1 = 1,\ \delta_2 = 2,\ \delta_3 = 0,\ \delta_1 = 1$

examine the complete controllability and observability properties as well as the modes of the system. Check the controllability and observability with respect to the possible pairs $(u_j,\ y_j)$. Then design an appropriate, decentralized static output feedback controller that yields satisfactory performance and provides an interpretation of the ensuing results.

4.8 Notes and References

In this chapter, three major problems related to decentralized structure of large-scale systems were addressed. The first problem was that of decentralized stabilization which amounts to finding a state or an output feedback gain guaranteeing that the closed-loop system has all its eigenvalues on the left half-plane. This is commonly known as feedback "stabilization." Alternatively, the closed-loop eigenvalues of a centralized controllable system may be preassigned through the state of output feedback. It is readily evident that the applications of these concepts in a decentralized fashion required certain extensions. In the foregoing sections, the problem of decentralized stabilization was mathematically formulated and some of the appropriate schemes for decentralized feedback stabilization were reviewed. The notions of "fixed modes," "fixed polynomials," and their role in dynamic compensation were disclosed.

The second problem addressed in this chapter was the decentralized "robust" control of large-scale linear systems with or without a known plant. The main thrust was the approach in [45]-[56] frequently known as "general servomechanism." The approach took advantage of the tuning regulators and dynamic compensators to design a feedback that both stabilized and regulated the system in a decentralized mode. The notion of "robust" feedback control was introduced when the control action continued to provide asymptotic stability and regulation under perturbation of plant parameters and matrices.

By the third problem, a near-optimal decentralized servomechanism controller was designed for a hierarchical interconnected system. The objectives of control design were to achieve satisfactory performance with respect to a prescribed linear-quadratic cost function and to be capable of rejecting unmeasurable external disturbances of known dynamics. The case when the system is subject to perturbation was explored. By making use of the information of every individual subsystem about the overall system, the control design was initialized. Since this information was inexact in practice, a procedure was presented to assess the degradation of the performance of the decentralized control system as a result of the erroneous information.

There are ample extensions to the developed approaches. One of these would be to examine the case of decentralized servomechanism problem for discrete-time systems. The literature lacks research results on this topic as virtually no publication material is available. Another extension is related to the tracking problem for exponentially increasing reference signals. Furthermore, since in practice the delay in the state or the input of the system is variable (may be uncertain), it would be of interest to find a bound on the delay that guarantees the closed-loop stability under the proposed linear-quadratic suboptimal decentralized controller. Finally for the case of arbitrary interaction topology, it is considered beneficial to develop decentralized control laws that yield reliable performance.

Chapter 5

Overlapping Control

In this Chapter, we continue our evaluation of the development of methods for tackling large-scale systems (LSS). In particular, the models of LSS summarized in the sequel are distinguished by the degree to which they reflect the internal structure of the overall system. Then we focus the concept of system decomposition and disclose the effective tool of overlapping decomposition . The process of actual tearing of the system may be performed from either conceptual or numerical reasons. Conceptual reasons correspond usually with the boundaries of physical subsystems. Numerical reasons require to develop a universal decomposition technique.

The units constituting a hierarchical structure are not completely independent but have to respond to data delivered by other units. Although this communication ensures that the global goals will eventually be satisfied, it prescribes the decision makers certain working regimes. Decentralization concerns the information structure of the decision making process. In decentralized decision making the decision units are completely independent or at least almost independent. That is, the network, which describes the information flow among the decision makers, can be divided into completely independent parts. The decision makers belonging to different subnetworks are completely separate from each other. Since such a complete division is possible only for specific problems, the term *decentralized* will also be used if the decision makers do communicate but this communication is restricted to certain time intervals or to a small part of the available information.

The outline of multilevel systems has shown that some coordination and, therefore, communication among the decision-making units is necessary if the overall goal is to be reached. In decentralized structures, such a coordination is impossible or restricted in accordance with the information exchange that is permitted. However, because of the simplifications in the practical implementation of coupled decision makers, which are gained from the absence of information links, decentralized structures are often used but reduce the quality of the solution.

5.1 Decomposition

The foregoing chapters taught us one fundamental lesson. As the amount of computation required to analyze and control a system grows faster than the size of the system, it is therefore considered beneficial to break down (decompose) the whole problem into smaller subproblems, to solve these subproblems separately, and then to combine (recompose) their solutions in order to get a global result for the original task. It must be emphasized that the subproblems are not independent. Some modification or coordination of the solutions of the subproblems is necessary in order to satisfy the interrelationships between the subproblems. The effort required to deal with the subproblems and their coordination can be allocated to various processors, which constitute a distributed computing system. Therefore, the concepts and techniques for reformulating a control problem as a set of interdependent subproblems and for solving these subproblems are referred to in this book as *coordinated control.*

5.1.1 Decoupled subsystems

The basis for the decomposition of the analytical or control problems is often provided by the internal structure of the process to be controlled. Accordingly, the process is not considered as a single object but as a compound of different interacting subsystems. Decomposition methods, generally speaking, exploit the system structure, which can be obtained from the building blocks of the process, or have to impose a structure for mathematical reasons. Let us consider the LSS described by the overall model:

$$\begin{aligned} \dot{\mathbf{x}}(t) &= \mathbf{A}\mathbf{x}(t) + \mathbf{B}\mathbf{u}(t), \quad \mathbf{x}(0) = \mathbf{x}_o \\ \mathbf{y}(t) &= \mathbf{C}\mathbf{x}(t) \end{aligned} \tag{5.1}$$

where $\mathbf{x}(t) \in \Re^n$ is the overall state vector, $\mathbf{u}(t) \in \Re^m$ is the control input and $\mathbf{y}(t) \in \Re^p$ is the overall output vector. The matrices $\mathbf{A} \in \Re^{n\times n}$, $\mathbf{B} \in \Re^{n\times m}$ and $\mathbf{C} \in \Re^{p\times m}$ are constants. In Chapter 3, we followed an *upward* approach and discussed ways to generate an overall system of the form (5.1) from its subsystem models. In what follows, the *downward* approach from the overall system (5.1) to the subsystem models is considered. The partition of the state vector $\mathbf{x}(t)$ into sub vectors $x_j(t),\ j = 1, ..., n_s$ yields

$$\begin{aligned} \dot{x}_j(t) &= A_j x_j(t) + \sum_{k=1, j\neq k}^{n_s} A_{jk} x_k(t) + \sum_{k=1}^{n_s} B_{jk} u_k(t) \\ y_j(t) &= \sum_{k=1, j\neq k}^{n_s} C_{jk} x_k(t) \end{aligned} \tag{5.2}$$

where the matrices $\mathbf{B}$ and $\mathbf{C}$ have been partitioned into blocks B_{jk} and $C_{jk},\ j = 1, ..., n_s,\ k = 1, ..., n_s$ according to the partition of $\mathbf{x}(t),\ \mathbf{u}(t)$ and $\mathbf{y}(t)$. Obviously, the subsystem state $x_j(t)$ depends on all inputs $u_j(t)$, and the subsystem output $y_j(t)$ on all states $x_j(t)$.

Since the ultimate goal is to get "weakly coupled" subsystems the partition of $\mathbf{x}(t)$ should be done in such a way that the dependencies of $y_j(t)$ on $u_k(t),\ j \neq k$ are zero or weak. Systems of the form

$$\begin{aligned} \dot{x}_j(t) &= A_j x_j(t) + \sum_{k=1, j\neq k}^{n_s} A_{jk} x_k(t) + B_j u_j(t) \\ y_j(t) &= \sum_{k=1, j\neq k}^{n_s} C_{jk} x_k(t), \quad j = 1, ..., n_s \end{aligned} \tag{5.3}$$

are often termed *input decentralized* and systems

$$\begin{aligned} \dot{x}_j(t) &= A_j x_j(t) + \sum_{k=1, j\neq k}^{n_s} A_{jk} x_k(t) + \sum_{k=1}^{n_s} B_{jk} u_k(t) \\ y_j(t) &= C_j x_j(t) \end{aligned} \tag{5.4}$$

are often termed *output decentralized* . By similarity,

$$\begin{aligned} \dot{x}_j(t) &= A_j x_j(t) + \sum_{k=1, j\neq k}^{n_s} A_{jk} x_k(t) + B_j u_j(t) \\ y_j(t) &= C_j x_(t) \end{aligned} \tag{5.5}$$

are often termed *input-output decentralized* .

The disjoint decomposition into input-decentralized, output decentralized or input-output decentralized subsystems has been proposed by [196]-[203] and [323] using different formats. In [310, 311], a decomposition method was given where the interactions of the resulting subsystems are lower than a prescribed threshold. In essence, when considering disjoint subsystems and considering (5.1)-(5.5), we are lead to the notion of ϵ-decomposition . The idea of ϵ-decomposition can be simply explained on the linear dynamic system (5.1) by defining

$$\begin{aligned} \mathbf{A} &= \mathbf{A}_D + \epsilon\ \mathbf{A}_C \\ \mathbf{A}_D &= Blockdiag\{A_1,\ A_2,\ \cdots,\ A_{n_s}\} \end{aligned} \tag{5.6}$$

the matrix $\mathbf{A}_C$ has all its elements smaller than one, and ϵ is a prescribed small number. The choice of ϵ influences on the strength of interconnections [200]. If each subsystem A_j is stable, then an appropriate choice of ϵ preserves the weak coupling property of the system and thereby the stability of the overall system. The increasing threshold of ϵ leads to the notion of nested ϵ-decomposition [324, 325].

5.1.2 Overlapping subsystems

The decomposition of the overall system into disjoint subsystems is not effective in situations where the subsystems are strongly coupled. It means that a given system has no ϵ-decomposition and one has to attempt alternative procedure. One such procedure is the *overlapping decomposition* in which the subsystems share some common parts. Basically, the overlapping subsystems may be weakly coupled although disjoint subsystems are not, that is, they may have an overlapping ϵ-decomposition [325].

A systematic way of overlapping decomposition means to expand the original system with strongly coupled subsystems into a larger dimensional systems with weakly coupled subsystems. There is a requirement of the relation between both systems. The solution of a large-dimensional system must include the solution of a lower dimensional original system. A circle of ideas, methods and algorithms devoted to overlapping decompositions has been formulated rigorously into a general mathematical framework called the inclusion principle [41, 324].

5.1.3 Transformations

Starting with the expansion of the original system (5.1) into the system

$$\begin{aligned} \dot{\tilde{\mathbf{x}}}(\mathbf{t}) &= \tilde{\mathbf{A}}\tilde{\mathbf{x}}(t) + \tilde{\mathbf{B}}\mathbf{u}(t), \quad \tilde{\mathbf{x}}(0) = \tilde{\mathbf{x}}_o \\ \mathbf{y}(t) &= \tilde{\mathbf{C}}\tilde{\mathbf{x}}(t) \end{aligned} \tag{5.7}$$

Formally, the systems (5.1) and (5.7) are related by some contraction transformation

$$\mathbf{x}(t) \;=\; \mathbf{T}^{\dagger}\,\tilde{\mathbf{x}}(t) \tag{5.8}$$

where

$$\mathbf{T}^{\dagger}\,\mathbf{T} = I \tag{5.9}$$

holds with the superscript $'\dagger'$ denoting the pseudo inverse of a rectangular matrix.

Definition 5.1.1 *A system (5.7) is said to include a system (5.1) if there exists an ordered pair of matrices $(\mathbf{T}, \mathbf{T}^{\dagger})$ such that relations (5.8) and (5.9) hold. The systems (5.1) or (5.7) are called contraction or expansion, respectively.*

It should be observed that the expansion of (5.1) into a new system (5.7) yields some parts of the original system (5.1) to appear more than once. Obviously f system (5.7) is decomposed into disjoint subsystems, then these parts of (5.1) belong simultaneously to two or more subsystems. That is, the subsystems *overlap*.

An expansion (5.7) can be found by using the matrices

$$\begin{aligned}\tilde{\mathbf{A}} &= \mathbf{T}\mathbf{A}\mathbf{T}^{\dagger} + \mathbf{M},\;\; \tilde{\mathbf{B}} = \mathbf{T}\mathbf{B} + \mathbf{N}\\ \tilde{\mathbf{C}} &= \mathbf{C}\mathbf{T}^{\dagger} + \mathbf{L}\end{aligned} \tag{5.10}$$

and by choosing appropriate matrices $\mathbf{M}$, $\mathbf{N}$ and $\mathbf{L}$. The following throe rem establishes the first basic result

Theorem 5.1.2 *The system (5.7) with relations (5.10) is an expansion of (5.1) if and only if the following conditions are satisfied*

$$\begin{aligned}\mathbf{T}^{\dagger}\mathbf{M}^{\ell}\mathbf{T} &= 0,\;\; \mathbf{T}^{\dagger}\mathbf{M}^{\ell-1}\mathbf{N} = 0,\\ \mathbf{L}\mathbf{M}^{\ell-1}\mathbf{T} &= 0,\;\; \mathbf{L}\mathbf{M}^{\ell-1}\mathbf{N} = 0\\ \ell &= 1, 2, ..., dim(\tilde{\mathbf{x}})\end{aligned} \tag{5.11}$$

Proof: The theorem can be readily derived by considering the standard relation

$$\mathbf{T}^{\dagger}\, exp(\tilde{\mathbf{A}}t)\, \mathbf{T} \;=\; exp(\mathbf{A}t)$$

along with the time series expansion of $exp(\tilde{\mathbf{A}}t)$ and $exp(\mathbf{A}t)$.

The method of investigating a given system (5.1) by considering the expansion (5.7) and inferring the results to the contraction (5.1) is called the *inclusion principle* . An important fact for the application of the inclusion principle is that the stability property of the system (5.1) is preserved in the expansion.

Theorem 5.1.3 *If the systems (5.1) and (5.7) are a contraction or an expansion, respectively, then the asymptotic stability of the system (5.7) implies the asymptotic stability of the system (5.1)*

5.1.4 Illustrative example 5.1

Consider the system (5.1) with partitioned state vector $\mathbf{x} = [\mathbf{x}_1^t,\ \mathbf{x}_2^t,\ \mathbf{x}_3^t]^t$ and structured matrices $\mathbf{A} = \mathbf{A}_{jk},\ \ \mathbf{B} = \mathbf{B}_{jk}$. An overlapping decomposition is given by $\tilde{\mathbf{x}}_1 = [\mathbf{x}_1^t,\ \mathbf{x}_2^t,\ \tilde{\mathbf{x}}_2 = [\mathbf{x}_2^t,\ \mathbf{x}_3^t]^t$ which satisfies (5.8) with

$$\mathbf{T} = \begin{bmatrix} I & 0 & 0 \\ 0 & I & 0 \\ 0 & I & 0 \\ 0 & 0 & I \end{bmatrix}, \quad \mathbf{T}^{\dagger} = \begin{bmatrix} I & 0 & 0 & 0 \\ 0 & 0.5I & 0.5I & 0 \\ 0 & 0 & 0 & I \end{bmatrix} \tag{5.12}$$

Simple computations using (5.10) yield:

$$\tilde{\mathbf{A}} = \begin{bmatrix} \mathbf{A}_{11} & \mathbf{A}_{12} & 0 & \mathbf{A}_{13} \\ \mathbf{A}_{21} & \mathbf{A}_{22} & 0 & \mathbf{A}_{23} \\ \mathbf{A}_{21} & 0 & \mathbf{A}_{22} & \mathbf{A}_{23} \\ \mathbf{A}_{31} & 0 & \mathbf{A}_{32} & \mathbf{A}_{33} \end{bmatrix}, \quad \mathbf{N} = 0$$

$$\tilde{\mathbf{B}} = \begin{bmatrix} \mathbf{B}_{11} & \mathbf{A}_{12} \\ \mathbf{B}_{21} & \mathbf{B}_{22} \\ \mathbf{B}_{21} & \mathbf{B}_{22} \\ \mathbf{B}_{31} & \mathbf{B}_{32} \end{bmatrix}, \quad \mathbf{M} = \begin{bmatrix} 0 & 0.5\mathbf{A}_{12} & -0.5\mathbf{A}_{12} & 0 \\ 0 & 0.5\mathbf{A}_{22} & -0.5\mathbf{A}_{22} & 0 \\ 0 & 0.5\mathbf{A}_{22} & -0.5\mathbf{A}_{22} & 0 \\ 0 & 0.5\mathbf{A}_{32} & -0.5\mathbf{A}_{32} & 0 \end{bmatrix}$$

Hence the isolated subsystems of the expansion (5.7) are

$$\dot{\tilde{\mathbf{x}}}_1(\mathbf{t}) = \begin{bmatrix} \mathbf{A}_{11} & \mathbf{A}_{12} \\ \mathbf{A}_{21} & \mathbf{A}_{22} \end{bmatrix} \tilde{\mathbf{x}}_1(t) + \begin{bmatrix} \mathbf{B}_{11} \\ \mathbf{B}_{21} \end{bmatrix} \mathbf{u}_1(t) \tag{5.13}$$

$$\dot{\tilde{\mathbf{x}}}_1(\mathbf{t}) = \begin{bmatrix} \mathbf{A}_{22} & \mathbf{A}_{23} \\ \mathbf{A}_{32} & \mathbf{A}_{33} \end{bmatrix} \tilde{\mathbf{x}}_2(t) + \begin{bmatrix} \mathbf{B}_{22} \\ \mathbf{B}_{32} \end{bmatrix} \mathbf{u}_2(t) \tag{5.14}$$

The overall system matrix has the form

$$\mathbf{A} = \begin{bmatrix} \mathbf{A}_{11} & & \mathbf{A}_{12} & \vdots & \mathbf{A}_{13} \\ & \cdots & \cdots & \cdots & \cdots \\ \mathbf{A}_{21} & \vdots & \mathbf{A}_{22} & \vdots & \mathbf{A}_{23} \\ \cdots & \cdots & \cdots & \vdots & \\ \mathbf{A}_{31} & \vdots & \mathbf{A}_{32} & & \mathbf{A}_{33} \end{bmatrix}$$

Little inspection shows that the original system (5.1) has been expanded in such a way that the state $\tilde{x}_2$ belongs to both new subsystems (5.13) and (5.14). Consequently, the matrix $\mathbf{A}_{22}$ is used twice. This explains the term *overlapping decomposition.* The overall system state space is not the direct sum of the subsystem state spaces. This decomposition is particularly useful for systems with $\mathbf{A}_{31} = 0$, $\mathbf{A}_{13} = 0$, $\mathbf{B}_{12} = 0$ and $\mathbf{B}_{21} = 0$, since these matrices are neglected when considering the decomposition (5.13) and (5.14).

It is fair to record that the *inclusion principle* was proposed in the early 1980s in the context of analysis and control of large scale systems [124]-[131] and [324]. Essentially, this principle establishes a mathematical framework for two dynamic systems with different dimensions, in which solutions of the system with larger-dimension include solutions of the system with smaller dimension.

5.2 Feedback Control Design

As we have noted earlier, the mathematical framework of the *inclusion principle* relies on the choice of appropriate linear transformations $\mathbf{M}$, $\mathbf{N}$, $\mathbf{L}$ between the inputs, states and outputs of both systems. These matrices, which are at the disposal of designer, are frequently called the complementary matrices [15]. Based thereon with the context of optimal control, the idea is to expand a system with overlapped components into a larger-dimensional system that appears decomposed into a number of disjoint subsystems. Then, decentralized controllers are designed for the expanded system and contracted to be implemented in the original system [12, 13, 124, 131, 129, 324].

The conditions given in previous works on the complementary matrices [124]-[130] to ensure the *inclusion principle* have a fundamental and implicit nature. Simply stated, they have the form of matrix products from which it is not easy to select specific values for the matrices. A few simple standard choices have been commonly used in practice, while the exploitation of the degree of freedom offered by the selection of the complementary matrices has been considered as one of interesting research issues, [128]. In this regard, a new characterization of the complementary matrices was recently presented in [15], which gave a more explicit way for their selection. It relies on introducing appropriate changes of basis in the expansion-contraction process as already suggested in [130]. In the sequel, we proceed along the same direction to provide results on overlapping decentralized state linear-quadratic optimal control.

5.2.1 Linear-quadratic control

For simplicity in exposition, we consider a pair of optimal control problems

$$\begin{aligned}\min_{u} J(u) &= \int_0^\infty (\mathbf{x}^t Q^* \mathbf{x} + \mathbf{u}^t R^* \mathbf{u}) dt, \\ s.t\, \mathbf{S} : \dot{\mathbf{x}}(t) &= \mathbf{A}\mathbf{x}(t) + \mathbf{B}\mathbf{u}(t) \end{aligned} \tag{5.15}$$

$$\begin{aligned}\min_{\tilde{u}} \tilde{J}(\tilde{\mathbf{u}}) &= \int_0^\infty (\tilde{\mathbf{x}}^t \tilde{Q}_* \tilde{\mathbf{x}} + \tilde{\mathbf{u}}^t \tilde{R}^* \tilde{\mathbf{u}}) dt, \\ s.t \tilde{\mathbf{S}} : \dot{\tilde{\mathbf{x}}}(\mathbf{t}) &= \tilde{\mathbf{A}}\tilde{\mathbf{x}}(t) + \tilde{\mathbf{B}}\tilde{\mathbf{u}}(t) \end{aligned} \tag{5.16}$$

where $\mathbf{x}(t) \in \Re^n,\ \mathbf{u}(t) \in \Re^m$ are the state and input of **S** at time $t \in \Re^+$ and $\tilde{\mathbf{x}}(t) \in \Re^{\tilde{n}},\ \tilde{\mathbf{u}}(t) \in \Re^{\tilde{m}}$ are those ones of $\tilde{\mathbf{S}}$. The matrices $\mathbf{A},\ \mathbf{B},\ Q^*,\ R^*$ and $\tilde{\mathbf{A}},\ \tilde{\mathbf{B}},\ \tilde{Q}^*,\ \tilde{R}^*$ are constant matrices of dimensions $n \times n,\ n \times m,\ n \times n,\ m \times m$, and $\tilde{n} \times \tilde{n},\ \tilde{n} \times \tilde{m},\ \tilde{n} \times \tilde{n},\ \tilde{m} \times \tilde{m}$, respectively. Matrices $Q^* \geq 0,\ R^* \geq 0,\ tildeR^* \geq 0$. Suppose that the dimensions of the state and input vectors $\mathbf{x}(t),\ \mathbf{u}(t)$ of **S** are smaller than (or at most equal to) those of $\tilde{\mathbf{x}},\ \tilde{\mathbf{u}}$ of $\tilde{\mathbf{S}}$. We use $\mathbf{x}(t; \mathbf{x}_0, \mathbf{u})$ to denote the state behavior of **S** for a fixed input $\mathbf{u}(t)$ and an initial state $\mathbf{x}(0) = \mathbf{x}_0$. Similar notation $\tilde{\mathbf{x}}(t; \tilde{\mathbf{x}_0}, \tilde{\mathbf{u}})$ is used for the state behavior of system $\tilde{\mathbf{S}}$.

In line of the preceding section, we can see that systems **S** and $\tilde{\mathbf{S}}$ are related by the transformations

$$\tilde{\mathbf{x}}(t) = V\mathbf{x}(t),\ \mathbf{x}(t) = U\tilde{\mathbf{x}}(t),\ \tilde{\mathbf{u}}(t) = R\mathbf{u}(t),\ \mathbf{u}(t) = Q\tilde{\mathbf{u}}(t)$$

where $V,\ U,\ Q$ and R are constant matrices of appropriate dimensions and full rank satisfying

$$UV = I_n,\ \ QR = I_m$$

where $I_n,\ I_m$ are identity matrices of indicated dimensions. The following definition provides a wide-sense statement of the *inclusion principle*:

Definition 5.2.1

1. *System $\tilde{\mathbf{S}}$ is said to include system* **S**, *that is $\tilde{\mathbf{S}} \supset \mathbf{S}$, if there exists a triplet $(U,\ V,\ R)$ such that, for any initial state $\mathbf{x}_0$ and any fixed input $\mathbf{u}(t)$ of* **S**, *the choice $\tilde{\mathbf{x}_0} = V\mathbf{x}_0$ and $\tilde{\mathbf{u}}\mathbf{t} = R\mathbf{u}(t)$ for all $t \geq 0$ of the initial state $\tilde{\mathbf{x}_0}$ and input $\tilde{\mathbf{u}}(t)$ of the system* **S** *implies $\mathbf{x}(t; \mathbf{x}_0, \mathbf{u}) = U\tilde{\mathbf{x}}(t; \tilde{\mathbf{x}_0}, \tilde{\mathbf{u}})$ for all $t \geq 0$.*
   If $\tilde{\mathbf{S}} \supset \mathbf{S}$, then $\tilde{\mathbf{S}}$ it is said to be an expansion of **S** *and* **S** *is called a contraction of $\tilde{\mathbf{S}}$.*

2. *The pair* $(\tilde{\mathbf{S}}; \tilde{J})$ *is said to include the pair* $(\mathbf{S}; J)$, *that is* $(\tilde{\mathbf{S}}; \tilde{J}) \supset (\mathbf{S}; J)$ *if*
$$\tilde{\mathbf{S}} \supset \mathbf{S}, \quad J(\mathbf{x}_0; \mathbf{u}) = \tilde{J}(\tilde{\mathbf{x}}_0; \tilde{\mathbf{u}}).$$
The matrices of $(\tilde{\mathbf{S}}; \tilde{J})$ *and* $(\mathbf{S}; J)$ *can be related via*
$$\begin{aligned} \tilde{\mathbf{A}} &= V\mathbf{A}U + M, \quad \tilde{\mathbf{B}} = V\mathbf{B}Q + N, \\ \tilde{Q^*} &= U_t Q^* U + M_{Q^*}, \quad \tilde{R^*} = Q_t R^* Q + N_{R^*} \end{aligned}$$
where M, N, M_{Q^*}, N_{R^*} *are constant complementary matrices of appropriate dimensions.*

The following preliminary results can be readily established:

Theorem 5.2.2 *System* $\tilde{\mathbf{S}}$ *is an expansion of system* $\mathbf{S}$ *if and only if* $UM^jV = 0$ *and* $UM^{j-1}NR = 0$ *for all* $j = 1, 2, ..., \tilde{n}$

Theorem 5.2.3 $\tilde{\mathbf{S}} \supset \mathbf{S}$ *if and only if there exists* $\bar{\mathbf{S}}$ *such that* $\tilde{\mathbf{S}} \supset \bar{\mathbf{S}} \supset \mathbf{S}$ *,where* $\bar{\mathbf{S}}$ *is a restriction (aggregation) of* $\tilde{\mathbf{S}}$ *and* $\mathbf{S}$ *is an aggregation (restriction) of* $\bar{\mathbf{S}}$.

The expansion and contraction between systems $\mathbf{S}$ and $\tilde{\mathbf{S}}$ can be illustrated in the form
$$\begin{aligned} &\mathbf{S} \rightarrow \tilde{\mathbf{S}} \rightarrow \mathbf{S} \\ &\Re^n \xrightarrow{V} \Re^{\tilde{n}} \xrightarrow{U} \Re^n \\ &\Re^m \xrightarrow{R} \Re^{\tilde{m}} \xrightarrow{Q} \Re^m \end{aligned} \tag{5.17}$$

It follows that for $(\tilde{\mathbf{S}}; \tilde{J})$ to be an expansion of $(\mathbf{S}; J)$, a proper choice of the matrices M, N, M_{Q^*}, N_{R^*} is required. The following theorem provides such a requirement [128]:

Theorem 5.2.4 *The pair* $(\tilde{\mathbf{S}}; \tilde{J}) \supset (\mathbf{S}; J)$ *if either*

1. $MV = 0, \quad NR = 0, \quad V^t M_{Q^*} V = 0, \quad R^t N_{R^*} R = 0$ *or*

2. $UM^jV = 0, \quad UM^{j-1}NR = 0, \quad M_{Q^*}M^{j-1}V = 0,$
$M_{Q^*}M^{j-1}NR = 0, \quad R^t N_{R^*} R = 0 \; \forall \; j = 1, 2, ..., \tilde{n}$

So far, all the statements and the foregoing analysis pertain to uncontrolled systems. The next definition and results are related to controlled systems:

Definition 5.2.5 *A control law* $\tilde{\mathbf{u}} = -\tilde{\mathbf{K}}\tilde{\mathbf{x}}$ *for* $\tilde{\mathbf{S}}$ *is contractible to the control law* $\mathbf{u} = -\mathbf{K}\mathbf{x}$ *for implementation to* $\mathbf{S}$ *if the choice* $\tilde{\mathbf{x}}_0 = V\mathbf{x}_0$ *and* $\tilde{\mathbf{u}} = R\mathbf{u}$ *implies* $\mathbf{K}\mathbf{x}(t; \mathbf{x}_0, \mathbf{u}) = Q\tilde{\mathbf{K}}\mathbf{x}(t; \mathbf{x}_0, \mathbf{u})$, *for all* $t \neq 0$, *for any initial state* $\mathbf{x}_0$ *and any fixed input* $\mathbf{u}(t)$ *of* $\mathbf{S}$.

Now suppose that $\tilde{\mathbf{K}} = R\mathbf{K}U + F$, where F denotes a complementary matrix. The conditions to satisfy **Definition** 5.2.5 are given by the following theorem whose proof is given in[128]:

Theorem 5.2.6 *A control law* $\tilde{\mathbf{u}} = -\tilde{\mathbf{K}}\tilde{\mathbf{x}}$ *for* $\tilde{\mathbf{S}}$ *is contractible to the control law* $\mathbf{u} = -\mathbf{K}\mathbf{x}$ *for* $\mathbf{S}$ *if and only if* $QFM^{j-1}V = 0$ *and* $QFM^{j-1}NR = 0$, *for all* $j = 1, 2, ..., \tilde{n}$.

An equivalent form of **Theorem** 5.2.6, which gives an explicit expression of matrix $\mathbf{K}$ from matrix $\tilde{\mathbf{K}}$, is summarized by the following theorem

Theorem 5.2.7 *A control law* $\tilde{\mathbf{u}} = -\tilde{\mathbf{K}}\tilde{\mathbf{x}}$ *for* $\tilde{\mathbf{S}}$ *is contractible to the control law* $\tilde{\mathbf{u}} = R\mathbf{u}$ *for* $\mathbf{S}$ *if and only if* $Q\tilde{\mathbf{K}}V = \mathbf{K}$, $Q\tilde{\mathbf{K}}M^{j-1}V = 0$ *for all* $j = 1, 2, ..., \tilde{n}$, *and* $Q\tilde{\mathbf{K}}M^{j-1}NR = 0$ *for all* $i = 1, 2, ..., \tilde{n}$

5.2.2 Change of basis

Since the Inclusion Principle does not depend on the specific basis used in the state, input and output spaces for both systems $\mathbf{S}$ and $\tilde{\mathbf{S}}$, we follow [15] and proceed to introduce convenient changes of basis in $\tilde{\mathbf{S}}$. Thus, scheme (5.17) can be modified into the form

$$\begin{array}{c} \mathbf{S} \to \tilde{\mathbf{S}} \to \overline{\tilde{\mathbf{S}}} \to \tilde{\mathbf{S}} \to \mathbf{S} \\ \Re^n \xrightarrow{V} \Re^{\tilde{n}} \xrightarrow{T_{\mathbf{A}}^{-1}} \overline{\Re}^{\tilde{n}} \xrightarrow{T_{\mathbf{A}}} \Re^{\tilde{n}} \xrightarrow{U} \Re^n \\ \Re^m \xrightarrow{R} \Re^{\tilde{m}} \xrightarrow{T_{\mathbf{B}}^{-1}} \overline{\Re}^{\tilde{m}} \xrightarrow{T_{\mathbf{B}}} \Re^{\tilde{m}} \xrightarrow{Q} \Re^m \end{array} \tag{5.18}$$

where now $\overline{\tilde{\mathbf{S}}}$ denotes the expanded system with the new basis. The idea of using changes of basis in the expansion-contraction process was already introduced in [130] to represent $\tilde{\mathbf{S}}$ in a canonical form. Now given V and R, we define $U = (V^tV)^{-1}V^t$, $Q = (R^tR)^{-1}R^t$ as their pseudo inverses, respectively, and consider

$$T_{\mathbf{A}} = (V\ W_{\mathbf{A}}), \qquad T_{\mathbf{B}} = (R\ W_{\mathbf{B}}) \tag{5.19}$$

where $W_{\mathbf{A}}$, $W_{\mathbf{B}}$ are chosen such that

$$Im(W_{\mathbf{A}}) = Ker(U), \quad Im(W_{\mathbf{B}}) = Ker(Q)$$

Using these transformations, it is not difficult to verify the following conditions

$$\begin{aligned}
\widetilde{UV} &= I_n, \quad \widetilde{VU} = \begin{pmatrix} I_n & 0 \\ 0 & 0 \end{pmatrix} \\
&and \\
widetildeQR &= I_m, \quad \widetilde{RQ} = \begin{pmatrix} I_m & 0 \\ 0 & 0 \end{pmatrix} \\
&where \\
\widetilde{V} &= T_{\mathbf{A}}^{-1} V = \begin{pmatrix} I_n \\ 0 \end{pmatrix} \\
\widetilde{U} &= UT_A = (I_n 0) \\
&and \\
\widetilde{R} &= T_{\mathbf{B}}^{-1} R = \begin{pmatrix} I_m \\ 0 \end{pmatrix}, \quad \widetilde{Q} = QT_B = (I_m 0)
\end{aligned}$$

In fact, these conditions will be crucial to obtain explicit block structures (with zero blocks) of the complementary matrices and further to give a general strategy for their selection.

5.2.3 Improved expansion-contraction

The expansion-contraction process will be developed for the system **S** having the following structure **S**:

$$\begin{bmatrix} \dot{x}_1 \\ \dot{x}_2 \\ \dot{x}_3 \end{bmatrix} = \begin{bmatrix} A_{11} & A_{12} & A_{13} \\ A_{21} & A_{22} & A_{23} \\ A_{31} & A_{32} & A_{33} \end{bmatrix} \begin{bmatrix} x_1 \\ x_2 \\ x_3 \end{bmatrix} + \begin{bmatrix} B_{11} & B_{12} & B_{13} \\ B_{21} & B_{22} & B_{23} \\ B_{31} & B_{32} & B_{33} \end{bmatrix} \begin{bmatrix} u_1 \\ u_2 \\ u_3 \end{bmatrix} \quad (5.20)$$

where $A_{ii}, B_{ii}, i = 1, 2, 3$, are $n_i \times n_i$ and $n_i \times m_i$ dimensional matrices, respectively. This system is composed of two subsystems with one overlapped part. This simple structure will help in smoothing the notation. The results obtained can be easily generalized for any number of interconnected overlapping subsystems. This structure has been extensively adopted as prototype in the literature within the Inclusion Principle. Consider the optimal control problem in the system $\widetilde{\overline{S}}$

$$\begin{aligned}
\min_{\widetilde{\mathbf{u}}} \widetilde{J}(\widetilde{\mathbf{x}}_0, \widetilde{\mathbf{u}}) &= \int_0^\infty (\widetilde{\mathbf{x}}^t \widetilde{Q}^* \widetilde{\mathbf{x}} + \widetilde{\mathbf{u}}^t \widetilde{R}^* \widetilde{\mathbf{u}}) dt \\
s.t\, \widetilde{\overline{S}} : \dot{\widetilde{\overline{x}}} &= \widetilde{\overline{A}} \widetilde{\mathbf{x}} + \widetilde{\overline{B}} \widetilde{\overline{u}}
\end{aligned} \quad (5.21)$$

where $\widetilde{\mathbf{x}}$ and $\widetilde{\mathbf{u}}$ are defined as

$$\widetilde{\mathbf{x}} = T_{\mathbf{A}}^{-1} V\mathbf{x} = \widetilde{V}\mathbf{x}, \;\; \widetilde{\mathbf{u}} = T_{\mathbf{B}}^{-1} Ru = \widetilde{R}\widetilde{\mathbf{u}}$$

In addition, $\widetilde{\mathbf{A}}, \widetilde{\mathbf{B}}, \widetilde{Q^*}$ and $\widetilde{R^*}$ are constant matrices of appropriate dimensions verifying

$$\begin{aligned} \widetilde{\mathbf{A}} &= \widetilde{V}\mathbf{A}\widetilde{U} + \widetilde{M}, \; \widetilde{\mathbf{B}} = \widetilde{V}\mathbf{B}\widetilde{Q} + \widetilde{N}, \\ \widetilde{Q^*} &= \widetilde{U}^t Q^* \widetilde{U} + \widetilde{M_Q}^*, \; \widetilde{R}^* = \widetilde{Q}^t R^* \widetilde{Q} + \widetilde{N}_R{}^* \end{aligned}$$

where the new complementary matrices are expressed as [15]:

$$\widetilde{M} = T_{\mathbf{A}}^{-1} M T_{\mathbf{A}}, \; \widetilde{N} = T_{\mathbf{A}}^{-1} N T_{\mathbf{B}}, \; \widetilde{M_Q}^* = T_{\mathbf{A}}^t M_Q{}^* T_{\mathbf{A}}, \; \widetilde{N}_R{}^* = T_{\mathbf{B}}^t N_R{}^* T_{\mathbf{B}}$$

Using these matrices, the conditions given by the *inclusion principle* as stated in **Theorem** 5.2.2 become

$$\widetilde{UM}^j \widetilde{V} = 0, \; \widetilde{UM}^{j-1} \widetilde{NR} = 0, \;\; \forall\, j = 1, 2, ..., \tilde{n}$$

Our immediate task is to analyze the form of matrices $\widetilde{M}, \; \widetilde{N}, \; \widetilde{M_Q}^*$, and $\widetilde{N}_R{}^*$. For this purpose, we consider for $j, k = 1, .., 4$ the complementary matrices

$$M = (M_{ij}), \; N = (N_{ij}), \; M_Q{}^* = (M_{Q}{}^*_{ij}), \; N_R{}^* = (N_{R}{}^*_{ij})$$

where all sub matrices have appropriate dimensions and $M_{Q}{}^*_{ij} = M^t_{Q^*_{ji}}, \; N_{R}{}^*_{ij} = N^t_{R^*_{ji}}$. To proceed further, we introduce the matrices

$$\begin{aligned} \widetilde{M} &= \begin{bmatrix} \widetilde{M}_{11} & \widetilde{M}_{12} \\ \widetilde{M}_{21} & \widetilde{M}_{22} \end{bmatrix}, \; \widetilde{N} = \begin{bmatrix} \widetilde{N}_{11} & \widetilde{N}_{12} \\ \widetilde{N}_{21} & \widetilde{N}_{22} \end{bmatrix}, \\ \widetilde{M_Q}^* &= \begin{bmatrix} \widetilde{M_Q}{}^*_{11} & \widetilde{M_Q}{}^*_{12} \\ \widetilde{M_Q}{}^*_{21} & \widetilde{M_Q}{}^*_{22} \end{bmatrix}, \; \widetilde{N}_R{}^* = \begin{bmatrix} \widetilde{N}_{R}{}^*_{11} & \widetilde{N}_{R}{}^*_{12} \\ \widetilde{N}_{R}{}^*_{21} & \widetilde{N}_{R}{}^*_{22} \end{bmatrix} \end{aligned}$$

where

$$\begin{aligned} \widetilde{M}_{11} &\in \Re^{n\times n}, \; \widetilde{M}_{22} \in \Re^{\tilde{n}-n\times\tilde{n}-n}, \; \widetilde{N}_{11} \in \Re^{n\times m}, \; \widetilde{N}_{22} \in \Re^{\tilde{n}-n\times\tilde{m}-m}, \\ \widetilde{M_Q}{}^*_{11} &\in \Re^{n\times n}, \; \widetilde{M_Q}{}^*_{22} \in \Re^{\tilde{n}-n\times\tilde{n}-n}, \; \widetilde{N}_{R}{}^*_{11} \in \Re^{m\times m}, \; \widetilde{N}_{R}{}^*_{22} \in \Re^{\tilde{m}-m\times\tilde{m}-m} \end{aligned}$$

The following lemmas summarize the main results

Lemma 5.2.8 *Consider system (5.15) and the corresponding expanded system (5.21) verifying the inclusion principle. It follows that*

$$\widetilde{M} = \begin{bmatrix} 0 & \widetilde{M}_{12} \\ \widetilde{M}_{21} & \widetilde{M}_{22} \end{bmatrix}$$

where $\widetilde{M}_{12}\widetilde{M}_{22}^{j-2}\widetilde{M}_{21} = 0, \; \forall\, j = 2, ..., \tilde{n}$

Lemma 5.2.9 *Consider system (5.15) and the corresponding expanded system (5.21) verifying the inclusion principle. It follows that*

$$\widetilde{N} = \begin{bmatrix} 0 & \widetilde{N}_{12} \\ \widetilde{N}_{21} & \widetilde{N}_{22} \end{bmatrix}$$

where $\widetilde{M}_{12}\widetilde{M}_{22}^{j-2}\widetilde{N}_{21} = 0, \ \forall\, j = 2, ..., \tilde{n}$

The proofs of **Lemmas** 5.2.8 and 5.2.9 can be established in line of [15]. Based thereon, we have the following result

Theorem 5.2.10 *Consider the optimal control problems (5.15) and (5.21). Then the pair* $(\widetilde{\mathbf{S}},\ \widetilde{J}) \supset (\mathbf{S},\ J)$ *if either of the following conditions hold*

1.

$$\begin{aligned} \widetilde{M} &= \begin{bmatrix} 0 & \widetilde{M}_{12} \\ 0 & \widetilde{M}_{22} \end{bmatrix}, \ \widetilde{N} = \begin{bmatrix} 0 & \widetilde{N}_{12} \\ 0 & \widetilde{N}_{22} \end{bmatrix}, \\ \widetilde{M_Q}^* &= \begin{bmatrix} 0 & \widetilde{M}_{Q12}^* \\ \widetilde{M}_{Q12}^{t\,*} & \widetilde{M}_{Q22}^* \end{bmatrix}, \ \widetilde{N_R}^* = \begin{bmatrix} 0 & \widetilde{N}_{R12}^* \\ \widetilde{N}_{R12}^{t\,*} & \widetilde{N}_{R22}^* \end{bmatrix} \end{aligned}$$

2.

$$\widetilde{M} = \begin{bmatrix} 0 & \widetilde{M}_{12} \\ \widetilde{M}_{21} & \widetilde{M}_{22} \end{bmatrix}, \ \widetilde{N} = \begin{bmatrix} 0 & \widetilde{N}_{12} \\ \widetilde{N}_{21} & \widetilde{N}_{22} \end{bmatrix}$$

such that $\widetilde{M}_{12}\widetilde{M}_{22}^{j-2}\widetilde{M}_{21} = 0$ *and* $\widetilde{M}_{12}\widetilde{M}_{22}^{j-2}\widetilde{N}_{21} = 0, \ \forall\, j = 2, ..., \tilde{n},$

$$\widetilde{M_Q}^* = \begin{bmatrix} 0 & 0 \\ 0 & \widetilde{M}_{Q22}^* \end{bmatrix}$$

such that $\widetilde{M}_{Q12}^*\widetilde{M}_{22}^{j-2}\widetilde{M}_{21} = 0, \ \forall\, j = 2, ..., \tilde{n}, \ \ \widetilde{M}_{Q12}^*\widetilde{M}_{22}^{i-2}\widetilde{N}_{21} = 0, \ \forall\, j = 2, ..., \tilde{n}+1,$ *and*

$$\widetilde{N_R}^* = \begin{bmatrix} 0 & \widetilde{N}_{R12}^* \\ \widetilde{N}_{R_{12}^*}^t & \widetilde{N}_{R22}^* \end{bmatrix}$$

Theorem 5.2.11 *The pair* $(\widetilde{\mathbf{S}},\ \widetilde{J}^{\circ}) \supset (\mathbf{S},\ J^{\circ})$ *if*

$$\widetilde{MV} = 0, \ R^* = (\tilde{Q}(\tilde{\tilde{R}}^*)^{-1}\tilde{Q}^t)^{-1}, \ \widetilde{N} = 0, \ \widetilde{V}^t\widetilde{M_Q}^*\widetilde{V} = 0$$

Proof: Let the Riccati matrices associated with the solutions of the problems (5.15) and (5.21) be P and $\widetilde{P}$, respectively. Then, the optimal costs are $J^{\circ}(\mathbf{x}_0 = \mathbf{x}_0^t P \mathbf{x}_0$ and $\widetilde{J}^{\circ} = \widetilde{\mathbf{x}}_0^t \widetilde{P} \widetilde{\mathbf{x}}_0$. It follows that the relation $P = \widetilde{V}^t \widetilde{P} V$ results directly from $\widetilde{\mathbf{x}}_0 = \widetilde{V}\mathbf{x}_0$ and $J^{\circ}(\mathbf{x}_0) = \widetilde{J^{\circ}}(\mathbf{x}_0)$. On substituting $\widetilde{\mathbf{A}} = \widetilde{V}\mathbf{A}\widetilde{U} + \widetilde{M}$ and $\widetilde{\mathbf{B}} = \widetilde{V}\mathbf{B}\widetilde{Q} + \widetilde{N}$ into the corresponding Riccati equation of problem (5.21) and $P = \widetilde{V}^t \widetilde{\widetilde{P}} V$ into the Riccati equation of problem (5.15) and comparing both equations, we reach the desired result.

Next, we proceed to determine the conditions under which a control law designed in the expanded system $\widetilde{\mathbf{S}}$ can be contracted to be implemented in the initial system $\mathbf{S}$. Considering the system (5.20), let K be the associated gain matrix. Denote F as the corresponding complementary matrix with the form $F = (F_{jk}),\ j, k = 1, ..., 4$. Further, define

$$\widetilde{F} = \begin{bmatrix} \widetilde{F}_{11} & \widetilde{F}_{12} \\ \widetilde{F}_{21} & \widetilde{F}_{22} \end{bmatrix}$$

where $\widetilde{F}_{11} \in \Re^{m \times n},\ \widetilde{F}_{22} \in \Re^{\tilde{m}-m \times \tilde{n}-n}$. It turns out that the gain matrix $\widetilde{\widetilde{K}}$ for the system $\widetilde{\mathbf{S}}$ has the form $\widetilde{K} = \widetilde{R}K\widetilde{U} + \widetilde{F},$ where $\widetilde{\widetilde{K}} = T_{\mathbf{B}}^{-1} K T_{\mathbf{A}}$ and $\widetilde{F} = T_{\mathbf{B}}^{-1} F T_{\mathbf{A}}$. By **Definition** 5.2.5, we have $\widetilde{\mathbf{u}} = -\widetilde{K}\widetilde{\mathbf{x}}$ is contractible to the control law $\mathbf{u} = -K\mathbf{x}$ if

$$K\mathbf{x}(t; \mathbf{x}_0, \mathbf{u}) = \widetilde{QK}\widetilde{\mathbf{x}}(t; \widetilde{V}\mathbf{x}_0, \widetilde{R}\mathbf{u})$$

Consequently, the following theorem follows from the preceding results:

Theorem 5.2.12 *A control law $\widetilde{\mathbf{u}} = -\widetilde{K}\widetilde{\mathbf{x}}$ designed in the expanded system $\widetilde{\mathbf{S}}$ is contractible to the control law $\mathbf{u} = -K\mathbf{x}$ of the system $\mathbf{S}$ if and only if*

$$\widetilde{F} = \begin{bmatrix} 0 & \widetilde{F}_{12} \\ \widetilde{F}_{21} & \widetilde{F}_{22} \end{bmatrix}$$

and $\widetilde{F}_{12}\widetilde{M_{22}^{j-2}}\widetilde{M}_{21} = 0,\ \widetilde{F}_{12}\widetilde{M_{22}^{j-2}}\widetilde{N}_{21} = 0,\ \forall\, j = 2, ..., \tilde{n} + 1.$

5.2.4 Particular selection

Recalling the results of example 5.1, and to make a practical use of the characterizations of $\widetilde{M},\ \widetilde{M}$ to enable the expansion-contraction process, we proceed by defining the specific transformations V and R to expand a given problem (5.15). Needless to stress that the choice of these expansion matrices is limited by the information structure constraints as it demands the preservation of the integrity of the local feedback and subsystems in overlapping decentralized control. Once V and R are chosen, the corresponding changes of basis $T_{\mathbf{A}}, T_{\mathbf{B}}$ are given by (5.19). It is straightforward to obtain the structure of the complementary matrices $M, N, \widetilde{M_Q}^*, \widetilde{N_R}^*$ and F from the corresponding expressions $\widetilde{M} = T_{\mathbf{A}}^{-1} M T_{\mathbf{A}}, \widetilde{N} = T_{\mathbf{A}}^{-1} N T_{\mathbf{B}}, \widetilde{M_Q}^* = T_{\mathbf{A}}^t M_Q{}^* T_{\mathbf{A}}, N_R{}^* = T_{\mathbf{B}}^t N_R{}^* T_{\mathbf{B}}$ and $\widetilde{F} = T_{\mathbf{B}}^{-1} F T_{\mathbf{A}}$. In conclusion, from the derived structure, the designer can select specific values of the elements of complementary matrices according to given specifications.

Let us illustrate this procedure for the following expansion transformation matrices:

$$V = \begin{bmatrix} I_{n1} & 0 & 0 \\ 0 & I_{n2} & 0 \\ 0 & I_{n2} & 0 \\ 0 & 0 & I_{n3} \end{bmatrix}, \quad R = \begin{bmatrix} I_{m1} & 0 & 0 \\ 0 & I_{m2} & 0 \\ 0 & I_{m2} & 0 \\ 0 & 0 & I_{m3} \end{bmatrix} \tag{5.22}$$

These transformations are chosen to help, in the manner of example 5.1, to an expanded system where state vector $\mathbf{x}_2$ and control vector $\mathbf{u}_2$ appear repeated in $\widetilde{\mathbf{x}}^t = (\mathbf{x}1^t, \mathbf{x}_2^t, \mathbf{x}_2^t, \mathbf{x}_3^t)$ and $\widetilde{\mathbf{u}}^t = (\mathbf{u}_1^t, \mathbf{u}_2^t, \mathbf{u}_2^t, \mathbf{u}_3^t$, respectively. According to (5.19), the changes of basis to define the system $\widetilde{\mathbf{S}}$ for matrices (5.22) are given by

$$T_{\mathbf{A}} = \begin{bmatrix} I_{n1} & 0 & 0 & 0 \\ 0 & I_{n2} & 0 & I_{n2} \\ 0 & I_{n2} & 0 & -I_{n2} \\ 0 & 0 & I_{n3} & 0 \end{bmatrix}, \quad T_{\mathbf{A}}^{-1} = \begin{bmatrix} I_{n1} & 0 & 0 & 0 \\ 0 & \frac{1}{2} I_{n2} & \frac{1}{2} I_{n2} & 0 \\ 0 & 0 & 0 & I_{n3} \\ 0 & \frac{1}{2} I_{n2} & -\frac{1}{2} I_{n2} & 0 \end{bmatrix} \tag{5.23}$$

In a similar way, we get $T_{\mathbf{B}}, T_{\mathbf{B}}^{-1}$. Then, by **Lemmas** 5.2.8 and 5.2.9 and using the relations $\widetilde{M} = T_{\mathbf{A}}^{-1} M T_{\mathbf{A}}$ and $\widetilde{N} = T_{\mathbf{A}}^{-1} N T_{\mathbf{B}}$, it is easy to obtain the following structure for complementary matrix M:

$$M = \begin{bmatrix} 0 & M_{12} & -M_{12} & 0 \\ M_{21} & M_{22} & M_{23} & M_{24} \\ -M_{21} & -(M_{22} + M_{23} + M_{33} & M_{33}) & -M_{24} \\ 0 & M_{42} & -M_{42} & 0 \end{bmatrix} \tag{5.24}$$

By the same token, we have a similar one for N, which, for all $j = 2, ..., \tilde{n}$, must verify

$$\begin{bmatrix} M_{12} \\ M_{23}+M_{33} \\ M_{42} \end{bmatrix} (M_{22}+M_{32})^{j-2} \begin{bmatrix} M_{21} & M_{22}+M_{23} & M_{24} \end{bmatrix} = 0$$

$$\begin{bmatrix} M_{12} \\ M_{23}+M_{33} \\ M_{42} \end{bmatrix} (M_{22}+M_{33})^{j-2} \begin{bmatrix} N_{21} & N_{22}+N_{23} & N_{24} \end{bmatrix} = 0 \quad (5.25)$$

The corresponding expanded system matrix $\tilde{A} = VAU + M$ is then given by

$$\tilde{A} = \begin{bmatrix} A_{11} & \frac{1}{2}A_{21}+M_{12} & \frac{1}{2}A_{21}-M_{12} & A_{13} \\ A_{21}+M_{21} & \frac{1}{2}A_{22}+M_{22}\frac{1}{2}A_{22}+M_{23} & A_{23}+M_{24} & \\ A_{21}-M_{21} & \frac{1}{2}A_{22}-(M_{22}+M_{23}+M_{33}) & \frac{1}{2}A_{22}+M_{33} & A_{23}-M_{24} \\ A_{31} & \frac{1}{2}A_{32}+M_{42} & \frac{1}{32}A_{21}-M_{42} & A_{33} \end{bmatrix}$$

A similar structure can be written for the expanded control matrix $\tilde{B} = VBQ + N$

Remark 5.2.13 *From (5.25), we may identify the following particular cases:*

1. *When* $M_{12} = 0,\ M_{23}+M_{33} = 0$ *and* $M_{42} = 0.$
2. *When* $M_{21} = 0,\ M_{22}+M_{23} = 0,\ M_{24} = 0,\ N_{21} = 0,\ N_{22}+N_{23} = 0$ *and* $N_{24} = 0$

From the above remark, the following special case arises:

1. Selecting $M_{22} + M_{33} = 0,$ it follows that conditions (5.25) hold for all $j > 2$. However, for j = 2, they are

$$\begin{bmatrix} M_{12} \\ M_{23}+M_{33} \\ M_{42} \end{bmatrix} \begin{bmatrix} M_{21} & M_{22}+M_{23} & M_{24} \end{bmatrix} = 0$$

$$\begin{bmatrix} M_{12} \\ M_{23}+M_{33} \\ M_{42} \end{bmatrix} \begin{bmatrix} N_{21} & N_{22}+N_{23} & N_{24} \end{bmatrix} = 0 \qquad (5.26)$$

2. $M_{23} + M_{33} = 0$ or $M_{22} + M_{23} = 0$, two sub cases are obtained:
subcase1: $M_{23} = -M_{33}$. Then, relations (5.26) are

$$\begin{bmatrix} M_{12} \\ 0 \\ M_{42} \end{bmatrix} \begin{bmatrix} M_{21} & M_{22} & M_{24} \end{bmatrix} = 0$$
$$\begin{bmatrix} M_{12} \\ 0 \\ M_{42} \end{bmatrix} \begin{bmatrix} N_{21} & N_{22} + N_{23} & N_{24} \end{bmatrix} = 0 \qquad (5.27)$$

subcase2: $M_{23} = -M_{22}$. Then, relations (5.26) are

$$\begin{bmatrix} M_{12} \\ M_{22} \\ M_{42} \end{bmatrix} \begin{bmatrix} M_{21} & 0 & M_{24} \end{bmatrix} = 0$$
$$\begin{bmatrix} M_{12} \\ M_{22} \\ M_{42} \end{bmatrix} \begin{bmatrix} N_{21} & N_{22} + N_{23} & N_{24} \end{bmatrix} = 0 \qquad (5.28)$$

5.2.5 Illustrative example 5.2

The objective is to illustrate the potential advantages of the complementary matrices for an overlapping decentralized state LQ optimal control design. For this purpose, we address problem (5.15) for system (5.20) with the following data

$$A = \begin{bmatrix} -2 & & 0 & 1 & \vdots & -2 \\ & \cdots & \cdots & \cdots & \cdots & \\ 0 & \vdots & 4 & -2 & \vdots & -1 \\ 0 & \vdots & -2 & -4 & \vdots & 0 \\ \cdots & \cdots & \cdots & \cdots & \cdots & \\ 1 & \vdots & 0 & 0 & & 2 \end{bmatrix},$$

$$B = \begin{bmatrix} 0 & & 1 & \vdots & 1 \\ & & \cdots & \cdots & \\ 1 & \vdots & 0 & \vdots & 1 \\ 1 & \vdots & 2 & \vdots & 1 \\ 1 & \vdots & 1 & & 0 \end{bmatrix},$$

$$Q^* = Blockdiag[1,\ 1,\ 1,\ 1], \quad R^* = Blockdiag[1,\ 1,\ 1] \qquad (5.29)$$

The overlapping decomposition is determined by dotted lines and note that the pair $(A,\ B)$ is controllable. For the purpose of illustration, the following procedure is applied

1. The pair $(\mathbf{S},\ J)$ is expanded to $(\tilde{\mathbf{S}},\ \tilde{J})$
2. A decentralized optimal control is designed for the decoupled expanded system in the form $\tilde{\mathbf{u}} = -\tilde{K}\tilde{\mathbf{u}}$ where $\tilde{K}$ is a block diagonal matrix
3. The designed control is then contracted to be implemented in the original system $\mathbf{S}$ as $\mathbf{u} = -K\mathbf{x}$, where $K = Q\tilde{K}V$ according to **Theorem** 5.2.7

For simplicity, we assume that $\mathbf{x}_0$ as a random variable uniformly distributed over the n-dimensional unit sphere. In this way, the expected value of the performance criterion is $\hat{J}^{\oplus} = \mathbf{trace}\{H\},$, where H satisfies the Lyapunov equation $(A-BK)^t H + H(A-BK) + K^t R^* K + Q^* = 0$. Four distinct cases will be treated with M being the design matrix and its sub optimality is analyzed. The computational results are summarized below:

$\mathbf{case 1}:\quad N \ = \ 0$ Following [128, 324] in evaluating the overlapping decomposition by using an Aggregation, we select matrix M as the most commonly used in the literature, that is:

$$M = \begin{bmatrix} 0 & 0 & 0 & 0 \\ A_{21} & \frac{1}{2}A_{22} & -\frac{1}{2}A_{22} & -A_{23} \\ -A_{21} & -\frac{1}{2}A_{22} & \frac{1}{2}A_{22} & A_{23} \\ 0 & 0 & 0 & 0 \end{bmatrix}$$
$$= \begin{bmatrix} 0 & 0 & 0 & 0 & 0 & 0 \\ 0 & 2 & -1 & -2 & 1 & 1 \\ 0 & -1 & -2 & 1 & 2 & 0 \\ 0 & -2 & 1 & 2 & -1 & -1 \\ 0 & 1 & 2 & -1 & -2 & 0 \\ 0 & 0 & 0 & 0 & 0 & 0 \end{bmatrix} \qquad (5.30)$$

and the corresponding expanded matrix

$$\tilde{A} = \begin{bmatrix} A_{11} & \frac{1}{2}A_{12} & \frac{1}{2}A_{12} & A_{13} \\ 2A_{21} & A_{22} & 0 & 0 \\ 0 & 0 & A_{22} & 2A_{23} \\ A_{31} & \frac{1}{2}A_{32} & \frac{1}{2}A_{32} & A_{33} \end{bmatrix}$$

$$= \begin{bmatrix} -2 & 0 & 0.5 & 0 & 0.5 & -2 \\ 0 & 4 & -2 & 0 & 0 & 0 \\ 0 & -2 & -4 & 0 & 0 & 0 \\ 0 & 0 & 0 & 4 & -2 & -2 \\ 0 & 0 & 0 & -2 & -4 & 0 \\ 1 & 0 & 0 & 0 & 0 & 2 \end{bmatrix} \tag{5.31}$$

Simple computation of the suboptimal performance index is $\hat{J}^{\oplus} = 13.63$.

case2: $M_Q{}^* = 0$ In this case overlapping decomposition is pursued by using a restriction. The typical choice of M is

$$M = \begin{bmatrix} 0 & \frac{1}{2}A_{12} & -\frac{1}{2}A_{12} & 0 \\ 0 & \frac{1}{2}A_{22} & -\frac{1}{2}A_{22} & 0 \\ 0 & -\frac{1}{2}A_{22} & \frac{1}{2}A_{22} & 0 \\ 0 & -\frac{1}{2}A_{32} & \frac{1}{2}A_{32} & 0 \end{bmatrix}$$

$$= \begin{bmatrix} 0 & 0 & 0.5 & 0 & -0.5 & 0 \\ 0 & 2 & -1 & -2 & 1 & 0 \\ 0 & -1 & -2 & 1 & 2 & 0 \\ 0 & -2 & 1 & 2 & -1 & 0 \\ 0 & 1 & 2 & -1 & -2 & 0 \\ 0 & 0 & 0 & 0 & 0 & 0 \end{bmatrix} \tag{5.32}$$

and the corresponding expanded matrix

$$\tilde{A} = \begin{bmatrix} A_{11} & A_{12} & 0 & A_{13} \\ A_{21} & A_{22} & 0 & A_{23} \\ A_{21} & 0 & A_{22} & A_{23} \\ A_{31} & 0 & A_{32} & A_{33} \end{bmatrix}$$

$$= \begin{bmatrix} -2 & 0 & 1 & 0 & 0 & 2 \\ 0 & 4 & -2 & 0 & 0 & -1 \\ 0 & -2 & -4 & 0 & 0 & 0 \\ 0 & 0 & 0 & 4 & -2 & -1 \\ 0 & 0 & 0 & -2 & -4 & 0 \\ 1 & 0 & 0 & 0 & 0 & 2 \end{bmatrix} \tag{5.33}$$

This yields the suboptimal performance index as $\hat{J}^{\oplus} = 15.44$.

case3: $\mathbf{N}_R{}^* = 0$ The overlapping decomposition is implemented by the procedure of the foregoing sections. To facilitate the procedure, we can select the complementary sub matrices of M as

$$\begin{aligned} M_{12} &= (1/2)A_{12} = [0,\ (1/2)],\ M_{21}^t = A_{21}^t = [0,\ 0], \\ M_{24} &= -A_{23} = [1,\ 0],\ M_{42} = -(1/2)A_{32} = [0,\ 0] \end{aligned}$$

If $M_{12}M_{22} = 0$, the relations (5.27) hold. Denote

$$M_{22} = \begin{bmatrix} m_{22} & m_{23} \\ 0 & 0 \end{bmatrix}$$

which means that we have the degree of freedom to select the values of m_{22} and m_{23}. To minimize $\hat{J}^{\oplus}$ with respect to m_{22} and m_{23}, the standard algorithm of steepest descent method is used solve the Lyapunov equation. This yields $m_{22} = -0.22,\ m_{23} = -0.82$. The complete matrices M and $\tilde{A}$ given by (5.24) and (5.26), respectively, are the following:

$$\begin{aligned} M &= \begin{bmatrix} 0 & 0 & 0.50 & 0 & -0.50 & 0 \\ 0 & -0.22 & -0.82 & -0.22 & -0.82 & 1 \\ 0 & 0 & 0 & 0 & 0 & 0 \\ 0 & 0.22 & 0.82 & 0.22 & 0.82 & -1 \\ 0 & 0 & 0 & 0 & 0 & 0 \\ 0 & 0 & 0 & 0 & 0 & 0 \end{bmatrix}, \\ \tilde{A} &= \begin{bmatrix} -2 & 0 & 1 & 0 & 0 & -2 \\ 0 & 1.78 & -1.82 & 1.78 & -1.82 & 0 \\ 0 & -1 & -2 & -1 & -2 & 0 \\ 0 & 2.22 & -0.18 & 2.22 & -0.18 & -2 \\ 0 & -1 & -2 & -1 & -2 & 0 \\ 1 & 0 & 0 & 0 & 0 & 2 \end{bmatrix} \end{aligned} \quad (5.34)$$

The computed suboptimal performance index is $\hat{J}^{\oplus} = 11.02$.

case4: $M_Q{}^* = 0$ By similarity, the overlapping decomposition is implemented by the procedure of the foregoing sections. In this case, the complementary sub matrices $M_{12},\ M_{21},\ M_{24}$ and M_{42} are the same as in **case3**. The sub matrix M_{22} can now be selected in the form

$$M_{22} = \begin{bmatrix} 0 & m_{23} \\ 0 & m_{33} \end{bmatrix}$$

so as to satisfy relations (5.28). In this case, the minimization algorithm gives $m_{23} = -0.41,\ m_{33} = 2.01$. In turn, M and $\tilde{A}$ are

$$M = \begin{bmatrix} 0 & 0 & 0.50 & 0 & -0.50 & 0 \\ 0 & 0 & -0.41 & 0 & 0.41 & 1 \\ 0 & 0 & 2.01 & 0 & -2.01 & 0 \\ 0 & 0 & -0.41 & 0 & 0.41 & -1 \\ 0 & 0 & 2.01 & 0 & -2.01 & 0 \\ 0 & 0 & 0 & 0 & 0 & 0 \end{bmatrix},$$

$$\tilde{A} = \begin{bmatrix} -2 & 0 & 1 & 0 & 0 & -2 \\ 0 & 2 & -1.41 & 2 & -0.59 & 0 \\ 0 & -1 & 0.01 & -1 & -4.01 & 0 \\ 0 & 2 & -1.41 & 2 & -0.59 & -2 \\ 0 & -1 & 0.01 & -1 & -4.01 & 0 \\ 1 & 0 & 0 & 0 & 0 & 2 \end{bmatrix} \tag{5.35}$$

The computed suboptimal performance index is $\hat{J}^{\oplus} = 10.48$.

It is of interest to note that the centralized optimal control for the initial system $\mathbf{S}$ in (5.15) is given by $\hat{J}^{\circ} = 9.88$. The sub optimality measure ranges from $16.194\% - 57.275\%$, which is better than those obtained via the usual aggregations or restrictions [15].

5.3 LMI-Based Overlapping Control

In this section we establish an alternative approach for the design of overlapping control, which avoids the main difficulties associated with direct expansion. The new element is to combine system expansion and linear matrix inequalities (LMI) in a way that bypasses the fixed mode problem, while placing only mild restrictions on the Lyapunov function. The material covered in this section follows the work of [352]. To formally introduce the alternative approach to overlapping, let us consider a linear system

$$\mathbf{S}:\ \dot{\mathbf{x}}(t) = \mathbf{A}\mathbf{x}(t) + \mathbf{B}\mathbf{u}(t) \tag{5.36}$$

where $\mathbf{x} \in \Re^n$ is the state , $\mathbf{u} \in \Re^m$ represents the control input and matrices $\mathbf{A}, \mathbf{B}$ have the following structure

$$\mathbf{A} = \begin{bmatrix} \mathbf{A}_{11} & & \mathbf{A}_{12} & \vdots & 0 \\ & \cdots & \cdots & \vdots & \cdots \\ \mathbf{A}_{21} & \vdots & \mathbf{A}_{22} & \vdots & \mathbf{A}_{23} \\ \cdots & \cdots & \cdots & \vdots & \\ 0 & \vdots & \mathbf{A}_{32} & & \mathbf{A}_{33} \end{bmatrix}, \mathbf{B} = \begin{bmatrix} \mathbf{B}_{11} & \vdots & 0 \\ & \vdots & \cdots \\ \mathbf{B}_{21} & \vdots & \mathbf{B}_{22} \\ \cdots & \vdots & \\ 0 & \vdots & \mathbf{B}_{32} \end{bmatrix} \quad (5.37)$$

where the overlapping decomposition is along the dotted lines, which implies that the state variables associated with $\mathbf{A}_{22}$ are available to both inputs. It is then natural to consider an overlapping control of the form $\mathbf{u}(t) = \mathbf{K}\mathbf{x}(t)$ where

$$\mathbf{K} = \begin{bmatrix} \mathbf{K}_{11} & & \mathbf{K}_{12} & \vdots & 0 \\ \cdots & \cdots & \cdots & \cdots & \cdots \\ 0 & \vdots & \mathbf{K}_{22} & & \mathbf{K}_{23} \end{bmatrix} \quad (5.38)$$

As we learned from the foregoing sections, overlapping control is not always related to the matrix structure identified in (5.37). In some relevant cases it can arise simply as a result of control information structure constraints, with no connection to the form of matrix $\mathbf{A}$. To illustrate this point, let us consider system (5.36) with the input matrix of the form

$$\mathbf{B} = \begin{bmatrix} \mathbf{B}_1 & 0 \\ \mathbf{B}_2 & 0 \end{bmatrix} \quad (5.39)$$

We consider further that $\mathbf{x}_1$ and $\mathbf{x}_2$ are the components of $\mathbf{x}_1$ that are available to inputs $\mathbf{u}_1$ and $\mathbf{u}_2$, and that $\mathbf{x}_1$ and $\mathbf{x}_2$ share a certain subset of state variables. It can be easily verified that such a system can always be permuted so that the new matrix $\mathbf{B}$ has the overlapping structure introduced in (5.37).

Remark 5.3.1 *At this stage, it is significant to distinguish between overlapping control laws that are induced by the structure of matrix* $\mathbf{A}$*, and those that are dictated by the availability of state information for control design. This distinction is essentially physical, and has no direct bearing on the mathematical treatment of the problem. From the standpoint of design, a more meaningful classification of overlapping control laws would be one that is based on the structure of matrix* $\mathbf{A}$.

We recall from the *inclusion principle* and the overlapping decomposition that system (5.36) in included in the

$$\widetilde{\mathbf{S}}: \ \dot{\widetilde{\mathbf{x}}}(t) = \widetilde{\mathbf{A}}\widetilde{\mathbf{x}}(t) + \widetilde{\mathbf{B}}\mathbf{u}(t) \tag{5.40}$$

such that

$$\begin{aligned} V\,\mathbf{A} &= \widetilde{\mathbf{A}}\,V, \ V\,\mathbf{B} = \widetilde{\mathbf{B}} \\ V &= \begin{bmatrix} I_1 & 0 & 0 \\ 0 & I_2 & 0 \\ 0 & I_2 & 0 \\ 0 & 0 & I_3 \end{bmatrix} \end{aligned} \tag{5.41}$$

where $I_1, \ I_2$ and I_3 represent identity matrices with dimensions corresponding to $\mathbf{A}_{11}, \ \mathbf{A}_{22}$ and $\mathbf{A}_{33}$, respectively.

Now consider that the expanded system $\widetilde{\mathbf{S}}$ can be stabilized with a decentralized control $\mathbf{u} = \widetilde{\mathbf{K}}_d\widetilde{\mathbf{x}}$, where

$$\widetilde{\mathbf{K}}_d = \begin{bmatrix} \widetilde{\mathbf{K}}_{11} & \widetilde{\mathbf{K}}_{12} & 0 & 0 \\ 0 & 0 & \widetilde{\mathbf{K}}_{23} & \widetilde{\mathbf{K}}_{24} \end{bmatrix} \tag{5.42}$$

It follows that the original system S is stabilized by $\mathbf{u} = \mathbf{K}\mathbf{x}$, where

$$\mathbf{K} = \widetilde{\mathbf{K}}_d\,V = \begin{bmatrix} \widetilde{\mathbf{K}}_{11} & \widetilde{\mathbf{K}}_{12} & 0 \\ 0 & \widetilde{\mathbf{K}}_{23} & \widetilde{\mathbf{K}}_{24} \end{bmatrix} \tag{5.43}$$

represents an overlapping control law.

Remark 5.3.2 *It is easy to realize that the problem of designing overlapping control can thus be formulated as a decentralized control problem in the expanded space. One of the main difficulties that arise in this context results from the fact that the pair* $(\widetilde{\mathbf{A}}, \ \widetilde{\mathbf{B}})$ *is inherently uncontrollable. It can be shown that the eigenvalues of* $\widetilde{\mathbf{A}}_{22}$ *represent fixed modes of the expansion* $\widetilde{\mathbf{S}}$, *so this method is not directly applicable to cases when matrix* $\widetilde{\mathbf{A}}_{22}$ *is unstable [324].*

5.3.1 Design procedure

In the sequel, we show that LMI provides a natural framework for the design of overlapping control. To see this, let us assume that there exists a gain matrix $\mathbf{K}$ that conforms to the structural constraint (5.38) such that the resulting closed-loop system

$$\mathbf{S}_K: \ \dot{\mathbf{x}}(t) = (\mathbf{A} + \mathbf{B}\mathbf{K})\mathbf{x}(t) \tag{5.44}$$

is guaranteed to be stable. By Lyapunov stability, it follows that there exists a matrix $0 < \mathcal{P}^t = \mathcal{P}$ such that

$$\mathbf{A}^t\mathcal{P} + \mathcal{P}\mathbf{A} + \mathcal{P}\mathbf{B}\mathbf{K} + \mathbf{K}^t\mathbf{B}^t\mathcal{P} \; < \; 0 \tag{5.45}$$

with $\mathbf{x}^t\mathcal{P}\mathbf{x}$ being the corresponding Lyapunov function. It is important to recognize that although (5.45) is actually a nonlinear matrix inequality in $\mathcal{P}$ and $\mathbf{K}$, it can be convexified by introducing matrices $\mathcal{Y} = \mathcal{P}^{-1}$ and $\mathbf{L} = \mathbf{K}\mathcal{Y}$. Based thereon, we reformulate (5.45) as an LMI problem in $\mathcal{Y}$ and $\mathbf{L}$:

Compute matrices $\mathcal{Y}$ and $\mathbf{L}$ such that

$$\begin{aligned} &\mathcal{Y} \; > \; 0, \\ &\mathcal{Y}\mathbf{A}^t + \mathbf{A}\mathcal{Y} + \mathbf{B}\mathbf{L} + \mathbf{L}^t\mathbf{B}^t \; < \; 0 \end{aligned} \tag{5.46}$$

Then, determine the gain matrix as $\mathbf{K} = \mathbf{L}\mathcal{Y}^{-1}$.

To ensure that the gain matrix $\mathbf{K}$ has the desired overlapping structure (5.38), it is necessary to look for matrices $\mathcal{Y}$ and $\mathbf{L}$ in the form

$$\mathbf{L} = \begin{bmatrix} \mathbf{L}_{11} & \mathbf{L}_{12} & 0 \\ 0 & \mathbf{L}_{22} & \mathbf{L}_{23} \end{bmatrix}, \; \mathcal{Y} = \begin{bmatrix} \mathcal{Y}_{11} & 0 & 0 \\ 0 & \mathcal{Y}_{22} & 0 \\ 0 & 0 & \mathcal{Y}_{33} \end{bmatrix} \tag{5.47}$$

since evidently this ensures that

$$\mathbf{K} = \begin{bmatrix} \mathbf{L}_{11}\mathcal{Y}_{11}^{-1} & \mathbf{L}_{12}\mathcal{Y}_{22}^{-1} & 0 \\ 0 & \mathbf{L}_{22}\mathcal{Y}_{22}^{-1} & \mathbf{L}_{23}\mathcal{Y}_{33}^{-1} \end{bmatrix} \tag{5.48}$$

The LMI design strategy is now lust rated by the following example

5.3.2 Illustrative example 5.3

Consider the system of the type (5.36) with

$$\mathbf{A} \; = \; \begin{bmatrix} 1 & & 4 & \vdots & 0 \\ \cdots & \cdots & \cdots & \cdots & \cdots \\ 1 & \vdots & 2 & \vdots & 2 \\ \cdots & \cdots & \cdots & \cdots & \\ 0 & \vdots & -2 & & 3 \end{bmatrix}, \; \mathbf{B} = \begin{bmatrix} 1 & \vdots & 0 \\ & \vdots & \cdots \\ 1 & \vdots & 0 \\ \cdots & \vdots & \\ 0 & \vdots & 1 \end{bmatrix}$$

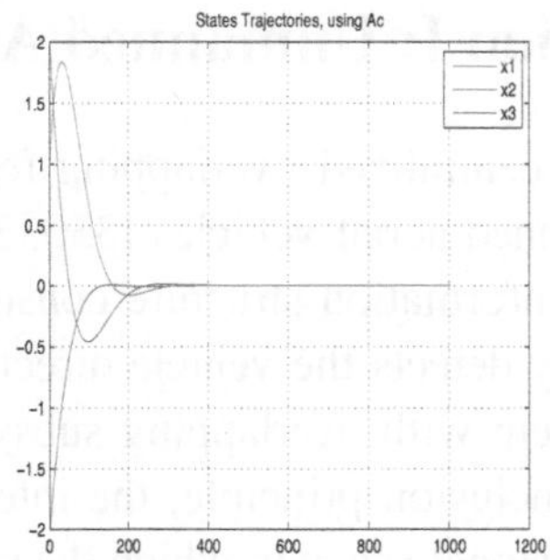

Figure 5.1: Closed-loop state trajectories

The overlapping decomposition is determined by dotted lines and note that the open-loop system is unstable with eigenvalues $\{-0.166,\ 3.08 \pm j1.59\}$. our objective will be to stabilize it using overlapping control. Using

$$V = \begin{bmatrix} 1 & 0 & 0 \\ 0 & 1 & 0 \\ 0 & 1 & 0 \\ 0 & 0 & 1 \end{bmatrix}$$

we get the expanded system $\widetilde{\mathbf{S}}$ as

$$\widetilde{\mathbf{A}} = \begin{bmatrix} 1 & 4 & \vdots & 0 & 0 \\ 1 & 2 & \vdots & 0 & 2 \\ \cdots & \cdots & \cdots & \cdots & \cdots \\ 1 & 0 & \vdots & 2 & 2 \\ 0 & 0 & \vdots & -2 & 3 \end{bmatrix}, \ \widetilde{\mathbf{B}} = \begin{bmatrix} 1 & \vdots & 0 \\ 1 & \vdots & 0 \\ \cdots & \cdots & \cdots \\ 1 & \vdots & 0 \\ 0 & \vdots & 1 \end{bmatrix}$$

which has an unstable fixed mode $A_{22} = 2$, so direct decentralized control design in the expanded space would fail. However, the LMI optimization (5.46) easily resolves this problem, yielding

$$\mathbf{K} = \begin{bmatrix} -1.2900 & -0.9344 & \vdots & 0 \\ 0 & \vdots & 0.4819 & -1.7227 \end{bmatrix}$$

and the corresponding closed-loop state trajectories are displayed in Fig. 5.1.

5.4 Application I: Unmanned Aerial Vehicles

In what follows, decentralized overlapping feedback laws are designed for a formation of unmanned aerial vehicles [350, 352]. The dynamic model of the formation with an information structure constraint in which each vehicle, except the leader, only detects the vehicle directly in front of it, is treated as an interconnected system with overlapping subsystems. Using the mathematical framework of the inclusion principle, the interconnected system is expanded into a higher dimensional space in which the subsystems appear to be disjoint. Then, at each subsystem, a static state feedback controller is designed to robustly stabilize the perturbed nominal dynamics of the subsystem.

5.4.1 A Kinematic model

Let us start with the following planar kinematic model for a single vehicle:

$$\begin{aligned} \dot{X} &= v \cos \psi, \\ \dot{Y} &= v \sin \psi, \\ \dot{\psi} &= \omega \end{aligned} \tag{5.49}$$

where X and Y denote rectangular coordinates, and ψ is the heading angle in the (X, Y) plane. The speed in the longitudinal direction (in body axes) v and angular turn rate ω are assumed to be the control inputs. It can easily be shown that the decoupling matrix of the input-state feedback linearization for the kinematic model (5.49) is singular. In order to deal with this problem, dynamic extension [308] is employed by considering speed v as a new state variable, and acceleration a as a new input variable. Now, the state and input variables are defined as

$$\zeta = \begin{bmatrix} \zeta_1 \\ \zeta_2 \\ \zeta_3 \\ \zeta_4 \end{bmatrix} = \begin{bmatrix} X \\ Y \\ \psi \\ v \end{bmatrix}, \quad \eta = \begin{bmatrix} \eta_1 \\ \eta_2 \end{bmatrix} = \begin{bmatrix} a \\ \omega \end{bmatrix} \tag{5.50}$$

Using (5.50) the kinematic model (5.49) can be rewritten as follows:

$$\begin{aligned} \dot{\zeta} &= f(\zeta)\ g(\zeta)\eta, \\ f(\zeta) &= \begin{bmatrix} \zeta_4 \cos(\zeta_3) \\ \zeta_4 \sin(\zeta_3) \\ 0 \\ 0 \end{bmatrix}, \quad g(\zeta) = \begin{bmatrix} 0 & 0 \\ 0 & 0 \\ 0 & 1 \\ 1 & 0 \end{bmatrix} \end{aligned} \tag{5.51}$$

At this point to facilitate feedback linearization, a change of state variables is introduced as

$$\mathbf{z} = T(\zeta) \longrightarrow \begin{bmatrix} \mathbf{z}_1 \\ \mathbf{z}_2 \\ \mathbf{z}_3 \\ \mathbf{z}_4 \end{bmatrix} = \begin{bmatrix} \zeta_1 \\ \zeta_2 \\ \zeta_4 \cos(\zeta_3) \\ \zeta_4 \sin(\zeta_3) \end{bmatrix} \tag{5.52}$$

and change of input variables, to define the new input $\mathbf{u} \in \Re^2$, $\eta = M(\zeta)u$ such that

$$M(\zeta) = \begin{bmatrix} \cos(\zeta_3) & \sin(\zeta_3) \\ -\sin(\zeta_3)/\zeta_4 & \cos(\zeta_3)/\zeta_4 \end{bmatrix} \tag{5.53}$$

It is not difficult to show that the transformations of (5.52) and (5.53) imply the following exact linearization of the nonlinear model (5.51)

$$\begin{aligned}
\dot{\mathbf{z}} &= \frac{\partial T}{\partial \zeta}\zeta \Rightarrow \\
\dot{\mathbf{z}} &= \begin{bmatrix} 0 & 0 & 1 & 0 \\ 0 & 0 & 0 & 1 \\ 0 & 0 & 0 & 0 \\ 0 & 0 & 0 & 0 \end{bmatrix} \mathbf{z} + \begin{bmatrix} 0 & 0 \\ 0 & 0 \\ 1 & 0 \\ 0 & 1 \end{bmatrix} \mathbf{z} \\
&= \mathbf{E}\,\mathbf{z} + \mathbf{F}\,\mathbf{z} \qquad (5.54) \\
\dot{\mathbf{z}} &= \begin{bmatrix} 0 & I_2 \\ 0 & 0 \end{bmatrix} \mathbf{z} + \begin{bmatrix} 0 \\ I_2 \end{bmatrix} \mathbf{z} \qquad (5.55)
\end{aligned}$$

with $z \in \Re^4$ and $u \in \Re^2$ being the state and input to the system.

5.4.2 A Leader-follower formation model

Our goal now is to develop a dynamic model for a leader-follower type of formation of vehicles. Let us introduce the following decomposition for the state variables of the *j*th vehicle in the formation of *q* vehicles,

$$\mathbf{z}_j = \begin{bmatrix} \mathbf{z}_j^A \\ \mathbf{z}_j^B \end{bmatrix} \in \Re^4, \ \mathbf{z}_j^A = \begin{bmatrix} \mathbf{z}_{j1} \\ \mathbf{z}_{j2} \end{bmatrix} \in \Re^2, \ \mathbf{z}_j^B = \begin{bmatrix} \mathbf{z}_{j3} \\ \mathbf{z}_{j4} \end{bmatrix} \in \Re^2 \tag{5.56}$$

We note that

$$\mathbf{z}_j^A \begin{bmatrix} X_j \\ Y_j \end{bmatrix}, \ \mathbf{z}_j^B = \begin{bmatrix} v_j \cos\psi_j \\ v_j \sin\psi_j \end{bmatrix}, \ j \in \{1, ..., q\} \tag{5.57}$$

We observe that $\mathbf{z}_j$ is split conveniently into two subvectors, where z_j^A includes position coordinates and z_j^B includes speed coordinates of the *j*th vehicle. The benefit gained by this decomposition to help in controlling the vehicles in a formation by controlling variables that represent distances between vehicles (not positions of the vehicles), and variables that represent speed coordinates for each independent vehicle. The control input for the *j*th vehicle as defined by (5.53), will be denoted as $u_j \in \Re^2$.

In the sequel, we consider a platoon of *r* vehicles and introduce the change of variables $e_1^B = z_1^B - v_{d1}$ for the leading vehicle and

$$\begin{aligned} e_j^A &= z_{j-1}^A - z_j^A - d_{j-1} \\ e_j^B &= z_j^B - v_{dj} \qquad j \in \{2, ..., r\} \end{aligned} \tag{5.58}$$

where $d_{j-1} \in \Re^2$ is a constant desired Euclidean distance between the $(j-1)st$ and *j*th vehicles, $j \in \{2, ..., r\}$, and $v_{dj} \in \Re^2$, represents the desired speed for the *j*th vehicle, $j \in \{1, ..., r\}$.. We note that for controlling distances between vehicles, position of the leading vehicle, that is, z_A^1 is not needed. Since the desired Euclidean distances between vehicles are assumed to be constant, the following assumption is necessary:

$$v_{dj} = v_d, \quad j \in \{1, ..., r\} \tag{5.59}$$

This means in order to achieve constant desired spacing in the formation, the desired speed for each vehicle must be the same. Then $\dot{e}_1^B = u_1$ for the leading vehicle and

$$\begin{aligned} \dot{e}_j^A &= e_{j-1}^B - e_j^B \\ \dot{e}_j^B &= u_j \quad j \in \{2, ..., r\} \end{aligned} \tag{5.60}$$

Notice that the goal is for the whole platoon, that is formation, to fly at constant desired speed v_d with desired spacing between vehicles, uniquely determined by desired Euclidean distances between successive vehicles equal to $d_j, j \in 1, ..., r-1$, which is accomplished if the system described by Eq.(5.60) is stable with respect to its zero equilibrium.

Due to symmetry, the procedure does not depend on the size of the platoon and therefore we take $r = 3$ and rewrite (5.60)in the compact form:

$$\begin{aligned} \begin{bmatrix} \dot{e}_1^B \\ \dot{e}_2^A \\ \dot{e}_2^B \\ \dot{e}_3^A \\ \dot{e}_3^B \end{bmatrix} &= \begin{bmatrix} 0 & 0 & 0 & 0 & 0 \\ I & 0 & -I & 0 & 0 \\ 0 & 0 & 0 & 0 & 0 \\ 0 & 0 & I & 0 & -I \\ 0 & 0 & 0 & 0 & 0 \end{bmatrix} \begin{bmatrix} e_1^{II} \\ e_2^{I} \\ e_2^{II} \\ e_3^{I} \\ e_3^{II} \end{bmatrix} + \begin{bmatrix} I & 0 & 0 \\ 0 & 0 & 0 \\ 0 & I & 0 \\ 0 & 0 & 0 \\ 0 & 0 & I \end{bmatrix} \begin{bmatrix} u_1 \\ u_2 \\ u_3 \end{bmatrix} \\ \dot{\mathbf{e}} &= \mathbf{Ae} + \mathbf{Bu} \end{aligned} \tag{5.61}$$

It is important to observe that system (5.60) (or 5.61) can be viewed as an interconnected system with subsystems having state variables defined

$$e_1 = e_1^B, \;\; e_j = \begin{bmatrix} e_j^A \\ e_j^B \end{bmatrix} \quad \forall\, j \in \{1, ..., r\}. \tag{5.62}$$

Applying the inclusion principle, it turns out that the expansion/contraction matrices for the state are given as

$$V = \begin{bmatrix} I & 0 & 0 & 0 & 0 \\ I & 0 & 0 & 0 & 0 \\ 0 & I & 0 & 0 & 0 \\ 0 & 0 & I & 0 & 0 \\ 0 & 0 & I & 0 & 0 \\ 0 & 0 & 0 & I & 0 \\ 0 & 0 & 0 & 0 & I \end{bmatrix}, \; U = \begin{bmatrix} \frac{1}{2}I & \frac{1}{2}I & 0 & 0 & 0 & 0 & 0 \\ 0 & 0 & I & 0 & 0 & 0 & 0 \\ 0 & 0 & 0 & \frac{1}{2}I & \frac{1}{2}I & 0 & 0 \\ 0 & 0 & 0 & 0 & 0 & I & 0 \\ 0 & 0 & 0 & 0 & 0 & 0 & I \end{bmatrix}$$

$$R = \begin{bmatrix} I & 0 & 0 \\ I & 0 & 0 \\ 0 & I & 0 \\ 0 & I & 0 \\ 0 & 0 & I \end{bmatrix}, \; Q = \begin{bmatrix} \frac{1}{2}I & \frac{1}{2}I & 0 & 0 & 0 \\ 0 & 0 & \frac{1}{2}I & \frac{1}{2}I & 0 \\ 0 & 0 & 0 & 0 & I \end{bmatrix}$$

and the stabilizing feedback gain matrix has the structure in the original space

$$\tilde{K}_M = \begin{bmatrix} \tilde{K}_1 & 0 & 0 & 0 & 0 \\ \tilde{K}_2 & \tilde{K}_3 & \tilde{K}_1 & 0 & 0 \\ 0 & 0 & \tilde{K}_2 & \tilde{K}_3 & \tilde{K}_1 \end{bmatrix} \tag{5.63}$$

Let us again consider the group of five vehicles flying in the formation as depicted in 5.2. Assume that the nominal speed v_d 5.59 is $[300, 0]$[ft/s], desired distances between vehicles (in absolute values) are all equal to $|d_{j-1}| \equiv d = [400, 400]^t$[ft] for both platoons and all *j* in 5.58, and consider external perturbations to be of sinusoidal functions with magnitudes equal to 10.

The design procedure is now applied to compute decentralized overlapping static feedback controllers. It was reported in [352] that typical simulation results for two different sets of initial conditions using superimposed snapshots of the formation at representative time instances, which are 40 nonuniform (depending on the nonuniform step size used in simulations) time intervals between $0 - 1.3$[s] are reproduced in Figs. 5.3 and 5.4. Position coordinates are given in feet. Horizontal distances between vehicles V_1 and V_2, and V_2 and V_3, corresponding to Fig. 5.2, for the set of initial conditions for the simulation presented in Fig. 5.3, are given in Figs. 5.5a and 5.5b, respectively. Time is in seconds and distances are in feet.

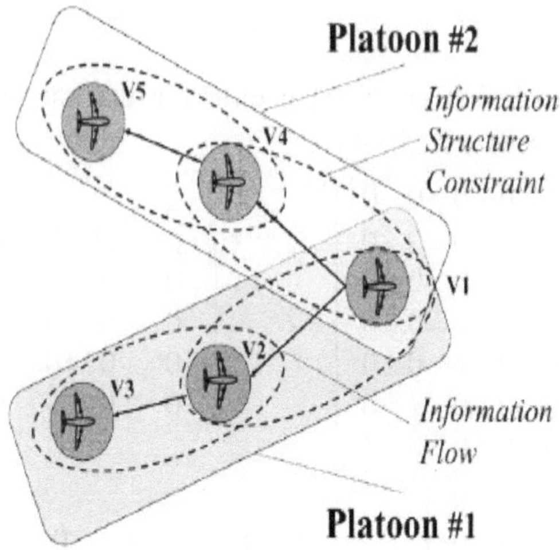

Figure 5.2: Leader-follower type formation with five vehicles and two platoons.

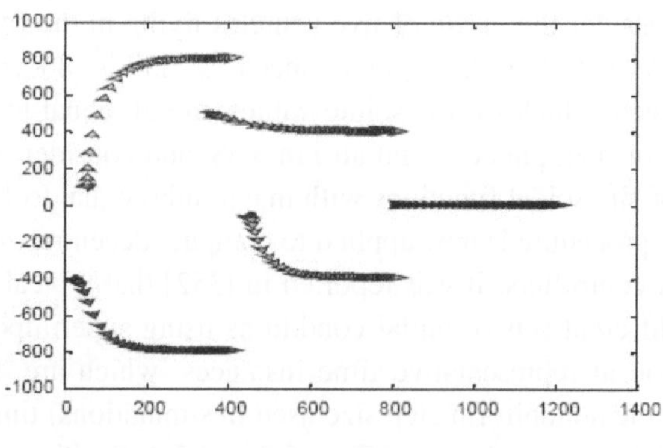

Figure 5.3: One possible formation of unmanned aerial vehicles

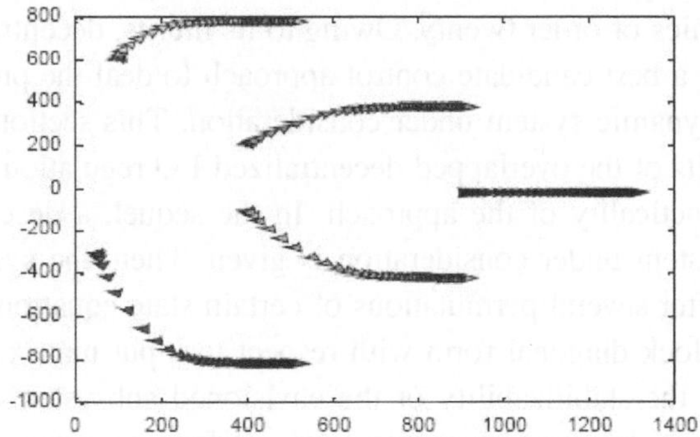

Figure 5.4: Another possible formation of unmanned aerial vehicles

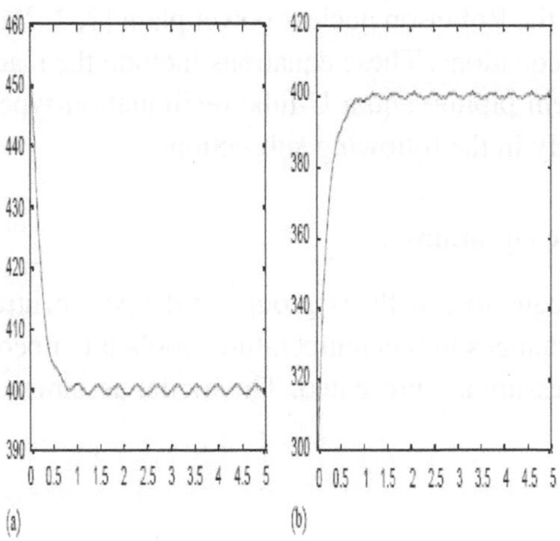

Figure 5.5: Horizontal distances between: (a) vehicles V_1 and V_2; (b) vehicles V_2 and V_3

5.5 Application II: Robinson Nuclear Power Plant

Robinson nuclear power plant considered in this report is a large sale system with dynamics of order twenty. Owing to its merits, decentralized control scheme appears a best candidate control approach to deal the problem of regulation of the dynamic system under consideration. This section provides the simulation results of the overlapped decentralized LQ regulation as a manifestation of the practicality of the approach. In the sequel, a description of the dynamics of system under consideration is given. Then, the system matrices are presented after several permutations of certain state equations to bring the system into a block diagonal form with respect to input matrix, as well as to assure, at least, the stabilizability of the envisioned subsystems. A complete process of system decomposition into four overlapping subsystems is given along with listing the important theorems of inclusion theory. Following this, Individual overlapped subsystems are then described with necessary transformation matrices. This completes the analysis stage. In the design stage, the LQR-based overlapping decentralized state feedback controller is constructed so as to guarantee an index of suboptimality mentioned. Simulation results obtained for open loop system, closed loop system with centrallized controller and closed loop system with overlappind decentralized controller are finally given.

The model for the Robinson nuclear power plant [**?**, **?**, **?**] consists of a set of linear differential equations. These equations include the reactor core, pressurizer, primary system piping, and a U-tube recirculation type steam generator, described separately in the following subsections.

Core point kinetic equations

The point kinetic equations with six groups of delayed neutrons and reactivity feedbacks due to changes in fuel temperature, coolant temperature and primary coolant system pressure are presented. The model assumes one fuel node and two coolant nodes.

$$
\begin{aligned}
\frac{d\delta P}{dt} &= -400\delta P + 0.0125\delta C_1 + 0.0305\delta C_2 + 0.111\delta C_3 \\
&+ 0.301\delta C_4 + 1.140\delta C_5 + 3.01\delta C_6 - 1781\delta T_f \\
&- 13700\delta T_{C1} - 13700\delta T_{C2} + 411\delta P_P + 10^6\delta\rho_{Rod} \\
\frac{d\delta C_1}{dt} &= 13.125\delta P - 0.0125\delta C_1 \\
\frac{d\delta C_2}{dt} &= 87.5\delta P - 0.0305\delta C_2 \\
\frac{d\delta C_3}{dt} &= 78.125\delta P - 0.111\delta C_3 \\
\frac{d\delta C_4}{dt} &= 158.125\delta P - 0.301\delta C_4 \\
\frac{d\delta C_5}{dt} &= 46.25\delta P - 1.140\delta C_5 \\
\frac{d\delta C_6}{dt} &= 16.875\delta P - 3.01\delta C_6 \\
\frac{d\delta T_f}{dt} &= 0.0756\delta P - 0.16466\delta T_f + 0.16466\delta T_{C1} \\
\frac{d\delta T_{C1}}{dt} &= 0.05707\delta T_f + 2.3832\delta T_{LP} - 2.4403\delta T_{C1} \\
\frac{d\delta T_{C2}}{dt} &= 0.05707\delta T_f + 2.3262\delta T_{C1} - 2.3832\delta T_{C2} \qquad (5.64)
\end{aligned}
$$

Pressurizer equations

The pressurizer model is based on mass, energy and volume balances with the assumption that saturation conditions always apply for steam-water mixture in the pressurizer.

$$
\begin{aligned}
\frac{d\delta P_P}{dt} &= 0.0207\delta T_f - 0.0207\delta T_{C1} + 0.0103\delta T_{C2} + 0.240\delta T_{UP} \\
&- 0.130\delta T_{IP} - 0.509\delta T_P + 0.634\delta T_m - 0.116\delta T_{OP} \\
&+ 0.121\delta T_{LP} - 0.279\delta T_{HL} + 0.0235\delta T_{CL} - 0.0062\delta Q
\end{aligned}
$$

Steam generator equations

The steam generator model with three regions: primary fluid, tube metal and secondary fluid.

$$
\begin{aligned}
\frac{d\delta T_P}{dt} &= 0.2238\delta T_{IP} - 0.76642\delta T_P + 0.53819\delta T_m \\
\frac{d\delta T_m}{dt} &= 3.07017\delta T_P - 5.3657\delta T_m + 0.33272\delta P_s \\
\frac{d\delta P_s}{dt} &= 1.349\delta T_m - 0.2034\delta P_s - 0.03843\delta W_{FW} \qquad (5.65)
\end{aligned}
$$

Piping equations

All piping sections are modeled as well mixed volumes.

$$
\begin{aligned}
\frac{d\delta T_{UP}}{dt} &= 0.33645\delta T_{C2} - 0.33645\delta T_{UP} \\
\frac{d\delta T_{HL}}{dt} &= 2.5\delta T_{UP} - 2.5\delta T_{HL} + 0.0016\delta W_P \\
\frac{d\delta T_{IP}}{dt} &= 1.45\delta T_{HL} - 1.45\delta T_{IP} \\
\frac{d\delta T_{OP}}{dt} &= 1.45\delta T_P - 1.45\delta T_{OP} \\
\frac{d\delta T_{CL}}{dt} &= 1.48\delta T_{OP} - 1.48\delta T_{CL} \\
\frac{d\delta T_{LP}}{dt} &= 0.516\delta T_{CL} - 0.516\delta T_{LP} \qquad (5.66)
\end{aligned}
$$

where δP is the deviation in reactor power from initial steady-state value, δC_i is the deviation of normalized precursor concentrations $i = 1, ..., 6$, δT_f is the deviation of fuel temperature in fuel node, δT_{C1} is the deviation of temperature in the first coolant node, δT_{C2} is the deviation of temperature in the second coolant node, δP_P is the deviation of primary system pressure, δT_P is the deviation of temperature of primary coolant node in the steam generator, δT_m is the deviation of steam generator tube metal temperature, δP_s is the deviation of steam pressure from its initial steady-state value, δT_{UP} is the deviation of the reactor upper plenum temperature, δT_{LP} is the deviation of the reactor lower plenum temperature, δT_{HL} is the deviation of hot leg temperature, δT_{IP} is the deviation of temperature of primary coolant in the steam generator or inlet plenum, δT_{OP} is the deviation of temperature of primary coolant in the steam generator or outlet plenum, δT_{CL} is the deviation of cold leg temperature, $\delta \rho_{rod}$ is the reactivity due to control rod movement, δQ is the rate of heat addition to the pressurizer fluid with electric heater, δW_{FW} is the deviation of feedwater flow rate in steam generator and δW_P is the deviation of primary water flow rate to the steam generator.

It must be noted that (5.64)-(5.66) describe the H. B. Robinson nuclear power plant. A state-space model is given by:

$$\dot{x} = Ax + Bu \tag{5.67}$$

where

$$\begin{aligned}
x &= [x_a \; x_b \; x_c] && (5.68)\\
x_a &= [\delta P \; \delta C_1 \; \delta C_2 \; \delta C_3 \; \delta C_4 \; \delta C_5 \; \delta C_6] && (5.69)\\
x_b &= [\delta T_f \; \delta T_{C1} \; \delta T_{C2} \; \delta P_P \; \delta T_m \; \delta T_p \; \delta P_s] && (5.70)\\
x_c &= [\delta T_{UP} \; \delta T_{HL} \; \delta T_{IP} \; \delta T_{OP} \; \delta T_{CL} \; \delta T_{LP}] && (5.71)\\
u &= [\delta \rho_{rod} \; \delta W_{FW} \; \delta W_P \; \delta Q] && (5.72)
\end{aligned}$$

Being of high dimensions, the system matrices A and B are given in partitioned forms as follows:

$$A = \begin{bmatrix} A_{11} & A_{12} & A_{13} \\ A_{21} & A_{22} & A_{23} \\ A_{31} & A_{32} & A_{33} \end{bmatrix} \tag{5.73}$$

where

$$A_{11} = \begin{bmatrix} -400 & 0.0125 & 0.0305 & 0.111 & 0.301 & 1.14 & 3.01 \\ 13.125 & -0.0125 & 0 & 0 & 0 & 0 & 0 \\ 87.5 & 0 & -0.0305 & 0 & 0 & 0 & 0 \\ 78.125 & 0 & 0 & -0.111 & 0 & 0 & 0 \\ 158.125 & 0 & 0 & 0 & -0.301 & 0 & 0 \\ 46.25 & 0 & 0 & 0 & 0 & -1.14 & 0 \\ 16.875 & 0 & 0 & 0 & 0 & 0 & -3.01 \end{bmatrix}$$

$$A_{12} = \begin{bmatrix} -1781 & -13700 & -13700 & 0.111 & 411 & 0 & 0 \\ 0 & 0 & 0 & 0 & 0 & 0 & 0 \\ 0 & 0 & 0 & 0 & 0 & 0 & 0 \\ 0 & 0 & 0 & 0 & 0 & 0 & 0 \\ 0 & 0 & 0 & 0 & 0 & 0 & 0 \\ 0 & 0 & 0 & 0 & 0 & 0 & 0 \\ 0 & 0 & 0 & 0 & 0 & 0 & 0 \end{bmatrix}$$

$$A_{13} = [0]_{6\times 7}, \quad A_{21} = \begin{bmatrix} 0.0756 & 0 & 0 & 0 & 0 & 0 & 0 \\ 0 & 0 & 0 & 0 & 0 & 0 & 0 \\ 0 & 0 & 0 & 0 & 0 & 0 & 0 \\ 0 & 0 & 0 & 0 & 0 & 0 & 0 \\ 0 & 0 & 0 & 0 & 0 & 0 & 0 \\ 0 & 0 & 0 & 0 & 0 & 0 & 0 \\ 0 & 0 & 0 & 0 & 0 & 0 & 0 \end{bmatrix}$$

$$A_{22} = \begin{bmatrix} -0.16466 & 0.16466 & 0 & 0 & 0 & 0 & 0 \\ 0.05707 & -24403 & 0 & 0 & 0 & 0 & 0 \\ 0.05707 & 23262 & -23832 & 0 & 0 & 0 & 0 \\ 0.0207 & -0.0207 & 0.0103 & 0 & 0.634 & -0.509 & 0 \\ 0 & 0 & 0 & 0 & -53657 & 307017 & 0.3372 \\ 0 & 0 & 0 & 0 & 0.53819 & -0.76642 & 0 \\ 0 & 0 & 0 & 0 & 1.349 & 0 & -0.2034 \end{bmatrix}$$

$$A_{23} = \begin{bmatrix} 0 & 0 & 0 & 0 & 0 & 0 \\ 0 & 0 & 0 & 0 & 0 & 0 \\ 0 & 0 & 0 & 0 & 0 & 2.3832 \\ 0.240 & -0.279 & -0.130 & -0.116 & 0.0235 & 0.121 \\ 0 & 0 & 0.2238 & 0 & 0 & 0 \\ 0 & 0 & 0 & 0 & 0 & 0 \\ 0 & 0 & 0 & 0 & 0 & 0 \end{bmatrix}$$

$$A_{31} = [0]_{6\times 7}, \quad A_{32} = \begin{bmatrix} 0 & 0 & 0.33645 & 0 & 0 & 0 & 0 \\ 0 & 0 & 0 & 0 & 0 & 0 & 0 \\ 0 & 0 & 0 & 0 & 0 & 0 & 0 \\ 0 & 0 & 0 & 0 & 0 & 0 & 0 \\ 0 & 0 & 0 & 0 & 0 & 1.45 & 0 \\ 0 & 0 & 0 & 0 & 0 & 0 & 0 \end{bmatrix}$$

$$A_{33} = \begin{bmatrix} -0.33645 & 0 & 0 & 0 & 0 & 0 \\ 2.5 & -2.5 & 0 & 0 & 0 & 0 \\ 0 & 1.45 & -1.45 & 0 & 0 & 0 \\ 0 & 0 & 0 & -1.45 & 0 & 0 \\ 0 & 0 & 0 & 1.48 & -1.48 & 0 \\ 0 & 0 & 0 & 0 & 0.516 & -1.516 \end{bmatrix}$$

$$B = \begin{bmatrix} B_1 & B_2 \end{bmatrix}^t$$

$$B_1 = \begin{bmatrix} 10^6 & 0 & 0 & 0 & 0 & 0 & 0 & 0 & 0 & 0 \\ 0 & 0 & 0 & 0 & 0 & 0 & 0 & 0 & 0 & 0 \\ 0 & 0 & 0 & 0 & 0 & 0 & 0 & 0 & 0 & 0 \\ 0 & 0 & 0 & 0 & 0 & 0 & 0 & 0 & 0 & 0 \end{bmatrix}^t$$

$$B_2 = \begin{bmatrix} 0 & 0 & 0 & 0 & 0 & 0 & 0 & 0 & 0 & 0 \\ 0 & 0 & 0 & -0.03843 & 0 & 0 & 0 & 0 & 0 & 0 \\ 0 & 0 & 0 & 0 & 0 & 0 & 0.0016 & 0 & 0 & 0 \\ -0.0062 & 0 & 0 & 0 & 0 & 0 & 0 & 0 & 0 & 0 \end{bmatrix}^t$$

5.5.1 System permutations

In order to successfully apply the decentralized control methodology, the input matrix B of a certain system should be in block-diagonal form. The system can then be decomposed into multiple subsystems with orders equal to the rows of the corresponding block of the input matrix B. It is obvious that in our case matrix B is not in the diagonal form. There are four inputs int system and only four non-zero elements in the input matrix B all appearing in different columns. As such, the B matrix can be transformed to block-diagonal form by a set of permutations. Although a transformation matrix can be found after all permutations, it contains sixteen all zero rows for apparent reasons, which therefore cannot be used as a transformation matrix for the whole system. After performing the permutations, the system described by (5.67) can be described as

$$\dot{\bar{x}} = \bar{A}\bar{x} + \bar{B}u \tag{5.74}$$

Since the columns of input matrix are not shuffled, the input vector remains unchanged, Whereas the re-arranged system state vector $\bar{x}$ and resultant matrices $\bar{A},\ \bar{B}$ after aforementioned permutations are mentioned in the followings.

$$\begin{aligned} \bar{x} &= [\bar{x}_1\ \bar{x}_2\ \bar{x}_3\ \bar{x}_4]^t \\ \bar{x}_1 &= [\delta P\ \delta C_1\ \delta C_2\ \delta C_3\ \delta C_4\ ] \\ \bar{x}_2 &= [\delta P_s\ \delta C_6\ \delta T_f\ \delta T_{C1}\ \delta T_{C2}] \\ \bar{x}_3 &= [\delta T_{HL}\ \delta P_P\ \delta T_p\ \delta C_5\ \delta T_f\ ] \\ \bar{x}_4 &= [\delta C_4\ \delta T_{IP}\ \delta T_{OP}\ \delta T_{CL}\ \delta T_{LP}] \end{aligned}$$

$$\bar{A} = \begin{bmatrix} \bar{A}_{11} & \bar{A}_{12} & \bar{A}_{13} & \bar{A}_{14} \\ \bar{A}_{21} & \bar{A}_{22} & \bar{A}_{23} & \bar{A}_{24} \\ \bar{A}_{31} & \bar{A}_{32} & \bar{A}_{33} & \bar{A}_{34} \\ \bar{A}_{41} & \bar{A}_{42} & \bar{A}_{43} & \bar{A}_{44} \end{bmatrix} \tag{5.75}$$

$$\bar{A}_{11} = \begin{bmatrix} -400.0000 & 0.0125 & 0.0305 & 0.1110 & 0.3010 \\ 13.1250 & -0.0125 & 0 & 0 & 0 \\ 87.5000 & 0 & -0.0305 & 0 & 0 \\ 78.1250 & 0 & 0 & -0.1110 & 0 \\ 158.1250 & 0 & 0 & 0 & -0.3010 \end{bmatrix}$$

$$\bar{A}_{12} = \begin{bmatrix} 0 & 30 & -1781 & -13700 & -13700 \\ 0 & 0 & 0 & 0 & 0 \\ 0 & 0 & 0 & 0 & 0 \\ 0 & 0 & 0 & 0 & 0 \\ 0 & 0 & 0 & 0 & 0 \end{bmatrix}$$

$$\bar{A}_{13} = \begin{bmatrix} 0 & 0 & 0 & 1.140 & 0 \\ 0 & 0 & 0 & 0 & 0 \\ 0 & 0 & 0 & 0 & 0 \\ 0 & 0 & 0 & 0 & 0 \\ 0 & 0 & 0 & 0 & 0 \end{bmatrix}, \ \bar{A}_{21} = \begin{bmatrix} 0 & 0 & 0 & 0 & 0 \\ 0 & 0 & 0 & 0 & 0 \\ 16.8750 & 0 & 0 & 0 & 0 \\ 0.0756 & 0 & 0 & 0 & 0 \\ 0 & 0 & 0 & 0 & 0 \end{bmatrix}$$

$$\bar{A}_{22} = \begin{bmatrix} -0.2034 & 0 & 0 & 0 & 0 \\ 0 & -3.0100 & 0 & 0 & 0 \\ 0 & 0 & -0.1647 & 0.1647 & 0 \\ 0 & 0 & 0.0571 & -2.4403 & 0 \\ 0 & 0 & 0.0571 & 2.3262 & -2.3832 \end{bmatrix}$$

$$\bar{A}_{23} = \begin{bmatrix} 0 & 1.3490 & 0 & 0 & 0 \\ 0 & 0 & 0 & 0 & 0 \\ 0 & 0 & 0 & 0 & 0 \\ 0 & 0 & 0 & 0 & 0 \\ 0 & 0 & 0 & 0 & 0 \end{bmatrix}, \ \bar{A}_{24} = \begin{bmatrix} 0 & 0 & 0 & 0 & 0 \\ 0 & 0 & 0 & 0 & 0 \\ 0 & 0 & 0 & 0 & 0 \\ 0 & 0 & 0 & 0 & 2.3832 \\ 0 & 0 & 0 & 0 & 0 \end{bmatrix}$$

$$\bar{A}_{31} = \begin{bmatrix} 0 & 0 & 0 & 0 & 0 \\ 0 & 0 & 0 & 0 & 0 \\ 0 & 0 & 0 & 0 & 0 \\ 46.25 & 0 & 0 & 0 & 0 \\ 0 & 0 & 0 & 0 & 0 \end{bmatrix}, \ \bar{A}_{32} = \begin{bmatrix} 0 & 0 & 0 & 0 & 0 \\ 0.3372 & 0 & 0 & 0 & 0 \\ 0 & 0 & 0 & 0 & 0 \\ 0 & 0 & 0 & 0 & 0 \\ 0 & 0 & 0 & 0 & 0.3365 \end{bmatrix}$$

$$\bar{A}_{33} = \begin{bmatrix} -2.5000 & 0 & 0 & 0 & 2.5000 \\ 0 & -5.3657 & 3.0702 & 0 & 0 \\ 0 & 0.5382 & -0.7664 & 0 & 0 \\ 0 & 0 & 0 & -1.1400 & 0 \\ 0 & 0 & 0 & 0 & -0.3365 \end{bmatrix}$$

$$\bar{A}_{34} = \begin{bmatrix} 0 & 0 & 0 & 0 & 0 \\ 0 & 0.2238 & 0 & 0 & 0 \\ 0 & 0 & 0 & 0 & 0 \\ 0 & 0 & 0 & 0 & 0 \\ 0 & 0 & 0 & 0 & 0 \end{bmatrix}$$

$$\bar{A}_{41} = [0]_{5\times 5}$$

$$\bar{A}_{42} = \begin{bmatrix} 0 & 0.027 & 0 & -0.0207 & 00103 \\ 0 & 0 & 0 & 0 & 0 \\ 0 & 0 & 0 & 0 & 0 \\ 0 & 0 & 0 & 0 & 0 \\ 0 & 0 & 0 & 0 & 0 \end{bmatrix}$$

$$\bar{A}_{43} = \begin{bmatrix} -0.2790 & 0.6340 & -0.5090 & 0 & 0.2400 \\ 1.4500 & 0 & 0 & 0 & 0 \\ 0 & 1.4500 & 0 & 0 & 0 \\ 0 & 0 & 0 & 0 & 0 \\ 0 & 0 & 0 & 0 & 0 \end{bmatrix}$$

$$\bar{A}_{44} = \begin{bmatrix} 0 & -0.1300 & -0.1160 & 0.0235 & 0.1210 \\ 0 & -1.4500 & 0 & 0 & 0 \\ 0 & 0 & -1.4500 & 0 & 0 \\ 0 & 0 & 1.4800 & -1.4800 & 0 \\ 0 & 0 & 0 & 0.5160 & -0.5160 \end{bmatrix}$$

$$\bar{B} = diag(\bar{B}_1, \bar{B}_2, \bar{B}_3, \bar{B}_4) \tag{5.76}$$

where

$$\bar{B}_1 = \begin{bmatrix} 10^6 \\ 0 \\ 0 \\ 0 \\ 0 \end{bmatrix}, \bar{B}_2 = \begin{bmatrix} -0.03843 \\ 0 \\ 0 \\ 0 \\ 0 \end{bmatrix},$$

$$\bar{B}_3 = \begin{bmatrix} 0.0016 \\ 0 \\ 0 \\ 0 \\ 0 \end{bmatrix}, \bar{B}_4 = \begin{bmatrix} -0.00062 \\ 0 \\ 0 \\ 0 \\ 0 \end{bmatrix}$$

The system (5.74) is decomposed in to four subsystems, and can be represented as in the followings:

$$\dot{\bar{x}}_i = \bar{A}_{ii}\bar{x}_i + \bar{B}_i u_i + H_i\bar{x} \tag{5.77}$$

for $i = 1, 2, ..4$. H_i represent the interconnections with other subsystems, such that

$$H_1 = \begin{bmatrix} 0 & \bar{A}_{12} & \bar{A}_{13} & \bar{A}_{14} \end{bmatrix} \tag{5.78}$$

$$H_2 = \begin{bmatrix} \bar{A}_{21} & 0 & \bar{A}_{23} & \bar{A}_{24} \end{bmatrix} \tag{5.79}$$

$$H_3 = \begin{bmatrix} \bar{A}_{31} & \bar{A}_{32} & 0 & \bar{A}_{34} \end{bmatrix} \quad (5.80)$$

$$H_4 = \begin{bmatrix} \bar{A}_{41} & \bar{A}_{42} & \bar{A}_{43} & 0 \end{bmatrix} \quad (5.81)$$

Remark: It can be obsereved by simple inspection that interconnection matrices H_i include fewer non zero elements with magnitudes comparable to those of corresponding subsystem state transition matrices $\bar{A}_{ii}$, thereby indicating the weak couplings.

Correctness check of the permutations

Since the system matrix is large in dimensions having four hundred elements, it is very difficult to check the correctness of the permutation by simply visual comparison. This correctness was checked by evaluating the eigen values of both A and $\bar{A}$ which are same. Thus the matrix pair $(\bar{A},\ \bar{B})$ represent the same system represented by matrix pair $(A,\ B)$

5.5.2 Centralized LQR design

In order to analyze the decentralized control scheme and compare its performance with the centralized one, first we design the centralized LQR controller. In LQR approach, an optimal state feedback control law $u^*(t)$ is obtained by minimizing the following quadratic cost functional J_l with respect to u(t), subject to the system dynamics.

$$J_l = \int_0^\infty [x(t)Qx(t) + u(t)Ru(t)]dt \quad (5.82)$$

where $Q \geq 0$ and $R > 0$ are weighting matrices with appropriate dimensions. Minimization of J_l in (5.82), in general, gives the following optimal control law:

$$u^*(t) = -K\ x(t) \quad (5.83)$$

where K is the optimal state feedback gain given as $K = R^{-1}B^tP$ with P being the solution of the ARE,

$$PA + A^tP - PBR^{-1}B^tP + Q = 0 \quad (5.84)$$

whereby it is assumed that initial state vector x_0 is known. Our objective here is illustrate the practicality of the application of decentralized control scheme. In what follows, we set the state and input weighting matrices as $Q = I_{20},\ R = I_{20}$ which corresponds to equal weighting for each subsystem.

In the centralized case, the state feedback matric K_c is computed as:

$$K_c = \left[\begin{array}{cccc} k_{c1} & k_{c2} & k_{c3} & k_{c4} \end{array} \right] \quad (5.85)$$

where

$$k_{c1} = \left[\begin{array}{ccccc} 0.0198 & 0.2248 & 0.8222 & 0.3926 & 0.5754 \\ -0.0000 & 0.0000 & 0.0001 & 0.0000 & 0.0000 \\ -0.0000 & 0.0000 & 0.0000 & 0.0000 & 0.0000 \\ -0.0000 & 0.0000 & 0.0000 & 0.0000 & 0.0000 \end{array} \right]$$

$$k_{c2} = \left[\begin{array}{ccccc} 0.0001 & 0.0441 & -0.0014 & -0.0137 & -0.0137 \\ -0.4886 & 0.0000 & -0.0818 & 0.0318 & 0.0319 \\ -0.0046 & 0.0000 & -0.0019 & 0.0012 & 0.0012 \\ -0.1311 & 0.0000 & -0.0494 & 0.0292 & 0.0291 \end{array} \right]$$

$$k_{c3} = \left[\begin{array}{ccccc} -0.0000 & 0.0000 & -0.0000 & 0.1313 & -0.0001 \\ 0.1110 & -0.2523 & -0.4349 & 0.0000 & 0.2505 \\ 0.0038 & -0.0032 & 0.0023 & 0.0000 & 0.0094 \\ 0.0869 & -0.0878 & 0.0377 & 0.0000 & 0.2244 \end{array} \right]$$

$$k_{c4} = \left[\begin{array}{ccccc} 0.0007 & -0.0000 & -0.0000 & 0.0000 & 0.0000 \\ -0.8127 & 0.0347 & 0.0329 & -0.0299 & -0.0466 \\ -0.0224 & 0.0017 & 0.0015 & -0.0003 & 0.0001 \\ -0.5823 & 0.0384 & 0.0366 & -0.0096 & -0.0013 \end{array} \right]$$

5.5.3 Decentralized LQR design

It is demonstrated earlier the nuclear system is decopmosed into four subsystems, represented by pairs

$$(\bar{A}_{ii}, \ \bar{B}_i), \ i = 1, 2, ..4)$$

For state feedback based regulation, all of the subsystem matrix pairs should be at least sterilizable. Pair $(\bar{A}_{11}, \ \bar{B}_1)$ is controllable, while the remaining three pairs are stabilizable. Our objective is to obtain a block diagonal state feedback gain matrix K_{dc} such that the closed loop system

$$\dot{\bar{x}} = (\bar{A} - \bar{B}k)\bar{x} \quad (5.86)$$

is stabilized. The block diagonal state feedback K is given by

$$K_{dc} = diag(K_1, K_2, K_3, K_4] \quad (5.87)$$

where $K_i, \imath = 1, 2, ..4$ are obtained such that closed loop subsystems

$$\dot{\bar{x}}_i = (\bar{A}_{ii} - \bar{B}_i k_i)\bar{x}_i \tag{5.88}$$

are stable. Notice that interconnections through H_i are ignored while designing K_i. If these interconnections are weak, then the block diagonal K will suffice the state feedback stabilization, thereby satisfying the notion of decentralized feedback control.

As in the centrallized case, the weighting matrices are chosen as $Q_i = I_5,\ R_i = I_5$. The LQR state feedback gain matrices K_i are computed as

$$\begin{aligned}
K_1 &= \begin{bmatrix} 0.9998 & 0.2249 & 0.8229 & 0.3935 & 0.5784 \end{bmatrix} \\
K_2 &= \begin{bmatrix} -0.0936 & 0 & 0 & 0 & 0 \end{bmatrix} \\
K_3 &= \begin{bmatrix} 0.0003200 & 0 & 0 & 0 & 0.000282 \end{bmatrix} \\
K_4 &= \begin{bmatrix} -1.0000 & 0.0893 & -0.0182 & -0.0963 & -0.2317 \end{bmatrix}
\end{aligned}$$

5.5.4 Multilevel control

In addition to subsystem controls, an additional compensatory control [322] can be provided at a higher level for the sake of getting close to the optimal centralized control. The additional state feedback gain K^g is given by:

$$K^g = (B^t B)^{-1} B^t H \tag{5.89}$$

The closed-loop dynamics with this additional gain are given by

$$\dot{\bar{x}} = [\bar{A} - \bar{B}(K_{dc} + K^g)]\, \bar{x}$$

5.5.5 Overlapping control

Since the system has four inputs, and therefore, to attain a maximum degree of decomposition, we decompose the system into four subsystems with certain degree of overlaps.To proceed, first we decompose the state vector $\bar{x}$ in to seven components as:

$$\bar{x} = \begin{bmatrix} \bar{x}_1 & \bar{x}_2 & \bar{x}_3 & \bar{x}_4 & \bar{x}_5 & \bar{x}_6 & \bar{x}_7 \end{bmatrix}^t \tag{5.90}$$

The corresponding dimensions of these components are $5, 2, 3, 2, 3, 2, 3$, totaling to 20. This partition of the state $\bar{x}$ induces a partition of the the matrix $\bar{A}$

as

$$\bar{A} = \begin{bmatrix} \bar{A}_{11} & \bar{A}_{12} & \bar{A}_{13} & \bar{A}_{14} & \bar{A}_{15} & \bar{A}_{16} & \bar{A}_{17} \\ \bar{A}_{21} & \bar{A}_{22} & \bar{A}_{23} & \bar{A}_{24} & \bar{A}_{25} & \bar{A}_{26} & \bar{A}_{27} \\ \bar{A}_{31} & \bar{A}_{32} & \bar{A}_{33} & \bar{A}_{34} & \bar{A}_{35} & \bar{A}_{36} & \bar{A}_{37} \\ \bar{A}_{41} & \bar{A}_{42} & \bar{A}_{43} & \bar{A}_{44} & \bar{A}_{45} & \bar{A}_{46} & \bar{A}_{47} \\ \bar{A}_{51} & \bar{A}_{52} & \bar{A}_{53} & \bar{A}_{55} & \bar{A}_{55} & \bar{A}_{56} & \bar{A}_{57} \\ \bar{A}_{61} & \bar{A}_{62} & \bar{A}_{63} & \bar{A}_{64} & \bar{A}_{65} & \bar{A}_{66} & \bar{A}_{67} \\ \bar{A}_{71} & \bar{A}_{72} & \bar{A}_{73} & \bar{A}_{74} & \bar{A}_{75} & \bar{A}_{76} & \bar{A}_{77} \end{bmatrix} \tag{5.91}$$

Where the submatrices have appropriate dimensions. The seven components of the state vector $\bar{x}$ are arranged to four overlapping components as follows:

$$\tilde{x}_1 = \begin{bmatrix} \bar{x}_1 & \bar{x}_2 \end{bmatrix}^t \tag{5.92}$$

$$\tilde{x}_2 = \begin{bmatrix} \bar{x}_2 & \bar{x}_3 & \bar{x}_4 \end{bmatrix}^t \tag{5.93}$$

$$\tilde{x}_3 = \begin{bmatrix} \bar{x}_4 & \bar{x}_5 & \bar{x}_6 \end{bmatrix}^t \tag{5.94}$$

$$\tilde{x}_4 = \begin{bmatrix} \bar{x}_6 & \bar{x}_7 \end{bmatrix}^t \tag{5.95}$$

These four overlapping state vectors components constitute a new state vector $\tilde{x} = \begin{bmatrix} \tilde{x}_1 & \tilde{x}_2 & \tilde{x}_3 & \tilde{x}_4 \end{bmatrix}^t$. The vector $\tilde{x}$ is related to $\bar{x}$ by a linear transformation $\tilde{x} = V\bar{x}$ where V is the $\tilde{n} \times n$ matrix

$$V = \begin{bmatrix} \mathbf{I}_5 & \mathbf{0}_{5\times2} & \mathbf{0}_{5\times3} & \mathbf{0}_{5\times2} & \mathbf{0}_{5\times3} & \mathbf{0}_{5\times2} & \mathbf{0}_{5\times3} \\ \mathbf{0}_{2\times5} & \mathbf{I}_2 & \mathbf{0}_{2\times3} & \mathbf{0}_{2\times2} & \mathbf{0}_{2\times3} & \mathbf{0}_{2\times2} & \mathbf{0}_{2\times3} \\ \mathbf{0}_{2\times5} & \mathbf{I}_2 & \mathbf{0}_{2\times3} & \mathbf{0}_{2\times2} & \mathbf{0}_{2\times3} & \mathbf{0}_{2\times2} & \mathbf{0}_{2\times3} \\ \mathbf{0}_{3\times5} & \mathbf{0}_{3\times2} & \mathbf{I}_3 & \mathbf{0}_{3\times2} & \mathbf{0}_{3\times3} & \mathbf{0}_{3\times2} & \mathbf{0}_{3\times3} \\ \mathbf{0}_{2\times5} & \mathbf{0}_{2\times2} & \mathbf{0}_{2\times3} & \mathbf{I}_2 & \mathbf{0}_{2\times3} & \mathbf{0}_{2\times2} & \mathbf{0}_{2\times3} \\ \mathbf{0}_{2\times5} & \mathbf{0}_{2\times2} & \mathbf{0}_{2\times3} & \mathbf{I}_2 & \mathbf{0}_{2\times3} & \mathbf{0}_{2\times2} & \mathbf{0}_{2\times3} \\ \mathbf{0}_{3\times5} & \mathbf{0}_{3\times2} & \mathbf{0}_{3\times3} & \mathbf{0}_{3\times2} & \mathbf{I}_3 & \mathbf{0}_{3\times3} & \mathbf{0}_{3\times2} \\ \mathbf{0}_{2\times5} & \mathbf{0}_{2\times2} & \mathbf{0}_{2\times3} & \mathbf{0}_{2\times2} & \mathbf{0}_{2\times3} & \mathbf{I}_2 & \mathbf{0}_{2\times3} \\ \mathbf{0}_{2\times5} & \mathbf{0}_{2\times3} & \mathbf{0}_{2\times3} & \mathbf{0}_{2\times2} & \mathbf{0}_{2\times3} & \mathbf{I}_2 & \mathbf{0}_{2\times3} \\ \mathbf{0}_{3\times5} & \mathbf{0}_{3\times2} & \mathbf{0}_{3\times3} & \mathbf{0}_{3\times2} & \mathbf{0}_{3\times3} & \mathbf{0}_{3\times2} & \mathbf{I}_3 \end{bmatrix} \tag{5.96}$$

and $\tilde{n} = 26$, hence V is a 27×20 matrix. I_i represent an identity matrix of order i and $\mathbf{0}_{i\times j}$ represent a $i \times j$ zero matrix with the state $\tilde{x}(t) \in \mathbf{R}^{\tilde{n}}$. If we introduce the following relations [324]

$$\begin{aligned} \tilde{A} = U\bar{A}V + M, \quad \tilde{B} = V\bar{B} + N \\ \tilde{Q} = U^tQU + M_Q, \quad \tilde{R} = R + N_R \end{aligned} \tag{5.97}$$

where we assume $UV = I$, and $\tilde{Q}$, Q, $\tilde{R}$ and R are the appropriate weighting matrices in LQR designs, then the inclusion conditions for $(\tilde{\mathbf{S}}, \tilde{\mathbf{J}})$ and $(\mathbf{S}, \mathbf{J})$ are formulated as follows.

The following theorem [324] is used hereafter
$(\tilde{\mathbf{S}}, \tilde{\mathbf{J}}) \supset (\mathbf{S}, \mathbf{J})$ if either

$$(i) \quad MV = 0, \quad N = 0, \quad V^t M_Q V = 0, \quad N_R = 0 \tag{5.98}$$

or

$$(ii) \quad UM^i V = 0, \quad M_Q M^{i-1} N = 0, \quad M_Q M^{i-1} V = 0 \\ UM^{i-1} N = 0, \quad N_R = 0, \quad \forall i \in \tilde{n} \tag{5.99}$$

Where $\tilde{\mathbf{J}}$ and $\mathbf{J}$ are the so called LQR performance indices for systems $\tilde{\mathbf{S}}$ and $\mathbf{S}$, respectively. We chose $\tilde{Q}$, Q, $\tilde{R}$ and R as identity matrices and the conditions $(i,\ ii)$ of the aforementioned theorem are verified by choosing

$$M = 0, \quad M_Q = 0, \quad N = 0 \ \ and \ \ N_R = 0 \tag{5.100}$$

With these conditions, the expanded system can be expressed as:

$$\tilde{\mathbf{S}} : \ \dot{\tilde{x}} = \tilde{A}\tilde{x} + \tilde{B}U \tag{5.101}$$

where

$$\tilde{A} = V\bar{A}U, \quad \tilde{B} = V\bar{B}$$

The matrix $U \in \mathbf{R}^{20\times 26}$ is defined using (5.97) as

$$U = \bar{A}^{-1} V^{\dagger} \tilde{A} \tag{5.102}$$

where $V^{\dagger}$ is the pseudo inverse of V. The $\tilde{A}$ is now expressed as

$$\tilde{A} = \begin{bmatrix} \bar{A}_{11} & \bar{A}_{12} & 0 & 0 & 0 & \bar{A}_{13} & \bar{A}_{14} & \bar{A}_{15} & \bar{A}_{16} & \bar{A}_{17} \\ \bar{A}_{21} & \bar{A}_{22} & 0 & 0 & 0 & \bar{A}_{23} & \bar{A}_{24} & \bar{A}_{25} & \bar{A}_{26} & \bar{A}_{27} \\ \bar{A}_{21} & 0 & \bar{A}_{22} & \bar{A}_{23} & \bar{A}_{24} & 0 & 0 & \bar{A}_{25} & \bar{A}_{26} & \bar{A}_{27} \\ \bar{A}_{31} & 0 & \bar{A}_{32} & \bar{A}_{33} & \bar{A}_{34} & 0 & 0 & \bar{A}_{35} & \bar{A}_{36} & \bar{A}_{37} \\ \bar{A}_{41} & 0 & \bar{A}_{42} & \bar{A}_{43} & \bar{A}_{44} & 0 & 0 & \bar{A}_{45} & \bar{A}_{46} & \bar{A}_{47} \\ \bar{A}_{41} & 0 & \bar{A}_{42} & \bar{A}_{43} & 0 & \bar{A}_{44} & \bar{A}_{45} & \bar{A}_{46} & 0 & \bar{A}_{47} \\ \bar{A}_{51} & 0 & \bar{A}_{52} & \bar{A}_{53} & 0 & \bar{A}_{54} & \bar{A}_{55} & \bar{A}_{56} & 0 & \bar{A}_{57} \\ \bar{A}_{61} & 0 & \bar{A}_{62} & \bar{A}_{63} & 0 & \bar{A}_{64} & \bar{A}_{65} & \bar{A}_{66} & 0 & \bar{A}_{67} \\ \bar{A}_{61} & 0 & \bar{A}_{62} & \bar{A}_{63} & 0 & \bar{A}_{64} & \bar{A}_{65} & 0 & \bar{A}_{66} & \bar{A}_{67} \\ \bar{A}_{71} & 0 & \bar{A}_{72} & \bar{A}_{73} & 0 & \bar{A}_{74} & \bar{A}_{75} & 0 & \bar{A}_{76} & \bar{A}_{77} \end{bmatrix} \tag{5.103}$$

The overlapping subsytems $\tilde{A}_1$, $\tilde{A}_2$, $\tilde{A}_3$ and $\tilde{A}_4$ are now described in the followings:

$$\tilde{A}_1 = \begin{bmatrix} \bar{A}_{11} & \bar{A}_{12} \\ \bar{A}_{21} & \bar{A}_{22} \end{bmatrix},$$

$$\bar{A}_{11} = \begin{bmatrix} -400.0000 & 0.0125 & 0.0305 & 0.1110 \\ 13.1250 & -0.0125 & 0 & 0 \\ 87.5000 & 0 & -0.0305 & 0 \\ 78.1250 & 0 & 0 & -0.1110 \end{bmatrix}$$

$$\bar{A}_{12} = \begin{bmatrix} 0.3010 & 0 & 3.0100 \\ 0 & 0 & 0 \\ 0 & 0 & 0 \\ 0 & 0 & 0 \end{bmatrix}$$

$$\bar{A}_{21} = \begin{bmatrix} 158.1250 & 0 & 0 & 0 \\ 0 & 0 & 0 & 0 \\ 16.8750 & 0 & 0 & 0 \end{bmatrix}$$

$$\bar{A}_{22} = \begin{bmatrix} -0.3010 & 0 & 0 \\ 0 & -0.2034 & 0 \\ 0 & 0 & -3.0100 \end{bmatrix} \tag{5.104}$$

$$\tilde{A}_2 = \begin{bmatrix} \bar{A}_{22} & \bar{A}_{23} & \bar{A}_{24} \\ \bar{A}_{32} & \bar{A}_{33} & \bar{A}_{34} \\ \bar{A}_{42} & \bar{A}_{43} & \bar{A}_{14} \end{bmatrix} \tag{5.105}$$

$$= \begin{bmatrix} -0.2034 & 0 & 0 & 0 & 0 & 0 & 1.3490 \\ 0 & -3.01 & 0 & 0 & 0 & 0 & 0 \\ 0 & 0 & -0.1647 & 0.1647 & 0 & 0 & 0 \\ 0 & 0 & 0.0571 & -2.4403 & 0 & 0 & 0 \\ 0 & 0 & 0.0571 & 2.3262 & -2.3832 & 0 & 0 \\ 0 & 0 & 0 & 0 & 0 & -2.5 & 0 \\ 0.3372 & 0 & 0 & 0 & 0 & 0 & -5.3657 \end{bmatrix}$$

$$\tilde{A}_3 = \begin{bmatrix} \bar{A}_{44} & \bar{A}_{45} & \bar{A}_{46} \\ \bar{A}_{54} & \bar{A}_{55} & \bar{A}_{56} \\ \bar{A}_{64} & \bar{A}_{65} & \bar{A}_{66} \end{bmatrix} \tag{5.106}$$

$$= \begin{bmatrix} -2.5000 & 0 & 0 & 0 & 2.5000 & 0 & 0 \\ 0 & -5.3657 & 3.0702 & 0 & 0 & 0 & 0.2238 \\ 0 & 0.5382 & -0.7664 & 0 & 0 & 0 & 0 \\ 0 & 0 & 0 & -1.1400 & 0 & 0 & 0 \\ 0 & 0 & 0 & 0 & -0.3365 & 0 & 0 \\ -0.2790 & 0.6340 & -0.5090 & 0 & 0.2400 & 0 & -0.1300 \\ 1.4500 & 0 & 0 & 0 & 0 & 0 & -1.4500 \end{bmatrix}$$

$$\tilde{A}_4 = \begin{bmatrix} \bar{A}_{66} & \bar{A}_{67} \\ \bar{A}_{76} & \bar{A}_{77} \end{bmatrix} \tag{5.107}$$

$$= \begin{bmatrix} 0 & -0.1300 & -0.1160 & 0.0235 & 0.1210 \\ 0 & -1.4500 & 0 & 0 & 0 \\ 0 & 0 & -1.4500 & 0 & 0 \\ 0 & 0 & 1.4800 & -1.48000 & \\ 0 & 0 & 0 & 0.5160 & -0.5160 \end{bmatrix}$$

The interconnections among the overlapped subsystems can be easily obtained by simple inspection of $\tilde{A}$ in (5.103). system. Interconnection matrices $H_i, i = 1, 2, 3, 4$ associated with each of the subsystem is given by

$$\tilde{H}_1 = \begin{bmatrix} 0 & 0 & 0 & 0 & 0 & \bar{A}_{13} & \bar{A}_{14} & \bar{A}_{15} & \bar{A}_{16} & \bar{A}_{17} \\ 0 & 0 & 0 & 0 & 0 & \bar{A}_{23} & \bar{A}_{24} & \bar{A}_{25} & \bar{A}_{26} & \bar{A}_{27} \end{bmatrix} \tag{5.108}$$

$$\tilde{H}_2 = \begin{bmatrix} \bar{A}_{21} & 0 & 0 & 0 & 0 & 0 & 0 & \bar{A}_{25} & \bar{A}_{26} & \bar{A}_{27} \\ \bar{A}_{31} & 0 & 0 & 0 & 0 & 0 & 0 & \bar{A}_{35} & \bar{A}_{36} & \bar{A}_{37} \\ \bar{A}_{41} & 0 & 0 & 0 & 0 & 0 & 0 & \bar{A}_{45} & \bar{A}_{46} & \bar{A}_{47} \end{bmatrix} \tag{5.109}$$

$$\tilde{H}_3 = \begin{bmatrix} \bar{A}_{41} & 0 & \bar{A}_{42} & \bar{A}_{43} & 0 & 0 & 0 & 0 & 0 & \bar{A}_{47} \\ \bar{A}_{51} & 0 & \bar{A}_{52} & \bar{A}_{53} & 0 & 0 & 0 & 0 & 0 & \bar{A}_{57} \\ \bar{A}_{61} & 0 & \bar{A}_{62} & \bar{A}_{63} & 0 & 0 & 0 & 0 & 0 & \bar{A}_{67} \end{bmatrix} \tag{5.110}$$

$$\tilde{H}_4 = \begin{bmatrix} \bar{A}_{61} & 0 & \bar{A}_{62} & \bar{A}_{63} & 0 & \bar{A}_{64} & \bar{A}_{65} & 0 & 0 & 0 \\ \bar{A}_{71} & 0 & \bar{A}_{72} & \bar{A}_{73} & 0 & \bar{A}_{74} & \bar{A}_{75} & 0 & 0 & 0 \end{bmatrix} \tag{5.111}$$

The expanded input matrix $\tilde{B}$ is given by:

$$\tilde{B} = \begin{bmatrix} \bar{B}_1 & 0 & 0 & 0 \\ 0 & 0 & 0 & 0 \\ 0 & \bar{B}_2 & 0 & 0 \\ 0 & 0 & 0 & 0 \\ 0 & 0 & \bar{B}_3 & 0 \\ 0 & 0 & 0 & 0 \\ 0 & 0 & 0 & \bar{B}_4 \end{bmatrix} \tag{5.112}$$

where $\bar{B}_i, \ i = 1, 2, 3, 4$ are given in (5.76)

5.5.6 Overlapping decentralized LQR design

Our objective is to design a decentralized state feedback LQR for the expanded system ($\tilde{\mathbf{S}}$ and then to contract that very LQR gain to apply to the actual system. In essence we have to choose a linear control law

$$u = -K\bar{x} \tag{5.113}$$

and to optimize $(\mathbf{S}, \mathbf{J})$ by choosing appropriately a control law

$$u = -\tilde{K}\tilde{x} \tag{5.114}$$

that optimizes $(\tilde{\mathbf{S}}, \tilde{\mathbf{J}})$ in the expanded space. This is only possible if the following contractibility conditions are satisfied.

5.5.7 Contractibility conditions

A control law $u = -\tilde{K}\tilde{x}$ is contractible to control law $u = -K\bar{x}$ if and only if [324]

$$FM^{i-1}V = 0, \quad FM^{i-1}N = 0 \forall i \in \tilde{n} \tag{5.115}$$

Where F is given by

$$\tilde{K} = KU + F \tag{5.116}$$

Since we have chosen $M = 0$ and $N = 0$, the aforementioned conditions are satisfied Siljak(1991) if

$$MV = 0, \quad N = 0 \tag{5.117}$$

then, the contracted K can be obtained as:

$$K = \tilde{K}V \tag{5.118}$$

Furthermore, note that all the four subsystems are stabilizable.

5.5.8 State feedback gain matrix

To determine the gain matrix K, first we found these gains for the overlapped subsystems and then contracted them back to form K. The gains $\tilde{K}_i$, $i = 1, 2, 3, 4$ are given by:

$$\begin{aligned}
\tilde{K}_1 &= \begin{bmatrix} 0.9998 & 0.2249 & 0.8228 & 0.3935 & 0.5782 & 0 & 0.0443 \end{bmatrix} \\
\tilde{K}_2 &= \begin{bmatrix} -0.1561 & 0 & 0.0000 & -0.0000 & 0.0000 & 0 & -0.0386 \end{bmatrix} \\
\tilde{K}_3 &= \begin{bmatrix} 0.1563 & -0.0862 & 0.3189 & 0 & 0.4451 & -1.0000 & 0.0766 \end{bmatrix} \\
\tilde{K}_4 &= \begin{bmatrix} -1.0000 & 0.0893 & -0.0182 & -0.0963 & -0.2317 \end{bmatrix}
\end{aligned}$$

The corresponding contracted state feedback gain matrix K is now given by:

$$K = \begin{bmatrix} \tilde{k}_{1_{(1-5)}} & \tilde{k}_{1_{(6-7)}} & \mathbf{0}_{1\times3} & \mathbf{0}_{1\times2} & \mathbf{0}_{1\times3} & \mathbf{0}_{1\times2} & \mathbf{0}_{1\times3} \\ \mathbf{0}_{1\times5} & \tilde{k}_{2_{(1-2)}} & \tilde{k}_{2_{(3-5)}} & \tilde{k}_{2_{(6-7)}} & \mathbf{0}_{1\times3} & \mathbf{0}_{1\times2} & \mathbf{0}_{1\times3} \\ \mathbf{0}_{1\times5} & \mathbf{0}_{1\times2} & \mathbf{0}_{1\times3} & \tilde{k}_{3_{(1-2)}} & \tilde{k}_{3_{(3-5)}} & \tilde{k}_{3_{(6-7)}} & \mathbf{0}_{1\times3} \\ \mathbf{0}_{1\times5} & \mathbf{0}_{1\times2} & \mathbf{0}_{1\times3} & \mathbf{0}_{1\times2} & \mathbf{0}_{1\times3} & \tilde{k}_{4_{(1-2)}} & \tilde{k}_{4_{(3-5)}} \end{bmatrix}$$

Where $\tilde{k}_{l_{(i-j)}}$ represent a component of the lth subsystem state feedback gain vector $\tilde{k}_l$ composed of elements i through j.

For example $\tilde{k}_{3_{(3-5)}}$ represent $[0.3189 \;\; 0 \;\; 0.4451]$ as can be verified from (**??**). Further more $\mathbf{0}_{1\times j}$ represent a zero row vector of length j, for example, $\mathbf{0}_{1\times3} = [0 \;\; 0 \;\; 0]$.

5.5.9 Simulation analysis

Figs (5.6) through (5.10) show the simulation results for centralized, and overlapped decentralized control schemes. Fig. (5.11) show the control efforts in the two cases. The centralized LQR is supposed to show the optimal results, theoretically, while in the other two cases, due to neglecting the interconnections in the designs of subsystem LQRs, the results are supposedly suboptimal. The level of sub-optimality for overlapped decentralized control is evaluated by the following index [324]:

$$\mu^* = \lambda_M^{-1}[H(V^t\tilde{P}_DV)^{-1}]$$

where $\tilde{P}_D$ is the solution of associated ARE and

$$\begin{aligned}
\hat{A}^t H + H\hat{A} &= -G_D \\
G_D &= Q_D + K_D^t R_D K_D
\end{aligned}$$

where subscript D indicates that these quantities correspond to block diagonal extended system $(\tilde{A}, \tilde{B})$. Also

$$\hat{A} = \bar{A} - \bar{B}K_D$$

When the overlapped decentralized controller applied to the system, it is found that

$$\mu^* = 0.00014614$$

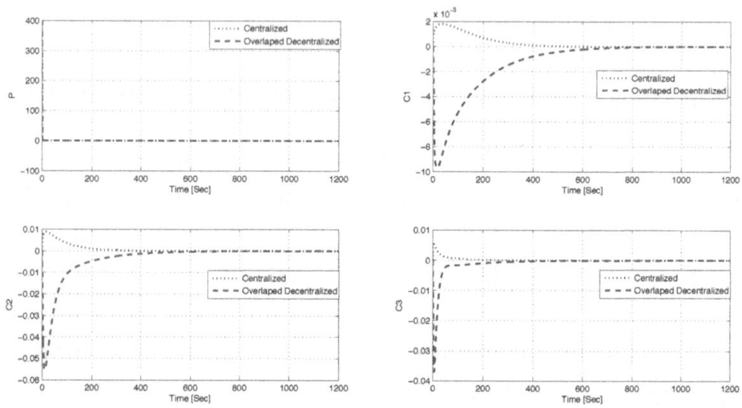

Figure 5.6: State trajectories δP, δC_1, δC_2 and δC_3 with different controllers

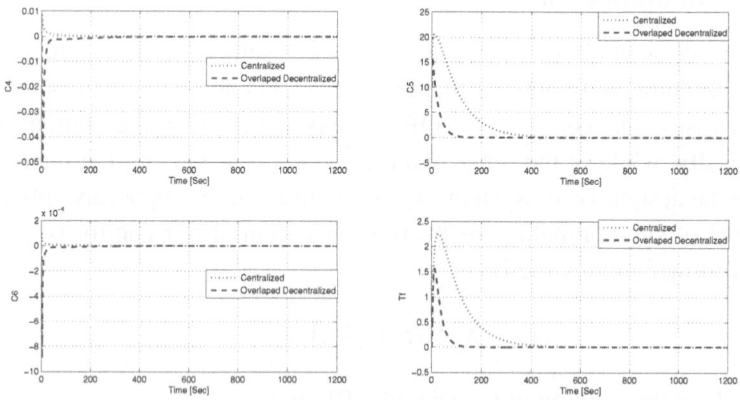

Figure 5.7: State trajectories δC_4, δP_s, δC_6 and δT_f with different controllers

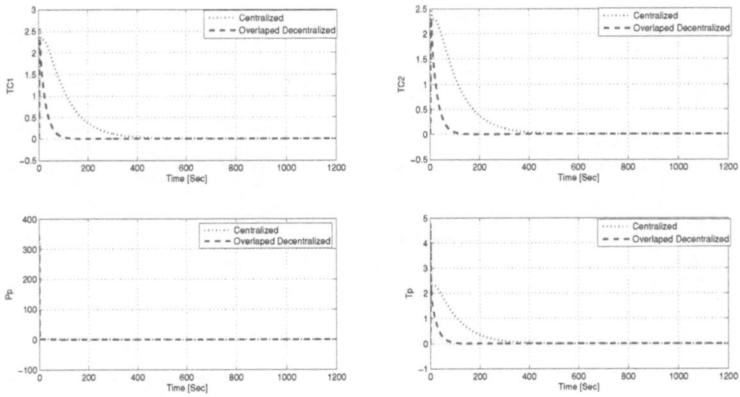

Figure 5.8: State trajectories δT_{C1}, δT_{C2}, δT_{HL} and δP_P with different controllers

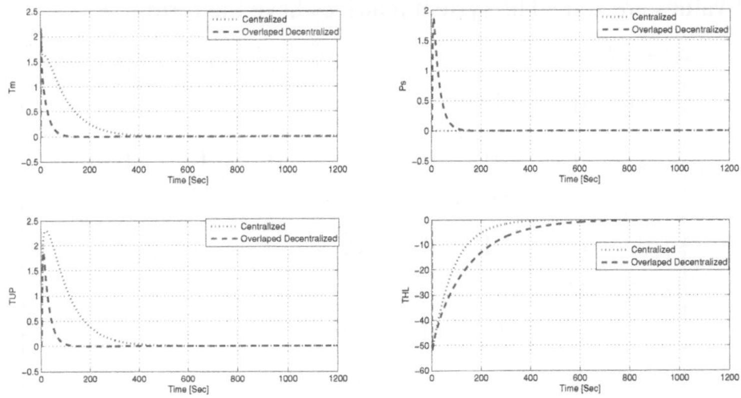

Figure 5.9: State trajectories δT_p, δC_5, δT_f and δC_4 with different controllers

5.6 Problem Set III

Problem III.1: Consider a mathematical model of a string of five high-speed moving vehicles [183]. The motion of each vehicle is the string is represented by two states: position and velocity. Under normal operating conditions, the dynamical model of the string can be written in term of the deviations from a desired separation distance between adjacent vehicles and the deviations from a desired nominal string velocity. In view of this, the equilibrium state and

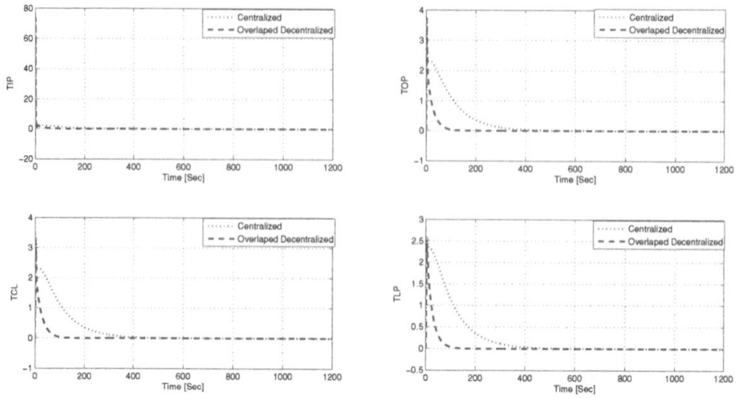

Figure 5.10: State trajectories $\delta T_{IP}, \delta T_{OP}, \delta T_{CL}$ and δT_{LP} with different controllers

control values are zero. The system matrices have the form:

$$
\mathbf{A} =
\left[\begin{array}{ccccccccc}
-1 & 0 & 0 \vdots & & & & & & \\
1 & 0 & -1 \vdots & 0 & 0 & 0 & 0 & 0 & 0 \\
 & & \cdots & \cdots & \cdots & & & & \\
0 & 0 & \vdots -1 \vdots & 0 & 0 \vdots & & & & \\
\cdots & \cdots & \cdots & & & & & & \\
 & & \vdots 1 & 0 & -1 \vdots & & & & \\
 & & & & \cdots & \cdots & \cdots & & \\
 & & \vdots 0 & 0 & \vdots -1 \vdots & 0 & 0 \vdots & & \\
 & & \cdots & \cdots & \cdots & & & & \\
 & & & & \vdots 1 & 0 & -1 \vdots & & \\
 & & & & & & \cdots & \cdots & \cdots \\
 & & & & 0 & 0 & \vdots -1 \vdots & 0 & 0 \\
 & & & & & & \vdots 1 & 0 & -1 \\
 & & & & & & \vdots 0 & 0 & -1
\end{array}\right],
$$

$$
\left[\begin{array}{ccc}
1 & 0 \vdots & \\
0 & 0 \vdots & \\
0 & 0 & \cdots \\
0 & 1 \vdots & 0 \vdots \\
\cdots & \cdots & \\
 & \vdots & 0 \vdots
\end{array}\right.
$$

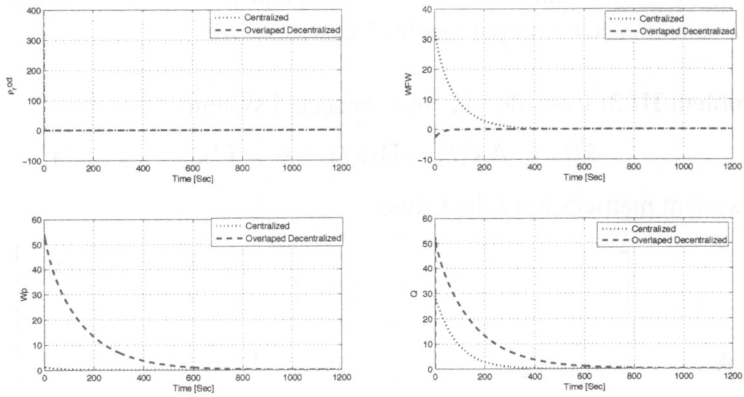

Figure 5.11: State trajectories $\delta\rho_{rod}$, δW_{FW}, δW_P and δQ with different controllers

where the overlapping subsystems are identified are substrings of two adjacent vehicles. Choose appropriate transformation matrices to generate the expanded system. Using suitable weighting matrices, compute the suboptimal control.

Problem III.2: Consider the interconnected system

$$\dot{\mathbf{x}}(t) = \mathbf{A}\mathbf{x}(t) + \mathbf{B}\mathbf{u}(t), \quad \mathbf{y} = \mathbf{C}\mathbf{x}$$

where system matrices have the values

$$\mathbf{A} = \left[\begin{array}{ccccc} 1 & & 3 & \vdots & 0 \\ & \cdots & \cdots & \cdots & \cdots \\ 0 & \vdots & -1 & \vdots & 0 \\ \cdots & \cdots & \cdots & \cdots & \\ 0 & \vdots & 3 & & 1 \end{array}\right], \quad B = \left[\begin{array}{ccccc} 1 & & 0 & \vdots & 0 \\ 0 & \vdots & 1 & \vdots & 0 \\ \cdots & \cdots & \cdots & \cdots & \\ 0 & \vdots & 0 & & 1 \end{array}\right],$$

$$\mathbf{C} = \left[\begin{array}{ccccc} 1 & & 0 & \vdots & 0 \\ & \cdots & \cdots & \cdots & \cdots \\ 0 & \vdots & 1 & \vdots & 0 \\ \cdots & \cdots & \cdots & \cdots & \\ 0 & \vdots & 0 & & 1 \end{array}\right]$$

where the dotted lines separate the overlapping subsystems. Design a decentralized overlapping controller to stabilize the system and to minimize a quadratic cost with weighting matrices $Q = [1 \;\; 0.5 \;\; 1]$ and $R = [1 \;\; 2 \;\; 1]$. Examine the cases of state and static output feedback cases.

Problem III.3: Consider the interconnected system

$$\dot{\mathbf{x}}(t) = \mathbf{A}\mathbf{x}(t) + \mathbf{B}\mathbf{u}(t), \quad \mathbf{y} = \mathbf{C}\mathbf{x}$$

where system matrices have the values

$$\mathbf{A} = \begin{bmatrix} 1 & & 4 & \vdots & 0 \\ & \cdots & \cdots & \cdots & \cdots \\ 1 & \vdots & 2 & \vdots & 2 \\ \cdots & \cdots & \cdots & \cdots & \\ 0 & \vdots & -2 & & 3 \end{bmatrix}, \; \mathbf{B} = \begin{bmatrix} 1 & \vdots & 0 \\ & \vdots & \cdots \\ 0 & \vdots & 0 \\ \cdots & \vdots & \\ 0 & \vdots & 1 \end{bmatrix}$$

where the dotted lines separate the overlapping subsystems. Design a decentralized overlapping controller to stabilize the system.

Problem III.4: Consider the following two linear systems

$$\begin{aligned} \mathbf{S} \quad &: \quad \dot{\mathbf{x}}(t) = \mathbf{A}\mathbf{x}(t) + \mathbf{B}\mathbf{u}(t), \; \mathbf{x}(0) = \mathbf{x}_o, \\ \mathbf{y} \quad &= \quad \mathbf{C}\mathbf{x}, \\ \tilde{\mathbf{S}} \quad &: \quad \dot{\tilde{\mathbf{x}}}(\mathbf{t}) = \tilde{\mathbf{A}}\tilde{\mathbf{x}}(\mathbf{t}) + \tilde{\mathbf{B}}\tilde{\mathbf{u}}(\mathbf{t}), \; \tilde{\mathbf{x}}(\mathbf{0}) = \tilde{\mathbf{x}}_{\mathbf{o}}, \\ \tilde{\mathbf{y}} \quad &= \quad \tilde{\mathbf{C}}\tilde{\mathbf{x}} \end{aligned}$$

where $\mathbf{x} \in \Re^n$, $\mathbf{u} \in \Re^m$, $\mathbf{y} \in \Re^p$, $\tilde{\mathbf{x}} \in \Re^{\tilde{n}}$, $\tilde{\mathbf{u}} \in \Re^{\tilde{m}}$, $\tilde{\mathbf{y}} \in \Re^{\tilde{p}}$ are the state, input and output vectors for systems $\mathbf{S}$ and $\tilde{\mathbf{S}}$ respectively. Consider the transformations

$$\begin{aligned} \mathbf{T} \; &: \; \Re^n \longrightarrow \Re^{\tilde{n}}, \; rank(\mathbf{T}) = n, \\ \mathbf{U} \; &: \; \Re^{\tilde{m}} \longrightarrow \Re^m, \; rank(\mathbf{U}) = m, \\ \mathbf{T}^\dagger \; &: \; \Re^{\tilde{n}} \longrightarrow \Re^n, \; \mathbf{T}^\dagger\mathbf{T} = I_n, \\ \mathbf{U}^\dagger \; &: \; \Re^m \longrightarrow \Re^{\tilde{m}}, \; rank(\mathbf{U}) = m \end{aligned}$$

such that $\tilde{\mathbf{x}}_o = \mathbf{T}\mathbf{x}_o, \;\; \mathbf{u} = \mathbf{U}\tilde{\mathbf{u}}$. Based thereon, establish that

$$\mathbf{TA} = \tilde{\mathbf{A}}\mathbf{T}, \;\; \mathbf{TBU} = \tilde{\mathbf{B}}$$

In this case, system $\tilde{\mathbf{S}}$ is called an extension of system $\mathbf{S}$. Proceed to deduce the following relations:

$$\tilde{\mathbf{A}} = \mathbf{TAT}^\dagger + \mathbf{E}, \;\; \tilde{\mathbf{B}} = \mathbf{TBU} + \mathbf{F}$$

and derive the conditions on the complementary matrices $\mathbf{E}$ and $\mathbf{F}$.

5.7 Notes and References

In this Chapter, the main concern has been on the decomposition and control of interconnected systems through coupled subystems. The main approach addressed has been the overlapping decomposition (OLD) and the associated overlapping control (OLC). Different types of into input-decentralized or output-decentralized subsystems has been proposed by [128]. Overlapping decomposition has been treated by [124]-[131] where emphasis has been placed on theoretical foundations of expansion and contraction relations. An extension of the inclusion principle to non-linear systems was given by [286] and generalizing this approach for overlapping inputs and outputs was reported in [120]. The notion of a non-classical information pattern has been discussed in connection with decentralized design in [185] and application of OLD to the power-frequency behavior of multiarea power systems was illustrated in [192].

A systematic extension to the class of complementary matrices to obtain a more flexible selection strategy has been developed in BakuleL1988 where the introduction of appropriate changes of basis in the expansion contraction process has been formalized. Overlapping decomposition and overlapping feedback control techniques have been developed in [19] for a cable-stayed bridge benchmark, in [350] for platoon of vehicles and in [352] for a formation of unmanned aerial vehicles.

Despite the major developments in deploying the concepts of overlapping and the inclusion principle, the issue of computational load and associated processing requirements has been addressed in comparison with other decentralized control approaches. It is hoped that this issue will attract the attention of control researchers.

Chapter 6

Interconnected Time-Delay systems

In this Chapter, we start our examination of the development of decentralized control for a class of interconnected continuous-time and discrete-time systems. The subsystems are time-delay plants subjected to convex-bounded parametric uncertainties and the interconnections are time-delay couplings. The major departure from the previous chapter lies in the analysis of delay-dependent stability criteria and the synthesis of delay-dependent feedback stabilization methods.

6.1 Introduction

In the sequel, we study the stability, robust stability and feedback stabilization problems of a class of interconnected systems. We start with linear continuous-time systems and then generalize the results to a class of nonlinear systems. Moreover, we extend the results to discrete-time systems. In all cases under consideration, the local subsystem has convex-bounded parametric uncertainties and time-varying delays within the local subsystems and across the interconnections. On one hand, linear time-delay systems represent a wide class of infinite-dimensional systems and are frequently encountered to describe propagation, transport phenomena and population dynamics in various engineering and physical applications [255]. On the other hand, large-scale interconnected system appear in a variety of engineering applications including power systems, large structures and manufacturing systems and for those applications, decentralized control schemes present a practical and effective means for designing control algorithms based on the individual subsystems [324]. Relevant research results on decentralized control of relevance to the present work can be found in [173, 288, 328].

Stability analysis and control design criteria for state-delay systems, on the other hand, can be broadly classified into two categories: delay-independent, which are applicable to delays of arbitrary size [255] and delay-dependent, which include information on the size of the delay, see [74]. Improved methods pertaining to the problems of determining robust stability criteria of uncertain time-delay systems have been recently reported [103] and the references cited therein. To reduce the level of design conservatism when dealing with time-varying delays, one has to select appropriate Lyapunov-Krasovskii functional (LKF) with moderate number of terms, avoid bounding methods and use parametrized relations and variables effectively to avoid redundancy. It appears from the existing results that general results pertaining to interconnected time-delay systems are few and restricted, see [174, 246, 253, 251, 252, 287] where most of the efforts were centered on matching conditions and were virtually delay-independent.

6.2 Linear Interconnected Systems: Continuous-Time

We consider a class of linear systems S structurally composed of n_s coupled subsystems $\mathsf{S_j}$ and modeled by the state-space model:

$$\begin{aligned} \dot{x}_j(t) &= A_j x_j(t) + A_{dj} x_j(t - \tau_j(t)) + B_j u_j(t) \\ &+ \sum_{k=1,k\neq j}^{n_s} E_{jk} x_k(t - \eta_{jk}(t)) + \Gamma_j w_j(t) \\ z_j(t) &= G_j x(t) + G_{dj} x(t - \tau_j(t)) + D_j u(t) + \Phi_j w_j(t) \end{aligned} \tag{6.1}$$

where for $j \in \{1, ..., n_s\}$, $x_j(t) \in \Re^{n_j}$ is the state vector; $u_j(t) \in \Re^{m_j}$ is the control input, $w_j(t) \in \Re^{q_j}$ is the disturbance input which belongs to $\mathcal{L}_2[0, \infty)$, $z_j(t) \in \Re^{q_j}$ is the controlled output and $\tau_j,\ \eta_{jk},\ j, k \in \{1, ..., n_s\}$ are unknown time-delay factors satisfying

$$\begin{aligned} 0 &\le \tau_j(t) \le \varrho_j, \quad \dot{\tau}_j(t) \le \mu_j, \\ 0 &\le \eta_{jk}(t) \le \varrho_{jk}, \quad \dot{\eta}_{jk}(t) \le \mu_{jk} \end{aligned} \tag{6.2}$$

where the bounds $\varrho_j,\ \varrho_{jk},\ \mu_j,\ \mu_{jk}$ are known constants in order to guarantee smooth growth of the state trajectories. The matrices $A_j \in \Re^{n_j \times n_j}$, $B_j \in \Re^{n_j \times m_j}$, $G_j \in \Re^{q_j \times n_j}$, $D_j \in \Re^{q_j \times m_j}$, $A_{dj} \in \Re^{n_j \times n_j}$, $\Phi_j \in \Re^{q_j \times q_j}$, $\Gamma_j \in \Re^{n_j \times q_j}$, $C_j \in \Re^{p_j \times n_j}$, $C_{dj} \in \Re^{p_j \times n_j}$, $F_j \in \Re^{p_j \times m_j}$, $\Psi_j \in \Re^{p_j \times q_j}$ are real and uncertain. The initial condition $\langle x_j(0), x_j(r)\langle = \langle x_{oj}, \phi_j\langle,\ j \in \{1, ..., n_s\}$ where $\phi_j(.) \in \mathcal{L}_2[-\tau_j^*, 0],\ j \in \{1, ..., n_s\}$. The inclusion of the term $A_{dj}x_j(t - \tau_j(t))$ is meant to emphasize the delay within the local subsystem.

The class of systems described by (6.1) subject to delay-pattern (6.2) is frequently encountered in modeling several physical systems and engineering applications including large space structures, multi-machine power systems, cold mills, transportation systems, water pollution management, to name a few [253, 252, 252, 324].

For the purpose of conducting stability analysis, we employ the Lyapunov-Krasovskii theory [93] and consider the Lyapunov-Krasovskii functional (LKF):

$$
\begin{aligned}
V(t) &= \sum_{j=1}^{n_s} V_j(t) \\
V_j(t) &= \Big[V_{oj}(t) + V_{aj}(t) + V_{mj}(t)) + V_{nj}(t)\Big] \\
V_{oj}(t) &= x_j^t(t)\mathcal{P}_j x_j(t), \\
V_{aj}(t) &= \int_{-\varrho_j}^{0}\int_{t+s}^{t} \dot{x}_j^t(\alpha)\mathcal{W}_j\dot{x}_j(\alpha)d\alpha\, ds, \\
V_{mj}(t) &= \int_{t-\tau_j(t)}^{t} x_j^t(s)\mathcal{Q}_j x_j(s)\, ds, \\
V_{nj}(t) &= \sum_{k=1,k\neq j}^{n_s}\int_{t-\eta_{jk}(t)}^{t} x_k^t(s)\mathcal{Z}_{jk}x_k(s)\, ds
\end{aligned} \qquad (6.3)
$$

where $0 < \mathcal{P}_j = \mathcal{P}_j^t,\ 0 < \mathcal{W}_{aj} = \mathcal{W}_{aj}^t,\ 0 < \mathcal{W}_{cj} = \mathcal{W}_{aj}^t,\ 0 < \mathcal{Q}_j = \mathcal{Q}_j^t,\ 0 < \mathcal{R}_j = \mathcal{R}_j^t,\ 0 < \mathcal{Z}_{jk} = \mathcal{Z}_{jk}^t,\ j, k \in \{1, ..., n_s\}$ are weighting matrices of appropriate dimensions.

Remark 6.2.1 *The first term in (6.3) is standard to nominal systems without delay while the second and third terms correspond to the delay-dependent conditions since they provide measures of the individual and difference signal energies during the delay-period (recall that $\int_{t-\varrho_j}^{t} \dot{x}_j(\alpha)d\alpha = x_j(t) - x_j(t - \varrho_j)$ by Leibniz-Newton formula) and the fourth term corresponds to the intra-connection delays. LKF composed of the first three terms has been considered in [74, 103, 261, 290] for single time-delay systems. Unlike the results on interconnected systems [251, 283, 287, 288, 328, 371], LKF (6.3) represents a new effective form and therefore the advantages and reduced conservatism afforded in [74, 103, 261, 290] is carried over hereafter.*

6.2.1 Local delay-dependent analysis

Local delay-dependent analysis In this section, we develop new criteria for LMI-based characterization of delay-dependent asymptotic stability and $\mathcal{L}_2$ gain analysis. The criteria includes some parameter matrices aims at expanding the range of applicability of the developed conditions. The following theorem establishes the main result subsystem S_j:

Theorem 6.2.2 *Given $\varrho_j > 0,\ \mu_j > 0,\ \varrho_{jk} > 0$ and $\mu_{jk} > 0,\ j,k = 1,...,n_s$. The family of subsystems $\{\mathsf{S}_j\}$ with $u_j(.) \equiv 0$ where S_j is described by (6.1) is delay-dependent asymptotically stable with $\mathcal{L}_2$-performance bound $\gamma_j,\ j = 1,...,n_s$ if there exist weighting matrices $\mathcal{P}_j,\ \mathcal{Q}_j,\ \mathcal{W}_j,\ \mathcal{Z}_{kj},\ k = 1,..n_s,$ parameter matrices $\Theta_j\ \Upsilon_j$ and a scalar $\gamma_j > 0,\ j = 1,...,n_s$ satisfying the following LMI*

$$\Xi_{\varrho j} = \begin{bmatrix} \Xi_{\varrho j1} & \Xi_{\varrho j2} \\ \bullet & \Xi_{\varrho j3} \end{bmatrix} < 0 \tag{6.4}$$

$$\Xi_{\varrho j1} = \begin{bmatrix} \Xi_{oj} & \Xi_{aj} & -\varrho_j\Theta_j & 0 \\ \bullet & -\Xi_{mj} & -\varrho_j\Upsilon_j & 0 \\ \bullet & \bullet & -\varrho_j\mathcal{W}_j & 0 \\ \bullet & \bullet & \bullet & -\Xi_{nj} \end{bmatrix}$$

$$\Xi_{\varrho j2} = \begin{bmatrix} \mathcal{P}_j\Gamma_j & G_j^t & \varrho_k A_j^t\mathcal{W}_k \\ 0 & G_{dj}^t & \varrho_k A_{d_j}^t\mathcal{W}_k \\ 0 & 0 & 0 \\ 0 & 0 & \varrho_k\sum_{k=1,k\neq j}^{n_s} E_{jk}\mathcal{W}_k \end{bmatrix}$$

$$\Xi_{\varrho j3} = \begin{bmatrix} -\gamma_j^2 I & \Phi_j^t & \varrho_k\Gamma_j^t\mathcal{W}_k \\ \bullet & -I & 0 \\ \bullet & \bullet & -\varrho_k\mathcal{W}_k \end{bmatrix} \tag{6.5}$$

where

$$\Xi_{oj} = \mathcal{P}_jA_j + A_j^t\mathcal{P}_j^t + \Theta_j + \Theta_j^t + \mathcal{Q}_j + \sum_{k=1}^{n_s}\mathcal{Z}_{kj} + (n_s-1)\mathcal{P}_j,$$

$$\Xi_{aj} = \mathcal{P}_jA_{dj} - \Theta_j + \Upsilon_j^t,$$

$$\Xi_{mj} = \Upsilon_j + \Upsilon_j^t + (1-\mu_j)\mathcal{Q}_j,$$

$$\Xi_{nj} = \sum_{k=1}^{n_s}(1-\mu_{kj})\mathcal{Z}_{kj} + \sum_{k=1,k\neq j}^{n_s} E_{kj}^t\mathcal{P}_kE_{kj} \tag{6.6}$$

Proof: It is readily seen from (6.5) using the delay bounds of (6.2) that there exists a scalar $\sigma_j > 0$ such that

$$\begin{bmatrix} \tilde{\Xi}_{\varrho j1} & \Xi_{\varrho j2} \\ \bullet & \Xi_{\varrho j3} \end{bmatrix} < 0 \tag{6.7}$$

with

$$\tilde{\Xi}_{\varrho j1} = \begin{bmatrix} \Xi_{oj} + \sigma_j I & \Xi_{aj} & -\Theta_j & 0 \\ \bullet & -\Xi_{mj} & -\Upsilon_j & 0 \\ \bullet & \bullet & -\varrho_j^{-1}\mathcal{W}_j & 0 \\ \bullet & \bullet & \bullet & -\Xi_{nj} \end{bmatrix} \tag{6.8}$$

Therefore, for all τ_j satisfying (6.2) we have

$$\Xi_{\sigma j} = \begin{bmatrix} \Xi_{\sigma j1} & \Xi_{\varrho j2} \\ \bullet & \Xi_{\varrho j3} \end{bmatrix} < 0 \tag{6.9}$$

$$\Xi_{\sigma j1} = \begin{bmatrix} \Xi_{oj} & \Xi_{aj} & -\tau_j\Theta_j & 0 \\ \bullet & -\Xi_{mj} & -\tau_j\Upsilon_j & 0 \\ \bullet & \bullet & -\tau_j\mathcal{W}_j & 0 \\ \bullet & \bullet & \bullet & -\Xi_{nj} \end{bmatrix} \tag{6.10}$$

Now to examine the internal stability , we need to compute the time-derivative of $V(t)$ along the solutions of (6.1) with $w(t) \equiv 0$. In terms of 6.3, we do this job component-wise as follows.

$$\begin{aligned} \dot{V}_{oj}(t) &= 2x_j^t\mathcal{P}_j\dot{x}_j = 2x_j^t\mathcal{P}_j[A_jx_j(t) + A_{dj}x_j(t-\tau_j(t))] \\ &+ 2x_j^t\mathcal{P}_j\sum_{k=1}^{n_s} E_{jk}x_k(t-\eta_{jk}(t))] \end{aligned} \tag{6.11}$$

Little algebra on (6.11) yields

$$\begin{aligned} \dot{V}_{oj}(t) &= 2x_j^t\mathcal{P}[A_j + A_{dj}]x_j(t) - 2x_j^t\mathcal{P}_jA_{dj}\int_{t-\tau_j(t)}^{t} \dot{x}_j(s)ds \\ &+ 2x_j^t\mathcal{P}_j\sum_{k=1}^{n_s} E_{jk}x_k(t-\eta_{jk}(t))]() \end{aligned} \tag{6.12}$$

Expanding terms in (6.12) gives

$$\begin{aligned}\dot{V}_{oj}(t) &= 2x_j^t\mathcal{P}_j[A_j + A_{dj}]x_j(t) \\ &+ 2x_j^t[\Theta_j - \mathcal{P}_jA_{dj}]\int_{t-\tau_j(t)}^{t} \dot{x}_j(s)ds \\ &+ 2x^t(t-\tau_j(t))\Upsilon_j\int_{t-\tau_j(t)}^{t} \dot{x}_j(s)ds - \Big[2x_j^t t\Theta_j\int_{t-\tau_j}^{t} \dot{x}_j(s)ds \\ &+ 2x^t(t-\tau_j(t))\Upsilon_j\int_{t-\tau_j(t)}^{t} \dot{x}_j(s)ds\Big] \\ &+ 2x_j^t\mathcal{P}_j\sum_{k=1}^{n_s} E_{jk}x_k(t-\eta_{jk}(t))] \end{aligned} \tag{6.13}$$

Further manipulation of (6.13) yields

$$\begin{aligned}\dot{V}_{oj}(t) &= \frac{1}{\tau_j(t)}\int_{t-\tau_j(t)}^{t} \Big[2x_j^t[\mathcal{P}_jA_j + \Theta_j]x_j \\ &+ 2x_j^t[\mathcal{P}_jA_{dj} - \Theta_j + \Upsilon_j^t]x_j(t-\tau_j(t)) \\ &- 2x_j^t(t-\tau_j(t))\Upsilon x(t-\tau_j) - 2x^t\tau_j\Theta\dot{x}_j(s) \\ &- 2x_j^t(t-\tau_j(t))\tau_j(t)\Upsilon_j\dot{x}_j(s) \\ &+ 2x_j^t\mathcal{P}_j\sum_{k=1,k\neq j}^{n_s} E_{jk}x_k(t-\eta_{jk}(t))]\Big]ds \end{aligned} \tag{6.14}$$

where $\Theta_j \in \Re^{n_j\times n_j}$ and $\Upsilon_j \in \Re^{n_j\times n_j}$ are appropriate relaxation matrices injected to facilitate the delay-dependence analysis. Next we have

$$\begin{aligned}\dot{V}_{aj}(t) &= \int_{-\varrho_j}^{0} [\dot{x}_j^t(t)\mathcal{W}_j\dot{x}_j(t) - \dot{x}_j^t(t+s)\mathcal{W}_j\dot{x}_j(t+s)]ds \\ &= \int_{t-\varrho_j}^{t} [\dot{x}_j^t(t)\mathcal{W}_j\dot{x}_j(t) - \dot{x}_j^t(s)\mathcal{W}_j\dot{x}_j(s)]d\ s \\ &= \varrho_j\ \dot{x}_j^t(t)\mathcal{W}_j\dot{x}_j(t) - \int_{t-\tau_j}^{t} \dot{x}_j^t(s)\mathcal{W}_j\dot{x}_j(s)ds \\ &- \int_{t-\varrho_j}^{t-\tau_j} \dot{x}_j^t(s)\mathcal{W}_j\dot{x}_j(s) \end{aligned} \tag{6.15}$$

Further manipulation of (6.15) yields

$$\begin{aligned}
\dot{V}_{aj}(t) &\leq \varrho_j \, \dot{x}_j^t(t)\mathcal{W}_j\dot{x}_j(t) - \int_{t-\tau_j}^{t} \dot{x}_j^t(s)\mathcal{W}_j\dot{x}_j(s)ds \\
&= \frac{1}{\tau_j(t)}\int_{t-\tau_j(t)}^{t} \Big[\varrho_j\dot{x}_j^t(t)\mathcal{W}_j\dot{x}_j(t) \\
&- \tau_j\dot{x}_j^t(s)\mathcal{W}_j\dot{x}_j(s)\Big] ds
\end{aligned} \tag{6.16}$$

Note that the term $T_j = \int_{t-\varrho_j}^{t-\tau_j} \dot{x}_j^t(s)\mathcal{W}_j\dot{x}_j(s)$ accounts for the enlarged time interval from $t - \varrho_j \to t$ to $t - \tau_j \to t$. It is obvious that $T_j > 0$ and hence expression (6.16) holds true without conservatism. There has been alternative route to handle T_j by employing extra parameter matrices and adding more identities [103]-[146]. Proceeding further, we get

$$\begin{aligned}
\dot{V}_{mj}(t) &= x_j^t(t)\mathcal{Q}_jx_j(t) \\
&- (1 - \dot{\tau}_j(t))\, x^t(t - \tau_j(t))\mathcal{Q}_jx_j(t - \tau_j(t)) \\
&\leq x_j^t(t)\mathcal{Q}_jx_j(t) \\
&- (1 - \mu_j)\, x_j^t(t - \tau_j(t))\mathcal{Q}_jx_j(t - \tau_j(t))
\end{aligned} \tag{6.17}$$

Little algebra on (6.17) yields

$$\begin{aligned}
\dot{V}_{mj}(t) &= \frac{1}{\tau_j(t)}\int_{t-\tau_j(t)}^{t} \Big[x_j^t(t)\mathcal{Q}_jx_j(t) \\
&- (1 - \mu_j)\, x_j^t(t - \tau_j(t))\mathcal{Q}_jx_j(t - \tau_j(t))\Big] ds
\end{aligned} \tag{6.18}$$

Also, we have

$$\begin{aligned}
\dot{V}_{nj}(t) &= \sum_{k=1,k\neq j}^{n_s} \Big[x_k^t(t)\mathcal{Z}_{jk}x_k(t) \\
&- (1 - \dot{\eta}_{jk}(t))x_k^t(t - \eta_{jk}(t))\mathcal{Z}_{jk}x_k(t - \eta_{jk}(t))\Big] \\
&\leq \sum_{k=1,k\neq j}^{n_s} \Big[x_k^t(t)\mathcal{Z}_{jk}x_k(t) \\
&- (1 - \mu_{jk})\, x_k^t(t - \eta_{jk}(t))\mathcal{Z}_{jk}x_k(t - \eta_{jk}(t))\Big]
\end{aligned} \tag{6.19}$$

For the class of interconnected systems (6.1), the following structural identity holds

$$\sum_{j=1}^{n_s}\sum_{k=1,k\neq j}^{n_s} x_k^t(t)\mathcal{Z}_{jk}x_k(t) = \sum_{j=1}^{n_s}\sum_{k=1,k\neq j}^{n_s} x_j^t(t)\mathcal{Z}_{kj}x_j(t) \tag{6.20}$$

Observe that the result of applying **Inequality** 3 of the appendix, we have

$$\begin{aligned}
& 2x_j^t\mathcal{P}_j \sum_{k=1,k\neq j}^{n_s} E_{jk}x_k(t-\eta_{jk}(t)) = \\
& \sum_{k=1,k\neq j}^{n_s} 2x_j^t\mathcal{P}_jE_{jk}x_k(t-\eta_{jk}(t)) = \\
& \sum_{k=1,k\neq j}^{n_s} 2[\mathcal{P}_jx_j]^t[E_{jk}x_k(t-\eta_{jk}(t))] \leq \\
& \sum_{k=1,k\neq j}^{n_s} [\mathcal{P}_jx_j]^t\mathcal{P}_j^{-1}[\mathcal{P}_jx_j] + \\
& \sum_{k=1,k\neq j}^{n_s} [E_{jk}x_k(t-\eta_{jk}(t))]^t\mathcal{P}_j[E_{jk}x_k(t-\eta_{jk}(t))] = \\
& (n_s-1)x_j^t\mathcal{P}_jx_j + \\
& \sum_{k=1,k\neq j}^{n_s} x_k(t-\eta_{jk}(t)E_{jk}^t\mathcal{P}_jE_{jk}x_k(t-\eta_{jk}(t))
\end{aligned} \tag{6.21}$$

which when incorporated with (6.19) yields

$$\begin{aligned}
& \sum_{j=1}^{n_s}\sum_{k=1,k\neq j}^{n_s} x_k(t-\eta_{jk}(t)E_{jk}^t\mathcal{P}_jE_{jk}x_k(t-\eta_{jk}(t)) = \\
& \sum_{j=1}^{n_s} x_j^t(t-\eta_{kj}(t))[\sum_{k=1,k\neq j}^{n_s} E_{jk}^t\mathcal{P}_jE_{jk}x_k(t-\eta_{kj}(t))
\end{aligned} \tag{6.22}$$

In addition, the following equality holds

$$\begin{aligned}
& \sum_{j=1}^{n_s}\{(\sum_{k=1,k\neq j}^{n_s} E_{jk}x_k(t-\eta_{jk}(t)))^t\varrho_j\mathcal{W}_j(\sum_{k=1,k\neq j}^{n_s} E_{jk}x_k(t-\eta_{jk}(t)))\} = \\
& \sum_{j=1}^{n_s}\{x_j^t(t-\eta_{kj}(t))(\sum_{k=1,k\neq j}^{n_s} E_{jk})^t\varrho_k\mathcal{W}_k(\sum_{k=1,k\neq j}^{n_s} E_{jk})x_j(t-\eta_{kj}(t))\}
\end{aligned} \tag{6.23}$$

By combining (6.3)-(6.22) and using (6.23) with Schur complements, we have

$$\dot{V}_j(t)|_{(6.1)} \leq \sum_{j=1}^{n_s} \left[\frac{1}{\tau_j} \int_{t-\tau_j}^{t} \chi_j^t(t,s)\, \overline{\Xi}_j\, \chi_j(t,s)\, ds \right] \tag{6.24}$$

$$\chi_j(t,s) = \left[x_j^t(t)\;\; x_j^t(t-\tau_j)\;\; \dot{x}_j^t(s)\;\; x_j^t(t-\eta_{kj}) \right]^t \tag{6.25}$$

where $\dot{V}_j(t)|_{(6.1)}$ defines the Lyapunov derivative along the solutions of system (6.1). Note that $\overline{\Xi}_j$ corresponds to $\Xi_{\sigma j}$ in with $G_j \equiv 0,\; G_{dj} \equiv 0,\; \Phi_j \equiv 0$. In view of (6.10), it follows that $\overline{\Xi}_j < 0$ and more importantly

$$\begin{aligned} &\dot{V}_j(t)|_{(6.1)} < \\ &\sum_{j=1}^{n_s} \left[\frac{1}{\tau_j(t)} \int_{t-\tau_j}^{t} \chi_j^t(t,s) diag[-\sigma_j,\, 0,\, 0,\, 0,\, 0,\, 0] \chi(t,s)_j\, ds \right] \\ &= \; -\sum_{j=1}^{n_s} \sigma_j\, \|x_j\|^2 \; < \; 0 \end{aligned} \tag{6.26}$$

This establishes the internal asymptotic stability.

Now consider the performance measure

$$J = \sum_{j=1}^{n_s} \int_0^\infty \left(z_j^t(s) z_j(s) - \gamma_j^2 w_j^t(s) w_j(s) \right) ds$$

For any $w_j(t) \in \mathcal{L}_2(0,\infty) \neq 0$ and zero initial condition $x_j(0) = 0$, we have

$$\begin{aligned} J \;&=\; \sum_{j=1}^{n_s} \int_0^\infty \left(z_j^t(s) z_j(s) - \gamma_j^2 w_j^t(s) w_j(s) \; + \dot{V}_j(t)|_{(6.1)} \right. \\ &- \left. \dot{V}_j(t)|_{(6.1)} \right) ds \\ &\leq \; \sum_{j=1}^{n_s} \int_0^\infty \left(z_j^t(s) z_j(s) - \gamma_j^2 w_j^t(s) w_j(s) \; + \dot{V}_j(t)|_{(6.1)} \right) ds \end{aligned}$$

Proceeding as before, we get

$$\begin{aligned} &\sum_{j=1}^{n_s} \left(z_j^t(s) z_j(s) - \gamma_j^2 w_j^t(s) w_j(s) \; + \dot{V}_j(t)|_{(6.1)} \right) \\ &= \sum_{j=1}^{n_s} \bar{\chi}_j^t(t,s)\, \widehat{\Xi}_j\, \bar{\chi}_j(t,s), \end{aligned}$$

$$\bar{\chi}_j(t,s) \;=\; \left[\; x_j^t(t)\;\; x_j^t(t-\tau_j)\;\; \dot{x}_j^t(s)\;\; w_j^t(s) \;\right]^t$$

where $\widehat{\Xi}_j$ corresponds to $\Xi_{\varrho j}$ in (6.5) by Schur complements. It is readily seen from (6.5) by Schur complements that

$$\sum_{j=1}^{n_s}\left(z_j^t(s)z_j(s) - \gamma_j^2 w_j^t(s)w_j(s) \; + \dot{V}_j(t)|_{(6.1)}\right) < 0$$

for arbitrary $s \in [t, \infty)$, which implies for any $w_j(t) \in \mathcal{L}_2(0,\infty) \neq 0$ that $J < 0$ leading to $\sum_{j=1}^{n_s} ||z_j(t)||_2 \; < \; \sum_{j=1}^{n_s} \gamma_j \; ||w(t)_j||_2$ and the proof is completed.

Remark 6.2.3 *It is significant to recognize that our methodology incorporates three weighting matrices $\mathcal{P}_j$, $\mathcal{Q}_j$, $\mathcal{W}_j$, and two parameter matrices Θ_j, Υ_j at the subsystem level to ensure least conservative delay-dependent stability results. This will eventually result in reduced computational requirements as evidenced by a simple comparison with the improved free-weighting matrices method of [103] in terms of two aspects. One aspect would be due to reduced computational load as evidenced by less number of manipulated variables and faster processing. Another aspect arises by noting that LMIs (6.5) subject to (6.6) theoretically cover the results of [146, 174, 186] as special cases. Furthermore, in the absence of delay ($A_{dj} \equiv 0$, $\mathcal{Q}_j \equiv 0$, $\mathcal{W}_j \equiv 0$, it is easy to infer that LMIs (6.5) subject to (6.6) will eventually reduce to a parametrized delay-independent criteria.*

6.2.2 Linear polytopic systems

Linear polytopic systems Suppose now that system (6.1) has the state-space model

$$\begin{aligned}
\dot{x}_j(t) &= A_{j\Delta}x_j(t) + A_{dj\Delta}x_j(t-\tau_j) + B_{j\Delta}u_j(t) \\
&+ \sum_{k=1}^{n_s} E_{jk\Delta}x_k(t-\eta_{jk}) + \Gamma_{j\Delta}w_j(t) \\
z_j(t) &= G_{j\Delta}x_j(t) + G_{dj\Delta}x_(t-\tau_j) + D_{j\Delta}u_j(t) \\
&+ \Phi_{j\Delta}w_j(t)
\end{aligned} \tag{6.27}$$

whose matrices containing uncertainties which belong to a real convex bounded polytopic model of the type

$$\begin{bmatrix} A_{j\Delta} & A_{dj\Delta} & B_{j\Delta} & E_{jk\Delta} & \Gamma_{j\Delta} \\ G_{j\Delta} & G_{dj\Delta} & D_{j\Delta} & 0 & \Phi_{j\Delta} \end{bmatrix} \in \Pi_\lambda$$

$$:= \left\{ \begin{bmatrix} A_{j\lambda} & A_{dj\lambda} & B_{j\lambda} & E_{jk\lambda} & \Gamma_{j\lambda} \\ G_{j\lambda} & G_{dj\lambda} & D_{j\lambda} & 0 & \Phi_{j\lambda} \end{bmatrix} = \right.$$

$$\left. \sum_{s=1}^{N} \lambda_s \begin{bmatrix} A_{js} & A_{djs} & B_{js} & E_{jks} & \Gamma_{js} \\ G_{js} & G_{djs} & D_{js} & 0 & \Phi_{js} \end{bmatrix}, \lambda_s \in \Lambda \right\} \tag{6.28}$$

where Λ is the unit simplex

$$\Lambda := \left\{ (\lambda_1, \cdots, \lambda_N) : \sum_{j=1}^{N} \lambda_j = 1 \, , \, \lambda_j \geq 0 \right\} \tag{6.29}$$

Define the vertex set $\mathcal{N} = \{1, ..., N\}$. We use $\{A_o, ..., \Phi_o\}$ to imply generic system matrices and $\{A_{oj}, ..., \Phi_{oj}, \; j \in \mathcal{N}\}$ to represent the respective values at the vertices. It is a straightforward task to show that the following result holds.

Theorem 6.2.4 *Given $\varrho_j > 0$ and $\mu_j > 0$. System (6.1) with polytopic representation (6.28)-(6.29) and $u_j(.) \equiv 0$ is delay-dependent asymptotically stable with $\mathcal{L}_2$-performance bound γ_j if there exist weighting matrices $\mathcal{P}_j, \; \mathcal{Q}_j, \; \mathcal{W}_j, \; \mathcal{Z}_{kj}, \; k = 1, ..n_s,$ parameter matrices Θ_j Υ_j and a scalar $\gamma_j > 0$ satisfying the following LMIs for $s = 1, ..., N$*

$$\Xi_{\varrho js} = \begin{bmatrix} \Xi_{ojs} & \Xi_{ajs} & -\varrho_j\Theta_j & 0 & \mathcal{P}_j\Gamma_j & G_j^t & \varrho_j A_{js}^t \mathcal{W}_j \\ \bullet & -\Xi_{mj} & -\varrho_j\Upsilon_j & 0 & 0 & G_{djs}^t & \varrho_j A_{djs}^t \mathcal{W}_k \\ \bullet & \bullet & -\varrho_j\mathcal{W}_k & 0 & 0 & 0 & \varrho_k \sum_{k=1,k\neq j}^{n_s} E_{kjs}\mathcal{W}_k \\ \bullet & \bullet & \bullet & -\Xi_{nj} & 0 & 0 & 0 \\ \bullet & \bullet & \bullet & \bullet & -\gamma_j^2 I & \Phi_j^t & \varrho_k \Gamma_j^t \mathcal{W}_k \\ \bullet & \bullet & \bullet & \bullet & \bullet & -I & 0 \\ \bullet & \bullet & \bullet & \bullet & \bullet & \bullet & -\varrho_k\mathcal{W}_k \end{bmatrix} < 0 \tag{6.30}$$

where

$$
\begin{aligned}
\Xi_{ojs} &= \mathcal{P}_j A_{js} + A_{js}^t \mathcal{P}_j^t + \Theta_j + \Theta_j^t + \mathcal{Q}_j \\
&+ \sum_{k=1}^{n_s} \mathcal{Z}_{kj} + (n_s - 1)\mathcal{P}_j, \\
\Xi_{ajs} &= \mathcal{P}_j A_{djs} - \Theta_j + \Upsilon_j^t, \\
\Xi_{mj} &= \Upsilon_j + \Upsilon_j^t + (1 - \mu_j)\mathcal{Q}_j, \\
\Xi_{nj} &= \sum_{k=1}^{n_s} (1 - \mu_{kj})\mathcal{Z}_{kj} + \sum_{k=1, k \neq j}^{n_s} E_{kjs}^t \mathcal{P}_j E_{kjs}
\end{aligned} \tag{6.31}
$$

Remark 6.2.5 *The optimal performance-level $\gamma_j,\ j = 1, .., n_s$ can be determined in case of decentralized stability by solving the following convex optimization problems :*

Problem C: Stability

$$
\begin{aligned}
& For\ \ j, k = 1, ..., n_s, \\
& \quad Given\ \ \varrho_j,\ \ \mu_j,\ \ \varrho_{jk},\ \ \mu_{jk}, \\
& \quad \min_{\mathcal{P}_j,\ \mathcal{Q}_j,\ \mathcal{W}_j,\ \mathcal{Z}_{kj},\ \Theta_j\ \Upsilon_j} \quad \gamma_j \\
& subject\ \ to\ \ LMI(6.5)
\end{aligned} \tag{6.32}
$$

6.2.3 Illustrative example 6.1

Consider an interconnected system composed of three subsystems, each is of the type (6.1) with the following coefficients

$$Subsystem\ 1:$$

$$A_1 = \begin{bmatrix} -2 & 0 \\ -2 & -1 \end{bmatrix},\ A_{d1} = \begin{bmatrix} -1 & 0 \\ -1 & 0 \end{bmatrix},\ \Gamma_1 = \begin{bmatrix} 0.2 \\ 0.2 \end{bmatrix},\ G_1 = \begin{bmatrix} 0.2 & 0.1 \end{bmatrix},$$

$$G_{d1} = \begin{bmatrix} -0.1 & 0 \end{bmatrix},\ \Phi_1 = 0.5$$

$$Subsystem\ 2:$$

$$A_2 = \begin{bmatrix} -1 & 0 \\ -1 & -4 \end{bmatrix},\ A_{d2} = \begin{bmatrix} 1 & 0 \\ -2 & -1 \end{bmatrix},\ \Gamma_2 = \begin{bmatrix} 0.1 \\ 0.3 \end{bmatrix},\ G_2 = \begin{bmatrix} 0.2 & 0.1 \end{bmatrix},$$

$$G_{d2} = \begin{bmatrix} 0.1 & 0 \end{bmatrix},\ \Phi_2 = 0.2$$

$$Subsystem\ 3:$$

$$A_3 = \begin{bmatrix} 0 & 1 \\ -1 & -2 \end{bmatrix},\ A_{d3} = \begin{bmatrix} 0 & 0 \\ 0 & -1 \end{bmatrix},\ \Gamma_3 = \begin{bmatrix} 0.1 \\ 0.5 \end{bmatrix},\ G_3 = \begin{bmatrix} 0.1 & -0.1 \end{bmatrix},$$

$$G_{d3} = \begin{bmatrix} -0.1 & 0 \end{bmatrix},\ \Phi_3 = 0.1$$

$$Couplings\ 1:$$

$$E_{12} = \begin{bmatrix} 1 & 0 \\ 1 & 0 \end{bmatrix},\ E_{13} = \begin{bmatrix} 0 & -1 \\ 0 & -1 \end{bmatrix},$$

$$Couplings\ 2:$$

$$E_{21} = \begin{bmatrix} -1 & -2 \\ 3 & 6 \end{bmatrix},\ E_{23} = \begin{bmatrix} -1 & 1 \\ 3 & -2 \end{bmatrix},$$

$$Couplings\ 3:$$

$$E_{31} = \begin{bmatrix} 1 & 2 \\ 1 & 2 \end{bmatrix},\ E_{32} = \begin{bmatrix} 0 & 0 \\ 0 & -1 \end{bmatrix}$$

Considering Problem C, it is found that the feasible solution is attained at

$$
\begin{aligned}
&Subsystem\ 1: \\
&\varrho_1 = 3,\ \mu_1 = 1.5,\ \varrho_{21} = 2,\ \mu_{21} = 0.8, \\
&\varrho_{31} = 2,\ \mu_{31} = 0.8,\ \gamma_1 = 2.2817 \\
&\mathcal{P}_1 = \begin{bmatrix} 2.2511 & 0.0020 \\ \bullet & 0.0362 \end{bmatrix},\ \mathcal{Q}_1 = \begin{bmatrix} 1.39501 & 0.0576 \\ \bullet & 20.0181 \end{bmatrix}, \\
&Subsystem\ 2: \\
&\varrho_2 = 32.5,\ \mu_2 = 1.3,\ \varrho_{12} = 1.5,\ \mu_{12} = 0.9, \\
&\varrho_{32} = 1.5,\ \mu_{32} = 0.9,\ \gamma_2 = 0.9930 \\
&\mathcal{P}_2 = \begin{bmatrix} 0.1386 & -0.0152 \\ \bullet & 0.0724 \end{bmatrix},\ \mathcal{Q}_2 = \begin{bmatrix} 5.9043 & 0.3326 \\ \bullet & 5.4613 \end{bmatrix}, \\
&Subsystem\ 3: \\
&\varrho_3 = 3,\ \mu_3 = 1.1,\ \varrho_{13} = 1.8,\ \mu_{13} = 0.75, \\
&\varrho_{23} = 1.8,\ \mu_{23} = 0.75,\ \gamma_3 = 1.7437 \\
&\mathcal{P}_3 = \begin{bmatrix} 1.8296 & 0.1004 \\ \bullet & 0.5422 \end{bmatrix},\ \mathcal{Q}_3 = \begin{bmatrix} 21.3065 & -2.5036 \\ \bullet & 28.6509 \end{bmatrix},
\end{aligned}
$$

Since $\mathcal{P}_j,\ \mathcal{Q}_j,\ > 0, j = 1, 2, 3$ then the conditions required by Theorem 6.2.2 are satisfied.

6.2.4 State-feedback stabilization

We now direct attention to decentralized state-feedback stabilization schemes. Applying the state-feedback control $u_j(t) = K_j x_j(t)$ to the linear system (6.1) and define $A_f = A_j + B_j K_j$ and $G_f = G_j + D_j K_j$. It then follows from Theorem 6.2.2 that the resulting closed-loop system is delay-dependent asymptotically stable with $\mathcal{L}_2$-performance bound γ if there exist matrices matrices $\mathcal{P}_j,\ \mathcal{Q}_j,\ \mathcal{W}_j,\ \mathcal{Z}_{kj},\ k = 1, ..n_s,$ parameter matrices $\Theta_j\ \Upsilon_j$ and a scalar

$\gamma_j > 0$ satisfying the following LMI

$$\widehat{\Xi}_{\varrho j} = \begin{bmatrix} \Xi_{sj} & \Xi_{aj} & -\varrho_j\Theta_j & 0 & \mathcal{P}_j\Gamma_j & G_f^t & \varrho_j A_f^t \mathcal{W}_k \\ \bullet & -\Xi_{mj} & -\varrho_j\Upsilon_j & 0 & 0 & G_{dj}^t & \varrho_j A_{d_j}^t \mathcal{W}_k \\ \bullet & \bullet & -\varrho_j\mathcal{W}_k & 0 & 0 & 0 & 0 \\ \bullet & \bullet & \bullet & -\Xi_{nj} & 0 & 0 & \varrho_k \sum_{k=1,k\neq j}^{n_s} E_{kj}\mathcal{W}_k \\ \bullet & \bullet & \bullet & \bullet & -\gamma_j^2 I & \Phi_j^t & \varrho_k\Gamma_j^t\mathcal{W}_k \\ \bullet & \bullet & \bullet & \bullet & \bullet & -I & 0 \\ \bullet & \bullet & \bullet & \bullet & \bullet & \bullet & -\varrho_k\mathcal{W}_k \end{bmatrix} < 0 \tag{6.33}$$

where

$$\Xi_{sj} = \mathcal{P}_j A_f + A_f^t \mathcal{P}_j + \Theta_j + \Theta_j^t + \mathcal{Q} + \sum_{k=1,k\neq j}^{n_s} \mathcal{Z}_{kj} + (n_s - 1)\mathcal{P}_j \tag{6.34}$$

and Ξ_{aj}, Ξ_{cj}, Ξ_{mj}, Ξ_{nj} are given by (6.6). The main decentralized feedback design results are established by the following theorems

Theorem 6.2.6 *Given the bounds $\varrho_j > 0$ and $\mu_j > 0$, the interconnected system (6.1) with decentralized controller $u_j(t) = K_j x_j(t)$ is delay-dependent asymptotically stable with $\mathcal{L}_2$-performance bound γ_j if there exist weighting matrices matrices $\mathcal{X}_j$, $\mathcal{Y}_j$, $\mathcal{M}_j$, $\{\Lambda_{kj}\}_{k=1}^{n_s}$, $\{\Psi_{rj}\}_1^3$, and a scalar $\gamma_j > 0$ satisfying the following LMI*

$$\Xi_{\varrho j} = \begin{bmatrix} \Pi_{oj} & \Pi_{aj} & -\varrho_j\Psi_{1j} & 0 & \Gamma_j & \Pi_{zj}^t & \varrho_k(\mathcal{X}_j A_j^t + \mathcal{Y}_j^t B_j^t) \\ \bullet & -\Pi_{mj} & -\varrho_k\Psi_{3j} & 0 & 0 & G_{dj}^t & \varrho_k \mathcal{X}_j A_{d_j}^t \\ \bullet & \bullet & -\varrho_j\Psi_{4j} & 0 & 0 & 0 & 0 \\ \bullet & \bullet & \bullet & -\Pi_{nj} & 0 & 0 & \sum_{k=1,k\neq j}^{n_s} \varrho_k \mathcal{X}_j E_{kj} \\ \bullet & \bullet & \bullet & \bullet & -\gamma_j^2 I & \Phi_j^t & \varrho_j\Gamma_j^t \\ \bullet & \bullet & \bullet & \bullet & \bullet & -I & 0 \\ \bullet & \bullet & \bullet & \bullet & \bullet & \bullet & -\varrho_k\mathcal{M}_k \end{bmatrix} < 0 \tag{6.35}$$

where

$$\begin{aligned}
\Pi_{oj} &= A_j\mathcal{X}_j + B_j\mathcal{Y}_j + \mathcal{X}_jA_j^t + \mathcal{Y}_j^tB_j^t + \Psi_{1j} + \Psi_{1j}^t \\
&+ \Psi_{2j} + \sum_{k=1,k\neq j}^{n_s} \Lambda_{kj} + (n_s-1)\mathcal{P}_j \\
\Pi_{aj} &= A_{dj}\mathcal{X}_j - \Psi_{1j} + \Psi_{3j}^t, \\
\Pi_{mj} &= \Psi_{3j} + \Psi_{3j}^t + (1-\mu_j)\Psi_{2j} - \Psi_{4j}^t, \\
\Pi_{nj} &= \sum_{k=1,k\neq j}^{n_s} (1-\mu_{kj})\Lambda_{kj} + \sum_{k=1,k\neq j}^{n_s} \mathcal{X}_jE_{kj}^t\mathcal{P}_kE_{kj}\mathcal{X}_j \\
\Pi_{zj} &= G_{js}\mathcal{X}_j + D_{js}\mathcal{Y}_j, \;\; \mathcal{M}_j = \mathcal{W}_j^{-1}
\end{aligned} \tag{6.36}$$

Moreover, the local gain matrix is given by $K_j = \mathcal{Y}_j\mathcal{X}_j^{-1}$.

Proof: Define $\mathcal{X}_j = \mathcal{P}_j^{-1}$ and apply the congruent transformation

$$diag[\mathcal{X}_j,\ \mathcal{X}_j,\ \mathcal{X}_j,\ \mathcal{X}_j,\ I,\ I,\ I]$$

to LMI (6.33) using the linearizations

$$\begin{aligned}
\mathcal{Y}_j &= K_j\mathcal{X}_j,\ \Psi_{1j} = \mathcal{X}_j\Theta_j\mathcal{X}_j,\ \Psi_{2j} = \mathcal{X}_j\mathcal{Q}_j\mathcal{X}_j, \\
\Psi_{3j} &= \mathcal{X}_j\Upsilon_j\mathcal{X}_j,\ \Omega_j = \alpha_j\mathcal{X}_jF_j^t,\ \Lambda_{kj} = \mathcal{X}_j\mathcal{Z}_{kj}\mathcal{X}_j, \\
\Psi_{4j} &= \mathcal{X}_j\mathcal{W}_j\mathcal{X}_j
\end{aligned}$$

we readily obtain LMI (6.35) by Schur complements.

Theorem 6.2.7 *Given* $\varrho_j > 0$ *and* $\mu_j > 0$. *System (6.1) with decentralized controller* $u_j(t) = K_jx_j(t)$ *and polytopic representation (6.28)-(6.29) is delay-dependent asymptotically stable with* $\mathcal{L}_2$*-performance bound* γ_j *if there exist weighting matrices matrices* $\mathcal{X}_j,\ \mathcal{Y}_j,\ \mathcal{M}_j,\ \{\Lambda_{kj}\}_{k=1}^{n_s},\ \{\Psi_{rj}\}_1^3$, *and a scalar* $\gamma_j > 0$ *satisfying the following LMIs for* $s = 1, ..., N$

$$\Xi_{\varrho js} = \begin{bmatrix}
\Pi_{ojs} & \Pi_{ajs} & -\varrho_j\Psi_{1j} & 0 & \Gamma_j & \Pi_{zjs}^t & \varrho_j(\mathcal{X}_jA_j^t + \mathcal{Y}_j^tB_j^t) \\
\bullet & -\Pi_{mj} & -\varrho_j\Psi_{3j} & 0 & 0 & G_{dj}^t & \varrho_k\mathcal{X}_jA_{d_js}^t \\
\bullet & \bullet & -\varrho_k\Psi_{4j} & 0 & 0 & 0 & 0 \\
\bullet & \bullet & \bullet & -\Xi_{nj} & 0 & 0 & \varrho_k\sum_{k=1,k\neq j}^{n_s}\varrho_k\mathcal{X}_jE_{kjs} \\
\bullet & \bullet & \bullet & \bullet & -\gamma_j^2I & \Phi_{js}^t & \varrho_k\mathcal{X}_j\Gamma_{js}^t \\
\bullet & \bullet & \bullet & \bullet & \bullet & -I & 0 \\
\bullet & \bullet & \bullet & \bullet & \bullet & \bullet & -\varrho_k\mathcal{M}_k
\end{bmatrix} < 0 \tag{6.37}$$

where

$$\begin{aligned}
\Pi_{ojs} &= A_j\mathcal{X}_j + B_j\mathcal{Y}_j + \mathcal{X}_jA_j^t + \mathcal{Y}_j^tB_j^t + \Psi_{1j} + \Psi_{1j}^t \\
&+ \Psi_{2j} + \sum_{k=1,k\neq j}^{n_s} \Lambda_{kj} + (n_s-1)\mathcal{X}_j \\
\Pi_{ajs} &= A_{djs}\mathcal{X}_j - \Psi_{1j} + \Psi_{3j}^t, \\
\Pi_{mj} &= \Psi_{3j} + \Psi_{3j}^t + (1-\mu_j)\Psi_{2j}, \\
\Pi_{zjs} &= G_{js}\mathcal{X}_j + D_{js}\mathcal{Y}_j + \sum_{k=1,k\neq j}^{n_s} \mathcal{X}_jE_{kjs}^t\mathcal{P}_kE_{kjs}\mathcal{X}_j
\end{aligned} \tag{6.38}$$

Moreover, the local gain matrix is given by $K_j = \mathcal{Y}_j\mathcal{X}_j^{-1}$.

Proof: Follows by parallel development to Theorems 6.2.4 and 6.2.6.

Remark 6.2.8 *Similarly, the optimal performance-level* $\gamma_j,\ j = 1,..,n_s$ *can be determined in case of decentralized stabilization by solving the following convex optimization problems:*

Problem D: $\mathcal{H}_\infty$ *Stabilization*

$$\begin{aligned}
&For\ \ j,k = 1,...,n_s, \\
&\quad Given\ \ \varrho_j,\ \ \mu_j,\ \ \varrho_{jk},\ \ \mu_{jk}, \\
&\qquad \min_{\mathcal{X}_j,\, \mathcal{Y}_j,\, \mathcal{M}_j,\, \Omega_j,\, \{\Lambda_{rj}\}_{r=1}^{n_s},\, \{\Psi_{rj}\}_1^4} \qquad \gamma_j \\
&subject\ \ to\ \ LMI(6.35)
\end{aligned} \tag{6.39}$$

6.2.5 Illustrative example 6.2

Consider the interconnected system of Example 6.1 with the additional coefficients

$$Subsystem \quad 1 \ :$$

$$B_1 = \begin{bmatrix} 1 \\ 2 \end{bmatrix},\ B_2 = \begin{bmatrix} 1 & 1 \\ -1 & 2 \end{bmatrix},\ B_3 = \begin{bmatrix} 2 \\ 1 \end{bmatrix},$$
$$D_1 = [0],\ D_2 = [0],\ D_3 = [0]$$

Considering Problem D, it is found that the feasible solution is attained at

$$\gamma_1 = 1.8011,\ K_1 = \begin{bmatrix} 1.1187 & 0.8962 \end{bmatrix},$$
$$\gamma_2 = 13.4931,\ K_2 = \begin{bmatrix} -8.8441 & 1.3237 \\ -3.9501 & -3.3300 \end{bmatrix},$$
$$\gamma_3 = 1.4746,\ K_3 = \begin{bmatrix} 0.6289 & -0.2550 \end{bmatrix}$$

6.3 Nonlinear Interconnected Systems

Extending on the foregoing section we consider, in this section, a class of interconnected continuous-time systems with Lipschitz-type nonlinearities and time-varying delays within the local subsystems and across the interconnections. In our analysis, we consider the time-delay factor as a differentiable time-varying function satisfying some bounding relations. We construct appropriate Lyapunov-Krasovskii functional in order to exhibit the delay-dependent dynamics. The developed methods for decentralized stability and stabilization deploy an "injection procedure" within the individual subsystems which eliminates the need for overbounding and utilizes smaller number of LMI decision variables. By this way, new and improved solutions are developed in terms of feasibility-testing of new parametrized linear matrix inequalities (LMIs).

6.3.1 Problem Statement

We consider a class of nonlinear systems S structurally composed of n_s coupled subsystems $\mathsf{S_j}$ and modeled by the state-space model:

$$\begin{aligned}\dot{x}_j(t) &= A_j x_j(t) + A_{dj} x_j(t-\tau_j) + B_j u(t) + f_j(x_j(t),t) + h_j(x_j(t-\tau_j),t) \\ &+ \sum_{k=1,k\neq j}^{n_s} E_{jk} x_k(t-\eta_{jk}) + \Gamma_j w_j(t) \\ z_j(t) &= G_j x(t) + G_{dj} x(t-\tau_j) + D_j u(t) + \Phi_j w_j(t) \end{aligned} \tag{6.40}$$

where for $j \in \{1,...,n_s\}$, $x_j(t) \in \Re^{n_j}$ is the state vector; $u_j(t) \in \Re^{m_j}$ is the control input, $w_j(t) \in \Re^{q_j}$ is the disturbance input which belongs to $\mathcal{L}_2[0,\infty)$, $z_j(t) \in \Re^{q_j}$ is the controlled output and $\tau_j,\ \eta_{jk},\ j,k \in \{1,...,n_s\}$ are unknown time-delay factors satisfying

$$0 \le \tau_j \le \varrho_j, \quad \dot{\tau}_j \le \mu_j, \quad 0 \le \eta_{jk} \le \varrho_{jk}, \quad \dot{\eta}_{jk} \le \mu_{jk} \tag{6.41}$$

where the bounds $\varrho_j,\ \varrho_{jk},\ \mu_j,\ \mu_{jk}$ are known constants in order to guarantee smooth growth of the state trajectories. The matrices $A_j \in \Re^{n_j\times n_j}$, $B_j \in \Re^{n_j\times m_j}$, $G_j \in \Re^{q_j\times n_j}$, $D_j \in \Re^{q_j\times m_j}$, $A_{dj} \in \Re^{n_j\times n_j}$, $\Phi_j \in \Re^{q_j\times q_j}$, $\Gamma_j \in \Re^{n_j\times q_j}$, $C_j \in \Re^{p_j\times n_j}$, $C_{dj} \in \Re^{p_j\times n_j}$, $F_j \in \Re^{p_j\times m_j}$, $\Psi_j \in \Re^{p_j\times q_j}$ are real and constants. The initial condition $\langle x_j(0), x_j(r)\langle = \langle x_{oj}, \phi_j\langle,\ j \in \{1,...,n_s\}$ where $\phi_j(.) \in \mathcal{L}_2[-\tau_j^*, 0],\ j \in \{1,...,n_s\}$. The inclusion of the term $A_{dj}x_j(t-\tau_j)$ is meant to emphasize the delay within the jth subsystem.

The unknown functions $f_j = f_j(x_j(t),t) \in \Re^{n_j}$, $h_j = h_j(x_j(t),t) \in \Re^{n_j}$ are vector-valued time-varying nonlinear perturbations with $f_j(0,t) =$

$0,\ h_j(0,t) = 0, \forall\ t$ and satisfy the following Lipschitz condition for all $(x_j,t), (\hat{x_j},t) \in \Re^{n_j} \times \Re$:

$$\begin{aligned} &||f_j(x_j(t),t) - f_j(\hat{x}_j(t),t)|| \leq \alpha_j\ ||F_j(x_j - \hat{x}_j)|| \\ &||h_j(x_j(t-\tau),t) - h_j(\hat{x}_j(t-\tau),t)|| \leq \\ &\beta_j\ ||H_j\ (x_j(t-\tau) - \hat{x}_j(t-\tau))|| \end{aligned} \tag{6.42}$$

for some constants $\alpha_j > 0\ \beta_j > 0$ and $F_j \in \Re^{n_j} \times \Re^{n_j},\ H_j \in \Re^{n_j} \times \Re^{n_j}$ are constant matrices. Note as a consequence of (6.42), we have

$$\begin{aligned} ||f_j(x_j(t),t)|| &\leq \alpha_j\ ||F_j\ x_j|| \\ ||h_j(x_j(t-\tau),t)|| &\leq \beta_j\ ||H_j\ x_j(t-\tau)|| \end{aligned} \tag{6.43}$$

Equivalently stated, condition (6.42) implies that

$$\begin{aligned} &\left[f_j^t(x_j(t),t) f_j(x_j(t),t) - \alpha_j^2 x_j^t(t) F_j^t F_j x_j(t)\right] \leq 0 \\ &\left[h_j^t(x(t-\tau),t) h_j(x_j(t-\tau),t) - \beta_j^2 x_j^t(t-\tau) H_j^t H_j x_j(t-\tau)\right] \leq 0 \end{aligned} \tag{6.44}$$

Remark 6.3.1 *It must be observed that system description (6.40) might bear some resemblance to dynamic models of neural networks in terms of interconnections and delays. However, careful considerations would reveal that system (6.40) is much more general since it allows delays within the subsystems as well as across the couping links. Additionally, the performance vector is permitted to depend on delays and both the states as well as output are subject to finite-energy disturbances and not constant inputs as in neural networks. Moreover, stabilization is the ultimate goal for large-scale systems, a feature which is not shared by neural networks. The tools of system stability are common and in the sequel we will provide an improved method for stability and state-feedback stabilization.*

6.3.2 Local delay-dependent analysis

Local delay-dependent analysis In this section, we develop new criteria for LMI-based characterization of delay-dependent asymptotic stability and $\mathcal{L}_2$ gain analysis. The criteria includes some parameter matrices aims at expanding the range of applicability of the developed conditions. The following theorem establishes the main result for S with subsystems $\mathsf{S_j},\ j = 1, ..., n_s$:

Theorem 6.3.2 *Given* $\varrho_j > 0,\ \mu_j > 0,\ \varrho_{jk} > 0$ *and* $\mu_{jk} > 0,\ j,k = 1, ..., n_s$. *The free global system* S *composed of the family of subsystems* $\{\mathsf{S}_j\}$

with $u_j(.) \equiv 0$ where $\mathsf{S_j}$ is described by (6.40) is delay-dependent asymptotically stable with $\mathcal{L}_2$-performance bound $\gamma_j,\ j = 1, ..., n_s$ if there exist weighting matrices $\mathcal{P}_j,\ \mathcal{Q}_j,\ \mathcal{W}_j$ $\mathcal{Z}_{kj},\ k = 1, ..n_s,$ parameter matrices $\Theta_j\ \Upsilon_j$ and scalars $\gamma_j > 0,\ \sigma_j > 0,\ \kappa_j > 0,\ j = 1, ..., n_s$ satisfying the following LMI

$$\Xi_{\varrho j} = \begin{bmatrix} \Xi_{\varrho j1} & \Xi_{\varrho j2} \\ \bullet & \Xi_{\varrho j3} \end{bmatrix} < 0 \tag{6.45}$$

with

$$\Xi_{\varrho j1} = \begin{bmatrix} \Xi_{oj} & \Xi_{aj} & -\varrho_j\Theta_j & 0 & \mathcal{P}_j & \mathcal{P}_j \\ \bullet & -\Xi_{mj} & -\varrho_j\Upsilon_j & 0 & 0 & 0 \\ \bullet & \bullet & -\varrho_j\mathcal{W}_j & 0 & 0 & 0 \\ \bullet & \bullet & \bullet & -\Xi_{nj} & 0 & 0 \\ \bullet & \bullet & \bullet & \bullet & -\sigma_j I & 0 \\ \bullet & \bullet & \bullet & \bullet & \bullet & -\kappa_j I \end{bmatrix},$$

$$\Xi_{\varrho j2} = \begin{bmatrix} \mathcal{P}_j\Gamma_j & G_j^t & \varrho_j A_j^t\mathcal{W}_k \\ 0 & G_{dj}^t & \varrho_j A_{d_j}^t\mathcal{W}_k \\ 0 & 0 & \varrho_k \sum_{k=1,k\neq j}^{n_s} E_{kj}^t\mathcal{W}_k \\ 0 & 0 & 0 \\ 0 & 0 & \mathcal{W}_k \\ 0 & 0 & \mathcal{W}_k \end{bmatrix},$$

$$\Xi_{\varrho j3} = \begin{bmatrix} -\gamma_j^2 I & \Phi_j^t & \varrho_j\Gamma_j^t\mathcal{W}_k \\ \bullet & -I & 0 \\ \bullet & \bullet & -\varrho_k\mathcal{W}_k \end{bmatrix} \tag{6.46}$$

where

$$\begin{aligned} \Xi_{oj} &= \mathcal{P}_j A_j + A_j^t\mathcal{P}_j^t + \Theta_j + \Theta_j^t + \mathcal{Q}_j + \sigma_j\alpha_j^2 F_j^t F_j + \sum_{k=1}^{n_s}\mathcal{Z}_{kj} + (n_s - 1)\mathcal{P}_j, \\ \Xi_{aj} &= \mathcal{P}_j A_{dj} - \Theta_j + \Upsilon_j^t, \\ \Xi_{mj} &= \Upsilon_j + \Upsilon_j^t + (1 - \mu_j)\mathcal{Q}_j - \kappa_j\beta_j^2 H_j^t H_j, \\ \Xi_{nj} &= \sum_{k=1}^{n_s}(1 - \mu_{kj})\mathcal{Z}_{kj} + \sum_{k=1,k\neq j}^{n_s} E_{kj}^t\mathcal{P}_k E_{kj} \end{aligned} \tag{6.47}$$

Proof: It is readily seen from (6.46) that there exists a scalar $\omega_j > 0$ such that

$$\bar{\Xi}_{\varrho j} = \begin{bmatrix} \bar{\Xi}_{\varrho j1} & \Xi_{\varrho j2} \\ \bullet & \Xi_{\varrho j3} \end{bmatrix} < 0 \tag{6.48}$$

$$\bar{\Xi}_{\varrho j1} = \begin{bmatrix} \Xi_{oj} + \omega_j I & \Xi_{aj} & -\Theta_j & \Xi_{cj} & \mathcal{P}_j & \mathcal{P}_j \\ \bullet & -\Xi_{mj} & -\Upsilon_j & 0 & 0 & 0 \\ \bullet & \bullet & -\varrho_j^{-1}\mathcal{W}_j & 0 & 0 & 0 \\ \bullet & \bullet & \bullet & -\Xi_{nj} & 0 & 0 \\ \bullet & \bullet & \bullet & \bullet & -\sigma_j I & 0 \\ \bullet & \bullet & \bullet & \bullet & \bullet & -\kappa_j I \end{bmatrix} \tag{6.49}$$

Therefore, for all τ_j satisfying (6.41) we have

$$\Xi_{\omega_j} = \begin{bmatrix} \tilde{\Xi}_{\varrho j1} & \Xi_{\varrho j2} \\ \bullet & \Xi_{\varrho j3} \end{bmatrix} < 0 \tag{6.50}$$

$$\tilde{\Xi}_{\varrho j1} = \begin{bmatrix} \Xi_{oj} & \Xi_{aj} & -\tau_j\Theta_j & \Xi_{cj} & \mathcal{P}_j & \mathcal{P}_j \\ \bullet & -\Xi_{mj} & -\tau_j\Upsilon_j & 0 & 0 & 0 \\ \bullet & \bullet & -\tau_j\mathcal{W}_j & 0 & 0 & 0 \\ \bullet & \bullet & \bullet & -\Xi_{nj} & 0 & 0 \\ \bullet & \bullet & \bullet & \bullet & -\sigma_j I & 0 \\ \bullet & \bullet & \bullet & \bullet & \bullet & -\kappa_j I \end{bmatrix} \tag{6.51}$$

Now consider the Lyapunov-Krasovskii functional (LKF) for the global system S:

$$\begin{aligned} V(t) &= \sum_{j=1}^{n_s} \Big[V_{oj}(t) + V_{aj}(t) + V_{mj}(t) + V_{nj}(t) \Big] \\ V_{oj}(t) &= x_j^t(t)\mathcal{P}_j x_j(t), \;\; V_{aj}(t) = \int_{-\varrho_j}^{0} \int_{t+s}^{t} \dot{x}_j^t(\alpha)\mathcal{W}_j\dot{x}_j(\alpha)d\alpha\, ds, \\ V_{mj}(t) &= \int_{t-\tau_j}^{t} x_j^t(s)\mathcal{Q}_j x_j(s)\, ds, \\ V_{nj}(t) &= \sum_{k=1,k\neq j}^{n_s} \int_{t-\eta_{jk}}^{t} x_k^t(s)\mathcal{Z}_{jk}x_k(s)\, ds \end{aligned} \tag{6.52}$$

where $0 < \mathcal{P}_j = \mathcal{P}_j^t,\ 0 < \mathcal{W}_j = \mathcal{W}_j^t,\ 0 < \mathcal{Q}_j = \mathcal{Q}_j^t,\ 0 < \mathcal{R}_j = \mathcal{R}_j^t,\ 0 < \mathcal{Z}_{jk} = \mathcal{Z}_{jk}^t,\ j,k \in \{1, ..., n_s\}$ are weighting matrices of appropriate dimensions. The first term in (6.52) is standard to nominal systems without delay while the second and third terms correspond to the delay-dependent conditions since they provide measures of the individual and difference signal energies during the delay-period [1] and the fourth term corresponds to the intra-connection delays. A straightforward computation gives the time-derivative of $V_j(t)$ along the solutions of (6.40) with $w(t) \equiv 0$ as:

$$\begin{aligned} \dot{V}_{oj}(t) &= 2x_j^t \mathcal{P}_j \dot{x}_j = 2x_j^t \mathcal{P}_j [A_j x(t) + A_{dj} x_j(t-\tau_j) + f_j + h_j] \\ &+ 2x_j^t \mathcal{P}_j \sum_{k=1}^{n_s} E_{jk} x_k(t-\eta_{jk})] \end{aligned} \tag{6.53}$$

Expanding terms of (6.53) results in

$$\begin{aligned} \dot{V}_{oj}(t) &= 2x_j^t \mathcal{P}[A_j x_j(t) + A_{dj}] x_j(t) - 2x_j^t \mathcal{P}_j A_{dj} \int_{t-\tau_j}^{t} \dot{x}_j(s) ds \\ &+ 2x_j^t \mathcal{P}_j [f_j + h_j] + 2x_j^t \mathcal{P}_j \sum_{k=1}^{n_s} E_{jk} x_k(t-\eta_{jk})] \\ &= 2x_j^t \mathcal{P}_j [A_j x_j(t) + A_{dj}] x_j(t) + 2x_j^t [\Theta_j - \mathcal{P}_j A_{dj}] \int_{t-\tau_j}^{t} \dot{x}_j(s) ds \\ &+ 2x^t(t-\tau_j) \Upsilon_j \int_{t-\tau_j}^{t} \dot{x}_j(s) ds \\ &- \left[2x_j^t t \Theta_j \int_{t-\tau_j}^{t} \dot{x}_j(s) ds + 2x^t(t-\tau_j) \Upsilon_j \int_{t-\tau_j}^{t} \dot{x}_j(s) ds \right] + 2x_j^t \mathcal{P}_j [f_j + \\ &+ 2x_j^t \mathcal{P}_j \sum_{k=1}^{n_s} E_{jk} x_k(t-\eta_{jk})] \end{aligned} \quad (6$$

Little algebra on (6.13) yields

$$\begin{aligned} \dot{V}_{oj}(t) &= \frac{1}{\tau_j} \int_{t-\tau_j}^{t} \Big[2x_j^t [\mathcal{P}_j A_j + \Theta_j] x_j + 2x_j^t [\mathcal{P}_j A_{dj} - \Theta_j + \Upsilon_j^t] x_j(t-\tau_j) \\ &- 2x_j^t (t-\tau_j) \Upsilon_j x_j(t-\tau_j) \\ &- 2x_j^t \tau_j \Theta_j \dot{x}_j(s) - 2x_j^t (t-\tau_j) \tau_j \Upsilon_j \dot{x}_j(s) + 2x_j^t \mathcal{P}_j [f_j + h_j] \\ &+ 2x_j^t \mathcal{P}_j \sum_{k=1}^{n_s} E_{jk} x_k(t-\eta_{jk})] \Big] ds \end{aligned} \tag{6.55}$$

where $\Theta_j \in \Re^{n_j \times n_j}$ and $\Upsilon_j \in \Re^{n_j \times n_j}$ are appropriate relaxation matrices

[1] Recall that $\int_{t-\varrho_j}^{t} \dot{x}_j(\alpha) d\alpha = x_j(t) - x_j(t-\varrho_j)$ by Leibniz-Newton formula

injected to facilitate the delay-dependence analysis. Proceeding further,

$$\begin{aligned}\dot{V}_{aj}(t) &= \int_{-\varrho_j}^{0} [\dot{x}_j^t(t)\mathcal{W}_j\dot{x}_j(t) - \dot{x}_j^t(t+s)\mathcal{W}_j\dot{x}_j(t+s)]ds \\ &= \int_{t-\varrho_j}^{t} [\dot{x}_j^t(t)\mathcal{W}_j\dot{x}_j(t) - \dot{x}_j^t(s)\mathcal{W}_j\dot{x}_j(s)]d\,s \end{aligned} \tag{6.56}$$

Little algebra on (6.56) yields

$$\begin{aligned}\dot{V}_{aj}(t) &= \varrho_j\,\dot{x}_j^t(t)\mathcal{W}_j\dot{x}_j(t) - \int_{t-\tau_j}^{t} \dot{x}_j^t(s)\mathcal{W}_j\dot{x}_j(s)ds - \int_{t-\varrho_j}^{t-\tau_j} \dot{x}_j^t(s)\mathcal{W}_j\dot{x}_j(s) \\ &\leq \varrho_j\,\dot{x}_j^t(t)\mathcal{W}_j\dot{x}_j(t) - \int_{t-\tau_j}^{t} \dot{x}_j^t(s)\mathcal{W}_j\dot{x}_j(s)ds \\ &= \frac{1}{\tau_j}\int_{t-\tau_j}^{t} \Big[\varrho_j\dot{x}_j^t(t)\mathcal{W}_j\dot{x}_j(t) - \tau_j\dot{x}_j^t(s)\mathcal{W}_j\dot{x}_j(s)\Big] \end{aligned} \tag{6.57}$$

Note that the term $T_j = \int_{t-\varrho_j}^{t-\tau_j} \dot{x}_j^t(s)\mathcal{W}_j\dot{x}_j(s)$ accounts for the enlarged time interval from $t-\varrho_j \to t$ to $t-\tau_j \to t$. It is readily evident that $T_j > 0$ and hence expression (6.57) holds true without conservatism. There has been alternative route to handle T_j by employing extra parameter matrices and adding more identities [103]-[146]. Next we have

$$\begin{aligned}\dot{V}_{mj}(t) &= x_j^t(t)\mathcal{Q}_jx_j(t) - (1-\dot{\tau}_j)\,x^t(t-\tau_j)\mathcal{Q}_jx_j(t-\tau_j) \\ &\leq x_j^t(t)\mathcal{Q}_jx_j(t) - (1-\mu_j)\,x_j^t(t-\tau_j)\mathcal{Q}_jx_j(t-\tau_j) \\ &= \frac{1}{\tau_j}\int_{t-\tau_j}^{t} \Big[x_j^t(t)\mathcal{Q}_jx_j(t) - (1-\mu_j)\,x_j^t(t-\tau_j)\mathcal{Q}_jx_j(t-\tau_j)\Big] \end{aligned} \tag{6.58}$$

Also, we obtain

$$\begin{aligned}\dot{V}_{nj}(t) &= \sum_{k=1,k\neq j}^{n_s} x_k^t(t)\mathcal{Z}_{jk}x_k(t) - \sum_{k=1,k\neq j}^{n_s} (1-\dot{\eta}_{jk})\,x_k^t(t-\eta_{jk})\mathcal{Z}_{jk}x_k(t-\eta_{jk}) \\ &\leq \sum_{k=1,k\neq j}^{n_s} x_k^t(t)\mathcal{Z}_{jk}x_k(t) - \sum_{k=1,k\neq j}^{n_s} (1-\mu_{jk})\,x_k^t(t-\eta_{jk})\mathcal{Z}_{jk}x_k(t-\eta_{jk}) \end{aligned} \tag{6.59}$$

For the class of interconnected systems (6.40), the following structural identity holds

$$\sum_{j=1}^{n_s}\sum_{k=1,k\neq j}^{n_s} x_k^t(t)\mathcal{Z}_{jk}x_k(t) = \sum_{j=1}^{n_s}\sum_{k=1,k\neq j}^{n_s} x_j^t(t)\mathcal{Z}_{kj}x_j(t) \tag{6.60}$$

It should be noted that relation (6.60) is employed to preserve the local variables with subsystem S_j thereby assuring decentralized computation structure. Taking into account (6.44) via the S-procedure [29], it follows that there exist scalars $\sigma_j > 0, \ \kappa_j > 0$ such that by combining (6.52)-(6.60) and Schur complements, we have

$$\dot{V}(t)|_{(6.40)} \leq \sum_{j=1}^{n_s} \Big[\frac{1}{\tau_j} \int_{t-\tau_j}^{t} \chi_j^t(t,s)\, \overline{\Xi}_j\, \chi_j(t,s)\, ds\Big] \tag{6.61}$$

$$\chi_j(t,s) = \big[x_j^t(t) \;\; x_j^t(t-\tau_j) \;\; \dot{x}_j^t(s) \;\; x_j^t(t-\eta_{kj}) \;\; f_j^t \;\; h_j^t\big]^t \tag{6.62}$$

where $\dot{V}(t)|_{(6.40)}$ defines the Lyapunov derivative along the solutions of the global system S with subsystem S_j given by (6.40). Note that $\overline{\Xi}_j$ corresponds to $\Xi_{\omega j}$ in (6.51) with $G_j \equiv 0, \ G_{dj} \equiv 0, \ \Gamma_j \equiv 0, \ \Phi_j \equiv 0$. In view of (6.51), it follows that $\overline{\Xi}_j < 0$ and more importantly

$$\begin{aligned} \dot{V}_j(t)|_{(6.40)} \ &< \ \sum_{j=1}^{n_s} \Big[\frac{1}{\tau_j} \int_{t-\tau_j}^{t} \chi_j^t(t,s) diag[-\omega_j, \ 0, \ 0, \ 0, \ 0, \ 0]\chi(t,s)_j\, ds\Big] \\ &= \ -\sum_{j=1}^{n_s} \omega_j\, ||x_j||^2 \ < \ 0 \end{aligned} \tag{6.63}$$

This establishes the internal asymptotic stability.

Consider the performance measure for the global system S

$$J = \sum_{j=1}^{n_s} \int_0^{\infty} \Big(z_j^t(s) z_j(s) - \gamma_j^2 w_j^t(s) w_j(s) \Big) ds$$

For any $w_j(t) \in \mathcal{L}_2(0,\infty) \neq 0$ and zero initial condition $x_j(0) = 0$, we have

$$\begin{aligned} J \ &= \ \sum_{j=1}^{n_s} \int_0^{\infty} \Big(z_j^t(s) z_j(s) - \gamma_j^2 w_j^t(s) w_j(s) \ + \dot{V}_j(t)|_{(6.40)} - \ \dot{V}_j(t)|_{(6.40)} \Big) ds \\ &\leq \ \sum_{j=1}^{n_s} \int_0^{\infty} \Big(z_j^t(s) z_j(s) - \gamma_j^2 w_j^t(s) w_j(s) \ + \dot{V}_j(t)|_{(6.40)} \Big) ds \end{aligned}$$

Proceeding as before, we get

$$\sum_{j=1}^{n_s} \Big(z_j^t(s) z_j(s) - \gamma_j^2 w_j^t(s) w_j(s) \ + \dot{V}_j(t)|_{(6.40)} \Big) = \sum_{j=1}^{n_s} \bar{\chi}_j^t(t,s)\, \widehat{\Xi}_j\, \bar{\chi}_j(t,s),$$

$$\bar{\chi}_j(t,s) = \big[\ x_j^t(t) \;\; x_j^t(t-\tau_j) \;\; \dot{x}_j^t(s) \;\; f_j^t \;\; h_j^t \;\; w_j^t(s)\ \big]^t$$

where $\widehat{\Xi}_j$ corresponds to $\Xi_{\varrho j}$ in (6.65) by Schur complements. It is readily seen from (6.65) by Schur complements that

$$\sum_{j=1}^{n_s} \left(z_j^t(s) z_j(s) - \gamma_j^2 w_j^t(s) w_j(s) \ + \dot{V}_j(t)|_{(6.40)} \right) < 0$$

for arbitrary $s \in [t, \infty)$, which implies for any $w_j(t) \in \mathcal{L}_2(0, \infty) \neq 0$ that $J < 0$ leading to $\sum_{j=1}^{n_s} ||z_j(t)||_2 \ < \ \sum_{j=1}^{n_s} \gamma_j \ ||w(t)_j||_2$ and the proof is completed.

Remark 6.3.3 *It is significant to recognize that our methodology incorporates three weighting matrices $\mathcal{P}_j$, $\mathcal{Q}_j$, $\mathcal{W}_j$, and two parameter matrices Θ_j, Υ_j at the subsystem level to ensure least conservative delay-dependent stability results. This will eventually result in reduced computational requirements as evidenced by a simple comparison with the improved free-weighting matrices method of [103]. The performance of the developed stability criteria is assessed in terms of the number of LMI variables L, the number of other unknown variables U and the basic storage requirements S. Based thereon, a comparison of the computational requirements of different methods is given in Table 1. It is readily evident from the developed method in this paper requires the least computational effort on the subsystems level, a significant feature that is naturally needed for large-scale interconnected systems.*

Method	L	U	S
Ref. [79]	4	4	$6n^2 + 2n$
Ref. [103]	5	9	$11.5n^2 + 2.5n$
Ref. [146]	11	2	$7.5n^2 + 5.5n$
Ref. [186]	7	9	$12.5n^2 + 3.5n$
Ref. [290]	5	6	$8.5n^2 + 2.5n$
Theorem 6.3.2	3	2	$3.5n^2 + 2n$

Table 6.1: Computational Requirements: A Comparison

6.3.3 State feedback stabilization

We now direct attention to decentralized state-feedback stabilization schemes. Applying the state-feedback control $u_j(t) = K_j x_j(t)$ to the nonlinear system (6.40) and define $A_f = A_j + B_j K_j$ and $G_f = G_j + D_j K_j$. It then follows from Theorem 6.3.2 that the resulting closed-loop system is delay-dependent asymptotically stable with $\mathcal{L}_2$-performance bound γ_j if there exist matrices matrices $\mathcal{P}_j,\ \mathcal{Q}_j,\ \mathcal{W}_j,\ \mathcal{Z}_{kj},\ k = 1, ..n_s,$ parameter matrices $\Theta_j\ \Upsilon_j$ and scalars $\gamma_j > 0,\ \sigma_j > 0,\ \kappa_j > 0$ satisfying the following LMI

$$\widehat{\Xi}_{\varrho j} = \begin{bmatrix} \widehat{\Xi}_{\varrho j1} & \widehat{\Xi}_{\varrho j2} \\ \bullet \widehat{\Xi}_{\varrho j3} & \end{bmatrix} < 0 \tag{6.64}$$

with

$$\widehat{\Xi}_{\varrho j1} = \begin{bmatrix} \Xi_{sj} & \Xi_{aj} & -\varrho_j\Theta_j & 0 & \mathcal{P}_j & \mathcal{P}_j \\ \bullet & -\Xi_{mj} & -\varrho_j\Upsilon_j & 0 & 0 & 0 \\ \bullet & \bullet & -\varrho_j\mathcal{W}_j & 0 & 0 & 0 \\ \bullet & \bullet & \bullet & -\Xi_{nj} & 0 & 0 \\ \bullet & \bullet & \bullet & \bullet & -\sigma_j I & 0 \\ \bullet & \bullet & \bullet & \bullet & \bullet & -\kappa_j I \end{bmatrix} \tag{6.65}$$

where

$$\begin{aligned} \Xi_{sj} &= \mathcal{P}_j A_f + A_f^t \mathcal{P}_j + \Theta_j + \Theta_j^t + \mathcal{Q} + \sigma_j \alpha_j^2 F_j^t F_j \\ &+ \sum_{k=1,k\neq j}^{n_s} \mathcal{Z}_{kj} + (n_s - 1)\mathcal{P}_j \end{aligned} \tag{6.66}$$

and $\Xi_{aj},\ \Xi_{cj},\ \Xi_{mj},\ \Xi_{nj}$ are given by (6.47). The main decentralized feedback design results are established by the following theorem

Theorem 6.3.4 *Given $\varrho_j > 0$ and $\mu_j > 0$. System (6.40) with decentralized controller $u_j(t) = K_j x_j(t)$ is delay-dependent asymptotically stable with $\mathcal{L}_2$-performance bound γ_j if there exist weighting matrices*

$$\mathcal{X}_j,\ \mathcal{Y}_j,\ \mathcal{M}_j,\ \Omega_j,\ \{\Lambda_{kj}\}_{k=1}^{n_s},\ \{\Psi_{rj}\}_1^4,\ \gamma_j > 0,\ \sigma_j > 0,\ \kappa_j > 0$$

satisfying the following LMI

$$\Xi_{\varrho j} = \begin{bmatrix} \Xi_{\varrho j1} & \Xi_{\varrho j2} \\ \bullet & \Xi_{\varrho j3} \end{bmatrix} < 0 \tag{6.67}$$

where

$$\Xi_{\varrho j1} = \begin{bmatrix} \Pi_{oj} & \Pi_{aj} & -\varrho_j\Psi_{1j} & 0 & I & I \\ \bullet & -\Pi_{mj} & -\varrho_j\Psi_{3j} & 0 & 0 & 0 \\ \bullet & \bullet & -\varrho_j\Pi_{wj} & 0 & 0 & 0 \\ \bullet & \bullet & \bullet & -\Pi_{nj} & 0 & 0 \\ \bullet & \bullet & \bullet & \bullet & -\sigma_j I & 0 \\ \bullet & \bullet & \bullet & \bullet & \bullet & -\kappa_j I \end{bmatrix}$$

$$\Xi_{\varrho j2} = \begin{bmatrix} \Gamma_j & \Pi_{zj}^t & \varrho_j(\mathcal{X}_j A_j^t + \mathcal{Y}_j^t B_j^t) & \Omega_j \\ 0 & G_{dj}^t & \varrho_k \mathcal{X}_j A_{d_j}^t & 0 \\ 0 & 0 & 0 & 0 \\ 0 & 0 & \sum_{k=1,k\neq j}^{n_s} \varrho_k \mathcal{X}_j E_{kj}^t & 0 \\ 0 & 0 & I & 0 \\ 0 & 0 & I & 0 \end{bmatrix}$$

$$\Xi_{\varrho j3} = \begin{bmatrix} -\gamma_j^2 I & \Phi_j^t & \varrho_k \mathcal{X}_j \Gamma_j^t & 0 \\ \bullet & -I & 0 & 0 \\ \bullet & \bullet & -\varrho_k \mathcal{M}_k & 0 \\ \bullet & \bullet & \bullet & -\sigma_j I \end{bmatrix} \tag{6.68}$$

with

$$\begin{aligned} \Pi_{oj} &= A_j\mathcal{X}_j + B_j\mathcal{Y}_j + \mathcal{X}_j A_j^t + \mathcal{Y}_j^t B_j^t + \Psi_{1j} + \Psi_{1j}^t + \Psi_{2j} \\ &+ \sum_{k=1,k\neq j}^{n_s} \Lambda_{kj} + (n_s - 1)\mathcal{X}_j, \\ \Pi_{aj} &= A_{dj}\mathcal{X}_j - \Psi_{1j} + \Psi_{3j}^t, \\ \Pi_{mj} &= \Psi_{3j} + \Psi_{3j}^t + (1-\mu_j)\Psi_{2j} - \Psi_{4j}^t, \\ \Pi_{nj} &= \sum_{k=1,k\neq j}^{n_s} (1-\mu_{kj})\Lambda_{kj} + \sum_{k=1,k\neq j}^{n_s} \mathcal{X}_j E_{kj}^t \mathcal{P}_k E_{kj}\mathcal{X}_j, \\ \Pi_{zj} &= G_{js}\mathcal{X}_j + D_{js}\mathcal{Y}_j,\ \mathcal{M}_j = \mathcal{W}_j^{-1} \end{aligned} \tag{6.69}$$

Moreover, the local gain matrix is given by $K_j = \mathcal{Y}_j\mathcal{X}_j^{-1}$.

Proof: Define $\mathcal{X}_j = \mathcal{P}_j^{-1}$ and apply the congruent transformation

$$diag[\mathcal{X}_j,\ \mathcal{X}_j,\ \mathcal{X}_j,\ \mathcal{X}_j,\ I,\ I,\ I,\ I,\ I]$$

to LMI (6.64) using the linearizations

$$\begin{aligned} \mathcal{Y}_j &= K_j\mathcal{X}_j,\ \Psi_{1j} = \mathcal{X}_j\Theta_j\mathcal{X}_j,\ \Psi_{2j} = \mathcal{X}_j\mathcal{Q}_j\mathcal{X}_j,\ \Psi_{3j} = \mathcal{X}_j\Upsilon_j\mathcal{X}_j, \\ \Omega_j &= \alpha_j\mathcal{X}_j F_j^t,\ \Lambda_{kj} = \mathcal{X}_j\mathcal{Z}_{kj}\mathcal{X}_j,\ \Psi_{4j} = \kappa_j\mathcal{X}_j\beta_j^2 H_j^t H_j\mathcal{X}_j \end{aligned}$$

we readily obtain LMI (6.67) by Schur complements.

Remark 6.3.5 *The optimal performance-level $\gamma_j, \;\; j = 1,..,n_s$ can be determined in case of decentralized stability and stabilization of interconnected nonlinear systems by solving the following convex optimization problems:*
Problem A: Stability:

$$\begin{aligned} & For \;\; j,k = 1,...,n_s, \\ & \quad Given \;\; \varrho_j, \;\; \mu_j, \;\; \varrho_{jk}, \;\; \mu_{jk}, \\ & \qquad \min_{\mathcal{P}_j,\, \mathcal{Q}_j,\, \mathcal{W}_j,\, \mathcal{Z}_{kj},\, \Theta_j\, \Upsilon_j,\, \sigma_j,\, \kappa_j} \qquad \gamma_j \\ & subject \;\; to \;\; LMI(6.65) \end{aligned} \tag{6.70}$$

Problem B: $\mathcal{H}_\infty$ Stabilization

$$\begin{aligned} & For \;\; j,k = 1,...,n_s, \\ & \quad Given \;\; \varrho_j, \;\; \mu_j, \;\; \varrho_{jk}, \;\; \mu_{jk}, \\ & \qquad \min_{\mathcal{X}_j,\, \mathcal{Y}_j,\, \mathcal{M}_j,\, \Omega_j,\, \{\Lambda_{rj}\}_{r=1}^{n_s},\, \{\Psi_{rj}\}_1^4,\, \sigma_j,\, \kappa_j} \qquad \gamma_j \\ & subject \;\; to \;\; LMI(6.67) \end{aligned} \tag{6.71}$$

6.3.4 Illustrative example 6.3

Consider an interconnected system composed of three subsystems, each is of the type (6.40) with the following coefficients

$Subsystem \quad 1 \quad :$

$$
\begin{aligned}
A_1 &= \begin{bmatrix} -2 & 0 \\ -2 & -1 \end{bmatrix}, \; \Gamma_1 = \begin{bmatrix} 0.2 \\ 0.2 \end{bmatrix}, \; B_1 = \begin{bmatrix} 1 \\ 1 \end{bmatrix}, \\
A_{d1} &= \begin{bmatrix} -1 & 0 \\ -1 & 0 \end{bmatrix} \\
G_1 &= \begin{bmatrix} 0.2 & 0.1 \end{bmatrix}, \; \Phi_1 = 0.5, \; \beta_1 = 2, \; \alpha_1 = 1, \\
G_{d1} &= \begin{bmatrix} -0.1 & 0 \end{bmatrix}, \\
F_1^t F_1 &= \begin{bmatrix} 0.04 & 0 \\ 0 & 0.04 \end{bmatrix}, \; H_1^t H_1 = \begin{bmatrix} 0.01 & 0 \\ 0 & 0.01 \end{bmatrix},
\end{aligned}
$$

$Couplings \quad 1 \quad :$

$$
E_{12} = \begin{bmatrix} 1 & 0 & 1 \\ 1 & 0 & 1 \end{bmatrix}, \; E_{13} = \begin{bmatrix} 0 & -1 \\ 0 & -1 \end{bmatrix},
$$

$Subsystem \quad 2 :$

$$
\begin{aligned}
A_2 &= \begin{bmatrix} -1 & 0 & 0 \\ -1 & -4 & -1 \\ 1 & 1 & 0 \end{bmatrix}, \; \Gamma_2 = \begin{bmatrix} 0.1 \\ 0.3 \\ 0.4 \end{bmatrix}, \; B_2^t = \begin{bmatrix} 1 & 1 & 0 \\ -1 & 2 & 1 \end{bmatrix}, \\
G_2 &= \begin{bmatrix} 0.2 & 0.1 & 0 \end{bmatrix}, \; \Phi_2 = 0.2, , \; \beta_2 = 1, \; \alpha_2 = 1, \\
G_{d2} &= \begin{bmatrix} 0.1 & 0 & -0.2 \end{bmatrix} \\
F_2^t F_2 &= \begin{bmatrix} 0.09 & 0 \\ 0 & 0.09 \end{bmatrix}, \; H_2^t H_2 = \begin{bmatrix} 0.01 & 0 \\ 0 & 0.01 \end{bmatrix}, \\
A_{d2} &= \begin{bmatrix} 1 & 0 & 0 \\ -2 & -1 & 0 \\ -1 & 0 & 0 \end{bmatrix},
\end{aligned}
$$

$Couplings \quad 2 \quad :$

$$
E_{21} = \begin{bmatrix} -1 & -2 \\ 3 & 6 \\ 1 & 2 \end{bmatrix}, \; E_{23} = \begin{bmatrix} -1 & 1 \\ 3 & -2 \\ 1 & -1 \end{bmatrix},
$$

$Subsystem \quad 3 \quad :$

$$
\begin{aligned}
A_3 &= \begin{bmatrix} 0 & 1 \\ -1 & -2 \end{bmatrix}, \; \Gamma_3 = \begin{bmatrix} 0.1 \\ 0.5 \end{bmatrix}, \; B_3 = \begin{bmatrix} 1 & 0 \\ 0 & 1 \end{bmatrix}, \\
A_{d3} &= \begin{bmatrix} 0 & 0 \\ 0 & -1 \end{bmatrix}, \\
G_3 &= \begin{bmatrix} 0.1 & -0.1 \end{bmatrix}, \; \Phi_3 = 0.1, \; \beta_1 = 1, \; \alpha_1 = 2, \\
G_{d3} &= \begin{bmatrix} -0.1 & 0 \end{bmatrix}, \\
F_3^t F_3 &= \begin{bmatrix} 0.01 & 0 \\ 0 & 0.01 \end{bmatrix}, \; H_1^t H_1 = \begin{bmatrix} 0.04 & 0 \\ 0 & 0.04 \end{bmatrix},
\end{aligned}
$$

Using the LMI-solver Scilab-5.1, it is found that the feasible solution of Problem A is attained at $\mu_1 = 1.8455,\ \varrho_1 = 0.895,\ \mu_2 = 1.9191,\ \varrho_2 = 0.8333,\ \mu_3 = 1.7784,\ \varrho_3 = 0.769,$ with disturbance attenuation levels $\gamma_1 = 3.5115,\ \gamma_2 = 3.1359,\ \gamma_3 = 3.6244$ thereby validating Theorem 6.3.2. Turning to Problem B, the feasible solution is summarized below

$$\varrho_1 = 1.154,\ \mu_1 = 1.015,\ K_1 = \left[\begin{array}{cc} 4.061 & 1.536 \end{array}\right],\ \gamma_1 = 3.3875,$$

$$\varrho_2 = 2.063,\ \mu_2 = 1.256,\ K_2 = \left[\begin{array}{ccc} 2.314 & 1.816 & 0.957 \\ -0.742 & 2.314 & 4.315 \end{array}\right],\ \gamma_2 = 3.2977,$$

$$\varrho_3 = 1.225,\ \mu_3 = 0.923,\ K_3 = \left[\begin{array}{cc} 4.498 & 0.155 \\ 0.163 & 3.147 \end{array}\right],\ \gamma_3 = 2.9938$$

6.4 Interconnected Discrete-Time Delay Systems

Large-scale interconnected systems can be found in such diverse fields as electrical power systems, space structures, manufacturing processes, transportation, and communication. An important motivation for the design of decentralized feedback control (DFC) schemes is that the information exchange between subsystems of a large-scale system is not needed; hence, the individual subsystem controllers are simple and use only locally available information. In this way, multiple separate controllers are articulated each of which has access to different measured information and has authority over different decision or actuation variables.

Decentralized control of large-scale systems has received considerable interest in the systems and control literature. It effectively exploits the information structure constraint commonly exists in many practical large-scale systems. Over the past few decades, a large body of literature has become available on this subject, see [57]-[406] and the references therein. The computational advantages of DFC approach have also recently attracted considerable attention, particularly in the context of parallel processing.

In recent years, a new systematic methodology has been proposed in [326, 348] based on linear matrix inequalities (LMIs) [29] and extended to continuous-time interconnected systems in [326, 327, 348, 398] and to time-delay systems in [28, 260, 261]. The appealing feature of these results is that the underlying problem is formulated as a convex optimization problem, which is designed to maximize the system robustness with respect to uncertainties.

The main objective of this section is to develop an LMI-based method for testing robust stability and designing feedback (state and output) schemes for robust decentralized stabilization of interconnected systems. We deal with systems composed of linear discrete-time subsystems coupled by static nonlinear interconnections satisfying quadratic constraints and subject to unknown-but-bounded state-delays. The developed solution is cast into the framework of convex optimization problem over LMIs. We then develop delay-dependent robust stability and feedback stabilization results. Feedback controllers are designed on the subsystem level to guarantee robust stability of the overall system and, in addition, maximize the bounds of unknown interconnection terms. By incorporating additive gain perturbations we establish new resilient feedback stabilization schemes for discrete-time delay systems.

6.4.1 Model description and preliminaries

A class of nonlinear interconnected discrete-time systems Σ with state-delay which is composed of N coupled subsystems $\Sigma_j,\ j = 1,..,N$ is represented by:

$$\begin{aligned} \Sigma: \ x(k+1) &= Ax(k) + Bu(k) + Dx(k-d(k)) + g(k, x(k), x(k-d(k))) \\ y(k) &= Cx(k) + Hu(k) \end{aligned} \qquad (6.72)$$

where for $k \in Z_+ := \{0, 1, ...\}$ and $x = (x_1^t, ..., x_N^t)^t \in \Re^n$, $n = \sum_{j=1}^N n_j$, $u = (u_1^t, ..., u_N^t)^t \in \Re^p$, $p = \sum_{j=1}^N p_j$, $y = (y_1^t, ..., y_N^t)^t \in \Re^q$, $q = \sum_{j=1}^N q_j$ being the state, control and output vectors of system Σ. The associated matrices are real constants and modeled as $A = diag\{A_1, .., A_N\}$, $A_j \in \Re^{n_j \times n_j}$, $B = diag\{B_1, .., B_N\}$, $B_j \in \Re^{n_j \times p_j}$, $D = diag\{D_1, .., D_N\}$, $D_j \in \Re^{n_j \times n_j}$, $C = diag\{C_1, .., C_N\}$, $C_j \in \Re^{q_j \times n_j}$, $H = diag\{H_1, .., H_N\}$, $H_j \in \Re^{q_j \times p_j}$. The function $g : Z_+ \times \Re^n \times \Re^n \to \Re^{n_j}$, $g(k, x(k), x(k-d(k))) = (g_1^t(k, x(k), x(k-d(k))), ..., g_N^t(k, x(k), x(k-d(k))))^t$ is a vector function piecewise-continuous in its arguments. In the sequel, we assume that this function is uncertain and the available information is that, in the domains of continuity $\mathbf{G} \subset Z_+ \times \Re^n \times \Re^n$, it satisfies the quadratic inequality

$$\begin{aligned} & g^t(k, x(k), x(k-d(k)))g(k, x(k), x(k-d(k))) \leq x^t(k)\widetilde{G}^t\widetilde{\Phi}^{-1}\widetilde{G}x(k) \\ + \ & x^t(k-d(k))\widetilde{G}_d^t\widetilde{\Psi}^{-1}\widetilde{G}_d x(k-d(k)) \end{aligned} \qquad (6.73)$$

where $\widetilde{G} = [\widetilde{G}_1^t, .., \widetilde{G}_N^t]^t$, $\widetilde{G}_j \in \Re^{r_j \times n}$, $\widetilde{G}_d = [\widetilde{G}_{d1}^t, .., \widetilde{G}_{dN}^t]^t$, $\widetilde{G}_{dj} \in \Re^{s_j \times n}$, $j = 1, .., N$ are constant matrices such that $g_j(k, 0, 0) = 0$ and $x = 0$ is an equilibrium of system (6.72) for $d(k) \geq 0$. Exploiting the structural form of system (6.72), a model of the jth subsystem Σ_j can be described by

$$\begin{aligned} \Sigma_j : \;\; x_j(k+1) &= A_j x_j(k) + B_j u_j(k) + D_j x_j(k - d_j(k)) \\ &+ \; g_j(k, x(k), x(k - d(k))) \\ y_j(k) &= C_j x_j(k) + H_j u_j(k) \end{aligned} \tag{6.74}$$

where $x_j(k) \in \Re^{n_j}$, $u_j(k) \in \Re^{p_j}$ and $y_j(k) \in \Re^{q_j}$ are the subsystem state, control input and measured output, respectively. The scalar $0 < d_j^* \leq d_j(k) \leq d_j^+$ represents the state-delay within subsystem Σ_j. The function $g_j : \mathsf{Z}_+ \times \Re^n \times \Re^n \rightarrow \Re^{n_j}$ is a piecewise-continuous vector function in its arguments and in line of (6.73) it satisfies the quadratic inequality

$$\begin{aligned} & g_j^t(k, x(k), x(k - d(k))) \; g_j(k, x(k), x(k - d(k))) \leq \phi_j^2 \; x^t(k) \widetilde{G}_j^t \widetilde{G}_j x(k) \\ & + \psi_j^2 \; x^t(k - d(k)) \widetilde{G}_{dj}^t \widetilde{G}_{dj} x(k - d(k)) \end{aligned} \tag{6.75}$$

where $\phi_j > 0, \; \psi_j > 0$ are bounding parameters such that

$$\begin{aligned} \widetilde{\Phi} &= Blockdiag\{\phi_1^2 I_{r_1}, .., \phi_{r_N}^2 I_{r_N}\}, \\ \widetilde{\Psi} &= Blockdiag\{\psi_1^2 I_{s_1}, .., \psi_{s_N}^2 I_{s_N}\} \end{aligned}$$

where I_{m_j} represents the $m_j \times m_j$ identity matrix. From (6.73) and (6.75), it is always possible to find matrices $\Phi, \; \Psi$ such that

$$\begin{aligned} & g^t(k, x(k), x(k - d(k))) g(k, x(k), x(k - d(k))) \leq x^t(k) G^t \Phi^{-1} G x(k) \\ + \;\; & x^t(k - d(k)) G_d^t \Psi^{-1} G_d x(k - d(k)) \end{aligned} \tag{6.76}$$

where

$$\begin{aligned} G &= Blockdiag\{G_1, .., G_N\}, \;\; G_j \in \Re^{r_j} \times n_j, \\ G_d &= Blockdiag\{G_{d1}, .., G_{dN}\}, \;\; G_{dj} \in \Re^{s_j} \times n_j, \\ \Phi &= Blockdiag\{\delta_1 I_{r_1}, .., \delta_N I_{r_N}\}, \; , \; \delta_j = \phi_j^{-2}, \\ \Psi &= Blockdiag\{\nu_1 I_{s_1}, .., \nu_N I_{s_N}\}, \;\; \nu_j = \psi_j^{-2} \end{aligned}$$

Remark 6.4.1 *These applications include cold rolling mills, decision-making processes and manufacturing systems. Related results for a class of discrete-time systems with time-varying delays can be found in [28] where delay-independent stability and stabilization conditions are derived. It should be stressed that although we consider only the case of single time-delay, extension to multiple time-delay systems can be easily attained using an augmentation procedure.*

Remark 6.4.2 *It should be observed that the interconnected system (6.72) has a decoupled structure for the individual subsystems and retains all the nonlinearities and uncertainties in the global term*

$$g(k, x(k), x(k - d(k))) = \sum_{j=1}^{N} g_j(k, x(k), x(k - d(k)))$$

Following [28], we note that the local nonlinear perturbations $g_j(.,.,.)$ *includes the class of norm-bounded uncertain systems in which*

$$g_j(k, x(k), x(k - d(k))) = \Delta A_j(x_j(k), k)\; x_j(k) \\ + \Delta A_j(x_j(k - d_j(k)), k)\; x_j(k - d_j(k))$$

where $\Delta A_j(x_j(k), k)$, $\Delta A_j(x_j(k - d_j(k)), k)$ *are norm bounded matrices. It is readily seen that the class of systems (6.72) casts more general classes satisfying the quadratic inequality bound (6.73). In addition, we do not impose any sort of matching conditions. More importantly, by defining* $x(k - d(k)) = [x_1^t(k - d_1(k)), ..., x_N^t(k - d_N(k))]^t$, *we allow the local subsystems to have different delay patterns.*

In the sequel, we let $\xi_j(k) = [x_j^t(k)\; x_j^t(k - d_j(k))\; g_j^t(k, x(k), x(k - d(k)))]^t$. Then (6.75) can be conveniently written as

$$\xi_j^t \begin{bmatrix} -\phi_j^2\; G_j^t G_j & 0 & 0 \\ \bullet & -\psi_j^2\; G_{dj}^t G_{dj} & 0 \\ \bullet & \bullet & I_j \end{bmatrix} \xi_j \leq 0 \tag{6.77}$$

The goal is to develop new tools for the analysis and design of a class of interconnected discrete-time state-delay systems with uncertain function of nonlinear perturbations by exploiting the decentralized information structure constraint. We seek to establish complete LMI-based procedures for the robust delay-dependent stability and feedback (state, output, resilient) stabilization by basing all the computations at the subsystem level.

6.4.2 Subsystem stability

Subsystem stability Our goal is to establish tractable conditions guaranteeing global asymptotic stability of the origin ($x = 0$) for all $g(k, x(k), x(k - d(k))) \in \mathbf{G}$. Let $\beta_j = (d_j^+ - d_j^* + 1)$ representing the number of samples within the delay range $d_j^* \leq d(k) \leq d_j^+$. The main result of subsystem stability is established by the following theorem

Theorem 6.4.3 *Given the delay sample number β_j. System (6.72) with $u \equiv 0$, is delay-dependent asymptotically stable for all nonlinear uncertainties satisfying (6.73) if there exist matrices $0 < \mathcal{Y}_j^t = \mathcal{Y}_j \in \Re^{n_j \times n_j}$, $0 < \mathcal{W}_j^t = \mathcal{W}_j \in \Re^{n_j \times n_j}$ and scalars $\delta_j > 0$ and $\nu_j > 0$ such that the following convex optimization problems for $j = 1, .., N$ are feasible*

$$\min_{\mathcal{Y}_j, \mathcal{W}_j, \delta_j, \nu_j} \quad \delta_j + \nu_j$$

$$subject\ to:$$

$$\mathcal{Y}_j > 0, \quad \mathcal{W}_j > 0$$

$$\begin{bmatrix} -\mathcal{Y}_j + \beta_j \mathcal{W}_j & 0 & 0 & \mathcal{Y}_j A_j^t & \mathcal{Y}_j A_j^t & \mathcal{Y}_j G_j^t \\ \bullet & -\mathcal{W}_j & \mathcal{Y}_j G_{dj}^t & \mathcal{Y}_j D_j^t & \mathcal{Y}_j D_j^t & 0 \\ \bullet & \bullet & -\nu_j I_j & 0 & 0 & 0 \\ \bullet & \bullet & \bullet & I_j - \mathcal{Y}_j & 0 & 0 \\ \bullet & \bullet & \bullet & \bullet & -\mathcal{Y}_j & 0 \\ \bullet & \bullet & \bullet & \bullet & \bullet & -\delta_j I_j \end{bmatrix} < 0 \quad (6.78)$$

Proof: Introduce the local Lyapunov-Krasovskii functional (LKF):

$$\begin{aligned} V_j(k) &= x_j^t(k)\mathcal{P}_j x_j(k) + \sum_{m=k-d_j(k)}^{k-1} x_j^t(m)\mathcal{Q}_j x_j(m) \\ &+ \sum_{s=2-d_j^+}^{1-d_j^*} \sum_{m=k+s-1}^{k-1} x_j^t(m)\mathcal{Q}_j x_j(m) \end{aligned} \quad (6.79)$$

where $0 < \mathcal{P}_j^t = \mathcal{P}_j$, $0 < \mathcal{Q}_j^t = \mathcal{Q}_j$ are weighting matrices of appropriate dimensions. A straightforward computation gives the first-difference of $\Delta V_j(k) = V_j(k+1) - V_j(k)$ along the solutions of (6.72) with $u_j(k) \equiv 0$ as:

$$\begin{aligned} \Delta V_j(k) &= [A_j x_j(k) + D_j x_j(k - d_j(k)) + g_j]^t \mathcal{P}_j \\ &\times [A_j x_j(k) + D_j x_j(k - d_j(k)) + g_j] \\ &- x_j^t(k)\mathcal{P}_j x_j(k) + x_j^t(k)\mathcal{Q}_j x_j(k) - x_j^t(k - d_j(k))\mathcal{Q}_j x_j(k - d_j(k)) \\ &+ \sum_{m=k+1-d_j(k+1)}^{k-1} x_j^t(m)\mathcal{Q}_j x_j(m) - \sum_{m=k+1-d_j(k)}^{k-1} x_j^t(m)\mathcal{Q}_j x_j(m) \\ &+ (d_j^+ - d_j^*)x_j^t(k)\mathcal{Q}_j x_j(k) - \sum_{m=k+1-d_j^+}^{k-d_j^*} x_j^t(m)\mathcal{Q}_j x_j(m) \end{aligned} \quad (6.80)$$

Since

$$\begin{aligned}
\sum_{m=k+1-d_j(k+1)}^{k-1} x_j^t(m)\mathcal{Q}_j x_j(m) &= \sum_{m=k+1-d_j^*}^{k-1} x_j^t(m)\mathcal{Q}_j x_j(m) \\
&+ \sum_{m=k+1-d_j(k+1)}^{k-d_j^*} x_j^t(m)\mathcal{Q}_j x_j(m) \\
&\leq \sum_{m=k+1-d_j(k)}^{k-1} x_j^t(m)\mathcal{Q}_j x_j(m) + \sum_{m=k+1-d_j^+}^{k-d_j^*} x_j^t(m)\mathcal{Q}_j x_j(m) \quad (6.81)
\end{aligned}$$

Then using (6.81) into (6.80) and manipulating, we reach

$$\begin{aligned}
\Delta V_j(k) &\leq [A_j x_j(k) + D_j x_j(k - d_j(k)) + g_j]^t \mathcal{P}_j \\
&\times [A_j x_j(k) + D_j x_j(k - d_j(k)) + g_j] \\
&+ x_j^t(k)[(d_j^+ - d_j^* + 1)\mathcal{Q}_j - \mathcal{P}_j] x_j(k) \\
&- x_j^t(k - d_j(k))\mathcal{Q}_j x_j(k - d_j(k)) \\
&= \xi_j^t(k)\, \Xi_j\, \xi_j(k) \qquad (6.82)
\end{aligned}$$

where

$$\begin{aligned}
\Xi_j &= \begin{bmatrix} A_j^t\mathcal{P}_j A_j + \beta_j\mathcal{Q}_j - \mathcal{P}_j & A_j^t\mathcal{P}_j D_j & A_j^t\mathcal{P}_j \\ \bullet & D_j^t\mathcal{P}_j D_j - \mathcal{Q}_j & D_j^t\mathcal{P}_j \\ \bullet & \bullet & \mathcal{P}_j \end{bmatrix}, \\
\xi_j(k) &= \begin{bmatrix} x_j^t(k) & x_j^t(k - d(k)) & g_j^t(k, x(k), x(k - d(k)))^t \end{bmatrix}^t \qquad (6.83)
\end{aligned}$$

The sufficient condition of stability $\Delta V_j(k) < 0$ implies that $\Xi_j < 0$. By resorting to the **S**-procedure [29], inequalities (6.77) and (6.82) can be rewritten together as

$$\begin{aligned}
&\mathcal{P}_j > 0,\ \omega_j \geq 0, \\
&\begin{bmatrix} \begin{array}{c} A_j^t\mathcal{P}_j A_j + \beta_j\mathcal{Q}_j \\ -\mathcal{P}_j + \omega_j\phi_j^2 G_j^t G_j \end{array} & A_j^t\mathcal{P}_j D_j & A_j^t\mathcal{P}_j \\ \bullet & \begin{array}{c} D_j^t\mathcal{P}_j D_j - \mathcal{Q}_j \\ +\omega_j\psi_j^2 G_{dj}^t G_{dj} \end{array} & D_j^t\mathcal{P}_j \\ \bullet & \bullet & \mathcal{P}_j - \omega_j I_j \end{bmatrix} < 0 \qquad (6.84)
\end{aligned}$$

which describes non-strict LMIs since $\omega_j \geq 0$. Recalling from [29] that minimization under non-strict LMIs corresponds to the same result as minimization under strict LMIs when both strict and non-strict LMI constraints are feasible. Moreover, if there is a solution for (6.84) for $\omega_j = 0,$ there will be also a solution for some $\omega_j > 0$ and sufficiently small $\phi_j,\ \psi_j$. Therefore, we safely replace $\omega_j \geq 0$ by $\omega_j > 0$. Equivalently, we may further rewrite (6.84) in the form

$$\bar{\mathcal{P}}_j > 0,$$
$$\begin{bmatrix} A_j^t\bar{\mathcal{P}}_jA_j + \beta_j\bar{\mathcal{Q}}_j - \\ \bar{\mathcal{P}}_j + \phi_j^2G_j^tG_j & A_j^t\bar{\mathcal{P}}_jD_j & A_j^t\bar{\mathcal{P}}_j \\ \bullet & D_j^t\bar{\mathcal{P}}_jD_j - \bar{\mathcal{Q}}_j \\ +\psi_j^2G_{dj}^tG_{dj} & D_j^t\bar{\mathcal{P}}_j \\ \bullet & \bullet & \bar{\mathcal{P}}_j - I_j \end{bmatrix} < 0 \qquad (6.85)$$

where $\bar{\mathcal{P}}_j = \omega_j^{-1}\ \mathcal{P}_j,\ \bar{\mathcal{Q}}_j = \omega_j^{-1}\ \mathcal{Q}_j$. Using the linearizations $\mathcal{Y}_j = \bar{\mathcal{P}}_j^{-1},\ \mathcal{W}_j = \bar{\mathcal{P}}_j^{-1}\bar{\mathcal{Q}}_j\bar{\mathcal{P}}_j^{-1}$ with $\delta_j = \phi_j^{-2}$ and $\nu_j = \psi_j^{-2}$, multiplying by ω_j then pre- and postmultiplying by $\bar{\mathcal{P}}_j^{-1}$ with some arrangement, we express (6.85) in the form (6.78). Robust stability of the nonlinear interconnected system (6.72) under the constraint(6.73) with maximal $\phi_j,\ \psi_j$ is thus established.

Remark 6.4.4 *Theorem 6.4.3 characterizes the local conditions for delay-dependent subsystem stability by resorting to a convenient LKF. The delay-dependence enters through the factor β_j for subsystem j. In the literature, there are more elaborate and efficient LKFs, see [78, 79], to achieve delay-dependent stability conditions for discrete-time systems by adding more appropriate terms. However, the selected LKF nicely paves the way to feedback stabilization by simple convex analysis within the interconnected framework. In addition, note that for all possible nonlinear perturbations $g(k,0,0) = 0$ and $x = 0$ is an equilibrium of system (6.72) for $d(k) \geq 0$. Therefore in the light of [348], the global asymptotic stable is guaranteed.*

In the next sections, we consider the decentralized feedback stabilization for interconnected system (6.72) within LMI-based formulation. We will be looking for a feedback controller which robustly stabilizes Σ. The main trust is to guarantee that local closed-loop subsystems is delay-dependent asymptotically stable for all possible nonlinear interconnections satisfying (6.75). In this way, the local controllers stabilize the linear part of Σ and, at the same time, maximize its tolerance to uncertain nonlinear interconnections and perturbations. We start with state feedback stabilization by assuming that all local state-variables are measurable and accessible to the local controllers.

6.4.3 State feedback design

State feedback design Now, we examine the application of a linear local feedback controller of the form

$$u_j(k) \;=\; K_j\, x_j(k) \tag{6.86}$$

to system (6.72), where $K_j \in \Re^{p_j \times n_j}$ is a constant gain matrix. Substituting (6.86) into (6.74) yields:

$$\begin{aligned}
\Sigma_s:\; x_j(k+1) &= A_{js}x_j(k) + D_j x_j(k-d_j(k)) + g_j(k, x(k), x(k-d(k))) \\
y_j(k) &= C_{js}x_j(k) \qquad (6.87)\\
A_{js} &= A_j + B_jK_j, \quad C_{js} = C_j + H_jK_j \qquad (6.88)
\end{aligned}$$

A direct application of **Theorem 6.4.3** leads to the following optimization problem

$$\min_{\mathcal{Y}_j, \mathcal{W}_j, \delta_j, \nu_j} \quad \delta_j + \nu_j$$

$$subject\ to:$$

$$\begin{bmatrix}
-\mathcal{Y}_j + \beta_j\mathcal{W}_j & 0 & 0 & \mathcal{Y}_jA^t_{js} & \mathcal{Y}_jA^t_{js} & \mathcal{Y}_jG^t_j \\
\bullet & -\mathcal{W}_j & \mathcal{Y}_jG^t_{dj} & \mathcal{Y}_jD^t_j & \mathcal{Y}_jD^t_j & 0 \\
\bullet & \bullet & -\nu_jI_j & 0 & 0 & 0 \\
\bullet & \bullet & \bullet & I_j - \mathcal{Y}_j & 0 & 0 \\
\bullet & \bullet & \bullet & \bullet & -\mathcal{Y}_j & 0 \\
\bullet & \bullet & \bullet & \bullet & \bullet & -\delta_jI_j
\end{bmatrix} < 0 \tag{6.89}$$

To put inequality (6.89) in an LMI setting, we introduce the variable $K_j\mathcal{Y}_j = \mathcal{M}_j$. This gives in turn the following LMIs for $j = 1, .., N$:

$$\min_{\mathcal{Y}_j, \mathcal{W}_j, \mathcal{M}_j, \delta_j, \nu_j} \quad \delta_j + \nu_j$$

$$subject\ to:$$

$$\begin{bmatrix}
-\mathcal{Y}_j + \beta_j\mathcal{W}_j & 0 & 0 & \mathcal{Y}_jA^t_j + \mathcal{M}_jB^t_j & \mathcal{Y}_jA^t_j + \mathcal{M}_jB^t_j & \mathcal{Y}_jG^t_j \\
\bullet & -\mathcal{W}_j & \mathcal{Y}_jG^t_{dj} & \mathcal{Y}_jD^t_j & \mathcal{Y}_jD^t_j & 0 \\
\bullet & \bullet & -\nu_jI_j & 0 & 0 & 0 \\
\bullet & \bullet & \bullet & I_j - \mathcal{Y}_j & 0 & 0 \\
\bullet & \bullet & \bullet & \bullet & -\mathcal{Y}_j & 0 \\
\bullet & \bullet & \bullet & \bullet & \bullet & -\delta_jI_j
\end{bmatrix}
< 0 \tag{6.90}$$

and thus the following theorem is established:

Theorem 6.4.5 *Given the delay sample number β_j. System (6.72) is robustly delay-dependent asymptotically stabilizable by control law (6.86) if there exist matrices $0 < \mathcal{Y}_j^t = \mathcal{Y}_j \in \Re^{n_j \times n_j}$, $0 < \mathcal{W}_j^t = \mathcal{W}_j \in \Re^{n_j \times n_j}$, $\mathcal{M}_j$ and scalars $\delta_j > 0,\ \nu_j > 0$ such that the convex optimization problem (6.90) has a feasible solution. The controller gain is given by $K_j = \mathcal{M}_j \mathcal{Y}_j^{-1}$.*

6.4.4 Bounded state feedback design

Bounded state feedback design Following the approach of [28, 348], we consider hereafter the case of local state feedback control with bounded gain matrix K_j of the form $K_j^t\, K_j \ <\ \kappa_j\, I$, with $\kappa_j > 0$. Since $K_j = \mathcal{M}_j \mathcal{Y}_j^{-1}$ this condition corresponds to the additional constraints on the component matrices $\mathcal{M}_j$ and $\mathcal{Y}_j^{-1}$ by setting $\mathcal{M}_j^t\, \mathcal{M}_j \ <\ \mu_j I,\ \ \mu_j > 0,\ \ Y^{-1}\ <\ \sigma_j\, I,\ \ \sigma_j > 0,\ j = 1, .., N$. In turn, these are equivalent to the LMIs

$$\begin{bmatrix} -\mu_j\, I_j & \mathcal{M}_j^t \\ \bullet & -I_j \end{bmatrix} < 0, \quad \begin{bmatrix} -\sigma_j\, I_j & I_j \\ \bullet & -\mathcal{Y}_j \end{bmatrix} < 0 \tag{6.91}$$

In a similar way, in order to guarantee desired values (ϕ_j, ψ_j) of the bounding factors (δ_j, ν_j), we recall that $\phi_j^{-2} = \delta_j,\ \psi_j^{-2} = \nu_j$. Thus we require

$$\delta_j \ -\ \frac{1}{\phi_j^2}\ <\ 0\ ,\ \nu_j\ -\ \frac{1}{\psi_j^2}\ <\ 0 \tag{6.92}$$

Incorporating the foregoing modifications into the present analysis leads to establishing the following convex optimization problem over LMIs for the local subsystem j:

$$\min_{\mathcal{Y}_j, \mathcal{W}_j, \mathcal{M}_j, \delta_j, \nu_j, \mu_j, \sigma_j} \quad \delta_j + \nu_j + \mu_j + \sigma_j$$

$$subject\ to:$$

$$\begin{bmatrix} -\mathcal{Y}_j + \beta_j \mathcal{W}_j & 0 & 0 & \mathcal{Y}_j A_j^t + \mathcal{M}_j B_j^t & \mathcal{Y}_j A_j^t + \mathcal{M}_j B_j^t & \mathcal{Y}_j G_j^t \\ \bullet & -\mathcal{W}_j & \mathcal{Y}_j G_{dj}^t & \mathcal{Y}_j D_j^t & \mathcal{Y}_j D_j^t & 0 \\ \bullet & \bullet & -\nu_j I_j & 0 & 0 & 0 \\ \bullet & \bullet & \bullet & I_j - \mathcal{Y}_j & 0 & 0 \\ \bullet & \bullet & \bullet & \bullet & -\mathcal{Y}_j & 0 \\ \bullet & \bullet & \bullet & \bullet & \bullet & -\delta_j I_j \end{bmatrix}$$

$$< 0$$

$$\delta_j \ -\ \frac{1}{\phi_j^2}\ <\ 0\ ,\ \nu_j\ -\ \frac{1}{\psi_j^2}\ <\ 0$$

$$\begin{bmatrix} -\mu_j\, I_j & \mathcal{M}_j^t \\ \bullet & -I_j \end{bmatrix} < 0, \quad \begin{bmatrix} -\sigma_j\, I_j & I_j \\ \bullet & -\mathcal{Y}_j \end{bmatrix} < 0 \tag{6.93}$$

Hence, the following theorem summarizes the main result

Theorem 6.4.6 *Given the delay sample number β_j and the bounds $\phi_j, \ \psi_j$. System (6.72) is robustly delay-dependent asymptotically stabilizable by control law (6.86) with constrained feedback gains and bounding factors if there exist matrices $0 < \mathcal{Y}_j^t = \mathcal{Y}_j \in \Re^{n_j \times n_j}, \ 0 < \mathcal{W}_j^t = \mathcal{W}_j \in \Re^{n_j \times n_j}, \ \mathcal{M}_j$ and scalars $\delta_j > 0, \ \nu_j > 0, \ \mu_j > 0, \ \sigma_j > 0$ such that the convex optimization problem (6.93) has a feasible solution. The controller gain is given by $K_j = \mathcal{M}_j \mathcal{Y}_j^{-1}$.*

6.4.5 Resilient state feedback design

Now, we now address the performance deterioration issue by considering that the actual linear local state-feedback controller has the following form:

$$u_j(k) \ = \ [K_j \ + \ \Delta K_j] \, x_j(k) \tag{6.94}$$

where $K_j \in \Re^{p_j \times n_j}$ is a constant gain matrix and ΔK_j is an additive gain perturbation matrix represented by

$$\Delta K_j \ = \ M_j \, \Delta_j \, N_j, \quad \Delta_j \in \mathbf{\Delta} := \{\Delta_j : \Delta_j^t \Delta_j \leq I\} \tag{6.95}$$

The application of control law (6.94) to system (6.74) yields the perturbed closed-loop system

$$\begin{aligned} \Sigma_p : \ x_j(k+1) &= (A_{js} + B_j \Delta K_j) x_j(k) + D_j x_j(k - d_j(k)) \\ &+ \ g_j(k, x(k), x(k - d(k))) \\ y_j(k) &= C_{js} x_j(k) \qquad\qquad\qquad\qquad\qquad\qquad (6.96) \\ A_{js} &= A_j + B_j K_j, \quad C_{js} = C_j + H_j K_j + M_j \, \Delta_j \, N_j \qquad (6.97) \end{aligned}$$

For simplicity in exposition, we let $\Upsilon_j = \mathcal{Y}_j A_{js}^t + \mathcal{Y}_j (B_j \Delta K_j)^t$. It follows by applying **Theorem 6.4.3** to system (6.97), we obtain the following convex problem

$$\min_{\mathcal{Y}_j, \mathcal{W}_j, \delta_j, \nu_j} \quad \delta_j + \nu_j$$

$$subject \ to :$$

$$\begin{bmatrix} -\mathcal{Y}_j + \beta_j \mathcal{W}_j & 0 & 0 & \Upsilon_j & \Upsilon_j & \mathcal{Y}_j G_j^t \\ \bullet & -\mathcal{W}_j & \mathcal{Y}_j G_{dj}^t & \mathcal{Y}_j D_j^t & \mathcal{Y}_j D_j^t & 0 \\ \bullet & \bullet & -\nu_j I_j & 0 & 0 & 0 \\ \bullet & \bullet & \bullet & I_j - \mathcal{Y}_j & 0 & 0 \\ \bullet & \bullet & \bullet & \bullet & -\mathcal{Y}_j & 0 \\ \bullet & \bullet & \bullet & \bullet & \bullet & -\delta_j I_j \end{bmatrix} < 0 \tag{6.98}$$

which we seek its feasibility over all possible perturbations $\Delta_j \in \mathbf{\Delta}$. In order to convexify inequality (6.98) and at the same time bypass the exhaustive search over the perturbation set $\mathbf{\Delta}$, we manipulate inequality of (6.98) with the aid of **Inequality 1** in the Appendix to reach

$$\begin{bmatrix} -\mathcal{Y}_j + \beta_j \mathcal{W}_j & 0 & 0 & \mathcal{Y}_j A_{js}^t & \mathcal{Y}_j A_{js}^t & \mathcal{Y}_j G_j^t \\ \bullet & -\mathcal{W}_j & \mathcal{Y}_j G_{dj}^t & \mathcal{Y}_j D_j^t & \mathcal{Y}_j D_j^t & 0 \\ \bullet & \bullet & -\nu_j I_j & 0 & 0 & 0 \\ \bullet & \bullet & \bullet & I_j - \mathcal{Y}_j & 0 & 0 \\ \bullet & \bullet & \bullet & \bullet & -\mathcal{Y}_j & 0 \\ \bullet & \bullet & \bullet & \bullet & \bullet & -\delta_j I_j \end{bmatrix}$$

$$+ \begin{bmatrix} 0 \\ 0 \\ 0 \\ B_j M_j \\ 0 \\ 0 \end{bmatrix} \Delta_j \begin{bmatrix} \mathcal{Y}_j N_j^t \\ 0 \\ 0 \\ 0 \\ 0 \\ 0 \end{bmatrix}^t +$$

$$\begin{bmatrix} \mathcal{Y}_j N_j^t \\ 0 \\ 0 \\ 0 \\ 0 \\ 0 \end{bmatrix} \Delta_j^t \begin{bmatrix} 0 \\ 0 \\ 0 \\ B_j M_j \\ 0 \\ 0 \end{bmatrix}^t + \begin{bmatrix} 0 \\ 0 \\ 0 \\ 0 \\ B_j M_j \\ 0 \end{bmatrix} \Delta_j \begin{bmatrix} \mathcal{Y}_j N_j^t \\ 0 \\ 0 \\ 0 \\ 0 \\ 0 \end{bmatrix}^t$$

$$+ \begin{bmatrix} \mathcal{Y}_j N_j^t \\ 0 \\ 0 \\ 0 \\ 0 \\ 0 \end{bmatrix} \Delta_j^t \begin{bmatrix} 0 \\ 0 \\ 0 \\ 0 \\ B_j M_j \\ 0 \end{bmatrix}^t \tag{6.99}$$

which can be overbounded as

$$\leq \begin{bmatrix} -\mathcal{Y}_j + \mathcal{W}_j & 0 & 0 & \mathcal{Y}_j A_{js}^t & \mathcal{Y}_j A_{js}^t & \mathcal{Y}_j G_j^t \\ \bullet & -\mathcal{W}_j & \mathcal{Y}_j G_{dj}^t & \mathcal{Y}_j D_j^t & \mathcal{Y}_j D_j^t & 0 \\ \bullet & \bullet & -\nu_j I_j & 0 & 0 & 0 \\ \bullet & \bullet & \bullet & I_j - \mathcal{Y}_j & 0 & 0 \\ \bullet & \bullet & \bullet & \bullet & -\mathcal{Y}_j & 0 \\ \bullet & \bullet & \bullet & \bullet & \bullet & -\delta_j I_j \end{bmatrix}$$

$$+ \quad \eta_{aj} \begin{bmatrix} 0 \\ 0 \\ 0 \\ B_j M_j \\ 0 \\ 0 \end{bmatrix} \begin{bmatrix} 0 \\ 0 \\ 0 \\ B_j M_j \\ 0 \\ 0 \end{bmatrix}^t +$$

$$\eta_{aj}^{-1} \begin{bmatrix} \mathcal{Y}_j N_j^t \\ 0 \\ 0 \\ 0 \\ 0 \\ 0 \end{bmatrix} \begin{bmatrix} \mathcal{Y}_j N_j^t \\ 0 \\ 0 \\ 0 \\ 0 \\ 0 \end{bmatrix}^t + \eta_{cj} \begin{bmatrix} 0 \\ 0 \\ 0 \\ 0 \\ B_j M_j \\ 0 \end{bmatrix} \begin{bmatrix} 0 \\ 0 \\ 0 \\ 0 \\ B_j M_j \\ 0 \end{bmatrix}^t$$

$$+ \quad \eta_{cj}^{-1} \begin{bmatrix} \mathcal{Y}_j N_j^t \\ 0 \\ 0 \\ 0 \\ 0 \\ 0 \end{bmatrix} \begin{bmatrix} \mathcal{Y}_j N_j^t \\ 0 \\ 0 \\ 0 \\ 0 \\ 0 \end{bmatrix}^t$$

$$:= \quad \begin{bmatrix} \Pi_{1j} & \Pi_{2j} \\ \bullet & \Pi_{3j} \end{bmatrix} \quad < \quad 0 \tag{6.100}$$

for some scalars $\eta_{aj} > 0, \ \eta_{cj} > 0,$ where

$$\begin{aligned}
\Pi_{1j} &= \begin{bmatrix} -\mathcal{Y}_j + \beta_j \mathcal{W}_j + \eta_{aj}^{-1} \mathcal{Y}_j N_j^t N_j \mathcal{Y}_j + \eta_{cj}^{-1} \mathcal{Y}_j N_j^t N_j \mathcal{Y}_j & 0 \\ \bullet & -\mathcal{W}_j \end{bmatrix} \\
\Pi_{2j} &= \begin{bmatrix} 0 & \mathcal{Y}_j A_j^t + \mathcal{Y}_j K_j^t B_j^t & \mathcal{Y}_j A_j^t + \mathcal{Y}_j K_j^t B_j^t & \mathcal{Y}_j G_j^t \\ \mathcal{Y}_j G_{dj}^t & \mathcal{Y}_j D_j^t & \mathcal{Y}_j D_j^t & 0 \end{bmatrix}, \\
\Pi_{3j} &= Blockdiag[\Pi_{31j}, \Pi_{32j}] \\
\Pi_{31j} &= Blockdiag[-\nu_j I_j, I_j - \mathcal{Y}_j + \eta_{aj} B_j M_j M_j^t B_j^t], \\
\Pi_{32j} &= Blockdiag[-\mathcal{Y}_j + \eta_{cj} B_j M_j M_j^t B_j^t, \ \ -\delta_j I_j]
\end{aligned} \tag{6.101}$$

Introducing the change of variables

$$K_j \mathcal{Y}_j = L_j, \ Z_{aj} = \eta_{aj} \mathcal{Y}_j N_j^t, \ Z_{cj} = \eta_{cj} \mathcal{Y}_j N_j^t, \ \bar{Z}_j = [Z_{aj}, \quad Z_{cj}]$$

This will eventually cast inequalities (6.100)-(6.101) into the following LMI format:

$$\begin{bmatrix} -\mathcal{Y}_j + \beta_j \mathcal{W}_j & 0 & \mathcal{S}_j^t \\ \bullet & -\mathcal{W}_j & 0 \\ \bullet & \bullet & -\mathcal{R}_j \end{bmatrix} \quad < \quad 0 \tag{6.102}$$

where

$$\begin{aligned}
\mathcal{S}_j &= \begin{bmatrix} 0 & \mathcal{Y}_j G_{dj}^t \\ \mathcal{Y}_j A_j^t + L_j^t B_j^t & \mathcal{Y}_j D_j^t \\ 0 & 0 \\ \mathcal{Y}_j A_j^t + L_j^t B_j^t & \mathcal{Y}_j D_j^t \\ 0 & 0 \\ \mathcal{Y}_j G_j^t & 0 \\ \bar{Z}^t & 0 \end{bmatrix}, \\
\mathcal{R}_j &= Blockdiag[\mathcal{R}_{1j} \quad \mathcal{R}_{2j}] > 0, \;\; \eta_j = Blockdiag[\eta_{aj} \quad \eta_{cj}] \\
\mathcal{R}_{1j} &= \begin{bmatrix} -\nu_j I_j & 0 & 0 \\ \bullet & I_j - \mathcal{Y}_j & B_j M_j \\ \bullet & \bullet & -\eta_{aj} I_j \end{bmatrix}, \\
\mathcal{R}_{2j} &= \begin{bmatrix} -\mathcal{Y}_j & B_j M_j & 0 & 0 \\ \bullet & -\eta_{cj} I_j & 0 & 0 \\ \bullet & \bullet & -\delta_j I_j & 0 \\ \bullet & \bullet & \bullet & -\eta_j \end{bmatrix}
\end{aligned} \tag{6.103}$$

The foregoing analysis has established the following theorem

Theorem 6.4.7 *Given the delay sample number β_j. System (6.72) is robustly delay-dependent asymptotically stabilizable by the actual control law (6.94) for all possible gain variations (6.95) if there exist matrices $0 < \mathcal{Y}_j^t = \mathcal{Y}_j \in \Re^{n_j \times n_j}$, $0 < \mathcal{W}_j^t = \mathcal{W}_j \in \Re^{n_j \times n_j}$, L_j and scalars $\delta_j > 0$, $\nu_j > 0$, η_{aj}, η_{cj}, Z_{aj}, Z_{cj} such that the following convex optimization problem*

$$\begin{aligned}
&\min_{\mathcal{Y}_j, \mathcal{W}_j, L_j, \delta_j, \nu_j, \eta_{aj}, \eta_{cj}, Z_{aj}, Z_{cj}} \quad \delta_j + \nu_j + \eta_{aj} + \eta_{cj} \\
&subject\ to: \\
&\mathcal{Y}_j > 0, \;\; \mathcal{W}_j > 0 \;\; and \;\; (6.102) - (6.103)
\end{aligned}$$

has a feasible solution. The controller gain is given by $K_j = L_j \mathcal{Y}_j^{-1}$

Had we taken the modifications made previously to constrain the local state feedback gains, we would have reached an alternative convex optimization problem over LMIs summarized by the following theorem.

Theorem 6.4.8 *Given the delay sample number β_j and the bounds ϕ_j, ψ_j. System (6.72) is robustly delay-dependent asymptotically stabilizable by the actual control law (6.94) for all possible gain variations (6.95) if there exist matrices $0 < \mathcal{Y}_j^t = \mathcal{Y}_j \in \Re^{n_j \times n_j}$, $0 < \mathcal{W}_j^t = \mathcal{W}_j \in \Re^{n_j \times n_j}$, L_j and scalars*

$\delta_j > 0,\ \nu_j > 0,\ \mu_j > 0,\ \sigma_j > 0,\ \eta_{aj},\ \eta_{cj},\ Z_{aj},\ Z_{cj}$ such that the following convex optimization problem

$$\begin{array}{l} \min\limits_{\mathcal{Y}_j,\mathcal{W}_j,L_j,\delta_j,\nu_j,\mu_j,\sigma_j,\eta_{aj},\eta_{cj},Z_{aj},Z_{cj}} \quad \delta_j + \nu_j + \eta_{aj} + \eta_{cj} + \mu_j + \sigma_j \\ subject\ to: \\ \mathcal{Y}_j > 0,\ \mathcal{W}_j > 0 \ \text{and}\ (6.102) - (6.103) \\ \delta_j - \dfrac{1}{\phi^2} < 0\ ,\ \nu_j - \dfrac{1}{\psi_j^2} < 0 \\ \begin{bmatrix} -\mu_j I & \mathcal{M}_j^t \\ \bullet & -I_j \end{bmatrix} < 0\ ,\ \begin{bmatrix} -\sigma_j I_j & I_j \\ \bullet & -\mathcal{Y}_j \end{bmatrix} < 0 \end{array}$$

has a feasible solution. The controller gain is given by $K_j = L_j\mathcal{Y}_j^{-1}$

Remark 6.4.9 *It should be observed that Theorems 6.4.5-6.4.8 provide a complete decoupled set of LMI-based state-feedback design algorithms for a class of interconnected discrete-time systems with unknown-but-bounded delays and quadratically-bounded nonlinearities and perturbations. It is significant to record that Theorem 6.4.8 is new addition to the resilient control theory [258, 389]. Indeed, these algorithms assume the accessibility of local states.*

In the next section, we drop the assumption on complete accessibility of local state-variables and consider the availability of local output measurements. We focus on the design of local dynamic output feedback stabilization scheme.

6.4.6 Output feedback stabilization

In this part, we consider the design of a local dynamic output feedback stabilization scheme of Luenberger-type:

$$\begin{aligned} z_j(k+1) &= A_j z_j(k) + B_j u_j(k) + K_{oj}[y_j(k) - w_j(k)] \\ w_j(k) &= C_j z_(k) + H_j u_j(k) \\ u_j(k) &= K_{cj} z_j(k) \end{aligned} \tag{6.104}$$

where the matrices $K_{oj} \in \Re^{n_j x q_j}$ and $K_{cj} \in \Re^{p_j \times n_j}$ are the local unknown observer and state-feedback gain matrices to be determined such that the local closed-loop controlled system achieves delay-dependent asymptotic stability.

We proceed by defining the state error $e_j(k) = x_j(k) - z_j(k)$ and obtain from (6.72) and (6.104), the local error system

$$\begin{aligned} e_j(k+1) &= [A_j - K_j C_j]e_j(k) + D_j x_j(k - d_j(k)) \\ &+ g_j(k, x(k), x(k - d(k))) \end{aligned} \tag{6.105}$$

By combining systems (6.72) and (6.105), we get the augmented system

$$\zeta_j(k+1) = \mathcal{A}_j\, \zeta_j(k) + \mathcal{D}_j\, \zeta_j(k - d_j(k)) + \bar{g}_j(k, x(k), x(k - d(k))) \qquad (6.106)$$

where

$$\begin{aligned}
\zeta_j(k) &= \begin{bmatrix} x_j(k) \\ e_j(k) \end{bmatrix} \in \Re^{2n_j}, \\
\bar{g}_j(k, x(k), x(k-d(k))) &= \begin{bmatrix} g_j(k, x(k), x(k-d(k))) \\ g_j(k, x(k), x(k-d(k))) \end{bmatrix} \\
\mathcal{A}_j &= \begin{bmatrix} A_j + B_j K_{cj} & -B_j K_{cj} \\ 0 & A_j - K_{oj} C_j \end{bmatrix}, \\
\mathcal{D}_j &= \begin{bmatrix} D_j & 0 \\ D_j & 0 \end{bmatrix}
\end{aligned} \qquad (6.107)$$

Let

$$\begin{aligned}
\bar{G}_j &= \begin{bmatrix} 2G_j & 0 \\ 0 & 0 \end{bmatrix}, \ \bar{G}_{dj} = \begin{bmatrix} 2G_{dj} & 0 \\ 0 & 0 \end{bmatrix} \\
\bar{\mathcal{Y}}_j &= \begin{bmatrix} \mathcal{Y}_{cj} & 0 \\ 0 & \mathcal{Y}_{oj} \end{bmatrix}, \ \bar{\mathcal{W}}_j = \begin{bmatrix} \mathcal{W}_{cj} & 0 \\ 0 & \mathcal{W}_{oj} \end{bmatrix}, \\
\hat{\nu}_j &= Blockdiag[\nu_{cj}, \nu_{oj}], \ \hat{\delta}_j = Blockdiag[\delta_{cj}, \delta_{oj}]
\end{aligned} \qquad (6.108)$$

where $\mathcal{Y}_{cj}, \mathcal{W}_{cj}, \mathcal{Y}_{oj}, \mathcal{W}_{oj} \in \Re^{n_j \times n_j}$ be symmetric and positive-definite matrices. It follows from Theorem 6.4.3 that the robust delay-dependent asymptotic stability of the augmented system (6.106) is guaranteed by the solution of the following convex optimization problem

$$\min_{\bar{\mathcal{Y}}_j, \bar{\mathcal{W}}_j, \hat{\nu}_j, \hat{\delta}_j} \quad \hat{\nu}_j + \hat{\delta}_j \qquad (6.109)$$

$$subject\ to:$$

$$\mathcal{Y}_{cj} > 0,\ \mathcal{Y}_{oj} > 0,\ \mathcal{W}_{cj} > 0,\ \mathcal{W}_{oj} > 0$$

$$\begin{bmatrix}
-\bar{\mathcal{Y}}_j + \beta_j \bar{\mathcal{W}}_j & 0 & 0 & \bar{\mathcal{Y}}_j \mathcal{A}_j^t & \bar{\mathcal{Y}}_j \mathcal{A}_j^t & \bar{\mathcal{Y}}_j \bar{G}_j^t \\
\bullet & -\bar{\mathcal{W}}_j & \bar{\mathcal{Y}}_j \bar{G}_{dj}^t & \bar{\mathcal{Y}}_j \mathcal{D}_j^t & \bar{\mathcal{Y}}_j \mathcal{D}_j^t & 0 \\
\bullet & \bullet & -\hat{\nu}_j I_j & 0 & 0 & 0 \\
\bullet & \bullet & \bullet & I_j - \bar{\mathcal{Y}}_j & 0 & 0 \\
\bullet & \bullet & \bullet & \bullet & -\bar{\mathcal{Y}}_j & 0 \\
\bullet & \bullet & \bullet & \bullet & \bullet & -\hat{\delta}_j I_j
\end{bmatrix} < 0 \qquad (6.110)$$

is feasible. Mathematical manipulations of inequality (6.110) using (6.107)-(6.108) lead to the following theorem

Theorem 6.4.10 *System (6.106) is robustly delay-dependent asymptotically stable if there exist matrices* $0 < \mathcal{Y}_{cj}^t = \mathcal{Y}_{cj} \in \Re^{n_j \times n_j}$, $0 < \mathcal{Y}_{oj}^t = \mathcal{Y}_{oj} \in \Re^{n_j \times n_j}$, $0 < \mathcal{W}_{cj}^t = \mathcal{W}_{cj} \in \Re^{n_j \times n_j}$, $0 < \mathcal{W}_{oj}^t = \mathcal{W}_{oj} \in \Re^{n_j \times n_j}$, $\mathcal{S}_{cj} \in \Re^{p_j \times n_j}$, $\mathcal{S}_{oj} \in \Re^{n_j \times n_j}$, $\mathcal{S}_{fj} \in \Re^{p_j \times n_j}$ *and scalars* $\delta_{cj} > 0$, $\nu_{cj} > 0$, $\delta_{oj} > 0$, $\nu_{oj} > 0$ *such that the following convex optimization problem*

$$\min_{\mathcal{Y}_{cj},\mathcal{Y}_{oj},\mathcal{W}_{cj},\mathcal{W}_{oj},\mathcal{S}_{cj},\mathcal{S}_{oj},\mathcal{S}_{fj},\delta_{cj},\delta_{oj},\nu_{cj},\nu_{oj}} \quad \delta_{cj} + \delta_{oj} + \nu_{cj} + \nu_{oj} \tag{6.111}$$

$$subject\ to:$$

$$\mathcal{Y}_{cj} > 0,\ \mathcal{Y}_{oj} > 0,\ \mathcal{W}_{cj} > 0,\ \mathcal{W}_{oj} > 0,\quad \begin{bmatrix} \Xi_{cj} & \Xi_{dj} \\ \bullet & \Xi_{oj} \end{bmatrix} < 0$$

$$\Xi_{cj} = \begin{bmatrix} \Xi_{cj1} & \Xi_{cj2} \end{bmatrix}$$

$$\Xi_{cj1} = \begin{bmatrix} -\mathcal{Y}_{cj} + \beta_j \mathcal{W}_{cj} & 0 & 0 \\ \bullet & -\mathcal{Y}_{oj} + \mathcal{W}_{oj} & 0 \\ \bullet & \bullet & -\mathcal{W}_{cj} \\ \bullet & \bullet & \bullet \\ \bullet & \bullet & \bullet \\ \bullet & \bullet & \bullet \end{bmatrix}$$

$$\Xi_{cj2} = \begin{bmatrix} 0 & 0 & 0 \\ 0 & 0 & 0 \\ 0 & 2\mathcal{Y}_{cj} G_{dj}^t & 0 \\ -\mathcal{W}_{oj} & 0 & 0 \\ \bullet & -\nu_{cj} I_j & 0 \\ \bullet & \bullet & -\nu_{oj} I_j \end{bmatrix}$$

$$\Xi_{dj} = \begin{bmatrix} \Xi_{dj1} & \Xi_{dj2} \end{bmatrix}$$

$$\Xi_{dj1} = \begin{bmatrix} \mathcal{Y}_{cj} A_j^t + \mathcal{S}_{cj}^t B_j^t & 0 & \mathcal{Y}_{cj} A_j^t + \mathcal{S}_{cj}^t B_j^t \\ -\mathcal{S}_{fj}^t B_j^t & \mathcal{Y}_{oj} A_j^t - \mathcal{S}_{oj}^t & -\mathcal{S}_{fj}^t B_j^t \\ \mathcal{Y}_{cj} D_j^t & \mathcal{Y}_{cj} D_j^t & \mathcal{Y}_{cj} D_j^t \\ 0 & 0 & 0 \\ 0 & 0 & 0 \\ 0 & 0 & 0 \end{bmatrix}$$

$$\Xi_{dj2} = \begin{bmatrix} 0 & 2\mathcal{Y}_{cj} G_j^t & 0 \\ \mathcal{Y}_{oj} A_j^t - \mathcal{S}_{oj}^t & 0 & 0 \\ \mathcal{Y}_{cj} D_j^t & 0 & 0 \\ 0 & 0 & 0 \\ 0 & 0 & 0 \\ 0 & 0 & 0 \end{bmatrix} \tag{6.112}$$

$$\begin{aligned}\Xi_{oj} &= Blockdiag\left[\ \Xi_{oj1}\quad \Xi_{oj2}\ \right]\\ \Xi_{oj1} &= Blockdiag\left[\ I_j - \mathcal{Y}_{cj}\quad I_j - \mathcal{Y}_{oj}\quad -\mathcal{Y}_{cj}\ \right]\\ \Xi_{oj2} &= Blockdiag\left[\ -\mathcal{Y}_{oj}\quad -\delta_{cj}I_j\quad -\delta_{oj}I_j\ \right]\end{aligned} \tag{6.113}$$

is feasible. Moreover, the gains are given by $K_{cj} = \mathcal{S}_{cj}\mathcal{Y}_{cj}^{-1}$, $K_{oj} = \mathcal{S}_{oj}\mathcal{Y}_{oj}^{-1}C_j^{\dagger}$, *where* $C_j^{\dagger} = C_j^t(C_jC_j^t)^{-1}$ *is the pseudo-inverse of* C_j.

Remark 6.4.11 *It can be argued that the block-diagonal form of the Lyapunov matrices (6.108) is not restrictive since it aims at developing convenient non-iterative computational algorithm while maintaining all the computations at the subsystem level. It has been widely used in dynamic output-feedback design [28, 346]. In the case of static output feedback the problem can be transformed into an LMI problem by a simple change of variables [82]. In the case of dynamic output feedback the problem becomes far more complex and in addition, a general dynamic output feedback design will eventually lead to nonlinear matrix inequalities.*

6.4.7 Resilient output feedback stabilization

Next we address the resilience problem of output feedback stabilization. For this purpose, we re-express the controller (6.104) into the form

$$\begin{aligned} z_j(k+1) &= A_jz_j(k) + B_ju_j(k) + [K_{oj} + \Delta K_{oj}][y_j(k) - w_j(k)]\\ w_j(k) &= C_jz_(k) + H_ju_j(k)\\ u_j(k) &= [K_{cj} + \Delta K_{cj}]z_j(k)\end{aligned} \tag{6.114}$$

where the matrices K_{oj} and K_{cj} are the local unknown gain parameters to be determined such that the local closed-loop controlled system achieves delay-dependent asymptotic stability and ΔK_{oj}, ΔK_{cj} are gain perturbations represented by

$$\begin{aligned}\Delta K_{oj} &= M_{oj}\Delta_{fj}N_{oj},\ \ \Delta K_{cj} = M_{cj}\Delta_{dj}N_{cj},\\ &\Delta_{fj},\ \Delta_{dj},\ \in \mathbf{\Delta} := \{\Delta_j : \Delta_j^t\Delta_j \le I\}\end{aligned} \tag{6.115}$$

Following the analysis above, the augmented system becomes

$$\begin{aligned}\zeta_j(k+1) &= [\mathcal{A}_j + \Delta\mathcal{A}_j]\ \zeta_j(k) + \mathcal{D}_j\ \zeta_j(k - d_j(k)) + \bar{g}_j(k, x(k), x(k-d(k)))\\ &= \bar{\mathcal{A}}_j\ \zeta_j(k) + \mathcal{D}_j\ \zeta_j(k - d_j(k)) + \bar{g}_j(k, x(k), x(k-d(k)))\end{aligned} \tag{6.116}$$

where $\zeta_j(k), \mathcal{A}_j, \mathcal{D}_j, \bar{g}_j(k, x(k), x(k-d(k)))$ are given by (6.107) and

$$\Delta\mathcal{A}_j = \left[\begin{array}{cc} B_j\Delta K_{cj} & -B_j\Delta K_{cj}\\ 0 & -\Delta K_{oj}C_j\end{array}\right] \tag{6.117}$$

In light of the foregoing procedures, it follows from Theorem 6.4.3 that the robust delay-dependent asymptotic stability of the augmented system (6.116) is guaranteed by the feasible solution of the following convex optimization problem

$$\min_{\bar{\mathcal{Y}}_j, \bar{\mathcal{W}}_j, \hat{\nu}_j, \hat{\delta}_j} \quad \hat{\nu}_j + \hat{\delta}_j \tag{6.118}$$

$$subject\ to:$$

$$\mathcal{Y}_{cj} > 0,\ \mathcal{Y}_{oj} > 0,\ \mathcal{W}_{cj} > 0,\ \mathcal{W}_{oj} > 0$$

$$\begin{bmatrix} -\bar{\mathcal{Y}}_j + \beta_j \bar{\mathcal{W}}_j & 0 & 0 & \bar{\mathcal{Y}}_j \bar{A}_j^t & \bar{\mathcal{Y}}_j \bar{A}_j^t & \bar{\mathcal{Y}}_j \bar{G}_j^t \\ \bullet & -\bar{\mathcal{W}}_j & \bar{\mathcal{Y}}_j \bar{G}_{dj}^t & \bar{\mathcal{Y}}_j \bar{\mathcal{D}}_j^t & \bar{\mathcal{Y}}_j \bar{\mathcal{D}}_j^t & 0 \\ \bullet & \bullet & -\hat{\nu}_j I_j & 0 & 0 & 0 \\ \bullet & \bullet & \bullet & I_j - \bar{\mathcal{Y}}_j & 0 & 0 \\ \bullet & \bullet & \bullet & \bullet & -\bar{\mathcal{Y}}_j & 0 \\ \bullet & \bullet & \bullet & \bullet & \bullet & -\hat{\delta}_j I_j \end{bmatrix} < 0 \tag{6.119}$$

In order to convexify (6.119) we invoke **Inequality 1** in the Appendix and manipulate using (6.113) to reach

$$\begin{aligned} &\begin{bmatrix} \Xi_{cj} & \Xi_{dj} \\ \bullet & \Xi_{oj} \end{bmatrix} + \Upsilon_{j1} \Delta_{dj}^t \Upsilon_{j2}^t + \Upsilon_{j2} \Delta_{dj} \Upsilon_{j1}^t + \\ &\Upsilon_{j3} \Delta_{fj}^t \Upsilon_{j4}^t + \Upsilon_{j4} \Delta_{fj} \Upsilon_{j3}^t \leq \\ &\begin{bmatrix} \Xi_{cj} & \Xi_{dj} \\ \bullet & \Xi_{oj} \end{bmatrix} + \eta_{jg} \Upsilon_{j1} \Upsilon_{j1}^t + \eta_{jg}^{-1} \Upsilon_{j2} \Upsilon_{j2}^t + \\ &\eta_{jf} \Upsilon_{j3} \Upsilon_{j3}^t + \eta_{jf}^{-1} \Upsilon_{j4} \Upsilon_{j4}^t \ < \ 0 \end{aligned} \tag{6.120}$$

for some scalars $\eta_{jg} > 0,\ \eta_{jf} > 0$ and where

$$\begin{aligned} \Upsilon_{j1}^t &= [N_{cj} \mathcal{Y}_{cj},\ -N_{cj} \mathcal{Y}_{oj},\ 0,\ 0,\ 0,\ 0,\ 0,\ 0,\ 0,\ 0,\ 0,\ 0] \\ \Upsilon_{j2}^t &= [0,\ 0,\ 0,\ 0,\ 0,\ 0,\ M_{cj}^t B_j^t,\ 0,\ M_{cj}^t B_j^t,\ 0,\ 0,\ 0] \\ \Upsilon_{j3}^t &= [0,\ -N_{oj} C_j \mathcal{Y}_{oj},\ 0,\ 0,\ 0,\ 0,\ 0,\ 0,\ 0,\ 0,\ 0,\ 0] \\ \Upsilon_{j4}^t &= [0,\ 0,\ 0,\ 0,\ 0,\ 0,\ 0,\ M_{oj}^t, 0,\ 0,\ 0,\ 0] \end{aligned} \tag{6.121}$$

Introducing the matrix notations

$$
\begin{aligned}
Z_{mj} &= \eta_{jg}\mathcal{Y}_{cj}N_{cj}^t \ , \ Z_{nj} = -\eta_{jg}\mathcal{Y}_{oj}N_{cj}^t \ , \ Z_{pj} = \eta_{jf}\mathcal{Y}_{cj}C_j^tN_{oj}^t \\
\Xi_{sj} &= \begin{bmatrix} Z_{mj} & 0 & 0 & 0 \\ -Z_{nj} & 0 & Z_{pj} & 0 \\ 0 & B_jM_{cj} & 0 & 0 \\ 0 & 0 & 0 & 0 \\ 0 & 0 & 0 & 0 \\ 0 & 0 & 0 & 0 \end{bmatrix}, \\
\Xi_{vj} &= Blockdiag\left[-\eta_{jg}, -\eta_{jg}, -\eta_{jf}, -\eta_{jf}\right] \qquad (6.122)
\end{aligned}
$$

The following theorem summarizes the main result about resilient output feedback stabilization

Theorem 6.4.12 *The augmented system (6.116) is robustly delay-dependent asymptoticly stable if there exist matrices* $0 < \mathcal{Y}_{cj}^t = \mathcal{Y}_{cj} \in \Re^{n_j \times n_j}$, $0 < \mathcal{Y}_{oj}^t = \mathcal{Y}_{oj} \in \Re^{n_j \times n_j}$, $0 < \mathcal{W}_{cj}^t = \mathcal{W}_{cj} \in \Re^{n_j \times n_j}$, $0 < \mathcal{W}_{oj}^t = \mathcal{W}_{oj} \in \Re^{n_j \times n_j}$, $\mathcal{S}_{cj} \in \Re^{p_j \times n_j}$, $\mathcal{S}_{oj} \in \Re^{n_j \times n_j}$, $\mathcal{S}_{fj} \in \Re^{p_j \times n_j}$, $Z_{mj} \in \Re^{n_j}$, $Z_{pj} \in \Re^{n_j}$, $Z_{nj} \in \Re^{n_j}$ *and scalars* $\delta_{cj} > 0$, $\nu_{cj} > 0$, $\delta_{oj} > 0$, $\nu_{oj} > 0$ *such that the following convex optimization problem*

$$
\min_{\mathcal{Y}_{cj},\mathcal{Y}_{oj},\mathcal{W}_{cj},\mathcal{W}_{oj},\mathcal{S}_{cj},\mathcal{S}_{oj},\mathcal{S}_{fj},Z_{mj},Z_{nj},Z_{pj},\delta_{cj},\delta_{oj},\nu_{cj},\nu_{oj}} \quad \delta_{cj} + \delta_{oj} + \nu_{cj} + \nu_{oj} \qquad (6.123)
$$

$$
subject\ to:
$$

$$
\mathcal{Y}_{cj} > 0,\ \mathcal{Y}_{oj} > 0,\ \mathcal{W}_{cj} > 0,\ \mathcal{W}_{oj} > 0, \quad \begin{bmatrix} \Xi_{cj} & \Xi_{dj} & \Xi_{sj} \\ \bullet & \Xi_{oj} & 0 \\ \bullet & \bullet & \Xi_{vj} \end{bmatrix} < 0 \qquad (6.124)
$$

is feasible. Moreover, the gains are given by $K_{cj} = \mathcal{S}_{cj}\mathcal{Y}_{cj}^{-1}$, $K_{oj} = \mathcal{S}_{oj}\mathcal{Y}_{oj}^{-1}C_j^{\dagger}$.

Remark 6.4.13 *It should be pointed out that the results reported here before represent major extension over [28, 348] since they either dealt with delay-free systems [348] or targeted independent stability and stabilization [28]. Here, we essentially dealt with systems composed of linear discrete-time subsystems coupled by static nonlinear interconnections satisfying quadratic constraints and subject to unknown-but-bounded state-delays. Subsequently we developed delay-dependent robust stability and feedback stabilization results. To attain decentralized computation, robust controllers based on state feedback and dynamic output feedback were designed on the subsystem level with guaranteed overall desired system performance and in addition, maximized the* *bounds of unknown interconnection terms. Moreover, we incorporated additive gain perturbations to establish new resilient feedback stabilization schemes for discrete-time delay systems.*

6.4.8 Illustrative example 6.4

To illustrate the usefulness of the results of the developed approach, let us consider the following interconnected time-delay system which is composed of three subsystems. With reference to (6.74), the data values are:

$$
\begin{aligned}
Subsystem\ 1 &: \\
A_1 &= \begin{bmatrix} 0.0 & 0.1 \\ 0.5 & -0.5 \end{bmatrix}, \; D_1 = \begin{bmatrix} 0.1 & 0.0 \\ 0.01 & 0.1 \end{bmatrix}, \\
B_1 &= \begin{bmatrix} 0 \\ 0.05 \end{bmatrix}, \; C_1^t = \begin{bmatrix} 0.4 \\ 0.6 \end{bmatrix}, \; H_1 = 1, \\
Subsystem\ 2 &: \\
A_2 &= \begin{bmatrix} 0 & 0 & 0.1 \\ 0.1 & -0.6 & 0.1 \\ 0.3 & 0.7 & 0.9 \end{bmatrix}, \; D_2 = \begin{bmatrix} 0.0 & 0.2 & 0.0 \\ 0.0 & 0.1 & -0.2 \\ 0.1 & -0.3 & -0.2 \end{bmatrix}, \\
B_2 &= \begin{bmatrix} 0.03 \\ 0 \\ 0.04 \end{bmatrix}, \; H_2 = 0.7, \; C_2 = \begin{bmatrix} 0.3 & 0.5 & 0.2 \end{bmatrix} \\
Subsystem\ 3 &: \\
A_3 &= \begin{bmatrix} 0.0 & 0.2 \\ -0.6 & 0.6 \end{bmatrix}, \; D_3 = \begin{bmatrix} 0.0 & 0.2 \\ 0.1 & -0.1 \end{bmatrix}, \\
B_3 &= \begin{bmatrix} 0.1 \\ 0.1 \end{bmatrix}, \; C_3^t = \begin{bmatrix} 0.5 \\ 0.5 \end{bmatrix}, \; H_3 = 0.5, \\
x_1(k) &= \begin{bmatrix} x_{11}(k) & x_{12}(k) \end{bmatrix}^t, \\
x_2(k) &= \begin{bmatrix} x_{21}(k) & x_{22}(k) & x_{23}(k) \end{bmatrix}^t, \\
x_3(k) &= \begin{bmatrix} x_{31}(k) & x_{32}(k) \end{bmatrix}^t
\end{aligned}
$$

In terms of the overall state vector $x(k) = \left[\begin{array}{ccc} x_1^t(k) & x_2^t(k) & x_3^t(k) \end{array}\right]^t \in \Re^7$, the subsystem interconnections are given by

$$\begin{aligned}
g_1(.,.,.) &= 0.1\, cos(x_{22})\, U_{2,5}\, x(k) + 0.15\, cos(x_{31})\, U_{2,5}\, x(k - d_1(k)), \\
2 &\leq d_1(k) \leq 6, \\
g_2(.,.,.) &= 0.3\, cos(x_{21})\, U_{3,5}\, x(k) + 0.2\, cos(x_{11})\, U_{3,5}\, x(k - d_2(k)), \\
3 &\leq d_2(k) \leq 7, \\
g_3(.,.,.) &= 0.2\, cos(x_{12})\, U_{2,5}\, x(k) + 0.25\, cos(x_{23})\, U_{2,5}\, x(k - d_1(k)), \\
3 &\leq d_3(k) \leq 8,
\end{aligned}$$

$$U_{2,5} = \left[\begin{array}{ccccccc} 1 & 1 & 1 & 1 & 1 & 1 & 1 \\ 1 & 1 & 1 & 1 & 1 & 1 & 1 \end{array}\right], \quad U_{3,5} = \left[\begin{array}{ccccccc} 1 & 1 & 1 & 1 & 1 & 1 & 1 \\ 1 & 1 & 1 & 1 & 1 & 1 & 1 \\ 1 & 1 & 1 & 1 & 1 & 1 & 1 \end{array}\right]$$

Using MATLAB-LMI solver, it is found that the feasible solution of Theorem 6.4.3 is given by

$$\delta_1 = 1.5365,\ \nu_1 = 1.3982,\ \mathcal{Y}_1 = \left[\begin{array}{cc} 1.629 & 0.315 \\ \bullet & 1.743 \end{array}\right],\ \mathcal{W}_1 = \left[\begin{array}{cc} 0.896 & 0.234 \\ \bullet & 1.137 \end{array}\right]$$

$$\delta_2 = 1.7025,\ \nu_2 = 1.6544,\ \mathcal{Y}_1 = \left[\begin{array}{ccc} 3.444 & 1.019 & 1.231 \\ \bullet & 2.381 & 1.123 \\ \bullet & \bullet & 4.075 \end{array}\right],$$

$$\mathcal{W}_2 = \left[\begin{array}{ccc} 4.112 & 1.156 & 1.765 \\ \bullet & 3.116 & 2.087 \\ \bullet & \bullet & 3.198 \end{array}\right],$$

$$\delta_3 = 1.4784,\ \nu_3 = 1.2986,\ \mathcal{Y}_1 = \left[\begin{array}{cc} 1.933 & 0.514 \\ \bullet & 1.688 \end{array}\right],$$

$$\mathcal{W}_3 = \left[\begin{array}{cc} 1.632 & 0.777 \\ \bullet & 1.729 \end{array}\right]$$

which illustrates the delay-dependent asymptotic stability. Proceeding to the feedback stabilization schemes, the feasible solution of Theorem 6.4.5 is summarized by

$$\begin{aligned}
\delta_1 &= 1.3614,\ \nu_1 = 1.8222,\ K_1 = \left[\begin{array}{cc} -29.2634 & -23.1445 \end{array}\right], \\
||K_1|| &= 37.3097 \\
\delta_2 &= 1.7025,\ \nu_2 = 1.6544,\ K_2 = \left[\begin{array}{ccc} -26.1974 & -27.3162 & -24.6767 \end{array}\right], \\
||K_2|| &= 45.1281 \\
\delta_3 &= 1.4784,\ \nu_3 = 1.2986,\ K_3 = \left[\begin{array}{cc} -25.2634 & -19.1445 \end{array}\right], \\
||K_3|| &= 31.6978
\end{aligned}$$

Observe that the gains $K_1, .., K_3$ are relatively high. Considering Theorem 6.4.6 and given the tolerance bounds $\phi_1^2 = 1.64$, $\psi_1^2 = 1.55$, $\phi_2^2 = 1.85$, $\psi_2^2 = 1.75$, $\phi_3^2 = 1.74$, $\psi_3^2 = 1.95$, the feasible solution is attained at

$$\begin{aligned}
\delta_1 &= 0.695,\ \nu_1 = 0.714,\ K_1 = \begin{bmatrix} -5.0732 & -3.6114 \end{bmatrix}, \\
||K_1|| &= 6.2273 \\
\delta_2 &= 0.632,\ \nu_2 = 0.644,\ K_2 = \begin{bmatrix} -3.1974 & -2.3162 & -2.9314 \end{bmatrix}, \\
||K_2|| &= 4.9174 \\
\delta_3 &= 0.708,\ \nu_3 = 0.647,\ K_3 = \begin{bmatrix} -2.2634 & -1.1215 \end{bmatrix}, \\
||K_3|| &= 4.9174
\end{aligned}$$

The great reduction in the norms of the local gain matrices is quite evident and supports the developed approach which aims at maximizing the tolerance to uncertain nonlinear interconnections and perturbations.

Admitting gain perturbations of the form (6.95) with $M_1 = [1.5\ \ 1.5]$, $N_1 = 0.3$, $M_2 = [2.5\ \ 2.5\ \ 2.5]$,
$N_2 = 0.2$, $M_3 = [2.0\ \ 2.0]$, $N_3 = 0.25$, we move to consider Theorem 6.4.7, numerical implementation of the LMI solver yields the following feasible solution 6.4.5 is summarized by

$$\begin{aligned}
\delta_1 &= 1.3614,\ \nu_1 = 1.8222,\ K_1 = \begin{bmatrix} -34.4265 & -28.5175 \end{bmatrix}, \\
||K_1|| &= 44.7038 \\
\delta_2 &= 1.7025,\ \nu_2 = 1.6544,\ K_2 = \begin{bmatrix} -29.1484 & -27.4444 & -28.7375 \end{bmatrix}, \\
||K_2|| &= 49.2815 \\
\delta_3 &= 1.4784,\ \nu_3 = 1.2986,\ K_3 = \begin{bmatrix} -30.2474 & -22.5354 \end{bmatrix}, \\
||K_3|| &= 37.7194
\end{aligned}$$

On the other hand, considering **Theorem** 6.4.8 and given the tolerance bounds $\phi_1^2 = 1.62$, $\psi_1^2 = 1.53$, $\phi_2^2 = 1.82$, $\psi_2^2 = 1.73$, $\phi_3^2 = 1.71$, $\psi_3^2 = 1.92$, the feasible solution is attained at

$$\begin{aligned}
\delta_1 &= 0.695,\ \nu_1 = 0.714,\ K_1 = \begin{bmatrix} -5.2132 & -3.6224 \end{bmatrix}, \\
||K_1|| &= 6.3482 \\
\delta_2 &= 0.632,\ \nu_2 = 0.644,\ K_2 = \begin{bmatrix} -3.2444 & -2.3354 & -2.9664 \end{bmatrix}, \\
||K_2|| &= 4.9774 \\
\delta_3 &= 0.708,\ \nu_3 = 0.647,\ K_3 = \begin{bmatrix} -2.4243 & -1.3235 \end{bmatrix}, \\
||K_3|| &= 4.9774
\end{aligned}$$

Once again, the great reduction in the norms of the local gain matrices is quite evident.

Turning to the output feedback stabilization, simulation results show that the feasible solution of **Theorem** 6.4.10 is given by

$$\begin{aligned}
\delta_{c1} &= 1.6674, \ \nu_{c1} = 1.7713, \ \delta_{o1} = 2.0165, \ \nu_{o1} = 2.1004, \\
K_{c1} &= \begin{bmatrix} -2.5953 & -2.4345 \end{bmatrix}, \ K_{o1} = \begin{bmatrix} 0.5030 \\ 0.0100 \end{bmatrix}, \\
\delta_{c2} &= 1.7732, \ \nu_{c2} = 1.8824, \ \delta_{o2} = 2.1365, \ \nu_{o2} = 2.1477, \\
K_{c2} &= \begin{bmatrix} -3.7598 & -1.7746 & 1.2324 \end{bmatrix}, \ K_{o2} = \begin{bmatrix} 0.4121 \\ 0.0322 \\ 0.0714 \end{bmatrix}, \\
\delta_{c3} &= 1.8235, \ \nu_{c3} = 1.8763, \ \delta_{o3} = 2.1125, \ \nu_{o3} = 2.1016, \\
K_{c3} &= \begin{bmatrix} -2.4423 & -1.9876 \end{bmatrix}, \ K_{o3} = \begin{bmatrix} 0.6112 \\ 0.1007 \end{bmatrix}
\end{aligned}$$

Finally, we attend to **Theorem** 6.4.12. By incorporating gain perturbations of the form (6.115) with

$$M_{o1} = [1.5 \ \ 1.5], \ N_{o1} = 0.3, \ M_{o2} = [2.5 \ \ 2.5 \ \ 2.5], \ N_{o2} = 0.2$$

$$M_{o3} = [2.0 \ \ 2.0], \ N_{o3} = 0.25, \ M_{c1}^t = [1.6 \ \ 1.6], \ N_{c1} = 0.4, \ M_{c2}^t = [2.7 \ \ 2.7 \ \ 2.7$$

$$N_{c2} = 0.2, \ M_{c3}^t = [2.1 \ \ 2.1], \ N_{c3} = 0.2$$

the feasible solution is given by

$$\begin{aligned}
\delta_{c1} &= 1.7784, \ \nu_{c1} = 1.8923, \ \delta_{o1} = 2.3355, \ \nu_{o1} = 2.2278, \\
K_{c1} &= \begin{bmatrix} -3.6643 & -2.7576 \end{bmatrix}, \ K_{o1} = \begin{bmatrix} 0.7104 \\ 0.1126 \end{bmatrix}, \\
\delta_{c2} &= 1.9810, \ \nu_{c2} = 2.0134, \ \delta_{o2} = 2.5122, \ \nu_{o2} = 2.4234, \\
K_{c2} &= \begin{bmatrix} -4.0283 & -1.6648 & 1.4684 \end{bmatrix}, \ K_{o2} = \begin{bmatrix} 0.7321 \\ 0.1323 \\ 0.1705 \end{bmatrix}, \\
\delta_{c3} &= 2.0135, \ \nu_{c3} = 1.9845, \ \delta_{o3} = 2.2675, \ \nu_{o3} = 2.5324, \\
K_{c3} &= \begin{bmatrix} -2.7322 & -1.8584 \end{bmatrix}, \ K_{o3} = \begin{bmatrix} 0.5514 \\ 0.2004 \end{bmatrix}
\end{aligned}$$

For the purpose of illustration, we plot in Fig. 6.1 through 6.4 typical closed-loop state-trajectories under the resilient output-feedback control strategy. The plotted results illustrate the smooth behavior of the system response and the

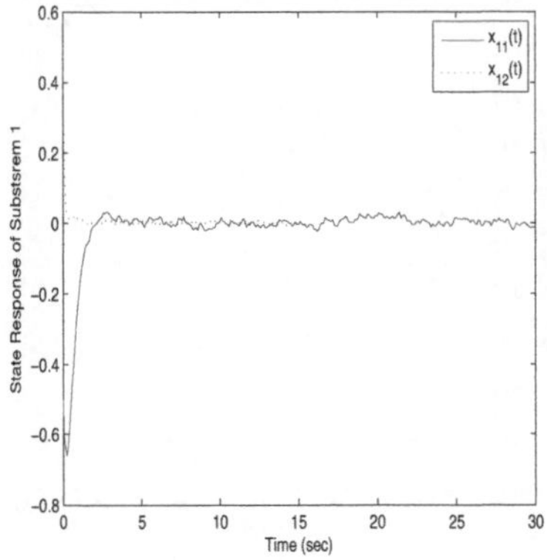

Figure 6.1: State trajectories of Subsystem 1

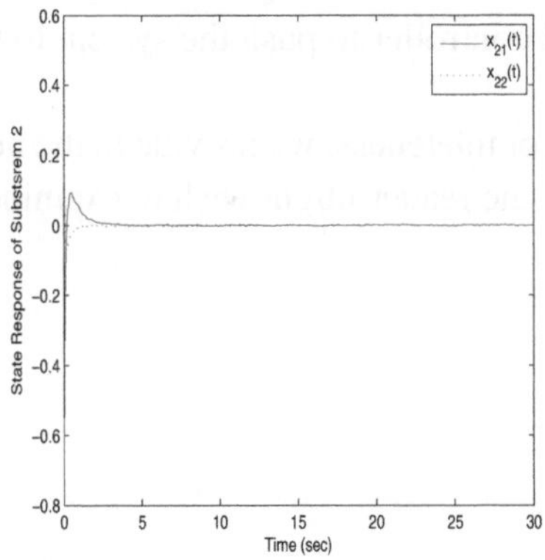

Figure 6.2: First two-state trajectories of Subsystem 2

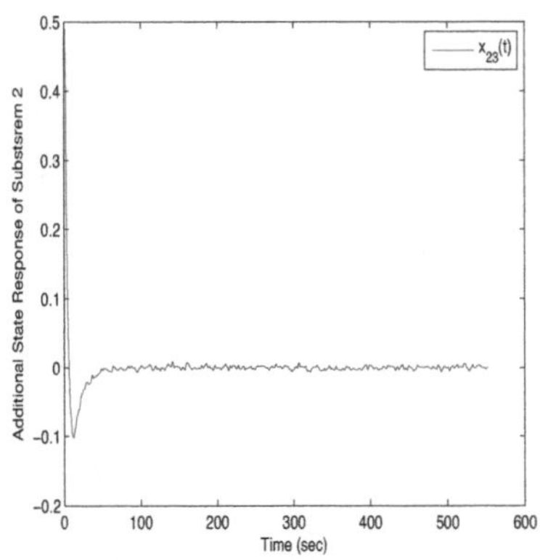

Figure 6.3: Third state trajectory of Subsystem 2

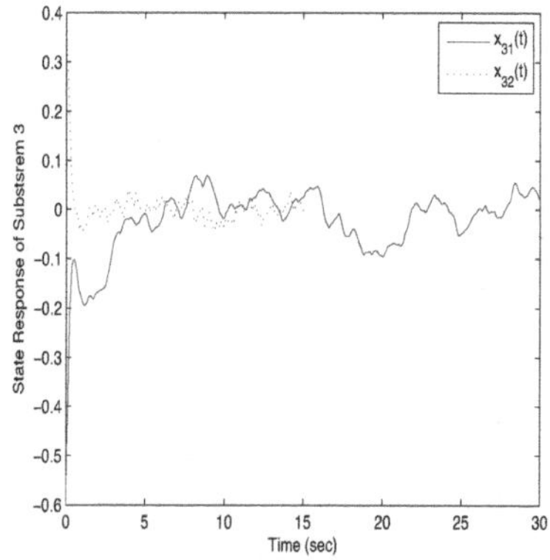

Figure 6.4: State trajectories of Subsystem 3

ability of the developed controller to push the system towards settlement along the zero level.

For the purpose of completeness, we provide in the next section some questions and problems that the reader might wish to examine.

6.5 Problem Set IV

Problem IV.1: Consider a large-scale system comprised of three subsystems and described by:

$$\begin{aligned}
A_1 &= \begin{bmatrix} -5 & -1 & 2 \\ -1 & 3 & -1 \\ -3 & 2 & -2 \end{bmatrix}, \; B_1 = \begin{bmatrix} 1 \\ -1 \\ 1 \end{bmatrix}, \; \Gamma_1 = \begin{bmatrix} 1 \\ -1 \\ 1 \end{bmatrix}, \\
A_{d1} &= \begin{bmatrix} 0.5 & 1 & 0 \\ 0 & -1 & 0 \\ -1 & 0 & -1 \end{bmatrix}, \; E_{12} = \begin{bmatrix} 0.1 \\ 0 \\ 0.4 \end{bmatrix}, \; E_{13} = \begin{bmatrix} 0.3 & 0.1 \\ 0 & -0.1 \\ -0.1 & 0 \end{bmatrix}, \\
G_{d1} &= \begin{bmatrix} 0.2 \\ 0 \\ -0.1 \end{bmatrix}, \; D_1 = 0, \; \Phi_1 = 0.8, \; G_1 = \begin{bmatrix} 0.1 \\ 0.3 \\ 0.2 \end{bmatrix}, \\
A_2 &= -3, \; B_2 = -3, \; \Gamma_2 = 0.4, \; A_{d2} = 0.5, \\
E_{21} &= \begin{bmatrix} -0.1 & 0 & 0.2 \end{bmatrix}, \; E_{23} = \begin{bmatrix} 0.1 & -0.1 \end{bmatrix}, \\
G_{d2} &= -0.3, \; D_2 = 0, \; \Phi_2 = 0.6, \; G_2 = 0.4, \\
A_3 &= \begin{bmatrix} -2 & 3 \\ 1 & 2 \end{bmatrix}, \; B_3 = \begin{bmatrix} 1 \\ 1 \end{bmatrix}, \; \Gamma_3 = \begin{bmatrix} -1 \\ 1 \end{bmatrix}, \\
A_{d3} &= \begin{bmatrix} 0.1 & 0 \\ 0 & -1 \end{bmatrix}, \; E_{31} = \begin{bmatrix} 0.1 & -0.3 & 0 \\ 0 & 0.4 & 0.1 \end{bmatrix}, \; E_{32} = \begin{bmatrix} 0.3 \\ -0.1 \end{bmatrix}, \\
G_{d3} &= \begin{bmatrix} 0.2 & 0 \\ 0 & -0.1 \end{bmatrix}, \; D_3 = 0, \; \Phi_3 = 0.4, \; G_3 = \begin{bmatrix} 0.1 & 0.3 \\ 0.2 & -0.1 \end{bmatrix}
\end{aligned}$$

Choose arbitrary bounds on the system delays and determine the subsystem stability and state feedback stabilization. Repeat for a different set of delay pattern and deduce the relevant results.

Problem IV.2: Consider a discrete-time LSS described by:

$$\begin{aligned}
x_1(k+1) &= \begin{bmatrix} 0 & 1 & 0 \\ 0 & 0 & 1 \\ -1 & 4 & 3 \end{bmatrix} x_1(k) + \begin{bmatrix} 0 \\ 0 \\ 1 \end{bmatrix} u_1(k) + f_1(x_k(k)), \\
x_2(k+1) &= \begin{bmatrix} 0 & 1 \\ 4 & 5 \end{bmatrix} x_2(k) + \begin{bmatrix} 0 \\ 1 \end{bmatrix} u_2(k) + f_2(x_k(k)), \\
f_1^t(x_k(k)) f_1(x_k(k)) &\leq \alpha_1 \, x_1^t(k) x_1(k), \\
f_2^t(x_k(k)) f_2(x_k(k)) &\leq \alpha_2 \, x_2^t(k) x_2(k)
\end{aligned}$$

where α_1 and α_2 are free parameters. Using appropriate quadratic Lyapunov function, derive sufficient conditions for stability and express them in LMI format. What is the maximal bounds for α_1 and α_2 that preserve the stability?

Problem IV.3: Consider a large-scale system comprised of three second-order subsystems and described by:

$$\begin{aligned}
\dot{x}_1(t) &= \begin{bmatrix} 1 & 1.5 \\ 0.3 & -2 \end{bmatrix} x_1(t) + \begin{bmatrix} 0.5\xi_1 & 0 \\ 0 & 0.1\xi_1 \end{bmatrix} x_2(t-0.25) \\
&+ \begin{bmatrix} 5 \\ 1 \end{bmatrix} \{u_1(t) + v_1(t)\}, \\
\dot{x}_2(t) &= \begin{bmatrix} -2 & 0 \\ -1 & -1 \end{bmatrix} x_1(t) + \begin{bmatrix} 0 & 0 \\ \xi_2 & \xi_3 \end{bmatrix} x_1(t-0.2) \\
&+ \begin{bmatrix} 0 & 0 \\ \xi_2 & 0 \end{bmatrix} x_3(t-0.1) \\
&+ \begin{bmatrix} 0 \\ 1 \end{bmatrix} \{\phi u_2(t) + v_2(t)\}, \\
\dot{x}_3(t) &= \begin{bmatrix} -1 & 0 \\ -1 & -2 \end{bmatrix} x_1(t) + \begin{bmatrix} 0 & 0 \\ \xi_4 & \xi_4 \end{bmatrix} x_2(t-0.1) \\
&+ \begin{bmatrix} 1 & 0 \\ 0 & 1 \end{bmatrix} \{u_3(t) + v_3(t)\}
\end{aligned}$$

where the different parameter bounds are given by

$$\begin{aligned}
||\xi_1|| &\leq 1, \;||v_1|| \leq 0.1, \;||\xi_2|| \leq 1, \;||\xi_3|| \leq 2, \;||v_2|| \leq 0.5, \\
||\xi_3|| &\leq 2, \;||v_3|| \leq 0.1, \;2 \leq \phi\, 4
\end{aligned}$$

The objective is to determine a decentralized controller such that the solutions of the overall system are uniformly ultimately bounded with respect to the bound $\varepsilon \leq 0.18$. To derive appropriate conditions, apply stability theory based on constructive use of Lyapunov function [152].

Problem IV.4: Consider two stands of cold rolling mill in tandem and described by:

$$\begin{aligned}
\dot{x}_1(t) &= \begin{bmatrix} -100 & 0 & 0 \\ 0 & -40 & 0 \\ 4.517 & -34.335 & -56.063 \end{bmatrix} x_1(t) \\
&+ \begin{bmatrix} 100 & 0 \\ 0 & 40 \\ 0 & 0 \end{bmatrix} u_1(t) + f_1(t,x,u), \\
\dot{x}_2(t) &= \begin{bmatrix} -100 & 0 & 0 \\ 0 & -40 & 0 \\ 6.428 & -35.982 & -100.295 \end{bmatrix} x_2(t) \\
&+ \begin{bmatrix} 100 & 0 \\ 0 & 40 \\ 0 & 0 \end{bmatrix} u_2(t) + f_2(t,x,u), \\
C_1 &= \begin{bmatrix} 0.455 & 0 & -5.589 \\ 0 & 0 & 1 \end{bmatrix}. \ C_2 = \begin{bmatrix} 0.491 & 0 & -6.931 \\ 0 & 0 & 1 \end{bmatrix}, \\
f_1^t(t,x,u)f_1(t,x,u) &\le \alpha_1 \, x^t(t)M_1x(t), \ \ f_2^t(t,x,u)f_2(t,x,u) \le \alpha_2 \, x^t(t)M_2x(t), \\
M_1 &= \begin{bmatrix} 0 & 0 & 0 & 0 & 0 & 0 \\ 0 & 0 & 0 & 0 & 0 & 0 \\ 0 & 0 & 0 & 0 & 0.1 & 0.1 \\ 0 & 0 & 0 & 0 & 0 & 0 \\ 0 & 0 & 0 & 0 & 0 & 0 \\ 0 & 0 & 0 & 0 & 0 & 0 \end{bmatrix}, \\
M_2 &= \begin{bmatrix} 0 & 0 & 0 & 0 & 0 & 0 \\ 0 & 0 & 0 & 0 & 0 & 0 \\ 0 & 0 & 0 & 0 & 0 & 0 \\ 0 & 0 & 0 & 0 & 0 & 0 \\ 0 & 0 & 0 & 0 & 0 & 0 \\ 0.1 & 0 & 0.1 & 0 & 0 & 0 \end{bmatrix}
\end{aligned}$$

where α_1 and α_2 are adjustable parameters. Using appropriate quadratic Lyapunov function, derive LMI-based sufficient conditions of asymptotic stability and static output feedback stabilization. What is the maximal bounds for α_1 and α_2 that preserve the stability?

Problem IV.5: Consider the following model of LSS with state-dependent delays

$$\begin{aligned} S_j: \quad \dot{x}_j(t) &= A_j x_j(t) + \Delta_j(x_j(t), t) + \sum_{k=1, k\neq j}^{n_s} A_{jk} x_k(t - \eta_{jk}(x_j(t), t)) \\ j &\in \{1, ..., n_s\} \end{aligned}$$

where $\eta_{jk}(x_j(t), t)$ are bounded continuous functions and the nonlinear parametric uncertainties are bounded in the form

$$||\Delta_j(x_j(t), t)|| \leq \beta_j ||x_j(t)||, \ \beta_j \ \in [0, \ \infty)$$

Apply the Lyapunov-Razumikhn theory to derive the sufficient conditions that guarantee the LSS is asymptotically stable independent of delay.

Problem IV.6: Consider the following model of LSS with parametric uncertainties:

$$\begin{aligned} S_j: \quad \dot{x}_j(t) &= [A_{jj} + \Delta A_{jj}(x(t), t)] x_j(t) \\ &+ \sum_{k=1, k\neq j}^{n_s} [A_{jk} + \Delta A_{jk}(x(t), t)] x_k(t) \\ &+ \sum_{k=1, k\neq j}^{n_s} [D_{jk} + \Delta D_{jk}(x(t - \eta_{jk}))] x_k(t - \eta_{jk}), \quad j \in \{1, ..., n_s\} \end{aligned}$$

where the nonlinear parametric uncertainties are bounded in the form

$$\begin{aligned} ||\Delta A_{jk}(x(t), t)|| &\leq \beta_{jk}, \ \ \beta_j \ \in [0, \ \infty), \\ ||\Delta D_{jk}(x(t - \eta_{jk}))|| &\leq \sigma_{jk}, \ \ \beta_j \ \in [0, \ \infty) \end{aligned}$$

Assume for simplicity that $\lambda(A_{jj}) \in \mathcal{C}^-, \ j \in \{1, ..., n_s\}$. Use the LKF

$$V(x) = \sum_{k=1}^{n_s} \left(x_j^t(t) P_j x_j(t) + \int_{t-\eta_{jk}}^{t} x_j^t(s) Q_j x_j(s) \, ds \right.$$

to derive the sufficient conditions that guarantee the LSS is asymptotically stable.

Problem IV.7: It is of interest to develop an overlapping decomposition procedure to deal with the case of time-delay systems with polytopic uncertainties. The objective is to derive appropriate expansion-contraction relations that allows the design of linear quadratic controller.

Problem IV.8: Extend the results of the overlapping control methods to the class of discrete-time systems

$$\begin{aligned}
\mathbf{S}: \quad x(k+1) &= \bar{A}(k)x(k) + \bar{B}(k)u(k) + \bar{A}_d(k)x(k-d) + \bar{D}(k)u(k-d), \\
y(k) &= Cx(k), \\
x(k) &= \varphi(k), -d \leq k \leq 0 \\
\bar{A}(k) &= A_o + \Delta A(k), \bar{B}(k) = B_o + \Delta B(k), \\
\bar{A}_d(k) &= A_{do} + \Delta A_d(k), \bar{D}(k) = D_o + \Delta D(k)
\end{aligned}$$

where

$$[\Delta A(k)\ \Delta B(k)\ \Delta A_d(k)\ \Delta D(k)] = H\ F(k)\ [E_1\ E_2\ E_3\ E_4],$$

where H, E_1, E_2, E_3, E_4 are are known constant real matrices of appropriate dimensions and $F^t(k)F(k) \leq I$.

6.6 Notes and References

A set of expansion-contraction relations extending the Inclusion Principle is proved for a class of linear continuous-time uncertain systems with state and control delays. Norm bounded arbitrarily time-varying uncertainties and a given point delay are considered. The resulting structural relations are easily extendable to polytopic systems with constant uncertainties. The presented inclusion relations are applied on the quadratic guaranteed cost control design. Conditions preserving the expansion-contraction relations for closed-loop systems including the equality of cost bounds have been proved. The guaranteed cost control design is performed using the LMI delay independent procedure in the expanded space and subsequently contracted into the original system. The results are specialized on the overlapping static output feedback design.

Chapter 7

Decentralized Reliable Control

In this chapter, we continue further into the decentralized-control techniques for interconnected systems, where we focus herein on methods for designing classes of reliable decentralized controllers to deal with possible actuator and/or sensor failures. Thus we study hereafter the problem of designing reliable decentralized feedback control for linear interconnected systems where we focus initially on subsystems with internal time-delays and additional time-delay couplings, under actuator and/or sensor failures. Then we specialize the result to delay-free subsystems. We equally treat continuous- and discrete-time system representations. The failures are described by a model that takes into consideration possible outages or partial failures in every single actuator/single sensor of each decentralized controller. The decentralized control design is performed through two steps. First, a decentralized stabilizing reliable feedback control set is derived at the subsystem level through the construction of appropriate Lyapunov-Krasovskii functional (LKF) and, second, a feasible linear matrix inequalities procedure is then established for the effective construction of the control set under different feedback schemes. Two schemes are considered: the first is based on state-measurement and the second utilizes static output-feedback. The decentralized feedback gains in both schemes are determined by convex optimization over linear matrix inequalities (LMIs).

7.1 Interconnected Continuous Systems

Large-scale interconnected systems appear in a variety of engineering applications including power systems, large structures, and manufacturing systems and for those applications, decentralized control schemes present a practical and effective means for designing control algorithms based on the individual subsystems [324]. Relevant research results on decentralized control of relevance to the present work can be found in [173, 288, 328]. There are fundamental issues arising quite frequently when designing feedback controllers for interconnected systems. The first major issue regards the construction of decentralized control schemes to confront the practical limitations in the number and the structure of the feedback loops. The second issue is due to the presence of uncertainties and/or time-varying delays both in the subsystems and in the interconnections. The third issue concerns the reliability of the control systems against different component failures. It becomes increasingly evident that the reliability of control systems in the presence of system component failures is of paramount importance for maintaining the critical functionality and survivability of many safety critical systems. For these systems, the overall reliability is enhanced not by using more reliable components, but by managing them in a way that the reliability of the overall system is greater than the reliability of its parts. This is the fundamental concept of fault-tolerant control and the reliable control problem. It is needless to stress that reliable operation is of prime importance in the case of interconnected dynamical systems since failures could occur independently in each subsystem or actuator channel in the form of total outage or partial degradation.

In the literature on designing reliable controllers for single systems, there is one approach based on the use of multiple redundant controls [161]. There is another direction that aims to design controls without redundancy by ensuring stability with some degree of performance for specified classes of admissible failures of particular control components [270]. Classical quadratic optimal control has been used for reliable design [108, 367, 401] to achieve stability and performance. Reliable optimal controllers have been designed by using $\mathcal{H}_2/\mathcal{H}_\infty$ tools in [83, 388] where a condition for decentralized and quadratic stabilizability was presented and subsequently used to provide a solution of $\mathcal{H}_2$-norm optimization. In [39], the integral constraints and the guaranteed cost control were applied to address robustness issues. In the reported papers, system interconnections were not considered and the classes of admissible failures were usually modeled as outages [83, 388]. This model considers the control set partitioned into one subset with the actuators whose failures are admissible in the control design and a complementary subset with the actuators that are assumed to keep a normal operation. The literature on designing reliable controllers for interconnected systems is quite limited [296] where an initial effort for decentralized reliable control was developed for a class of interconnected systems.

7.1.1 Problem description

We consider a class of linear systems S structurally composed of n_s coupled subsystems $\mathsf{S_j}$ and the model of the jth subsystem is described by the state-space representation:

$$\begin{aligned} \dot{x}_j(t) &= A_j x_j(t) + A_{dj} x_j(t - \tau_j(t)) + B_j u_j(t) + c_j(t) + \Omega_j w_j(t) \\ z_j(t) &= G_j x_j(t) + G_{dj} x_j(t - \tau_j(t)) + D_j u_j(t) + B_j w_j(t) \\ y_j(t) &= C_j x_j(t) + C_{dj} x_j(t - \tau_j(t)) \end{aligned} \quad (7.1)$$

where for $j \in \{1, ..., n_s\}$, $x_j(t) \in \Re^{n_j}$ is the state vector, $u_j(t) \in \Re^{m_j}$ is the control input, $y_j(t) \in \Re^{p_j}$ is the measured output, $w_j(t) \in \Re^{q_j}$ is the disturbance input which belongs to $\mathcal{L}_2[0, \infty)$, $z_j(t) \in \Re^{q_j}$ is the performance output. The matrices $A_j \in \Re^{n_j \times n_j}$, $B_j \in \Re^{n_j \times m_j}$, $D_j \in \Re^{q_j \times m_j}$, $A_{dj} \in \Re^{n_j \times n_j}$, $B_j \in \Re^{q_j \times q_j}$, $\Omega_j \in \Re^{n_j \times q_j}$, $C_j \in \Re^{p_j \times n_j}$, $C_{dj} \in \Re^{p_j \times n_j}$, $G_j \in \Re^{q_j \times n_j}$, $G_{dj} \in \Re^{q_j \times n_j}$ are real and constants. The initial condition is $B_j \in \mathcal{L}_2[-\tau_j^*, 0]$, $j \in \{1, ..., n_s\}$. The interactions between subsystem $\mathsf{S_j}$ and the other subsystems are summarized by the coupling vector $c_j(t) \in \Re^{n_j}$ where

$$c_j(t) = \sum_{k=1, k \neq j}^{n_s} F_{jk} x_k(t) + \sum_{k=1, k \neq j}^{n_s} E_{jk} x_k(t - \eta_{jk}(t)) \quad (7.2)$$

The matrices $F_{jk} \in \Re^{n_j \times n_k}$, $E_{jk} \in \Re^{n_j \times n_k}$ are real and constants. The factors τ_j, η_{jk}, $j, k \in \{1, ..., n_s\}$ are unknown time-delay factors satisfying

$$\begin{aligned} 0 &\leq \tau_j(t) \leq \varrho_j, \quad \dot{\tau}_j(t) \leq \mu_j, \\ 0 &\leq \eta_{jk}(t) \leq \varrho_{jk}, \quad \dot{\eta}_{jk}(t) \leq \mu_{jk} \end{aligned} \quad (7.3)$$

where the bounds ϱ_j, ϱ_{jk}, μ_j, μ_{jk} are known constants in order to guarantee smooth growth of the state trajectories. The inclusion of the terms $A_{dj} x_j(t - \tau_j(t))$, $E_{jk} x_k(t - \eta_{jk}(t))$ is meant to emphasize the delay within each subsystem (local delay) and among the subsystems (coupling delay), respectively. A block-diagram representation of the subsystem model (7.1) is depicted in Figure 7.1. It should be observed that the class of systems described by (7.1)-(7.2) subject to delay-pattern (7.3) is frequently encountered in modeling several physical systems and engineering applications including large space structures, multimachine power systems, cold mills, transportation systems, and water pollution management, to name a few [252, 253, 324].

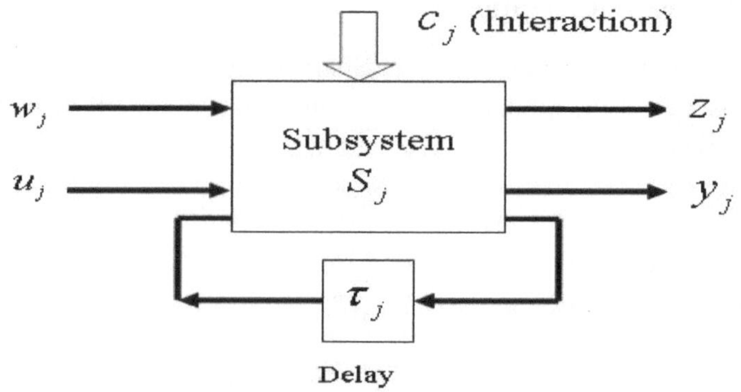

Figure 7.1: Subsystem model

7.1.2 Actuator failure model

In the sequel, we consider that system (7.1) is assumed to operate under failures described by extension of the representation [39] to the decentralized setting treated hereafter. Such representation allows independent outage or partial degradation in any single actuator of every decentralized controller. In this way, a rather general scenario is defined for the design of reliable decentralized control structure for a class of interconnected systems simultaneously facing uncertainties and failures. Let $u_j^f \in \Re^{f_j}$ denote the vector of signals from the f_j actuators that control the jth subsystem. We consider the following failure model:

$$u_j^f(t) = \Sigma_j u_j(t) + \beta_j(u_j) \tag{7.4}$$

where $0 < \Sigma_j \in \Re^{f_j \times f_j} = diag(\sigma_{j1}, ..., \sigma_{jf_j})$ and the function $\beta_j(u_j) = [\beta_{j1}(u_{j1}), ..., \beta_{jf_j}(u_{jf_j})]^t$ is uncertain ans satisfies, for each j,

$$\beta_{jk}^2 \leq \gamma_{jk}^2 \, u_{jk}^2, \quad k = 1, ..., f_j, \;\; \gamma_{jk}^2 \geq 0 \tag{7.5}$$

When (7.5) holds, then

$$||\beta_j(x_j)||^2 \leq ||\Gamma_j u_j||^2, \quad j = 1, ..., n_s \tag{7.6}$$

where $0 \leq \Gamma_j = diag(\gamma_{j1}, ..., \gamma_{jf_j}) \in \Re^{f_j \times f_j}$. The value of $\sigma_{jk}, \; k = 1, .., f_j$ represents the percentage of failure in the actuator j controlling subsystem S_j. With this representation, each subsystem actuator can fail independently. Note that the condition $\sigma_{jk} = 1, \; \gamma_{jk} = 0$ represents the normal case for the kth actuator of the jth subsystem $(u_{jk}^f(t) = u_{jk}(t))$. When this situation holds for all k, we get that $\Sigma_j = I_{f_j}$ and $\Gamma_j = 0$ which in turn clarifies the normal case in the jth channel ($u_j^f(t) = u_j(t)$). In the particular case $\sigma_{jk} = \gamma_{jk}$, (7.4) and (7.5) covers the outage case ($u_{jk}^f = 0$) since $\beta_{jk} = -\sigma_{jk}u_{jk}$ satisfies (7.5). Alternatively, the case $\beta_j(u_j) = -\sigma_j u_j$ discloses the outage of the whole controller of the jth system. Indeed, there are different cases that would reveal partial failures or partial degradations of the actuators.

Our objective now is to study two main problems:

1. *The first problem is with the decentralized reliable stabilization by developing decentralized feedback controllers (based on state, static output, and dynamic output) in the presence of failures described by (7.4)-(7.5) and deriving a feasibility testing at the subsystem level so as to guarantee the overall system asymptotic stability with a prescribed performance measure.*

2. *The second problem deals with the design of resilient decentralized controllers guaranteeing reliable stabilization by developing resilient decentralized feedback controllers that takes into consideration additive gain perturbations while ensuring that the overall closed-loop system is asymptotically stable with a prescribed performance measure.*

7.1.3 State-feedback reliable design

In this section, we develop new criteria for LMI-based characterization of decentralized reliable stabilization by local state feedback. The criteria include some parameter matrices aimed at expanding the range of applicability of the developed conditions. Using the local state feedback $u_j(t) = K_j x_j(t),\ j = 1, ..., N$, the local closed-loop subsystem under failure becomes

$$\begin{aligned}
\dot{x}_j(t) &= \widehat{A}_j x_j(t) + A_{dj} x_j(t - \tau_j(t)) + B_j \beta_j(u_j) + c_j(t) + \Omega_j w_j(t) \\
z_j(t) &= \widehat{G}_j x_j(t) + G_{dj} x_j(t - \tau_j(t)) + D_j \beta_j(u_j) + B_j w_j(t) \\
c_j(t) &= \sum_{k=1, k\neq j}^{n_s} F_{jk} x_k(t) + \sum_{k=1, k\neq j}^{n_s} E_{jk} x_k(t - \eta_{jk}(t)) \\
\widehat{A}_j &= A_j + B_j \Sigma_j K_j, \quad \widehat{G}_j = G_j + D_j \Sigma_j K_j
\end{aligned} \tag{7.7}$$

The following theorem establishes the main design result for subsystem S_j where in the subsections that follow, asymptotic stability conditions and the reliable decentralized feedback gains are derived.

Theorem 7.1.1 *Given the bounds $\varrho_j > 0,\ \mu_j > 0,\ \varrho_{jk} > 0,\ \mu_{jk} > 0,\ j,k = 1,...,n_s$ and matrix $\mathcal{W}_j$, then the family of subsystems $\{\mathsf{S}_j\}$ where S_j is described by (7.7) is delay-dependent asymptotically stabilizable by decentralized controller $u_j(t) = K_j x_j(t),\ j = 1,...,n_s$ with $\mathcal{L}_2$-performance bound $\gamma_j,\ j = 1,...,n_s$ if there exist positive-definite matrices $\mathcal{X}_j,\ \mathcal{Y}_j,\ \mathcal{W}_k,\ \{\Lambda_{kj}\}_{k=1}^{n_s},\ \{\Psi_{rj}\}_{r=1}^{4},$ satisfying the following LMIs for $k,j = 1,...,n_s$:*

$$\widetilde{\Pi}_j = \begin{bmatrix} \widetilde{\Pi}_{aj} & \widetilde{\Pi}_{cj} \\ \bullet & \widetilde{\Pi}_{ej} \end{bmatrix} < 0$$

$$\widetilde{\Pi}_{aj} = \begin{bmatrix} \Pi_{oj} & \Pi_{aj} & -\varrho_j\Psi_{1j} & 0 & B_j \\ \bullet & -\Pi_{mj} & -\varrho_j\Psi_{3j} & 0 & 0 \\ \bullet & \bullet & -\varrho_j\Psi_{4j} & 0 & 0 \\ \bullet & \bullet & \bullet & -\Pi_{nj} & 0 \\ \bullet & \bullet & \bullet & \bullet & -I_j \end{bmatrix},$$

$$\widetilde{\Pi}_{cj} = \begin{bmatrix} \Omega_j & \Pi_{zj}^t & \mathcal{Y}_j^t\Gamma_j^t & \varrho_k(\mathcal{X}_j A_j^t + \mathcal{Y}_j^t\Sigma_j^t B_j^t)\mathcal{W}_k \\ 0 & G_{dj}^t & 0 & \varrho_k\mathcal{X}_j A_{dj}^t\mathcal{W}_k \\ 0 & 0 & 0 & 0 \\ 0 & 0 & 0 & \varrho_k\sum_{k=1}^{n_s}\mathcal{X}_j E_{kj}\mathcal{W}_k \\ 0 & 0 & 0 & 0 \end{bmatrix},$$

$$\widetilde{\Pi}_{ej} = \begin{bmatrix} -\gamma_j^2 I_j & B_j^t & 0 & \varrho_k\Gamma_j^t\mathcal{W}_k \\ \bullet & -I_j & 0 & 0 \\ \bullet & \bullet & -I_j & 0 \\ \bullet & \bullet & \bullet & -\varrho_k\mathcal{W}_k \end{bmatrix} \tag{7.8}$$

where

$$\begin{aligned} \Pi_{oj} &= [A_j + \sum_{k=1}^{n_s} F_{kj}]\mathcal{X}_j + B_j\Sigma_j\mathcal{Y}_j + \mathcal{X}_j[A_j + \sum_{k=1}^{n_s} F_{kj}]^t \\ &+ \mathcal{Y}_j^t\Sigma_j^t B_j^t + \Psi_{1j} + \Psi_{1j}^t + \Psi_{2j} + \sum_{k=1,k\neq j}^{n_s} \Lambda_{kj} + (n_s - 1)\mathcal{X}_j \\ \Pi_{aj} &= A_{dj}\mathcal{X}_j - \Psi_{1j} + \Psi_{3j}^t,\ \ \Pi_{zj} = G_j\mathcal{X}_j + D_j\Sigma_j\mathcal{Y}_j, \\ \Pi_{nj} &= \sum_{k=1,k\neq j}^{n_s} (1 - \mu_{kj})\Lambda_{kj} + \sum_{k=1,k\neq j}^{n_s} \mathcal{X}_j E_{kj}^t \mathcal{P}_k E_{kj}\mathcal{X}_j, \\ \Pi_{mj} &= \Psi_{3j} + \Psi_{3j}^t + (1 - \mu_j)\Psi_{2j} - \Psi_{4j}^t + \Psi_{4js}^t \end{aligned} \tag{7.9}$$

Moreover, the local gain matrix is given by $K_j = \mathcal{Y}_j\mathcal{X}_j^{-1}$.

Proof: Consider the LKF:

$$\begin{aligned}
V(t) &= \sum_{j=1}^{n_s} V_j(t), \;\; V_j(t) = V_{oj}(t) + V_{aj}(t) + V_{mj}(t) + V_{nj}(t), \\
V_{oj}(t) &= x_j^t(t)\mathcal{P}_j x_j(t), \;\; V_{aj}(t) = \int_{-\varrho_j}^{0}\int_{t+s}^{t} \dot{x}_j^t(\alpha)\mathcal{W}_j\dot{x}_j(\alpha)d\alpha\, ds, \\
V_{mj}(t) &= \int_{t-\tau_j(t)}^{t} x_j^t(s)\mathcal{Q}_j x_j(s)\, ds, \\
V_{nj}(t) &= \sum_{k=1,k\neq j}^{n_s}\int_{t-\eta_{jk}(t)}^{t} x_k^t(s)\mathcal{Z}_{jk}x_k(s)\, ds \qquad (7.10)
\end{aligned}$$

where $0 < \mathcal{P}_j = \mathcal{P}_j^t$, $0 < \mathcal{W}_j = \mathcal{W}_j^t$, $0 < \mathcal{Q}_j = \mathcal{Q}_j^t$, $0 < \mathcal{Z}_{jk} = \mathcal{Z}_{jk}^t$, $j,k \in \{1,...,n_s\}$ are weighting matrices of appropriate dimensions. Observe that $V(t)$ has been used before in Chapter 6. Now by using the failure model inequality (7.6) with $u_j(t) = K_j x_j(t)$, we have

$$\beta_j^t(u_j)\beta_j(u_j) \;\leq\; x_j^t K_j^t \Gamma_j^t \Gamma_j K_j x_j \qquad (7.11)$$

For the time being, we consider that the gains K_j are specified. A straightforward computation gives the time-derivative of $V(t)$ along the solutions of (7.7) with $w(t) \equiv 0$ as:

$$\begin{aligned}
\dot{V}_{oj}(t) &= 2x_j^t\mathcal{P}_j\dot{x}_j = 2x_j^t\mathcal{P}_j[\widehat{A}_j x_j(t) + A_{dj}x_j(t-\tau_j(t)) + B_j\beta_j(u_j)] \\
&+ 2x_j^t\mathcal{P}_j\sum_{k=1}^{n_s}F_{jk}x_k(t) + 2x_j^t\mathcal{P}_j\sum_{k=1}^{n_s}E_{jk}x_k(t-\eta_{jk}(t)) \qquad (7.12)
\end{aligned}$$

Standard manipulations of (7.12) yield

$$\begin{aligned}
\dot{V}_{oj}(t) &= 2x_j^t\mathcal{P}_j[\widehat{A}_j + A_{dj}]x_j(t) + 2x_j^t[\Theta_j - \mathcal{P}_j A_{dj}]\int_{t-\tau_j(t)}^{t}\dot{x}_j(s)ds \\
&+ 2x_j^t\mathcal{P}_j B_j\beta_j(u_j) \\
&+ 2x^t(t-\tau_j(t))\Upsilon_j\int_{t-\tau_j(t)}^{t}\dot{x}_j(s)ds - \Big[2x_j^t t\Theta_j\int_{t-\tau_j}^{t}\dot{x}_j(s)ds \\
&+ 2x^t(t-\tau_j(t))\Upsilon_j\int_{t-\tau_j(t)}^{t}\dot{x}_j(s)ds\Big] \\
&+ 2x_j^t\mathcal{P}_j\sum_{k=1}^{n_s}F_{jk}x_k(t) + 2x_j^t\mathcal{P}_j\sum_{k=1}^{n_s}E_{jk}x_k(t-\eta_{jk}(t)) \qquad (7.13)
\end{aligned}$$

Little algebra on (7.13) yields

$$\dot{V}_{oj}(t) = \frac{1}{\tau_j(t)} \int_{t-\tau_j(t)}^{t} \Bigg[2x_j^t[\mathcal{P}_j \widehat{A}_j + \Theta_j]x_j$$
$$+2x_j^t[\mathcal{P}_j A_{dj} - \Theta_j + \Upsilon_j^t]x_j(t-\tau_j(t)) - 2x_j^t(t-\tau_j(t))\Upsilon_j x(t-\tau_j)$$
$$-2x^t \tau_j \Theta \dot{x}_j(s) - 2x_j^t(t-\tau_j(t))\tau_j(t)\Upsilon_j \dot{x}_j(s)$$
$$+2x_j^t \mathcal{P}_j B_j \beta_j(u_j) + 2x_j^t \mathcal{P}_j \sum_{k=1}^{n_s} F_{jk} x_k(t)$$
$$+2x_j^t \mathcal{P}_j \sum_{k=1}^{n_s} E_{jk} x_k(t-\eta_{jk}(t)) \Bigg] ds \tag{7.14}$$

where $\Theta_j \in \Re^{n_j \times n_j}$ and $\Upsilon_j \in \Re^{n_j \times n_j}$ are appropriate relaxation matrices injected to facilitate the delay-dependence analysis.

$$\dot{V}_{aj}(t) = \int_{t-\varrho_j}^{t} [\dot{x}_j^t(t)\mathcal{W}_j \dot{x}_j(t) - \dot{x}_j^t(s)\mathcal{W}_j \dot{x}_j(s)]d\ s$$
$$= \varrho_j\ \dot{x}_j^t(t)\mathcal{W}_j \dot{x}_j(t) - \int_{t-\tau_j}^{t} \dot{x}_j^t(s)\mathcal{W}_j \dot{x}_j(s)ds$$
$$- \int_{t-\varrho_j}^{t-\tau_j} \dot{x}_j^t(s)\mathcal{W}_j \dot{x}_j(s) \tag{7.15}$$

Little algebra on (7.15) yields

$$\dot{V}_{aj}(t) \le \varrho_j\ \dot{x}_j^t(t)\mathcal{W}_j \dot{x}_j(t) - \int_{t-\tau_j}^{t} \dot{x}_j^t(s)\mathcal{W}_j \dot{x}_j(s)ds$$
$$= \frac{1}{\tau_j(t)} \int_{t-\tau_j(t)}^{t} \Big[\varrho_j\ \dot{x}_j^t(t)\mathcal{W}_j \dot{x}_j(t) - \tau_j \dot{x}_j^t(s)\mathcal{W}_j \dot{x}_j(s) \Big] ds \tag{7.16}$$

Note that the term $T_j = \int_{t-\varrho_j}^{t-\tau_j} \dot{x}_j^t(s)\mathcal{W}_j \dot{x}_j(s)$ accounts for the enlarged time interval from $t-\varrho_j \rightarrow t$ to $t-\tau_j \rightarrow t$. It is obvious that $T_j > 0$ and hence expression (7.16) holds true without conservatism. There has been an alternative route to handle T_j by employing extra parameter matrices and adding more identities [103] and [146]. Also,

$$\begin{aligned}
\dot{V}_{mj}(t) &= x_j^t(t)\mathcal{Q}_j x_j(t) - (1-\dot{\tau}_j(t))\, x^t(t-\tau_j(t))\mathcal{Q}_j x_j(t-\tau_j(t)) \\
&\leq x_j^t(t)\mathcal{Q}_j x_j(t) - (1-\mu_j)\, x_j^t(t-\tau_j(t))\mathcal{Q}_j x_j(t-\tau_j(t)) \\
&= \frac{1}{\tau_j(t)} \int_{t-\tau_j(t)}^{t} \Big[x_j^t(t)\mathcal{Q}_j x_j(t) \\
&- (1-\mu_j)\, x_j^t(t-\tau_j(t))\mathcal{Q}_j x_j(t-\tau_j(t)) \Big] ds \qquad (7.17)
\end{aligned}$$

Also, we have

$$\begin{aligned}
\dot{V}_{oj}(t)\dot{V}_{nj}(t) &= \sum_{k=1,k\neq j}^{n_s} \\
&\times \Big[x_k^t(t)\mathcal{Z}_{jk} x_k(t) \\
&- (1-\dot{\eta}_{jk}(t)) x_k^t(t-\eta_{jk}(t))\mathcal{Z}_{jk} x_k(t-\eta_{jk}(t)) \Big] \\
&\leq \sum_{k=1,k\neq j}^{n_s} \Big[x_k^t(t)\mathcal{Z}_{jk} x_k(t) \\
&- (1-\mu_{jk})\, x_k^t(t-\eta_{jk}(t))\mathcal{Z}_{jk} x_k(t-\eta_{jk}(t)) \Big] \qquad (7.18)
\end{aligned}$$

For the class of interconnected systems (7.1), the following structural identity holds:

$$\sum_{j=1}^{n_s} \sum_{k=1,k\neq j}^{n_s} x_k^t(t)\mathcal{Z}_{jk} x_k(t) = \sum_{j=1}^{n_s} \sum_{k=1,k\neq j}^{n_s} x_j^t(t)\mathcal{Z}_{kj} x_j(t) \qquad (7.19)$$

By combining (7.10)-(7.19) and using Schur complements, we have

$$\dot{V}_j(t)|_{(7.7)} \leq \sum_{j=1}^{n_s} \left[\frac{1}{\tau_j} \int_{t-\tau_j}^{t} \chi_j^t(t,s)\, \Xi_j\, \chi_j(t,s)\, ds \right] \qquad (7.20)$$

where

$$\chi_j(t,s) = \begin{bmatrix} x_j^t(t) & x_j^t(t-\tau_j) & \dot{x}_j^t(s) & x_j^t(t-\eta_{kj}) & \beta^t(x_j) \end{bmatrix}^t$$

$$
\begin{aligned}
\overline{\Xi}_j &= \begin{bmatrix} \Xi_{oj} & \Xi_{aj} & -\varrho_j\Theta_j & \Xi_{cj} & \mathcal{P}_jB_j & \varrho_j\widehat{A}_j^t\mathcal{W}_j \\ \bullet & -\Xi_{mj} & -\varrho_j\Upsilon_j & 0 & 0 & \varrho_kA_{d_j}^t\mathcal{W}_k \\ \bullet & \bullet & -\varrho_k\mathcal{W}_k & 0 & 0 & 0 \\ \bullet & \bullet & \bullet & -\Xi_{nj} & 0 & \varrho_k\sum_{k=1}^{n_s}E_{kj}\mathcal{W}_k \\ \bullet & \bullet & \bullet & \bullet & -I_j & 0 \\ \bullet & \bullet & \bullet & \bullet & \bullet & -\varrho_k\mathcal{W}_k \end{bmatrix} \\
\Xi_{oj} &= \mathcal{P}_j[\widehat{A}_j + \sum_{k=1}^{n_s} F_{kj}] + [\widehat{A}_j + \sum_{k=1}^{n_s} F_{kj}]^t\mathcal{P}_j + \Theta_j + \Theta_j^t + \mathcal{Q}_j \\
&+ \sum_{k=1}^{n_s} \mathcal{Z}_{kj} + K_j^t\Gamma_j^t\Gamma_jK_j + (n_s-1)\mathcal{P}_j, \\
\Xi_{aj} &= \mathcal{P}_jA_{dj} - \Theta_j + \Upsilon_j^t, \; \Xi_{mj} = \Upsilon_j + \Upsilon_j^t + (1-\mu_j)\mathcal{Q}_j, \\
\Xi_{nj} &= \sum_{k=1}^{n_s}(1-\mu_{kj})\mathcal{Z}_{kj} + \sum_{k=1,k\neq j}^{n_s} E_{kj}^t\mathcal{P}_kE_{kj}
\end{aligned} \tag{7.21}
$$

where $\dot{V}_j(t)|_{(7.7)}$ defines the Lyapunov derivative along the solutions of system (7.1). Internal stability requirement $\dot{V}_j(t)|_{(7.7)} < 0$ implies that $\overline{\Xi}_j < 0$. It is readily seen from (7.21) using the delay bounds of (7.3) that there exists a scalar $\sigma_j > 0$ such that

$$
\begin{bmatrix} \Xi_{oj}+\sigma_jI_j & \Xi_{aj} & -\Theta_j & 0 & \mathcal{P}_jB_j & \varrho_k\widehat{A}_j^t\mathcal{W}_k \\ \bullet & -\Xi_{mj} & -\Upsilon_j & 0 & 0 & \varrho_kA_{d_j}^t\mathcal{W}_k \\ \bullet & \bullet & -\varrho_j^{-1}\mathcal{W}_j & 0 & 0 & 0 \\ \bullet & \bullet & \bullet & -\Xi_{nj} & 0 & \varrho_k\sum_{k=1}^{n_s}E_{kj}\mathcal{W}_k \\ \bullet & \bullet & \bullet & \bullet & -I_j & 0 \\ \bullet & \bullet & \bullet & \bullet & \bullet & -\varrho_k\mathcal{W}_k \end{bmatrix} < 0 \tag{7.22}
$$

Therefore, for all τ_j satisfying (7.3) we have

$$
\begin{bmatrix} \Xi_{oj}+\sigma_jI_j & \Xi_{aj} & -\tau_j\Theta_j & 0 & \mathcal{P}_jB_j & \varrho_k\widehat{A}_j^t\mathcal{W}_k \\ \bullet & -\Xi_{mj} & -\tau_j\Upsilon_j & 0 & 0 & \varrho_kA_{d_j}^t\mathcal{W}_k \\ \bullet & \bullet & -\tau_j\mathcal{W}_j & 0 & 0 & 0 \\ \bullet & \bullet & \bullet & -\Xi_{nj} & 0 & \varrho_k\sum_{k=1}^{n_s}E_{kj}\mathcal{W}_k \\ \bullet & \bullet & \bullet & \bullet & -I_j & 0 \\ \bullet & \bullet & \bullet & \bullet & \bullet & -\varrho_k\mathcal{W}_k \end{bmatrix} < 0 \tag{7.23}
$$

and hence

$$\dot{V}_j(t)|_{(7.7)} \;<\; \sum_{j=1}^{n_s}\left[\frac{1}{\tau_j(t)}\int_{t-\tau_j}^{t} \chi_j^t(t,s) diag[-\sigma_j,\ 0,\ 0,\ 0,\ 0,\ 0]\chi_j(t,s)\, ds\right]$$
$$= \; -\sum_{j=1}^{n_s} \sigma_j\ ||x_j||^2 \;<\; 0 \qquad (7.24)$$

We continue further and consider the $\mathcal{L}_2 - gain$ performance measure

$$J = \sum_{j=1}^{n_s}\int_0^{\infty}\left(z_j^t(s)z_j(s) - \gamma_j^2 w_j^t(s)w_j(s)\right)ds$$

For any $w_j(t) \in \mathcal{L}_2(0,\infty) \neq 0$ with zero initial condition $x_j(0) = 0$ hence $V(0) = 0$, we have

$$J \;=\; \sum_{j=1}^{n_s}\int_0^{\infty}\left(z_j^t(s)z_j(s) - \gamma_j^2 w_j^t(s)w_j(s) + \dot{V}_j(t)|_{(7.1)} - V_j(\infty)\right)ds$$
$$\leq \; \sum_{j=1}^{n_s}\int_0^{\infty}\left(z_j^t(s)z_j(s) - \gamma_j^2 w_j^t(s)w_j(s) + \dot{V}_j(t)|_{(7.1)}\right)ds$$

Proceeding as before, we make use of (7.24) to get

$$\sum_{j=1}^{n_s}\left(z_j^t(s)z_j(s) - \gamma_j^2 w_j^t(s)w_j(s) + \dot{V}_j(t)|_{(7.1)}\right) =$$
$$\sum_{j=1}^{n_s} \bar{\chi}_j^t(t,s)\ \widehat{\Xi}_j\ \bar{\chi}_j(t,s) \qquad (7.25)$$

where

$$\bar{\chi}_j(t,s) = \left[\ x_j^t(t)\quad x_j^t(t-\tau_j)\quad \dot{x}_j^t(s)\quad x_j^t(t-\eta_{kj})\quad \beta^t(x_j)\quad w_j^t(s)\ \right]^t$$

$$\widehat{\Xi}_j =$$
$$\begin{bmatrix} \Xi_{oj} & \Xi_{aj} & -\varrho_j\Theta_j & \Xi_{cj} & \mathcal{P}_jB_j & \mathcal{P}_j\Omega_j & \widehat{G}_j^t & \varrho_j\widehat{A}_j^t\mathcal{W}_j \\ \bullet & -\Xi_{mj} & -\varrho_j\Upsilon_j & 0 & 0 & 0 & G_{dj}^t & \varrho_jA_{d_j}^t\mathcal{W}_j \\ \bullet & \bullet & -\varrho_j\mathcal{W}_j & 0 & 0 & 0 & 0 & 0 \\ \bullet & \bullet & \bullet & -\Xi_{nj} & 0 & 0 & 0 & \varrho_j\sum_{k=1}^{n_s}E_{kj}\mathcal{W}_j \\ \bullet & \bullet & \bullet & \bullet & -I_j & 0 & 0 & 0 \\ \bullet & \bullet & \bullet & \bullet & \bullet & -\gamma_j^2I_j & B_j^t & \varrho_j\Gamma_j^t\mathcal{W}_j \\ \bullet & \bullet & \bullet & \bullet & \bullet & \bullet & -I_j & 0 \\ \bullet & \bullet & \bullet & \bullet & \bullet & \bullet & \bullet & -\varrho_j\mathcal{W}_j \end{bmatrix}$$

It is readily seen that when $\widehat{\Xi}_j < 0$ the condition

$$\sum_{j=1}^{n_s} \left(z_j^t(s) z_j(s) - \gamma_j^2 w_j^t(s) w_j(s) + \dot{V}_j(t)|_{(7.1)} \right) < 0$$

for arbitrary $s \in [t, \infty)$. This in turn implies that for any $w_j(t) \in \mathcal{L}_2(0, \infty) \neq 0$ we have

$$J < 0$$

leading to $\sum_{j=1}^{n_s} ||z_j(t)||_2 < \sum_{j=1}^{n_s} \gamma_j \; ||w(t)_j||_2$, which assures the desired performance.

7.1.4 Feedback gains

The immediate task now is to determine the decentralized gain matrices K_j. Applying the congruent transformation $T = diag[\mathcal{X}_j,\ \mathcal{X}_j,\ \mathcal{X}_j,\ \mathcal{X}_j,\ I_j,\ I_j,\ I_j,\ I_j]$, $\mathcal{X}_j = \mathcal{P}_j^{-1}$ to $\widehat{\Xi}_j$ with Schur complements and using the linearizations

$$\begin{aligned} \mathcal{Y}_j &= K_j \mathcal{X}_j,\ \Psi_{1j} = \mathcal{X}_j \Theta_j \mathcal{X}_j,\ \Psi_{2j} = \mathcal{X}_j \mathcal{Q}_j \mathcal{X}_j, \\ \Psi_{3j} &= \mathcal{X}_j \Upsilon_j \mathcal{X}_j,\ \Lambda_{kj} = \mathcal{X}_j \mathcal{Z}_{kj} \mathcal{X}_j,\ \Psi_{4j} = \mathcal{X}_j \mathcal{W}_j \mathcal{X}_j \end{aligned}$$

we readily obtain LMI (7.8) by Schur complements and therefore the proof is completed.

Remark 7.1.2 *It is significant to recognize that the methodology of this section incorporates four weighting matrices $\mathcal{P}_j$, $\mathcal{Q}_j$, $\mathcal{W}_j$, $\mathcal{Z}_{jk}$ and two parameter matrices Θ_j, Υ_j at the subsystem level in order to ensure least conservative delay-dependent stability results. This will eventually result in reduced computational requirements as evidenced by a simple comparison with the improved free-weighting matrices method of [103] for single time-delay systems in terms of two aspects. One aspect would be due to reduced computational load as evidenced by less number of manipulated variables and faster processing. Another aspect arises by noting that LMIs (7.8) subject to (7.9) for $n_s = 1$ theoretically cover the results of [146, 174] as special cases. Furthermore, in the absence of delay ($A_{dj} \equiv 0$, $\mathcal{Q}_j \equiv 0$, $\mathcal{W}_j \equiv 0$), it is easy to infer that LMIs (7.8) subject to (7.9) for $n_s = 1$ will eventually reduce to parameterized delay-independent criteria. LKF composed of the first three terms has been considered in [74, 103, 261, 290] for single time-delay systems. In comparison with the reported results on interconnected systems [251, 283, 287, 288, 328, 371], LKF (7.10) represents a new effective form and therefore the advantages and reduced conservatism afforded in [74, 103, 261, 290] are carried over herein.*

Remark 7.1.3 *The optimal performance-level* $\gamma_j,\ j = 1,..,n_s$ *for the interconnected system can be determined for decentralized reliable state feedback by solving the following convex optimization problem:*
Problem A:

$$\begin{array}{ll} For\ j,k = 1,...,n_s, & Given\ \varrho_j,\ \mu_j,\ \varrho_{jk},\ \mu_{jk},\ \mathcal{W}_j \\ \min\limits_{\mathcal{X}_j,\ \mathcal{Y}_j,\ \{\Lambda_{kj}\}_{k=1}^{n_s},\ \{\Psi_{rj}\}_{r=1}^{3}} & \gamma_j \\ subject\ to\ LMI(7.8) & \end{array} \tag{7.26}$$

7.1.5 Interconnected uncertain systems

Suppose now that the interconnected system (7.1) undergoes parametric uncertainties. One convenient representation would be the state-space model

$$\begin{aligned} \dot{x}_j(t) &= A_{j\Delta}x_j(t) + A_{dj\Delta}x_j(t-\tau_j) + B_{j\Delta}u_j(t) + c_j(k) + \Omega_{j\Delta}w_j(t) \\ z_j(t) &= G_{j\Delta}x_j(t) + G_{dj\Delta}x_(t-\tau_j) + D_{j\Delta}u_j(t) + B_{j\Delta}w_j(t) \\ y_j(t) &= C_{j\Delta}x_j(t) + C_{dj\Delta}x_(t-\tau_j) \\ c_j(t) &= \sum_{k=1,k\neq j}^{n_s} F_{jk\Delta}x_k(t) + \sum_{k=1,k\neq j}^{n_s} E_{jk\Delta}x_k(t-\eta_{jk}(t)) \end{aligned} \tag{7.27}$$

whose matrices containing uncertainties which belong to a real convex bounded polytopic model of the type

$$\begin{aligned} &\begin{bmatrix} A_{j\Delta} & A_{dj\Delta} & B_{j\Delta} & \Omega_{j\Delta} \\ G_{j\Delta} & G_{dj\Delta} & D_{j\Delta} & B_{j\Delta} \\ C_{j\Delta} & C_{dj\Delta} & E_{jk\Delta} & F_{jk\Delta} \end{bmatrix} \in \Pi_\lambda := \left\{ \begin{bmatrix} A_{j\lambda} & A_{dj\lambda} & B_{j\lambda} & \Omega_{j\lambda} \\ G_{j\lambda} & G_{dj\lambda} & D_{j\lambda} & B_{j\lambda} \\ C_{j\lambda} & C_{dj\lambda} & E_{jk\lambda} & F_{jk\lambda} \end{bmatrix} \right. \\ &= \left. \sum_{s=1}^{N} \lambda_s \begin{bmatrix} A_{js} & A_{djs} & B_{js} & \Omega_{js} \\ G_{js} & G_{djs} & D_{js} & B_{js} \\ C_{js} & C_{djs} & E_{jks} & F_{jks} \end{bmatrix}, \lambda_s \in \Lambda \right\} \end{aligned} \tag{7.28}$$

where Λ is the unit simplex

$$\Lambda := \left\{ (\lambda_1, \cdots, \lambda_N) : \sum_{j=1}^{N} \lambda_j = 1\ ,\ \lambda_j \geq 0 \right\} \tag{7.29}$$

Define the vertex set $\mathcal{N} = \{1, ..., N\}$. We use $\{A_{jo}, ..., B_{jo}\}$ to imply generic system matrices and $\{A_{js}, ..., B_{js},\ s \in \mathcal{N}\}$ to represent the respective values at the vertices. It is a straightforward task to show that the following result holds.

Theorem 7.1.4 *Given the bounds* $\varrho_j > 0,\ \mu_j > 0,\ \varrho_{jk} > 0,\ \mu_{jk} > 0,\ j,k = 1,...,n_s$ *and matrix* $\mathcal{W}_j$*, then the family of subsystems* $\{\mathsf{S}_j\}$ *with polytopic representation (7.28)-(7.29) where* S_j *is described by (7.2) is delay-dependent asymptotically stabilizable by decentralized controller* $u_j(t) = K_j x_j(t),\ j,k = 1,...,n_s$ *with* $\mathcal{L}_2$*-performance bound* $\gamma_j,\ j = 1,...,n_s$ *if there exist positive-definite matrices* $\mathcal{X}_j,\ \mathcal{Y}_j,\ \mathcal{W}_k,\ \{\Lambda_{kj}\}_{k=1}^{n_s},\ \{\Psi_{rj}\}_{r=1}^{3}$*, satisfying the following LMIs for* $j = 1,...,n_s,\ s = 1,...,N$

$$\widetilde{\Pi}_{js} = \begin{bmatrix} \widetilde{\Pi}_{ajs} & \widetilde{\Pi}_{cjs} \\ \bullet & \widetilde{\Pi}_{ejs} \end{bmatrix} < 0 \tag{7.30}$$

where

$$\widetilde{\Pi}_{ajs} = \begin{bmatrix} \Pi_{ojs} & \Pi_{ajs} & -\varrho_j\Psi_{1js} & \Pi_{cjs} & B_{js} \\ \bullet & -\Pi_{mjs} & -\varrho_j\Psi_{3js} & 0 & 0 \\ \bullet & \bullet & -\varrho_j\Psi_{4js} & 0 & 0 \\ \bullet & \bullet & \bullet & -\Pi_{njs} & 0 \\ \bullet & \bullet & \bullet & \bullet & -I_j \end{bmatrix},$$

$$\widetilde{\Pi}_{cjs} = \begin{bmatrix} \Omega_{js} & \Pi_{zjs}^t & \mathcal{Y}_j^t\Gamma_{js}^t & \varrho_k(\mathcal{X}_j A_{js}^t + \mathcal{Y}_j^t\Sigma_j^t B_{js}^t)\mathcal{W}_k \\ 0 & G_{djs}^t & 0 & \varrho_k\mathcal{X}_j A_{djs}^t\mathcal{W}_k \\ 0 & 0 & 0 & 0 \\ 0 & 0 & 0 & \varrho_k\sum_{k=1}^{n_s}\mathcal{X}_j E_{kjs}\mathcal{W}_k \\ 0 & 0 & 0 & 0 \end{bmatrix},$$

$$\widetilde{\Pi}_{ejs} = \begin{bmatrix} -\gamma_j^2 I_j & B_j^t & 0 & \varrho_k\Gamma_j^t\mathcal{W}_k \\ \bullet & -I_j & 0 & 0 \\ \bullet & \bullet & -I_j & 0 \\ \bullet & \bullet & \bullet & -\varrho_k\mathcal{W}_k \end{bmatrix} \tag{7.31}$$

with

$$\begin{aligned}
\Pi_{ojs} &= [A_{js} + \sum_{k=1}^{n_s} F_{kjs}]\mathcal{X}_j + B_{js}\Sigma_j\mathcal{Y}_j + \mathcal{X}_j[A_{js} + \sum_{k=1}^{n_s} F_{kjs}]^t + \mathcal{Y}_j^t\Sigma_j^t B_{js}^t \\
&+ \Psi_{1js} + \Psi_{1js}^t + \Psi_{2js} + \sum_{k=1}^{n_s}\Lambda_{kjs} + (n_s - 1)\mathcal{X}_j \\
\Pi_{ajs} &= A_{djs}\mathcal{X}_j - \Psi_{1js} + \Psi_{3js}^t,\quad \Pi_{zj} = G_{js}\mathcal{X}_j + D_{js}\Sigma_j\mathcal{Y}_j, \\
\Pi_{njs} &= \sum_{k=1,k\neq j}^{n_s} (1 - \mu_{kj})\Lambda_{kjs} + \sum_{k=1,k\neq j}^{n_s} \mathcal{X}_j E_{kjs}^t\mathcal{P}_k E_{kjs}\mathcal{X}_j, \\
\Pi_{mj} &= \Psi_{3js} + \Psi_{3js}^t + (1 - \mu_j)\Psi_{2js} + \Psi_{4js}^t
\end{aligned} \tag{7.32}$$

Moreover, the local gain matrix is given by $K_j = \mathcal{Y}_j\mathcal{X}_j^{-1}$.

7.1.6 Static output-feedback reliable design

In this section, we develop new criteria for LMI-based characterization of decentralized reliable stabilization by local static output feedback. The criteria include some parameter matrices aims at expanding the range of applicability of the developed conditions. To facilitate further development, we consider the case where the set of output matrices $C_j,\ j = 1, ..., n_s$ is assumed to be of full row rank.

Using the local static output-feedback control $u_j(t) = K_{oj} y_j(t),\ j = 1, ..., N$, the local closed-loop subsystem dynamics under failure become

$$\begin{aligned}
\dot{x}_j(t) &= \mathcal{A}_j x_j(t) + \mathcal{A}_{dj} x_j(t - \tau_j(t)) + B_j \beta_j(u_j) + c_j(t) + \Omega_j w_j(t) \\
z_j(t) &= \mathcal{G}_j x_j(t) + \mathcal{G}_{dj} x_j(t - \tau_j(t)) + D_j \beta_j(u_j) + B_j w_j(t) \\
c_j(t) &= \sum_{k=1}^{n_s} F_{jk} x_k(t) + \sum_{k=1}^{n_s} E_{jk} x_k(t - \eta_{jk}(t)) \\
\mathcal{A}_j &= A_j + B_j \Sigma_j K_{oj} C_j,\ \ \mathcal{A}_{dj} = A_{dj} + B_j \Sigma_j K_{oj} C_j, \\
\mathcal{G}_j &= G_j + D_j \Sigma_j K_{oj} C_j,\ \mathcal{G}_{dj} = G_{dj} + D_j \Sigma_j K_{oj} C_{dj}
\end{aligned} \tag{7.33}$$

Proceeding further, we adopt the Lyapunov functional (7.10) to the resulting closed-loop system (7.33). As a consequence, system (7.33) is asymptotically stable with $\mathcal{L}_2$-performance bound γ_j if there exist matrices $\overline{\mathcal{P}}_j,\ \overline{\mathcal{Q}}_j,\ \overline{\mathcal{W}}_j,\ \overline{\mathcal{Z}}_{kj},\ j, k = 1, ..n_s$, parameter matrices $\overline{\Theta}_j\ \overline{\Upsilon}_j$ satisfying the following LMI:

$$\begin{bmatrix}
\Xi_{oj} & \Xi_{aj} & -\varrho_j \overline{\Theta}_j & 0 & \overline{\mathcal{P}}_j B_j & \overline{\mathcal{P}}_j \Omega_j & \mathcal{G}_j^t & \varrho_k \mathcal{A}_j^t \overline{\mathcal{W}}_k \\
\bullet & -\Xi_{mj} & -\varrho_j \overline{\Upsilon}_j & 0 & 0 & 0 & \mathcal{G}_{dj}^t & \varrho_k \mathcal{A}_{d_j}^t \overline{\mathcal{W}}_k \\
\bullet & \bullet & -\varrho_j \overline{\mathcal{W}}_j & 0 & 0 & 0 & 0 & 0 \\
\bullet & \bullet & \bullet & -\Xi_{nj} & 0 & 0 & 0 & \varrho_k \sum_{k=1}^{n_s} E_{kj} \overline{\mathcal{W}}_k \\
\bullet & \bullet & \bullet & \bullet & -I_j & 0 & 0 & 0 \\
\bullet & \bullet & \bullet & \bullet & \bullet & -\gamma_j^2 I_j & B_j^t & \varrho_k \Gamma_j^t \overline{\mathcal{W}}_k \\
\bullet & \bullet & \bullet & \bullet & \bullet & \bullet & -I_j & 0 \\
\bullet & \bullet & \bullet & \bullet & \bullet & \bullet & \bullet & -\varrho_k \overline{\mathcal{W}}_k
\end{bmatrix} < 0 \tag{7.34}$$

where

$$\begin{aligned}
\Xi_{oj} &= \overline{\mathcal{P}}_j [\mathcal{A}_j + \sum_{k=1}^{n_s} F_{kj}] + [\mathcal{A}_j + \sum_{k=1}^{n_s} F_{kj}]^t \overline{\mathcal{P}}_j + \overline{\Theta}_j + \overline{\Theta}_j^t + \overline{\mathcal{Q}}_j + \sum_{k=1}^{n_s} \overline{\mathcal{Z}}_{kj} \\
&+ C_j^t K_{oj}^t \Gamma_j^t \Gamma_j K_{oj} C_j + (n_s - 1) \mathcal{P}_j,
\end{aligned}$$

$$
\begin{aligned}
\Xi_{aj} &= \overline{\mathcal{P}}_j \mathcal{A}_{dj} - \overline{\Theta}_j + \overline{\Upsilon}_j^t + C_j^t K_{oj}^t \Gamma_j^t \Gamma_j K_{oj} C_{dj}, \\
\Xi_{mj} &= \overline{\Upsilon}_j + \overline{\Upsilon}_j^t + (1-\mu_j)\overline{\mathcal{Q}}_j + C_{dj}^t K_{oj}^t \Gamma_j^t \Gamma_j K_{oj} C_{dj}, \\
\Xi_{nj} &= \sum_{k=1, k\neq j}^{n_s} (1-\mu_{kj}) \mathcal{Z}_{kj} + \sum_{k=1, k\neq j}^{n_s} E_{kj}^t \mathcal{P}_j E_{kj} \qquad (7.35)
\end{aligned}
$$

The following theorem establishes the main design result for subsystem $\mathsf{S_j}$.

Theorem 7.1.5 *Given the bounds $\varrho_j > 0,\ \mu_j > 0,\ \varrho_{jk} > 0,\ \mu_{jk} > 0,\ j,k = 1,...,n_s$ and matrix $\mathcal{W}_j$, then the family of subsystems $\{\mathsf{S}_j\}$ where $\mathsf{S_j}$ is described by (7.7) is delay-dependent asymptotically stabilizable by decentralized static output-feedback controller $u_j(t) = K_{oj} y_j(t),\ j = 1,...,N$ with $\mathcal{L}_2$-performance bound $\gamma_j,\ j = 1,...,n_s$ if there exist positive-definite matrices $\mathcal{X}_j,\ \mathcal{Y}_j,\ \mathcal{W}_k,\ \{\Lambda_{kj}\}_{k=1}^{n_s}$,*
$\{\Psi_{rj}\}_{r=1}^{6}$, satisfying the following LMIs for $j,k = 1,...,n_s$:

$$
\widetilde{\Pi}_j = \begin{bmatrix} \widetilde{\Pi}_{aj} & \widetilde{\Pi}_{cj} \\ \bullet & \widetilde{\Pi}_{ej} \end{bmatrix} < 0 \qquad (7.36)
$$

$$
\widetilde{\Pi}_{aj} = \begin{bmatrix} \Pi_{oj} & \Pi_{aj} & -\varrho_j \Psi_{1j} & \Pi_{cj} & B_j \\ \bullet & -\Pi_{mj} & -\varrho_j \Psi_{3j} & 0 & 0 \\ \bullet & \bullet & -\varrho_j \Psi_{4j} & 0 & 0 \\ \bullet & \bullet & \bullet & -\Pi_{nj} & 0 \\ \bullet & \bullet & \bullet & \bullet & -I_j \end{bmatrix},
$$

$$
\widetilde{\Pi}_{cj} = \begin{bmatrix} \Omega_j & \Pi_{zj}^t & \mathcal{Y}_j^t \Gamma_j^t & \varrho_k (\mathcal{X}_j A_j^t + \mathcal{Y}_j^t \Sigma_j^t B_j^t) \mathcal{W}_k \\ 0 & \Pi_{xj}^t & \Psi_{6j}^t \Gamma_j^t & \varrho_k \mathcal{X}_j A_{dj}^t \mathcal{W}_k \\ 0 & 0 & 0 & 0 \\ 0 & 0 & 0 & \varrho_k \sum_{k=1}^{n_s} \mathcal{X}_j E_{kj} \mathcal{W}_k \\ 0 & 0 & 0 & 0 \end{bmatrix},
$$

$$
\widetilde{\Pi}_{ej} = \begin{bmatrix} -\gamma_j^2 I_j & B_j^t & 0 & \varrho_k \Gamma_j^t \mathcal{W}_k \\ \bullet & -I_j & 0 & 0 \\ \bullet & \bullet & -I_j & 0 \\ \bullet & \bullet & \bullet & -\varrho_k \mathcal{W}_k \end{bmatrix} \qquad (7.37)
$$

where

$$
\begin{aligned}
\Pi_{oj} &= [A_j + \sum_{k=1}^{n_s} F_{kj}]\mathcal{X}_j + B_j \Sigma_j \mathcal{Y}_j + \mathcal{X}_j [A_j + \sum_{k=1}^{n_s} F_{kj}]^t + \mathcal{Y}_j^t \Sigma_j^t B_j^t \\
&+ \Psi_{1j} + \Psi_{1j}^t + \Psi_{2j} + \sum_{k=1}^{n_s} \Lambda_{kj} + (n_s - 1)\mathcal{X}_j
\end{aligned}
$$

$$
\begin{aligned}
\Pi_{aj} &= A_{dj}\mathcal{X}_j - \Psi_{1j} + \Psi_{3j}^t, \;\; \Pi_{mj} = \Psi_{3j} + \Psi_{3j}^t + (1-\mu_j)\Psi_{2j} \\
\Pi_{nj} &= \sum_{k=1}^{n_s}(1-\mu_{kj})\Lambda_{kj} + + \sum_{k=1, k\neq j}^{n_s} \mathcal{X}_j E_{kjs}^t \mathcal{P}_k E_{kjs}\mathcal{X}_j, \\
\Pi_{xj} &= G_{dj}\mathcal{X}_j + D_j\Sigma_j\Psi_{6j}, \;\; \Pi_{zj} = G_j\mathcal{X}_j + D_j\Sigma_j\mathcal{Y}_j \qquad (7.38)
\end{aligned}
$$

Moreover, the local gain matrix is given by $K_j = \mathcal{Y}_j\mathcal{X}_j^{-1}C_j^\dagger$.

Proof: Applying the congruent transformation

$$T = diag[\mathcal{X}_j,\; \mathcal{X}_j,\; \mathcal{X}_j,\; \mathcal{X}_j,\; I_j,\; I_j,\; I_j,\; I_j],\; \mathcal{X}_j = \overline{\mathcal{P}}_j^{-1}$$

to LMI (7.34) with Schur complements and using the linearizations

$$
\begin{aligned}
\mathcal{Y}_j &= K_{oj}C_j\mathcal{X}_j,\; \Psi_{1j} = \mathcal{X}_j\overline{\Theta}_j\mathcal{X}_j,\; \Psi_{2j} = \mathcal{X}_j\overline{\mathcal{Q}}_j\mathcal{X}_j,\; \Psi_{3j} = \mathcal{X}_j\overline{\Upsilon}_j\mathcal{X}_j \\
\Lambda_{kj} &= \mathcal{X}_j\overline{\mathcal{Z}}_{kj}\mathcal{X}_j,\; \Psi_{4j} = \mathcal{X}_j\overline{\mathcal{W}}_j\mathcal{X}_j,\; \Psi_{5j} = \mathcal{X}_jC_{dj}^tK_{oj}^t\Gamma_j^t,\; \Psi_{6j} = K_{oj}^tC_{dj}^t\mathcal{X}_j
\end{aligned}
$$

we readily obtain LMI (7.36) by Schur complements and therefore the proof is completed.

Remark 7.1.6 *Similarly, the optimal performance-level* $\gamma_j,\; j = 1,..,n_s$ *can be determined in case of decentralized static output-feedback stabilization by solving the following convex optimization problems*:
Problem B: *Static Output-Feedback Stabilization*

$$
\begin{aligned}
&For\;\; j,k = 1,...,n_s, \;\; Given\;\; \varrho_j,\; \mu_j,\; \varrho_{jk},\; \mu_{jk}, \\
&\min_{\mathcal{X}_j,\, \mathcal{Y}_j,\, \mathcal{M}_j,\, \Omega_j,\, \{\Lambda_{rj}\}_{r=1}^{n_s},\, \{\Psi_{rj}\}_1^6} \quad \gamma_j \\
&subject\;\; to \quad LMI(7.36) \qquad (7.39)
\end{aligned}
$$

7.1.7 Illustrative example 7.1

Consider an interconnected system composed of three subsystems, each of the type (7.1) with the following subsystem coefficients:

$$
\begin{aligned}
& Subsystem\; 1: \\
& A_1 = \begin{bmatrix} -2 & 0 \\ -2 & -1 \end{bmatrix},\; \Omega_1 = \begin{bmatrix} 0.2 \\ 0.2 \end{bmatrix},\; B_1 = \begin{bmatrix} 1 \\ 2 \end{bmatrix}, \\
G_1 &= \begin{bmatrix} 0.2 & 0.1 \end{bmatrix}, \\
& A_{d1} = \begin{bmatrix} -1 & 0 \\ -1 & 0 \end{bmatrix},\; G_{d1} = \begin{bmatrix} -0.1 & 0 \end{bmatrix},\; B_1 = 0.5,\; D_1 = [0.1] \\
& Subsystem\; 2: \\
& A_c = \begin{bmatrix} -1 & 0 \\ -1 & -4 \end{bmatrix},\; \Omega_c = \begin{bmatrix} 0.1 \\ 0.3 \end{bmatrix},\; B_2 = \begin{bmatrix} 1 & 1 \\ -1 & 2 \end{bmatrix},
\end{aligned}
$$

$$G_2 = \begin{bmatrix} 0.2 & 0.1 \end{bmatrix},$$

$$A_{d2} = \begin{bmatrix} 1 & 0 \\ -2 & -1 \end{bmatrix}, \; G_{d2} = \begin{bmatrix} 0.1 & 0 \end{bmatrix}, \; B_2 = 0.2, \; D_2 = [0.2 \;\; 0.4]$$

$Subsystem \; 3:$

$$A_3 = \begin{bmatrix} 0 & 1 \\ -1 & -2 \end{bmatrix}, \; \Omega_3 = \begin{bmatrix} 0.1 \\ 0.5 \end{bmatrix}, \; B_3 = \begin{bmatrix} 2 \\ 1 \end{bmatrix},$$

$$G_3 = \begin{bmatrix} 0.1 & -0.1 \end{bmatrix},$$

$$A_{d3} = \begin{bmatrix} 0 & 0 \\ 0 & -1 \end{bmatrix}, \; G_{d3} = \begin{bmatrix} -0.1 & 0 \end{bmatrix}, \; B_3 = 0.1, \; D_3 = [0.3]$$

and the coupling pattern

$Couplings \; 1:$

$$E_{12} = \begin{bmatrix} 1 & 0 \\ 1 & 0 \end{bmatrix}, \; E_{13} = \begin{bmatrix} 0 & -1 \\ 0 & -1 \end{bmatrix}, \; F_{12} = \begin{bmatrix} 0.2 & -0.1 \\ 0 & 0.2 \end{bmatrix},$$

$$F_{13} = \begin{bmatrix} 0 & -0.4 \\ 0.1 & -0.5 \end{bmatrix},$$

$Couplings \; 2:$

$$E_{21} = \begin{bmatrix} -1 & -2 \\ 3 & 6 \end{bmatrix}, \; E_{23} = \begin{bmatrix} -1 & 1 \\ 3 & -2 \end{bmatrix}, \; F_{21} = \begin{bmatrix} 0 & -0.5 \\ 0.8 & -0.5 \end{bmatrix},$$

$$F_{23} = \begin{bmatrix} -0.7 & 0.6 \\ 0 & -0.2 \end{bmatrix},$$

$Couplings \; 3:$

$$E_{31} = \begin{bmatrix} 1 & 2 \\ 1 & 2 \end{bmatrix}, \; E_{32} = \begin{bmatrix} 0 & 0 \\ 0 & -1 \end{bmatrix},$$

$$F_{31} = \begin{bmatrix} 0.7 & 0 \\ 0 & 0.8 \end{bmatrix}, \; F_{32} = \begin{bmatrix} 0 & 0 \\ 0 & -1 \end{bmatrix}$$

In order to illustrate the effectiveness of our reliable control algorithm, we initially consider the nominal control design in which no actuator failures will occur. Then, by using the data values $\Sigma_1 = 1, \; \Sigma_c = 1, \; \Sigma_3 = 1, \; \Gamma_1 = 0, \; \Gamma_c = 0, \; \Gamma_3 = 0$ and solving **Problem A**, it is found that the feasible solution yields the following control gain matrices and performance levels:

$$\begin{aligned} \gamma_1 &= 1.7874, \; K_1 = \begin{bmatrix} 1.2551 & -2.3786 \end{bmatrix}, \\ \gamma_c &= 3.1742, \; K_2 = \begin{bmatrix} -12.9437 & 2.0436 \\ -9.4414 & -8.0502 \end{bmatrix}, \\ \gamma_3 &= 2.0297, \; K_3 = \begin{bmatrix} -0.0540 & 1.0240 \end{bmatrix} \end{aligned} \tag{7.40}$$

Next, we study different failure scenarios. The first scenario is described by the following model:

$$\begin{aligned} u_1^f(t) &= 0.6\, u_1(t) + \beta_1(u_1), \;\; \beta_1^2(u_1) \le 0.01\, u_1^2, \\ u_2^f(t) &= 0.8\, u_2(t) + \beta_c(u_2), \;\; \beta_2^2(u_2) \le 0.09\, u_2^2, \\ u_3^f(t) &= 0.4\, u_3(t) + \beta_3(u_3), \;\; \beta_3^2(u_3) \le 0.04\, u_3^2 \end{aligned} \tag{7.41}$$

which corresponds to allowing a failure of the order of 60% in the actuator of system S_j with an error of the order of 10%. In a similar way, the tolerances allowed for other actuators are interpreted in the same way. Thus we have $\Sigma_1 = 0.6,\ \Gamma_1 = 0.1,\ \Sigma_c = 0.8,\ \Gamma_c = 0.3,\ \Sigma_3 = 0.4,\ \Gamma_3 = 0.2$ and upon solving **Problem A** again, the resulting feasible solution is summarized by

$$\begin{aligned} \gamma_1 &= 1.9492,\ K_1 = \begin{bmatrix} 2.4945 & -4.6355 \end{bmatrix}, \\ \gamma_c &= 3.7842,\ K_2 = \begin{bmatrix} -18.3947 & 3.3116 \\ -14.5364 & -12.5472 \end{bmatrix}, \\ \gamma_3 &= 2.4123,\ K_3 = \begin{bmatrix} -0.1177 & 2.1270 \end{bmatrix} \end{aligned}$$

Next consider the static output feedback design with the additional coefficients

$$C_1 = [0.2 \quad 0],\ \ C_1 = [0.6 \quad 0.4],\ \ C_1 = [0.5 \quad 0]$$

Considering **Problem C**, it is found that the feasible solution is attained at

$$\begin{aligned} \gamma_1 &= 1.8011,\ K_1 = \begin{bmatrix} 1.1187 & 0.8962 \end{bmatrix}, \\ \gamma_c &= 13.4931,\ K_2 = \begin{bmatrix} -8.8441 & 1.3237 \\ -3.9501 & -3.3300 \end{bmatrix}, \\ \gamma_3 &= 1.4746,\ K_3 = \begin{bmatrix} 0.6289 & -0.2550 \end{bmatrix} \end{aligned} \tag{7.42}$$

Using numerical simulation, three typical experiments were performed for three operational modes.

1) *Ideal mode*, where the nominal controllers (7.40) were implemented in an operation without failures.

2) *Failure mode*, where the nominal controllers (7.40) are implemented in an operation under the failures described in (7.41).

3) *Reliable control failure mode*, where the reliable controllers (7.42) are implemented in an operation under the failures described in (7.41).

Simulation results of the first state trajectories of the three subsystems are displayed in Figures 7.2 through 7.4 for the prescribed modes of operation. The trajectories of the state-feedback control input for the first subsystem are

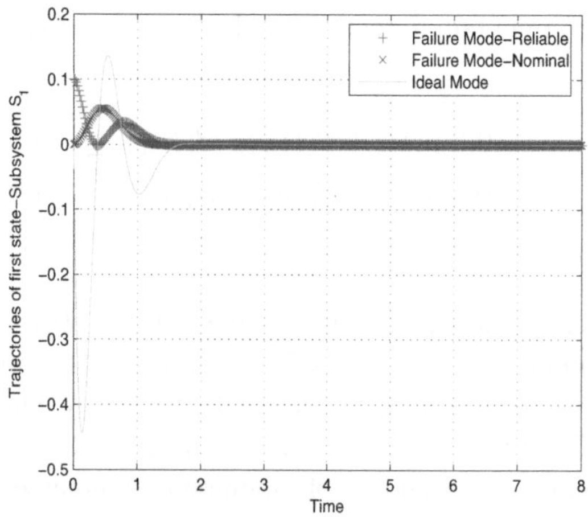

Figure 7.2: Trajectories of first state-Subsystem 1

depicted in Figure 7.5. In the case of static output feedback, the trajectories of the output of the first subsystem are depicted in Figure 7.6 (under the three modes) and the corresponding trajectories for the output-feedback control signal are displayed in Figure 7.7.

It is observed that the nominal control clearly deteriorates its ideal performance when operating under failures. When using the reliable control, we notice that the state response results drastically improved in spite of working under the same failures, being very close to the one obtained in the ideal case. As illustrated by Figures 7.5 and 7.7, the control effort in this case has a bigger initial magnitude than for the nominal cases, quickly approaching the ideal control signals. These results illustrate the satisfactory behavior of the proposed reliable control scheme.

7.2 Application to Multi-Area Power Systems

It becomes increasingly apparent that large-scale systems manifest the real world and appear in different applications such as power systems, digital communication networks, economic systems, and urban traffic networks. An integral feature of these systems is that they are composed of a set of small interconnected subsystems [224]; due to technical and/or economical reasons it is generally quite difficult to incorporate feedback from all the subsystems into the controller design. These difficulties motivate the development of decentralized control theory where each subsystem is controlled independently based

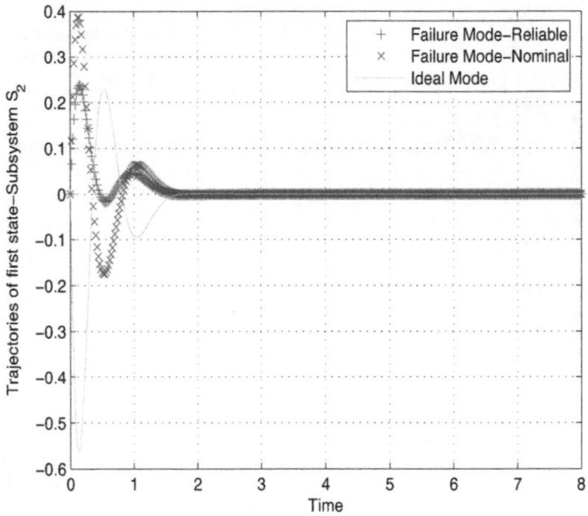

Figure 7.3: Trajectories of first state-Subsystem 2

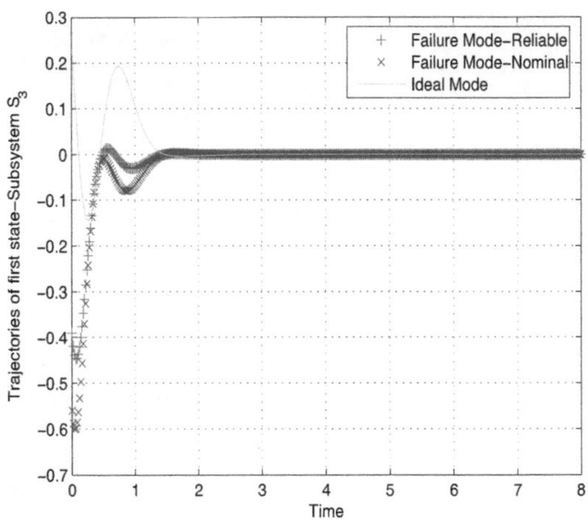

Figure 7.4: Trajectories of first state-Subsystem 3

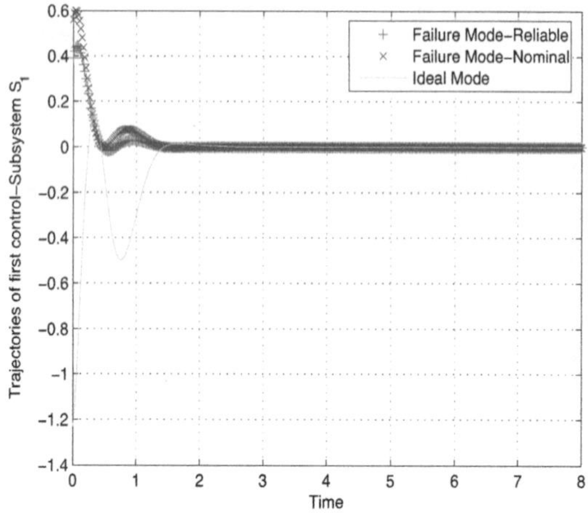

Figure 7.5: Trajectories of control input-Subsystem 1

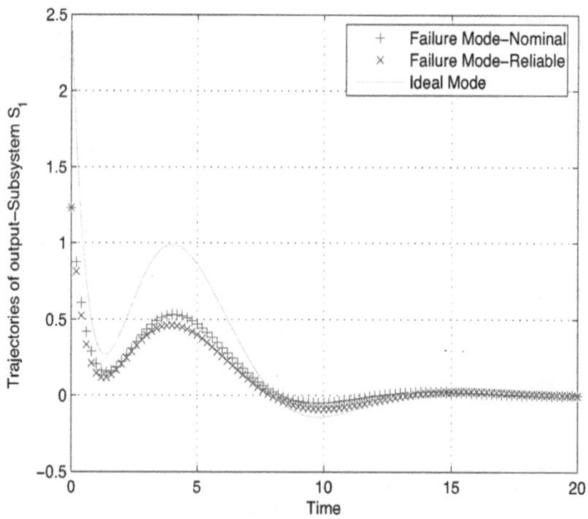

Figure 7.6: Trajectories of output-Subsystem 1

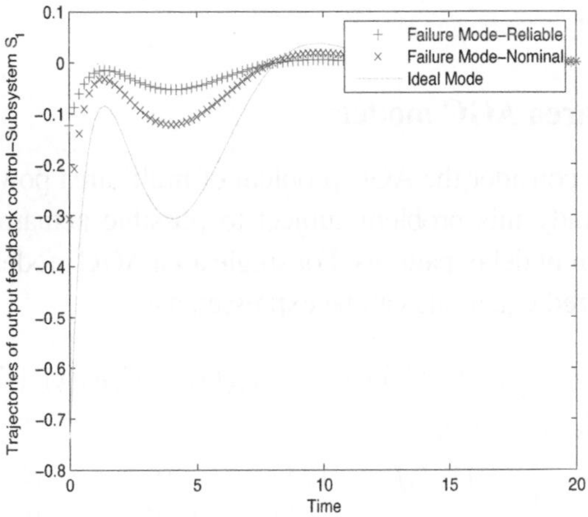

Figure 7.7: Trajectories of output feedback control-Subsystem 1

on locally available information [324]. Relevant research results on decentralized control of relevance to this paper can be found in [173, 288, 328]. Among the fundamental issues that arise quite frequently when designing feedback controllers for interconnected systems is the reliability of the control systems against different component failures. It becomes increasingly evident that the reliability of control systems in the presence of system component failures is of paramount importance for maintaining the critical functionality and survivability of many safety critical systems.

On another research front, various advanced control technologies have been applied to excitation and steam valving controllers of power systems including differential geometric tools [33, 190, 256]. It has been shown that the dynamics of power systems can be exactly linearized by employing nonlinear feedback and a state transformation. In the load-frequency control function, it is necessary to keep the system frequency and the inter-area tie-line power as near to the scheduled values as possible through control action. Considering the automatic-generation control (AGC) function, it is required to design the controls at the governor terminals while taking into account possible delays and control actuator failures.

In what follows, the AGC problem of multi-area power systems subject to actuator failures is translated into an equivalent problem of decentralized reliable feedback control for a class of linear interconnected continuous-time systems having internal subsystem time-delays and additional time-delay couplings under a class of control failures. The failure model takes into consideration possible outages or partial failures in every single actuator of each decentralized controller.

7.2.1 Single-area AGC model

In the sequel, we consider the AGC problem of multi-area power systems. Our objective is to study this problem subject to possible actuator failures while considering different delay patterns. For single area AGC model including ACE delay, the linearized equations can be expressed as:

$$\begin{aligned}
\dot{x}_j(t) &= A_j x_j(t) + A_{dj} x_j(t - \tau_j(t)) + B_j u_j(t) + \Gamma_j \Delta P_{dj} \\
y_j(t) &= C_j x_j(t), \\
A_j &= \begin{bmatrix} -D_j/M_j & 1/M_j & 0 & 0 \\ 0 & -1/T_{ej} & 1/T_{ej} & 0 \\ -1/R_j T_{gj} & 0 & -1/T_{gj} & -1/T_{gj} \\ K_j & 0 & 0 & 0 \end{bmatrix}
\end{aligned} \tag{7.43}$$

and

$$\begin{aligned}
B_j &= \begin{bmatrix} 0 \\ 0 \\ 1/T_{gj} \\ 0 \end{bmatrix}, \; C_j^t = \begin{bmatrix} 1 \\ 0 \\ 1 \\ 1 \end{bmatrix}, \; \Gamma_j^t = \begin{bmatrix} -1/M_j \\ 0 \\ 0 \\ 0 \end{bmatrix} \\
A_{dj} &= \begin{bmatrix} 0 & 0 & 0 & 0 \\ 0 & 0 & 0 & 0 \\ 0 & 0 & 0 & 1/T_{gj} \\ 0 & 0 & 0 & 0 \end{bmatrix}
\end{aligned} \tag{7.44}$$

with $x_j = [\Delta f_j, \;\; \Delta P_{mj}, \;\; \Delta P_{vj}, \;\; \Delta E_j]^t$, u_j is the control at the governor terminal, ΔP_{dj} is the load disturbance, D_j is a load damping constant, M_j is the inertia constant, T_{ej} is the turbine time-constant, T_{gj} is the governor time-constant, R_j is the speed regulation due to governor action, K_j is the integral control gain, Δf_j is the incremental frequency deviation, ΔP_{mj} is the incremental generator output change, and ΔP_{vj} is the incremental governor valve position change. Observe that A_{dj} reflects the effect of delayed ACE signal and the structure of matrix C_j implies that the output simply tracks the frequency deviation and the integral control error (ACE) deviation. In the case of a linearized multi-area power system composed of n_s power areas, the model (7.43) is then modified to include the area couplings and associated delays into the state-space representation:

$$\begin{aligned}\dot{x}_j(t) &= A_j x_j(t) + A_{dj} x_j(t - \tau_j(t)) + B_j u_j(t) \\ &+ \sum_{k=1,k\neq j}^{n_s} F_{jk} x_k(t) + \sum_{k=1,k\neq j}^{n_s} E_{jk} x_k(t - \eta_{jk}(t))\end{aligned} \tag{7.45}$$

where the matrices $F_{jk} \in \Re^{n_j \times n_k}$, $E_{jk} \in \Re^{n_j \times n_k}$ are real and constants. The factors τ_j, η_{jk}, $j, k \in \{1, ..., n_s\}$ are unknown time-delay factors satisfying

$$0 \leq \tau_j(t) \leq \varrho_j, \; 0 \leq \eta_{jk}(t) \leq \varrho_{jk} \tag{7.46}$$

where the bounds ϱ_j, η_{jk} are known constants in order to guarantee smooth growth of the state trajectories. The inclusion of the term $E_{jk} x_k(t - \eta_{jk}(t))$ is meant to emphasize the communication and telemetry delays and the delays among the power areas.

In the sequel, we consider that system (7.45) is assumed to operate under failures described by extension of the representation [39] to the decentralized setting treated hereafter. Such representation allows independent outage or partial degradation in any single actuator of every decentralized controller. In this way, a rather general scenario is defined for the design of a reliable decentralized control structure for a class of interconnected systems simultaneously facing uncertainties and failures. Our objective is to apply the decentralized reliable stabilization method by developing decentralized state feedback controllers in the presence of failures described in **Theorem 7.1.1** and deriving a feasibility testing at the subsystem level so as to guarantee the overall system asymptotic stability with a prescribed performance measure.

7.2.2 Simulation results

In the following, the developed decentralized reliable control method is demonstrated on a three-area power system model. For purposes of AGC, area one is modeled by two generators while the other two areas have single generator equivalents. Standard simplified models are used [328]. In order to illustrate the effectiveness of our reliable control algorithm, we initially consider the nominal control design in which no actuator failures will occur. Then, by using the data values $\Sigma_1 = 1$, $\Sigma_c = 1$, $\Sigma_3 = 1$, $\Pi_1 = 0$, $\Pi_2 = 0$, $\Pi_3 = 0$, it is found that the feasible solution of **Theorem 7.1.1** yields the following control gain matrices and performance levels:

$$\begin{aligned}
\gamma_1 &= 3.0297, \\
K_1 &= \begin{bmatrix} 1.2551 & -0.1421 & 0.0011 & -2.3786 \end{bmatrix}, \\
\gamma_c &= 1.9117, \\
K_1 &= \begin{bmatrix} -1.9437 & 2.0436 & -0.0541 & 1.0242 \end{bmatrix}, \\
\gamma_3 &= 1.9117, \\
K_1 &= \begin{bmatrix} 1.9437 & 2.0436 & -0.0541 & 1.0242 \end{bmatrix}
\end{aligned}$$

Next, we study different failure scenarios. The first scenario is described by the following model:

$$u_1^f(t) = 0.6\, u_1(t) + \beta_1(u_1), \;\; \beta_1^2(u_1) \le 0.01\, u_1^2, \tag{7.47}$$

$$u_2^f(t) = 0.8\, u_2(t) + \beta_c(u_2), \;\; \beta_2^2(u_2) \le 0.09\, u_2^2$$

$$u_3^f(t) = 0.4\, u_3(t) + \beta_3(u_3), \;\; \beta_3^2(u_3) \le 0.04\, u_3^2 \tag{7.48}$$

which corresponds to allowing a failure of the order of 60% in the actuator of system S_j with an error of the order of 10%. In a similar way, the tolerances allowed for other actuators are interpreted in the same way. Thus we have $\Sigma_1 = 0.6,\ \Pi_1 = 0.1,\ \Pi_2 = 0.8,\ \Pi_2 = 0.3,\ \Sigma_3 = 0.4,\ \Pi_3 = 0.2$ and **Theorem 7.1.1** was solved again. Using numerical simulation, three typical experiments were performed for three operational modes.

1) *Ideal mode*, where the nominal AGC controllers are implemented in an operation without failures.

2) *Failure mode*, where the nominal AGC controllers are implemented in an operation under the failures described in (7.48).

3) *Reliable control failure mode*, where the reliable AGC controllers are implemented in an operation under the failures described earlier.

Simulation results of the frequency deviations of the three power areas are displayed Figures 7.8 through 7.10 for the prescribed modes of operation. The trajectories of the state-feedback AGC control input for the first power area are depicted in Figure 7.11.

It is observed that the nominal AGC control clearly deteriorates its ideal performance when operating under failures. When using the reliable AGC control, we notice that the state response results drastically improved in spite of working under the same failures, being very close to the one obtained in the ideal case.

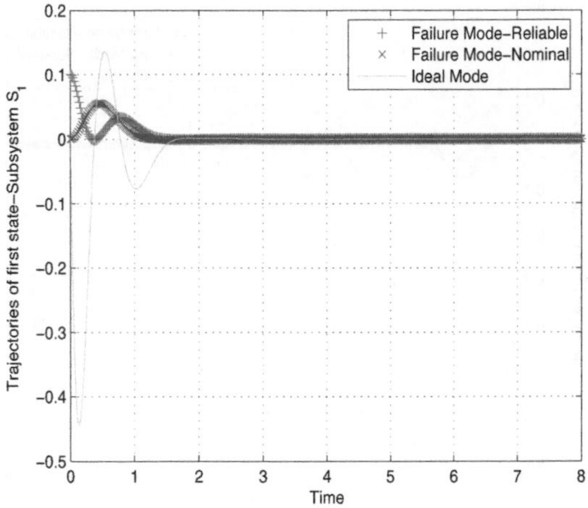

Figure 7.8: Trajectories of frequency deviation (pu)-Area 1

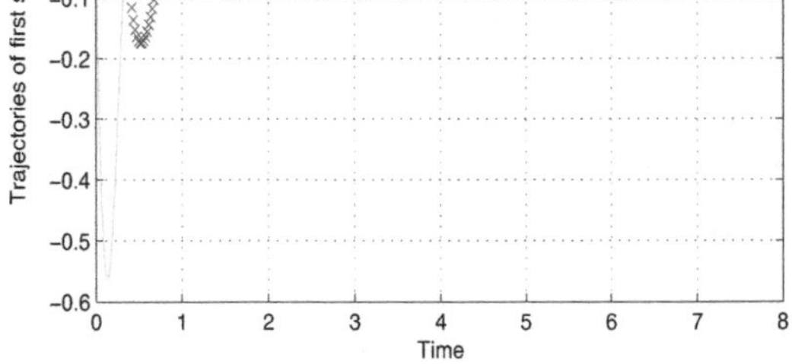

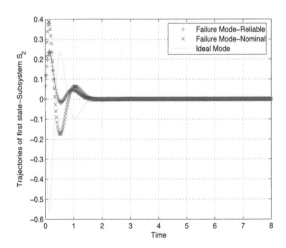

Figure 7.9: Trajectories of frequency deviation (pu)-Area 2

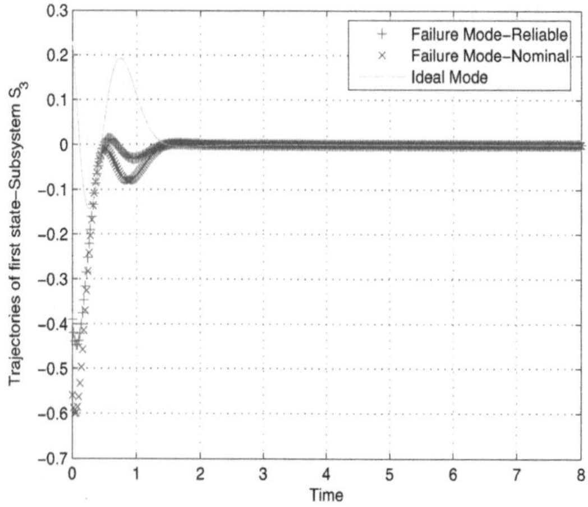

Figure 7.10: Trajectories of frequency deviation (pu)-Area 3

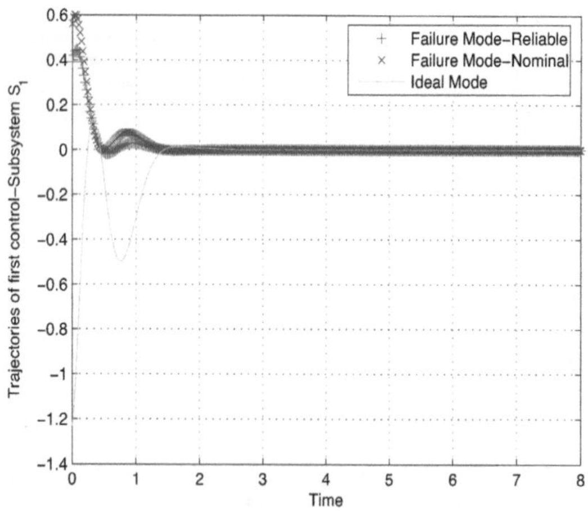

Figure 7.11: Trajectories of AGC control input-Area 1

7.2.3 Reliable control against sensor failures

In what follows, we direct attenuation to the reliable decentralized feedback stabilization problem under a class of sensor failures. These failures are described by a model that takes into consideration possible outages or partial failures in every sensor of each decentralized control loop. Our development follows the results of the foregoing section with appropriate modifications.

We consider a class of linear systems S structurally composed of n_s coupled subsystems S_j and the model of the jth subsystem is described by the state-space representation:

$$\begin{aligned} \dot{x}_j(t) &= A_j x_j(t) + A_{dj} x_j(t - \tau_j(t)) + B_j u_j(t) + c_j(t) \\ y_j(t) &= C_j x_j(t) + C_{dj} x_j(t - \tau_j(t)) \end{aligned} \tag{7.49}$$

where for $j \in \{1, ..., n_s\}$, $x_j(t) \in \Re^{n_j}$ is the state vector, $u_j(t) \in \Re^{m_j}$ is the control input, $y_j(t) \in \Re^{p_j}$ is the measured output, $w_j(t) \in \Re^{q_j}$ is the disturbance input which belongs to $\mathcal{L}_2[0, \infty)$, $z_j(t) \in \Re^{q_j}$ is the performance output. The matrices $A_j \in \Re^{n_j \times n_j}$, $B_j \in \Re^{n_j \times m_j}$, $D_j \in \Re^{q_j \times m_j}$, $A_{dj} \in \Re^{n_j \times n_j}$, $B_j \in \Re^{q_j \times q_j}$, $\Omega_j \in \Re^{n_j \times q_j}$, $C_j \in \Re^{p_j \times n_j}$, $C_{dj} \in \Re^{p_j \times n_j}$, $G_j \in \Re^{q_j \times n_j}$, $G_{dj} \in \Re^{q_j \times n_j}$ are real and constants. The initial condition $\langle x_j(0), x_j(r) \rangle = \langle x_{oj}, B_j \rangle$, $j \in \{1, ..., n_s\}$ where $B_j(.) \in \mathcal{L}_2[-\tau_j^*, 0]$, $j \in \{1, ..., n_s\}$. The interactions between subsystem S_j and the other subsystems is summarized by the coupling vector $c_j(t) \in \Re^{n_j}$ where

$$c_j^t(t) c_j(t) \le \alpha_j x_j^t(t) E_j^t E_j x_j(t) + \beta_j x_j^t(t - \tau_j(t)) F_j^t F_j x_j(t - \tau_j(t)) \tag{7.50}$$

where α_j, β_j are adjustable parameters and the matrices $F_j \in \Re^{n_j \times n_j}$, $F_j \in \Re^{n_j \times n_{kj}}$ are real and constants. The factors τ_j, $j \in \{1, ..., n_s\}$ are unknown time-delays satisfying

$$0 \;\le\; \tau_j(t) \;\le\; \varrho_j, \quad \dot{\tau}_j(t) \;\le\; \mu_j \tag{7.51}$$

where the bounds ϱ_j, μ_j are known constants in order to guarantee smooth growth of the state trajectories. It must be emphasized that inequality (7.50) is guaranteed to satisfy the interconnection patterns (7.2) when taking over all subsystems under the condition (7.19). Similar inequalities have been used in the foregoing chapters to classify the information structures.

Consider that system (7.49) is operating under sensor failures and our approach goes in parallel with the case of actuator failure. We extend the representation of [39] to a decentralized setting since such representation allows independent outage or partial degradation in any single sensor of every decentralized observation channel. In this way, a rather general scenario is defined for the design of reliable decentralized control structure for a class of interconnected systems facing simultaneous uncertainties and sensor failures. Proceeding further, we let $y_j^f \in \Re^{f_j}$ denote the vector of signals from the f_j sensor of the jth subsystem. In the sequel, we consider the following failure model:

$$y_j^f(t) \;=\; \Pi_j y_j(t) + \theta_j(y_j) \tag{7.52}$$

where $0 < \Pi_j \in \Re^{f_j \times f_j} = diag(\pi_{j1}, ..., \pi_{jf_j})$ and the function

$$\theta_j(y_j) = [\theta_{j1}(y_{j1}), ..., \theta_{jf_j}(y_{jf_j})]^t$$

is uncertain and satisfies, for each j,

$$\theta_{jk}^2 \;\leq\; \lambda_{jk}^2\, u_{jk}^2, \quad k = 1, ..., f_j, \;\; \lambda_{jk}^2 \;\geq\; 0 \tag{7.53}$$

When (7.53) holds, then

$$||\theta_j(x_j)||^2 \;\leq\; ||\Lambda_j u_j||^2, \quad j = 1, ..., n_s \tag{7.54}$$

where $0 \leq \Lambda_j = diag(\lambda_{j1}, ..., \lambda_{jf_j}) \in \Re^{f_j \times f_j}$. The value of $\pi_{jk}, \; k = 1, .., f_j$ represents the percentage of failure in the sensor j controlling subsystem S_j. With this representation, each subsystem sensor can fail independently. Note that the condition $\lambda_{jk} = 1, \; \lambda_{jk} = 0$ represents the normal case for the kth sensor of the jth subsystem $(y_{jk}^f(t) = y_{jk}(t))$. When this situation holds for all k, we get that $\Pi_j = I_{f_j}$ and $\Lambda_j = 0$ which in turn clarifies the normal case in the jth channel $(y_j^f(t) = y_j(t))$. In the particular case $\pi_{jk} = \lambda_{jk}$, (7.52) and (7.53) cover the outage case $(y_{jk}^f = 0)$ since $\theta_{jk} = -\pi_{jk} y_{jk}$ satisfies (7.53). Alternatively, the case $\theta_j(y_j) = -\pi_j u_j$ discloses the outage of the observation channel of the jth system. Indeed, there are different cases that would reveal partial failures or partial degradations of the sensors.

Our objective hereafter is to study two main problems:

1. *The first problem is with the decentralized reliable stabilization by developing decentralized feedback controllers (based on state, static output, and dynamic output) in the presence of sensor failures described by (7.52)-(7.53) and deriving a feasibility testing at the subsystem level so as to guarantee the overall system asymptotic stability with a prescribed performance measure.*

2. *The second problem deals with the resilient decentralized reliable stabilization by developing resilient decentralized feedback controllers that take into consideration additive gain perturbations while ensuring that the overall closed-loop system is asymptotically stable with a prescribed performance measure.*

7.2.4 State-feedback reliable design

New criteria for LMI-based characterization of decentralized reliable stabilization are developed using local state feedback. The criteria include some parameter matrices aimed at expanding the range of applicability of the developed conditions. Using the local state feedback $u_j(t) = K_j x_j(t),\ j = 1, ..., N$, the local closed-loop subsystem under failure becomes

$$\begin{aligned} \dot{x}_j(t) &= \widehat{A}_j x_j(t) + A_{dj} x_j(t - \tau_j(t)) + B_j \beta_j(u_j) + c_j(t) + \Omega_j w_j(t) \\ z_j(t) &= \widehat{G}_j x_j(t) + G_{dj} x_j(t - \tau_j(t)) + D_j \beta_j(u_j) + B_j w_j(t) \\ \widehat{A}_j &= A_j + B_j \Sigma_j K_j, \quad \widehat{G}_j = G_j + D_j \Sigma_j K_j \end{aligned} \tag{7.55}$$

The following theorem establishes the main design result for subsystem $\mathsf{S_j}$ where in the subsections that follow, asymptotic stability conditions and the reliable decentralized feedback gains are derived.

Theorem 7.2.1 *Given the bounds $\varrho_j > 0,\ \mu_j > 0,\ \varrho_{jk} > 0,\ \mu_{jk} > 0,\ j, k = 1, ..., n_s$ and matrix $\mathcal{W}_j$, then the family of subsystems $\{\mathsf{S}_j\}$ where $\mathsf{S_j}$ is described by (7.55) is delay-dependent asymptotically stabilizable by decentralized controller $u_j(t) = K_j x_j(t),\ j = 1, ..., n_s$ with $\mathcal{L}_2$-performance bound $\gamma_j,\ j = 1, ..., n_s$ if there exist positive-definite matrices $\mathcal{X}_j,\ \mathcal{Y}_j,\ \{\Lambda_{kj}\}_{k=1}^{n_s},\ \{\Psi_{rj}\}_{r=1}^{4},$ satisfying the following LMIs for $j = 1, ..., n_s$:*

$$\widetilde{\Pi}_j = \begin{bmatrix} \widetilde{\Pi}_{j1} & \widetilde{\Pi}_{j2} \\ \bullet & \widetilde{\Pi}_{j3} \end{bmatrix} < 0 \tag{7.56}$$

$$\widetilde{\Pi}_{j1} = \begin{bmatrix} \Pi_{oj} & \Pi_{aj} & -\varrho_j \Psi_{1j} & \Pi_{cj} & B_j \\ \bullet & -\Pi_{mj} & -\varrho_j \Psi_{3j} & 0 & 0 \\ \bullet & \bullet & -\varrho_j \Psi_{4j} & 0 & 0 \\ \bullet & \bullet & \bullet & -\Pi_{nj} & 0 \\ \bullet & \bullet & \bullet & \bullet & -I_j \end{bmatrix}$$

$$\widetilde{\Pi}_{j2} = \begin{bmatrix} \Omega_j & \Pi_{zj}^t & \mathcal{Y}_j^t \Gamma_j^t & \varrho_j (\mathcal{X}_j A_j^t + \mathcal{Y}_j^t \Sigma_j^t B_j^t) \mathcal{W}_j \\ 0 & G_{dj}^t & 0 & \varrho_k \mathcal{X}_j A_{dj}^t \mathcal{W}_k \\ 0 & 0 & 0 & 0 \\ 0 & 0 & 0 & \varrho_k \sum_{k=1}^{n_s} \mathcal{X}_j E_{kj} \mathcal{W}_k \\ 0 & 0 & 0 & 0 \end{bmatrix}$$

$$\widetilde{\Pi}_{j3} = \begin{bmatrix} -\gamma_j^2 I_j & B_j^t & 0 & \varrho_k \Gamma_j^t \mathcal{W}_k \\ \bullet & -I_j & 0 & 0 \\ \bullet & \bullet & -I_j & 0 \\ \bullet & \bullet & \bullet & -\varrho_k \mathcal{W}_k \end{bmatrix} \tag{7.57}$$

where

$$
\begin{aligned}
\Pi_{oj} &= [A_j + \sum_{k=1}^{n_s} F_{kj}]\mathcal{X}_j + B_j\Sigma_j\mathcal{Y}_j + \mathcal{X}_j[A_j + \sum_{k=1}^{n_s} F_{kj}]^t \\
&+ \mathcal{Y}_j^t\Sigma_j^t B_j^t + \Psi_{1j} + \Psi_{1j}^t + \Psi_{2j} + \sum_{k=1}^{n_s} \Lambda_{kj} \\
\Pi_{aj} &= A_{dj}\mathcal{X}_j - \Psi_{1j} + \Psi_{3j}^t, \ \Pi_{mj} = \Psi_{3j} + \Psi_{3j}^t + (1-\mu_j)\Psi_{2j}, \\
\Pi_{nj} &= \sum_{k=1,k\neq j}^{n_s} (1-\mu_{kj})\Lambda_{kj}, \ \Pi_{cj} = \sum_{k=1,k\neq j}^{n_s} E_{kj}\mathcal{X}_j, \\
\Pi_{zj} &= G_j\mathcal{X}_j + D_j\Sigma_j\mathcal{Y}_j
\end{aligned} \tag{7.58}
$$

Moreover, the local gain matrix is given by $K_j = \mathcal{Y}_j\mathcal{X}_j^{-1}$.

Proof: Consider the LKF:

$$
\begin{aligned}
V(t) &= \sum_{j=1}^{n_s} V_j(t), \ \ V_j(t) = V_{oj}(t) + V_{aj}(t) + V_{mj}(t) + V_{nj}(t), \\
V_{oj}(t) &= x_j^t(t)\mathcal{P}_j x_j(t), \ \ V_{aj}(t) = \int_{-\varrho_j}^{0}\int_{t+s}^{t} \dot{x}_j^t(\alpha)\mathcal{W}_j\dot{x}_j(\alpha)d\alpha\, ds, \\
V_{mj}(t) &= \int_{t-\tau_j(t)}^{t} x_j^t(s)\mathcal{Q}_j x_j(s)\, ds, \\
V_{nj}(t) &= \sum_{k=1,k\neq j}^{n_s} \int_{t-\eta_{jk}(t)}^{t} x_k^t(s)\mathcal{Z}_{jk}x_k(s)\, ds
\end{aligned} \tag{7.59}
$$

where $0 < \mathcal{P}_j = \mathcal{P}_j^t, \ 0 < \mathcal{W}_j = \mathcal{W}_j^t, \ 0 < \mathcal{Q}_j = \mathcal{Q}_j^t, \ 0 < \mathcal{Z}_{jk} = \mathcal{Z}_{jk}^t, \ j,k \in \{1,...,n_s\}$ are weighting matrices of appropriate dimensions. Observe that by using the failure model inequality (7.53) with $u_j(t) = K_j x_j(t)$, we have

$$
\beta_j^t(x_j)\beta_j(x_j) \ \leq \ x_j^t K_j^t \Gamma_j^t \Gamma_j K_j x_j \tag{7.60}
$$

For the time being, we consider that the gains K_j are specified. A straightforward computation gives the time derivative of $V(t)$ along the solutions of (7.55) with $w(t) \equiv 0$ as:

$$
\begin{aligned}
\dot{V}_{oj}(t) &= 2x_j^t\mathcal{P}_j\dot{x}_j = 2x_j^t\mathcal{P}_j[\widehat{A}_j x_j(t) + A_{dj}x_j(t-\tau_j(t)) + B_j\beta_j(u_j)] \\
&+ 2x_j^t\mathcal{P}_j\sum_{k=1}^{n_s} F_{jk}x_k(t) + 2x_j^t\mathcal{P}_j\sum_{k=1}^{n_s} E_{jk}x_k(t-\eta_{jk}(t))
\end{aligned} \tag{7.61}
$$

Some manipulations lead to

$$\begin{aligned}
\dot{V}_{oj}(t) &= = 2x_j^t \mathcal{P}_j [\widehat{A}_j + A_{dj}] x_j(t) \\
&+ 2x_j^t [\Theta_j - \mathcal{P}_j A_{dj}] \int_{t-\tau_j(t)}^{t} \dot{x}_j(s) ds + 2x_j^t \mathcal{P}_j B_j \beta_j(u_j) \\
&+ 2x^t(t-\tau_j(t)) \Upsilon_j \int_{t-\tau_j(t)}^{t} \dot{x}_j(s) ds \\
&- \left[2x_j^t t \Theta_j \int_{t-\tau_j}^{t} \dot{x}_j(s) ds + 2x^t(t-\tau_j(t)) \Upsilon_j \int_{t-\tau_j(t)}^{t} \dot{x}_j(s) ds \right] \\
&+ +2x_j^t \mathcal{P}_j \sum_{k=1, k\neq j}^{n_s} F_{jk} x_k(t) + 2x_j^t \mathcal{P}_j \sum_{k=1, k\neq j}^{n_s} E_{jk} x_k(t - \eta_{jk}(t))
\end{aligned} \tag{7.62}$$

Little algebra gives

$$\begin{aligned}
\dot{V}_{oj}(t) &= \frac{1}{\tau_j(t)} \int_{t-\tau_j(t)}^{t} \left[2x_j^t [\mathcal{P}_j \widehat{A}_j + \Theta_j] x_j \right. \\
&+ 2x_j^t [\mathcal{P}_j A_{dj} - \Theta_j + \Upsilon_j^t] x_j(t - \tau_j(t)) \\
&- 2x_j^t (t - \tau_j(t)) \Upsilon x(t - \tau_j) - 2x^t \tau_j \Theta \dot{x}_j(s) \\
&- 2x_j^t (t - \tau_j(t)) \tau_j(t) \Upsilon_j \dot{x}_j(s) + 2x_j^t \mathcal{P}_j B_j \beta_j(u_j) \\
&+ \left. 2x_j^t \mathcal{P}_j \sum_{k=1, k\neq j}^{n_s} F_{jk} x_k(t) + 2x_j^t \mathcal{P}_j \sum_{k=1, k\neq j}^{n_s} E_{jk} x_k(t - \eta_{jk}(t)) \right] ds
\end{aligned} \tag{7.63}$$

where $\Theta_j \in \Re^{n_j \times n_j}$ and $\Upsilon_j \in \Re^{n_j \times n_j}$ are appropriate relaxation matrices injected to facilitate the delay-dependence analysis.

$$\begin{aligned}
\dot{V}_{aj}(t) &= \int_{t-\varrho_j}^{t} [\dot{x}_j^t(t) \mathcal{W}_j \dot{x}_j(t) - \dot{x}_j^t(s) \mathcal{W}_j \dot{x}_j(s)] d\, s \\
&= \varrho_j \dot{x}_j^t(t) \mathcal{W}_j \dot{x}_j(t) - \int_{t-\tau_j}^{t} \dot{x}_j^t(s) \mathcal{W}_j \dot{x}_j(s) ds - \int_{t-\varrho_j}^{t-\tau_j} \dot{x}_j^t(s) \mathcal{W}_j \dot{x}_j(s) \\
&\leq \varrho_j\, \dot{x}_j^t(t) \mathcal{W}_j \dot{x}_j(t) - \int_{t-\tau_j}^{t} \dot{x}_j^t(s) \mathcal{W}_j \dot{x}_j(s) ds \\
&= \frac{1}{\tau_j(t)} \int_{t-\tau_j(t)}^{t} \left[\varrho_j\, \dot{x}_j^t(t) \mathcal{W}_j \dot{x}_j(t) - \tau_j \dot{x}_j^t(s) \mathcal{W}_j \dot{x}_j(s) \right] ds
\end{aligned} \tag{7.64}$$

Note that the term $T_j = \int_{t-\varrho_j}^{t-\tau_j} \dot{x}_j^t(s)\mathcal{W}_j\dot{x}_j(s)$ accounts for the enlarged time interval from $t - \varrho_j \rightarrow t$ to $t - \tau_j \rightarrow t$ and it is obvious that $T_j > 0$. Next, we have

$$\begin{aligned}
\dot{V}_{mj}(t) &= x_j^t(t)\mathcal{Q}_j x_j(t) - (1 - \dot{\tau}_j(t))\, x^t(t - \tau_j(t))\mathcal{Q}_j x_j(t - \tau_j(t)) \\
&\leq x_j^t(t)\mathcal{Q}_j x_j(t) - (1 - \mu_j)\, x_j^t(t - \tau_j(t))\mathcal{Q}_j x_j(t - \tau_j(t)) \\
&= \frac{1}{\tau_j(t)} \int_{t-\tau_j(t)}^{t} \Big[x_j^t(t)\mathcal{Q}_j x_j(t) \\
&- (1 - \mu_j)\, x_j^t(t - \tau_j(t))\mathcal{Q}_j x_j(t - \tau_j(t)) \Big] ds
\end{aligned} \tag{7.65}$$

Also, we have

$$\begin{aligned}
&\dot{V}_{nj}(t) \\
&= \sum_{k=1,k\neq j}^{n_s} \Big[x_k^t(t)\mathcal{Z}_{jk} x_k(t) - (1 - \dot{\eta}_{jk}(t)) x_k^t(t - \eta_{jk}(t))\mathcal{Z}_{jk} x_k(t - \eta_{jk}(t)) \Big] \\
&\leq \sum_{k=1,k\neq j}^{n_s} \Big[x_k^t(t)\mathcal{Z}_{jk} x_k(t) - (1 - \mu_{jk})\, x_k^t(t - \eta_{jk}(t))\mathcal{Z}_{jk} x_k(t - \eta_{jk}(t)) \Big]
\end{aligned} \tag{7.66}$$

For the class of interconnected systems (7.49), the following structural identity holds:

$$\sum_{j=1}^{n_s} \sum_{k=1,k\neq j}^{n_s} x_k^t(t)\mathcal{Z}_{jk} x_k(t) = \sum_{j=1}^{n_s} \sum_{k=1,k\neq j}^{n_s} x_j^t(t)\mathcal{Z}_{kj} x_j(t) \tag{7.67}$$

By combining (7.59)-(7.67) with the aid of Schur complements, we have

$$\begin{aligned}
\dot{V}_j(t)|_{(7.55)} &\leq \sum_{j=1}^{n_s} \left[\frac{1}{\tau_j} \int_{t-\tau_j}^{t} \chi_j^t(t,s)\; \Xi_j\; \chi_j(t,s)\; ds \right] \\
\chi_j(t,s) &= \left[x_j^t(t)\;\; x_j^t(t-\tau_j)\;\; \dot{x}_j^t(s)\;\; x_j^t(t-\eta_{kj})\;\; \beta^t(x_j) \right]^t
\end{aligned} \tag{7.68}$$

$$\overline{\Xi}_j = \begin{bmatrix} \Xi_{oj} & \Xi_{aj} & -\varrho_j\Theta_j & \Xi_{cj} & \mathcal{P}_jB_j & \varrho_j\widehat{A}_j^t\mathcal{W}_j \\ \bullet & -\Xi_{mj} & -\varrho_j\Upsilon_j & 0 & 0 & \varrho_jA_{d_j}^t\mathcal{W}_j \\ \bullet & \bullet & -\varrho_j\mathcal{W}_j & 0 & 0 & 0 \\ \bullet & \bullet & \bullet & -\Xi_{nj} & 0 & \varrho_j\sum_{k=1}^{n_s}E_{kj}\mathcal{W}_j \\ \bullet & \bullet & \bullet & \bullet & -I_j & 0 \\ \bullet & \bullet & \bullet & \bullet & \bullet & -\varrho_j\mathcal{W}_j \end{bmatrix}$$

$$\begin{aligned} \Xi_{oj} &= \mathcal{P}_j[\widehat{A}_j + \sum_{k=1}^{n_s}F_{kj}] + [\widehat{A}_j + \sum_{k=1}^{n_s}F_{kj}]^t\mathcal{P}_j \\ &+ \Theta_j + \Theta_j^t + \mathcal{Q}_j + \sum_{k=1}^{n_s}\mathcal{Z}_{kj} + K_j^t\Gamma_j^t\Gamma_jK_j, \\ \Xi_{aj} &= \mathcal{P}_jA_{dj} - \Theta_j + \Upsilon_j^t,\ \Xi_{cj} = \mathcal{P}_j\sum_{k=1}^{n_s}E_{kj} \\ \Xi_{mj} &= \Upsilon_j + \Upsilon_j^t + (1-\mu_j)\mathcal{Q}_j,\ \Xi_{nj} = \sum_{k=1}^{n_s}(1-\mu_{kj})\mathcal{Z}_{kj} \end{aligned} \tag{7.69}$$

where $\dot{V}_j(t)|_{(7.55)}$ defines the Lyapunov derivative along the solutions of system (7.49). Internal stability requirement $\dot{V}_j(t)|_{(7.55)} < 0$ implies that $\overline{\Xi}_j < 0$. It is readily seen from (7.68) using the delay bounds of (7.51) that there exists a scalar $\sigma_j > 0$ such that

$$\begin{bmatrix} \Xi_{oj}+\sigma_jI_j & \Xi_{aj} & -\Theta_j & \Xi_{cj} & \mathcal{P}_jB_j & \varrho_j\widehat{A}_j^t\mathcal{W}_j \\ \bullet & -\Xi_{mj} & -\Upsilon_j & 0 & 0 & \varrho_jA_{d_j}^t\mathcal{W}_j \\ \bullet & \bullet & -\varrho_j^{-1}\mathcal{W}_j & 0 & 0 & 0 \\ \bullet & \bullet & \bullet & -\Xi_{nj} & 0 & \varrho_j\sum_{k=1}^{n_s}E_{kj}\mathcal{W}_j \\ \bullet & \bullet & \bullet & \bullet & -I_j & 0 \\ \bullet & \bullet & \bullet & \bullet & \bullet & -\varrho_j\mathcal{W}_j \end{bmatrix} < 0 \tag{7.70}$$

Therefore, for all τ_j satisfying (7.51) we have

$$\begin{bmatrix} \Xi_{oj}+\sigma_jI_j & \Xi_{aj} & -\tau_j\Theta_j & \Xi_{cj} & \mathcal{P}_jB_j & \varrho_j\widehat{A}_j^t\mathcal{W}_j \\ \bullet & -\Xi_{mj} & -\tau_j\Upsilon_j & 0 & 0 & \varrho_jA_{d_j}^t\mathcal{W}_j \\ \bullet & \bullet & -\tau_j\mathcal{W}_j & 0 & 0 & 0 \\ \bullet & \bullet & \bullet & -\Xi_{nj} & 0 & \varrho_j\sum_{k=1}^{n_s}E_{kj}\mathcal{W}_j \\ \bullet & \bullet & \bullet & \bullet & -I_j & 0 \\ \bullet & \bullet & \bullet & \bullet & \bullet & -\varrho_j\mathcal{W}_j \end{bmatrix} < 0 \tag{7.71}$$

and hence

$$\dot{V}_j(t)|_{(7.55)} < \sum_{j=1}^{n_s}\Big[\frac{1}{\tau_j(t)}\int_{t-\tau_j}^{t} \chi_j^t(t,s) diag[-\sigma_j,\ 0,\ 0,\ 0,\ 0,\ 0]\chi_j(t,s)\, ds\Big]$$
$$= -\sum_{j=1}^{n_s}\sigma_j\ ||x_j||^2 < 0 \qquad (7.72)$$

Consider the $\mathcal{L}_2 - gain$ performance measure

$$J = \sum_{j=1}^{n_s}\int_0^{\infty}\Big(z_j^t(s)z_j(s) - \gamma_j^2 w_j^t(s)w_j(s)\Big)ds$$

For any $w_j(t) \in \mathcal{L}_2(0,\infty) \neq 0$ with zero initial condition $x_j(0) = 0$ hence $V(0) = 0$, we have

$$J = \sum_{j=1}^{n_s}\int_0^{\infty}\Big(z_j^t(s)z_j(s) - \gamma_j^2 w_j^t(s)w_j(s) + \dot{V}_j(t)|_{(7.55)} - V_j(\infty)\Big)ds$$
$$\leq \sum_{j=1}^{n_s}\int_0^{\infty}\Big(z_j^t(s)z_j(s) - \gamma_j^2 w_j^t(s)w_j(s) + \dot{V}_j(t)|_{(7.55)}\Big)ds$$

Proceeding as before, we make use of (7.68) to get

$$\sum_{j=1}^{n_s}\Big(z_j^t(s)z_j(s) - \gamma_j^2 w_j^t(s)w_j(s) + \dot{V}_j(t)|_{(7.55)}\Big)$$
$$= \sum_{j=1}^{n_s}\bar{\chi}_j^t(t,s)\ \widehat{\Xi}_j\ \bar{\chi}_j(t,s) \qquad (7.73)$$

where

$$\bar{\chi}_j(t,s) = \big[\ x_j^t(t)\quad x_j^t(t-\tau_j)\quad \dot{x}_j^t(s)\quad x_j^t(t-\eta_{kj})\quad \beta^t(x_j)\quad w_j^t(s)\ \big]^t$$
$$\widehat{\Xi}_j =$$
$$\begin{bmatrix} \Xi_{oj} & \Xi_{aj} & -\varrho_j\Theta_j & \Xi_{cj} & \mathcal{P}_jB_j & \mathcal{P}_j\Omega_j & \widehat{G}_j^t & \varrho_j\widehat{A}_j^t\mathcal{W}_j \\ \bullet & -\Xi_{mj} & -\varrho_j\Upsilon_j & 0 & 0 & 0 & G_{dj}^t & \varrho_jA_{d_j}^t\mathcal{W}_j \\ \bullet & \bullet & -\varrho_j\mathcal{W}_j & 0 & 0 & 0 & 0 & 0 \\ \bullet & \bullet & \bullet & -\Xi_{nj} & 0 & 0 & 0 & \varrho_j\sum_{k=1}^{n_s}E_{kj}\mathcal{W}_j \\ \bullet & \bullet & \bullet & \bullet & -I_j & 0 & 0 & 0 \\ \bullet & \bullet & \bullet & \bullet & \bullet & -\gamma_j^2I_j & B_j^t & \varrho_j\Gamma_j^t\mathcal{W}_j \\ \bullet & \bullet & \bullet & \bullet & \bullet & \bullet & -I_j & 0 \\ \bullet & \bullet & \bullet & \bullet & \bullet & \bullet & \bullet & -\varrho_j\mathcal{W}_j \end{bmatrix}$$

It is readily seen that when $\widehat{\Xi}_j < 0$ the condition

$$\sum_{j=1}^{n_s} \left(z_j^t(s) z_j(s) - \gamma_j^2 w_j^t(s) w_j(s) \ + \dot{V}_j(t)|_{(7.55)} \right) < 0$$

for arbitrary $s \in [t, \infty)$, which implies for any $w_j(t) \in \mathcal{L}_2(0, \infty) \neq 0$ that $J < 0$ leading to $\sum_{j=1}^{n_s} ||z_j(t)||_2 \ < \ \sum_{j=1}^{n_s} \gamma_j \ ||w(t)_j||_2$, which assures the desired performance.

The immediate task now is to determine the decentralized gain matrices K_j. Applying the congruent transformation

$$T = diag[\mathcal{X}_j, \ \mathcal{X}_j, \ \mathcal{X}_j, \ \mathcal{X}_j, \ I_j, \ I_j, \ I_j, \ I_j], \ \mathcal{X}_j = \mathcal{P}_j^{-1}$$

to $\overline{\Xi}_j$ with Schur complements and using the linearizations

$$\begin{aligned} \mathcal{Y}_j &= K_j \mathcal{X}_j, \ \Psi_{1j} = \mathcal{X}_j \Theta_j \mathcal{X}_j, \ \Psi_{2j} = \mathcal{X}_j \mathcal{Q}_j \mathcal{X}_j, \\ \Psi_{3j} &= \mathcal{X}_j \Upsilon_j \mathcal{X}_j, \ \Lambda_{kj} = \mathcal{X}_j \mathcal{Z}_{kj} \mathcal{X}_j, \ \Psi_{4j} = \mathcal{X}_j \mathcal{W}_j \mathcal{X}_j \end{aligned}$$

we readily obtain LMI (7.56) by Schur complements and therefore the proof is completed.

Remark 7.2.2 *It is significant to recognize that our methodology incorporates four weighting matrices $\mathcal{P}_j, \ \mathcal{Q}_j, \ \mathcal{W}_j, \ \mathcal{Z}_{jk}$ and two parameter matrices $\Theta_j, \ \Upsilon_j$ at the subsystem level in order to ensure least conservative delay-dependent stability results. This will eventually result in reduced computational requirements as evidenced by a simple comparison with the improved free-weighting matrices method of [103] for single time-delay systems in terms of two aspects. One aspect would be due to reduced computational load as evidenced by a less number of manipulated variables and faster processing. Another aspect arises by noting that LMIs (7.77) subject to (7.58) for $n_s = 1$ theoretically cover the results of [146, 174, 186] as special cases. Furthermore, in the absence of delay ($A_{dj} \equiv 0, \ \mathcal{Q}_j \equiv 0, \ \mathcal{W}_j \equiv 0$), it is easy to infer that LMIs (7.77) subject to (7.58) for $n_s = 1$ will eventually reduce to a parameterized delay-independent criteria. LKF composed of the first three terms has been considered in [74, 103, 261, 290] for single time-delay systems. In comparison with the reported results on interconnected systems [251, 283, 287, 288, 328, 371], LKF (7.59) represents a new effective form and therefore the advantages and reduced conservatism afforded in [74, 103, 261, 290] are carried over hereafter.*

Remark 7.2.3 *The optimal performance-level $\gamma_j,\ j = 1,..,n_s$ for the interconnected system can be determined for decentralized reliable state feedback by solving the following convex optimization problem:*
Problem A

$$\begin{aligned} & For\ \ j,k = 1,...,n_s,\ \ Given\ \ \varrho_j,\ \ \mu_j,\ \ \varrho_{jk},\ \ \mu_{jk},\ \ \mathcal{W}_j \\ & \min_{\mathcal{X}_j,\ \mathcal{Y}_j,\ \{\Lambda_{kj}\}_{k=1}^{n_s},\ \{\Psi_{rj}\}_{r=1}^{3}} \quad \gamma_j \\ & subject\ \ to\ \ LMI(7.77) \end{aligned} \tag{7.74}$$

7.2.5 Dynamic output-feedback reliable design

In this section, we develop new criteria for LMI-based characterization of decentralized reliable stabilization by local static output feedback. The criteria include some parameter matrices aimed at expanding the range of applicability of the developed conditions. To facilitate further development, we consider the case where the set of output matrices $C_j,\ j = 1,...,n_s$ is assumed to be of full row rank. Using the local dynamic output-feedback controller:

$$\begin{aligned} \dot{\xi}_j(t) &= A_j\xi_j(t) + B_j u_j(t) + L_j[y_j^f(t) - C_j\xi_j(t)] \\ u_j(t) &= K_j\xi_j(t) \end{aligned} \tag{7.75}$$

where $L_j,\ \ K_j$ are the unknown gain matrices. By defining the state error $e_j(t) = x_j(t) - \xi_j(t)$, then we obtain the closed-loop system as:

$$\begin{aligned} \dot{x}_j(t) &= A_j x_j(t) + B_j K_j x_j(t) + A_{dj} x_j(t - \tau_j(t)) - B_j K_j e_j(t) + c_j(t) \\ \dot{e}_j(t) &= (A_j - L_j C_j)e_j(t) + A_{dj} x_j(t - \tau_j(t)) \\ &+ L_j(F_j - I)C_j x_j(t) + L_j C_{dj} x_j(t - \tau_j(t)) + c_j(t) \\ &+ L_j\theta_j(C_j x_j(t) + C_{dj} x_j(t - \tau_j(t))) \end{aligned} \tag{7.76}$$

The following theorem establishes the main design result for subsystem S_j.

Theorem 7.2.4 *Given the bounds $\varrho_j > 0,\ \mu_j > 0,\ j,k = 1,...,n_s$ and matrix $\mathcal{W}_j$, then the family of subsystems $\{S_j\}$ where S_j is described by (7.76) is delay-dependent asymptotically stabilizable by decentralized controller $u_j(t) = K_j x_j(t),\ j = 1,...,n_s$ with $\mathcal{L}_2$-performance bound $\gamma_j,\ j = 1,...,n_s$ if there exist positive-definite matrices $\mathcal{X}_j,\ \mathcal{Y}_j,\ \{\Lambda_{kj}\}_{k=1}^{n_s},\ \{\Psi_{rj}\}_{r=1}^{4},$ satisfying the following LMIs for $j = 1,...,n_s$:*

$$\widetilde{\Pi}_j = \begin{bmatrix} \Pi_{1j} & N_j & \Pi_{2j} & \Pi_{3j} \\ \bullet & -\rho\mathcal{W}_j & 0 & 0 \\ \bullet & \bullet & -\rho\mathcal{W}_j & 0 \\ \bullet & \bullet & \bullet & -1 \end{bmatrix} \tag{7.77}$$

where

$$\Pi_{1j} = \begin{bmatrix} E_1 & E_2 & E_3 & E_4 \\ \bullet & E_5 & E_6 & E_7 \\ \bullet & \bullet & E_8 & E_9 \\ \bullet & \bullet & \bullet & E_{10} \end{bmatrix} \tag{7.78}$$

$$\begin{aligned} E_1 &= P_j A_{cj} + A_{cj}^t P_j + Q_j + R_j + N_{1j} + N_{1j}^t + \alpha_j \\ E_2 &= P_j A_{dj}, \ E_3 = N_{1j}, \quad E_4 = Y_j(\Sigma_j - I)C_j - B_j K_j \\ E_5 &= -(1-\mu_j)Q_j - 2N_{2j} - 2N_{2j}^t + \beta_j, \quad E_6 = N_{2j} \\ E_7 &= A_{dj}^t + C_{dj}^t F_j Y_j^t, \quad E_8 = -R_j, \ A_{cj} = A_j + B_j K_j \\ E_9 &= 0, \quad E_{10} = S_j A_j + A_j^t S_j - Y_j C_j - C_j^t Y_j^t \end{aligned} \tag{7.79}$$

$$\Pi_{2j} = \begin{bmatrix} \rho_j(A_j^t + B_j K_j)\mathcal{W}_j \\ \rho_j A_{dj}^t \mathcal{W}_j \\ 0 \\ -\rho_j K_j^t B_j^t \mathcal{W}_j \\ \varrho_j \mathcal{W}_j \end{bmatrix}, \quad \Pi_{3j} = \begin{bmatrix} C_j^t \Gamma_j^t Y_j^t \\ C_{dj}^t \Gamma_j^t Y_j^t \\ 0 \\ S_j \\ 0 \end{bmatrix}, \quad N_j = \begin{bmatrix} N_{1j} \\ N_{2j} \\ 0 \\ 0 \\ 0 \end{bmatrix}$$

Moreover, the local gain matrix is given by $L_j = S_j^{-1} Y_j$.

Proof: Consider the LKF:

$$\begin{aligned} V(t) &= \sum_{j=1}^{n_s} V_j(t), \ \ V_j(t) = V_{1j}(t) + V_{2j}(t) + V_{3j}(t)) + V_{4j}(t) + V_{5j}(t), \\ V_{1j}(t) &= x_j^t(t) P_j x_j(t), \ \ V_{2j}(t) = \int_{-\varrho_j}^{0} \int_{t+s}^{t} \dot{x}_j^t(\alpha) \mathcal{W}_j \dot{x}_j(\alpha) d\alpha \, ds, \\ V_{3j}(t) &= \int_{t-\tau_j(t)}^{t} x_j^t(s) Q_j x_j(s) \, ds, \\ V_{4j}(t) &= \int_{t-\rho_j}^{t} x_j^t(s) R_j x_j(s) \, ds, \ \ V_{5j}(t) = e_j^t(t) S_j e_j(t) \end{aligned} \tag{7.80}$$

where $0 < P_j = P_j^t, \ 0 < \mathcal{W}_j = \mathcal{W}_j^t, \ 0 < Q_j = Q_j^t, \ , \ j,k \in \{1, ..., n_s\}$ are weighting matrices of appropriate dimensions. Straightforward calculations yield

$$\begin{aligned} \dot{V}_{1j}(t) &= 2x_j^t \mathcal{P}_j \dot{x}_j = 2x_j^t \mathcal{P}_j [A_j x_j(t) + K_j B_j x_j(t) \\ &+ A_{dj} x_j(t - \tau_j(t)) - K_j B_j e_j(t) + c_j(t)] \\ \dot{V}_{2j}(t) &= \int_{t-\varrho_j}^{t} [\dot{x}_j^t(t) \mathcal{W}_j \dot{x}_j(t) - \dot{x}_j^t(s) \mathcal{W}_j \dot{x}_j(s)] d\, s \end{aligned} \tag{7.81}$$

and

$$\begin{aligned}
\dot{V}_{3j}(t) &= x_j^t(t)\mathcal{Q}_j x_j(t) - (1-\dot{\tau}_j(t))\, x_j^t(t-\tau_j(t))\mathcal{Q}_j x_j(t-\tau_j(t)) \\
&\leq x_j^t(t)\mathcal{Q}_j x_j(t) - (1-\mu_j)\, x_j^t(t-\tau_j(t))\mathcal{Q}_j x_j(t-\tau_j(t)) \quad (7.82) \\
\dot{V}_{4j}(t) &= x_j^t(t)R_j x_j(t) - x^t(t-\rho_j)R_j x_j(t-\rho_j \\
&= 2e_j^t S_j((A_j - L_jC_j)e_j + (A_{dj} + L_jC_{dj})x_j(t-\tau_j(t)) \\
&+ L_j(F_j - I)C_j x_j(t) \quad (7.83) \\
&\quad +c_j(t) + L_j\beta_j(C_jx_j(t) + C_{dj}x_j(t-\tau_j(t))) \\
&\leq 2e_j^t S_j[(A_j - L_jC_j)e_j + (A_{dj} + L_jC_{dj})x_j(t-\tau_j(t)) \\
&+ L_j(F_j - I)C_j x_j(t) + c_j(t)] \\
&\quad +e_j^t(t)S_jS_je_j(t) + (C_jx_j(t) \\
&+ C_{dj}x_j(t-\tau_j(t)))^t\Gamma^t Y_j^t Y_j \Gamma(C_jx_j(t) + C_{dj}x_j(t-\tau_j(t)))
\end{aligned} \tag{7.84}$$

Then by adding the following zero value terms

$$\begin{aligned}
&\left[\begin{array}{cc} x_j^t(t) & x_j^t(t-\tau_j(t)) \end{array}\right]\left[\begin{array}{c} 2N_{1j} \\ 2N_{2j} \end{array}\right][x_j(t) - x_j(t-\tau_j(t)) \\
&- \int_{t-\tau_j(t)}^{t} \dot{x}_j(s)ds] \\
&\left[\begin{array}{cc} x_j^t(t) & x_j^t(t-\tau_j(t)) \end{array}\right]\left[\begin{array}{c} -N_{1j} \\ -N_{2j} \end{array}\right][x_j(t) \\
&- x_j(t-\rho_j) - \int_{t-\rho_j}^{t} \dot{x}_j(s)ds]
\end{aligned} \tag{7.85}$$

Algebraic manipulations lead to

$$\begin{aligned}
&x_j^t(t)[N_{1j} + N_{1j}^t]x_j(t) + 2x_j^t(t)[-2N_{1j} + N_{2j}^t]x_j(t-\tau(t)) \\
&+2x_j^t(t)[N_{1j}]x_j(t-\varrho) + 2x_j^t(t-\tau(t))[-2N_{2j} - 2N_{2j}^t]x_j(t-\tau(t)) \\
&+2x_j^t(t-\tau(t))[N_{2j}]x_j(t-\varrho) \\
&-2\,\xi_j^t(t)2N_j\int_{t-\tau(t)}^{t}\dot{x}_j(s)ds - 2\,\xi_j^t(t)(N_j)\int_{t-\varrho}^{t}\dot{x}_j(s)ds = 0
\end{aligned} \tag{7.86}$$

where $\xi_j^t(t) = [x_j^t(t)\;\; x_j^t(t-\tau_j(t))\;\; x_j^t(t-\rho_j)\;\; e_j(t)]$

Since

$$\begin{aligned}
&\varrho\xi_j^t(t)\, N_j\, \mathcal{W}_j^{-1}\, N_j^t\, \xi_j(t) - \tau(t)\xi_j^t(t)\, N_j\, \mathcal{W}_j^{-1}\, N_j^t\, \xi_j(t) \\
&-(\varrho - \tau(t))\xi_j^t(t)\, N_j\, \mathcal{W}_j^{-1}\, N_j^t\, \xi_j(t) \\
&-2\, \xi_j^t(t)N_j \int_{t-\tau(t)}^{t} \dot{x}_j(s)ds + 2\, \xi_j^t(t)N_j \int_{t-\varrho}^{t-\tau(t)} \dot{x}_j(s)ds \\
&+ \int_{t-\varrho}^{t-\tau(t)} \dot{x}_j^t(s)\mathcal{W}_j\dot{x}(s)_j ds \\
&+ \int_{t-\tau(t)}^{t} \dot{x}_j^t(s)\mathcal{W}_j\dot{x}_j(s)ds \qquad (7.87)\\
&= \xi_j^t(t)\Big\{ \varrho N_j\mathcal{W}_j^{-1}N_j^t \Big\}\xi_j(t) - \int_{t-\tau_j(t)}^{t} [\xi_j^t(t)N_j \\
&+\dot{x}_j^t\mathcal{W}_j]\mathcal{W}_c^{-1}[\xi_j(t)^tN_j + \dot{x}_j^t(s)\mathcal{W}_j]^t ds \\
&- \int_{t-\rho_j}^{t-\tau_j(t)} [-\xi_j^t(t)N_j + \dot{x}_j^t(s)\mathcal{W}_j]\mathcal{W}_j^{-1}[-\xi_j^t(t)N_j + \dot{x}_j^t(s)\mathcal{W}_j]^t ds
\end{aligned}$$

Applying the congruent transformation

$$T = diag[\mathcal{X}_j,\ \mathcal{X}_j,\ \mathcal{X}_j,\ \mathcal{X}_j,\ I_j,\ I_j,\ I_j,\ I_j],\ \mathcal{X}_j = \overline{\mathcal{P}}_j^{-1}$$

with Schur complements and using the linearizations

$$\begin{aligned}
\mathcal{Y}_j &= K_{oj}C_j\mathcal{X}_j,\ \Psi_{1j} = \mathcal{X}_j\overline{\Theta}_j\mathcal{X}_j,\ \Psi_{2j} = \mathcal{X}_j\overline{\mathcal{Q}}_j\mathcal{X}_j,\ \Psi_{3j} = \mathcal{X}_j\overline{\Upsilon}_j\mathcal{X}_j \\
\Lambda_{kj} &= \mathcal{X}_j\overline{\mathcal{Z}}_{kj}\mathcal{X}_j,\ \Psi_{4j} = \mathcal{X}_j\overline{\mathcal{W}}_j\mathcal{X}_j,\ \Psi_{5j} = \mathcal{X}_jC_{dj}^tK_{oj}^t\Gamma_j^t,\ \Psi_{6j} = K_{oj}^tC_{dj}^t\mathcal{X}_j
\end{aligned}$$

we readily obtain LMI (7.77) by Schur complements and therefore the proof is completed.

7.2.6 Illustrative example 7.2

$$A_1 = \begin{bmatrix} -4.93 & -1.01 & 0 & 0 \\ -3.2 & -5.3 & -12.8 & 0 \\ 6.4 & 0.347 & -32.5 & -1.04 \\ 0 & 0.833 & 11.0 & -3.96 \end{bmatrix}, \quad B_j = \begin{bmatrix} 1 & 0 \\ 0 & 1 \\ 0 & 0 \\ 0 & 0 \end{bmatrix},$$

$$A_{d1} = \begin{bmatrix} 2.92 & 0 & 0 & 0 \\ 0 & 2.92 & 0 & 0 \\ 0 & 0 & 2.87 & 0 \\ 0 & 0 & 0 & 2.724 \end{bmatrix}, \quad F_{12} = \begin{bmatrix} 1 & 0 & 0 & 0 \\ 0 & 1 & 0 & 0 \\ 0 & 0 & 1 & 0 \\ 0 & 0 & 0 & 1 \end{bmatrix},$$

$$A_2 = \begin{bmatrix} -4.0 & -2.01 & 0 & 1 \\ -3.2 & -4.3 & -1.8 & 0 \\ 0 & 1.347 & -32.5 & -1.04 \\ 0 & 1 & 11.0 & -4 \end{bmatrix}, \quad B_j = \begin{bmatrix} 1 & 0 \\ 0 & 1 \\ 0 & 0 \\ 0 & 0 \end{bmatrix},$$

$$A_{d2} = \begin{bmatrix} 1.92 & 0 & 1 & 0 \\ 0 & 2 & 1 & 0 \\ 0 & 0 & 2.87 & 0 \\ 0.5 & 0 & 0 & 1.5 \end{bmatrix}, \quad F_{21} = \begin{bmatrix} 1 & 0 & 0 & 0 \\ 0 & 1 & 0 & 0 \\ 0 & 0 & 1 & 0 \\ 0 & 0 & 0 & 1 \end{bmatrix},$$

$$C_{o2} = \begin{bmatrix} 1 & 10 & 0 & 0 \\ 0 & 0 & 10 & 10 \end{bmatrix} \quad C_{do2} = \begin{bmatrix} 2 & 1 & 0 & 0 \\ 0 & 0 & 1 & 3 \end{bmatrix}$$

It is found that the feasible solution of LMI (7.77) yields

$$K_{c1} = \begin{bmatrix} -34.5094 & -7.2128 & 14.8074 & -12.4102 \\ 3.0766 & -15.0809 & -29.8711 & -18.4153 \end{bmatrix}$$

$$L_1 = \begin{bmatrix} -2.5426 & 4.1353 \\ 3.9288 & -0.3947 \\ -1.0332 & 1.3764 \\ -0.1978 & 10.9388 \end{bmatrix}$$

$$K_{c2} = \begin{bmatrix} -24.4560 & -4.9061 & -52.4436 & -130.5408 \\ 17.5392 & -12.7566 & 79.5070 & 157.7008 \end{bmatrix}$$

$$L_2 = \begin{bmatrix} -2.1034 & 2.9994 \\ 3.9288 & 0.7478 \\ 0.6681 & 0.6586 \\ 3.2221 & 10.9388 \end{bmatrix}$$

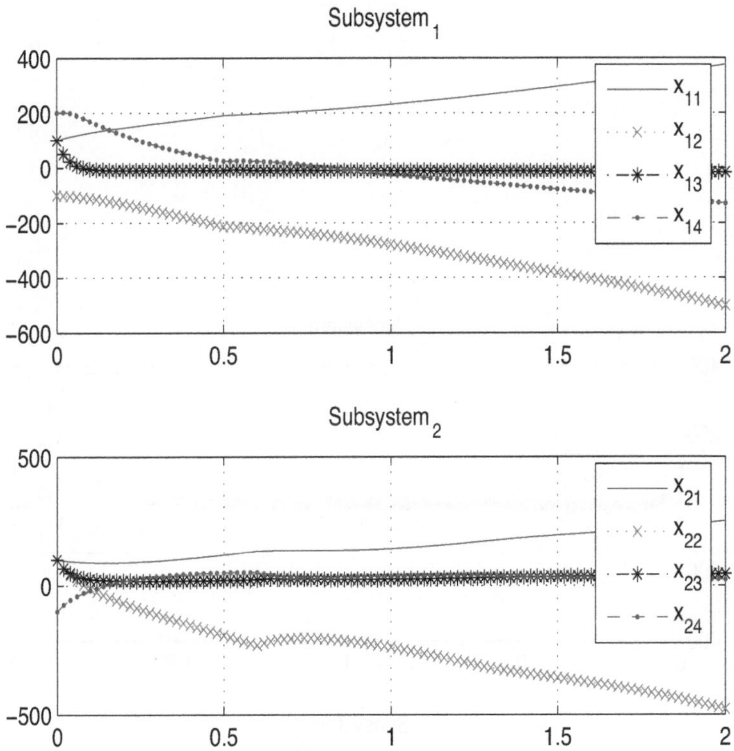

Figure 7.12: Failure mode response

Next, to study the effect of sensor failures, we consider a scenario described by the following model:

$$\begin{aligned} y_1^f(t) &= 0.6\, y_1(t) + \theta_1(y_1), \;\; \theta_1^2(y_1) \leq 0.01\, y_1^2, \\ y_2^f(t) &= 0.7\, y_2(t) + \theta_c(y_2), \;\; \theta_2^2(y_2) \leq 0.02\, y_2^2 \end{aligned} \tag{7.88}$$

which corresponds to allowing a failure of the order of 60% in the sensor of system S_1 with an error of the order of 10% and a failure of the order of 70% in the sensor of system S_2 with an error of the order of 20%. Using numerical simulation, typical experiments were performed for two operational modes.

1) *Failure mode*, where failures occurred without implementing controls.

2) *Reliable control mode*, where the reliable output-feedback controllers are implemented in an operation under the failures described in (7.88).

From the plotted graphs, we record the following observations. First, the failure-mode response in Figure 7.12 clearly shows that the system under this operational mode is unstable. Second, injecting the reliable controller yields the

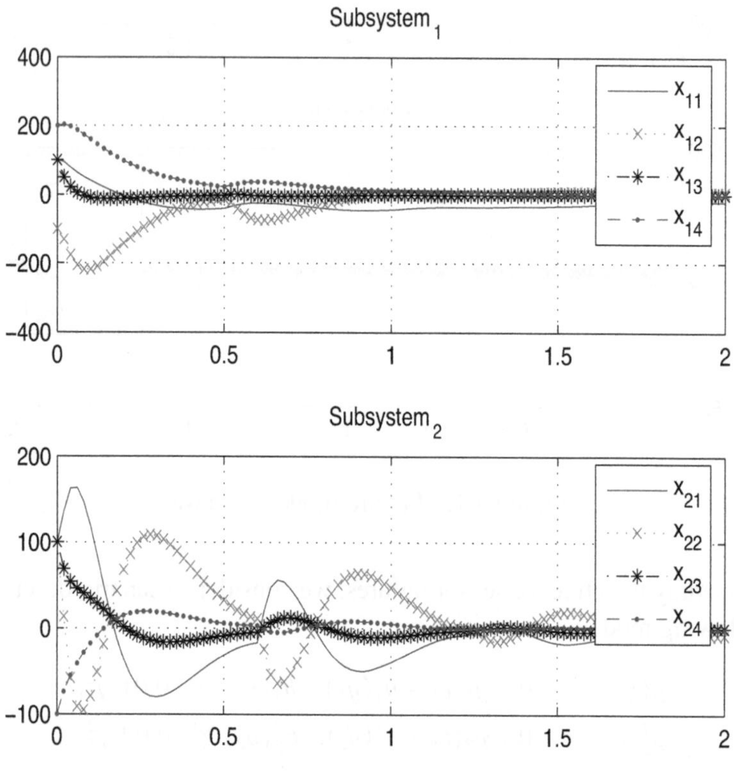

Figure 7.13: Reliable control mode response

closed-loop response shown in Figure 7.13. Here the impact of the controller is evidently quite effective in deriving the system states toward the stable level. These results illustrate the satisfactory behavior of the proposed reliable control scheme.

7.3 Interconnected Discrete Delay Systems

In this section, we study the reliable decentralized feedback stabilization problem of a class of linear interconnected discrete-time systems with interval delays under a class of control failures. These failures are described by a model that takes into consideration possible outages or partial failures in every single actuator of each decentralized controller. The local subsystem has convex-bounded parametric uncertainties and time-varying delays within the local subsystems and across the interconnections. We construct appropriate Lyapunov-Krasovskii functional in order to exhibit the delay-dependent dynamics. We develop reliable decentralized feedback stabilization methods which guarantee overall delay-dependent asymptotic stability with an ℓ_2 gain smaller that a prescribed constant level. The solution is expressed in terms of convex optimization over new parametrized linear matrix inequalities (LMIs) which can be conveniently solved by interior-point minimization methods. The developed results are tested on a representative example.

7.3.1 Problem formulation

We consider a class of linear systems S structurally composed of n_s coupled subsystems $\mathsf{S_j}$ and the model of the jth subsystem is described by the state-space representation:

$$\begin{aligned} x_j(k+1) &= A_j x_j(k) + D_j x_j(k-d_j(k)) + B_j u_j(k) \\ &+ c_j(k) + \Omega_j w_j(k) \quad (7.89) \\ z_j(k) &= G_j x_j(k) + H_j x_j(k-d_j(k)) + L_j u_j(k) \\ &+ B_j w_j(k) \quad (7.90) \end{aligned}$$

where for $j \in \{1, ..., n_s\}$, $x_j(k) \in \Re^{n_j}$ is the state vector, $u_j(k) \in \Re^{m_j}$ is the control input, $w_j(k) \in \Re^{q_j}$ is the disturbance input which belongs to $\ell_2[0, \infty)$, $z_j(k) \in \Re^{q_j}$ is the performance output, and $c_j(k) \in \Re^{n_j}$ is the coupling vector. The matrices $A_j \in \Re^{n_j \times n_j}$, $B_j \in \Re^{n_j \times m_j}$, $D_j \in \Re^{q_j \times m_j}$, $B_j \in \Re^{q_j \times q_j}$, $\Omega_j \in \Re^{n_j \times q_j}$, $L_j \in \Re^{q_j \times m_j}$, $G_j \in \Re^{q_j \times n_j}$, $H_j \in \Re^{q_j \times n_j}$, $F_{jk} \in \Re^{n_j \times n_k}$, and $E_{jk} \in \Re^{n_j \times n_k}$ are real and constants. The initial condition $\langle x_j(0), x_j(r)\rangle = \langle x_{oj}, B_j\rangle$, $j \in \{1, ..., n_s\}$ where $B_j(.) \in \ell_2[-\tau_j^*, 0]$, $j \in \{1, ..., n_s\}$. In the sequel, we treat $c_j(k)$ as a piecewise-continuous vector function in its arguments and it satisfies the quadratic inequality

$$c_j^t(k, ., ,)\ c_j(k, ., .) \leq B_j\ x_j^t(k) F_j^t F_j x_j(k) + \psi_j\ x_j^t(k - \eta_j(k)) E_{dj}^t E_{dj} x_j(k - \eta_j(k)) \tag{7.91}$$

where $B_j > 0$, $\psi_j > 0$ are adjustable bounding parameters. The factors τ_j, η_{jk}, $j, k \in \{1, ..., n_s\}$ are unknown time-delay factors satisfying

$$0 < d_j^- \leq d_j(k) \leq d_j^+,\ 0 < \eta_j^- \leq \eta_j(k) \leq \eta_j^+ \tag{7.92}$$

where the bounds d_j^-, d_j^+, η_{jk}^-, η_{jk}^+ are known constants in order to guarantee smooth growth of the state trajectories. Note in (7.89) and (7.91) that the delay within each subsystem (local delay) and among the subsystems (coupling delay), respectively, are emphasized. The class of systems described by (7.89)-(7.90) subject to delay-pattern (7.92) is frequently encountered in modeling several physical systems and engineering applications including large space structures, multi-machine power systems, cold mills, transportation systems, and water pollution management, to name a few [252, 253, 324].

7.3.2 Failure model

In the sequel, we follow the treatment of interconnected continuous-time systems and the reader will notice exact matching between the continuous-time domain and the discrete-time domain. Thus, we consider that system (7.90) is assumed to operate under failures described by extension of the representation [39] to the decentralized setting treated hereafter. Such representation allows independent outage or partial degradation in any single actuator of every decentralized controller. In this way, a rather general scenario is defined for the design of reliable decentralized control structure for a class of interconnected systems simultaneously facing uncertainties and failures. Let $u_j^f \in \Re^{f_j}$ denote the vector of signals from the f_j actuators that control the jth subsystem. We consider the following failure model:

$$u_j^f(k) = \Sigma_j u_j(k) + \beta_j(u_j) \tag{7.93}$$

where $0 < \Sigma_j \in \Re^{f_j \times f_j} = diag(\sigma_{j1}, ..., \sigma_{jf_j})$ and the function $\beta_j(u_j) = [\beta_{j1}(u_{j1}), ..., \beta_{jf_j}(u_{jf_j})]^t$ is uncertain and satisfies, for each j,

$$\beta_{jk}^2 \leq \gamma_{jk}^2\ u_{jk}^2, \quad k = 1, ..., f_j,\ \gamma_{jk}^2 \geq 0 \tag{7.94}$$

When (7.94) holds, then

$$||\beta_j(u_j)||^2 \leq ||\Gamma_j u_j||^2, \quad j = 1, ..., n_s \tag{7.95}$$

where $0 \leq \Gamma_j = diag(\gamma_{j1}, ..., \gamma_{jf_j}) \in \Re^{f_j \times f_j}$. The value of $\sigma_{jk},\ k = 1, .., f_j$ represents the percentage of failure in the actuator j controlling subsystem $\mathsf{S_j}$. With this representation, each subsystem actuator can fail independently. Note that the condition $\sigma_{jk} = 1,\ \gamma_{jk} = 0$ represents the normal case for the kth actuator of the jth subsystem ($u^f_{jk}(t) = u_{jk}(t)$). When this situation holds for all k, we get that $\Sigma_j = I_{f_j}$ and $\Gamma_j = 0$ which in turn clarifies the normal case in the jth channel ($u^f_j(k) = u_j(k)$). Of the particular case $\sigma_{jk} = \gamma_{jk}$, (7.93) and (7.94) cover the outage case ($u^f_{jk} = 0$) since $\beta_{jk} = -\sigma_{jk}u_{jk}$ satisfies (7.94). Alternatively, the case $\beta_j(u_j) = -\sigma_j u_j$ discloses the outage of the whole controller of the jth system. Indeed, there are different cases that would reveal partial failures or partial degradations of the actuators.

Our objective is to study two main problems: the first problem is with the decentralized reliable stabilization by developing decentralized feedback controllers (based on state, static output, and dynamic output) in the presence of failures described by (7.93)-(7.94) and deriving a feasibility testing at the subsystem level so as to guarantee the overall system asymptotic stability with a prescribed performance measure. The second problem deals with the resilient decentralized reliable stabilization by developing resilient decentralized feedback controllers that take into consideration additive gain perturbations while ensuring that the overall closed-loop system is asymptotically stable with a prescribed performance measure.

7.3.3 State-feedback reliable design

In this section, we develop new criteria for LMI-based characterization of decentralized reliable stabilization by local state feedback. The criteria include some parameter matrices aimed at expanding the range of applicability of the developed conditions. Using the local state feedback $u_j(k) = K_j x_j(k),\ j = 1, ..., N$, the local closed-loop subsystem under failure becomes

$$\begin{aligned} x_j(k+1) &= \widehat{A}_j x_j(k) + D_j x_j(k - d_j(k)) + B_j \beta_j(u_j) \\ &+ c_j(k) + \Omega_j w_j(k) \end{aligned} \tag{7.96}$$

$$\begin{aligned} z_j(k) &= \widehat{G}_j x_j(k) + H_j x_j(k - d_j(k)) + L_j \beta_j(u_j) \\ &+ B_j w_j(k) \end{aligned} \tag{7.97}$$

$$\widehat{A}_j = A_j + B_j \Sigma_j K_j,\ \widehat{G}_j = G_j + L_j \Sigma_j K_j \tag{7.98}$$

Observe that by using the failure model inequality (7.95) with $u_j(t) = K_j x_j(t)$, we have

$$\beta_j^t(x_j)\beta_j(x_j) \leq x_j^t K_j^t \Gamma_j^t \Gamma_j K_j x_j \tag{7.99}$$

For the time being, we consider that the gains K_j are specified. Let $d_j^* = (d_j^+ - d_j^- + 1)$, $\eta_j^* = (\eta_j^+ - \eta_j^- + 1)$ represent the respective number of samples. The following theorem establishes the asymptotic stability conditions of the feedback design for subsystem $\mathsf{S_j}$.

Theorem 7.3.1 *Given the bounds $d_j^+ > 0,\ d_j^- > 0,\ \eta_j^+ > 0,\ \eta_j^- > 0,\ j = 1, ..., N$. The global system S with subsystem $\mathsf{S_j}$ given by (7.89)-(7.90) is delay-dependent asymptotically stable with an $\ell_2-gain < \gamma_j$ if there exist weighting matrices $0 < \mathcal{P}_j,\ 0 < \mathcal{Q}_j,\ 0 < \mathcal{Z}_{jm},\ \forall j = 1, ..., n_s,\ m = 1, ..., n_s$ and scalars $\gamma_j > 0$ satisfying the following LMIs:*

$$\widehat{\Xi}_j = \begin{bmatrix} \Xi_{aj} & \Xi_{cj} & \widehat{\Xi}_{zj} \\ \bullet & -\widehat{\Xi}_{oj} & \widehat{\Xi}_{wj} \\ \bullet & \bullet & -I_j \end{bmatrix} < 0 \tag{7.100}$$

where

$$\begin{aligned} \widehat{\Xi}_{aj} &= \begin{bmatrix} \Xi_{1j} & 0 & 0 & 0 \\ \bullet & -\mathcal{Q}_j & 0 & 0 \\ \bullet & \bullet & -I_j & 0 \\ \bullet & \bullet & \bullet & -I_j \end{bmatrix}, \\ \Xi_{cj} &= \begin{bmatrix} 0 & 0 & \widehat{A}_j^t \mathcal{P}_j & K_j^t \Gamma_j^t & F_j^t & 0 \\ 0 & 0 & D_j^t \mathcal{P}_j & 0 & 0 & E_{dj}^t \\ 0 & 0 & B_j^t \mathcal{P}_j & 0 & 0 & 0 \\ 0 & 0 & \mathcal{P}_j & 0 & 0 & 0 \end{bmatrix} \end{aligned} \tag{7.101}$$

with

$$\begin{aligned} \widehat{\Xi}_{oj} &= \begin{bmatrix} \widehat{\Xi}_{o1j} & 0 \\ \bullet & \Xi_{o2j} \end{bmatrix}, \\ \widehat{\Xi}_{o1j} &= \begin{bmatrix} -\gamma^2 I_j & 0 & \Omega_j^t \mathcal{P}_j \\ \bullet & -\sum_{k=1,k\neq j}^{n_s} \mathcal{Z}_{jk} & 0 \\ \bullet & \bullet & -\mathcal{P}_j \end{bmatrix}, \\ \Xi_{o2j} &= \begin{bmatrix} -I_j & 0 & 0 \\ \bullet & -B_j I & 0 \\ \bullet & \bullet & -\psi_j I \end{bmatrix} \end{aligned} \tag{7.102}$$

and

$$\widehat{\Xi}_{zj} = \begin{bmatrix} \widehat{G}_j^t \\ H_j^t \\ L_j^t \\ 0 \end{bmatrix}, \widehat{\Xi}_{wj} = \begin{bmatrix} B_j^t \\ 0 \\ 0 \\ 0 \\ 0 \\ 0 \end{bmatrix},$$

$$\Xi_{1j} = -\mathcal{P}_j + d_j^* \mathcal{Q}_j + \sum_{k=1,k\neq j}^{n_s} \mathcal{Z}_{jk} \qquad (7.103)$$

Proof: Introduce the LKF:

$$\begin{aligned} V(k) &= \sum_{j=1}^{n_s} V_j(k), \ V_j(k) = x_j^t(k)\mathcal{P}_j x_j(k) \\ &+ \sum_{m=k-d_j(k)}^{k-1} x_j^t(m)\mathcal{Q}_j x_j(m) \\ &+ \sum_{s=2-d_j^+}^{1-d_j^-} \sum_{m=k+s-1}^{k-1} x_j^t(m)\mathcal{Q}_j x_j(m) \\ &+ \sum_{k=1,k\neq j}^{n_s} \sum_{m=k-\eta_{jk}(k)}^{k-1} x_k^t(m)\mathcal{Z}_{jk} x_k(m) \\ &+ \sum_{k=1,k\neq j}^{n_s} \sum_{s=2-\eta_{jk}^+}^{1-\eta_{jk}^-} \sum_{m=k+s-1}^{k-1} x_k^t(m)\mathcal{Z}_{jk} x_k(m) \end{aligned} \qquad (7.104)$$

where $0 < \mathcal{P}_j, \ 0 < \mathcal{Q}_j, \ 0 < \mathcal{Z}_{jk}$ are weighting matrices of appropriate dimensions. A straightforward computation gives the first difference of $\Delta V(k) = V(k+1) - V(k)$ along the solutions of (7.96) as:

$$\begin{aligned} \Delta V(k) &= \sum_{j=1}^{n_s} \Delta V_j(k) \\ \Delta V_j(k) &= [\widehat{A}_j x_j(k) + D_j x_j(k - d_j(k)) + B_j \beta_j(u_j) \\ &+ c_j(k) + \Omega_j w_j(k)]^t \mathcal{P}_j [\widehat{A}_j x_j(k) + D_j x_j(k - d_j(k)) \end{aligned}$$

$$
\begin{aligned}
&+B_j\beta_j(u_j) + c_j(k) + \Omega_j w_j(k)] \\
&-x_j^t(k)\mathcal{P}_j x_j(k) + x_j^t(k)\mathcal{Q}_j x_j(k) \\
&-x_j^t(k - d_j(k))\mathcal{Q}_j x_j(k - d_j(k)) \\
&+ \sum_{k=1,k\neq j}^{n_s} x_k^t(k)\mathcal{Z}_{jk} x_k(k) \\
&- \sum_{k=1,k\neq j}^{n_s} x_k^t(k - \eta_j(k))\mathcal{Z}_{jk} x_k(k - \eta_j(k)) \\
&+ \sum_{m=k+1-d_j(k+1)}^{k-1} x_j^t(m)\mathcal{Q}_j x_j(m) \\
&- \sum_{m=k+1-d_j(k)}^{k-1} x_j^t(m)\mathcal{Q}_j x_j(m) \\
&+(d_j^+ - d_j^-) x_j^t(k)\mathcal{Q}_j x_j(k) \\
&- \sum_{m=k+1-d_j^+}^{k-d_j^*} x_j^t(m)\mathcal{Q}_j x_j(m) \\
&+ \sum_{k=1,k\neq j}^{n_s} \sum_{m=k+1-\eta_k(k+1)}^{k-1} x_k^t(m)\mathcal{Z}_{jk} x_k(m) \\
&- \sum_{k=1,k\neq j}^{n_s} \sum_{m=k+1-\eta_k(k)}^{k-1} x_k^t(m)\mathcal{Z}_{jk} x_k(m) \\
&+ \sum_{k=1,k\neq j}^{n_s} (\eta_j^+ - \eta_k^-) x_k^t(m)\mathcal{Z}_{jk} x_k(m) \\
&- \sum_{k=1,k\neq j}^{n_s} \sum_{m=k+1-\eta_k^+}^{k-\eta_k^-} x_k^t(m)\mathcal{Z}_{jk} x_k(m) \qquad (7.105)
\end{aligned}
$$

Since

$$\sum_{m=k+1-d_j(k+1)}^{k-1} x_j^t(m)\mathcal{Q}_j x_j(m) = \sum_{m=k+1-d_j^-}^{k-1} x_j^t(m)\mathcal{Q}_j x_j(m)$$

$$+ \sum_{m=k+1-d_j(k+1)}^{k-d_j^-} x_j^t(m)\mathcal{Q}_j x_j(m)$$

$$\leq \sum_{m=k+1-d_j(k)}^{k-1} x_j^t(m)\mathcal{Q}_j x_j(m) + \sum_{m=k+1-d_j^+}^{k-d_j^-} x_j^t(m)\mathcal{Q}_j x_j(m) \quad (7.106)$$

In addition, we have

$$\sum_{k=1,k\neq j}^{n_s} \sum_{m=k+1-\eta_k(k+1)}^{k-1} x_k^t(m)\mathcal{Z}_{jk} x_k(m)$$

$$= \sum_{k=1,k\neq j}^{n_s} \sum_{m=k+1-\eta_k^-}^{k-1} x_k^t(m)\mathcal{Z}_{jk} x_k(m)$$

$$+ \sum_{k=1,k\neq j}^{n_s} \sum_{m=k+1-\eta_k(k+1)}^{k-\eta_k^-} x_k^t(m)\mathcal{Z}_{jk} x_k(m)$$

$$\leq \sum_{k=1,k\neq j}^{n_s} \sum_{m=k+1-\eta_j(k)}^{k-1} x_k^t(m)\mathcal{Z}_{jk} x_k(m)$$

$$+ \sum_{k=1,k\neq j}^{n_s} \sum_{m=k+1-\eta_j^+}^{k-\eta_j^-} x_k^t(m)\mathcal{Z}_{jk} x_k(m) \quad (7.107)$$

Then using (7.106) and (7.107) into (7.105) and manipulating, we reach

$$\begin{aligned}
\Delta V_j(k) &\leq [\widehat{A}_j x_j(k) + D_j x_j(k-d_j(k)) + B_j\beta_j(u_j) \\
&+c_j(k) + \Omega_j w_j(k)]^t \mathcal{P}_j[\widehat{A}_j x_j(k) + D_j x_j(k-d_j(k)) \\
&+B_j\beta_j(u_j) + c_j(k) + \Omega_j w_j(k)] \\
&+x_j^t(k)[d_j^* \mathcal{Q}_j - \mathcal{P}_j]x_j(k) \\
&-x_j^t(k-d_j(k))\mathcal{Q}_j x_j(k-d_j(k)) \\
&+\sum_{k=1,k\neq j}^{n_s} \eta_k^* x_k^t(k)\mathcal{Z}_{jk}x_k(k) \\
&-\sum_{k=1,k\neq j}^{n_s} x_k^t(k-\eta_j(k))\mathcal{Z}_{jk}x_k(k-\eta_j(k))
\end{aligned} \tag{7.108}$$

For the class of interconnected systems (7.1), the following structural identity holds:

$$\sum_{j=1}^{n_s}\sum_{k=1,k\neq j}^{n_s} x_k^t(m)\mathcal{Z}_{jk}x_k(m) = \sum_{j=1}^{n_s}\sum_{k=1,k\neq j}^{n_s} x_j^t(m)\mathcal{Z}_{kj}x_j(m)$$

In terms of the vectors

$$\xi_{1j}(k) = [x_j^t(k),\ x_j^t(k-d_j(k)),\ \beta_j(k),\ c_j^t(k)]^t$$

$$\xi_{2j}(k) = [\omega_j^t(k),\ \sum_{m=1,m\neq j}^{n_s} x_m^t(k-\eta_m(k))]^t$$

$$\xi_j(k) = [\xi_{1j}^t(k),\ \xi_{2j}^t(k)]^t$$

we combine (7.99)-(7.108) with algebraic manipulations using the constraint inequalities (7.91) and (7.95) and Schur complements [29] to arrive at

$$\Delta V(k) = \sum_{j=1}^{n_s}\Delta V_j(k) = \sum_{j=1}^{n_s}\xi_j^t(k)\ \Xi_j\ \xi_j(k) \tag{7.109}$$

where Ξ_j corresponds to $\widehat{\Xi}_j$ in (7.100) with $\omega_j(k) \equiv 0,\ G_j \equiv 0,\ H_j \equiv 0,\ L_j \equiv 0,\ B \equiv 0,\ \gamma_j \equiv 0$. It is known that the sufficient condition of subsystem internal stability is $\Delta V(k) < 0$ when $\omega_j(k) \equiv 0,\ G_j \equiv 0,\ H_j \equiv 0,\ L_j \equiv 0,\ B \equiv 0,\ \gamma_j \equiv 0$. This implies that $\Xi_j < 0$ under the same requirements.

Next, consider the performance measure

$$J = \sum_{j=1}^{n_s} \sum_{j=0}^{\infty} \left(z_j^t(k) z_j(k) - \gamma^2 \omega_j^t(k) \omega_j(k) \right)$$

For any $\omega_j(k) \in \ell_2(0,\infty) \neq 0$ and zero initial condition $x_{jo} = 0$ (hence $V_j(0) = 0$), we have

$$\begin{aligned} J &= \sum_{j=1}^{n_s} \sum_{j=0}^{\infty} \left(z_j^t(k) z_j(k) - \gamma^2 \omega_j^t(k) \omega_j(k) \ + \Delta V_j(k)|_{(7.89)} \right) \\ &- \sum_{j=1}^{n_s} \sum_{j=0}^{\infty} \Delta V_j(k)|_{(7.89)} \\ &= \sum_{j=1}^{n_s} \sum_{j=0}^{\infty} \left(z_j^t(k) z_j(k) - \gamma^2 \omega_j^t(k) \omega_j(k) \ + \Delta V_j(k)|_{(7.89)} \right) \\ &- V_j(\infty) \\ &\leq \sum_{j=1}^{n_s} \sum_{j=0}^{\infty} \left(z_j^t(k) z_j(k) - \gamma^2 \omega_j^t(k) \omega_j(k) \right. \\ &+ \left. \Delta V_j(k)|_{(7.89)} \right) \end{aligned} \tag{7.110}$$

where $\Delta V_j(k)|_{(7.89)}$ defines the Lyapunov difference along the solutions of system (7.89) with $u_j(k) \equiv 0$. On considering (7.90), (7.100), (7.103), and (7.110), it can easily be shown by algebraic manipulations that

$$\begin{aligned} & z_j^t(k) z_j(k) - \gamma^2 \omega_j^t(k) \omega_j(k) \ + \Delta V_j(k)|_{(7.89)} = \\ & \chi_j^t(k) \ \widehat{\Xi}_j \ \chi_j^t(k) \end{aligned} \tag{7.111}$$

where $\widehat{\Xi}_j$ is given in (7.100) by Schur complements [29]. It is readily seen that

$$z_j^t(k) z_j(k) - \gamma^2 \omega_j^t(k) \omega_j(k) \ + \Delta V_j(k)|_{(7.89)} < 0$$

for arbitrary $j \in [0, \infty)$, which implies for any $\omega_j(k) \in \ell_2(0,\infty) \neq 0$ that $J < 0$. This eventually leads to $||z_j(k)||_2 \ < \ \gamma \ ||\omega_j(k)||_2$ and hence the proof is completed. The next theorem provides expression for reliable decentralized feedback gains.

Theorem 7.3.2 *Given the bounds* $d_j^+ > 0, \ d_j^- > 0, \ \eta_j^+ > 0, \ \eta_j^- > 0, \ j = 1, ..., n_s$. *The closed-loop subsystem* $\mathsf{S_j}$ *(7.96)-(7.98) is delay-dependent asymptotically stable with an* $\ell_2 - gain < \gamma_j$ *if there exist weighting matrices*

$0 < \mathcal{X}_j,\ \mathcal{Y}_j,\ 0 < \Psi_j,\ 0 < \Lambda_{jm},\ j = 1, ..., n_s,\ m = 1, ..., n_s$ *and scalars* $B_j > 0,\ \psi_j > 0,\ \gamma_j > 0$ *satisfying the following LMIs:*

$$\widehat{\Pi}_j = \begin{bmatrix} \Pi_{aj} & \Pi_{cj} & \widehat{\Pi}_{zj} \\ \bullet & -\widehat{\Pi}_{oj} & \widehat{\Pi}_{wj} \\ \bullet & \bullet & -I_j \end{bmatrix} < 0 \tag{7.112}$$

where

$$\begin{aligned}
\widehat{\Pi}_{aj} &= \begin{bmatrix} \Pi_{1j} & 0 & 0 & 0 \\ \bullet & -\mathcal{Q}_j & 0 & 0 \\ \bullet & \bullet & -I_j & 0 \\ \bullet & \bullet & \bullet & -I_j \end{bmatrix}, \\
\Pi_{cj} &= \begin{bmatrix} 0 & 0 & \Pi_{2j} & \mathcal{Y}_j^t \Gamma_j^t & \mathcal{X}_j F_j^t & 0 \\ 0 & 0 & \mathcal{X} D_j^t & 0 & 0 & \mathcal{X}_j E_{dj}^t \\ 0 & 0 & B_j^t & 0 & 0 & 0 \\ 0 & 0 & I_j & 0 & 0 & 0 \end{bmatrix}, \\
\widehat{\Pi}_{oj} &= \begin{bmatrix} \widehat{\Pi}_{o1j} & 0 \\ \bullet & \Pi_{o2j} \end{bmatrix}, \\
\widehat{\Pi}_{o1j} &= \begin{bmatrix} -\gamma^2 I_j & 0 & \Omega_j^t \\ \bullet & -\sum_{m=1, m\neq j}^{n_s} \Lambda_{jm} & 0 \\ \bullet & \bullet & -\mathcal{X}_j \end{bmatrix},
\end{aligned}$$

$$\begin{aligned}
\Pi_{o2j} &= \begin{bmatrix} -I_j & 0 & 0 \\ \bullet & -B_j I & 0 \\ \bullet & \bullet & -\psi_j I \end{bmatrix}, \\
\widehat{\Pi}_{zj} &= \begin{bmatrix} \Pi_{3j} \\ \mathcal{X}_j H_j^t \\ L_j^t \\ 0 \end{bmatrix}, \quad \widehat{\Xi}_{wj} = \begin{bmatrix} B_j^t \\ 0 \\ 0 \\ 0 \\ 0 \\ 0 \end{bmatrix}
\end{aligned} \tag{7.113}$$

$$\begin{aligned}
\Pi_{1j} &= -\mathcal{X}_j + d_j^* \Psi_j + \sum_{m=1, m\neq j}^{n_s} \Lambda_{jm} \\
\Pi_{2j} &= \mathcal{X}_j A_j^t + \mathcal{Y}_j^t \Sigma_j^t B_j^t, \\
\Pi_{3j} &= \mathcal{X}_j G_j^t + \mathcal{Y}_j^t \Sigma_j^t L_j^t
\end{aligned} \tag{7.114}$$

Moreover, the local gain matrix is given by $K_j = \mathcal{Y}_j \mathcal{X}_j^{-1}$.

Proof: Applying the congruent transformation

$$T = diag[\mathcal{X}_j,\ \mathcal{X}_j,\ I_j,\ I_j,\ I_j,\ \mathcal{X}_j,\ I_j,\ I_j,\ I_j,\ I_j,\ I_j]$$

with $\mathcal{X}_j = \mathcal{P}_j^{-1}$ to LMI (7.100) with Schur complements and using the linearizations

$$\mathcal{Y}_j = K_j\mathcal{X}_j,\ \Psi_j = \mathcal{X}_j\mathcal{Q}_j\mathcal{X}_j,\ \Lambda_{mj} = \mathcal{X}_j\mathcal{Z}_{mj}\mathcal{X}_j$$

we readily obtain LMI (7.112) by Schur complements and therefore the proof is completed.

Remark 7.3.3 *The optimal performance-level* $\gamma_j,\ j = 1,..,n_s$ *for the interconnected system can be determined for decentralized reliable state feedback by solving the following convex optimization problem:*
Problem A

$$\begin{aligned} & For\ \ j,m = 1,...,n_s, \\ & \quad Given\ \ d_j^+ > 0,\ d_j^- > 0,\ \eta_j^+ > 0,\ \eta_j^- > 0 \\ & \quad \min_{\mathcal{X}_j,\ \mathcal{Y}_j,\ \Psi_j,\ \{\Lambda_{jm}\}_{m=1}^{n_s}} \quad \gamma_j \\ & \quad subject\ to\ LMI(7.112) \end{aligned} \tag{7.115}$$

7.3.4 Illustrative example 7.3

To illustrate the usefulness of the results of the developed approach, let us consider the following interconnected time-delay system which is composed of three subsystems. With reference to (7.96)-(7.98), the data values are:

$$Subsystem\ 1:$$

$$A_1 = \begin{bmatrix} 0.0 & 0.1 \\ 0.5 & -0.5 \end{bmatrix},\ D_1 = \begin{bmatrix} 0.1 & 0.0 \\ 0.01 & 0.1 \end{bmatrix},$$

$$B_1 = \begin{bmatrix} 0 \\ 0.05 \end{bmatrix},\ \Omega_1 = diag\begin{bmatrix} 0.4 & 0.6 \end{bmatrix},$$

$$F_1 = \begin{bmatrix} 0 & 0.1 \\ 0.01 & 0.02 \end{bmatrix},\ E_{d1} = \begin{bmatrix} 0.03 & 0.1 \\ 0 & 0.04 \end{bmatrix}$$

$$Subsystem\ 2:$$

$$A_c = \begin{bmatrix} 0 & 0 & 0.1 \\ 0.1 & -0.6 & 0.1 \\ 0.3 & 0.7 & 0.9 \end{bmatrix}, \ D_2 = \begin{bmatrix} 0.0 & 0.2 & 0.0 \\ 0.0 & 0.1 & -0.2 \\ 0.1 & -0.3 & -0.2 \end{bmatrix},$$

$$B_2 = \begin{bmatrix} 0.03 \\ 0 \\ 0.04 \end{bmatrix}, \ E_{d2} = \begin{bmatrix} 0.05 & 0 & 0.01 \\ 0 & 0.01 & 0 \\ 0.01 & 0.04 & 0.02 \end{bmatrix},$$

$$F_2 = \begin{bmatrix} 0 & 0.1 & 0 \\ 0.01 & 0 & 0.02 \\ 0.03 & 0.04 & 0 \end{bmatrix}, \ \Omega_c = diag \begin{bmatrix} 0.3 & 0.5 & 0.7 \end{bmatrix}$$

$$Subsystem\ 3:$$

$$A_3 = \begin{bmatrix} 0.0 & 0.2 \\ -0.6 & 0.6 \end{bmatrix}, \ D_3 = \begin{bmatrix} 0.0 & 0.2 \\ 0.1 & -0.1 \end{bmatrix},$$

$$B_3 = \begin{bmatrix} 0.1 \\ 0.1 \end{bmatrix}, \ \Omega_3 = diag \begin{bmatrix} 0.5 & 0.5 \end{bmatrix},$$

$$F_3 = \begin{bmatrix} 0.02 & 0.01 \\ 0.01 & 0.03 \end{bmatrix}, \ E_{d3} = \begin{bmatrix} 0.04 & 0.05 \\ 0.1 & 0.04 \end{bmatrix}$$

In order to illustrate the effectiveness of our reliable control algorithm, we initially consider the nominal control design in which no actuator failures will occur. Then, by using the data values $\Sigma_1 = 1, \ \Sigma_c = 1, \ \Sigma_3 = 1, \ \Gamma_1 = 0, \ \Gamma_c = 0, \ \Gamma_3 = 0$ and solving **Problem A** using MATLAB®-LMI solver, it is found that the feasible solution of **Theorem 7.3.2** yields the following control gain matrices and performance levels:

$$\begin{aligned} \gamma_1 &= 1.7874, \ K_1 = \begin{bmatrix} -5.2132 & -3.6224 \end{bmatrix}, \\ \gamma_c &= 3.1742, \ K_2 = \begin{bmatrix} -3.2444 & -2.3354 & -2.9664 \end{bmatrix}, \\ \gamma_3 &= 2.0297, \ K_3 = \begin{bmatrix} -2.4243 & -1.3235 \end{bmatrix} \end{aligned} \tag{7.116}$$

Next, we study different failure scenarios. The first scenario is described by the following model:

$$\begin{aligned} u_1^f(k) &= 0.6\, u_1(k) + \beta_1(u_1), \ \beta_1^2(u_1) \le 0.01\, u_1^2, \\ u_2^f(k) &= 0.8\, u_2(k) + \beta_c(u_2), \ \beta_2^2(u_2) \le 0.09\, u_2^2, \\ u_3^f(k) &= 0.4\, u_3(k) + \beta_3(u_3), \ \beta_3^2(u_3) \le 0.04\, u_3^2 \end{aligned} \tag{7.117}$$

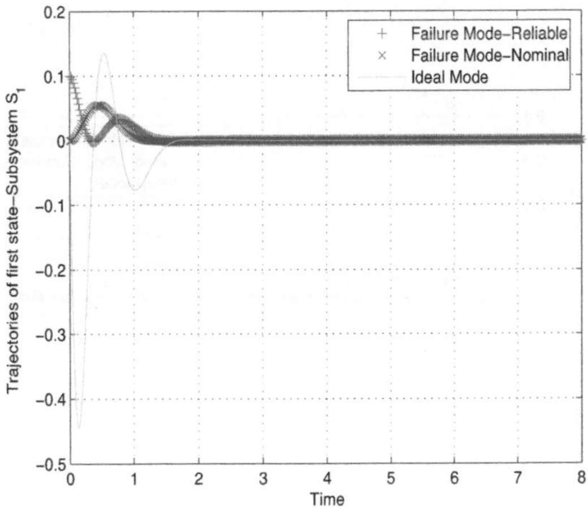

Figure 7.14: Trajectories of first state-Subsystem 1

which corresponds to allowing a failure of the order of 60% in the actuator of system S_j with an error of the order of 10%. In a similar way, the tolerances allowed for other actuators are interpreted in the same way. Thus we have $\Sigma_1 = 0.6,\ \Gamma_1 = 0.1,\ \Sigma_c = 0.8,\ \Gamma_c = 0.3,\ \Sigma_3 = 0.4,\ \Gamma_3 = 0.2$ and upon solving **Problem A** again, the resulting feasible solution is summarized by

$$\begin{aligned} \gamma_1 &= 1.9492,\ K_1 = \left[\begin{array}{cc} 2.4945 & -4.6355 \end{array}\right], \\ \gamma_c &= 3.7842,\ K_2 = \left[\begin{array}{ccc} -1.3947 & -4.5364 & -2.5472 \end{array}\right], \\ \gamma_3 &= 2.4123,\ K_3 = \left[\begin{array}{cc} -0.1177 & 2.1270 \end{array}\right] \end{aligned} \tag{7.118}$$

Using numerical simulation, three typical experiments were performed for three operational modes.

1) *Ideal mode*, where the feedback controllers (7.116) were implemented in an operation without failures.

2) *Failure mode*, where the feedback controllers (7.116) were implemented in an operation under the failures described in (7.117).

3) *Reliable control failure mode*, where the reliable controllers (7.118) were implemented in an operation under the failures described in (7.117). Simulation results of the first state trajectories of the three subsystems are displayed in Figures 7.14 through 7.16 for the prescribed modes of operation. The trajectories of the state-feedback control input for the first subsystem are depicted in Figure 7.17. It is observed that the nominal control clearly deteriorates its ideal performance when operating under failures. When using the reliable control, we notice that the state response results drastically improved in spite of work-

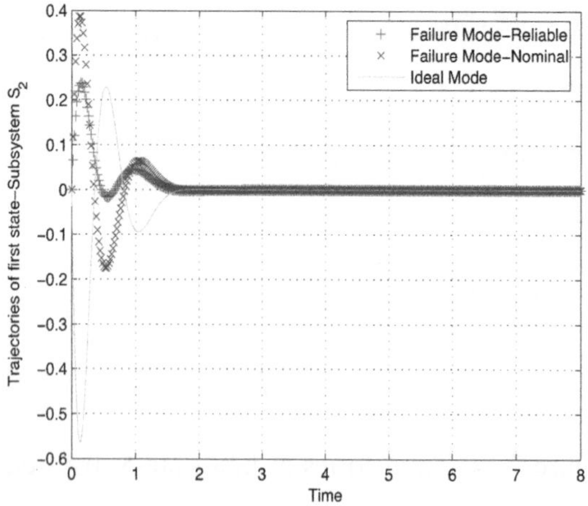

Figure 7.15: Trajectories of first state-Subsystem 2

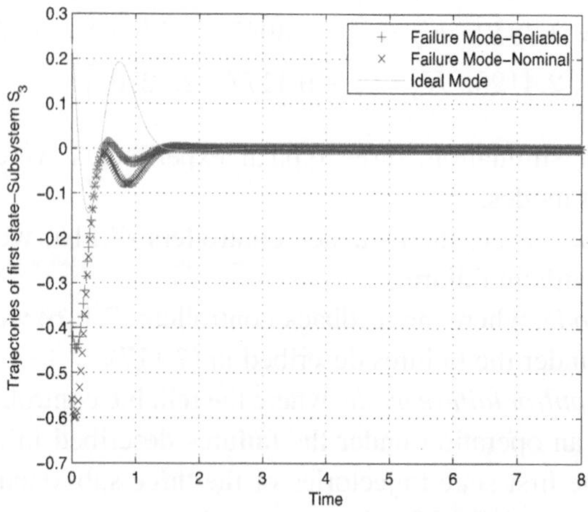

Figure 7.16: Trajectories of first state-Subsystem 3

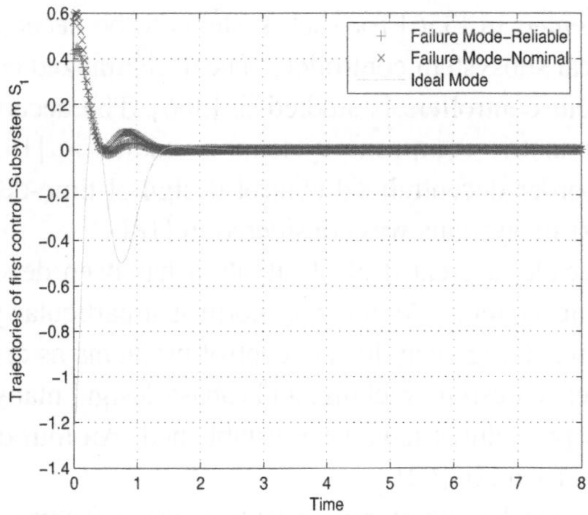

Figure 7.17: Trajectories of control input-Subsystem 1

ing under the same failures, being very close to the one obtained in the ideal case. As Figure 7.17 shows, the control effort in this case has a bigger initial magnitude than for the nominal cases, quickly approaching the ideal control signals. These results illustrate the satisfactory behavior of the proposed reliable control scheme.

7.4 Reliable Control of Symmetric Composite Systems

Symmetric composite systems are those composed of identical subsystems that are symmetrically interconnected. The motivation for studying this class of systems is due to its very diverse application areas, such as in electric power systems, industrial manipulators, and computer networks [107, 191, 356]. In recent years there has been a great interest in studying symmetric composite systems. It is shown that many analysis and synthesis problems for symmetric composite systems can be simplified because of their special structure. In [191] the state-space model of symmetric composite systems is proposed and some fundamental properties of the systems are investigated. For centralized control systems, the output regulation for symmetric composite systems is treated in [180]. The model reduction problem was considered in [160]. Methods of $\mathcal{H}_2$ and $\mathcal{H}_\infty$ optimal control were studied in [107]. In [383], the reliable control of such systems was analyzed. For the decentralized control of symmetric composite systems [191], it is proved that the system has no decentralized fixed modes if and only if it is completely controllable and observable. A sufficient condition is presented in [356] for such systems to be decentralized stabilizable using identical subsystem controllers. The decentralized control using two kinds of subsystem controllers is studied in [366]. The decentralized control for uncertain symmetric composite systems is studied in [14, 366]. Furthermore, a reduced-order decentralized control design of time-delayed uncertain symmetric composite systems was considered in [18].

In the last decade, a great deal of attention has been devoted to the $\mathcal{H}_\infty$ control of dynamic systems. As the $\mathcal{H}_\infty$-norm is a particularly useful performance measure in solving such diverse control problems as disturbance rejection, model reference design, tracking, and robust design, many corresponding important design procedures have been established. An introduction of some standard results is found in [62].

Sometimes, control systems may result in unsatisfactory performance or even instability in the event of control component failures. Recently, in [366], the design of reliable control systems was considered. The resulting control systems provide guaranteed stability and satisfy an $\mathcal{H}_\infty$-norm disturbance attenuation bound not only when all control components are operational, but also in the case of control-channel outages in the systems. The outages were restricted to occur within a preselected subset of available measurements or control inputs. In [385], the reliable control is studied using redundant controller using a simple design approach.

In [117], the state feedback fault-tolerant decentralized $\mathcal{H}_\infty$ control problem for symmetric composite systems is studied. By using the special structure of the systems, a state feedback decentralized $\mathcal{H}_\infty$ control law is constructed by design state feedback $\mathcal{H}_\infty$ controllers for two modified subsystems. Since the full state of the control systems are generally not available in practice, to study the more general output feedback problem is of great importance.

This section is concerned with the dynamic output feedback decentralized $\mathcal{H}_\infty$ control and reliability analysis of symmetric composite systems. First, it is shown that the dynamic output feedback decentralized $\mathcal{H}_\infty$ control problem is equivalent to a simultaneous $\mathcal{H}_\infty$ control problem of two modified subsystems. A design procedure based on the simultaneous $\mathcal{H}_\infty$ control method given in [34] is presented. The reliability considered in this paper concerns the largest number of control inputs or measurements failures that will keep the closed-loop system stable and maintain the required level of performance. By exploiting the special structure of the systems, it is shown that the computation of the poles and the $\mathcal{H}_\infty$-norm of the resulting closed-loop system when control-channel outages occur can be reduced to the computation of the poles and the $\mathcal{H}_\infty$-norm of three auxiliary systems. The order of one auxiliary system is twice that of any isolated subsystem, while the orders of the other two auxiliary systems are equal to that of any isolated subsystem. Thus the tolerance to control input failures and measurement failures can be tested more efficiently.

7.4.1 Problem formulation

The symmetric composite system under consideration consists of N subsystems; the *jth* subsystem is described by

$$\begin{aligned}
\dot{x}_j &= A_o x_j + \sum_{k=1,k\neq j}^{N} A_c x_k + \Gamma w_j + \sum_{k=1,k\neq j}^{N} \Phi w_k + B u_j \\
z_j &= C x_j + D u_j \\
y_j &= G x_j + F w_j
\end{aligned} \qquad (7.119)$$

where $j = 1, 2, \ldots, N$ and $x_j \in \Re^n, u_j \in \Re^m, w_j \in \Re^r, z_j \in \Re^s, y_j \in \Re^p$ are the state, control input, exogenous input, penalty and measured variables, respectively, and $A_o \in \Re^{nxn}, A_c \in \Re^{nxn}, \Gamma \in \Re^{nxr}, \Phi \in \Re^{nxr}, B \in \Re^{nxm}, C \in \Re^{sxn}, D \in \Re^{sxm}, G \in \Re^{pxn}, F \in \Re^{pxr}$.

Then the overall system is given by

$$\begin{aligned}
\dot{x} &= A_s x + \Gamma_s w + B_s u \\
z &= C_s x + D_s u \\
y &= G_s x + F_s w
\end{aligned} \qquad (7.120)$$

where

$$\begin{aligned}
x &= \left[x_1^t, \ldots, x_N^t\right], \; u = \left[u_1^t, \ldots, u_N^t\right], \; w = \left[u_1^t, \ldots, w_N^t\right], \\
z &= \left[z_1^t, \ldots, z_N^t\right], \; y = \left[y_1^t, \ldots, y_N^t\right]
\end{aligned}$$

and

$$\begin{aligned}
A_s &\in \Re^{Nn\times Nn}, \; \Gamma_s \in \Re^{Nn\times Nr}, \; B_s \in \Re^{Nn\times Nm}, \; C_s \in \Re^{Ns\times Nn}, \\
D_s &\in \Re^{Ns\times Nm}, \; G_s \in \Re^{Np\times Nn}, \; F_s \in \Re^{Np\times Nr}
\end{aligned}$$

have the following structure:

$$\begin{aligned}
A_s &= \begin{bmatrix} A_o & A_c & \cdots & A_c \\ A_c & A_o & \cdots & A_c \\ \vdots & \vdots & \ddots & \vdots \\ A_c & A_c & \cdots & A_o \end{bmatrix}, \quad \Gamma_s = \begin{bmatrix} \Gamma & \Phi & \cdots & \Phi \\ \Phi & \Gamma & \cdots & \Phi \\ \vdots & \vdots & \ddots & \vdots \\ \Phi & \Phi & \vdots & \Gamma \end{bmatrix}, \\
B_s &= Blockdiag[B, \cdots, B], \quad C_s = Blockdiag[C, \cdots, C], \\
D_s &= Blockdiag[D, \cdots, D], \quad G_s = Blockdiag[G, \cdots, G], \\
F_s &= Blockdiag[F, \cdots, F]
\end{aligned}$$

Remark 7.4.1 *It is quite evident from the structure of matrices A_s, ..., F_s that the individual subsystems have similar dynamics and the coupling pattern is similar too among different subsystems. This is the basis for the designation of symmetric composite systems.*

In what follows, we examine the decentralized H_∞ control problem that is considered based on dynamic output feedback and formally described by:

Given $\gamma > 0$, find a decentralized dynamic output feedback controller

$$\left\{ \begin{array}{l} \zeta_j = F_c\zeta_j + G_c y_j \\ u_j = H_c\zeta_j \end{array} \right. \tag{7.121}$$

where $\zeta_j \in \Re^v$, $j = 1, \cdots, N$, such that
(C1): *The matrix*

$$\begin{bmatrix} A_s & B_s\mathsf{H} \\ \mathsf{G}C_2 & \mathsf{F} \end{bmatrix},$$

is stable, where

$$\begin{aligned} \mathsf{H} &= \textit{Block diag}[H_c, \cdots, H_c],\ \mathsf{G} = \textit{Block diag}[G_c, \cdots, G_c], \\ \mathsf{F} &= \textit{Block diag}[F_c, \cdots, F_c] \end{aligned}$$

(C2): *The transfer function matrix $T^c(s)$ of the closed-loop system,*

$$\left\{ \begin{array}{l} \dot{x}_j = A_s x + B_s\mathsf{H}\zeta + B_s w \\ \dot{\zeta} = \mathsf{HF}\zeta + \mathsf{G}G_s x + \mathsf{G}F_s w \\ z = C_s x + D_s\mathsf{H}\zeta \end{array} \right.$$

where $\zeta = [\zeta_1^t, \cdots, \zeta_N^t]^t$, satisfies $\|T^c(s)\| < \gamma$.

Remark 7.4.2 *It should be emphasized that the feedback controllers (7.121) are identical for all subsystems, thereby exploiting the structural properties of system (7.120) to reduce the complexity of the controller design.*

We now examine the reliability of the closed-loop system composed of system (7.120) and decentralized controller of the form (7.121). To achieve our goal, we will study the tolerance to control-channel outages, which include control input failures and measurement failures. In this course, we consider the *jth* control input failure takes the form $u_j = 0$ and the *jth* measurement failure is modeled as $y_k = 0,\ j, k \in \{1, \cdots, N\}$.

Following [117], for a positive integer p, we denote

$$m_k = [1\ v_k\ v_k^2\ \cdots v_k^{p-1}]^t \quad k = 1, 2, \cdots, p$$

where $v_k = \exp\left(2\Pi(k-1)\sqrt{1}/p\right),\ k = 1, 2, \cdots, p$. This implies that v_k is a root of the equation $v_p = 1$. For simplicity in exposition, we denote

$$r_1 = m_1 = [1\ 1\ \cdots\ 1]^t, \quad r_{\frac{p}{2}} + 1 = m_{\frac{p}{2}} + 1$$

if p is an even number. Let $t = \frac{p+1}{2}$ if p is odd and $t = \frac{p}{2}$ if p is even. Then for $j = 2, 3, \cdots, t$, we define

$$r_j = \frac{1}{\sqrt{2}}(m_j + m_{p+2-j}), r_{p+2-j} = \frac{\sqrt{-1}}{\sqrt{2}}(m_j - m_{p+2-j})$$

Finally, introduce

$$R_p = \frac{1}{\sqrt{p}}[r_1\ r_2\ \cdots\ r_p] \tag{7.122}$$

$$T_{pj} = R_p \otimes I_j \tag{7.123}$$

where $\otimes$ denotes the Kronecker product.

7.4.2 Decentralized $\mathcal{H}_\infty$ control

To simplify the notations, we denote

$$A_\alpha = A_o + (N-1)A_c, \quad A_\beta = A_o - A_c, \tag{7.124}$$

$$\Gamma_\alpha = \Gamma + (N-1)\Phi, \quad \Gamma_\beta = \Gamma - \Phi \tag{7.125}$$

then from (7.123) and the results of [117], there exist appropriate transformations

$$T_{Nn},\ T_{Nm},\ T_{Np},\ T_{Nr},\ T_{Ns}$$

leading to

$$\begin{aligned}
T_{Nn}^{-1} A_s T_{Nn} &= \text{diag}[A_\alpha, A_\beta, \cdots, A_\beta], \\
T_{Nn}^{-1} \Gamma_s T_{Nr} &= \text{diag}[\Gamma_\alpha,\ \Gamma_\beta, \cdots, \Gamma_\beta], \\
T_{Nn}^{-1} B_s T_{Nm} &= Blockdiag[B, \cdots, B], \\
T_{Ns}^{-1} C_s T_{Nn} &= Blockdiag[C, \cdots, C], \\
T_{Np}^{-1} G_s T_{Nn} &= \text{diag}[G, \cdots, G], \\
T_{Ns}^{-1} D_s T_{Nm} &= Blockdiag[D, \cdots, D], \\
T_{Np}^{-1} F_s T^{Nr} &= \text{diag}[F, \cdots, F]
\end{aligned} \tag{7.126}$$

and further we denote

$$\begin{aligned} \mathsf{A}^c &= \begin{bmatrix} A_s & B_s\mathsf{H} \\ G_cG_s & \mathsf{H} \end{bmatrix}, \quad \mathsf{A}^c_\alpha = \begin{bmatrix} A_\alpha & BH_c \\ G_cG & F_c \end{bmatrix}, \qquad (7.127) \\ \mathsf{A}^c_\beta &= \begin{bmatrix} A_\beta & BH_1c \\ G_cG & F_c \end{bmatrix}, \\ W &= sI - A^c,\ W_\alpha = sI - A^c_\alpha,\ , W_\beta = sI - A_c \qquad (7.128) \end{aligned}$$

Then the following lemma holds [117].

Lemma 7.4.3 *There exists a permutation matrix P such that*

$$\mathsf{P}^{-1} \begin{bmatrix} T_{Nn}^{-1} & 0 \\ 0 & T_{Nv}^{-1} \end{bmatrix} W^{-1} \begin{bmatrix} T_{Nn} & 0 \\ 0 & T_{Nv} \end{bmatrix} \mathsf{P} = diag \left[W_\alpha^{-1},\ W_\beta^{-1}, \cdots, W_\beta^{-1} \right]$$

Proof 7.4.4 *Let* P *be the permutation matrix such that*

$$\begin{aligned} &\mathsf{P}^{-1} \begin{bmatrix} diag[A_\alpha, A_\beta, \cdots, A_\beta] & diag[BH_c, \cdots, BH_c] \\ diag[G_cG, \cdots, G_cG] & diag[F_c, \cdots, F_c] \end{bmatrix} \mathsf{P} \\ &= diag \left\{ \begin{bmatrix} A_\alpha & BH_c \\ G_cG & F_c \end{bmatrix}, \begin{bmatrix} A_\beta & BH_c \\ G_cG & F_c \end{bmatrix}, \cdots, \begin{bmatrix} A_\beta & BH_c \\ G_cG & F_c \end{bmatrix} \right\} \end{aligned}$$

Then from (7.126), we have

$$\begin{aligned} &\mathsf{P}^{-1} \begin{bmatrix} T_{Nn}^{-1} & 0 \\ 0 & T_{Nv}^{-1} \end{bmatrix} W^{-1} \begin{bmatrix} T_{Nn} & 0 \\ 0 & T_{Nv} \end{bmatrix} \mathsf{P} \\ &= \left\{ \mathsf{P}^{-1} \begin{bmatrix} T_{Nn}^{-1} & 0 \\ 0 & T_{Nv}^{-1} \end{bmatrix} W \begin{bmatrix} T_{Nn} & 0 \\ 0T_{Nv} & \end{bmatrix} \mathsf{P} \right\}^{-1} \\ &= \mathsf{P}^{-1} diag[sI - A_\alpha, sI - A_\beta, \cdots, sI - A_\beta] \\ &\quad diag[-G_1G, \cdots, -G_1G] \\ &\quad diag[-BH_c, \cdots, -BH_c] \\ &\quad diag[sI - F_c, \cdots, sI - F_c]\mathsf{P} \\ &diag \begin{bmatrix} sI - A_\alpha & -BH_c \\ -G_cG & sI - F_c \end{bmatrix}, \begin{bmatrix} sI - A_\beta & -BH_c \\ -G_cG & sI - F_c \end{bmatrix}, \\ &\quad , \cdots, \begin{bmatrix} sI - A_\beta & -BH_c \\ -G_cG & sI - F_c \end{bmatrix} \\ &\quad diag \left[W_\alpha^{-1}, W_\beta^{-1}, \cdots, W_\beta^{-1} \right] \end{aligned}$$

Some manipulations yield the desired results and thus the proof is completed.

Exploiting the special structure of system (7.120), the the decentralized $\mathcal{H}_\infty$ control (DHC) problem under consideration will be shown to be simplified to a simultaneous $\mathcal{H}_\infty$ control problem for two modified subsystems. The main result is established by the following theorem.

Theorem 7.4.5 *A decentralized dynamic output feedback controller of the form (7.121) satisfies conditions* (**C1**) *and* (**C2**) *if and only if it satisfies the following two conditions:*
(C3): A_α^c *and* A_β^c *are stable.*
(C4): $\|T_\alpha^c(s)\|_\infty \leq \gamma$ *and* $\|T_\beta^c(s)\|_\infty \leq \gamma$
where

$$\begin{aligned} T_\alpha^c(s) &= [C \quad DH_c]W_\alpha^{-1}\begin{bmatrix}\Gamma_\alpha \\ G_c\mathsf{F}\end{bmatrix} \\ &\textit{and} \\ T_\beta^c(s) &= [C \quad DH_c]W_\beta^{-1}\begin{bmatrix}\Gamma_\beta \\ G_c\mathsf{F}\end{bmatrix} \end{aligned} \qquad (7.129)$$

Proof 7.4.6 (**C1**) $\Leftrightarrow$ (**C3**)*: From (7.126), we have*

$$\begin{aligned} & spec\left\{\begin{bmatrix} A_s & B_s\mathsf{H} \\ GG_s & \mathsf{F}\end{bmatrix}\right\} \\ = \; & spec\left\{\begin{bmatrix} T_{Nn}^{-1} & 0 \\ 0 & T_{Nv}^{-1}\end{bmatrix}\begin{bmatrix} A_s & B_s\mathsf{H} \\ GG_s & \mathsf{F}\end{bmatrix}\begin{bmatrix} T_{Nn} & 0 \\ 0 & T_{Nv}\end{bmatrix}\right\} \\ = \; & spec\left\{\begin{bmatrix} T_{Nn}^{-1}A_sT_{Nn} & T_{Nn}^{-1}B_sT_{Nm}T_{Nm}^{-1}\mathsf{H}T_{Nv} \\ T_{Nv}^{-1}GT_{Np}T_{Np}^{-1}G_sT_{Nn} & T_{Nv}^{-1}\mathsf{F}T_{Nv}\end{bmatrix}\right\} \\ = \; & spec\left\{\begin{bmatrix} diag[A_\alpha, A_beta, \cdots, A_\beta] & diag[BH_c, \cdots, BH_c] \\ diag[G_cG, \cdots, G_cG] & diag[F_c, \cdots, F_c]\end{bmatrix}\right\} \\ = \; & spec\left\{Block\ diag\left\{\begin{bmatrix} A_\alpha & BH_1 \\ G_cG & F_c\end{bmatrix}, \begin{bmatrix} A_\beta & BH_c \\ G_cG & F1\end{bmatrix}, \cdots,\right.\right. \\ & \left.\left.\begin{bmatrix} A_\beta & BH_c \\ G_cG & F_c\end{bmatrix}\right\}\right\} \end{aligned}$$

Hence (**C1**) *holds if and only if* (**C3**) *holds.*

(**C2**)$\Longleftrightarrow$(**C4**)*: From (7.123), it is easy to see that T_{pj} is an orthogonal matrix for integers $p \geq 2$ and $j \geq 1$. Since $T^c(s) = [C_s \quad D_s\mathsf{H}]W^{-1}\begin{bmatrix}\Gamma_s \\ GF_s\end{bmatrix}$, premultiplication or postmultiplication of $T^c(s)$ by orthogonal matrices will* *leave the H_∞-norm intact. Hence we have*

$$
\begin{aligned}
\|T^c(s)\|_\infty &= \|T_{Ns}^{-1}T^c(s)T_{Nr}\|_\infty \\
&= \|T_{Ns}^{-1}[C_s \quad D_s\mathsf{H}]\begin{bmatrix} T_{Nn} & 0 \\ 0 & T_{Nv}\end{bmatrix}\mathsf{P}\mathsf{P}^{-1}\begin{bmatrix} T_{Nn}^{-1} & 0 \\ 0 & T_{Nv}^{-1}\end{bmatrix}W^{-1} \\
&\quad \times \begin{bmatrix} T_{Nn} & 0 \\ 0 & T_{Nv}\end{bmatrix}\mathsf{P}\mathsf{P}^{-1}\begin{bmatrix} T_{Nn}^{-1} & 0 \\ 0 & T_{Nv}^{-1}\end{bmatrix}\begin{bmatrix}\Gamma_s \\ GF_s\end{bmatrix}T_{Nr}\|_\infty
\end{aligned}
$$

Then from ***Lemma 7.4.3*** *we have*

$$
\begin{aligned}
\|T^c(s)\|_\infty &= \|[Block\ diag[C,\cdots,C] \quad Block\ diag[DH_c,\cdots,DH_c]]\mathsf{P} \\
&\quad \times Blockdiag[W_\alpha^{-1},W_\beta^{-1},\cdots,W_\beta^{-1}]\mathsf{P}^{-1} \\
&\quad \times \begin{bmatrix} Blockdiag[\Gamma_\alpha,\Gamma_\beta,\cdots,\Gamma_\beta] \\ Blockdiag[G_c\mathsf{F},\cdots,G_c\mathsf{F}]\end{bmatrix}\|_\infty \\
&= \|\{Blockdiag[C \quad DH_1],\cdots,[C \quad DH_1]\} \\
&\quad \times [W_\alpha^{-1},W_\beta^{-1},\cdots,W_\beta^{-1}] \\
&\quad \times Blockdiag\left\{\begin{bmatrix}\Gamma_\alpha \\ G_c\mathsf{F}\end{bmatrix},\begin{bmatrix}\Gamma_\beta \\ G_c\mathsf{F}\end{bmatrix},\cdots,\begin{bmatrix}\Gamma_\beta \\ G_c\mathsf{F}\end{bmatrix}\right\}\|_\infty \\
&= \|Blockdiag[T_\alpha^c(s),T_\beta^c(s),\cdots,T_\beta^c(s)]\|_\infty \\
&= \max\{\|T_\alpha^c(s)\|_\infty,\|T_\beta^c(s)\|_\infty\}
\end{aligned}
$$

Thus **(C2)** *holds if and only if* **(C4)** *holds.*

Remark 7.4.7 *It should be noted that* ***Theorem 7.4.5*** *establishes that the DHC problem under consideration is equivalent to the simultaneous $\mathcal{H}_\infty$ control problem of two modified subsystems*

$$
\left\{
\begin{aligned}
\dot{x}_\alpha &= A_\alpha x_j + \Gamma_{s\alpha}w_j + Bu_j \\
z_\alpha &= Cx_\alpha + Du \\
y &= Gx_\alpha + Fw_\alpha \\
\dot{x}_\beta &= A_\beta x_j + \Gamma_{s\beta}w_j + Bu_j \\
z_\beta &= Cx_\alpha + Du \\
y &= Gx_\beta + Fw_\beta
\end{aligned}
\right.
\tag{7.130}
$$

using the same controller

$$
\left\{
\begin{aligned}
\dot{\eta} &= F_{c\eta} + G_c y \\
u &= H_{c\eta}
\end{aligned}
\right.
\tag{7.131}
$$

7.4.3 Controller design

Initially, we make some the standard assumptions:

1. $(A_\alpha, \Gamma_\alpha)$, (A_α, B), (A_β, Γ_β), (A_β, B) are all stabilizable,
2. (C, A_α), (G, A_α), (C, A_β), (G, A_β) are all detectable,
3. $D^t[C \quad D] = [0 \quad I]$,
4. $\begin{bmatrix} \Gamma_\alpha \\ F \end{bmatrix} F^t = \begin{bmatrix} 0 \\ I \end{bmatrix}, \begin{bmatrix} \Gamma_\beta \\ F \end{bmatrix} F^t = \begin{bmatrix} 0 \\ I \end{bmatrix}.$

It should be pointed out that the assumptions are equivalent to those usually applied to (7.120) in a standard $\mathcal{H}_\infty$ synthesis problem and that lead to the following well-known result [62].

Lemma 7.4.8 *For $\gamma > 0$ and $i = \alpha, \beta$, there exists an admissible controller such that $\|T_j^c(s)\|_\infty < \gamma$ if and only if the following three conditions hold:*
(S1) *there exists a $X_{i\infty} \geq 0$, such that*

$$A_j^t X_{j\infty} + X_{j\infty} A_j + X_{j\infty}(\gamma^{-2}\Gamma_{sj}\Gamma_{sj}^t - BB^t)X_{i\infty} + C^tC = 0 \qquad (7.132)$$

(S2) *there exists a $Y_{j\infty} \geq 0$, such that*

$$A_j Y_{j\infty} + Y_{j\infty} A_j^t + Y_{j\infty}(\gamma^{-2}C^tC - G^tG)Y_{j\infty} + \Gamma_{sj}\Gamma_{sj}^t = 0 \qquad (7.133)$$

(S3) $\rho(Y_{j\infty}X_{j\infty}) < \gamma^2$.
Defining

$$\begin{aligned} F_{j\infty} &= B^t X_{j\infty} \\ L_{j\infty} &= (I - \gamma^{-2}Y_{j\infty}X_{j\infty})^{-1}Y_{j\infty}G^t \\ \hat{A}_{j\infty} &= A_j + \gamma^{-2}\Gamma_{j\infty}\Gamma_{j\infty}^t X_{j\infty} - BF_{j\infty} - L_{j\infty}G \\ Z_{j\infty} &= (I - \gamma^{-2}Y_{j\infty}X_{j\infty}) \end{aligned} \qquad (7.134)$$

then all admissible controllers $K_{j\infty}(Q_j(s))$ resulting in $\|T_j^c(s)\|_\infty < \gamma$ are parameterized as in Figure 7.18, where

$$M_{j\infty}(s) \leftrightarrow \begin{bmatrix} \hat{A}_{j\infty} & L_{j\infty} & Z_{i\infty}B \\ -F_{i\infty} & 0 & I \\ -G & I & 0 \end{bmatrix}$$

and $Q_j(s)$ is a stable real-rational transfer function satisfying $\|Q_j(s)\|_\infty < \gamma$.

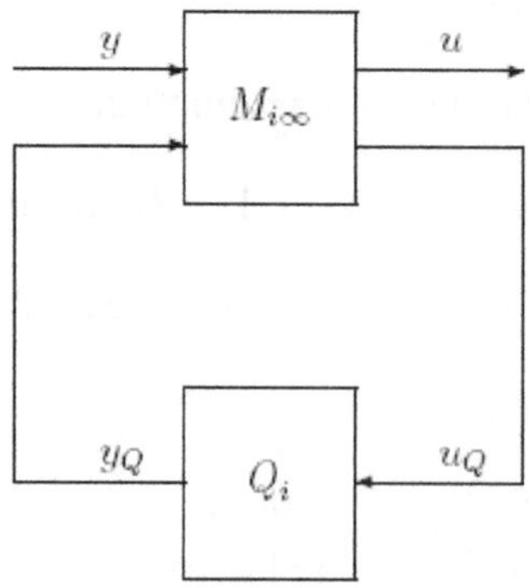

Figure 7.18: Modified system

From **Theorem 7.4.5** and **Lemma 7.4.8**, the necessary conditions for the DHC problem to have a solution are that conditions **(S1)**-**(S3)** hold for $j = \alpha, \beta$. And if these conditions hold, the problem is to find $Q_\alpha(s)$ and $Q_\beta(s)$ such that

$$\|Q_\alpha(s)\|_\infty < \gamma$$

and

$$K_{\alpha\infty}(Q_\alpha(s)) = K_{\beta\infty}(Q_\beta(s))$$

Such a problem was considered in [34] and based on the ensuing results, we have the following theorem.

Theorem 7.4.9 *For $\gamma \geq 0$, if there exist two matrices $N_\alpha \geq 0, N_\beta \geq 0$, such that*

$$\begin{aligned} N_\alpha \tilde{A}_\alpha^t + \tilde{A}_\alpha N_\alpha + N_\alpha \tilde{E}_\alpha N_\alpha + \tilde{R}_\alpha \tilde{\Re}_\alpha^t = 0, \\ N_\beta \tilde{A}_\beta^t + \tilde{A}_\beta N_\beta + N_\beta \tilde{E}_\beta N_\beta + \tilde{R}_\beta \tilde{\Re}_\beta^t = 0 \end{aligned} \tag{7.135}$$

where

$$\begin{aligned} \tilde{A}_\alpha &= \begin{bmatrix} \hat{A}_{\alpha\infty} + Z_{\alpha\infty} B F_{\alpha\infty} & Z_{\alpha\infty} B F_{\beta\infty} - L_{\alpha\infty} G \\ 0 & \hat{A}_{\beta\infty} + L_{\beta\infty} G \end{bmatrix} \\ \tilde{A}_\beta &= \begin{bmatrix} \hat{A}_{\alpha\infty} + L_{\alpha\infty} G & 0 \\ Z_{\beta\infty} B F_{\alpha\infty} - L_{\beta\infty} G & \hat{A}_{\beta\infty} + Z_{\beta\infty} B F_{\beta\infty} \end{bmatrix} \end{aligned}$$

$$\tilde{E}_\alpha = \tilde{E}_\beta = \begin{bmatrix} \gamma^{-2}F^t_{\alpha\infty}F_{\alpha\infty} - G^tG & \gamma^{-2}F^t_{\alpha\infty}F_{\beta\infty} - C^t_{21}G \\ \gamma^{-2}F^t_{\beta\infty}F_{\alpha\infty} - G^tG & \gamma^{-2}F^t_{\beta\infty}F_{\beta\infty} - G^tG \end{bmatrix}$$

$$\tilde{R}_\alpha = \begin{bmatrix} \frac{1}{2}L_{\alpha\infty} \\ \frac{1}{2}L_{\beta\infty} \end{bmatrix} + N_\beta \begin{bmatrix} G^t \\ G^t \end{bmatrix} \tag{7.136}$$

$$\tilde{R}_\beta = \begin{bmatrix} \frac{1}{2}L_{\alpha\infty} \\ -\frac{1}{2}L_{\beta\infty} \end{bmatrix} + N_\alpha \begin{bmatrix} G^t \\ G^t \end{bmatrix} \tag{7.137}$$

then the controller

$$\begin{cases} \dot{\zeta}_j = (\hat{A}^- \hat{L}\bar{C})\zeta_j + \hat{L}y_j \\ u_j = \hat{F}\zeta_j \end{cases} \tag{7.138}$$

is a solution of the DHC problem under consideration, where $j = 1, \cdots, N$,

$$\hat{A} = \begin{bmatrix} \hat{A}_{\alpha\infty} + L_{\alpha\infty}G & 0 \\ 0 & A_{\beta\infty} + L_{\beta\infty}G \end{bmatrix},$$

$$\hat{L} = \begin{bmatrix} \frac{1}{2}L_{\alpha\infty} \\ -\frac{1}{2}L_{\beta\infty} \end{bmatrix} + (N_\alpha + N_\beta) \begin{bmatrix} G^t \\ G^t \end{bmatrix},$$

$$\hat{F} = [-F_{\alpha\infty} \quad - F_{\beta\infty}],$$

$$\bar{C} = [G \quad G]$$

Proof 7.4.10 *It follows from [34], if (7.135) holds, then the controller*

$$\begin{cases} \dot{\eta} = (\hat{A} - \hat{L}\bar{C})\eta + \hat{L}y \\ u = \hat{F}\eta \end{cases}$$

is a simultaneous $\mathcal{H}_\infty$ *suboptimal controller guaranteeing* $\|T^c_\alpha(s)\|_\infty < \gamma$ *and* $\|T^c_\beta(s)\|_\infty < \gamma$. *The desired results then follow from* ***Theorem 7.4.5***.

Remark 7.4.11 *It was demonstrated in [34] that the parameter-embedding method may be used to solve the two equations in (7.135). Alternatively, we may use an iterative method as follows:*
First let $N^{(0)}_\alpha = N^{(0)}_\beta = 0$ *in (7.136) and (7.137), then the two equations in (7.135) can be solved. Use the obtained solutions* $N^{(1)}_\alpha$ *and* $N^{(1)}_\beta$ *to calculate* $\tilde{R}_\alpha$ *and* $\tilde{R}_\beta$ *and solve for* $N^{(2)}_\alpha$ *and* $N^{(2)}_\beta$. *The process is continued until* $N^{(k)}_\alpha$ *and* $N^{(k)}_\beta$ *converge.*

7.4.4 Illustrative example 7.4

Consider the voltage/reactive power behavior of a multimachine power system consisting of several synchronous machines including their PI-voltage controllers, which feed the load through a distribution net [191] with a system model given by

$$\begin{aligned}
\dot{x}_j &= \begin{bmatrix} -2.51 & -0.16 \\ 2.55 & 0 \end{bmatrix} x_j + \sum_{k=1,k\neq i}^{N} \begin{bmatrix} -0.065 & 0 \\ -0.0027 & 0 \end{bmatrix} x_k + \begin{bmatrix} 0.9 \\ -1 \end{bmatrix} u_j \\
&+ \begin{bmatrix} 0 & 0.02 \\ 0 & 0.01 \end{bmatrix} w_j + \sum_{k=1,k\neq j}^{N} \begin{bmatrix} 0 & 0.01 \\ 0 & 0.01 \end{bmatrix} w_k \qquad (7.139) \\
z_j &= \begin{bmatrix} 2 & 0.2 \\ 0 & 0 \end{bmatrix} x_j + \begin{bmatrix} 0 \\ 1 \end{bmatrix} u_j \\
y_j &= [2.54 \quad 0] x_j + [1 \quad 0] w_j \quad (j = 1, \cdots, 20) \qquad (7.140)
\end{aligned}$$

Assume that $\gamma = 0.62$, it is easy to test that assumptions (i)-(iv) hold and hence **Theorem 7.4.9** can be applied to design the controller. Using the iterative method suggested in **Remark 7.4.11** to solve the two equations in (7.135), we obtain the decentralized $\mathcal{H}_\infty$ controller

$$\begin{aligned}
\zeta_j &= \begin{bmatrix} -2.4451 & -0.1597 & -0.0003 & 0 \\ 2.5527 & -0.0002 & 0.0002 & 0 \\ -0.0247 & 0 & -3.6915 & -0.0484 \\ 0.1767 & 0 & 2.6478 & -0.0400 \end{bmatrix} \zeta_j + \begin{bmatrix} 0.0001 \\ -0.0001 \\ 0.0097 \\ 0.0696 \end{bmatrix} y_j \\
u_j &= [0.0002 \quad 0.0002 \quad 0.0549 \quad 0.0788] \zeta_j \qquad (7.141)
\end{aligned}$$

where $j = 1, \ldots, 20$. The resulting closed-loop system is stable and satisfies the performance requirement. In fact, we have

$$|T_\alpha^c(s)\|_\infty = 0.6143 < \gamma, \quad |T_\beta^c(s)\|_\infty = 0.0613 < \gamma$$

7.4.5 Illustrative example 7.5

Consider an open-loop unstable symmetric composite system composed of $N = 20$ subsystems given by

$$\begin{aligned}
\dot{x}_j &= \begin{bmatrix} -2 & -6 \\ 5 & 0 \end{bmatrix} x_j + \sum_{k=1,k\neq i}^{N} \begin{bmatrix} -0.500 & 0.1 \\ -0.002 & 0.1 \end{bmatrix} x_k + \begin{bmatrix} 2 \\ -1 \end{bmatrix} u_j \\
&+ \begin{bmatrix} 0 & 0.02 \\ 0 & 0.01 \end{bmatrix} w_j + \sum_{k=1,k\neq i}^{N} \begin{bmatrix} 0 & 0.01 \\ 0 & 0.01 \end{bmatrix} w_k \qquad (7.142)
\end{aligned}$$

$$
\begin{aligned}
z_j &= \begin{bmatrix} 2 & 0.2 \\ 0 & 0 \end{bmatrix} x_j + \begin{bmatrix} 0 \\ 1 \end{bmatrix} u_j \\
y_j &= [2 \quad 3] x_j + [1 \quad 0] w_j \quad (i = 1, \ldots, 20)
\end{aligned}
$$

Letting $\gamma = 10$, it is easy to test that assumptions (1)-(4) hold and hence **Theorem 7.4.9** can be applied to design the controller. Once again, using the iterative method suggested in **Remark 7.4.11** to solve the two equations in (7.135), we obtain the decentralized $\mathcal{H}_\infty$ controller

$$
\begin{cases}
\dot{\zeta}_j = \begin{bmatrix} -10.2465 & 0.3761 & 0.1692 & 0.2538 \\ 3.8375 & -1.0802 & -0.5880 & -0.8821 \\ 0.0086 & 0.0130 & -4.1214 & -4.8844 \\ 0.0077 & 0.0115 & 6.3247 & -0.6898 \end{bmatrix} \zeta_j + \begin{bmatrix} -0.0846 \\ 0.2940 \\ -0.0043 \\ -0.0038 \end{bmatrix} y_j \\
u_j = [0.5402 \quad 2.1067 \quad -1.3150 \quad 0.6013] \zeta_j
\end{cases}
$$

where $j = 1, \cdots, 20$. To test the effectiveness of the controller, we compute spec(A^c_α), spec (A^c_β), $\|T^c_\alpha(s)\|_\infty$, $\|T^c_\beta(s)\|_\infty$ and obtain

$$
\begin{aligned}
&\text{spec}(A^c_\alpha), \\
&\{-9.7727, \ -10.4685, \ -0.1682, \ -0.5199, \ -2.4043 \pm 5.2885i\} \subset C^-, \\
&\text{spec}(A^c_\beta), \\
&\{-10.3945, \ -0.2453, \ -1.1333 \pm 5.4437i, \ -2.4157 \pm 5.2913i\} \subset C^-, \\
&\|T^c_\alpha(s)\|_\infty = 4.9213 < \gamma, \quad \|T^c_\beta(s)\|_\infty = 2.4335 < \gamma
\end{aligned}
$$

7.4.6 Reliability analysis

In this section, we consider the reliability of the closed-loop system composed of system (7.120) and decentralized controller of the form (7.121). Here the reliability means the ability to tolerate control channel failures such as control-input channel failure or sensor measurement channel failures. For integer $1 \leq \ell \leq N-1$, we denote

$$
\begin{aligned}
U_\ell &= A_o + (\ell - 1) A_c, \quad \tilde{U}_\ell = \sqrt{\ell(N - l)} A_c \\
V_\ell &= \Gamma + (\ell - 1)\Phi, \quad \tilde{V}_\ell = \sqrt{\ell(N - l)} \Phi
\end{aligned}
$$

7.4.7 Control input failures

Since the subsystems (7.119) are all identical, without loss of generality, we may assume that the first l control inputs fail. Then the system matrix of the resulting closed-loop system is

$$
A^c_\ell = \begin{bmatrix} A & (B_s H)_\ell \\ G G_s & F \end{bmatrix}
$$

where

$$(B_sH)_\ell = \text{diag}\left[\overbrace{0,\ldots,0}^{\ell},\overbrace{BH_c,\ldots,BH_c}^{N-\ell}\right]$$

The transfer function matrix of the resulting closed-loop system is

$$T_\ell^c(s) = [C_s \quad (D_sH)_\ell](sI - A_\ell^c)^{-1}\begin{bmatrix}\Gamma_s\\ GF_s\end{bmatrix}$$

where

$$(D_sH)_\ell = \text{diag}\left[\overbrace{0,\ldots,0}^{\ell},\overbrace{DH_c,\ldots,DH_c}^{N-\ell}\right]$$

Using a similar method as in the proof of Theorem 8.4.12 in [117], the following theorem can be established.

Theorem 7.4.12 *When ℓ control inputs fail, the poles of the resulting closed-loop system are*

$$spec(A_\ell^c) = \begin{cases} spec(\bar{A}_l^c) \cup spec(A_\beta^c) \quad (\ell = 1) \\ spec(\bar{A}_\ell^c) \cup spec(\bar{A}_\beta^c) \cup spec(\bar{A}_\beta^c) \quad (2 \le \ell \le N-2) \\ spec(\bar{A}_\ell^c) \cup spec(\bar{A}_\beta^c) \quad (\ell = N-1) \end{cases}$$

where

$$spec(\bar{A}_\ell^c) = \begin{bmatrix} U_\ell & \tilde{U}_\ell & 0 & 0 \\ \tilde{U}_\ell & U_{N-\ell} & 0 & BH_c \\ G_cG & 0 & F_c & 0 \\ 0 & G_cG & 0 & F_c \end{bmatrix}, \bar{A}_\beta^c = \begin{bmatrix} A_\beta & 0 \\ G_cG & Fc \end{bmatrix}$$

and A_β^c is defined in (7.127). The transfer function matrix of the resulting closed-loop system is

$$\begin{aligned} T_\ell^c(s) &= \begin{bmatrix} T_{\ell s} & 0 \\ 0T_{(N-\ell)s} & \end{bmatrix} diag\left[\bar{T}_\ell^c(s),\overbrace{\bar{T}_\beta^c(s),\ldots,\bar{T}_\beta^c(s)}^{\ell-1},\overbrace{T_\beta^c(s),\ldots,T_\beta^c(s)}^{N-\ell-1}\right] \\ &\times \begin{bmatrix} T_{\ell r}^{-1} & 0 \\ 0 & T_{(N-\ell)r}^{-1} \end{bmatrix} \end{aligned} \tag{7.143}$$

where

$$\begin{aligned} \bar{T}_\ell^c(s) &= \begin{bmatrix} C & 0 & 0 & 0 \\ 0 & C & 0 & DH_c \end{bmatrix}(sI - \bar{A}_\ell^c)^{-1}\begin{bmatrix} V_\ell & \tilde{V}_\ell \\ \tilde{V}_\ell & V_{N-\ell} \\ G_cF & 0 \\ 0 & G_cF \end{bmatrix}, \\ \bar{T}_\beta^c(s) &= [C \quad 0](sI - \bar{A}_\beta^c)^{-1}\begin{bmatrix}\Gamma_{s\beta}\\ G_cF\end{bmatrix} \end{aligned}$$

and $T^c_\beta(s)$ *is defined in (7.129). Moreover, if* A^c_ℓ *is stable, the* $\mathcal{H}_\infty$*-norm of the transfer function matrix is*

$$\|T^c_\ell(s)\|_\infty = \begin{cases} max\left\{\|\bar{T}^c_\ell(s)\|_\infty, \|T^c_\beta(s)\|_\infty\right\} & (\ell = 1) \\ max\left\{\|\bar{T}^c_\ell(s)\|_\infty, \|T^c_\beta(s)\|_\infty, \|T^c_\beta(s)\|_\infty\right\} & (2 \le \ell \le N-2) \\ max\left\{\|\bar{T}^c_\ell(s)\|_\infty, \|T^c_\beta(s)\|_\infty\right\} & (\ell - N - 1) \end{cases}$$

7.4.8 General failures

In practice, a particular control-channel failure may have three possibilities: control-input failure only, measurement failure only, and simultaneous control-input and measurement failure. The following theorem shows that the above three possibilities can be considered together.

Theorem 7.4.13 *For* $1 \le \ell \le N-1$, *the poles of the resulting closed-loop system for any l control-channel failures are the same. Moreover, the transfer function matrices are also identical.*

Proof 7.4.14 *Suppose there are* $1 \le \ell \le N-1$ *control-channel failures in which* ℓ_1 *control-channels with only control-input failures,* ℓ_2 *control-channels with only measurement failures, and* ℓ_3 *control-channels with both control-input and measurement failures, thus* $\ell_1 + \ell_2 + \ell_3 = l$. *The system matrix and the transfer function matrix of the resulting closed-loop system are*

$$\bar{A}^c_\ell = \begin{bmatrix} A & B_s H_* \\ G_* G_s & F \end{bmatrix}$$

and

$$\bar{T}^c_\ell(s) = [C_s \quad D_s H_*](sI - \hat{A}^c_\ell)^{-1} \begin{bmatrix} \Gamma_s \\ G_* F_s \end{bmatrix}$$

respectively, where

$$\begin{aligned} G_* &= diag\left[\overbrace{G_c, \ldots, G_c}^{\ell_2}, \overbrace{0, \ldots, 0}^{\ell_2}, \overbrace{0, \ldots, 0}^{\ell_3}, \overbrace{G_c, \ldots, G_c}^{N-l}\right] \\ H_* &= diag\left[\overbrace{0, \ldots, 0}^{\ell_2}, \overbrace{H_c, \ldots, H_c}^{\ell_2}, \overbrace{0, \ldots, 0}^{\ell_3}, \overbrace{H_c, \ldots, H_c}^{N-l}\right] \end{aligned}$$

When $sI - F$ *is invertible, we have*

$$\begin{aligned} |sI - \hat{A}^c_\ell| &= \begin{vmatrix} sI - A & -B_2 H_* \\ -G_* G_s & sI - F \end{vmatrix} \\ &= |sI - F| \, |(sI - A) - B_s H_* (sI - F)^{-1} G_* G_s| \end{aligned}$$

Since

$$(sI - F)^{-1} = diag[(sI - F_c)^{-1}, \ldots, (sI - F_c)^{-1}]$$

we have

$$B_s H_*(sI - F)^{-1} G_* G_s$$
$$= diag\left[\overbrace{0, \ldots, 0,}^{\ell} \overbrace{B}\, H_c(sI - F_c)^{-1} G_c G, \ldots, B H_c(sI - F_c)^{-1} G_c G^{N-\ell}\right]$$

which is only affected by $\ell = \ell_1 + \ell_2 + \ell_3$. Therefore

$$spec(\hat{A}_l^c) = spec(A_l^c)$$

Since system (7.120) has the particular symmetric structure and the decentralized controller (7.121) has identical subsystem controllers, measurement failures and control-input failures have the same effect on the closed-loop transfer function matrix [366], that is,

$$\hat{T}_\ell^c(s) = T_\ell^c(s)$$

which concludes the proof.

Remark 7.4.15 *Since ℓ measurement failures is a special case of l control-channel failures, the results of ℓ measurement failures are the same as that of ℓ control-input failures.*

Remark 7.4.16 *It can be easily seen from (7.143) that the necessary condition for A_ℓ^c to be stable ($\ell \geq 1$) is $spec(F_c) \subset C^-$. In other words, $spec(F_c) \subset C^-$ is necessary for the closed-loop system to endure at least one control-channel failure.*

Remark 7.4.17 *From **Theorems 7.4.12** and **7.4.13**, one can determine the reliability of the controller by simply computing the poles and the $\mathcal{H}_\infty$-norm of at most three lower order systems. Noting that there is no conservativeness introduced in the simplification, thus the resulting reliability is also exact.*

Remark 7.4.18 *The control-channel failure considered in this paper means the control channel of particular subsystems completely fail. If each subsystem is a multiple-input system and the same part of the input fails for ℓ subsystems, similar reliability analysis results can be obtained. However, if a different part of the input fails in different subsystems, the result in this paper cannot be applied because the symmetric structure of the whole system will no longer hold. Certainly the reliability analysis method suitable for an arbitrary system could still be applied but the analysis will not be so simple.*

7.4.9 Illustrative example 7.6

Consider the closed-loop system composed of system (7.140) and the controller (7.141) designed in the previous section. For $1 \leq \ell \leq 19$ control-channel failures, **Theorem 7.4.12** is used to compute $\|T_\ell^c(s)\|_\infty$. The results are summarized in Table 7.1 ($\ell = 0$ refers to the closed-loop system with no control-channel failure). Following the foregoing analysis, we take $\gamma = 0.62$. Table 7.1 shows that for $1 \leq \ell \leq 16$, $\|T_\ell^c(s)\|_\infty < \gamma$ while $\|T_{17}^c(s)\|_\infty > \gamma$. Hence the closed-loop system will maintain its stability and with the norm of the transfer function matrix less than γ when $\ell < 16$ control-channel failures occur.

Table 7.1: Summary of Results for Example 7.6 ($\gamma = 0.62$)

ℓ	$\|T_\ell^c(s)\|_\infty$	ℓ	$\|T_\ell^c(s)\|_\infty$	ℓ	$\|T_\ell^C(s)\|_\infty$	ℓ	$\|T_\ell^C(s)\|_\infty$
0	0.6143	5	0.6160	10	0.6177	15	0.6195
1	0.6146	6	0.6164	11	0.6181	16	0.6198
2	0.6150	7	0.6167	12	0.6184	17	0.6201
3	0.6153	8	0.6170	13	0.6188	18	0.6205
4	0.6157	9	0.6174	14	0.6191	19	0.6208

7.4.10 Illustrative example 7.7

Consider the closed-loop system derived in **Illustrative example 7.5**. For $1 \leq \ell \leq 19$ control-channel failures, **Theorem 7.4.12** is used to compute spec(A_ℓ^c) and $\|T_\ell^c(s)\|_\infty$. The results are summarized in Table 7.2 ($\ell = 0$ and $\ell = 20$ refer to the closed-loop system with no control-channel failure and the open-loop system, respectively).

Following the foregoing analysis, we take $\gamma = 10$; Table 7.2 shows that for $1 \leq \ell \leq 8$, A_ℓ^c is stable and $\|T_\ell^c(s)\|_\infty > \gamma$ while $\|T_g^c(s)\|_\infty > \gamma$. Therefore, the closed-loop system will maintain its stability and with the norm of the transfer function matrix less than γ when $\ell < 9$ control-channel failures occur. Moreover, the closed-loop system becomes unstable when $\ell \geq 14$ control-channel failures occur.

7.4.11 Illustrative example 7.8

Consider the voltage/reactive power behavior of a multimachine power system. The overall system consists of several synchronous machines including their PI-voltage controller, which feeds the load through a distribution net [191].

Table 7.2: Summary of Results for Example 7.7 ($\gamma = 10$)

ℓ	A^c_ℓ	$\|T^c_\ell(s)\|_\infty$
0	stable	4.9213
1	stable	5.1988
2	stable	5.5178
3	stable	5.8893
4	stable	6.3286
5	stable	6.8579
6	stable	7.5104
7	stable	8.3385
8	stable	9.4292
9	stable	10.9398
10	stable	13.1864
11	stable	16.9131
12	stable	24.3899
13	stable	47.4609
≥ 14	unstable	−−

The system can be modeled by

$$\dot{x}_i = \begin{bmatrix} -2.51 & -0.16 \\ 2.55 & 0 \end{bmatrix} x_i + \sum_{k=1,k\neq i}^{N} \begin{bmatrix} -0.065 & 0 \\ -0.0027 & 0 \end{bmatrix} x_k$$

$$+ \begin{bmatrix} 0.9 \\ -1 \end{bmatrix} u_i + \begin{bmatrix} 0.2 \\ 0.1 \end{bmatrix} + \sum_{k=1,k\neq i}^{N} \begin{bmatrix} 0.1 \\ 0.1 \end{bmatrix} w_k$$

$$z_i = [2.540]x_i + u_i, \quad i = 1, 2, \ldots, N$$

Taking $N = 20$ and computing directly, we have

$$A_s = \begin{bmatrix} -2.445 & -0.16 \\ 2.5527 & 0 \end{bmatrix}$$

$$A_o = \begin{bmatrix} -3.745 & -0.16 \\ 2.4987 & 0 \end{bmatrix}$$

$$G_s = \begin{bmatrix} 0.1 \\ 0 \end{bmatrix}$$

$$G_o = \begin{bmatrix} 2.1 \\ 2 \end{bmatrix}$$

Suppose $\gamma = 0.8$, we choose $\alpha = 0.4, \epsilon = 0.0002$. Solving the Riccati equations (7.135) and (7.136), we have

$$P_o = \begin{bmatrix} 0.001 & 0.0011 \\ 0.0011 & 0.0014 \end{bmatrix}$$
$$P_s = \begin{bmatrix} 0.000694 & 0.000625 \\ 0.000625 & 0.0006025 \end{bmatrix}$$

By testing, we know that inequalities **Theorem 7.4.5** do not hold; we try by choosing

$$K_1 = -R_1^{-1}(B_1^T P_o + D_1^T C_1)$$

and obtain $K_1 = [-2.5397, 0 : 0003]$. From **Theorem 7.2.12,** we get $\text{spec}(A_c) = 4.552, -0.179, -5.8942, -0.1368 \subset C^-$, and $\|T\|_\infty = 0 : 0083 < \gamma$. Thus the decentralized H_∞ control law can be chosen as

$$u_i = K_1 x_i = [-2.5397, 0.0003]x_i, \quad i = 1, \ldots, N$$

For $l = 1, 2, 3, 4$, **Theorems 7.4.12** and **7.4.13** are used to compute spec (A_{cl}) and $\|T_l\|_\infty$. The results are summarized in Table 7.3.

Since for l = 1, 2, 3, spec$(A_{cl}) \subset C^-$ and $\|T_l\|_\infty < \gamma$,but $\|T_4\|_\infty > \gamma$, hence $l_0 = 4$. As a result, the closed-loop system will maintain its stability and the transfer matrix will satisfy $\|T_l\|_\infty \leq \gamma$ when less than four subsystem controllers fail.

Table 7.3: Summary of Results

l	spec (A_{cl})	$\|T_l\|_\infty$
1	-4.552, -0.179, 5.8517, -2.3089, -0.802, -0.1352	0.4501
2	-4.552, -0.179, -2.2646, -0.1804, -5.8082, -2.3541, -0.1801, -0.1336	0.6368
3	-4.552, -0.179, -2.2646, -0.1804, -5.7635, -2.4005, -0.18, -0.1321	0.7803
4	-4.552, -0.179, -2.2646, -0.1804, -5.7176, -2.448, -0.1799, -0.1305	0

7.5 Problem Set V

Problem V.1: Consider the stabilization problem of subsystems with collocated sensors and actuators, which are described by

$$\begin{aligned} M_j \ddot{q}_j + D_j \dot{q}_j + K_j q_j &= L_j u_j \\ y_j &= L_j^t q_j, \quad j = 1, 2, \ldots n_s \end{aligned}$$

where $q_j \in \Re^n_j$, $u_j \in \Re^r_j$, and $y_j \in \Re^r_j$ are the displacement (translational displacement and rotational angle), control input (force and torque), and measured displacement output of the jth subsystem, respectively. The mass matrix M_j is positive-definite while the damping and stiffness matrices D_j and K_j are positive-semidefinite such that

$$rank[D_j \quad K_j] = rank[D_j] = rank[K_j] < n_j$$

The matrix L_j is defined by the locations and directions of actuators and the matrix L^t_j expresses the locations and directions of sensors. Assume that the rigid modes of each subsystem are controllable and observable. Establish conditions under which the use of a dynamic displacement feedback controller at the subsystem level

$$\begin{aligned} \dot{x}_j &= -R_j x_j + S_j y_j \\ u_j &= S^t_j x_j - K_j y_j \end{aligned}$$

where $x_j \in \Re^p_j$, $p_j \geq r_j$, $R_j > 0$ and the matrix S_j is of full rank, guarantees the closed-loop subsystem stability. Study the reliability of the closed-loop system when the jth local subsystem controller fails and justify your results.

Problem V.2: Consider the multicontroller problem of linear time-invariant systems of the form

$$\begin{aligned} \dot{x} &= Ax + \sum_{j=1}^{q} Bu_j + Gw_o, \quad q > 1 \\ y_j &= Cx + w_j, \quad j = 1, 2, ...q, \\ z &= [x^t H^t \; u^t_1 \; \cdots \; u^t_q]^t \end{aligned}$$

where $x \in \Re^n$ is the state, y_j, $j = 1, 2, ...q$ are the measured outputs, z is an output to be regulated, w_j, $j = 0, 1, 2, ...q$ are the square-integrable disturbances, and u_j, $j = 1, 2, ...q$ are the control inputs. We seek to design q identical controllers for the system, generated by the dynamic form

$$\begin{aligned} \dot{\xi}_j &= A_c \xi_j + L y_j \\ u_j &= K \, \xi_j \;\; j = 1, 2, ...q \end{aligned}$$

where K is the feedback gain, L is the observer gain, K_w is the disturbance estimate gain, and $A_c = A + BK + GK_w - LC$. It is required to propose a method to design the gains K, L, and K_w so that for any p controller failures

$(0 \leq p \leq q-1)$, the resulting closed-loop system is internally stable and the $\mathcal{H}_\infty$-norm of the closed-loop transfer function matrix is bounded by some prescribed constant $\beta > 0$. Illustrate your method on the following data:

$$A = \begin{bmatrix} -2 & 1 & 1 & 1 \\ 3 & 0 & 0 & 2 \\ -1 & 0 & -2 & -3 \\ -2 & -1 & 2 & -1 \end{bmatrix}, \ G = \begin{bmatrix} 0.5 \\ 0 \\ 0 \\ 0 \end{bmatrix}, \ B = \begin{bmatrix} 0 \\ 0 \\ 0 \\ 1 \end{bmatrix},$$
$$H = \begin{bmatrix} 0 & 0 & 0.5 & 0 \end{bmatrix}, \ C = \begin{bmatrix} 1 & 0 & 0 & 0 \end{bmatrix}, \ q = 2$$

and provide some interpretations.

Problem V.3: A linear state-delayed system is described by

$$\begin{aligned} \dot{x}_j(t) &= A_j x_j(t) + A_{dj} x_j(t - \tau_j(t)) + B_j u_j + \sum_{k=1, k \neq j}^{n_s} F_{jk} x_k(t) \\ z_j(t) &= G_j x_j(t) + \Omega_j w_j(t), \ \ j = 1, ..., n_s, \\ x_j(t) &= \phi_j(t), \ \ t \in [-\tau_j, 0] \end{aligned}$$

where $x_j \in \Re^{n_j}$ is the state, $z_j \in \Re^{p_j}$ is the measurement output, $u_j \in \Re^{m_j}$ is the control input, $w_j, \ j = 0, 1, 2, ...q$ is the unexpected sensor signal (sensor fault), τ_j is the constant delay, and $\phi_j(t) \in \Re^{n_j}$ is the initial state vector. Consider an observer-based controller of the form

$$\begin{aligned} \dot{\xi}_j &= A_j \xi_j + A_{dj} \xi_j(t - \tau_j(t)) + B_j(u_j) + K_j(z_j - G_j \xi_j) \\ u_j &= K F_j \ \xi_j \end{aligned}$$

where $\xi_j \in \Re^{n_j}$ is the state estimate, K_j is the observer gain, and F_j is the state-feedback gain. Generate the augmented closed-loop dynamics in terms of the augmented state-vector $\zeta = [x^t \ \ x^t - \xi^t]^t$ and establish conditions of stability when $w_j = 0$. To study the performance of the closed-loop observer-based control system, we consider adding appropriate control signals $B_{fj} u_j(t - \eta)$ and $D_{fj} u_j(t - \eta)$ to the input and output channels, respectively. It is required to develop complete analysis and control synthesis for the reliable observer-based control system.

Problem V.4: Consider the multicontroller problem of linear time-invariant systems of the form

$$\begin{aligned}\dot{x} &= Ax + \sum_{j=1}^{M} Bu_j + Gw_o, \quad M > 1 \\ y_j &= C_j x + w_j, \quad j = 1, 2, ...M, \\ z &= [x^t H^t \ u_1^t \ \cdots \ u_M^t]^t\end{aligned}$$

where $x \in \Re^n$ is the state, $y_j, \ j = 1, 2, ...M$ are the measured outputs, z is an output to be regulated, $w_j, \ j = 0, 1, 2, ...M$ are the square-integrable disturbances, and $u_j, \ j = 1, 2, ...M$ are the control inputs. For the measured output $y_j, \ j = 1, 2, ...M,$ let y_j^F be the signal from the sensor that has failed. Then the following sensor failure model is used:

$$y_j^F = \alpha_{sj} y_j \ + \ \alpha_{soj} \beta_{sj}^F$$

where β_{sj}^F is a bounded disturbance, and $\alpha_{sj}, \ \alpha_{soj}$ satisfy

$$0 \leq \alpha_{sjm} \leq \alpha_{sj} \leq \alpha_{sjM}, \ \ 0 \leq \alpha_{soj} \leq \alpha_{sojM}$$

for $j = 1, 2, ...M$. In a similar way, let u_j^F be the signal from the actuator that has failed. The actuator failure model has the form

$$u_j^F = \alpha_{aj} u_j \ + \ \alpha_{aoj} \beta_{aj}^F$$

where β_{aj}^F is a bounded disturbance, and $\alpha_{aj}, \ \alpha_{aoj}$ satisfy

$$0 \leq \alpha_{ajm} \leq \alpha_{aj} \leq \alpha_{ajM}, \ \ 0 \leq \alpha_{aoj} \leq \alpha_{aojM}$$

for $j = 1, 2, ...M$ and where $\alpha_{sojM} \geq 1$ and $\alpha_{aojM} \geq 1$.

It is desired to design a controller of the following form:

$$\begin{aligned}\dot{\zeta} &= F\zeta \ + \ Ly \\ u &= K \ \zeta \quad \ j = 1, 2, ...M\end{aligned}$$

such that the resulting closed-loop system is asymptotically stable and with an $\mathcal{H}_\infty$-norm bound not only when all control components are operational, but also in the case of some sensor and actuator failures. Examine the developed conditions under different operational conditions.

7.6 Notes and References

In the first part of this chapter, the decentralized reliable feedback stabilization problem was discussed for a class of linear interconnection of continuous time-delay plants subjected to convex-bounded parametric uncertainties and coupled with time-delay interconnections. In the problem set-up, either actuator or sensor failures have been described by a model that considers possible outage or partial failures in every actuator/sensor of each decentralized controller. We first establish a sufficient condition for the existence of a decentralized reliable control scheme. This control, based on either state or static output-feedback, has been effectively constructed by means of a feasible LMI optimization problem. A key point in the control design has been the formulation of an LMI characterization, which uses assumptions over model uncertainties to remove the parameter dependence of the control characterization. This part is further complemented by developing the complete counterpart results of the decentralized reliable feedback stabilization problem for a class of linear interconnection of discrete-time-delay plants.

In the second part of this chapter, we moved to examine the dynamic output feedback decentralized $\mathcal{H}_\infty$ control and reliability analysis of symmetric composite systems. By using the structural properties of the systems, we provided a simple method to design its dynamic output feedback decentralized H1 controller. Moreover, the reliability of the controller can be easily tested by computing the poles and the $\mathcal{H}_\infty$-norm of systems of possibly much lower orders. Though we presented simple controller design and reliability analysis methods for symmetric composite systems, the reliability analysis can only be conducted after the controller design. In practice, the method proposed in this chapter can be applied in an iterative way. That is, if the reliability of the designed system is not satisfactory, then we can increase the performance requirement for the nominal case design and use our method to check the reliability again.

The area of decentralized reliable control is attactive and much more is anticipated to deal with different models of actuator/sensor faults.

Chapter 8

Decentralized Resilient Control

In this chapter, we continue probing further into the decentralized-control techniques for interconnected systems, where we focus herein on methods for designing resilient controllers to accommodate both parametric uncertainties and controller gain perturbations. We address two approaches: the first approach is an extension of Chapter 6 on time-delay systems and the second approach is an extension of the overlapping methods discussed in Chapter 5.

8.1 Introduction

In this paper, we develop robust decentralized delay-dependent stability and resilient feedback stabilization methods for a class of linear interconnected continuous-time systems. The subsystems are subjected to convex-bounded parametric uncertainties while time-varying delays occur within the local subsystems and across the interconnections and additive feedback gain perturbations are allowed. In this way, our control design offers decentralized structure and possesses robustness with respect to both parametric uncertainties and gain perturbations. For related results on resilient control, the reader is referred to [258]-[266] where it is shown to provide a framework of extended robustness properties. In this chapter, we construct appropriate Lyapunov-Krasovskii functional in order to exhibit the delay-dependent dynamics. The developed methods for decentralized stability and stabilization deploy an "injection procedure" within the individual subsystems which eliminates the need for overbounding and utilizes a smaller number of LMI decision variables. By this way, improved and less conservative solutions to the robust decentralized delay-dependent stability and resilient feedback stabilization problems in terms of feasibility testing of new parameterized linear matrix inequalities (LMIs) are developed. In our analysis, we consider the time-delay factor as a differentiable time-varying function satisfying some bounding relations and derive the solution criteria for nominal and polytopic models. Then, resilient decentralized feedback stabilization methods are provided based on state measurements, static output feedback, and dynamic output feedback by allowing additive gain perturbations while guaranteeing that the corresponding closed-loop family of individual subsystems enjoys the robust decentralized delay-dependent stability with an $\mathcal{L}_2$ gain smaller than a prescribed constant level.

8.2 Problem Statement

We consider a class of linear systems S structurally composed of n_s coupled subsystems S_j and modeled by the state-space model:

$$\begin{aligned}
\dot{x}_j(t) &= A_j x_j(t) + A_{dj} x_j(t-\tau_j(t)) + B_j u_j(t) + c_j(t) + \Gamma_j w_j(t) \\
z_j(t) &= G_j x_j(t) + G_{dj} x_j(t-\tau_j(t)) + D_j u_j(t) + \Phi_j w_j(t) \\
y_j(t) &= C_j x_j(t) + C_{dj} x_j(t-\tau_j(t)) \\
c_j(t) &= \sum_{k=1,k\neq j}^{n_s} F_{jk} x_k(t) + \sum_{k=1,k\neq j}^{n_s} E_{jk} x_k(t-\eta_{jk}(t))
\end{aligned} \tag{8.1}$$

where for $j \in \{1,...,n_s\}$, $x_j(t) \in \Re^{n_j}$ is the state vector, $u_j(t) \in \Re^{m_j}$ is the control input, $y_j(t) \in \Re^{p_j}$ is the measured output, $w_j(t) \in \Re^{q_j}$ is the disturbance input which belongs to $\mathcal{L}_2[0,\infty)$, $z_j(t) \in \Re^{q_j}$ is the performance output, $c_j(t) \in \Re^{n_j}$ is the coupling vector, and τ_j, η_{jk}, $j,k \in \{1,...,n_s\}$ are unknown time-delay factors satisfying

$$\begin{aligned}
0 &\leq \tau_j(t) \leq \varrho_j, \quad \dot{\tau}_j(t) \leq \mu_j, \\
0 &\leq \eta_{jk}(t) \leq \varrho_{jk}, \quad \dot{\eta}_{jk}(t) \leq \mu_{jk}
\end{aligned} \tag{8.2}$$

where the bounds ϱ_j, ϱ_{jk}, μ_j, μ_{jk} are known constants in order to guarantee smooth growth of the state trajectories. The matrices $A_j \in \Re^{n_j \times n_j}$, $B_j \in \Re^{n_j \times m_j}$, $D_j \in \Re^{q_j \times m_j}$, $A_{dj} \in \Re^{n_j \times n_j}$, $\Phi_j \in \Re^{q_j \times q_j}$, $\Gamma_j \in \Re^{n_j \times q_j}$, $C_j \in \Re^{p_j \times n_j}$, $C_{dj} \in \Re^{p_j \times n_j}$, $G_j \in \Re^{q_j \times n_j}$, $G_{dj} \in \Re^{q_j \times n_j}$, $F_{jk} \in \Re^{n_j \times n_k}$, $E_{jk} \in \Re^{n_j \times n_k}$ are real and constants.

The initial condition $\langle x_j(0), x_j(r) \rangle = \langle x_{oj}, \phi_j \rangle$, $j \in \{1,...,n_s\}$ where $\phi_j(.) \in \mathcal{L}_2[-\tau_j^*, 0]$, $j \in \{1,...,n_s\}$. The inclusion of the terms $A_{dj} x_j(t-\tau_j(t))$, $E_{jk} x_k(t-\eta_{jk}(t))$ is meant to emphasize the delay within each subsystem (local delay) and among the subsystems (coupling delay), respectively; see Figure 8.1. The class of systems described by (8.1) subject to delay-pattern

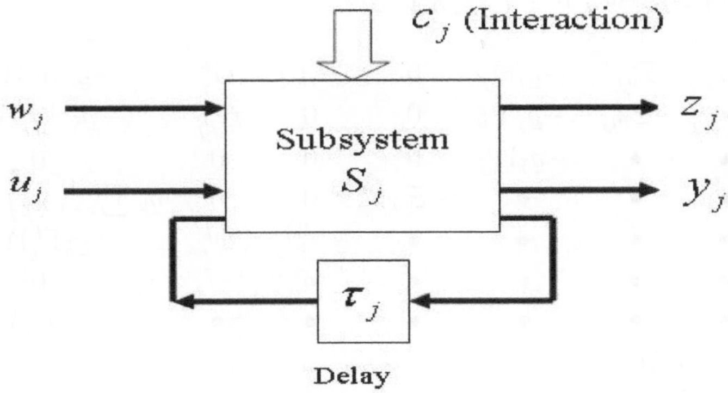

Figure 8.1: Subsystem model

(8.2) is frequently encountered in modeling several physical systems and engineering applications including large space structures, multimachine power systems, cold mills, transportation systems, and water pollution management, to name a few [252, 253, 324]. Our objective in this chapter is to study two main problems: the first problem is the decentralized delay-dependent asymptotic stability by deriving a feasibility testing at the subsystem level so as to guarantee the overall system asymptotic stability. The second problem deals with the resilient decentralized feedback stabilization by developing decentralized feedback controllers (based on state, static output, and dynamic output) that takes into consideration additive gain perturbations while ensuring that the overall closed-loop system is delay-dependent asymptotically stable.

The following theorem is a variant of results from Chapter 6.

Theorem 8.2.1 *Given* $\varrho_j > 0,\ \mu_j > 0,\ \varrho_{jk} > 0,$ *and* $\mu_{jk} > 0,\ j,k = 1,...,n_s$. *The family of subsystems* $\{\mathsf{S}_j\}$ *with* $u_j(.) \equiv 0$ *where* S_j *is described by (8.1) is delay-dependent asymptotically stable with* $\mathcal{L}_2$-*performance bound* $\gamma_j, j = 1,...,n_s$ *if there exist positive-definite matrices* $\mathcal{P}_j,\ \mathcal{Q}_j,\ \mathcal{W}_j,\ \mathcal{Z}_{kj},\ k,j = 1,..n_s,$ *parameter matrices* $\Theta_j\ \Upsilon_j$ *satisfying the* *following LMIs for* $j, k = 1, ..., n_s$:

$$\begin{bmatrix} \Xi_{oj} & \Xi_{aj} & -\varrho_j\Theta_j & 0 & \mathcal{P}_j\Gamma_j & G_j^t & \varrho_k A_j^t \mathcal{W}_k \\ \bullet & -\Xi_{mj} & -\varrho_j\Upsilon_j & 0 & 0 & G_{dj}^t & \varrho_k A_{d_j}^t \mathcal{W}_k \\ \bullet & \bullet & -\varrho_j\mathcal{W}_j & 0 & 0 & 0 & 0 \\ \bullet & \bullet & \bullet & -\Xi_{nj} & 0 & 0 & \varrho_k \sum_{k=1,k\neq j}^{n_s} E_{kj}\mathcal{W}_k \\ \bullet & \bullet & \bullet & \bullet & -\gamma_j^2 I_j & \Phi_j^t & \varrho_k \Gamma_j^t \mathcal{W}_k \\ \bullet & \bullet & \bullet & \bullet & \bullet & -I_j & 0 \\ \bullet & \bullet & \bullet & \bullet & \bullet & \bullet & -\varrho_k\mathcal{W}_k \end{bmatrix} < 0 \tag{8.3}$$

where

$$\begin{aligned} \Xi_{oj} &= \mathcal{P}_j[A_j + \sum_{k=1}^{n_s} F_{kj}] + [A_j + \sum_{k=1}^{n_s} F_{kj}]^t\mathcal{P}_j^t + \Theta_j + \Theta_j^t \\ &+ \mathcal{Q}_j + \sum_{k=1,k\neq j}^{n_s} \mathcal{Z}_{kj} + (n_s - 1)\mathcal{P}_j, \\ \Xi_{mj} &= \Upsilon_j + \Upsilon_j^t + (1 - \mu_j)\mathcal{Q}_j, \ \ \Xi_{aj} = \mathcal{P}_j A_{dj} - \Theta_j + \Upsilon_j^t, \\ \Xi_{nj} &= \sum_{k=1,k\neq j}^{n_s} (1 - \mu_{kj})\mathcal{Z}_{kj} + \sum_{k=1,k\neq j}^{n_s} E_{kj}^t \mathcal{P}_k E_{kj} \end{aligned} \tag{8.4}$$

In the foregoing section, we considered the design of feedback controllers under the condition that the implementations of the gains do not undergo any perturbations or parameter tolerances. In what follows, we extend the design results to the case when the feedback gain perturbations are incorporated into the analysis. This design approach is often called *resilient control theory*; see [258] for further reading.

8.3 Resilient Decentralized Stabilization

We now direct attention to resilient decentralized feedback stabilization schemes that take into account possible feedback gain variations. Essentially, our approach builds upon the framework of resilient control theory [258] and extends the results of [105, 174]. We start with the state feedback case.

8.3.1 Resilient state feedback

Consider the resilient state feedback control

$$\begin{aligned} u_j(t) &= [K_j + \Delta K_j(t)]x_j(t), \;\; \Delta K_j(t) = M_j \Delta_j(t) N_j, \\ & \Delta_j^t(t)\Delta_j(t) \leq I, \;\; \forall\, t \end{aligned} \tag{8.5}$$

where $\Delta K_j(t)$ represent additive gain perturbations with $M_j,\ N_j$ being constant matrices. Applying controller (8.5) to the linear system (8.1) yields the following closed-loop system and matrices:

$$\begin{aligned} \dot{x}_j(t) &= A_{fj}x_j(t) + A_{dj}x_j(t-\tau_j(t)) + c_j(k) + \Gamma_j w_j(t) \\ z_j(t) &= G_{fj}x(t) + G_{dj}x(t-\tau_j(t)) + \Phi_j w_j(t) \\ y_j(t) &= C_j x_j(t) + C_{dj}x_j(t-\tau_j(t)) \\ c_j(t) &= \sum_{k=1,k\neq j}^{n_s} F_{jk}x_k(t) + \sum_{k=1,k\neq j}^{n_s} E_{jk}x_k(t-\eta_{jk}(t)) \\ A_{fj} &= A_j + B_jK_j + B_j\Delta K_j, \;\; G_{fj} = G_j + D_jK_j + D_j\Delta K_j \end{aligned} \tag{8.6}$$

Then it follows from **Theorem 8.2.1** that the resulting closed-loop system (8.6) is delay-dependent asymptotically stable with $\mathcal{L}_2$-performance bound γ if there exist matrices $\mathcal{P}_j,\ \mathcal{Q}_j,\ \mathcal{W}_j,\ \mathcal{W}_k,\ \mathcal{Z}_{kj},\ j,k = 1,..n_s,$ parameter matrices $\Theta_j\ \Upsilon_j$, and a scalar $\gamma_j > 0$ satisfying the following LMI:

$$\begin{bmatrix} \Xi_{fj} & \Xi_{aj} & -\varrho_j\Theta_j & \Xi_{cj} & \mathcal{P}_j\Gamma_j & G_{fj}^t & \varrho_k A_{fj}^t \mathcal{W}_k \\ \bullet & -\Xi_{mj} & -\varrho_j\Upsilon_j & 0 & 0 & G_{dj}^t & \varrho_k A_{dj}^t \mathcal{W}_k \\ \bullet & \bullet & -\varrho_j\mathcal{W}_j & 0 & 0 & 0 & 0 \\ \bullet & \bullet & \bullet & -\Xi_{nj} & 0 & 0 & \varrho_k \sum_{k=1,k\neq j}^{n_s} E_{kj}\mathcal{W}_k \\ \bullet & \bullet & \bullet & \bullet & -\gamma_j^2 I_j & \Phi_j^t & \varrho_k\Gamma_j^t\mathcal{W}_k \\ \bullet & \bullet & \bullet & \bullet & \bullet & -I_j & 0 \\ \bullet & \bullet & \bullet & \bullet & \bullet & \bullet & -\varrho_k\mathcal{W}_k \end{bmatrix} < 0 \tag{8.7}$$

where

$$\begin{aligned} \Xi_{fj} &= \mathcal{P}_j[A_{fj} + \sum_{k=1}^{n_s} F_{kj}] + [A_{fj} + \sum_{k=1}^{n_s} F_{kj}]^t \mathcal{P}_j^t + \Theta_j + \Theta_j^t \\ &+ \mathcal{Q}_j + \sum_{k=1,k\neq j}^{n_s} \mathcal{Z}_{kj} + (n_s - 1)\mathcal{P}_j \end{aligned} \tag{8.8}$$

and Ξ_{aj}, Ξ_{cj}, Ξ_{mj}, Ξ_{nj} are given by (8.4). Application of **Inequality 1** from the Appendix to LMI (8.7) using (8.5) yields

$$\begin{bmatrix} \Xi_{kj} & \Xi_{aj} & -\varrho_j\Theta_j & \Xi_{cj} & \mathcal{P}_j\Gamma_j & G^t_{kj} & \varrho_j A^t_{kj}\mathcal{W}_j \\ \bullet & -\Xi_{mj} & -\varrho_j\Upsilon_j & 0 & 0 & G^t_{dj} & \varrho_j A^t_{dj}\mathcal{W}_j \\ \bullet & \bullet & -\varrho_j\mathcal{W}_j & 0 & 0 & 0 & 0 \\ \bullet & \bullet & \bullet & -\Xi_{nj} & 0 & 0 & \varrho_j\sum_{k=1}^{n_s} E_{kj}\mathcal{W}_j \\ \bullet & \bullet & \bullet & \bullet & -\gamma_j^2 I_j & \Phi^t_j & \varrho_j\Gamma^t_j\mathcal{W}_j \\ \bullet & \bullet & \bullet & \bullet & \bullet & -I_j & 0 \\ \bullet & \bullet & \bullet & \bullet & \bullet & \bullet & -\varrho_j\mathcal{W}_j \end{bmatrix}$$

$$+\varepsilon_j \begin{bmatrix} \mathcal{P}_j B_j M_j \\ 0 \\ 0 \\ 0 \\ 0 \\ D_j M_j \\ \varrho_j\mathcal{W}_j B_j M_j \end{bmatrix}\begin{bmatrix} \mathcal{P}_j B_j M_j \\ 0 \\ 0 \\ 0 \\ 0 \\ D_j M_j \\ \varrho_j\mathcal{W}_j B_j M_j \end{bmatrix}^t + \varepsilon_j^{-1}\begin{bmatrix} N^t_j \\ 0 \\ 0 \\ 0 \\ 0 \\ 0 \\ 0 \end{bmatrix}\begin{bmatrix} N^t_j \\ 0 \\ 0 \\ 0 \\ 0 \\ 0 \\ 0 \end{bmatrix}^t < 0 \quad (8.9)$$

$$\Xi_{kj} = \mathcal{P}_j[A_{kj} + \sum_{k=1}^{n_s} F_{kj}] + [A_{kj} + \sum_{k=1}^{n_s} F_{kj}]^t\mathcal{P}^t_j + \Theta_j + \Theta^t_j +$$

$$\mathcal{Q}_j + + \sum_{k=1,k\neq j}^{n_s} \mathcal{Z}_{kj} + (n_s - 1)\mathcal{P}_j$$

$$G_{kj} = G_j + D_j K_j, \quad A_{kj} = A_j + B_j K_j \quad (8.10)$$

for some $\varepsilon_j > 0$. The main decentralized resilient feedback design results are established by the following theorems.

Theorem 8.3.1 *Given the bounds $\varrho_j > 0$, $\mu_j > 0$ and matrix $\mathcal{W}_j$, the interconnected system (8.1) with decentralized resilient controller (8.5) is delay-dependent asymptotically stable with $\mathcal{L}_2$-performance bound γ_j if there exist positive-definite matrices $\mathcal{X}_j$, $\mathcal{Y}_j$, $\{\Lambda_{kj}\}_{k=1}^{n_s}$, $\{\Psi_{rj}\}_{r=1}^{4}$, and a scalar $\varepsilon_j > 0$ satisfying the following LMIs for $j = 1, ..., n_s$:*

$$\begin{bmatrix} \Xi_{1j} & \Xi_{2j} \\ \bullet & \Xi_{3j} \end{bmatrix} < 0 \quad (8.11)$$

$$\Xi_{1j} = \begin{bmatrix} \Xi_{1aj} & \Xi_{1cj} \\ \bullet & \Xi_{1sj} \end{bmatrix}$$

$$\Xi_{1aj} = \begin{bmatrix} \Pi_{oj} & \Pi_{aj} & -\varrho_j \Psi_{1j} & 0 \\ \bullet & -\Pi_{mj} & -\varrho_j \Psi_{3j} & 0 \\ \bullet & \bullet & -\varrho_j \Psi_{4j} & 0 \\ \bullet & \bullet & \bullet & -\Pi_{nj} \end{bmatrix}$$

$$\Xi_{1cj} = \begin{bmatrix} \Gamma_j & \Pi_{zj}^t & \varrho_k (\mathcal{X}_j A_j^t + \mathcal{Y}_j^t B_j^t) \mathcal{W}_k \\ 0 & G_{dj}^t & \varrho_k \mathcal{X}_j A_{dj}^t \mathcal{W}_k \\ 0 & 0 & 0 \\ 0 & 0 & \varrho_k \sum_{k=1, k \neq j}^{n_s} \mathcal{X}_j E_{kj} \mathcal{W}_k \end{bmatrix}$$

$$\Xi_{1sj} = \begin{bmatrix} -\gamma_j^2 I_j & \Phi_j^t & \varrho_k \Gamma_j^t \mathcal{W}_k \\ \bullet & -I_j & 0 \\ \bullet & \bullet & -\varrho_k \mathcal{W}_k \end{bmatrix}$$

$$\Xi_{2j} = \begin{bmatrix} \varepsilon_j B_j M_j & \mathcal{X}_j N_j^t \\ 0 & 0 \\ 0 & 0 \\ 0 & 0 \\ 0 & 0 \\ \varepsilon_j D_j M_j & 0 \\ \varepsilon_j \varrho_j \mathcal{W}_j B_j M_j & 0 \end{bmatrix}, \quad \Xi_{3j} = \begin{bmatrix} -\varepsilon_j I_j & 0 \\ \bullet & -\varepsilon_j I_j \end{bmatrix} \tag{8.12}$$

where

$$\begin{aligned} \Pi_{oj} &= [A_j + \sum_{k=1}^{n_s} F_{kj}] \mathcal{X}_j + B_j \mathcal{Y}_j + \mathcal{X}_j [A_j + \sum_{k=1}^{n_s} F_{kj}]^t + \mathcal{Y}_j^t B_j^t + \Psi_{1j} \\ &+ \Psi_{1j}^t + \Psi_{2j} + \sum_{k=1, k \neq j}^{n_s} \Lambda_{kj} + +(n_s - 1) \mathcal{X}_j \\ \Pi_{aj} &= A_{dj} \mathcal{X}_j - \Psi_{1j} + \Psi_{3j}^t, \;\; \Pi_{mj} = \Psi_{3j} + \Psi_{3j}^t + (1 - \mu_j) \Psi_{2j} + \Psi_{4j}^t, \\ \Pi_{nj} &= \sum_{k=1, k \neq j}^{n_s} (1 - \mu_{kj}) \Lambda_{kj} + \sum_{k=1, k \neq j}^{n_s} \mathcal{X}_j E_{kj}^t \mathcal{P}_k E_{kj} \mathcal{X}_j, \\ \Pi_{zj} &= G_{js} \mathcal{X}_j + D_{js} \mathcal{Y}_j \end{aligned} \tag{8.13}$$

Moreover, the local gain matrix is given by $K_j = \mathcal{Y}_j \mathcal{X}_j^{-1}$.

Proof: Applying the congruent transformation

$$Blockdiag[\mathcal{X}_j, \mathcal{X}_j, \mathcal{X}_j, \mathcal{X}_j, I_j, I_j, I_j, I_j, I_j], \; \mathcal{X}_j = \mathcal{P}_j^{-1}$$

to LMI (8.9) Schur complements and using the linearizations

$$\begin{aligned}\mathcal{Y}_j &= K_j\mathcal{X}_j,\ \Psi_{1j} = \mathcal{X}_j\Theta_j\mathcal{X}_j,\ \Psi_{2j} = \mathcal{X}_j\mathcal{Q}_j\mathcal{X}_j,\\ \Psi_{3j} &= \mathcal{X}_j\Upsilon_j\mathcal{X}_j,\ \Omega_j = \alpha_j\mathcal{X}_jF_j^t,\ \Lambda_{kj} = \mathcal{X}_j\mathcal{Z}_{kj}\mathcal{X}_j,\ \Psi_{4j} = \mathcal{X}_j\mathcal{W}_j\mathcal{X}_j\end{aligned}$$

we readily obtain LMI (8.11) by Schur complements.

Theorem 8.3.2 *Given the bounds $\varrho_j > 0,\ \mu_j > 0$ and matrix $\mathcal{W}_j$, the interconnected system (8.1) with decentralized controller $u_j(t) = K_jx_j(t)$ and polytopic representation (6.28)-(6.29) is delay-dependent asymptotically stable with $\mathcal{L}_2$-performance bound γ_j if there exist weighting matrices $\mathcal{X}_j$, $\mathcal{Y}_j$, $\{\Lambda_{kj}\}_{k=1}^{n_s}$, $\{\Psi_{rj}\}_{r=1}^{3}$ satisfying the following LMIs for $j = 1, ..., n_s$, $s = 1, ..., N$:*

$$\begin{bmatrix} \Xi_{1js} & \Xi_{2js} \\ \bullet & \Xi_{3j} \end{bmatrix} < 0 \tag{8.14}$$

$$\Xi_{1js} = \begin{bmatrix} \Xi_{ajs} & \Xi_{cjs} \\ \bullet & \Xi_{ejs} \end{bmatrix}$$

$$\Xi_{ajs} = \begin{bmatrix} \Pi_{ojs} & \Pi_{ajs} & -\varrho_j\Psi_{1j} & 0 \\ \bullet & -\Pi_{mj} & -\varrho_j\Psi_{3j} & 0 \\ \bullet & \bullet & -\varrho_j\Psi_{4j} & 0 \\ \bullet & \bullet & \bullet & -\Pi_{nj} \end{bmatrix}$$

$$\Xi_{cjs} = \begin{bmatrix} \Gamma_{js} & \Pi_{zjs}^t & \varrho_k(\mathcal{X}_jA_{js}^t + \mathcal{Y}_j^tB_{js}^t)\mathcal{W}_k \\ 0 & G_{djs}^t & \varrho_k\mathcal{X}_jA_{djs}^t\mathcal{W}_k \\ 0 & 0 & 0 \\ 0 & 0 & \varrho_k\sum_{k=1,k\neq j}^{n_s}\mathcal{X}_jE_{kjs}\mathcal{W}_k \end{bmatrix}$$

$$\Xi_{ejs} = \begin{bmatrix} -\gamma_j^2I_j & \Phi_j^t & \varrho_j\Gamma_j^t\mathcal{W}_j \\ \bullet & -I_j & 0 \\ & & \\ bullet & \bullet & -\varrho_k\mathcal{W}_k \end{bmatrix}$$

$$\Xi_{2js} = \begin{bmatrix} \varepsilon_jB_{js}M_j & \mathcal{X}_jN_j^t \\ 0 & 0 \\ 0 & 0 \\ 0 & 0 \\ 0 & 0 \\ \varepsilon_jD_{js}M_j & 0 \\ \varepsilon_j\varrho_j\mathcal{W}_jB_{js}M_j & 0 \end{bmatrix},\ \Xi_{3j} = \begin{bmatrix} -\varepsilon_jI & 0 \\ \bullet & -\varepsilon_jI \end{bmatrix} \tag{8.15}$$

where

$$\begin{aligned}
\Pi_{ojs} &= [A_{js} + \sum_{k=1}^{n_s} F_{kjs}]\mathcal{X}_j + B_{js}\mathcal{Y}_j + \mathcal{X}_j[A_{js} + \sum_{k=1}^{n_s} F_{kjs}]^t + \mathcal{Y}_j^t B_{js}^t \\
&+ \Psi_{1j} + \Psi_{1j}^t + \Psi_{2j} + \sum_{k=1,k\neq j}^{n_s} \Lambda_{kjs} + (n_s - 1)\mathcal{X}_j \\
\Pi_{ajs} &= A_{dj}\mathcal{X}_j - \Psi_{1j} + \Psi_{3j}^t, \; \Pi_{mj} = \Psi_{3j} + \Psi_{3j}^t + (1 - \mu_j)\Psi_{2j}, \\
\Pi_{nj} &= \sum_{k=1,k\neq j}^{n_s} (1 - \mu_{kj})\Lambda_{kjs} + \sum_{k=1}^{n_s} \mathcal{X}_j E_{kjs}^t \mathcal{P}_k E_{kjs}\mathcal{X}_j, \\
\Pi_{zjs} &= G_{js}\mathcal{X}_j + D_{js}\mathcal{Y}_j
\end{aligned} \tag{8.16}$$

Moreover, the local gain matrix is given by $K_j = \mathcal{Y}_j \mathcal{X}_j^{-1}$.

Proof: Follows by parallel development to **Theorems 6.2.6** and **6.2.7**.

Remark 8.3.3 *Similarly, the optimal performance-level* $\gamma_j, \; j = 1, .., n_s$ *can be determined in case of decentralized resilient state feedback stabilization by solving the following convex optimization problems:*

Problem A: Resilient state feedback stabilization

$$\begin{aligned}
& For \;\; j,k = 1,...,n_s, \quad Given \;\; \varrho_j, \;\; \mu_j, \;\; \varrho_{jk}, \;\; \mu_{jk}, \\
& \min_{\mathcal{X}_j,\, \mathcal{Y}_j,\, \mathcal{M}_j,\, \Omega_j,\, \{\Lambda_{rj}\}_{r=1}^{ns},\, \{\Psi_{rj}\}_1^4} \quad \gamma_j \\
& subject \;\; to \;\; LMI(8.12)
\end{aligned} \tag{8.17}$$

Next, we move to the static output feedback case. To facilitate further development, we consider the case where the set of output matrices $C_j, \; j = 1, ..., n_s$ are assumed to be of full row rank.

8.3.2 Resilient static output feedback

Consider the resilient static output feedback control

$$\begin{aligned}
u_j(t) &= [K_{oj} + \Delta K_{oj}(t)]y_j(t), \;\; \Delta K_{oj}(t) = M_{oj}\Delta_j(t)N_{oj}, \\
& \Delta_j^t(t)\Delta_j(t) \leq I, \quad \forall \, t
\end{aligned} \tag{8.18}$$

where $\Delta K_{oj}(t)$ represent additive gain perturbations with $M_{oj}, \;\; N_{oj}$ being constant matrices. Using controller (8.18) and the output equation, the closed-loop subsystem dynamics can be written as

$$\begin{aligned}
\dot{x}_j(t) &= A_j x_j(t) + A_{dj} x_j(t - \tau_j(t)) + c_j(t) + \Gamma_j w_j(t) \\
&+ B_j (K_{oj} + \Delta K_{oj})[C_j x(t) + C_{dj} x(t - \tau_j(t))] \\
&= A_{gj} x_j(t) + A_{dgj} x_j(t - \tau_j(t)) + c_j(t) + \Gamma_j w_j(t) \quad (8.19) \\
z_j(t) &= G_j x_j(t) + G_{dj} x_j(t - \tau_j(t)) + \Phi_j w_j(t) \\
&+ D_j (K_{oj} + \Delta K_{oj})[C_j x(t) + C_{dj} x(t - \tau_j(t))] \\
&= G_{gj} x_j(t) + G_{dgj} x_j(t - \tau_j(t)) + \Phi_j w_j(t) \quad (8.20) \\
c_j(t) &= \sum_{k=1, k \neq j}^{n_s} F_{jk} x_k(t) + \sum_{k=1, k \neq j}^{n_s} E_{jk} x_k(t - \eta_{jk}(t)) \quad (8.21)
\end{aligned}$$

where

$$\begin{aligned}
A_{gj} &= A_j + B_j K_{oj} C_j + B_j \Delta K_{oj} C_j = \hat{A}_{gj} + B_j \Delta K_{oj} C_j \\
A_{dgj} &= A_{dj} + B_j K_{oj} C_{dj} + B_j \Delta K_{oj} C_{dj} = \hat{A}_{dgj} + B_j \Delta K_{oj} C_{dj} \\
G_{gj} &= G_j + D_j K_{oj} C_j + D_j \Delta K_{oj} C_j = \hat{G}_{gj} + D_j \Delta K_{oj} C_j \\
G_{dgj} &= G_{dj} + D_j K_{oj} C_{dj} + D_j \Delta K_{oj} C_{dj} = \hat{G}_{dgj} + D_j \Delta K_{oj} C_{dj}
\end{aligned} \quad (8.22)$$

Proceeding further, we apply **Theorem 8.2.1** to the resulting closed-loop system (8.22). As a consequence, system (8.22) is delay-dependent asymptotically stable with $\mathcal{L}_2$-performance bound γ if there exist matrices

$$\mathcal{P}_j,\ \mathcal{Q}_j,\ \mathcal{W}_j,\ \mathcal{W}_k,\ \mathcal{Z}_{kj},\ j, k = 1, ..n_s$$

and parameter matrices Θ_j Υ_j and a scalar $\gamma_j > 0$ satisfying the following LMI:

$$\begin{bmatrix}
\Xi_{gj} & \Xi_{aj} & -\varrho_j \Theta_j & \Xi_{cj} & \mathcal{P}_j \Gamma_j & G_{gj}^t & \varrho_j A_{gj}^t \mathcal{W}_j \\
\bullet & -\Xi_{mj} & -\varrho_j \Upsilon_j & 0 & 0 & G_{dgj}^t & \varrho_j A_{dgj}^t \mathcal{W}_j \\
\bullet & \bullet & -\varrho_j \mathcal{W}_j & 0 & 0 & 0 & 0 \\
\bullet & \bullet & \bullet & -\Xi_{nj} & 0 & 0 & \varrho_j \sum_{k=1}^{n_s} E_{kj} \mathcal{W}_j \\
\bullet & \bullet & \bullet & \bullet & -\gamma_j^2 I_j & \Phi_j^t & \varrho_j \Gamma_j^t \mathcal{W}_j \\
\bullet & \bullet & \bullet & \bullet & \bullet & -I_j & 0 \\
\bullet & \bullet & \bullet & \bullet & \bullet & \bullet & -\varrho_j \mathcal{W}_j
\end{bmatrix} < 0 \quad (8.23)$$

where

$$\begin{aligned}
\Xi_{gj} &= \mathcal{P}_j[\hat{A}_{gj} + \sum_{k=1,k\neq j}^{n_s} F_{kj} + B_j \Delta K_{oj} C_j] \\
&+ \Theta_j + \Theta_j^t + \mathcal{Q}_j + [\hat{A}_{gj} + \sum_{k=1,k\neq j}^{n_s} F_{kj} + B_j \Delta K_{oj} C_j]^t \mathcal{P}_j \\
&+ \sum_{k=1,k\neq j}^{n_s} \mathcal{Z}_{kj} + (n_s - 1)\mathcal{P}_j \\
&= \mathcal{P}_j[\hat{A}_{gj} + \sum_{k=1}^{n_s} F_{kj} + B_j \Delta K_{oj} C_j] + [\hat{A}_{gj} + \sum_{k=1}^{n_s} F_{kj} + B_j \Delta K_{oj} C_j]^t \mathcal{P}_j \\
&+ \Theta_j + \Theta_j^t + \mathcal{Q}_j + \sum_{k=1,k\neq j}^{n_s} \mathcal{Z}_{kj} + (n_s - 1)\mathcal{P}_j \\
&+ \mathcal{P}_j B_j \Delta K_{oj} C_j + C_j^t \Delta K_{oj}^t B_j^t \mathcal{P}_j \\
&= \hat{\Xi}_{gj} + \mathcal{P}_j B_j \Delta K_{oj} C_j + C_j^t \Delta K_{oj}^t B_j^t \mathcal{P}_j
\end{aligned} \tag{8.24}$$

and Ξ_{aj}, Ξ_{cj}, Ξ_{mj}, Ξ_{nj} are given by (8.4). Algebraic manipulation of (8.23) using (8.22) yields

$$\begin{bmatrix}
\hat{\Xi}_{gj} & \Xi_{aj} & -\varrho_j \Theta_j & \Xi_{cj} & \mathcal{P}_j \Gamma_j & \hat{G}_{gj}^t & \varrho_k \hat{A}_{gj}^t \mathcal{W}_k \\
\bullet & -\Xi_{mj} & -\varrho_j \Upsilon_j & 0 & 0 & \hat{G}_{dgj}^t & \varrho_k \hat{A}_{dgj}^t \mathcal{W}_k \\
\bullet & \bullet & -\varrho_j \mathcal{W}_j & 0 & 0 & 0 & 0 \\
\bullet & \bullet & \bullet & -\Xi_{nj} & 0 & 0 & \varrho_k \sum_{k=1,k\neq j}^{n_s} E_{kj} \mathcal{W}_k \\
\bullet & \bullet & \bullet & \bullet & -\gamma_j^2 I_j & \Phi_j^t & \varrho_k \Gamma_j^t \mathcal{W}_k \\
\bullet & \bullet & \bullet & \bullet & \bullet & -I_j & 0 \\
\bullet & \bullet & \bullet & \bullet & \bullet & \bullet & -\varrho_k \mathcal{W}_k
\end{bmatrix}$$

$$+ \begin{bmatrix} \mathcal{P}_j B_j M_{oj} \\ 0 \\ 0 \\ 0 \\ 0 \\ D_j M_{oj} \\ \varrho_j \mathcal{W}_j B_j M_{oj} \end{bmatrix} \Delta_j \begin{bmatrix} C_j^t N_{oj}^t \\ 0 \\ 0 \\ 0 \\ 0 \\ 0 \\ 0 \end{bmatrix}^t + \begin{bmatrix} C_j^t N_{oj}^t \\ 0 \\ 0 \\ 0 \\ 0 \\ 0 \\ 0 \end{bmatrix} \Delta_j^t \begin{bmatrix} \mathcal{P}_j B_j M_{oj} \\ 0 \\ 0 \\ 0 \\ 0 \\ D_j M_{oj} \\ \varrho_j \mathcal{W}_j B_j M_{oj} \end{bmatrix}^t$$

$$+\begin{bmatrix} 0 \\ 0 \\ 0 \\ 0 \\ 0 \\ D_j M_{oj} \\ \varrho_j \mathcal{W}_j B_j M_{oj} \end{bmatrix} \Delta_j \begin{bmatrix} 0 \\ C_{dj}^t N_{oj}^t \\ 0 \\ 0 \\ 0 \\ 0 \\ 0 \end{bmatrix}^t + \begin{bmatrix} 0 \\ C_{dj}^t N_{oj}^t \\ 0 \\ 0 \\ 0 \\ 0 \\ 0 \end{bmatrix} \Delta_j^t \begin{bmatrix} 0 \\ 0 \\ 0 \\ 0 \\ 0 \\ D_j M_{oj} \\ \varrho_j \mathcal{W}_j B_j M_{oj} \end{bmatrix}^t$$
$$< 0 \tag{8.25}$$

With the aid of **Inequality 1** of the Appendix, expression (8.25) becomes

$$\leq \begin{bmatrix} \hat{\Xi}_{gj} & \Xi_{aj} & -\varrho_j \Theta_j & \Xi_{cj} & \mathcal{P}_j \Gamma_j & \hat{G}_{gj}^t & \varrho_k \hat{A}_{gj}^t \mathcal{W}_k \\ \bullet & -\Xi_{mj} & -\varrho_j \Upsilon_j & 0 & 0 & \hat{G}_{dgj}^t & \varrho_k \hat{A}_{dgj}^t \mathcal{W}_k \\ \bullet & \bullet & -\varrho_j \mathcal{W}_j & 0 & 0 & 0 & 0 \\ \bullet & \bullet & \bullet & -\Xi_{nj} & 0 & 0 & \varrho_k \sum_{k=1,k\neq j}^{n_s} E_{kj} \mathcal{W}_k \\ \bullet & \bullet & \bullet & \bullet & -\gamma_j^2 I & \Phi_j^t & \varrho_k \Gamma_j^t \mathcal{W}_k \\ \bullet & \bullet & \bullet & \bullet & \bullet & -I & 0 \\ \bullet & \bullet & \bullet & \bullet & \bullet & \bullet & -\varrho_k \mathcal{W}_k \end{bmatrix}$$
$$+\varepsilon_j \begin{bmatrix} \mathcal{P}_j B_j M_{oj} \\ 0 \\ 0 \\ 0 \\ 0 \\ D_j M_{oj} \\ \varrho_j \mathcal{W}_j B_j M_{oj} \end{bmatrix} \begin{bmatrix} \mathcal{P}_j B_j M_{oj} \\ 0 \\ 0 \\ 0 \\ 0 \\ D_j M_{oj} \\ \varrho_j \mathcal{W}_j B_j M_{oj} \end{bmatrix}^t + \varepsilon_j^{-1} \begin{bmatrix} C_j^t N_{oj}^t \\ 0 \\ 0 \\ 0 \\ 0 \\ 0 \\ 0 \end{bmatrix} \begin{bmatrix} C_j^t N_{oj}^t \\ 0 \\ 0 \\ 0 \\ 0 \\ 0 \\ 0 \end{bmatrix}^t$$
$$+\varphi_j \begin{bmatrix} 0 \\ 0 \\ 0 \\ 0 \\ 0 \\ D_j M_{oj} \\ \varrho_j \mathcal{W}_j B_j M_{oj} \end{bmatrix} \begin{bmatrix} 0 \\ 0 \\ 0 \\ 0 \\ 0 \\ D_j M_{oj} \\ \varrho_j \mathcal{W}_j B_j M_{oj} \end{bmatrix}^t + \varphi_j^{-1} \begin{bmatrix} 0 \\ C_{dj}^t N_{oj}^t \\ 0 \\ 0 \\ 0 \\ 0 \\ 0 \end{bmatrix} \begin{bmatrix} 0 \\ C_{dj}^t N_{oj}^t \\ 0 \\ 0 \\ 0 \\ 0 \\ 0 \end{bmatrix}^t$$
$$< 0 \tag{8.26}$$

for some $\varepsilon_j > 0,\ \varphi_j > 0$. By Schur complements, we express the last five terms in (8.26) into the form

$$\begin{bmatrix} \Xi_{4j} & \Xi_{5j} \\ \bullet & \Xi_{6j} \end{bmatrix} < 0 \tag{8.27}$$

where

$$\Xi_{4j} = \begin{bmatrix} \Xi_{41j} & \Xi_{42j} \\ \bullet & \Xi_{43j} \end{bmatrix} \quad (8.28)$$

where

$$\Xi_{41j} = \begin{bmatrix} \hat{\Xi}_{gj} & \Xi_{aj} & -\varrho_j \Theta_j & 0 \\ \bullet & -\Xi_{mj} & -\varrho_j \Upsilon_j & 0 \\ \bullet & \bullet & -\varrho_j \mathcal{W}_j & 0 \\ \bullet & \bullet & \bullet & -\Xi_{nj} \end{bmatrix}$$

$$\Xi_{42j} = \begin{bmatrix} \mathcal{P}_j \Gamma_j & \hat{G}_{gj}^t & \varrho_j \hat{A}_{gj}^t \mathcal{W}_j \\ 0 & \hat{G}_{dgj}^t & \varrho_k \hat{A}_{dgj}^t \mathcal{W}_k \\ 0 & 0 & 0 \\ 0 & 0 & \varrho_k \sum_{k=1,k\neq j}^{n_s} E_{kj} \mathcal{W}_k \end{bmatrix}$$

$$\Xi_{43j} = \begin{bmatrix} -\gamma_j^2 I_j & \Phi_j^t & \varrho_k \Gamma_j^t \mathcal{W}_k \\ \bullet & -I_j & 0 \\ \bullet & \bullet & -\varrho_k \mathcal{W}_k \end{bmatrix} \quad (8.29)$$

and

$$\Xi_{5j} = \begin{bmatrix} \varepsilon_j \mathcal{P}_j B_j M_{oj} & C_j^t N_{oj}^t & 0 & 0 \\ 0 & 0 & C_{dj}^t N_{oj}^t & 0 \\ 0 & 0 & 0 & 0 \\ 0 & 0 & 0 & 0 \\ 0 & 0 & 0 & 0 \\ \varepsilon_j D_j M_{oj} & 0 & \varphi_j D_j M_{oj} & 0 \\ \varepsilon_j \varrho_j \mathcal{W}_j B_j M_{oj} & 0 & \varphi_j \varrho_j \mathcal{W}_j B_j M_{oj} & 0 \end{bmatrix}$$

$$\Xi_{6j} = \begin{bmatrix} -\varepsilon_j I_j & 0 & 0 & 0 \\ \bullet & -\varepsilon_j I_j & 0 & 0 \\ \bullet & \bullet & -\varphi_j I_j & 0 \\ \bullet & \bullet & \bullet & -\varphi_j I_j \end{bmatrix}$$

$$\Xi_{6j} = \begin{bmatrix} -\varepsilon_j I_j & 0 & 0 & 0 \\ \bullet & -\varepsilon_j I_j & 0 & 0 \\ \bullet & \bullet & -\varphi_j I_j & 0 \\ \bullet & \bullet & \bullet & -\varphi_j I_j \end{bmatrix} \quad (8.30)$$

The main decentralized resilient feedback design results are established by the following theorem.

Theorem 8.3.4 *Given the bounds $\varrho_j > 0,\ \mu_j > 0$ and matrix $\mathcal{W}_j,\ j = 1, ..., n_s$, the interconnected system (8.1) with decentralized resilient controller (8.19) is delay-dependent asymptotically stable with $\mathcal{L}_2$-performance bound γ_j if there exist weighting matrices $\mathcal{X}_j,\ \mathcal{G}_j,\ \mathcal{R}_j,\ \{\Lambda_{kj}\}_{k=1}^{n_s},\ \{\Psi_{rj}\}_1^7$, and scalar $\gamma_j > 0,\ \varepsilon_j > 0$ satisfying the following LMI:*

$$\begin{bmatrix} \Xi_{7j} & \Xi_{8j} \\ \bullet & \Xi_{6j} \end{bmatrix} < 0 \tag{8.31}$$

$$C_j \mathcal{X}_j = \mathcal{G}_j C_j \tag{8.32}$$

where

$$\begin{aligned}
\Xi_{7j} &= \begin{bmatrix} \Xi_{71j} & \Xi_{72j} \\ \bullet & \Xi_{73j} \end{bmatrix} \\
\Xi_{71j} &= \begin{bmatrix} \Pi_{oj} & \Pi_{aj} & -\varrho_j \Psi_{1j} & 0 \\ \bullet & -\Pi_{mj} & -\varrho_j \Psi_{3j} & 0 \\ \bullet & \bullet & -\varrho_j \Psi_{4j} & 0 \\ \bullet & \bullet & \bullet & -\Pi_{nj} \end{bmatrix} \\
\Xi_{72j} &= \begin{bmatrix} \Gamma_j & \mathcal{X}_j G_j^t + \Psi_{5j} & \varrho_k (\mathcal{X}_j A_j^t + C_j^t \mathcal{R}_j^t B_j^t) \mathcal{W}_k \\ 0 & \mathcal{X}_j G_{dj}^t + \Psi_{6j} & \varrho_k (\mathcal{X}_j A_{dj}^t + \Psi_{7j}) \mathcal{W}_k \\ 0 & 0 & 0 \\ 0 & 0 & \varrho_k \sum_{k=1,k\neq j}^{n_s} \mathcal{X}_j E_{kj} \mathcal{W}_k \end{bmatrix} \\
\Xi_{73j} &= \begin{bmatrix} -\gamma_j^2 I_j & \Phi_j^t & \varrho_k \Gamma_j^t \mathcal{W}_k \\ \bullet & -I_j & 0 \\ \bullet & \bullet & -\varrho_k \mathcal{W}_k \end{bmatrix} \\
\Xi_{8j} &= \begin{bmatrix} \varepsilon_j B_j M_{oj} & \mathcal{X}_j C_j^t N_{oj}^t & 0 & 0 \\ 0 & 0 & C_{dj}^t N_{oj}^t & 0 \\ 0 & 0 & 0 & 0 \\ 0 & 0 & 0 & 0 \\ 0 & 0 & 0 & 0 \\ \varepsilon_j D_j M_{oj} & 0 & \varphi_j D_j M_{oj} & 0 \\ \varepsilon_j \varrho_j \mathcal{W}_j B_j M_{oj} & 0 & \varphi_j \varrho_j \mathcal{W}_j B_j M_{oj} & 0 \end{bmatrix} \qquad (8.33) \\
\Pi_{oj} &= [A_j + \sum_{k=1,k\neq j}^{n_s} F_{kj}] \mathcal{X}_j + B_j \mathcal{R}_j C_j + \mathcal{X}_j [A_j \\
&+ \sum_{k=1,k\neq j}^{n_s} F_{kj}]^t + C_j^t \mathcal{R}_j^t B_j^t \\
&+ \Psi_{1j} + \Psi_{1j}^t + \Psi_{2j} + \sum_{k=1,k\neq j}^{n_s} \Lambda_{kj} + (n_s - 1) \mathcal{X}_j
\end{aligned}$$

$$\begin{aligned}
\Pi_{aj} &= A_{dj}\mathcal{X}_j - \Psi_{1j} + \Psi_{3j}^t, \; \Pi_{mj} = \Psi_{3j} + \Psi_{3j}^t + (1-\mu_j)\Psi_{2j}, \\
\Pi_{zj} &= D_{js}\mathcal{Y}_j \\
\Pi_{nj} &= \sum_{k=1,k\neq j}^{n_s} (1-\mu_{kj})\Lambda_{kj} + \sum_{k=1,k\neq j}^{n_s} E_{kj}\mathcal{X}_j E_{kj}^t \mathcal{P}_k E_{kj}\mathcal{X}_j \qquad (8.34)
\end{aligned}$$

Moreover, the local gain matrix is given by $K_{oj} = \mathcal{R}_j \mathcal{G}_j^{-1}$.

Proof: Introduce the relation $C_j\mathcal{X}_j = \mathcal{G}_j C_j$ and let $K_{oj}\mathcal{G}_j = \mathcal{R}_j$. Applying the congruent transformation $diag[\mathcal{X}_j,\ \mathcal{X}_j,\ \mathcal{X}_j,\ \mathcal{X}_j,\ I_j,\ I_j,\ I_j,\ I_j,\ I_j],\ \mathcal{X}_j = \mathcal{P}_j^{-1}$ to LMI (8.31) with Schur complements and using the linearizations

$$\begin{aligned}
\Psi_{1j} &= \mathcal{X}_j\Theta_j\mathcal{X}_j,\ \Psi_{2j} = \mathcal{X}_j\mathcal{Q}_j\mathcal{X}_j,\ \Psi_{3j} = \mathcal{X}_j\Upsilon_j\mathcal{X}_j,\ \Omega_j = \alpha_j\mathcal{X}_j F_j^t, \\
\Lambda_{kj} &= \mathcal{X}_j\mathcal{Z}_{kj}\mathcal{X}_j,\ \Psi_{7j} = \mathcal{X}_j C_{dj}^t K_{oj}^t B_j^t, \\
\Psi_{4j} &= \mathcal{X}_j\mathcal{W}_j\mathcal{X}_j,\ \Psi_{5j} = \mathcal{X}_j C_j^t K_{oj}^t D_j^t,\ \Psi_{6j} = \mathcal{X}_j C_{dj}^t K_{oj}^t D_j^t
\end{aligned}$$

we readily obtain LMI (8.33) by Schur complements.

We note that the presence of matrix equality in (8.32) renders the computations of (8.31)-(8.33) using MATLAB®-LMI Toolbox or similar LMI solver rather costly. Therefore one is encouraged to convert (8.31)-(8.32) into true LMIs. With this in mind, we recall that the use of singular value decomposition (SVD) can express the output matrix C_j in the form

$$C_j = \mathcal{U}_j\ [\Lambda_{pj},\ \ 0]\ \mathcal{V}_j^t \qquad (8.35)$$

where $\mathcal{U}_j \in \Re^{p_j \times p_j}$, $\mathcal{V}_j \in \Re^{n_j \times n_j}$ are unitary matrices and $\Lambda_{pi} \in \Re^{p_j \times p_j}$ is a diagonal matrix with positive diagonal elements in decreasing order. The conversion to LMIs can now be accomplished by the following theorem [258].

Theorem 8.3.5 *Given a matrix* $C_j \in \Re^{p_j \times n_j}$, $rank[C_j] = p_j$ *and let* $0 < \mathcal{X}_j = \mathcal{X}_j^t \in \Re^{n_j \times n_j}$. *Then there exists a matrix* $0 < \mathcal{G}_j \in \Re^{p_j \times p_j}$ *such that*

$$C_j\mathcal{X}_j = \mathcal{G}_j C_j \qquad (8.36)$$

if and only if

$$\mathcal{X}_j = \mathcal{V}_j \begin{bmatrix} \mathcal{X}_{ju} & 0 \\ \bullet & \mathcal{X}_{jv} \end{bmatrix} \mathcal{V}_j^t,\ \ \mathcal{X}_{iu} \in \Re^{p_j \times p_j},\ \ \mathcal{X}_{iv} \in \Re^{(n_j - p_j)\times(n_j - p_j)} \qquad (8.37)$$

It is significant to observe that **Theorem 8.3.5** substitutes the matrix equation (8.32) by structural selection of the matrix variable $\mathcal{X}_j$. Incorporating this result into **Theorem 8.3.5**, we have thus established the following design result.

Theorem 8.3.6 *Given the bounds $\varrho_j > 0,\ \mu_j > 0$ and matrix $\mathcal{W}_j$, the interconnected system (8.1) with decentralized resilient controller (8.19) with output matrix C_j having the SVD form (8.35) is delay-dependent asymptotically stable with $\mathcal{L}_2$-performance bound γ_j if there exist weighting matrices*

$$\mathcal{X}_j,\ \mathcal{G}_j,\ \mathcal{R}_j,\ \{\Lambda_{kj}\}_{k=1}^{n_s},\ \{\Psi_{rj}\}_1^7$$

and scalars $\varepsilon_j > 0,\ \varphi_j > 0$ satisfying the following LMI:

$$\begin{bmatrix} \Xi_{7j} & \Xi_{8j} \\ \bullet & \Xi_{6j} \end{bmatrix} < 0 \tag{8.38}$$

where $\Xi_{7j},\ \Xi_{8j},\ \Pi_{oj},\ \Pi_{aj},\ \Pi_{nj},\ \Pi_{mj},\ \Pi_{cj}$ are given by (8.34). Moreover, the local gain matrix is given by $K_{oj} = \mathcal{R}_j \mathcal{U}_j \Lambda_{pj} \mathcal{X}_{ju}^{-1} \Lambda_{pj}^{-1} \mathcal{U}_j^{-1}$.

Remark 8.3.7 *Similarly, the optimal performance-level $\gamma_j,\ j = 1,..,n_s$ can be determined in case of decentralized resilient static output-feedback stabilization by solving the following convex optimization problems:*

Problem B: Resilient static output feedback stabilization

$$\begin{aligned} &For\ \ j,k = 1,...,n_s, \quad Given\ \ \varrho_j,\ \mu_j,\ \varrho_{jk},\ \mu_{jk}, \\ &\min_{\mathcal{X}_j,\ \mathcal{Y}_j,\ \mathcal{M}_j,\ \Omega_j,\ \{\Lambda_{rj}\}_{r=1}^{n_s},\ \{\Psi_{rj}\}_1^4} \quad \gamma_j \\ &subject\ \ to \quad LMI(8.38) \end{aligned} \tag{8.39}$$

Finally, we deal with the dynamic output feedback case.

8.3.3 Resilient dynamic output feedback

Consider the resilient dynamic output feedback control

$$\begin{aligned} \dot{\hat{x}}_j(t) &= A_j \hat{x}_j(t) + B_j u_j(t) + [K_{oj} + \Delta K_{oj}(t)][y_j(t) - C_j \hat{x}_j(t)] \\ u_j(t) &= [K_{cj} + \Delta K_{cj}(t)]\hat{x}_j(t), \end{aligned} \tag{8.40}$$

$$\begin{aligned} \Delta K_{oj}(t) &= M_{oj}\Delta_j(t)N_{oj},\ \ \Delta K_{cj}(t) = M_{cj}\Delta_j(t)N_{cj}, \\ &\Delta_j^t(t)\Delta_j(t) \le I,\ \ \forall\, t \end{aligned} \tag{8.41}$$

where $\Delta K_{oj}(t),\ \Delta K_{cj}(t)$ represent additive observer-gain and controller-gain perturbations with $M_{oj},\ N_{oj},\ M_{cj},\ N_{cj}$ being constant matrices. Applying the dynamic controller (8.40) to the linear system (8.1), we obtain the closed-loop system and associated matrices

$$
\begin{aligned}
e_j(t) &= \left[\begin{array}{cc} x_j^t(t) & x_j^t(t) - \hat{x}_j^t(t) \end{array} \right]^t && (8.42) \\
\dot{e}_j(t) &= [\mathcal{A}_j + \Delta\mathcal{A}_j]e_j(t) + [\mathcal{A}_{dj} + \Delta\mathcal{A}_{dj}]e_j(t - \tau_j(t)) + \hat{c}_j(t) + \hat{\Gamma}_j w_j(t) \\
z_j(t) &= [\hat{G}_j + \Delta\hat{G}_j]e_j(t) + \hat{G}_{dj} e(t - \tau_j(t)) + \Phi_j w_j(t) \\
c_j(t) &= \sum_{k=1}^{n_s} \hat{F}_{jk} e_k(t) + \sum_{k=1}^{n_s} \hat{E}_{jk} e_k(t - \eta_{jk}(t)) && (8.43)
\end{aligned}
$$

where

$$
\begin{aligned}
\mathcal{A}_j &= \left[\begin{array}{cc} A_j + B_j K_{cj} & -B_j K_{cj} \\ 0 & A_j - K_{oj} C_j \end{array} \right], \; \Delta\mathcal{A}_{dj} = \left[\begin{array}{cc} 0 & 0 \\ -\Delta K_{oj} C_{dj} & 0 \end{array} \right], \\
\mathcal{A}_{dj} &= \left[\begin{array}{cc} A_{dj} & 0 \\ 0 & A_{dj} - K_{oj} C_{dj} \end{array} \right], \; \Delta\mathcal{A}_j = \left[\begin{array}{cc} B_j \Delta K_{cj} & -B_j \Delta K_{cj} \\ 0 & -\Delta K_{oj} C_j \end{array} \right], \\
\hat{G}_j &= \left[\begin{array}{cc} G_j + D_j K_{cj} & -D_j K_{cj} \end{array} \right], \; \hat{G}_{dj} = \left[\begin{array}{cc} G_{dj} & 0 \end{array} \right], \\
\hat{c}_j(t) &= \left[\begin{array}{c} c_j(t) \\ c_j(t) \end{array} \right], \; \hat{\Gamma}_j = \left[\begin{array}{c} \Gamma_j \\ \Gamma_j \end{array} \right], \; \hat{F}_{jk} = \left[\begin{array}{cc} F_{jk} & 0 \\ F_{jk} & 0 \end{array} \right], \\
\hat{E}_{jk} &= \left[\begin{array}{cc} E_{jk} & 0 \\ E_{jk} & 0 \end{array} \right], \\
\Delta\hat{G}_j &= \left[\begin{array}{cc} D_j \Delta K_{cj}(t) & -D_j \Delta K_{cj}(t) \end{array} \right] && (8.44)
\end{aligned}
$$

Then it follows from **Theorem 8.2.1** that the resulting closed-loop system (8.6) is delay-dependent asymptotically stable with $\mathcal{L}_2$-performance bound γ if there exist matrices $\widehat{\mathcal{P}}_j$, $\widehat{\mathcal{Q}}_j$, $\widehat{\mathcal{W}}_j$, $\widehat{\mathcal{Z}}_{kj}$, $k = 1, ..n_s$, parameter matrices $\widehat{\Theta}_j$ $\widehat{\Upsilon}_j$, and a scalar $\gamma_j > 0$ satisfying the following LMI:

$$
\left[\begin{array}{cc} \widehat{\Xi}_{s1j} & \widehat{\Xi}_{s2j} \\ \bullet & \widehat{\Xi}s3j \end{array} \right] < 0 \qquad (8.45)
$$

with

$$\widehat{\Xi}_{s1j} = \begin{bmatrix} \widehat{\Xi}_{oj} & \widehat{\Xi}_{aj} & -\varrho_j\widehat{\Theta}_j & 0 \\ \bullet & -\widehat{\Xi}_{mj} & -\varrho_j\widehat{\Upsilon}_j & 0 \\ \bullet & \bullet & -\varrho_j\widehat{\mathcal{W}}_j & 0 \\ \bullet & \bullet & \bullet & -\widehat{\Xi}_{nj} \end{bmatrix}$$

$$\widehat{\Xi}_{s2j} = \begin{bmatrix} \widehat{\mathcal{P}}_j\hat{\Gamma}_j & \hat{G}_j^t + \Delta\hat{G}_j^t & \varrho_k[\mathcal{A}_j + \Delta\mathcal{A}_j]^t\widehat{\mathcal{W}}_k \\ 0 & \hat{G}_{dj}^t & \varrho_k[\mathcal{A}_{dj} + \Delta\mathcal{A}_{dj}]^t\widehat{\mathcal{W}}_k \\ 0 & 0 & 0 \\ 0 & 0 & \varrho_k\sum_{k=1,k\neq j}^{n_s}\hat{E}_{kj}\widehat{\mathcal{W}}_k \end{bmatrix} < 0$$

$$\widehat{\Xi}_{s3j} = \begin{bmatrix} -\gamma_j^2 I_j & \Phi_j^t & \varrho_k\hat{\Gamma}_j^t\widehat{\mathcal{W}}_k \\ \bullet & -I_j & 0 \\ \bullet & \bullet & -\varrho_k\widehat{\mathcal{W}}_k \end{bmatrix} < 0 \tag{8.46}$$

where

$$\begin{aligned} \widehat{\Xi}_{oj} &= \widehat{\mathcal{P}}_j[\mathcal{A}_j + \Delta\mathcal{A}_j + \sum_{k=1}^{n_s}\hat{F}_{kj}] + [\mathcal{A}_j + \Delta\mathcal{A}_j + \sum_{k=1}^{n_s}\hat{F}_{kj}]^t\widehat{\mathcal{P}}_j \\ &+ \widehat{\Theta}_j + \widehat{\Theta}_j^t + \widehat{\mathcal{Q}}_j + \sum_{k=1,k\neq j}^{n_s}\widehat{\mathcal{Z}}_{kj} + (n_s-1)\widehat{\mathcal{P}}_j \\ &= \overline{\Xi}_{oj} + \widehat{\mathcal{P}}_j\Delta\mathcal{A}_j + \Delta\mathcal{A}_j^t\widehat{\mathcal{P}}_j \\ \widehat{\Xi}_{mj} &= \widehat{\Upsilon}_j + \widehat{\Upsilon}_j^t + (1-\mu_j)\widehat{\mathcal{Q}}_j, \\ \widehat{\Xi}_{nj} &= \sum_{k=1}^{n_s}(1-\mu_{kj})\widehat{\mathcal{Z}}_{kj} + \sum_{k=1}^{n_s}\hat{E}_{kj}^t\widehat{\mathcal{P}}_j\hat{E}_{kj}, \\ \widehat{\Xi}_{aj} &= \widehat{\mathcal{P}}_j[\mathcal{A}_{dj} + \Delta\mathcal{A}_{dj}] - \widehat{\Theta}_j + \widehat{\Upsilon}_j^t = \overline{\Xi}_{aj} + \widehat{\mathcal{P}}_j\Delta\mathcal{A}_{dj} \end{aligned} \tag{8.47}$$

along with the following block matrices:

$$\begin{aligned} \widehat{\mathcal{P}}_j &= \begin{bmatrix} \mathcal{P}_{cj} & 0 \\ \bullet & \mathcal{P}_{oj} \end{bmatrix}, \widehat{\mathcal{Q}}_j = \begin{bmatrix} \mathcal{Q}_{cj} & 0 \\ \bullet & \mathcal{Q}_{oj} \end{bmatrix}, \\ \widehat{\mathcal{W}}_j &= \begin{bmatrix} \mathcal{W}_{cj} & 0 \\ \bullet & \mathcal{W}_{oj} \end{bmatrix}, \widehat{\mathcal{Z}}_{jk} = \begin{bmatrix} \mathcal{Z}_{cjk} & 0 \\ \bullet & \mathcal{Z}_{ojk} \end{bmatrix}, \\ \widehat{\Theta}_j &= \begin{bmatrix} \Theta_{cj} & 0 \\ \bullet & \Theta_{oj} \end{bmatrix}, \widehat{\Upsilon}_j = \begin{bmatrix} \Upsilon_{cj} & 0 \\ \bullet & \Upsilon_{oj} \end{bmatrix} \end{aligned} \tag{8.48}$$

Manipulating LMI (8.45) using (8.47)-(8.48), it yields

$$
\begin{bmatrix}
\overline{\Xi}_{oj} & \overline{\Xi}_{aj} & -\varrho_j\widehat{\Theta}_j & \widehat{\Xi}cj & \widehat{\mathcal{P}}_j\hat{\Gamma}_j & \hat{G}_j^t & \varrho_k\mathcal{A}_j^t\widehat{\mathcal{W}}_k \\
\bullet & -\widehat{\Xi}_{mj} & -\varrho_j\widehat{\Upsilon}_j & 0 & 0 & \hat{G}_{dj}^t & \varrho_k\mathcal{A}_{dj}^t\widehat{\mathcal{W}}_k \\
\bullet & \bullet & -\varrho_j\widehat{\mathcal{W}}_j & 0 & 0 & 0 & 0 \\
\bullet & \bullet & \bullet & -\widehat{\Xi}_{nj} & 0 & 0 & \varrho_k\sum_{k=1,k\neq j}^{n_s}\hat{E}_{kj}\widehat{\mathcal{W}}_k \\
\bullet & \bullet & \bullet & \bullet & -\gamma_j^2 I_j & \Phi_j^t & \varrho_k\hat{\Gamma}_j^t\widehat{\mathcal{W}}_k \\
\bullet & \bullet & \bullet & \bullet & \bullet & -I_j & 0 \\
\bullet & \bullet & \bullet & \bullet & \bullet & \bullet & -\varrho_k\widehat{\mathcal{W}}_k
\end{bmatrix}
$$

$$
+ \begin{bmatrix} H_{aj} \\ 0 \\ 0 \\ 0 \\ 0 \\ D_jM_{cj} \\ H_{cj} \end{bmatrix} \Delta_j \begin{bmatrix} H_{oj} \\ 0 \\ 0 \\ 0 \\ 0 \\ 0 \\ O_j \end{bmatrix}^t + \begin{bmatrix} H_{oj} \\ 0 \\ 0 \\ 0 \\ 0 \\ 0 \\ O_j \end{bmatrix} \Delta_j^t \begin{bmatrix} H_{aj} \\ 0 \\ 0 \\ 0 \\ 0 \\ D_jM_{cj} \\ H_{wj} \end{bmatrix}^t
$$

$$
+ \begin{bmatrix} H_{sj} \\ 0 \\ 0 \\ 0 \\ 0 \\ D_jM_{cj} \\ H_{wj} \end{bmatrix} \Delta_j \begin{bmatrix} H_{vj} \\ 0 \\ 0 \\ 0 \\ 0 \\ 0 \\ O_j \end{bmatrix}^t + \begin{bmatrix} H_{vj} \\ 0 \\ 0 \\ 0 \\ 0 \\ 0 \\ O_j \end{bmatrix} \Delta_j^t \begin{bmatrix} H_{sj} \\ 0 \\ 0 \\ 0 \\ 0 \\ D_jM_{cj} \\ H_{wj} \end{bmatrix}^t
$$

$$
+ \begin{bmatrix} -H_{sj} \\ 0 \\ 0 \\ 0 \\ 0 \\ 0 \\ -H_{wj} \end{bmatrix} \Delta_j \begin{bmatrix} H_{ej} \\ 0 \\ 0 \\ 0 \\ 0 \\ 0 \\ O_j \end{bmatrix}^t + \begin{bmatrix} H_{ej} \\ 0 \\ 0 \\ 0 \\ 0 \\ 0 \\ O_j \end{bmatrix} \Delta_j^t \begin{bmatrix} -H_{sj} \\ 0 \\ 0 \\ 0 \\ 0 \\ 0 \\ -H_{wj} \end{bmatrix}^t
$$

$$
< 0 \qquad (8.49)
$$

Applying **Inequality 1** of the Appendix to LMI (8.49), it becomes

$$
\begin{aligned}
&\leq \begin{bmatrix} \Xi_{oj} & \Xi_{aj} & -\varrho_j\widehat{\Theta}_j & \widehat{\Xi}_{cj} & \widehat{\mathcal{P}}_j\hat{\Gamma}_j & \hat{G}_j^t & \varrho_k\mathcal{A}_j^t\widehat{\mathcal{W}}_k \\ \bullet & -\widehat{\Xi}_{mj} & -\varrho_j\widehat{\Upsilon}_j & 0 & 0 & \hat{G}_{dj}^t & \varrho_k\mathcal{A}_{dj}^t\widehat{\mathcal{W}}_k \\ \bullet & \bullet & -\varrho_j\widehat{\mathcal{W}}_j & 0 & 0 & 0 & 0 \\ \bullet & \bullet & \bullet & -\widehat{\Xi}_{nj} & 0 & 0 & \varrho_k\sum_{k=1,k\neq j}^{n_s}\hat{E}_{kj}\widehat{\mathcal{W}}_k \\ \bullet & \bullet & \bullet & \bullet & -\gamma_j^2 I_j & \Phi_j^t & \varrho_k\hat{\Gamma}_j^t\widehat{\mathcal{W}}_k \\ \bullet & \bullet & \bullet & \bullet & \bullet & -I_j & 0 \\ \bullet & \bullet & \bullet & \bullet & \bullet & \bullet & -\varrho_k\widehat{\mathcal{W}}_k \end{bmatrix} \\
&+\varepsilon_j \begin{bmatrix} H_{aj} \\ 0 \\ 0 \\ 0 \\ 0 \\ D_jM_{cj} \\ H_{cj} \end{bmatrix} \begin{bmatrix} H_{aj} \\ 0 \\ 0 \\ 0 \\ 0 \\ D_jM_{cj} \\ H_{cj} \end{bmatrix}^t + \varepsilon_j^{-1} \begin{bmatrix} H_{oj} \\ 0 \\ 0 \\ 0 \\ 0 \\ 0 \\ O_j \end{bmatrix} \begin{bmatrix} H_{oj} \\ 0 \\ 0 \\ 0 \\ 0 \\ 0 \\ O_j \end{bmatrix}^t \\
&+\varphi_j \begin{bmatrix} H_{sj} \\ 0 \\ 0 \\ 0 \\ 0 \\ D_jM_{cj} \\ H_{wj} \end{bmatrix} \begin{bmatrix} H_{sj} \\ 0 \\ 0 \\ 0 \\ 0 \\ D_jM_{cj} \\ H_{wj} \end{bmatrix}^t + \varphi_j^{-1} \begin{bmatrix} H_{vj} \\ 0 \\ 0 \\ 0 \\ 0 \\ 0 \\ O_j \end{bmatrix} \begin{bmatrix} H_{vj} \\ 0 \\ 0 \\ 0 \\ 0 \\ 0 \\ O_j \end{bmatrix}^t \\
&+\psi_j \begin{bmatrix} -H_{sj} \\ 0 \\ 0 \\ 0 \\ 0 \\ D_jM_{cj} \\ -H_{wj} \end{bmatrix} \begin{bmatrix} -H_{sj} \\ 0 \\ 0 \\ 0 \\ 0 \\ D_jM_{cj} \\ -H_{wj} \end{bmatrix}^t + \psi_j^{-1} \begin{bmatrix} H_{ej} \\ 0 \\ 0 \\ 0 \\ 0 \\ 0 \\ O_j \end{bmatrix} \begin{bmatrix} H_{ej} \\ 0 \\ 0 \\ 0 \\ 0 \\ 0 \\ O_j \end{bmatrix}^t \\
&< 0 \qquad (8.50)
\end{aligned}
$$

where

$$
\begin{aligned}
H_{aj} &= \begin{bmatrix} \mathcal{P}_{cj}B_jM_{cj} \\ 0 \end{bmatrix}, \; H_{oj} = \begin{bmatrix} N_{cj}^t \\ -N_{cj}^t \end{bmatrix}, \; H_{cj} = \begin{bmatrix} \varrho_j\mathcal{W}_{cj}B_jM_{cj} \\ 0 \end{bmatrix}, \\
H_{sj} &= \begin{bmatrix} 0 \\ \mathcal{P}_{oj}M_{oj} \end{bmatrix}, \; H_{ej} = \begin{bmatrix} C_{dj}^tN_{oj}^t \\ 0 \end{bmatrix}, \; O_j = \begin{bmatrix} 0 \\ 0 \end{bmatrix}, \\
H_{vj} &= \begin{bmatrix} C_{dj}^tN_{oj}^t \\ 0 \end{bmatrix}, \; H_{wj} = \begin{bmatrix} 0 \\ \varrho_j\mathcal{W}_{oj}M_{oj} \end{bmatrix} \qquad (8.51)
\end{aligned}
$$

for some $\varepsilon_j > 0, \ \varphi_j > 0, \ \psi_j > 0$. Inequality (8.50) with the help of Schur complements can be written as

$$\begin{bmatrix} \overline{\Xi}_{1j} & \overline{\Xi}_{2j} \\ \bullet & \widehat{\Xi}_{3j} \end{bmatrix} < 0 \tag{8.52}$$

where

$$\overline{\Xi}_{1j} = \begin{bmatrix} \overline{\Xi}_{oj} & \overline{\Xi}_{aj} & -\varrho_j \widehat{\Theta}_j & 0 \\ \bullet & -\widehat{\Xi}_{mj} & -\varrho_j \widehat{\Upsilon}_j & 0 \\ \bullet & \bullet & -\varrho_j \widehat{\mathcal{W}}_j & 0 \\ \bullet & \bullet & \bullet & -\widehat{\Xi}_{nj} \end{bmatrix},$$

$$\overline{\Xi}_{2j} = \begin{bmatrix} \widehat{\mathcal{P}}_j \hat{\Gamma}_j & \hat{G}_j^t & \varrho_k \mathcal{A}_j^t \widehat{\mathcal{W}}_k & \Xi_{sj} \\ 0 & \hat{G}_{dj}^t & \varrho_k \mathcal{A}_{dj}^t \widehat{\mathcal{W}}_k & 0 \\ 0 & 0 & 0 & 0 \\ 0 & 0 & \varrho_j \sum_{k=1,k\neq j}^{n_s} \hat{E}_{kj} \widehat{\mathcal{W}}_k & 0 \end{bmatrix},$$

$$\overline{\Xi}_{3j} = \begin{bmatrix} -\gamma_j^2 I_j & \Phi_j^t & \varrho_k \hat{\Gamma}_j^t \widehat{\mathcal{W}}_k & 0 \\ \bullet & -I_j & 0 & 0 \\ \bullet & \bullet & -\varrho_k \widehat{\mathcal{W}}_k & 0 \\ \bullet & \bullet & \bullet & -\Xi_{vj} \end{bmatrix} \tag{8.53}$$

and

$$\Xi_{sj} = \begin{bmatrix} \varepsilon_j H_{aj} & H_{oj} & \varphi_j H_{sj} & H_{vj} & -\psi_j H_{sj} & H_{ej} \\ 0 & 0 & 0 & 0 & 0 & 0 \\ 0 & 0 & 0 & 0 & 0 & 0 \\ 0 & 0 & 0 & 0 & 0 & 0 \\ 0 & 0 & 0 & 0 & 0 & 0 \\ \varepsilon_j D_j M_{cj} & 0 & \varphi_j D_j M_{cj} & 0 & \psi_j D_j M_{cj} & 0 \\ \varepsilon_j H_{cj} & O_j & \varphi_j H_{wj} & O_j & -\psi_j H_{wj} & O_j \end{bmatrix}$$

$$\Xi_{vj} = Blockdiag \begin{bmatrix} \varepsilon_j I_j & \varepsilon_j I_j & \varphi_j I_j & \varphi_j I_j & \psi_j I_j & \psi_j I_j \end{bmatrix} \tag{8.54}$$

Our immediate task is to determine the observer and controller gain matrices K_{oj}, K_{cj} and we do this by convexification of inequality (8.52). The main decentralized resilient feedback design results are summarized by the following theorems.

Theorem 8.3.8 *Given the bounds $\varrho_j > 0, \ \mu_j > 0$ and matrices $\mathcal{W}_{cj}, \ \mathcal{W}_{oj}$, the interconnected system (8.1) with decentralized resilient observer-based controller (8.40) is delay-dependent asymptotically stable* *with* $\mathcal{L}_2$*-performance bound* γ_j *if there exist positive-definite matrices* $\mathcal{X}_{cj}$, $\mathcal{X}_{oj}$, $\mathcal{Y}_{cj}$, $\mathcal{Y}_{oj}$, $\mathcal{Y}_{mj}$, $\mathcal{Y}_{vj}$, $\mathcal{Y}_{wj}$, $\{\Lambda_{kj1}\}_{k=1}^{n_s}$, $\{\Lambda_{kj1}\}_{k=1}^{n_s}$, $\{\Psi_{rsj}\}_{r=1}^{4}$, $s = 1, 2$, *and scalars* $\varepsilon_j > 0$, $\varphi_j > 0$, $\psi_j > 0$ *satisfying the following LMIs for* $j = 1, ..., n_s$:

$$\begin{bmatrix} \widetilde{\Xi}_{d1j} & \widetilde{\Xi}_{d2j} \\ \bullet & \Xi_{vj} \end{bmatrix} < 0 \tag{8.55}$$

$$\widetilde{\Xi}_{d1j} = \begin{bmatrix} \widetilde{\Pi}_{oj} & \widetilde{\Pi}_{aj} & -\varrho_j\widetilde{\Psi}_{1j} & \widetilde{\Pi}_{cj} & \hat{\Gamma}_j & \widetilde{\Pi}_{zj}^t & \varrho_j\widetilde{\Pi}_{vj}^t \\ \bullet & -\widetilde{\Pi}_{mj} & -\varrho_j\widetilde{\Psi}_{3j} & 0 & 0 & \hat{G}_{dj}^t & \varrho_j\widetilde{\Pi}_{wj}^t \\ \bullet & \bullet & -\varrho_j\widetilde{\Psi}_{4j} & 0 & 0 & 0 & 0 \\ \bullet & \bullet & \bullet & -\widetilde{\Pi}_{nj} & 0 & 0 & \varrho_j\widetilde{\Pi}_{sj}^t \\ \bullet & \bullet & \bullet & \bullet & -\gamma_j^2 I_j & \Phi_j^t & \varrho_j\widetilde{\Pi}_{rj}^t \\ \bullet & \bullet & \bullet & \bullet & \bullet & -I_j & 0 \\ \bullet & \bullet & \bullet & \bullet & \bullet & \bullet & -\varrho_j\widetilde{\mathcal{W}}_j \end{bmatrix}$$

$$\widetilde{\Pi}_{d2j} = \begin{bmatrix} \varepsilon_j\widetilde{H}_{aj} & \widetilde{H}_{oj} & \varphi_j\widetilde{H}_{sj} & \widetilde{H}_{vj} & -\psi_j\widetilde{H}_{sj} & \widetilde{H}_{ej} \\ 0 & 0 & 0 & 0 & 0 & 0 \\ 0 & 0 & 0 & 0 & 0 & 0 \\ 0 & 0 & 0 & 0 & 0 & 0 \\ 0 & 0 & 0 & 0 & 0 & 0 \\ \varepsilon_j D_j M_{cj} & 0 & \varphi_j D_j M_{cj} & 0 & \psi_j D_j M_{cj} & 0 \\ \varepsilon_j\widetilde{H}_{cj} & O_j & \varphi_j\widetilde{H}_{wj} & O_j & -\psi_j\widetilde{H}_{wj} & O_j \end{bmatrix} \tag{8.56}$$

where

$$\widetilde{\Pi}_{oj} = \begin{bmatrix} \widetilde{\Pi}_{o1j} & \widetilde{\Pi}_{o2j} \\ \bullet & \widetilde{\Pi}_{o3j} \end{bmatrix},$$

$$\widetilde{\Pi}_{aj} = \begin{bmatrix} A_{dj}\mathcal{X}_{cj} - \Psi_{11j} + \Psi_{31j}^t & 0 \\ \bullet & A_{dj}\mathcal{X}_{oj} - \mathcal{Y}_{vj} - \Psi_{12j} + \Psi_{32j}^t \end{bmatrix}$$

$$\widetilde{\Pi}_{o1j} = [A_j + \sum_{k=1,k\neq j}^{n_s} F_{kj}]\mathcal{X}_{cj} + B_j\mathcal{Y}_{cj} + \mathcal{X}_{cj}[A_j + \sum_{k=1,k\neq j}^{n_s} F_{kj}]^t + \mathcal{Y}_{cj}^t B_j^t$$

$$+ \Psi_{11j} + \Psi_{11j}^t + \Psi_{21j} + \sum_{k=1,k\neq j}^{n_s} \Lambda_{kj1} + (n_s - 1)\mathcal{X}_{cj}$$

$$\widetilde{\Pi}_{o2j} = -\mathcal{Y}_{mj} + \mathcal{X}_{cj}\sum_{k=1,k\neq j}^{n_s} F_{kj}^t, \; \widetilde{\Pi}_{zj} = [G_j\mathcal{X}_{cj} + \mathcal{Y}_j \quad -\mathcal{Y}_{wj}]$$

$$\begin{aligned}
\widetilde{\Pi}_{o3j} &= A_j\mathcal{X}_{oj} + \mathcal{Y}_{oj} + \mathcal{X}_{oj}A_j^t + \mathcal{Y}_{oj}^t + \Psi_{12j} + \Psi_{12j}^t + \Psi_{22j} \\
&+ \sum_{k=1,k\neq j}^{n_s} \Lambda_{kj2} + (n_s-1)\mathcal{X}_{oj} \\
\widetilde{\Pi}_{mj} &= \begin{bmatrix} \Psi_{31j} + \Psi_{31j}^t + (1-\mu_j)\Psi_{21j} & 0 \\ \bullet & \Psi_{32j} + \widetilde{\Psi}_{32j}^t + (1-\mu_j)\Psi_{22j} \end{bmatrix} \\
\widetilde{\Pi}_{nj} &= Blockdiag[\widetilde{\Pi}_{n1j} \quad \widetilde{\Pi}_{n2j}], \\
\widetilde{\Pi}_{n1j} &= \sum_{k=1,k\neq j}^{n_s} (1-\mu_{kj})\Lambda_{kj1} + \sum_{k=1,k\neq j}^{n_s} \mathcal{X}_{cj}E_{kj}^t\mathcal{P}_{cj}E_{kj}\mathcal{X}_{cj}, \\
\widetilde{\Pi}_{n2j} &= \sum_{k=1,k\neq j}^{n_s} (1-\mu_{kj})\Lambda_{kj2} + \sum_{k=1,k\neq j}^{n_s} \mathcal{X}_{oj}E_{kj}^t\mathcal{P}_{oj}E_{kj}\mathcal{X}_{oj}, \\
\widetilde{\Psi}_{1j} &= \begin{bmatrix} \Psi_{11j} & 0 \\ \bullet & \Psi_{12j} \end{bmatrix}, \ \widetilde{\Psi}_{2j} = \begin{bmatrix} \Psi_{21j} & 0 \\ \bullet & \Psi_{22j} \end{bmatrix}, \ \widetilde{\Psi}_{3j} = \begin{bmatrix} \Psi_{31j} & 0 \\ \bullet & \Psi_{32j} \end{bmatrix}, \\
\widehat{\Lambda}_{kj} &= \begin{bmatrix} \Lambda_{kj1} & 0 \\ \bullet & \Lambda_{kj2} \end{bmatrix}, \ \widetilde{\Psi}_{4j} = \begin{bmatrix} \Psi_{41j} & 0 \\ \bullet & \Psi_{42j} \end{bmatrix} \\
\widetilde{\Pi}_{vj}^t &= \begin{bmatrix} (\mathcal{X}_{cj}A_j^t + \mathcal{Y}_{cj}^tB_j^t)\mathcal{W}_{cj} & 0 \\ -\mathcal{Y}_{mj}^t\mathcal{W}_{cj} & (\mathcal{X}_{oj}A_j^t - \mathcal{Y}_{oj}^t)\mathcal{W}_{oj} \end{bmatrix}, \\
\widetilde{\Pi}_{rj}^t &= [\Gamma_j^t\mathcal{W}_{cj} \quad \Gamma_j^t\mathcal{W}_{oj}], \ \widetilde{H}_{ej} = \begin{bmatrix} \mathcal{X}_{cj}C_{dj}^tN_{oj}^t \\ 0 \end{bmatrix} \\
\widetilde{\Pi}_{wj}^t &= \begin{bmatrix} \mathcal{X}_{cj}A_{dj}^t\mathcal{W}_{cj} & 0 \\ 0 & \mathcal{X}_{oj}A_{dj}^t - \mathcal{Y}_{vj}^t)\mathcal{W}_{oj} \end{bmatrix} \\
\widetilde{H}_{aj} &= \begin{bmatrix} B_jM_{cj} \\ 0 \end{bmatrix}, \ \widetilde{H}_{oj} = \begin{bmatrix} \mathcal{X}_{cj}N_{cj}^t \\ -\mathcal{X}_{oj}N_{cj}^t \end{bmatrix}, \\
\widetilde{H}_{cj} &= \begin{bmatrix} \varrho_j\mathcal{W}_{cj}B_jM_{cj} \\ 0 \end{bmatrix}, \ \widetilde{H}_{sj} = \begin{bmatrix} 0 \\ M_{oj} \end{bmatrix}, \\
\widetilde{H}_{vj} &= \begin{bmatrix} \mathcal{X}_{cj}C_{dj}^tN_{oj}^t \\ 0 \end{bmatrix}, \ \widetilde{H}_{wj} = \begin{bmatrix} 0 \\ \varrho_j\mathcal{W}_{oj}M_{oj} \end{bmatrix}, \\
\widetilde{\Pi}_{sj}^t &= \begin{bmatrix} \mathcal{X}_{cj}\sum_{k=1}^{n_s}E_{kj}^t\mathcal{W}_{cj} & 0 \\ \mathcal{X}_{oj}\sum_{k=1}^{n_s}E_{kj}^t\mathcal{W}_{cj} & 0 \end{bmatrix} \qquad (8.57)
\end{aligned}$$

Moreover, the local gain matrix is given by $K_{cj} = \mathcal{Y}_{cj}\mathcal{X}_{cj}^{-1}$, $K_{oj} = \mathcal{Y}_{oj}\mathcal{X}_{oj}^{-1}C_j^{\dagger}$.

Proof: Applying the congruent transformation

$$Blockdiag[\widehat{\mathcal{X}}_j, \ \widehat{\mathcal{X}}_j, \ \widehat{\mathcal{X}}_j, \ \widehat{\mathcal{X}}_j, \ I_j, \ I_j, \ \mathcal{T}_j], \ \widehat{\mathcal{X}}_j = \widehat{\mathcal{P}}_j^{-1}$$

with

$$\widehat{\mathcal{X}}_j = diag[\mathcal{X}_{cj} \quad \mathcal{X}_{oj}],\ \mathcal{T}_j = diag[I_j,\ I_j,\ I_j,\ I_j,\ I_j,\ I_j\ ]$$

to LMI (8.52) with (8.54) Schur complements and using the linearizations

$$\begin{aligned}
\mathcal{Y}_{cj} &= K_{cj}\mathcal{X}_{cj},\ \mathcal{Y}_{oj} = K_{oj}C_j\mathcal{X}_{oj},\ \mathcal{Y}_{mj} = B_jK_{cj}\mathcal{X}_{oj}, \\
\widetilde{\Psi}_{1j} &= \widehat{\mathcal{X}}_j\widehat{\Theta}_j\widehat{\mathcal{X}}_j = \begin{bmatrix} \Psi_{11j} & 0 \\ \bullet & \Psi_{12j} \end{bmatrix},\ \widetilde{\Psi}_{2j} = \widehat{\mathcal{X}}_j\widehat{\mathcal{Q}}_j\widehat{\mathcal{X}}_j = \begin{bmatrix} \Psi_{21j} & 0 \\ \bullet & \Psi_{22j} \end{bmatrix}, \\
\widetilde{\Psi}_{3j} &= \widehat{\mathcal{X}}_j\widehat{\Upsilon}_j\widehat{\mathcal{X}}_j = \begin{bmatrix} \Psi_{31j} & 0 \\ \bullet & \Psi_{32j} \end{bmatrix},\ \widehat{\Lambda}_{kj} = \widehat{\mathcal{X}}_j\widehat{\mathcal{Z}}_{kj}\widehat{\mathcal{X}}_j = \begin{bmatrix} \Lambda_{kj1} & 0 \\ \bullet & \Lambda_{kj2} \end{bmatrix}, \\
\mathcal{Y}_{vj} &= K_{oj}C_{dj}\mathcal{X}_{oj},\ \mathcal{Y}_{wj} = D_jK_{cj}\mathcal{X}_{oj}
\end{aligned}$$

we readily obtain LMI (8.56) with (8.57) by Schur complements.

Remark 8.3.9 *The optimal performance-level* $\gamma_j,\ j = 1,..,n_s$ *can be determined in case of decentralized resilient dynamic output feedback stabilization by solving the following convex optimization problems:*

Problem C: Resilient dynamic output feedback stabilization

$$\begin{aligned}
&For\ \ j,k = 1,...,n_s,\quad Given\ \ \varrho_j,\ \ \mu_j,\ \ \varrho_{jk},\ \ \mu_{jk}, \\
&\min_{\mathcal{X}_j,\ \mathcal{Y}_j,\ \mathcal{M}_j,\ \Omega_j,\ \{\Lambda_{rj}\}_{r=1}^{n_s},\ \{\Psi_{rj}\}_1^4} \quad \gamma_j \\
&subject\ \ to \quad LMI\ (8.56)\ with\ \ (8.57)
\end{aligned} \tag{8.58}$$

Remark 8.3.10 *It is important to note that the methodology developed in this chapter established robust decentralized criteria for delay-dependent stability by deploying an injection procedure and resilient feedback stabilization schemes applicable to a class of linear interconnected continuous time-delay systems. These criteria are LMI-based to handle local subsystems that are subjected to convex-bounded parametric uncertainties and additive feedback gain perturbations while allowing time-varying delays to occur within the local subsystems and across the interconnections. Our control design offers efficient decentralized control structures and possesses superior robustness properties with respect to both parametric uncertainties and gain perturbations. Note also in the nominal case (without parametric uncertainties and gain perturbations) and for delay-free systems, our results are an improved version of [105, 174]. This in turn enhances the utility and effectiveness of our approach.*

8.3.4 Illustrative example 8.1

Consider an interconnected system composed of three subsystems; each is of the type (8.1) with the following coefficients:

$$Subsystem\ 1:$$

$$A_1 = \begin{bmatrix} -2 & 0 \\ -2 & -1 \end{bmatrix}, \Gamma_1 = \begin{bmatrix} 0.2 \\ 0.2 \end{bmatrix}, G_1 = \begin{bmatrix} 0.2 & 0.1 \end{bmatrix},$$

$$A_{d1} = \begin{bmatrix} -1 & 0 \\ -1 & 0 \end{bmatrix}, G_{d1} = \begin{bmatrix} -0.1 & 0 \end{bmatrix}, \Phi_1 = 0.5$$

$$Couplings\ 1:$$

$$E_{12} = \begin{bmatrix} 1 & 0 \\ 1 & 0 \end{bmatrix}, E_{13} = \begin{bmatrix} 0 & -1 \\ 0 & -1 \end{bmatrix}, F_{12} = \begin{bmatrix} 0.2 & -0.1 \\ 0 & 0.2 \end{bmatrix},$$

$$F_{13} = \begin{bmatrix} 0 & -0.4 \\ 0.1 & -0.5 \end{bmatrix}$$

$$Subsystem\ 2:$$

$$A_2 = \begin{bmatrix} -1 & 0 \\ -1 & -4 \end{bmatrix}, \Gamma_2 = \begin{bmatrix} 0.1 \\ 0.3 \end{bmatrix}, G_2 = \begin{bmatrix} 0.2 & 0.1 \end{bmatrix},$$

$$A_{d2} = \begin{bmatrix} 1 & 0 \\ -2 & -1 \end{bmatrix}, G_{d2} = \begin{bmatrix} 0.1 & 0 \end{bmatrix}, \Phi_2 = 0.2$$

$$Couplings\ 2:$$

$$E_{21} = \begin{bmatrix} -1 & -2 \\ 3 & 6 \end{bmatrix}, E_{23} = \begin{bmatrix} -1 & 1 \\ 3 & -2 \end{bmatrix}, F_{21} = \begin{bmatrix} 0 & -0.5 \\ 0.8 & -0.5 \end{bmatrix},$$

$$F_{23} = \begin{bmatrix} -0.7 & 0.6 \\ 0 & -0.2 \end{bmatrix}$$

$$Subsystem\ 3:$$

$$A_3 = \begin{bmatrix} 0 & 1 \\ -1 & -2 \end{bmatrix}, \Gamma_3 = \begin{bmatrix} 0.1 \\ 0.5 \end{bmatrix}, G_3 = \begin{bmatrix} 0.1 & -0.1 \end{bmatrix},$$

$$A_{d3} = \begin{bmatrix} 0 & 0 \\ 0 & -1 \end{bmatrix}, G_{d3} = \begin{bmatrix} -0.1 & 0 \end{bmatrix}, \Phi_3 = 0.1$$

$$Couplings\ 3:$$

$$E_{31} = \begin{bmatrix} 1 & 2 \\ 1 & 2 \end{bmatrix}, E_{32} = \begin{bmatrix} 0 & 0 \\ 0 & -1 \end{bmatrix}, F_{31} = \begin{bmatrix} 0.7 & 0 \\ 0 & 0.8 \end{bmatrix},$$

$$F_{32} = \begin{bmatrix} 0 & 0 \\ 0 & -1 \end{bmatrix}$$

Considering **Theorem 8.2.1**, it is found that the feasible solution is attained at

$$
\begin{aligned}
&Subsystem\ 1: \\
&\varrho_1 = 3,\ \mu_1 = 1.5,\ \varrho_{21} = 2,\ \mu_{21} = 0.8,\ \varrho_{31} = 2,\ \mu_{31} = 0.8, \\
&\gamma_1 = 2.2817 \\
&\mathcal{P}_1 = \left[\begin{array}{cc} 2.0612 & 0.0023 \\ \bullet & 0.0322 \end{array}\right],\ \mathcal{Q}_1 = \left[\begin{array}{cc} 14.1238 & -1.3752 \\ \bullet & 19.5388 \end{array}\right] \\
&Subsystem\ 2: \\
&\varrho_2 = 2.5,\ \mu_2 = 1.3,\ \varrho_{12} = 1.5,\ \mu_{12} = 0.9,\ \varrho_{32} = 1.5,\ \mu_{32} = 0.9, \\
&\gamma_2 = 0.9958 \\
&\mathcal{P}_2 = \left[\begin{array}{cc} 0.1706 & -0.0281 \\ \bullet & 0.0817 \end{array}\right],\ \mathcal{Q}_2 = \left[\begin{array}{cc} 5.8316 & 0.4328 \\ \bullet & 5.4059 \end{array}\right] \\
&Subsystem\ 3: \\
&\varrho_3 = 3,\ \mu_3 = 1.1,\ \varrho_{13} = 1.8,\ \mu_{13} = 0.75,\ \varrho_{23} = 1.8,\ \mu_{23} = 0.75, \\
&\gamma_3 = 1.0142 \\
&\mathcal{P}_3 = \left[\begin{array}{cc} 0.5784 & 0.0662 \\ \bullet & 0.1153 \end{array}\right],\ \mathcal{Q}_3 = \left[\begin{array}{cc} 16.7851 & -0.5316 \\ \bullet & 7.2042 \end{array}\right]
\end{aligned}
$$

Since $\mathcal{P}_j,\ \mathcal{Q}_j,\ > 0, j = 1, 2, 3$ then the conditions required by **Theorem 8.2.1** are satisfied.

8.3.5 Illustrative example 8.2

Consider the interconnected system of Example 8.2 with the additional coefficients

$$
\begin{aligned}
&B_1 = \left[\begin{array}{c} 1 \\ 2 \end{array}\right],\ B_2 = \left[\begin{array}{cc} 1 & 1 \\ -1 & 2 \end{array}\right],\ B_3 = \left[\begin{array}{c} 2 \\ 1 \end{array}\right],\ D_1 = [0.1], \\
&D_2 = [0.2 \quad 0.4],\ D_3 = [0.3]
\end{aligned}
$$

Considering **Problem A**, it is found that the feasible solution is attained at

$$
\begin{aligned}
&\gamma_1 = 3.9292,\ K_1 = \left[\begin{array}{cc} 1.2551 & -2.3786 \end{array}\right],\ \gamma_3 = 7.0397, \\
&K_3 = \left[\begin{array}{cc} -0.0540 & 1.0240 \end{array}\right], \\
&\gamma_2 = 13.1742,\ K_2 = \left[\begin{array}{cc} -12.9437 & 2.0436 \\ -9.4414 & -8.0502 \end{array}\right]
\end{aligned}
$$

8.3.6 Illustrative example 8.3

Consider the interconnected system of Example 8.1 with the additional coefficients

$$C_1 = [0.2 \quad 0], \;\; C_2 = [0.6 \quad 0.4], \;\; C_3 = [0.5 \quad 0]$$

Considering **Problem B**, it is found that the feasible solution is attained at

$$\begin{aligned}
\gamma_1 &= 18.8905, \;\; K_1 = 5.9245, \;\; \gamma_3 = 37.0898, \;\; K_3 = 0.8416, \\
\gamma_2 &= 22.7111, \;\; K_2 = \begin{bmatrix} -0.5448 \\ 1.6161 \end{bmatrix}
\end{aligned}$$

8.3.7 Illustrative example 8.4

Consider the interconnected system of Example 8.1 with the additional coefficients

$$C_2 = \begin{bmatrix} 0.6 & 0.4 \\ 0.1 & 0 \end{bmatrix}, \;\; G_2 = \begin{bmatrix} 0.2 & 0.1 \\ 0 & 0 \end{bmatrix}, \;\; G_{d2} = \begin{bmatrix} 0.1 & 0 \\ 0 & 0 \end{bmatrix},$$
$$C_{d2} = \begin{bmatrix} 0.6 & 0.3 \\ 0.2 & 0 \end{bmatrix}, \;\; D_2 = \begin{bmatrix} 0.2 & 0.4 \\ 0.1 & 0.2 \end{bmatrix}$$

Considering **Problem C**, it is found that the feasible solution is attained at

$$\gamma_1 = 13.3086, \;\; K_{c1} = \begin{bmatrix} 0.9070 & -0.0722 \end{bmatrix}, \;\; K_{o1} = \begin{bmatrix} 8.8571 \\ 7.8083 \end{bmatrix},$$
$$\gamma_3 = 6.9512, \;\; K_{c3} = \begin{bmatrix} 0.0270 & 0.4458 \end{bmatrix}, \;\; K_{o3} = \begin{bmatrix} -1.4133 \\ 1.6144 \end{bmatrix},$$
$$\gamma_2 = 3.2553, \;\; K_{c2} = \begin{bmatrix} 0.1575 & -1.1567 \\ 0.2612 & 1.1867 \end{bmatrix}, \;\; K_{o2} = \begin{bmatrix} 0.2890 & -8.9054 \\ 9.9524 & 54.4992 \end{bmatrix}$$

The closed-loop state trajectories under the action of state feedback, static output feedback, and dynamic output feedback, which correspond to solutions of **Problems B, C**, and **D**, respectively, are depicted in Figures 8.2 through 8.4. From the plotted graphs, it is quite clear that all of the generated feedback controls guarantee regulation to the zero level. In addition, the dynamic output feedback control yields smoother trajectories in comparison to the other feedback controls. This is generally expected since the dynamic output feedback control employs more degrees of freedom. In order to demonstrate the effectiveness of the developed decentralized control strategies, the closed-loop state trajectories under the action of regular (without gain perturbations) and

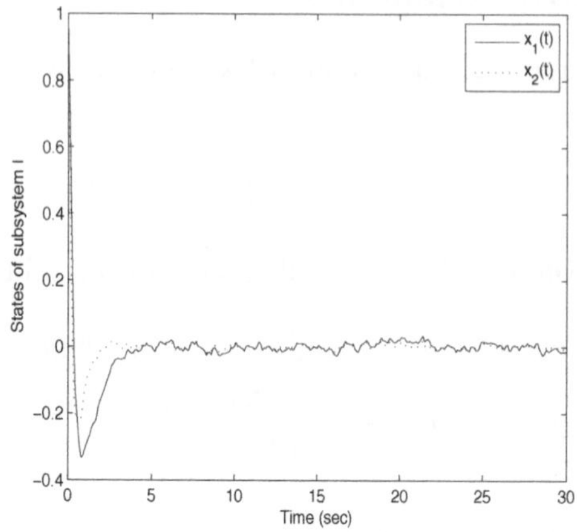

Figure 8.2: State trajectories of state feedback control

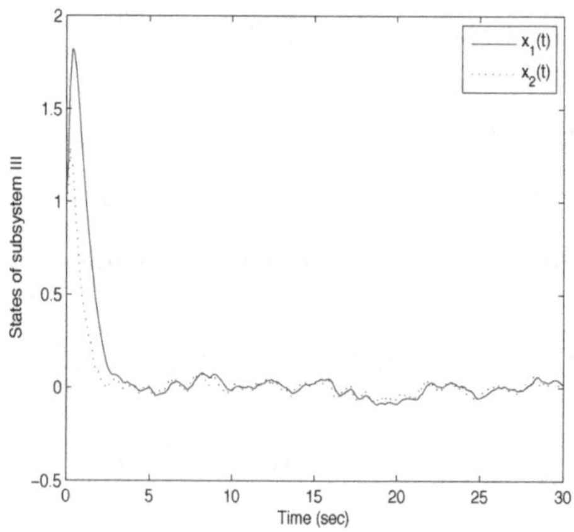

Figure 8.3: State trajectories of static output feedback control

resilient (with gain perturbations) state feedback controls are depicted in Figures 8.5 though 8.10. From the plotted graphs, it is quite clear that all of the generated feedback controls guarantee regulation to the zero level. The effect of resilience is significant in the initial period before settling to the desired level. This enhances the need of resilient control in decentralized systems to accommodate the possible gain perturbations.

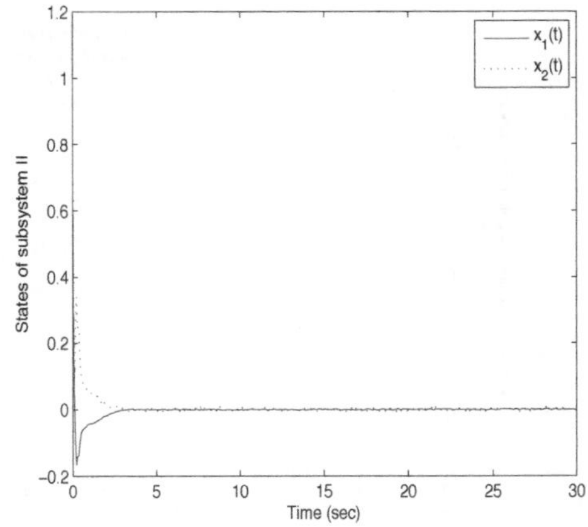

Figure 8.4: State trajectories of dynamic output feedback control

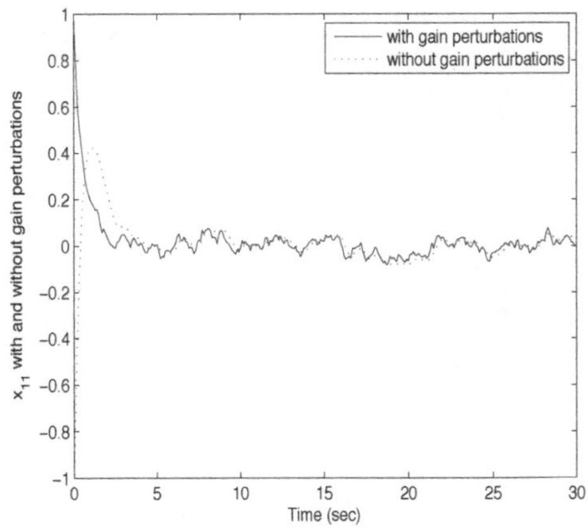

Figure 8.5: Trajectories of x_{11} with regular and resilient control

8.4 Resilient Overlapping Control

Standard assumption on designed controllers is that they can be implemented exactly into real world systems. In practice, control laws designed using theoretical methods and simulations are implemented imprecisely because of various reasons such as finite word length in any digital system, round-off errors in the imprecision inherent in analog systems, or the need for additional tuning of parameters in the final controller implementation. The controller designed

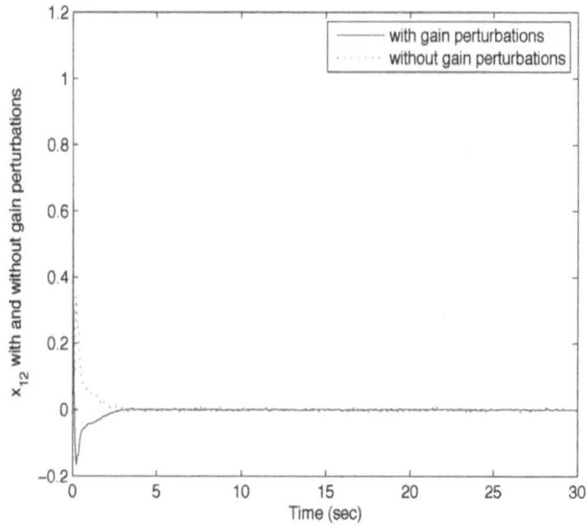

Figure 8.6: Trajectories of x_{12} with regular and resilient control

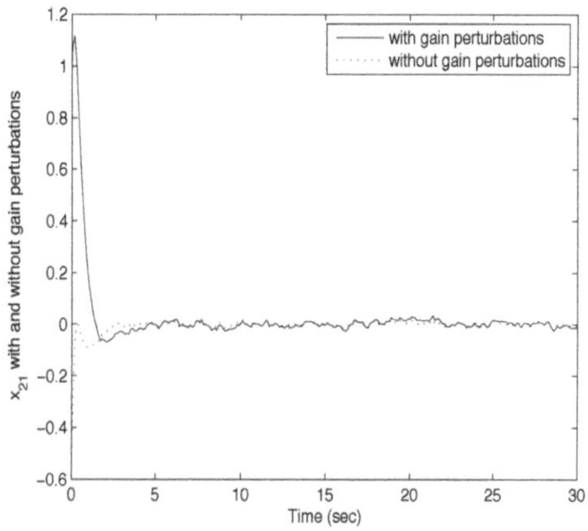

Figure 8.7: Trajectories of x_{21} with regular and resilient control

for uncertain plants may be sufficiently robust against system parameters, but the controller parameters themselves may be sensitive to relatively small perturbations and could even destabilize a closed-loop system. The importance of fragility, i.e., high sensitivity of controller parameters on its very small changes, is underlined when considering large-scale complex systems controlled by low cost local controllers. Such control systems are generally characterized by uncertainties, information structure constraints, delays, and high dimensionality. This situation naturally motivates the development of new effective control de-

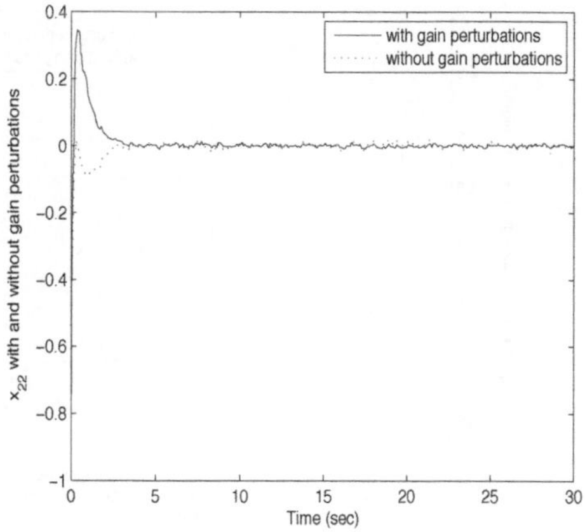

Figure 8.8: Trajectories of x_{22} with regular and resilient control

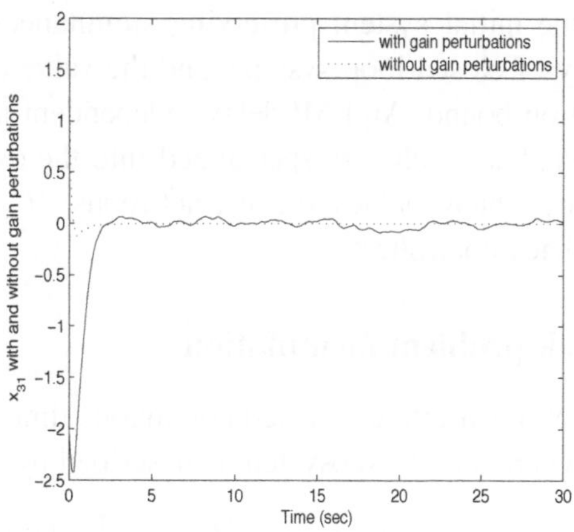

Figure 8.9: Trajectories of x_{31} with regular and resilient control

sign methods taking into account particular features of these systems including implementation aspects of controller parameter uncertainties.

In this section, the expansion-contraction relations, developed earlier within the inclusion principle, are applied for a class of continuous-time state-delayed uncertain systems when considering $\mathcal{H}_\infty$ state memoryless control with additive controller uncertainty. All uncertainties are supposed to be time-varying norm bounded. The main contribution is the derivation of conditions under which a resilient $\mathcal{H}_\infty$ control law designed in the expanded space is

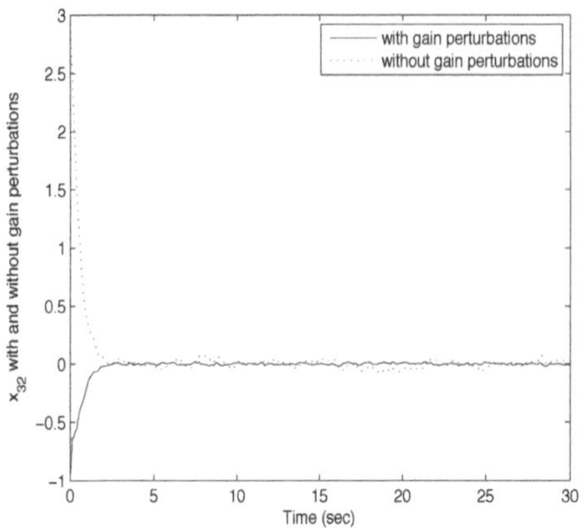

Figure 8.10: Trajectories of x_{32} with regular and resilient control

contracted into the initial system preserving simultaneously both the robust quadratic stability of closed-loop systems and the value of the $\mathcal{H}_\infty$-norm disturbance attenuation bound. An LMI delay-independent procedure is supplied for control design. The results are specialized into the overlapping decentralized control setting which enables to construct robust $\mathcal{H}_\infty$ resilient block tridiagonal state feedback controllers.

8.4.1 Feedback problem formulation

Consider a class of linear interconnected continuous-time uncertain time-delay (UTD) systems where the jth subsystem is described by

$$\begin{aligned} S_j: \dot{x}_j(t) &= [A_j + \Delta A_j(t)]x_j(t) + [B_j + \Delta B_j(t)]u_j(t) \\ &+ [A_{dj} + \Delta A_{dj}(t)]x_j(t-d_j) + \sum_{k=1,k\neq j}^{n_s} Z_{jk}x_k(t) + \Gamma_j w_j(t), \\ x_j(t_0) &= \phi_j(t_0),\ -d_j \leq t_0 \leq 0 \qquad (8.59) \\ z_j(t) &= Cx_j(t) + Du_j(t) \qquad (8.60) \end{aligned}$$

where for subsystem j , $x_j(t) \in \Re^{n_j}$ is the state, $u_j(t) \in \Re^{m_j}$ is the control input, $d_j > 0$ is the delay factor, $w_j(t) \in L_2^p[0,\infty)$ is the disturbance input, $z_j(t) \in \Re_j^q$ is the controlled output, and $\phi_j(t_0)$ is a given continuous initial function. The term $\sum_{k=1,k\neq j}^{n_s} Z_{jk}x_k(t)$ represents a static-type coupling among subsystems which implies the coupling transfer is faster than the isolated subsystem dynamics. The matrices $A_j,\ B_j,\ A_{dj},\ \Gamma_j,\ C_j,\ D_j$ are known real constant of appropriate dimensions and $\Delta A_j(t), \Delta B_j(t), \ \Delta A_{dj}(t)$ are real-valued matrices of time-varying, norm-bounded uncertain parameters of the form

$$\begin{aligned} \Delta A_j(t) &= H_{Aj} F_j(t) E_{Aj}, \quad \Delta A_{dj}(t) = H_{dj} F_j(t) E_{dj}, \\ \Delta B_j(t) &= H_{Bj} F_j(t) E_{Bj} \end{aligned} \tag{8.61}$$

where $H_{Aj}, \ H_{dj}, \ H_{Bj}, \ E_{Aj}, \ E_{Bj}, \ E_{dj}$ are known real constant matrices of appropriate dimensions and $F_j(t) \in \Re^{s_j \times r_j}$ are unknown real time-varying matrices with Lebesgue measurable elements satisfying the condition $F_j^t(t) F_j(t) \leq I_j$. From now onward, we let

$$\begin{aligned} \bar{A}_j(t) &= A_j + \Delta A_j(t), \quad \bar{B}_j(t) = B_j + \Delta B_j(t), \\ \bar{A}_{dj}(t) &= A_{dj} + \Delta A_{dj}(t) \end{aligned} \tag{8.62}$$

System (8.60) can be expressed into the form

$$\begin{aligned} S_j \colon \dot{x}_j(t) &= \bar{A}_j(t) x(t) + \bar{B}_j(t) u_j(t) + \bar{A}_{dj}(t) x(t - d_j) \\ &+ \sum_{k=1, k \neq j}^{n_s} Z_{jk} x_k(t) + \Gamma_j w_j(t), \\ z_j(t) &= C_j x_j(t) + D_j u_j(t), \\ x_j(t_0) &= \phi_j(t_0), \ -d_j \leq t_0 \leq 0 \end{aligned} \tag{8.63}$$

From (8.60)-(8.64), the interconnected system is described by

$$\begin{aligned} S \colon \dot{x}(t) &= \bar{\mathsf{A}}(t) x(t) + \bar{\mathsf{B}}(t) u(t) + \bar{\mathsf{A}}_d(t) x(t - d) + \Gamma w(t), \\ z(t) &= \mathsf{C} x(t) + \mathsf{D} u(t) \end{aligned} \tag{8.64}$$

where

$$\begin{aligned} x &= \begin{bmatrix} x_1 & x_2 & \cdots & x_{n_s} \end{bmatrix}, \ u = \begin{bmatrix} u_1 & u_2 & \cdots & u_{n_s} \end{bmatrix}, \\ \bar{\mathsf{B}} &= Blockdiag\{\bar{B}_1, \ ..., \ \bar{B}_{n_s}\}, \ \bar{\mathsf{A}}_d(t) = Blockdiag\{\bar{A}_{d1}, \ ..., \ \bar{A}_{dn_s}\}, \\ \bar{\mathsf{C}}(t) &= Blockdiag\{C_1, \ ..., \ C_{n_s}\}, \ \bar{\mathsf{D}} = Blockdiag\{D_1, \ ..., \ D_{n_s}\}, \\ \bar{\mathsf{A}}(t) &= \begin{bmatrix} \bar{A}_1 & Z_{12} & \cdots & Z_{1n_s} \\ Z_{21} & \bar{A}_2 & \cdots & Z_{2n_s} \\ \vdots & \vdots & \vdots & \vdots \\ Z_{n_s 1} & Z_{n_s 2} & \cdots & \bar{A}_{n_s} \end{bmatrix} \end{aligned} \tag{8.65}$$

Now, consider a local resilient state feedback controller in the form

$$u_j(t) = [K_j + \Delta K_j(t)] x_j(t) = \bar{K}_j(t) x_j(t) \tag{8.66}$$

for system (8.60), where $K_j \in \Re^{m_j \times n_j}$. The gain matrix perturbation $\Delta K_j(t)$ satisfies

$$\Delta K_j(t) = H_{kj} F_{kj}(t) E_{kj} \tag{8.67}$$

where $F_{kj}^t(t) F_{kj}(t) \leq I_j$ with $F_{kj} \in \Re^{s_j \times r_j}$ being an unknown real time-varying matrix with Lebesgue measurable elements. $H_{kj},\ E_{kj}$ are known constant matrices of appropriate dimensions. The overall resilient state feedback controller takes the form

$$\begin{aligned} u(t) &= [\mathsf{K} + \Delta K(t)]x(t) = \bar{\mathsf{K}}(t)x(t) \qquad (8.68)\\ \bar{\mathsf{K}} &= Blockdiag[\bar{K}_1,\ ...,\ \bar{K}_{n_s}],\ \mathsf{K} = Blockdiag[K_1,\ ...,\ K_{n_s}],\\ \Delta K &= Blockdiag[\Delta K_1,\ ...,\ \Delta K_{n_s}] \end{aligned}$$

Applying controller (8.68) to system (8.60), we get the resulting overall closed-loop system in the form

$$\begin{aligned} \mathbf{S}_c: \quad \dot{x}(t) &= [\mathsf{A} + \mathsf{AK} + H_B F(t) E_B \mathsf{K} + H_A F(t) E_A + \mathsf{B} H_k F_k(t) E_k \\ &+ H_B F(t) E_B H_k F_k(t) E_k] x(t) + \sum_{k=1,k\neq j}^{n_s} Z_{jk} x_k(t - h_{jk}) \\ &+ [\mathsf{A}_d + H_d F(t) E_d] x(t-d) + \mathsf{Z}(t) x(t-h) + \Gamma w(t) \\ &= \mathsf{A}_p(t) x(t) + \mathsf{A}_q(t) x(t-d) + \mathsf{Z}(t) x(t-h) + \Gamma w(t), \\ z(t) &= [\mathsf{C} + \mathsf{DK} + \mathsf{D} H_k F_k(t) E_k] x(t) \qquad (8.69) \end{aligned}$$

where

$$H_A = \{H_{Aj}\},\ H_B = \{H_{Bj}\},\ H_k = \{H_{kj}\},\ H_d = \{H_{dj}\},$$
$$E_A = \{E_{Aj}\},\ E_k = \{E_{kj}\},\ E_B = \{E_{Bj}\},\ E_d = \{E_{dj}\}$$

are the block-structure of matrices in (8.61) and (8.67). Now to examine the robust stability, we consider the overall Lyapunov-Krasovskii functional (LKF):

$$V(x,t) = x^t(t)\mathsf{P}x(t) + \int_{t-d}^{t} x^t(\sigma)[I + E_d^t E_d]x(\sigma)d\sigma \tag{8.70}$$

where $\mathsf{P} = \{P_j\},\ P_j \in \Re^{n_j \times n_j},\ (I + E_d^t E_d) \in \Re^{n \times n}$ are positive-definite symmetric matrices. The following results can derived using arguments from basic time-delay theory [255].

Lemma 8.4.1 *System (8.60) with $u(t) = 0, w(t) = 0$ is robustly quadratically stable if there exist matrices $0 < \mathsf{P}^t = \mathsf{P}$ such that the family of inequality*

$$\begin{aligned} &\mathsf{PA} + \mathsf{A}^t\mathsf{P} + \mathsf{P}(H_A H_A^t + H_d H_d^t + \mathsf{A}_d \mathsf{A}_d^t)\mathsf{P} + \\ &\mathsf{PZP} + E_A^t E_A + E_d^t E_{dj} + I \end{aligned} \tag{8.71}$$

for all admissible uncertainties (8.61).

Lemma 8.4.2 *System (8.60) is robustly quadratically stabilizable if there exist a matrix* $0 < P^t = P$ *and a gain matrix* K *given by (8.68) such that*

$$\begin{aligned} &\mathsf{P}(\mathsf{A}+\mathsf{B}\mathsf{K})+(\mathsf{A}+\mathsf{B}\mathsf{K})^t\mathsf{P}+\varepsilon\mathsf{K}^tE_B^tE_B\mathsf{K}+\\ &\mathsf{P}(H_AH_A^t+H_dH_d^t+(1+\frac{1}{\varepsilon})H_BH_B^t+A_dA_d^t+BH_kH_k^tB^t)\mathsf{P}+\\ &E_A^tE_A+E_d^tE_d+(1+\alpha)E_k^tE_K+I<0 \end{aligned} \quad (8.72)$$

where $\alpha = \|H_k^tE_B^tE_BH_K\|$ *for all* $\varepsilon > 0$ *and all admissible uncertainties (8.61) and (8.67).*

To extend these results further, we introduce the following definitions.

Definition 8.4.3 *Given system (8.60) and a scalar* $\gamma > 0$*. This system is robustly quadratically stable with* $\mathcal{H}_\infty$*-norm bound* γ *if it is robustly quadratically stable in light of* **Lemma 8.4.1** *and under zero initial conditions* $||z(t)||_2 \leq \gamma||w(t)||_2$ *holds for any* $w(t) \neq 0$ *and for all admissible uncertainties (8.61).*

Definition 8.4.4 *Given system (8.60) and a scalar* $\gamma > 0$*. This system is robustly quadratically stabilizable with* $\mathcal{H}_\infty$*-norm bound* γ *if there exist a matrix* $0 < P^t = P$ *and a gain matrix* K *given by (8.68) such that*

$$\begin{aligned} &\mathsf{P}(\mathsf{A}+\mathsf{B}\mathsf{K})+(\mathsf{A}+\mathsf{B}\mathsf{K})^t\mathsf{P}\\ +\;&\mathsf{P}(H_AH_A^t+H_dH_d^t+\mathsf{A}_d\mathsf{A}_d^t+\gamma^{-2}\Gamma\Gamma^t)\mathsf{P}\\ +\;&\varepsilon\mathsf{K}^tE_B^tE_B\mathsf{K}+\left(1+\frac{1}{\varrho}\right)(\mathsf{C}+\mathsf{D}\mathsf{K})^t(\mathsf{C}+\mathsf{D}\mathsf{K})\\ +\;&(1+\varrho)\sigma E_k^tE_k+E_A^tE_A+E_d^tE_d+(1+\alpha)E_k^tE_K+I<0 \end{aligned} \quad (8.73)$$

hold for all admissible uncertainties (8.61) and (8.67), where $\sigma = \|H_k^t\mathsf{D}^t\mathsf{D}H_k\|$.

8.4.2 LMI control design

Since the delay factor is considered constant in the present formulation, the best we hope for is a delay-independent procedure [255]. It is included as an effective control design tool.

Theorem 8.4.5 *Consider the linear UTD system (8.60) and constant* $\gamma > 0$*. If for some positive real values* ε, ϱ, *there exist matrices* Q *and* Y *satisfying the* *following LMI,*

$$\begin{bmatrix} \Pi & \mathsf{Q}\Phi^t & \mathsf{Q}E_k^t & \mathsf{Y}^tE_B^t & \Psi^t \\ \bullet & -I & 0 & 0 & 0 \\ \bullet & \bullet & -[\frac{\sigma^{-1}}{1+\varrho}]I & 0 & 0 \\ \bullet & \bullet & \bullet & -[\frac{1}{\varepsilon}]I & 0 \\ \bullet & \bullet & \bullet & \bullet & -[\frac{\varrho}{1+\varrho}]I \end{bmatrix} < 0 \tag{8.74}$$

where

$$\begin{aligned} \Pi &= \mathsf{AQ} + \mathsf{QA}^t + \mathsf{BY} + \mathsf{Y}^t\mathsf{Q}^t \\ &+ H_AH_A^t + H_dH_d^t + \mathsf{A}_d\mathsf{A}_d^t + \gamma^{-2}\Gamma\Gamma^t, \\ \Phi^t &= [E_A^t \quad E_d^t \quad (1+\sigma)^{1/2}E_k^t \quad I] \\ \Psi^t &= \mathsf{QQC}^t + \mathsf{Y}^t\mathsf{D}^t \end{aligned} \tag{8.75}$$

then there exists a memoryless state feedback controller (8.68) such that the resulting closed-loop system (8.69) is robustly quadratically stable with $\mathcal{H}_\infty$-norm bound γ. Moreover, the gain matrix K is given by $\mathsf{K} = \mathsf{YQ}^{-1}$.

8.4.3 Application of the inclusion principle

We now proceed to apply the inclusion principle discussed in Chapter 5. We consider a new larger-dimension system, similar to system (8.60), in the form

$$\begin{aligned} \tilde{S}: \dot{\tilde{x}}(t) &= [\tilde{\mathsf{A}} + \Delta\tilde{\mathsf{A}}(t)]\tilde{x}(t) + [\tilde{\mathsf{B}} + \Delta\tilde{\mathsf{B}}(t)]u(t) \\ &+ [\tilde{\mathsf{A}}_d + \Delta\tilde{\mathsf{A}}_d(t)]\tilde{x}(t-d) + \tilde{\Gamma}w(t), \end{aligned} \tag{8.76}$$

$$\tilde{z}(t) = \tilde{\mathsf{C}}\tilde{x}(t) + \mathsf{A}u(t), \tag{8.77}$$

$$\tilde{x}(t_0) = \tilde{\phi}(t_0), \; -d \le t_0 \le 0$$

where $\tilde{x}(t) \in \Re^{\tilde{n}}$ is the state, $u(t) \in \Re^m$ is the control input, $d > 0$ is the delay time, $w(t) \in L_2^p[0,\infty)$ is the disturbance input, $\tilde{z}(t) \in R^q$ is the controlled output and $\tilde{\phi}(t_0)$ is a continuous initial function. According to the inclusion principle, we consider $n \le \tilde{n}, q \le \tilde{q}$. The conditions (8.61) and the notation (8.62) for system **S** are analogous for system $\tilde{S}$, while considering now all matrices with tilde ($\tilde{}$). The corresponding resilient state feedback controller for the system $\tilde{S}$ is given by

$$u(t) = [\tilde{\mathsf{K}} + \Delta\tilde{\mathsf{K}}(t)]\tilde{x}(t) \tag{8.78}$$

where $\tilde{\mathsf{K}} \in \Re^{m\times\tilde{n}}$. The uncertain gain matrix $\Delta\tilde{\mathsf{K}}(t)$ satisfies

$$\Delta\tilde{\mathsf{K}}(t) = \tilde{H}_k\tilde{F}_k(t)\tilde{E}_k \tag{8.79}$$

where $\tilde{F}_k^t(t)\tilde{F}_k(t) \leq I,\ \tilde{F}_k \in \Re^{s\times r}$ is an unknown real time-varying matrix with Lebesgue measurable elements and $\tilde{H}_k, \tilde{E}_k$ are known constant matrices of appropriate dimensions.

The performance measure of the expanded system $\tilde{S}$ is now assessed by

$$\|\tilde{z}(t)\|_2 \leq \tilde{\gamma}\|w(t)\|_2 \tag{8.80}$$

provided that the corresponding expanded closed-loop system is robustly quadratically stable with zero initial condition.

Now let $x(t) = x(t;\phi(t_0), u(t), w(t)), \tilde{x}(t) = \tilde{x}(t;\tilde{\phi}(t_0), u(t), w(t))$ be the solutions of (8.60) and (8.77) for initial functions $\phi(t_0)$ and $\tilde{\phi}(t_0)$, given inputs $u(t)$ and disturbance inputs $w(t)$, respectively. Applying the standard relations of the inclusion principle, it means that the systems S and $\tilde{S}$ are related by the following linear transformations:

$$\tilde{x}(t) = \mathsf{V}x(t), \quad x(t) = \mathsf{U}\tilde{x}(t) \tag{8.81}$$

where V and its pseudo-inverse matrix $\mathsf{U} = (\mathsf{V}^t\mathsf{V})^{-1}\mathsf{V}^t$ are constant full-rank matrices of appropriate dimensions [324]. Therefore, the following definitions are recalled.

Definition 8.4.6 *A system $\tilde{S}$ includes the system S, denoted by $\tilde{S} \supset S$, if there exists a pair of constant matrices (U, V) such that $\mathsf{UV} = I_n$, and for any initial function $\phi(t_0)$, any fixed input $u(t)$ and any disturbance input $w(t)$ of S, it follows that*

$$x(t;\phi(t_0), u(t), w(t)) = U\tilde{x}(t; V\phi(t_0), u(t), w(t)),\ \forall t$$

Definition 8.4.7 *A pair $(\tilde{S},\ \tilde{\gamma})$ is an expansion of $(S,\ \gamma)$, denoted by $(\tilde{S},\ \tilde{\gamma}) \supset (S,\ \gamma)$, if $\tilde{S} \supset S$ and $\gamma = \tilde{\gamma}$.*

Definition 8.4.8 *A controller $u(t) = [\tilde{\mathsf{K}} + \Delta\mathsf{K}(t)]\tilde{x}(t)$ for $\tilde{S}$ is contractible to $u(t) = [\mathsf{K} + \Delta\mathsf{K}(t)]x(t)$ for S if the choice $\tilde{\phi}(t_0) = \mathsf{V}\phi(t_0)$ implies*

$$[\mathsf{K} + \Delta\mathsf{K}(t)]x(t;\phi(t_0), u(t), w(t)) = [\tilde{\mathsf{K}} + \Delta\tilde{\mathsf{K}}(t)]\tilde{x}(t; \mathsf{V}\phi(t_0), u(t), w(t))$$

for all t, any initial function $\phi(t_0)$, any fixed input u(t), and any disturbance input $w(t)$ of S.

Remark 8.4.9 *As we learned from Chapter 6, the inclusion principle can be used for the analysis and control design of different classes of dynamic systems with different objectives. In general, a dynamic system is expanded to obtain* *another larger dimension system containing all information about the initial system. Then, the controller design is usually performed for the expanded system and consequently contracted into the original system. This approach is quite effective mainly when considering decentralized controller design for the expanded system without shared parts. In the sequel, we consider that the gain matrix* $\tilde{\mathsf{K}}$ *appearing in the controller* $u(t) = [\tilde{\mathsf{K}} + \Delta\tilde{\mathsf{K}}(t)]\tilde{x}(t)$ *is not the expanded gain matrix of* K *but a "free" matrix designed for the system* $\tilde{S}$ *with given* $\tilde{\gamma}$.

Suppose we are given a pair of matrices $(\mathsf{U},\ \mathsf{V})$. Then, the matrices $\tilde{\mathsf{A}}, \Delta\tilde{\mathsf{A}}(t), \Delta\mathsf{B}, \Delta\tilde{\mathsf{B}}(t), \tilde{\mathsf{A}}_d, \Delta\tilde{\mathsf{A}}_d(t), \tilde{\Gamma}$ and $\tilde{\mathsf{C}}$ can be described as follows:

$$\begin{aligned} \tilde{\mathsf{A}} = \mathsf{VAU} + \mathsf{M}, \quad \Delta\tilde{\mathsf{A}}(t) &= \mathsf{V}\Delta\mathsf{A}(t)\mathsf{U}, \\ \tilde{\mathsf{B}} = \mathsf{VB} + \mathsf{N}, \quad \Delta\tilde{\mathsf{B}}(t) &= \mathsf{V}\Delta\mathsf{B}(t), \\ \tilde{\mathsf{A}}_d = \mathsf{VA}_d\mathsf{U} + \mathsf{M}_d, \quad \Delta\tilde{\mathsf{A}}_d(t) &= \mathsf{V}\Delta\mathsf{A}_d(t)\mathsf{U}, \qquad (8.82) \\ \tilde{\Gamma} = \mathsf{V}\Gamma + \mathsf{R}, \quad \tilde{\mathsf{C}} &= \mathsf{CU} + \mathsf{L} \qquad (8.83) \end{aligned}$$

where $\mathsf{M},\ \mathsf{N},\ \mathsf{M}_d, \mathsf{R}$, and L are called *complementary matrices*. The transformations $(\mathsf{U},\ \mathsf{V})$ can be selected a priori to define structural relations between the state variables in both systems S and $\tilde{S}$. The choice of the complementary matrices then provides degrees of freedom to obtain different expanded spaces with desirable properties [15, 16].

For system (8.60) with positive numbers $\gamma,\ \tilde{\gamma}$, consider an expanded system $\tilde{S}$ described by (8.77) via the relations (8.83) such that $\tilde{S} \supset S$ holds. The immediate objective is to derive conditions under which $(\tilde{S}_C, \tilde{\gamma}) \supset (S_C, \gamma)$ using robust quadratic stability with $\mathcal{H}_\infty$-norm bound γ. At this stage, we recall that the necessity of overlapping structure is dictated by structural constraints on the system and there is no principal need to expand also the controlled output. Thus, we can put $z(t) = \tilde{z}(t)$ without any restriction. The same applies to the disturbance input. As a result, we can introduce the equality on the H_∞-norm bound γ, that is, $\gamma = \tilde{\gamma}$. Additionally, some conditions on the complementary matrices (8.83) must be imposed on $(\tilde{S}, \tilde{\gamma})$ to be an expansion of (S, γ) as per **Definition 8.4.7**. This is summarized by the following theorem, the proof of which can be derived following [17].

Theorem 8.4.10 *Consider the systems (8.60) and (8.77) with (8.80). A pair* $(\tilde{S}, \tilde{\gamma})$ *includes the pair* $(S,\ \gamma)$ *if and only if*

$$\begin{aligned} \mathsf{U}\tilde{\Phi}(t,0)\mathsf{V} &= \Phi(t,0), \quad \mathsf{U}\Phi(t,s)\mathsf{M}_d\mathsf{V} = 0, \\ \mathsf{U}\tilde{\Phi}(t,s)\mathsf{N} &= 0, \quad \mathsf{U}\Phi(t,s)\mathsf{R} = 0, \qquad (8.84) \\ \mathsf{LV} &= 0 \end{aligned}$$

hold for all t and s.

Remark 8.4.11 *It is significant to note that obtaining an explicit solution of a time-varying system is a difficult problem. Therefore, effort should be focused on conditions under which $(\tilde{S}, \tilde{\gamma}) \supset (S, \gamma)$ holds but without any necessity to compute the transition matrices.*

Equivalent conditions to those of **Theorem 8.4.10** expressed in terms of complementary matrices are provided by the following theorem.

Theorem 8.4.12 *Consider the systems (8.60) and (8.77) with (8.80). A pair $(\tilde{S}, \tilde{\gamma})$ includes the pair (S, γ) if and only if*

$$\begin{aligned} \mathsf{UM}^j\mathsf{V} &= 0, \quad \mathsf{UM}^{j-1}\mathsf{M}_d\mathsf{V} = 0, \\ \mathsf{UM}^{j-1}\mathsf{N} &= 0, \quad \mathsf{UM}^{j-1}\mathsf{R} = 0, \\ \mathsf{LV} &= 0 \end{aligned} \tag{8.85}$$

hold for all $k = 1, 2, ..., \tilde{n}$.

Proof 8.4.13 *Consider the transition matrix $\Phi(t, 0)$ of $\tilde{\tilde{\mathsf{A}}}$. Following the notation given in (3), $\tilde{\tilde{\mathsf{A}}} = \tilde{\mathsf{A}} + \Delta\tilde{\mathsf{A}}(\mathsf{t})$ represents the state matrix of the expanded space $\tilde{S}$ as a function of two variables defined by the Peano-Baker series [17]:*

$$\begin{aligned} \Phi(t,0) = I \quad &+ \int_0^t \tilde{\tilde{\mathsf{A}}}(\sigma_1)\mathsf{d}\sigma_1 + \int_0^t \tilde{\tilde{\mathsf{A}}}(\sigma_1) \int_0^{\sigma_1} \tilde{\tilde{\mathsf{A}}}(\sigma_2)\mathsf{d}\sigma_2\mathsf{d}\sigma_1 \\ &+ \int_0^t \tilde{\tilde{\mathsf{A}}}(\sigma_1) \int_0^{\sigma_1} \tilde{\tilde{\mathsf{A}}}(\sigma_2) \int_0^{\sigma_2} \tilde{\tilde{\mathsf{A}}}(\sigma_3)\mathsf{d}\sigma_3\mathsf{d}\sigma_2\mathsf{d}\sigma + ... \end{aligned} \tag{8.86}$$

In the light of (8.83),

$$\begin{aligned} \tilde{\tilde{\mathsf{A}}}(\sigma_\mathsf{i}) &= \tilde{\mathsf{A}} + \Delta\tilde{\mathsf{A}}(\sigma_i), \\ &= \mathsf{VAU} + \mathsf{M} + \mathsf{V}\Delta\mathsf{A}(\sigma_j)\mathsf{U}, \quad \forall j = 1, 2, ... \end{aligned}$$

*From **Theorem 8.4.10**, pre- and postmultiplying both sides of $\tilde{\Phi}(t, 0)$ by A and A, respectively, we can prove that $\mathsf{U}\tilde{\Phi}(t, 0)\mathsf{V} = \Phi(t, 0)$ is equivalent to $\mathsf{UM}^j\mathsf{V} = 0$ for $j = 1, 2, ..., \tilde{n}$. In a similar way, $\mathsf{U}\tilde{\Phi}(t, s)\mathsf{M}_d\mathsf{V} = 0$ is equivalent to $\mathsf{UM}^{j-1}\mathsf{M}_d\mathsf{V} = 0$ for $j = 1, 2, ..., \tilde{n}$. The condition $\mathsf{U}\tilde{\Phi}(t, s)\mathsf{N} = 0$ is equivalent to $\mathsf{UM}^{j-1}\mathsf{N} = 0$ and $\mathsf{U}\tilde{\Phi}(t, s)\mathsf{R} = 0$ is equivalent to $\mathsf{UM}^{j-1}\mathsf{R} = 0$, for all $j = 1, 2, ..., \tilde{n}$. The requirement $\mathsf{LV} = 0$ can be obtained by imposing $\gamma_* = \tilde{\gamma}$, which completes the proof.*

It is evident that **Theorem 8.4.12** paves the way to derive expanded systems satisfying the inclusion principle with the same $\mathcal{H}_\infty$ performance attenuation bound γ without an exact knowledge of transition matrices. The following corollary stands out.

Corollary 8.4.14 *Consider the systems (8.60) and (8.77) with (8.80). A pair $(\tilde{S}, \tilde{\gamma})$ includes the pair (S, γ) if $\mathsf{LV} = 0$ and*

$$\begin{aligned} \mathsf{MV} &= 0, \quad \mathsf{M}_d\mathsf{V} = 0, \quad \mathsf{N} = 0, \quad \mathsf{R} = 0 \text{ or} \\ \mathsf{UM} &= 0, \quad \mathsf{UM}_d = 0, \quad \mathsf{UN} = 0, \quad \mathsf{UR} = 0 \end{aligned} \tag{8.87}$$

Remark 8.4.15 *If $\mathsf{M}_d = 0,\ \mathsf{R} = 0$ in (8.87), then 8.87 corresponds to particular cases within the inclusion principle called restrictions and aggregations, respectively [324].*

Recall that **Definition 8.4.8** provides the conditions under which a control law designed in the expanded system $\tilde{S}$ can be contracted and implemented into the initial system **S**.These requirements, however, do not guarantee that the closed-loop system $\tilde{S}_c$ includes the closed-loop system S_c in the sense of the *inclusion principle*, that is, $\tilde{S}_c \supset S_c$. Now consider conditions which include also $\mathcal{H}_\infty$ performance attenuation bounds γ. The ensuing results are presented by the following theorem using complementary matrices.

Theorem 8.4.16 *Consider the systems (8.60) and (8.77) with (8.80) such that $\tilde{S} \supset S$. Suppose that $u(t) = [\tilde{\mathsf{K}} + \Delta\tilde{\mathsf{K}}(t)]\tilde{x}(t)$ is a contractible control law designed in $\tilde{S}$. If $\mathsf{MV} = 0,\ \mathsf{M}_d\mathsf{V} = 0,\ \mathsf{N} = 0,\ \mathsf{R} = 0$, and $\mathsf{LV} = 0$, then $(\tilde{S}_c, \tilde{\gamma}) \supset (S_c, \gamma)$.*

Proof 8.4.17 *Suppose $\tilde{S}_c \supset S_c$ and consider the resilient control $u(t) = [\tilde{\mathsf{K}} + \Delta\tilde{\mathsf{K}}(t)]\tilde{x}(t)$ a contractible law designed in $\tilde{S}$. The corresponding closed-loop expanded system is given as follows:*

$$\begin{aligned} \tilde{S}_c\colon \dot{\tilde{x}}(t) &= [\tilde{\mathsf{A}} + \tilde{\mathsf{A}}\tilde{\mathsf{A}} + \Delta\tilde{\mathsf{A}}(t) + \Delta\mathsf{A}(t)\tilde{\mathsf{A}} + \tilde{\mathsf{A}}\Delta\mathsf{A}(t) \\ &+ \Delta\tilde{\mathsf{A}}(t)\Delta\tilde{\mathsf{A}}(t)]\tilde{x}(t) + [\tilde{\mathsf{A}}_d + \Delta\tilde{\mathsf{A}}_d(t)]\tilde{x}(t-d) \\ &+ \tilde{\Gamma}w(t) \\ &= \tilde{\mathsf{A}}_p(t)\tilde{x}(t) + \tilde{\mathsf{A}}_q(t)\tilde{x}(t-d) + \tilde{\Gamma}w(t), \\ \tilde{z}(t) &= [\tilde{\mathsf{A}} + \mathsf{A}[\tilde{\mathsf{A}} + \Delta\tilde{\mathsf{A}}(t)]]\tilde{x}(t) \end{aligned} \tag{8.88}$$

It follows from **Definition 8.4.8** that the contracted state feedback controller $u(t) = [\mathsf{K} + \Delta\mathsf{K}(t)]x(t)$ has the form $u(t) = [\tilde{\mathsf{K}} + \Delta\tilde{\mathsf{K}}(t)]\mathsf{V}x(t)$. Consider the relation between the state matrices of the closed-loop systems S_C and $\tilde{S}_C$ given in (8.69) and (8.88), respectively. It follows that

$$\begin{aligned}
\tilde{\mathsf{A}}_p(t) &= \mathsf{VA}_p(t)\mathsf{U} + \mathsf{M}_p \Longrightarrow \\
\mathsf{M}_p &= \mathsf{M} + \mathsf{N}[\tilde{\mathsf{K}} + \Delta\tilde{\mathsf{K}}(t)] + \mathsf{VB}[\tilde{K} + \Delta\tilde{K}(t)] - \mathsf{VB}[\tilde{K} + \Delta\tilde{K}(t)]\mathsf{VU} \\
&+ \mathsf{V}\Delta\mathsf{B}(t)[\tilde{\mathsf{K}} + \Delta\tilde{\mathsf{K}}(t)](I - \mathsf{VU})
\end{aligned}$$

where M_p is a complementary matrix, will be determined shortly. Since we seek $\tilde{S} \supset S_C$, the condition $\mathsf{UM}_p^j\mathsf{V} = 0,\ j = 1, 2, ..., \tilde{n}$, must be satisfied. Imposing this condition and using (8.83), it can be established that $\mathsf{MV} = 0,\ \mathsf{N} = 0$ is a sufficient condition so that $\mathsf{UM}_p^j\mathsf{V} = 0$ holds for all $j = 1, 2, ..., \tilde{n}$. Similarly, the relation $\tilde{\mathsf{A}}_q(t) = \mathsf{VA}_q(t)\mathsf{U} + \mathsf{AM}_q$ implies $\mathsf{M}_q = \mathsf{M}_d$. In order to verify the inclusion principle, we impose $\mathsf{UM}_j^d\mathsf{V} = 0$ $\mathsf{M}_d\mathsf{V} = 0$ to reach that $\mathsf{M}_d\mathsf{V} = 0$ is a sufficient condition to satisfy for all $j = 1, 2, ..., \tilde{n}$. The condition $\mathsf{U}[\tilde{\Gamma}w(t)] = \Gamma w(t)$ is equivalent to $\mathsf{UR} = 0$ in terms of complementary matrices and consequently $\mathsf{R} = 0$ is a sufficient condition. Finally, $\mathsf{LV} = 0$ is the same requirement as given in **Theorem 8.4.12** when imposing $\gamma = \tilde{\gamma}$.

The particular case 1) satisfies **Theorem 8.4.16** under which $(\tilde{S}_c, \tilde{\gamma}) \supset (S_c, \gamma)$. Suppose that a controller $u(t) = [\tilde{\mathsf{K}} + \Delta\tilde{\mathsf{K}}(t)]\tilde{x}(t)$ designed in the expanded system $\tilde{S}$ satisfies the inequality in **Definition 8.4.4**. This implies that the robust quadratic stability with $\mathcal{H}_\infty$-norm bound γ is guaranteed for the closed-loop system (8.88). Note that the controller design is performed with a given $\tilde{\gamma}$.

It is necessary to establish whether the condition $\tilde{\gamma} = \gamma$ is satisfied for the closed-loop system (8.69) with the contracted controller $u(t) = [\mathsf{K} + \Delta\mathsf{K}(t)]x(t)$ implemented into the initial system **S**. This condition is provided by the following theorem.

Theorem 8.4.18 *Consider the systems (8.60) and (8.77) with (8.80). Suppose that*

$$\mathsf{MV} = 0, \quad \mathsf{M}_d\mathsf{V} = 0, \quad \mathsf{N} = 0, \quad \mathsf{R} = 0, \quad \mathsf{LV} = 0$$

hold. Also, suppose that there exist a matrix $0 < \tilde{\mathsf{P}}^t = \tilde{\mathsf{P}}$ *and a gain matrix* $\tilde{K}$ *such that the inequality*

$$\begin{aligned}
&\tilde{\mathsf{P}}(\tilde{\mathsf{A}} + \tilde{\mathsf{B}}\tilde{\mathsf{K}}) + (\tilde{\mathsf{A}} + \tilde{\mathsf{B}}\tilde{\mathsf{K}})^t\tilde{\mathsf{P}} + \\
&\tilde{\mathsf{P}}(\mathsf{V}[H_A H_A^t + H_d H_d^t + \mathsf{A}_d\mathsf{A}_d^t]\mathsf{U} + \gamma^{-2}\tilde{\Gamma}\tilde{\Gamma}^t)\tilde{\mathsf{P}} + \\
&\varepsilon\tilde{\mathsf{K}}^t\tilde{E}_B^t\tilde{E}_B\tilde{\mathsf{K}} + \left(1 + \frac{1}{\varrho}\right)(\tilde{\mathsf{C}} + \tilde{\mathsf{D}}\tilde{\mathsf{K}})^t(\tilde{\mathsf{C}} + \tilde{\mathsf{D}}\tilde{\mathsf{K}}) + \\
&\tilde{\sigma}(1 + \varrho)(E_k^t E_k + E_A^t E_A + E_d^t E_d + (1 + \alpha)E_k^t E_K + I) < 0 \quad (8.89)
\end{aligned}$$

holds for $\varepsilon > 0, \ \varrho > 0$ *and all admissible uncertainties, where* $\tilde{\sigma} = \|\tilde{H}_k^t D^t D \tilde{H}_k\|$. *Then, there exist a positive-definite symmetric matrix* $\mathcal{P} = \mathsf{V}^t \tilde{\mathsf{P}} \mathsf{V}$ *and a gain matrix* $\mathsf{K} = \tilde{\mathsf{K}}\mathsf{V}$ *such that the robust stability with* $\mathcal{H}_\infty$ *performance attenuation bound* γ *is guaranteed for closed-loop system (8.69). Moreover,* $\gamma = \tilde{\gamma}$.

Proof 8.4.19 *Consider* $u(t) = [\tilde{\mathsf{K}} + \Delta\tilde{\mathsf{K}}(t)]\tilde{x}(t)$ *a contractible controller designed for the system* $\tilde{S}$. *Suppose the gain matrix* $\tilde{\mathsf{K}}$ *together with a matrix* $0 < \tilde{\mathsf{P}}^t = \tilde{\mathsf{P}}$ *satisfying (8.89). Using (8.83) into (8.89), then inequality (8.89) implies inequality (8.73) by* ***Definition 8.4.4*** *when matrix* $0 < \mathsf{P}^t = \mathsf{P}$ *in the form* $\mathsf{P} = \mathsf{V}^t\tilde{\mathsf{P}}\mathsf{V}, \ E_B = \tilde{E}_B, \ E_k = \tilde{E}_k\mathsf{V}$ *and* $H_k = \tilde{H}_k\mathsf{V}$. *Hence, the robust quadratic stability with* $\mathcal{H}_\infty$*-norm bound* γ *is guaranteed for closed-loop system (8.69) using the contracted controller* $u(t) = [\mathsf{K} + \Delta\mathsf{K}(t)]x(t)x(t) = [\tilde{\mathsf{K}} + \Delta\tilde{\mathsf{K}}(t)]\mathsf{V}x(t)$ *and consequently* $\gamma = \tilde{\gamma}$.

8.4.4 Overlapping state feedback

Information structure constraints on the state feedback gain matrices include different practically important structures corresponding with the sparsity forms of gain matrices well known in the theory of sparse matrices. These particular forms are a block diagonal form, a block tridiagonal form, and a double-bordered block diagonal form corresponding with decentralized, overlapping, and gain matrices, respectively. Generally, their control design may be effectively performed using LMI approach [329].

To demonstrate this approach, consider two overlapping subsystems with the structure of matrices $\mathsf{A}, \ \Delta\mathsf{A}(t), \ \mathsf{A}_d, \ \Delta\mathsf{A}_d(t)$ and $\mathsf{B}, \ \Delta\mathsf{B}(t), \ \Gamma$, respectively, in the form:

$$\begin{bmatrix} * & * & * \\ * & * & * \\ * & * & * \end{bmatrix}, \quad \begin{bmatrix} * & \vdots & * \\ * & \vdots & * \\ * & \vdots & * \end{bmatrix} \tag{8.90}$$

where $\mathsf{A}_{jj}, \ \Delta\mathsf{A}_{jj}(t), \ \mathsf{A}_{d_\mathsf{A}}, \ \Delta\mathsf{A}_{d_{jj}}(t) \in \Re^{n_j \times n_j}$ for $j = 1,2,3,$ and $\mathsf{B}_{jk}, \ \Delta\mathsf{B}_{ij}(t) \in \Re^{n_j \times m_j}, \ \Gamma_{jk} \in \Re^{n_j \times p_j}$ for $j = 1,2,3$. The dimensions of the components of the vector $x^t(t) = [x_1^t(t), x_2^t(t), x_3^t(t)]$ are n_1, n_2, n_3, respectively, and satisfy $n_1 + n_2 + n_3 = n$. The partition of $u^t(t) = [u_1^t(t), u_2^t(t)]$ has two components of dimensions m_1, m_2 such that $m_1 + m_2 = m$. A standard particular selection of the matrix V is selected as follows:

$$\begin{bmatrix} I_{n_1} & 0 & 0 \\ 0 & I_{n_2} & 0 \\ 0 & I_{n_2} & 0 \\ 0 & 0 & I_{n_3} \end{bmatrix} \tag{8.91}$$

As it has been shown before, this transformation leads naturally to an expanded system where the state vector x_2 appears repeated in $\tilde{x}^t(t) = [x_1^t(t), x_2^t(t), x_2^t(t), x_3^t(t)]$. The expanded controller has a block diagonal form with two subblocks of dimensions $m_1 \times (n_1 + n_2)$ and $m_2 \times (n_2 + n_3)$ as follows:

$$\tilde{\mathsf{K}}_d = \begin{bmatrix} \tilde{\mathsf{K}}_{11} & \tilde{\mathsf{K}}_{12} & 0 & 0 \\ 0 & 0 & \tilde{\mathsf{K}}_{23} & \tilde{\mathsf{K}}_{24} \end{bmatrix} \tag{8.92}$$

The corresponding contracted gain matrix has a block tridiagonal form as follows:

$$\mathsf{K}_{td} = \begin{bmatrix} \tilde{\mathsf{K}}_{11} & \tilde{\mathsf{K}}_{12} & 0 \\ 0 & \tilde{\mathsf{K}}_{23} & \tilde{\mathsf{K}}_{24} \end{bmatrix} \tag{8.93}$$

The design of overlapping controllers depends on the structure of matrices B, $\Delta\mathsf{B}(t)$. As has been demonstrated in [352], Type I corresponds with all nonzero elements of all input matrices in (8.90), while Type II corresponds with all elements $-(*)_{21} = 0$ and $(*)_{22} = 0$. The LMI control design for Type I can be performed directly on the original system. Type II requires performing the LMI control design in the expanded space because the direct design usually leads to infeasibility [329].

To simplify the control design for the Type II case, we denote P_{td} a block-tridiagonal matrix with the dimensions of its blocks corresponding with the dimensions of overlapping subsystems. To proceed, we introduce the following.

Definition 8.4.20 *Consider the system (8.60). The controller $u_{td}x(t) = [\mathsf{K}_{td} + \Delta\mathsf{K}_{td}(t)]x(t)$ with block-tridiagonal matrices K_{td} and $\Delta\mathsf{K}_{td}(t)$ is a tridiagonal-resilient $\mathcal{H}_\infty$ state feedback controller if there exist a matrix $0 < \mathsf{P}_{td}^t = \mathsf{P}_{td}$ and a gain matrix K_{td} such that the inequality*

$$\begin{aligned} &\mathsf{P}_{td}(\mathsf{A} + \mathsf{B}\mathsf{K}_{td}) + (\mathsf{A} + \mathsf{B}\mathsf{K}_{td})^t\mathsf{P}_{td} + \\ &\mathsf{P}_{td}(H_A H_A^t + H_d H_d^t + (1 + \frac{1}{\varepsilon})H_B H_B^t + A_d A_d^t + \mathsf{B}H_k H_k^t \mathsf{B}^t \\ &+\gamma^{-2}\Gamma\Gamma^t)\mathsf{P}_{td} + \\ &\varepsilon \mathsf{K}_{td}^t E_B^t E_B \mathsf{K}_{td} + (1 + \frac{1}{\varrho})(\mathsf{C} + \mathsf{D}\mathsf{K}_{td})^t(\mathsf{C} + \mathsf{D}\mathsf{K}_{td}) + \\ &(1 + \varrho)\sigma E_k^t E_k + E_A^t E_A + E_d^t E_d + (1 + \sigma)E_k^t E_K + I < 0 \end{aligned} \tag{8.94}$$

holds for all admissible uncertainties (8.61) and (8.67), where $\sigma = \|H_k^t \mathsf{D}^t \mathsf{D} H_k\|$.

The robust quadratic stability with $\mathcal{H}_\infty$-norm bound γ of the closed-loop system (8.69) is guaranteed. The following result can be easily established as a particular case of **Theorem 8.4.18**:

Theorem 8.4.21 *Consider systems (8.60) and (8.77). Suppose that*

$$\begin{aligned} \mathsf{MV} &= 0, \quad \mathsf{M}_d\mathsf{V} = 0, \quad \mathsf{N} = 0, \\ \mathsf{LV} &= 0, \quad \mathsf{R} = 0 \end{aligned} \tag{8.95}$$

holds. Consider the subsystem structure (8.90) and the transformation matrix (8.91). If there exist a matrix $\tilde{\mathsf{P}}_D > 0$ *and a gain matrix* $\tilde{\mathsf{K}}_D$ *satisfying (8.89), then*

$$u_{td}x(t) = [\mathsf{K}_{td} + \Delta\mathsf{K}_{td}(t)]x(t) = [\tilde{\mathsf{K}}_D + \Delta\mathsf{K}_D(t)]\mathsf{V}x(t)$$

is a tridiagonal-resilient $\mathcal{H}_\infty$ *state feedback controller with the matrix* $\mathsf{P}_{td} = \mathsf{V}^t\tilde{\mathsf{P}}_D\mathsf{V} > 0$ *for the system* **S** *satisfying* $\gamma = \tilde{\gamma}$.

8.4.5 Illustrative example 8.5

Consider the interconnected system with the numerical coefficients

$$A = \begin{bmatrix} -2 & 0 & -1 & 1 \\ -1 & 0 & 2 & 0 \\ 0 & -2 & -1 & 0 \\ 1 & 0 & 0 & -1 \end{bmatrix}, \quad B = \begin{bmatrix} 0.5 & 0 \\ 0.3 & 0.4 \\ 0 & 0.4 \\ 0 & 0.1 \end{bmatrix},$$

$$A_d = \begin{bmatrix} -0.2 & 0 & 0 & 0 \\ 0 & 0.2 & 0.1 & 0 \\ 0 & 0.1 & 0 & 0 \\ 0 & 0 & 0 & 0.2 \end{bmatrix}, \quad \Gamma = \begin{bmatrix} 0.3 & 0.2 \\ 0.4 & 0.3 \\ 0.2 & 0.2 \\ 0.1 & 0 \end{bmatrix},$$

$$C = \begin{bmatrix} 0.1 & 0 & 0 & 0 \\ 0 & 0.1 & -0.1 & 0 \\ 0 & 0 & 0 & 0.1 \end{bmatrix}, \quad H_A = \begin{bmatrix} 0.2 & 0 \\ 0.4 & 0 \\ 0.2 & 0.1 \\ 0.1 & 0 \end{bmatrix},$$

$$H_B^t = \begin{bmatrix} 0.1 & 0 \\ 0 & 0.1 \\ 0.2 & 0 \\ 0.1 & 0.1 \end{bmatrix}, \; H_d^t = \begin{bmatrix} 0 & 0.2 \\ 0 & 0.4 \\ 0.1 & 0.2 \\ 0 & 0.2 \end{bmatrix},$$

$$E_A = \begin{bmatrix} 0.1 & 0 & 0.1 & 0 \\ 0 & 0 & 0 & 0.1 \end{bmatrix}, \; E_B = \begin{bmatrix} 0.1 & 0.1 \\ 0 & 0.1 \end{bmatrix},$$

$$E_d = \begin{bmatrix} 0 & 0.1 & 0 & 0 \\ 0 & 0.1 & 0 & 0.1 \end{bmatrix}$$

The overlapped subsystems are

$$A_{22} = \begin{bmatrix} 0 & 2 \\ -2 & 1 \end{bmatrix}, \; C_{22} = \begin{bmatrix} 0.1 & -0.1 \end{bmatrix}$$

in the matrices A and C, respectively. The remaining overlapped subsystems corresponding to the matrices $\Delta A(t)\ A_1$ and $\Delta A_1(t)$ are also 2×2 dimensional blocks. Following [352], the transformation matrices are taken as

$$\mathsf{V} = \begin{bmatrix} 1 & 0 & 0 & 0 \\ 0 & 1 & 0 & 0 \\ 0 & 0 & 1 & 0 \\ 0 & 1 & 0 & 0 \\ 0 & 0 & 1 & 0 \\ 0 & 0 & 0 & 1 \end{bmatrix}, \quad \mathsf{M} = \begin{bmatrix} 1 & 0 & 0 \\ 0 & 1 & 0 \\ 0 & 1 & 0 \\ 0 & 0 & 1 \end{bmatrix}$$

We select $\mathsf{N} = 0,\ \mathsf{M}_d = 0,\ \mathsf{R} = 0$ and evaluate the remaining complementary matrices as

$$\mathsf{M} = \begin{bmatrix} 0 & 0 & -0.5 & 0 & 0.5 & 0 \\ 0 & 0 & 1 & 0 & -1 & 0 \\ 0 & -10 & -0.5 & 1 & 0.5 & 0 \\ 0 & 0 & -1 & 0 & 1 & 0 \\ 0 & 1 & 0.5 & -1 & -0.5 & 0 \\ 0 & 0 & 0 & 0 & 0 & 0 \end{bmatrix},$$

$$\mathsf{L} = \begin{bmatrix} 0 & 0 & 0 & 0 & 0 & 0 \\ 0 & 0.05 & -0.05 & -0.05 & 0.05 & 0 \\ 0 & -0.05 & 0.05 & 0.05 & -0.05 & 0 \\ 0 & 0 & 0 & 0 & 0 & 0 \end{bmatrix}$$

which satisfy the required relations (8.95). Application of **Theorem 8.4.21** yields the contracted and centralized gain matrices which are

$$\mathsf{K}_D = \begin{bmatrix} -0.7356 & -0.6244 & 0 \\ 0 & 0.1198 & 0.4763 \end{bmatrix},$$
$$\mathsf{K}_C = \begin{bmatrix} -4.8886 & -2.2419 & -1.3895 \\ 4.8957 & 0.8587 & -1.1244 \end{bmatrix}$$

8.5 Resilient Stabilization for Discrete Systems

Power systems, transportation systems, manufacturing processes, communication networks, and economic systems are some of the examples of large-scale interconnected systems [224]. Conventional centralized control methodology of such systems requires means of exchanging information between the subsystems for the controller implementation, and, therefore, sufficiently large communication bandwidth is needed for transferring information between the subsystems. This makes centralized controllers when applied to large-scale interconnected systems impractical and uneconomical. In addition, often, there are no means for subsystem information exchange, which prevents the application of centralized control. To overcome this, decentralized control methodology has been developed, which only uses local information available at each subsystem level for the controller implementation. In this way, multiple separate controllers are articulated each of which has access to different measured information and has authority over different decision or actuation variables. Hence, in real applications, decentralized controllers are simpler and more practical than centralized controllers. An exhaustive list of publications on the subject of decentralized control of large-scale interconnected systems is given in [57, 324]. A survey of early results can be found in [194].

Decentralized feedback control (DFC) of large-scale systems has received considerable interest in the systems and control literature. It effectively exploits the information structure constraint commonly existing in many practical large-scale systems. Over the past few decades, a large body of literature has become available on this subject; see [82]-[406] and the references therein. The computational advantages of DFC the approach have also recently attracted considerable attention, particularly in the context of parallel processing.

Decentralized robust controller design in the presence of uncertainties can be found in [88, 248]. Most of the proposed decentralized control strategies assume that subsystem states are available for feedback implementation (see, for example, [327, 396, 400]). However, in real applications, the availability of the states of each subsystem cannot be guaranteed. This motivated the development of decentralized output feedback controllers, which incorporate local observers to estimate the states of the subsystems (see, for example, [346, 406]). In recent years, a new systematic methodology has been proposed in [348] based on linear matrix inequalities (LMIs) [29] and extended to continuous-time interconnected systems in [327, 348, 398] and to time-delay systems in [28]. The appealing feature of these results is that the underlying problem is formulated as a convex optimization problem, which is designed to maximize the system robustness with respect to uncertainties.

In this section, we consider the stabilization problem of a class of large-scale interconnected discrete-time systems with unknown nonlinear interconnections and gain perturbations. The interconnection of each subsystem satisfies a quadratic constraint bound as, for example, in [28] and extending the results of [289, 348]. We develop a resilient control scheme[1] to cope with the gain perturbations. The main objective is to develop an LMI-based method for testing robust stability and designing robust feedback schemes for decentralized stabilization of interconnected systems. Our contribution is to extend the work in [28, 348] further and generalize their results in different ways. First, we deal with systems composed of linear discrete-time subsystems coupled by static nonlinear interconnections satisfying quadratic constraints. The developed solution is cast into the framework of convex optimization problems over LMIs. Second, we develop robust asymptotic stability and feedback stabilization results. Output feedback controllers are designed on the subsystem level to guarantee robust stability of the overall system and, in addition, maximize the bounds of unknown interconnection terms. Third, by incorporating additive gain perturbations we establish new resilient feedback stabilization schemes for discrete-time delay systems. To illustrate the application of the developed schemes we include a numerical example of three interconnected time-delay systems and provide detailed results of implementation.

8.5.1 Problem statement

We consider a class of nonlinear interconnected discrete-time systems Σ composed of N coupled subsystems $\Sigma_j,\ j = 1, .., N$ and represented by:

$$\begin{aligned} \Sigma:\ x(k+1) &= Ax(k) + Bu(k) + g(k, x(k)) \\ y(k) &= Cx(k) \end{aligned} \tag{8.96}$$

where for $k \in Z_+ := \{0, 1, ...\}$ and $x = (x_1^t, ..., x_N^t)^t \in \Re^n$, $n = \sum_{j=1}^N n_j$, $u = (u_1^t, ..., u_N^t)^t \in \Re^p$, $p = \sum_{j=1}^N p_j$, $y = (y_1^t, ..., y_N^t)^t \in$

[1]Throughout the chapter, we use the term "resilient" as introduced in [258] corresponding to "non-fragile" used in [60, 70].

$\Re^q$, $q = \sum_{j=1}^N q_j$ being the state, control, and output vectors of the interconnected system Σ. The associated matrices are real constants and modeled as $A = diag\{A_1, .., A_N\}$, $A_j \in \Re^{n_j \times n_j}$, $B = diag\{\Gamma, .., B_N\}$, $B_j \in \Re^{n_j \times p_j}$, $C = diag\{C_1, .., C_N\}$, $C_j \in \Re^{q_j \times n_j}$. The function $g : \mathsf{Z}_+ \times \Re^n \times \Re^n \rightarrow \Re^{n_j}$, $g(k, x(k)) = (g_1^t(k, x(k)), .., g_N^t(k, x(k)))^t$ is a vector function piecewise-continuous in its arguments. In the sequel, we assume that this function is uncertain and the available information is that, in the domains of continuity $\mathbf{G}$, it satisfies the quadratic inequality

$$g^t(k, x(k))g(k, x(k)) \leq x^t(k)\widetilde{G}^t\widetilde{\Phi}^{-1}\widetilde{G}x(k) \quad (8.97)$$

where $\widetilde{G} = [\widetilde{G}_1^t, .., \widetilde{G}_N^t]^t$, $\widetilde{G}_j \in \Re^{r_j \times n_j}$ are constant matrices such that $g_j(k, 0, 0) = 0$ and $x = 0$ is an equilibrium of system (8.96).

Exploiting the structural form of system (8.96), a model of the jth subsystem Σ_j can be described by

$$\begin{aligned} \Sigma_j : \; x_j(k+1) &= A_j x_j(k) + B_j u_j(k) + g_j(k, x(k)) \\ y_j(k) &= C_j x_j(k) \end{aligned} \quad (8.98)$$

where $x_j(k) \in \Re^{n_j}$, $u_j(k) \in \Re^{p_j}$, and $y_j(k) \in \Re^{q_j}$ are the subsystem state, control input, and measured output, respectively. The function $g_j : \mathsf{Z}_+ \times \Re^n \times \Re^n \rightarrow \Re^{n_j}$ is a piecewise-continuous vector function in its arguments and in line of (8.97) it satisfies the quadratic inequality

$$g_j^t(k, x(k)) \; g_j(k, x(k)) \leq \phi_j^2 \; x^t(k)\widetilde{G}_j^t\widetilde{G}_j x(k) \quad (8.99)$$

where $\phi_j > 0$ are bounding parameters such that $\widetilde{\Phi} = diag\{\phi_1^{-2} I_{r_1}, .., \phi_N^{-2} I_{r_N}\}$ where I_{m_j} represents the $m_j \times m_j$ identity matrix. From (8.97) and (8.99), it is always possible to find matrices Φ such that

$$g^t(k, x(k))g(k, x(k)) \;\; \leq \;\; x^t(k)G^t\Phi^{-1}Gx(k) \quad (8.100)$$

where $G = diag\{G_1, .., G_N\}$, $\Phi = diag\{\delta_1 I_{r_1}, .., \delta_N I_{r_N}\}$, and $\delta_j = \phi_j^{-2}$.

Our aim in this chapter is to develop new tools for the analysis and design of a class of interconnected discrete-time systems with uncertain function of nonlinear perturbations by exploiting the decentralized information structure constraint. We seek to establish complete LMI-based procedures for the robust stability and feedback (output, resilient) stabilization by basing all the computations at the subsystem level.

8.5.2 Subsystem stability

Our goal is to establish tractable conditions guaranteeing global asymptotic stability of the origin ($x = 0$) for all $g(k, x(k)) \in \mathbf{G}$. The main result of subsystem stability is established by the following theorem.

Theorem 8.5.1 *System (8.96) with $u \equiv 0$ is asymptotically stable for all nonlinear uncertainties satisfying (8.97) if there exist matrices $0 < \mathcal{Y}_j^t = \mathcal{Y}_j \in \Re^{n_j \times n_j}$ and scalars $\delta_j > 0$ such that the following convex optimization problem is feasible:*

$$\min_{\mathcal{Y}_j, \delta_j} \sum_{j=1}^{N} \delta_j$$
$$subject\ to: \ \mathcal{Y}_1 > 0, ..., \mathcal{Y}_N > 0,$$
$$diag\Big(\Gamma_1, ..., \Gamma_N\Big) < 0, \quad \Gamma_j = \begin{bmatrix} -\mathcal{Y}_j & \mathcal{Y}_j A_j^t & \mathcal{Y}_j A_j^t & \mathcal{Y}_j G_j^t \\ \bullet & I_j - \mathcal{Y}_j & 0 & 0 \\ \bullet & \bullet & -\mathcal{Y}_j & 0 \\ \bullet & \bullet & \bullet & -\delta_j I_j \end{bmatrix} \tag{8.101}$$

Proof: Introduce the Lyapunov functional (LF):

$$V(k) = \sum_{j=1}^{N} V_j(k) = \sum_{j=1}^{N} \Big[x_j^t(k) \mathcal{P}_j x_j(k) \Big] \tag{8.102}$$

where $0 < \mathcal{P}_j^t = \mathcal{P}_j \in \Re^{n_j \times n_j}$ are weighting matrices for $j = 1, .., N$. A straightforward computation gives the first-difference of $\Delta V(k) = V(k+1) - V(k)$ along the solutions of (8.96) with $u_j(k) \equiv 0,\ j = 1, .., N$ as:

$$\begin{aligned} \Delta V(k) &= \sum_{j=1}^{N} [A_j x_j(k) + g_j]^t \mathcal{P}_j [A_j x_j(k) + g_j] - \sum_{j=1}^{N} x_j^t(k) \mathcal{P}_j x_j(k) \\ &= \xi^t(k)\ \Xi\ \xi(k) \end{aligned} \tag{8.103}$$

where

$$\begin{aligned} \Xi_j &= diag(\Xi_1, ..., \Xi_N),\ \Xi_j = \begin{bmatrix} A_j^t \mathcal{P}_j A_j - \mathcal{P}_j & A_j^t \mathcal{P}_j \\ \bullet & \mathcal{P}_j \end{bmatrix}, \\ \xi(k) &= [\xi_1^t(k), ..., \xi_N^t(k)]^t,\ \ \xi_j^t(k) = \begin{bmatrix} x_j^t(k) & g_j^t(k, x(k)) \end{bmatrix} \end{aligned} \tag{8.104}$$

Recalling that (8.99) can be conveniently written as

$$\xi^t \, diag\left(\begin{bmatrix} -\phi_1^2 \, G_1^t G_1 & 0 \\ \bullet & I_1 \end{bmatrix}, ..., \begin{bmatrix} -\phi_N^2 \, G_N^t G_N & 0 \\ \bullet & I_N \end{bmatrix} \right) \xi \; \leq 0 \tag{8.105}$$

Then the sufficient condition of stability $\Delta V(k) < 0$ implies that $\Xi < 0$. By resorting to the **S**-procedure, inequalities (8.103) and (8.105) can be rewritten together as

$$\begin{aligned} & \mathcal{P}_1 > 0, ..., \mathcal{P}_N > 0, \; \omega_1 \geq 0, ..., \omega_N \geq 0, \\ & diag\Big(\Pi_1, ..., \Pi_N \Big) < 0, \\ \Pi_j \; = \; & \begin{bmatrix} A_j^t \mathcal{P}_j A_j - \mathcal{P}_j + \omega_j \phi_j^2 G_j^t G_j & A_j^t \mathcal{P}_j \\ \bullet & \mathcal{P}_j - \omega_j I_j \end{bmatrix} \end{aligned} \tag{8.106}$$

which describes nonstrict LMIs since $\omega_j \geq 0$. Recall from [29] that minimization under nonstrict LMIs corresponds to the same result as minimization under strict LMIs when both strict and nonstrict LMI constraints are feasible. Moreover, if there is a solution for (8.106) for $\omega_j = 0$, there will be also a solution for some $\omega_j > 0$ and sufficiently small ϕ_j. Therefore, we safely replace $\omega_j \geq 0$ by $\omega_j > 0$. Equivalently, we may further rewrite (8.106) in the form

$$\begin{aligned} & \bar{\mathcal{P}}_1 \; > \; 0, ..., \bar{\mathcal{P}}_N \; > \; 0, \\ & diag\Big(\bar{\Pi}_1, ..., \bar{\Pi}_N \Big) < 0, \\ \bar{\Pi}_j \; = \; & \begin{bmatrix} A_j^t \bar{\mathcal{P}}_j A_j - \bar{\mathcal{P}}_j + \phi_j^2 G_j^t G_j & A_j^t \bar{\mathcal{P}}_j \\ \bullet & \bar{\mathcal{P}}_j - I_j \end{bmatrix} \end{aligned} \tag{8.107}$$

where $\bar{\mathcal{P}}_j \; = \; \omega_j^{-1} \, \mathcal{P}_j$. Using the change of variable $\mathcal{Y}_j = \bar{\mathcal{P}}_j^{-1}$ with $\delta_j = \phi_j^{-2}$, multiplying by ω_j, then pre- and postmultiplying by $\bar{\mathcal{P}}_j$ with some arrangement, we express (8.107) using Schur complements in the form (8.101). Robust stability of the nonlinear interconnected system (8.96) under the constraint (8.97) with maximal ϕ_j is thus established.

Remark 8.5.2 *Note that for all possible nonlinear perturbations $g(k, 0) \; = \; 0$, the origin $x = 0$ is an equilibrium of system (8.96) and therefore in the light of [348], the overall asymptotic stable is guaranteed. It is significant to observe that* **Theorem 8.5.1** *yields a block-diagonal structure and provides an*

LMI-based characterization of the overall system stability in terms of the local subsystem asymptotic stability. In implementation, the local LMI problems could be solved in parallel since they virtually decoupled and the overall minimal solution is attained by minimizing the sum of the local tolerances

In what follows, we consider the decentralized feedback stabilization for interconnected system (8.96) within LMI-based formulation. We will be looking for a feedback controller that robustly stabilizes Σ. The main trust is to guarantee that local closed-loop subsystems are asymptotically stable for all possible nonlinear interconnections satisfying (8.99). In this way, the local controllers stabilize the linear part of Σ and, at the same time, maximize its tolerance to uncertain nonlinear interconnections and perturbations.

8.5.3 State feedback design

Now we examine the application of a linear local feedback controller of the form

$$u_j(k) = K_j\, x_j(k) \tag{8.108}$$

to system (8.96), where $K_j \in \Re^{p_j \times n_j}$ is a constant gain matrix. Substituting (8.108) into (8.98) yields:

$$\begin{aligned} \Sigma_s: \; x_j(k+1) &= [A_j + B_j K_j] x_j(k) + g_j(k, x(k)) \\ &= A_{js} x_j(k) + g_j(k, x(k)) \qquad (8.109) \\ y_j(k) &= C_j x(k) \qquad (8.110) \end{aligned}$$

A direct application of **Theorem 8.5.1** leads to the following optimization problem:

$$\begin{aligned} &\min_{\mathcal{Y}_j, \delta_j} \sum_{j=1}^{N} \delta_j \\ &subject\ to: \; \mathcal{Y}_1 > 0, ..., \mathcal{Y}_N > 0, \\ &diag\Big(\Gamma_{1s}, ..., \Gamma_{Ns}\Big) < 0, \quad \Gamma_{js} = \begin{bmatrix} -\mathcal{Y}_j & \mathcal{Y}_j A_{js}^t & \mathcal{Y}_j A_{js}^t & \mathcal{Y}_j G_j^t \\ \bullet & I_j - \mathcal{Y}_j & 0 & 0 \\ \bullet & \bullet & -\mathcal{Y}_j & 0 \\ \bullet & \bullet & \bullet & -\delta_j I_j \end{bmatrix} \end{aligned} \tag{8.111}$$

To put inequality (8.111) in a proper LMI setting, we need to separate the local gain from the local Lyapunov matrix. This is achieved by introducing the variable $K_j\mathcal{Y}_j = \mathcal{M}_j$. This in turn gives the following convex minimization problem over LMIs for:

$$\min_{\mathcal{Y}_j,\delta_j} \sum_{j=1}^{N} \delta_j$$
$$subject\ to:\ \mathcal{Y}_1 > 0, ..., \mathcal{Y}_N > 0,$$
$$Blockdiag\Big(\Gamma_{1s}, ..., \Gamma_{Ns}\Big) < 0,$$
$$\Gamma_{js} = \begin{bmatrix} -\mathcal{Y}_j & \mathcal{Y}_j A_j^t + \mathcal{M}_j B_j^t & \mathcal{Y}_j A_j^t + \mathcal{M}_j B_j^t & \mathcal{Y}_j G_j^t \\ \bullet & I_j - \mathcal{Y}_j & 0 & 0 \\ \bullet & \bullet & -\mathcal{Y}_j & 0 \\ \bullet & \bullet & \bullet & -\delta_j I_j \end{bmatrix} \tag{8.112}$$

and thus the following theorem is established.

Theorem 8.5.3 *System (8.96) is robustly asymptotically stabilizable by the decentralized control law (8.108) if there exist matrices* $0 < \mathcal{Y}_j^t = \mathcal{Y}_j \in \Re^{n_j \times n_j}$, $\mathcal{M}_j \in \Re^{p_j \times n_j}$ *and scalars* $\delta_j > 0$ *such that the convex optimization problem (8.112) has a feasible solution. The controller gain is given by* $K_j = \mathcal{M}_j \mathcal{Y}_j^{-1}$.

8.5.4 Bounded state feedback design

Following the approach of [28, 348], we consider hereafter the case of local state feedback control with bounded gain matrix K_j of the form $K_j^t\ K_j < \kappa_j\ I$, with $\kappa_j > 0$. Since $K_j = \mathcal{M}_j \mathcal{Y}_j^{-1}$ this condition corresponds to the additional constraints on the component matrices $\mathcal{M}_j$ and $\mathcal{Y}_j^{-1}$ by setting $\mathcal{M}_j^t\ \mathcal{M}_j < \mu_j I,\ \ \mu_j > 0,\ \ \mathcal{Y}_j^{-1} < \sigma_j\ I,\ \ \sigma_j > 0,\ j = 1, .., N$. In turn, these are equivalent to the LMIs

$$\begin{bmatrix} -\mu_j\ I_j & \mathcal{M}_j^t \\ \bullet & -I_j \end{bmatrix} < 0, \quad \begin{bmatrix} -\sigma_j\ I_j & I_j \\ \bullet & -\mathcal{Y}_j \end{bmatrix} < 0 \tag{8.113}$$

In a similar way, in order to guarantee desired values (ϕ_j) of the bounding factors (δ_j), we recall that $\phi_j^{-2} = \delta_j$. Thus we require

$$\delta_j - \frac{1}{\phi_j^2} < 0 \tag{8.114}$$

Incorporating the foregoing modifications into the present analysis leads to establishing the following convex optimization problem over LMIs for the local subsystem j:

$$\begin{aligned}
&\min_{\mathcal{Y}_j,\delta_j} \sum_{j=1}^{N} \delta_j \\
&subject\ to: \mathcal{Y}_1 > 0, ..., \mathcal{Y}_N > 0, \\
&diag\Big(\Gamma_{1s}, ..., \Gamma_{Ns}\Big) < 0, \qquad (8.115) \\
&\Gamma_{js} = \begin{bmatrix} -\mathcal{Y}_j & \mathcal{Y}_j A_j^t + \mathcal{M}_j B_j^t & \mathcal{Y}_j A_j^t + \mathcal{M}_j B_j^t & \mathcal{Y}_j G_j^t \\ \bullet & I_j - \mathcal{Y}_j & 0 & 0 \\ \bullet & \bullet & -\mathcal{Y}_j & 0 \\ \bullet & \bullet & \bullet & -\delta_j I_j \end{bmatrix}, \\
&diag\Big(\delta_1 - \frac{1}{\phi_1^2}, ..., \delta_N - \frac{1}{\phi_N^2}\Big) < 0, \\
&diag\Big(\begin{bmatrix} -\mu_1 I_1 & \mathcal{M}_1^t \\ \bullet & -I_1 \end{bmatrix}, ..., \begin{bmatrix} -\mu_N I_N & \mathcal{M}_N^t \\ \bullet & -I_N \end{bmatrix}\Big) < 0, \\
&diag\Big(\begin{bmatrix} -\sigma_1 I_1 & I_1 \\ \bullet & -\mathcal{Y}_1 \end{bmatrix}, ..., \begin{bmatrix} -\sigma_N I_N & I_N \\ \bullet & -\mathcal{Y}_N \end{bmatrix}\Big) < 0
\end{aligned}$$

Hence, the following theorem summarizes the main result.

Theorem 8.5.4 *Given the bounds ϕ_j, σ_j, μ_j, system (8.96) is robustly asymptotically stabilizable by control law (8.108) with constrained feedback gains and bounding factors if there exist matrices $0 < \mathcal{Y}_j^t = \mathcal{Y}_j \in \Re^{n_j \times n_j}$, $\mathcal{M}_j \in \Re^{p_j \times n_j}$ and scalars $\delta_j > 0$, $\mu_j > 0$, $\sigma_j > 0$ such that the convex optimization problem (8.115) has a feasible solution. The controller gain is given by $K_j = \mathcal{M}_j \mathcal{Y}_j^{-1}$.*

8.5.5 Resilient state feedback design

Now we address the performance deterioration issue in controller implementation [258] by considering that the actual linear local state feedback controller has the following perturbed form:

$$u_j(k) = [K_j + \Delta K_j]\, x_j(k) \qquad (8.116)$$

where $K_j \in \Re^{p_j \times n_j}$ is a constant gain matrix and ΔK_j is an additive gain perturbation matrix represented by

$$\Delta K_j = M_j\, \Delta_j\, N_j, \quad \Delta_j \in \mathbf{\Delta} := \{\Delta_j : \Delta_j^t \Delta_j \le I\} \qquad (8.117)$$

The application of control law (8.116) to system (8.98) yields the perturbed closed-loop system

$$\begin{aligned} \Sigma_p: \; x_j(k+1) &= (A_{js} + B_j \Delta K_j) x_j(k) + g_j(k, x(k)) \\ y_j(k) &= C_j x(k) \end{aligned} \tag{8.118}$$

For simplicity in exposition, we let $\Upsilon_j = \mathcal{Y}_j A_{js}^t + \mathcal{Y}_j (B_j \Delta K_j)^t$. It follows by applying **Theorem 8.5.1** to system (8.118), we obtain the following convex problem:

$$\begin{aligned} & \min_{\mathcal{Y}_j, \delta_j} \sum_{j=1}^{N} \delta_j \\ & subject\ to: \ \mathcal{Y}_1 > 0, ..., \mathcal{Y}_N > 0, \\ & Blockdiag\left(\Gamma_1, ..., \Gamma_N\right) < 0, \\ & \Gamma_j = \begin{bmatrix} -\mathcal{Y}_j & \Upsilon_j & \Upsilon_j & \mathcal{Y}_j G_j^t \\ \bullet & I_j - \mathcal{Y}_j & 0 & 0 \\ \bullet & \bullet & -\mathcal{Y}_j & 0 \\ \bullet & \bullet & \bullet & -\delta_j I_j \end{bmatrix} \end{aligned} \tag{8.119}$$

which we seek its feasibility over all possible perturbations $\Delta_j \in \mathbf{\Delta}$. In order to convexify inequality (8.119) and at the same time bypass the exhaustive search over the perturbation set $\mathbf{\Delta}$, we exploit the diagonal structure of inequality (8.119) and manipulate the jth-block (corresponding to subsystem j) with the aid of **Inequality 1** of the Appendix to reach

$$\begin{bmatrix} -\mathcal{Y}_j & \mathcal{Y}_j A_{js}^t & \mathcal{Y}_j A_{js}^t & \mathcal{Y}_j C_j^t \\ \bullet & I_j - \mathcal{Y}_j & 0 & 0 \\ \bullet & \bullet & -\mathcal{Y}_j & 0 \\ \bullet & \bullet & \bullet & -\delta_j I_j \end{bmatrix} + \begin{bmatrix} 0 \\ B_j M_j \\ B_j M_j \\ 0 \end{bmatrix} \Delta_j \begin{bmatrix} \mathcal{Y}_j N_j^t \\ 0 \\ 0 \\ 0 \end{bmatrix}^t +$$

$$\begin{bmatrix} \mathcal{Y}_j N_j^t \\ 0 \\ 0 \\ 0 \end{bmatrix} \Delta_j^t \begin{bmatrix} 0 \\ B_j M_j \\ B_j M_j \\ 0 \end{bmatrix}^t \leq$$

$$\begin{bmatrix} -\mathcal{Y}_j & \mathcal{Y}_j A_{js}^t & \mathcal{Y}_j A_{js}^t & \mathcal{Y}_j C_j^t \\ \bullet & I_j - \mathcal{Y}_j & 0 & 0 \\ \bullet & \bullet & -\mathcal{Y}_j & 0 \\ \bullet & \bullet & \bullet & -\delta_j I_j \end{bmatrix} + \eta_j \begin{bmatrix} 0 \\ B_j M_j \\ B_j M_j \\ 0 \end{bmatrix} \begin{bmatrix} 0 \\ B_j M_j \\ B_j M_j \\ 0 \end{bmatrix}^t +$$

$$\eta_j^{-1} \begin{bmatrix} \mathcal{Y}_j N_j^t \\ 0 \\ 0 \\ 0 \end{bmatrix} \begin{bmatrix} \mathcal{Y}_j N_j^t \\ 0 \\ 0 \\ 0 \end{bmatrix}^t < 0 \tag{8.120}$$

for some scalars $\eta_j > 0$. By Schur complements, inequality (8.120) becomes

$$\begin{bmatrix} -\mathcal{Y}_j & \mathcal{Y}_j A_{js}^t & \mathcal{Y}_j A_{js}^t & \mathcal{Y}_j C_j^t & 0 & \mathcal{Y}_j N_j^t \\ \bullet & I_j - \mathcal{Y}_j & 0 & 0 & \eta_j B_j M_j & 0 \\ \bullet & \bullet & -\mathcal{Y}_j & 0 & \eta_j B_j M_j & 0 \\ \bullet & \bullet & \bullet & -\delta_j I_j & 0 & 0 \\ \bullet & \bullet & \bullet & \bullet & -\eta_j I_j & \\ \bullet & \bullet & \bullet & \bullet & \bullet & -\eta_j I_j \end{bmatrix} < 0 \tag{8.121}$$

Introducing the change of variables $K_j \mathcal{Y}_j = \mathcal{M}_j$, inequality (8.121) becomes

$$\Theta_j = \begin{bmatrix} -\mathcal{Y}_j & \mathcal{Y}_j A_j^t + \mathcal{M}_j B_j^t & \mathcal{Y}_j A_j^t + \mathcal{M}_j B_j^t & \mathcal{Y}_j C_j^t & 0 & \mathcal{Y}_j N_j^t \\ \bullet & I_j - \mathcal{Y}_j & 0 & 0 & \eta_j B_j M_j & 0 \\ \bullet & \bullet & -\mathcal{Y}_j & 0 & \eta_j B_j M_j & 0 \\ \bullet & \bullet & \bullet & -\delta_j I_j & 0 & 0 \\ \bullet & \bullet & \bullet & \bullet & -\eta_j I_j & \\ \bullet & \bullet & \bullet & \bullet & \bullet & -\eta_j I_j \end{bmatrix} < 0 \tag{8.122}$$

The foregoing analysis has established the following theorem.

Theorem 8.5.5 *System (8.96) is robustly asymptotically stabilizable by the actual control law (8.116) for all possible gain variations (8.117) if there exist matrices* $0 < \mathcal{Y}_j^t = \mathcal{Y}_j \in \Re^{n_j \times n_j}$, $\mathcal{M}_j \in \Re^{p_j \times n_j}$ *and scalars* $\delta_j > 0,\ \eta_j$ *such that the following convex optimization problem*

$$\min_{\mathcal{Y}_j, \mathcal{M}_j, \delta_j, \eta_j} \sum_{j=1}^{N} \delta_j + \eta_j$$

$$subject\ to:\ \mathcal{Y}_1 > 0, ..., \mathcal{Y}_N > 0, \mathcal{M}_1, ..., \mathcal{M}_N$$

$$Blockdiag\Big(\Theta_1, ..., \Theta_N\Big) < 0$$

has a feasible solution where Θ_j *is given by (8.122). The controller gain is given by* $K_j = \mathcal{M}_j \mathcal{Y}_j^{-1}$.

Had we taken the modifications made previously to constrain the local state feedback gains, we would have reached an alternative convex optimization problem over LMIs summarized by the following theorem.

Theorem 8.5.6 *Given the bounds $\phi_j,\ \sigma_j,\ \mu_j$, system (8.96) is robustly asymptotically stabilizable by the actual control law (8.116) for all possible gain variations (8.117) if there exist matrices $0 < \mathcal{Y}_j^t = \mathcal{Y}_j \in \Re^{n_j \times n_j},\ \mathcal{M}_j \in \Re^{p_j \times n_j}$ and scalars $\delta_j > 0,\ \mu_j > 0,\ \sigma_j > 0,\ \eta_j$ such that the following convex optimization problem*

$$
\begin{aligned}
&\min_{\mathcal{Y}_j,\mathcal{M}_j,\delta_j} \sum_{j=1}^{N} \delta_j + \eta_j\mu_j + \sigma_j \\
&subject\ to:\ \mathcal{Y}_1 > 0, ..., \mathcal{Y}_N > 0, \mathcal{M}_1, ..., \mathcal{M}_N \\
&\qquad diag\left(\Theta_1, ..., \Theta_N\right) < 0, \\
&diag\left(\delta_1 - \frac{1}{\phi_1^2}, ..., \delta_N - \frac{1}{\phi_N^2}\right) < 0, \\
&diag\left(\begin{bmatrix} -\mu_1 I_1 & \mathcal{M}_1^t \\ \bullet & -I_1 \end{bmatrix}, ..., \begin{bmatrix} -\mu_N I_N & \mathcal{M}_N^t \\ \bullet & -I_N \end{bmatrix}\right) < 0, \\
&diag\left(\begin{bmatrix} -\sigma_1 I_1 & I_1 \\ \bullet & -\mathcal{Y}_1 \end{bmatrix}, ..., \begin{bmatrix} -\sigma_N I_N & I_N \\ \bullet & -\mathcal{Y}_N \end{bmatrix}\right) < 0
\end{aligned}
$$

has a feasible solution. The controller gain is given by $K_j = \mathcal{M}_j\mathcal{Y}_j^{-1}$.

Remark 8.5.7 *It should be observed that* ***Theorems 8.5.3-8.5.6*** *provide a complete decoupled set of LMI-based state feedback design algorithms for a class of interconnected discrete-time systems and quadratically bounded nonlinearities and perturbations. It is significant to record that* ***Theorem 8.5.6*** *is a new addition to the resilient control theory [258, 389]. Indeed, these algorithms assume the accessibility of local states.*

In the sequel, we drop the assumption on complete accessibility of local state variables and consider the availability of local output measurements. We focus on the design of a local dynamic output feedback stabilization scheme. To facilitate further development, we consider the case where the set of output matrices $C_j,\ j = 1, ..., n_s$ are assumed to be of full row rank. We shall be looking for a general linear time-invariant dynamic controller that obeys the decentralized information structure constraint requiring that each subsystem is controlled using only its own local output.

8.5.6 Output feedback stabilization

In this part, we consider the design of a local dynamic output feedback stabilization scheme:

$$\begin{aligned} z_j(k+1) &= L_j z_j(k) + F_j y_j(k) \\ u_j(k) &= K_{cj} z_j(k) + K_{oj} y_j(k) \end{aligned} \tag{8.123}$$

where $z_j(k) \in \Re^{m_j}$ is the local observer state such that for the interconnected system $z = (z_1^t, ..., z_N^t)^t \in \Re^m$, $m = \sum_{j=1}^N m_j$, $L = diag\{L_1, ..., L_N\}$, $F = diag\{F_1, ..., F_N\}$, $K_c = diag\{K_{c1}, ..., K_{cN}\}$, $K_o = diag\{K_{o1}, ..., K_{oN}\}$. The matrices $L_j, \in \Re^{m_j \times m_j}$, $F_j, \in \Re^{m_j \times q_j}$, $K_{cj}, \in \Re^{p_j \times m_j}$, $K_{oj}, \in \Re^{p_j \times q_j}$ are unknown gains to be determined. In the sequel, we let

$$\mathcal{K}_j = \begin{bmatrix} L_j & F_j \\ K_{cj} & K_{oj} \end{bmatrix} \in \Re^{m_j + p_j} \times \Re^{m_j + q_j} \tag{8.124}$$

to represent the local control parameter matrix of subsystem j to be determined such that the overall closed-loop controlled system achieves asymptotic stability.

We proceed, in line of the foregoing analysis, by appending the subsystem dynamics (8.98) and output feedback controller (8.123), to get the local augmented system

$$\begin{aligned} \zeta_j(k+1) &= \mathcal{A}_j\, \zeta_j(k) + \bar{g}_j(k, \zeta(k)) \\ \xi_j(k) &= \mathcal{C}_j\, \zeta_j(k) \end{aligned} \tag{8.125}$$

where

$$\begin{aligned} \zeta_j(k) &= \begin{bmatrix} x_j(k) \\ z_j(k) \end{bmatrix} \in \Re^{n_j + m_j}, \ \bar{g}_j(k, \zeta(k)) = \begin{bmatrix} g_j(k, x(k)) \\ 0 \end{bmatrix} \\ \mathcal{A}_j &= \begin{bmatrix} A_j + B_j K_{oj} C_j & B_j K_{cj} \\ F_j C_j & L_j \end{bmatrix}, \ \mathcal{C}_j = \begin{bmatrix} C_j & 0 \end{bmatrix} \end{aligned} \tag{8.126}$$

Note in light of (8.100), we have

$$\bar{g}^t(k, \zeta(k)) \bar{g}(k, \zeta(k)) \ \leq \ \zeta^t(k) \widehat{G}^t \Phi^{-1} \widehat{G} \zeta(k) \tag{8.127}$$

where $\widehat{G} = diag[\widehat{G}_1, ..., \widehat{G}_N]$, $\widehat{G}_j = [G_j \quad 0]$. Introduce the block matrix

$$\bar{\mathcal{Y}}_j = \begin{bmatrix} \mathcal{Y}_{cj} & 0 \\ 0 & \mathcal{Y}_{oj} \end{bmatrix} \tag{8.128}$$

where $\mathcal{Y}_{cj} \in \Re^{n_j \times n_j}$ and $\mathcal{Y}_{oj} \in \Re^{m_j \times m_j}$ are symmetric and positive-definite matrices. It follows from **Theorem 8.5.1** that the robust asymptotic stability of the augmented system (8.125) guaranteed by the solution of the following convex optimization problem

$$\min_{\bar{\mathcal{Y}}_j, \delta_j} \sum_{j=1}^{N} \delta_j$$

$$subject\ to: \ \mathcal{Y}_{c1} > 0, \ , ..., \mathcal{Y}_{cN} > 0, \ \mathcal{Y}_{o1} > 0, \ , ..., \mathcal{Y}_{oN} > 0,$$

$$diag\left(\widetilde{\Gamma}_1, ..., \widetilde{\Gamma}_N\right) < 0, \quad \widetilde{\Gamma}_j = \begin{bmatrix} -\bar{\mathcal{Y}}_j & \bar{\mathcal{Y}}_j \mathcal{A}_j^t & \bar{\mathcal{Y}}_j \mathcal{A}_j^t & \bar{\mathcal{Y}}_j \widehat{G}_j^t \\ \bullet & I_j - \bar{\mathcal{Y}}_j & 0 & 0 \\ \bullet & \bullet & -\bar{\mathcal{Y}}_j & 0 \\ \bullet & \bullet & \bullet & -\delta_j I_j \end{bmatrix} < 0 \tag{8.129}$$

is feasible. Algebraic manipulations of inequality (8.129) using (8.126)-(8.128) and basing the analysis on the subsystem level lead to the following theorem.

Theorem 8.5.8 *System (8.125) is robustly asymptotically stable if there exist matrices* $0 < \mathcal{Y}_{cj}^t = \mathcal{Y}_{cj} \in \Re^{n_j \times n_j}$, $0 < \mathcal{Y}_{oj}^t = \mathcal{Y}_{oj} \in \Re^{n_j \times n_j}$, $\mathcal{S}_{cj} \in \Re^{p_j \times n_j}$, $\mathcal{S}_{oj} \in \Re^{m_j \times m_j}$, $\mathcal{R}_{cj} \in \Re^{m_j \times n_j}$, $\mathcal{R}_{oj} \in \Re^{p_j \times m_j}$ *and scalars* $\delta_j > 0$ *such that the following convex optimization problem*

$$\min_{\mathcal{Y}_{cj}, \mathcal{Y}_{oj}, \mathcal{S}_{cj}, \mathcal{S}_{oj}, \mathcal{R}_{oj}, \mathcal{R}_{cj}, \delta_j} \sum_{j=1}^{N} \delta_j$$

$$subject\ to: \ \mathcal{Y}_{c1} > 0, \ , ..., \mathcal{Y}_{cN} > 0, \ \mathcal{Y}_{o1} > 0, \ , ..., \mathcal{Y}_{oN} > 0,$$

$$Blockdiag\left(\begin{bmatrix} \Xi_{c1} & \Xi_{d1} \\ \bullet & \Xi_{o1} \end{bmatrix}, ..., \begin{bmatrix} \Xi_{cN} & \Xi_{dN} \\ \bullet & \Xi_{oN} \end{bmatrix} \right) < 0 \tag{8.130}$$

where for $j = 1, ..., N$

$$\Xi_{cj} = \begin{bmatrix} -\mathcal{Y}_{cj} & 0 & \mathcal{Y}_{cj} A_j^t + \mathcal{S}_{cj}^t B_j^t & \mathcal{R}_{cj}^t \\ \bullet & -\mathcal{Y}_{oj} & \mathcal{R}_{oj}^t B_j^t & \mathcal{S}_{oj}^t \\ \bullet & \bullet & I_j - \mathcal{Y}_{cj} & 0 \\ \bullet & \bullet & \bullet & I_j - \mathcal{Y}_{oj} \end{bmatrix},$$

$$\Xi_{dj} = \begin{bmatrix} \mathcal{Y}_{cj} A_j^t + \mathcal{S}_{cj}^t B_j^t & \mathcal{R}_{cj}^t & \mathcal{Y}_{cj} G_j^t \\ \mathcal{R}_{oj}^t B_j^t & \mathcal{S}_{oj}^t & 0 \\ 0 & 0 & 0 \end{bmatrix},$$

$$\Xi_{oj} = \begin{bmatrix} -\mathcal{Y}_{cj} & 0 & 0 \\ \bullet & -\mathcal{Y}_{oj} & 0 \\ \bullet & \bullet & -\delta_j I_j \end{bmatrix} \tag{8.131}$$

is feasible. Moreover, the gain matrices are given by

$$\mathcal{K}_j = \begin{bmatrix} \mathcal{S}_{oj} & \mathcal{R}_{cj} \\ \mathcal{R}_{oj} & \mathcal{S}_{cj} \end{bmatrix} \begin{bmatrix} \mathcal{Y}_{oj} & 0 \\ 0 & \mathcal{Y}_{cj} \end{bmatrix}^{-1} \begin{bmatrix} I_j & 0 \\ 0 & C_j^{\dagger} \end{bmatrix}$$

Proof: Introducing the change of variables:

$$\begin{aligned} \mathcal{S}_{oj} &= L_j \mathcal{Y}_{oj},\ \mathcal{R}_{oj} = K_{oj} \mathcal{Y}_{oj},\ \mathcal{R}_{cj} = F_j C_j \mathcal{Y}_{cj}, && (8.132) \\ \mathcal{S}_{cj} &= K_{cj} C_j \mathcal{Y}_{cj},\ \mathcal{R}_{sj} = F_j \mathcal{Y}_{cj} && (8.133) \end{aligned}$$

and expanding the LMI (8.129) using (8.128), we readily obtain LMI (8.130) with (8.131).

Remark 8.5.9 *In the literature, it is reported that general forms of Lyapunov matrices $\mathcal{Y}_j$ lead to nonlinear matrix inequalities. Therefore, the selection of the block-diagonal form of the Lyapunov matrices (8.128) becomes common in control system design and has been widely used in dynamic output feedback design [28, 346]. In our approach, this selection yields systematic design procedure that aims at developing convenient noniterative computational algorithm while maintaining all the computations at the subsystem level.*

8.5.7 Resilient output feedback stabilization

Next we address the resilience problem of output feedback stabilization. For this purpose, we re-express the controller (8.123) into the form

$$\begin{aligned} z_j(k+1) &= L_j z_j(k) + F_j y_j(k) \\ u_j(k) &= [K_{cj} + \Delta K_{cj}] z_j(k) + [K_{oj} + \Delta K_{oj}] y_j(k) && (8.134) \end{aligned}$$

where the matrices ΔK_{oj}, ΔK_{cj} are gain perturbations represented by

$$\begin{aligned} \Delta K_{oj} &= M_{oj} \Delta_j N_{oj},\ \ \Delta K_{cj} = M_{cj} \Delta_j N_{cj}, \\ \Delta_j &\in \mathbf{\Delta} := \{\Delta_j : \Delta_j^t \Delta_j \leq I\} && (8.135) \end{aligned}$$

Following the foregoing analysis, the augmented local subsystem becomes

$$\begin{aligned} \zeta_j(k+1) &= [\mathcal{A}_j + \Delta \mathcal{A}_j]\, \zeta_j(k) + \bar{g}_j(k, \zeta(k)) \\ &= \bar{\mathcal{A}}_j\, \zeta_j(k) + \bar{g}_j(k, \zeta(k)) \\ \xi_j(k) &= \mathcal{C}_j\, \zeta_j(k) && (8.136) \end{aligned}$$

where

$$\Delta \mathcal{A}_j = \begin{bmatrix} B_j \Delta K_{oj} C_j & B_j \Delta K_{cj} \\ 0 & 0 \end{bmatrix} \qquad (8.137)$$

where $\zeta_j(k), \mathcal{A}_j, \mathcal{C}_j$ are given by (8.126). In light of the procedure developed earlier, it follows from **Theorem 8.5.1** that the robust asymptotic stability of the augmented system (8.136) is guaranteed by the feasible solution of the following convex optimization problem:

$$\min_{\bar{\mathcal{Y}}_j, \delta_j} \sum_{j=1}^{N} \delta_j$$

$$subject\ to: \ \mathcal{Y}_{c1} > 0, \ ,..., \mathcal{Y}_{cN} > 0, \ \mathcal{Y}_{o1} > 0, \ ,..., \mathcal{Y}_{oN} > 0,$$

$$diag\left(\widetilde{\Gamma}_1, ..., \widetilde{\Gamma}_N\right) < 0,$$

$$\widetilde{\Gamma}_j = \begin{bmatrix} -\bar{\mathcal{Y}}_j & \bar{\mathcal{Y}}_j \bar{\mathcal{A}}^t_{\underline{j}} & \bar{\mathcal{Y}}_j \bar{\mathcal{A}}^t_j & \bar{\mathcal{Y}}_j \widehat{\mathcal{G}}^t_j \\ \bullet & I_j - \mathcal{Y}_j & 0 & 0 \\ \bullet & \bullet & -\bar{\mathcal{Y}}_j & 0 \\ \bullet & \bullet & \bullet & -\delta_j I_j \end{bmatrix} < 0 \tag{8.138}$$

In order to convexify (8.138) we invoke **Inequality 1** of the Appendix while basing the analysis on the subsystem level and manipulate using (8.135) to reach

$$\begin{bmatrix} -\bar{\mathcal{Y}}_j & \bar{\mathcal{Y}}_j \mathcal{A}^t_{\underline{j}} & \bar{\mathcal{Y}}_j \mathcal{A}^t_j & \bar{\mathcal{Y}}_j \mathcal{C}^t_j \\ \bullet & I_j - \mathcal{Y}_j & 0 & 0 \\ \bullet & \bullet & -\bar{\mathcal{Y}}_j & 0 \\ \bullet & \bullet & \bullet & -\delta_j I_j \end{bmatrix} +$$

$$\begin{bmatrix} \sigma_{1j} \\ U \\ U \\ 0 \end{bmatrix} \Delta^t \begin{bmatrix} U \\ \sigma_{2j} \\ \sigma_{2j} \\ 0 \end{bmatrix}^t + \begin{bmatrix} U \\ \sigma_{2j} \\ \sigma_{2j} \\ 0 \end{bmatrix} \Delta \begin{bmatrix} \sigma_{1j} \\ U \\ U \\ 0 \end{bmatrix}^t$$

$$\begin{bmatrix} \sigma_{3j} \\ U \\ U \\ 0 \end{bmatrix} \Delta^t \begin{bmatrix} U \\ \sigma_{4j} \\ \sigma_{4j} \\ 0 \end{bmatrix}^t + \begin{bmatrix} U \\ \sigma_{4j} \\ \sigma_{4j} \\ 0 \end{bmatrix} \Delta \begin{bmatrix} \sigma_{3j} \\ U \\ U \\ 0 \end{bmatrix}^t$$

$$\leq \begin{bmatrix} -\bar{\mathcal{Y}}_j & \bar{\mathcal{Y}}_j \mathcal{A}_j^t & \bar{\mathcal{Y}}_j \mathcal{A}_j^t & \bar{\mathcal{Y}}_j \mathcal{C}_j^t \\ \bullet & I_j - \bar{\mathcal{Y}}_j & 0 & 0 \\ \bullet & \bullet & -\bar{\mathcal{Y}}_j & 0 \\ \bullet & \bullet & \bullet & -\delta_j I_j \end{bmatrix} +$$

$$\varepsilon_j^{-1} \begin{bmatrix} U \\ \sigma_{2j} \\ \sigma_{2j} \\ 0 \end{bmatrix} \begin{bmatrix} U \\ \sigma_{2j} \\ \sigma_{2j} \\ 0 \end{bmatrix}^t + \varepsilon_j \begin{bmatrix} \sigma_{1j} \\ U \\ U \\ 0 \end{bmatrix} \begin{bmatrix} \sigma_{1j} \\ U \\ U \\ 0 \end{bmatrix}^t +$$

$$\varphi_j^{-1} \begin{bmatrix} U \\ \sigma_{4j} \\ \sigma_{4j} \\ 0 \end{bmatrix} \begin{bmatrix} U \\ \sigma_{4j} \\ \sigma_{4j} \\ 0 \end{bmatrix}^t +$$

$$\varphi_j \begin{bmatrix} \sigma_{3j} \\ U \\ U \\ 0 \end{bmatrix} \begin{bmatrix} \sigma_{3j} \\ U \\ U \\ 0 \end{bmatrix}^t < 0 \tag{8.139}$$

where

$$\begin{aligned} \sigma_{1j} &= \begin{bmatrix} C_j^t N_{oj}^t \\ 0 \end{bmatrix}, \ \sigma_{2j} = \begin{bmatrix} \mathcal{Y}_{cj} B_j M_{oj} \\ 0 \end{bmatrix}, \ \sigma_{3j} = \begin{bmatrix} 0 \\ N_{cj}^t \end{bmatrix}, \\ \sigma_{4j} &= \begin{bmatrix} \mathcal{Y}_{cj} B_j M_{cj} \\ 0 \end{bmatrix}, \ U = \begin{bmatrix} 0 \\ 0 \end{bmatrix} \end{aligned} \tag{8.140}$$

for some $\varepsilon_j > 0, \ \varphi_j > 0$. Inequality (8.139) with the help of Schur complements can be written as

$$\begin{bmatrix} -\bar{\mathcal{Y}}_j & \bar{\mathcal{Y}}_j \mathcal{A}_j^t & \bar{\mathcal{Y}}_j \mathcal{A}_j^t & \bar{\mathcal{Y}}_j \mathcal{C}_j^t & \Xi_{vj} \\ \bullet & I_j - \bar{\mathcal{Y}}_j & 0 & 0 & 0 \\ \bullet & \bullet & -\bar{\mathcal{Y}}_j & 0 & 0 \\ \bullet & \bullet & \bullet & -\delta_j I_j & 0 \\ \bullet & \bullet & \bullet & \bullet & -\Xi_{wj} \end{bmatrix} < 0 \tag{8.141}$$

with

$$\Xi_{vj} = \begin{bmatrix} U & \varepsilon_j\sigma_{1j} & U & \varepsilon_j\sigma_{3j} \\ \sigma_{2j} & U & \sigma_{4j} & U \\ \sigma_{2j} & U & \sigma_{4j} & U \\ 0 & 0 & 0 & 0 \end{bmatrix},$$

$$\Xi_{wj} = \begin{bmatrix} \varepsilon_j I_j & 0 & 0 & 0 \\ \bullet & \varphi_j I_j & 0 & 0 \\ \bullet & \bullet & \varepsilon_j I_j & 0 \\ \bullet & \bullet & \bullet & \varphi_j I_j \end{bmatrix} \tag{8.142}$$

Our immediate task is to determine the feedback controller gain matrices

$$L_j,\ F_j,\ K_{oj},\ K_{cj}$$

and we do this by extending on the convexification procedure of **Theorem 8.5.8**. The desired result is summarized by the following theorem.

Theorem 8.5.10 *System (8.136) is robustly asymptotically stable if there exist matrices* $0 < \mathcal{Y}_{cj}^t = \mathcal{Y}_{cj} \in \Re^{n_j \times n_j}$, $0 < \mathcal{Y}_{oj}^t = \mathcal{Y}_{oj} \in \Re^{n_j \times n_j}$, $\mathcal{S}_{cj} \in \Re^{p_j \times n_j}$, $\mathcal{S}_{oj} \in \Re^{m_j \times m_j}$, $\mathcal{R}_{cj} \in \Re^{m_j \times n_j}$, $\mathcal{R}_{oj} \in \Re^{p_j \times m_j}$, $\mathcal{R}_{sj} \in \Re^{m_j \times n_j}$, $\mathcal{M}_{oj} \in \Re^{n_j \times n_j}$ *and scalars* $\delta_j > 0,\ \nu_j > 0,\ \varepsilon_j > 0,\ \varphi_j > 0$ *such that the following convex optimization problem*

$$\min_{\mathcal{Y}_{cj},\mathcal{Y}_{oj},\mathcal{S}_{cj},\mathcal{R}_{oj},\mathcal{R}_{cj},\mathcal{R}_{sj},\mathcal{M}_{oj},\delta_j,\varepsilon_j,\varphi_j} \quad \sum_{j=1}^{N} \delta_j + \varepsilon_j + \varphi_j \tag{8.143}$$

$$subject\ to:\ \mathcal{Y}_{c1} > 0,\ ,...,\mathcal{Y}_{cN} > 0,\ \mathcal{Y}_{o1} > 0,\ ,...,\mathcal{Y}_{oN} > 0,$$

$$diag\left(\begin{bmatrix} \Xi_{c1} & \Xi_{d1} & \Xi_{v1} \\ \bullet & \Xi_{o1} & 0 \\ \bullet & \bullet & -\Xi_{w1} \end{bmatrix}, ..., \begin{bmatrix} \Xi_{cN} & \Xi_{dN} & \Xi_{vN} \\ \bullet & \Xi_{oN} & 0 \\ \bullet & \bullet & -\Xi_{wN} \end{bmatrix} \right) < 0 \tag{8.144}$$

is feasible, where $\Xi_{cj},\ \Xi_{dj},\ \Xi_{oj}$ *are as in (8.131) and* $\Xi_{vj},\ \Xi_{wj}$ *are given by (8.142). Moreover, the gain matrices are given by*

$$\mathcal{K}_j = \begin{bmatrix} \mathcal{S}_{oj} & \mathcal{R}_{cj} \\ \mathcal{R}_{oj} & \mathcal{S}_{cj} \end{bmatrix} \begin{bmatrix} \mathcal{Y}_{oj} & 0 \\ 0 & \mathcal{Y}_{cj} \end{bmatrix}^{-1} \begin{bmatrix} I_j & 0 \\ 0 & C_j^\dagger \end{bmatrix}$$

Proof: By introducing the change of variables (8.133) and expanding the LMI (8.141) using (8.128), we readily obtain LMI (8.144).

8.5.8 Illustrative example 8.6

To illustrate the usefulness of the developed resilience approach, let us consider the following interconnected system which is composed of three subsystems. With reference to (8.98), the data values are:

$$
\begin{aligned}
&\textit{Subsystem 1}: \\
A_1 &= \begin{bmatrix} 0.9 & 0.5 \\ 0.8 & 0.1 \end{bmatrix}, \Gamma = \begin{bmatrix} 1 \\ 0.5 \end{bmatrix}, C_1 = \begin{bmatrix} 0 & 1 \end{bmatrix} \\
&\textit{Subsystem 2}: \\
A_2 &= \begin{bmatrix} 0 & 0 & 0.1 \\ 0.1 & -0.6 & 0.1 \\ 0.3 & 0.7 & 0.9 \end{bmatrix}, B_2 = \begin{bmatrix} 0.3 \\ 0 \\ 0.4 \end{bmatrix}, \\
C_2 &= \begin{bmatrix} 1 & 0 & 0 \end{bmatrix} \\
&\textit{Subsystem 3}: \\
A_3 &= \begin{bmatrix} 0.3 & 0.2 \\ -0.6 & 0.6 \end{bmatrix}, B_3 = \begin{bmatrix} 0.5 \\ 0.7 \end{bmatrix}, C_3 = \begin{bmatrix} 0.5 & 0 \end{bmatrix}, \\
x_1(k) &= \begin{bmatrix} x_{11}(k) & x_{12}(k) \end{bmatrix}^t, \\
x_2(k) &= \begin{bmatrix} x_{21}(k) & x_{22}(k) & x_{23}(k) \end{bmatrix}^t, \\
x_3(k) &= \begin{bmatrix} x_{31}(k) & x_{32}(k) \end{bmatrix}^t
\end{aligned}
$$

In terms of the overall state vector $x(k) = \begin{bmatrix} x_1^t(k) & x_2^t(k) & x_3^t(k) \end{bmatrix}^t \in \Re^7$, the interconnection is given by

$$
g(k, x(k)) = \left[\begin{array}{cc|ccc|cc}
0.1 & -0.1 & 0 & 0.2 & 0 & 0.1 & 0 \\
0.1 & 0.3 & 1 & 0 & 0 & 0 & 1 \\
\hline
-0.1 & 0.1 & 0 & -0.2 & 0 & 0.1 & 0 \\
0 & -0.1 & 0 & 0 & 0.3 & 0 & 0.1 \\
0 & 0 & -0.3 & 0.4 & 0.3 & 0 & 0.2 \\
\hline
0.1 & -0.1 & 0.1 & -0.1 & 0 & 0.2 & -0.1 \\
0.1 & -0.1 & 0 & 0.2 & 0 & 0.4 & 0.5
\end{array}\right] e(k, x(k))\, x(k)
$$

where $e(k, x(k)) : \Re^8 \longrightarrow [0, 1]$ represents a normalized interconnection parameter and the dashed (horizontal and vertical) lines correspond to the three subsystems.

The objective here is to compute a decentralized control law that would connectively stabilize the system for all values of $e(k, x(k)) \in [0, 1]$ based on state or output feedback schemes. Using MATLAB-LMI solver, it is found that the feasible solution of **Theorem 8.5.1** is given by

$$\delta_1 = 1.6535, \ \mathcal{Y}_1 = \begin{bmatrix} 1.711 & 0.318 \\ \bullet & 1.844 \end{bmatrix},$$
$$\delta_2 = 1.7335, \ \mathcal{Y}_2 = \begin{bmatrix} 4.252 & 1.121 & 1.441 \\ \bullet & 2.824 & 1.327 \\ \bullet & \bullet & 4.225 \end{bmatrix},$$
$$\delta_3 = 1.4964, \ \mathcal{Y}_3 = \begin{bmatrix} 2.223 & 0.625 \\ \bullet & 1.914 \end{bmatrix}$$

which illustrates the asymptotic stability of the individual subsystems and hence the overall system.

For the output feedback stabilization design, we used second-order dynamic controllers ($m_1 = m_2 = m_3 = 2$). The ensuing simulation results show that the feasible solution of **Theorem 8.5.8** is given by

$$\begin{aligned}
\delta_1 &= 1.6764, \\
L_1 &= \begin{bmatrix} -0.0123 & 0.0043 \\ 0.0046 & -0.0134 \end{bmatrix}, \ F_1 = \begin{bmatrix} 0.1167 & 0.3914 \\ -0.1015 & -0.3208 \end{bmatrix}, \\
K_{c1} &= \begin{bmatrix} 0.1109 & -0.1103 \end{bmatrix}, \\
K_{o1} &= \begin{bmatrix} -0.3829 & -0.2531 \end{bmatrix}, \\
\delta_2 &= 1.7864, \\
L_2 &= \begin{bmatrix} -0.0424 & 0.0118 \\ 0.0216 & -0.0209 \end{bmatrix}, \ F_2 = \begin{bmatrix} 0.1203 & 0.4024 & 0.2247 \\ -0.1276 & -0.2958 & -0.4216 \end{bmatrix}, \\
K_{c2} &= \begin{bmatrix} -0.1007 & -0.2125 \end{bmatrix}, \ K_{o2} = \begin{bmatrix} -0.2471 & -0.3667 & -0.0845 \end{bmatrix}, \\
\delta_3 &= 1.8795, \\
L_3 &= \begin{bmatrix} -0.0583 & 0.0029 \\ -0.0189 & -0.2601 \end{bmatrix}, \ F_3 = \begin{bmatrix} 0.2088 & 0.4314 \\ -0.1023 & -0.2678 \end{bmatrix}, \\
K_{c3} &= \begin{bmatrix} -0.1243 & -0.3108 \end{bmatrix}, \\
K_{o3} &= \begin{bmatrix} -0.5104 & -0.3017 \end{bmatrix}
\end{aligned}$$

Turning to **Theorem 8.5.8**, by incorporating gain perturbations of the form (8.135) with $M_{o1} = [0.5 \ \ 0.5]$, $N_{o1} = 0.8$, $M_{o2} = [1.5 \ \ 1.5 \ \ 2.5]$, $N_{o2} = 0.6$, $M_{o3} = [1.0 \ \ 1.0]$, $N_{o3} = 0.5$, $M_{c1}^t = [1.2 \ \ 1.8]$, $N_{c1} = 0.6$, $M_{c2}^t = [2.0 \ \ 2.5 \ \ 2.0]$, $N_{c2} = 0.4$, $M_{c3}^t = [1.5 \ \ 2.5]$, $N_{c3} = 0.1$, the feasible solution is given by

$$\begin{aligned}
\delta_1 &= 1.9814, \\
L_1 &= \begin{bmatrix} -0.0245 & 0.0173 \\ 0.0136 & -0.0323 \end{bmatrix}, \; F_1 = \begin{bmatrix} 0.1143 & 0.3764 \\ -0.1015 & -0.4218 \end{bmatrix}, \\
K_{c1} &= \begin{bmatrix} 0.1109 & -0.1103 \end{bmatrix}, \\
K_{o1} &= \begin{bmatrix} -0.3779 & -0.2620 \end{bmatrix}, \\
\delta_2 &= 2.0456, \\
L_2 &= \begin{bmatrix} -0.0451 & 0.0129 \\ 0.0334 & -0.0212 \end{bmatrix}, \; F_2 = \begin{bmatrix} -0.1269 & -0.2768 & -0.4213 \\ 0.1204 & 0.4033 & 0.2239 \end{bmatrix}, \\
K_{c2} &= \begin{bmatrix} -0.1013 & -0.2131 \end{bmatrix}, \; K_{o2} = \begin{bmatrix} -0.2471 & -0.3427 & -0.0785 \end{bmatrix}, \\
\delta_3 &= 2.1135, \\
L_3 &= \begin{bmatrix} -0.0578 & 0.0114 \\ -0.0189 & -0.2501 \end{bmatrix}, \; F_3 = \begin{bmatrix} 0.2098 & 0.4294 \\ -0.1023 & -0.2678 \end{bmatrix}, \\
K_{c3} &= \begin{bmatrix} -0.1243 & -0.3108 \end{bmatrix}, \\
K_{o3} &= \begin{bmatrix} -0.5105 & -0.3016 \end{bmatrix}
\end{aligned}$$

The obtained results indicated slightly higher tolerance levels δ_j in the case of resilient feedback control which is expected to accommodate for the effects of gain perturbations.

8.6 Problem Set VI

Problem VI.1: Consider the following two linear systems

$$\begin{aligned}
\mathbf{S} &: \quad \dot{\mathbf{x}}(t) = \mathbf{A}\mathbf{x}(t) + \mathbf{B}\mathbf{u}(t), \; \mathbf{x}(0) = \mathbf{x}_o, \\
\mathbf{y} &= \mathbf{C}\mathbf{x}, \\
\tilde{\mathbf{S}} &: \quad \dot{\tilde{\mathbf{x}}}(\mathbf{t}) = \tilde{\mathbf{A}}\tilde{\mathbf{x}}(\mathbf{t}) + \tilde{\mathbf{B}}\tilde{\mathbf{u}}(\mathbf{t}), \; \tilde{\mathbf{x}}(\mathbf{0}) = \tilde{\mathbf{x}}_\mathbf{o}, \\
\tilde{\mathbf{y}} &= \tilde{\mathbf{C}}\tilde{\mathbf{x}}
\end{aligned}$$

where $\mathbf{x} \in \Re^n$, $\mathbf{u} \in \Re^m$, $\mathbf{y} \in \Re^p$, $\tilde{\mathbf{x}} \in \Re^{\tilde{n}}$, $\tilde{\mathbf{u}} \in \Re^{\tilde{m}}$, $\tilde{\mathbf{y}} \in \Re^{\tilde{p}}$ are the state, input, and output vectors for systems $\mathbf{S}$ and $\tilde{\mathbf{S}}$, respectively. Consider the transformations

$$\begin{aligned}
\mathbf{T} &: \; \Re^n \longrightarrow \Re^{\tilde{n}}, \; rank(\mathbf{T}) = n, \\
\mathbf{U} &: \; \Re^{\tilde{m}} \longrightarrow \Re^m, \; rank(\mathbf{U}) = m, \\
\mathbf{T}^\dagger &: \; \Re^{\tilde{n}} \longrightarrow \Re^n, \; \mathbf{T}^\dagger\mathbf{T} = I_n, \\
\mathbf{U}^\dagger &: \; \Re^m \longrightarrow \Re^{\tilde{m}}, \; rank(\mathbf{U}) = m
\end{aligned}$$

such that $\tilde{\mathbf{x}}_o = \mathbf{T}\mathbf{x}_o, \quad \mathbf{u} = \mathbf{U}\tilde{\mathbf{u}}$. Based thereon, establish that

$$\mathbf{T}\mathbf{A} = \tilde{\mathbf{A}}\mathbf{T}, \quad \mathbf{T}\mathbf{B}\mathbf{U} = \tilde{\mathbf{B}}$$

In this case, system $\tilde{\mathbf{S}}$ is called an extension of system $\mathbf{S}$. Consider the linear feedback stabilizing controls:

$$\mathbf{u} = \mathbf{K}\mathbf{x}, \quad \tilde{\mathbf{u}} = \tilde{\mathbf{K}}\tilde{\mathbf{u}}$$

such that $\tilde{\mathbf{A}} + \tilde{\mathbf{B}}\tilde{\mathbf{K}}$ is asymptotically stable and the following relation

$$\mathbf{K}\mathbf{x}(t; \mathbf{x}_o, \mathbf{U}\tilde{\mathbf{u}} = \mathbf{U}\tilde{\mathbf{K}}\tilde{\mathbf{x}}(t; \mathbf{T}\mathbf{x}_o, \tilde{\mathbf{u}})$$

holds. Verify that the condition

$$\mathbf{K} = \mathbf{U}\tilde{\mathbf{K}}\mathbf{T}$$

holds.

Problem VI.2: Extending on **Problem VI.1**, consider the following cost functions

$$\begin{aligned} \mathbf{S} \quad &: \quad J = \int_0^\infty (\mathbf{x}^t\mathbf{Q}\mathbf{x} + \mathbf{u}^t\mathbf{R}\mathbf{u})\, dt, \\ \tilde{\mathbf{S}} \quad &: \quad \tilde{J} = \int_0^\infty (\tilde{\mathbf{x}}^t\tilde{\mathbf{Q}}\tilde{\mathbf{x}} + \tilde{u}^t\tilde{\mathbf{R}}\tilde{\mathbf{u}})\, dt \end{aligned}$$

Verify in light of the contractability conditions that

$$J(\mathbf{x}_o, \mathbf{K}\mathbf{x}) = \tilde{J}(\mathbf{T}\mathbf{x}_o, \tilde{\mathbf{K}}\tilde{\mathbf{x}})$$

Hence derive the relations

$$\tilde{\mathbf{Q}} = (\mathbf{T}^\dagger)^t\mathbf{Q}\mathbf{T}^\dagger + \tilde{\mathbf{E}}, \quad \tilde{\mathbf{R}} = \mathbf{U}^t\mathbf{R}\mathbf{U} + \tilde{\mathbf{F}}$$

Extend these results to resilient overlapping control.

Problem VI.3: For a class of uncertain linear interconnected system of the type (8.60)-(8.62), consider a local resilient state feedback controller

$$u_j(t) = K_j[I_j + \Delta K_j(t)]x_j(t), \; \Delta K_j(t) = H_{kj}F_{kj}(t)E_{kj}$$

Examine the robust stability with respect to the overall Lyapunov-Krasovskii functional (LKF):

$$V(x,t) = x^t(t)\mathsf{P}x(t) + \int_{t-d}^t x^t(\sigma)[I + E_d^t E_d]x(\sigma)d\sigma$$

where $\mathsf{P} = \{P_j\},\ P_j \in \Re^{n_j \times n_j},\ (I + E_d^t E_d) \in \Re^{n \times n}$ are positive-definite symmetric matrices. Then, or otherwise, derive an LMI-based condition for robust stabilizability.

Problem VI.4: Consider an uncertain linear composite system comprised of n_s subsystems, where the jth subsystem is described by

$$\begin{aligned} S_j: \quad \dot{x}_j(t) &= [A + \Delta A_j(t)]x_j(t) + [A_d + \Delta A_{dj}(t)]x_j(t - d_j) \\ &+ \ Bu_j(t) + z_j(t), \\ y_j(t) &= Cx_j(t), \ \ x_j(t) = \phi_j(t), \ \ \forall\, t \in [-d_j,\, 0], \ \ j \in \{1, ..., n_s\}, \\ z_j(t) &= \sum_{k=1}^{n_s} L_{jk} y_{xk} \ + \ L_{djk} y_{dk}, \\ y_{xk} &= C_y x_j, \ \ y_{dk} = C_{dy} x_j(t - d_j) \end{aligned}$$

where the interconnection matrices have the following structure:

$$\begin{aligned} L_{jj} &= 0, \ \ L_{jk} = L_q + \Delta L_{qjk}(t), \\ L_{djj} &= 0, \ \ L_{djk} = L_{dq} + \Delta L_{dqjk}(t), \ \ j \neq k \end{aligned}$$

and the nonlinear parametric uncertainties are given by

$$\begin{aligned} \Delta A_j(t) &= D_a \Delta_{aj} E_a, \ \ \Delta A_{dj}(t) = D_{da} \Delta_{daj} E_{da}, \\ \Delta L_{qjk}(t) &= D_\ell \Delta_{\ell j} E_\ell, \ \ \Delta L_{dqjk}(t) = D_{d\ell} \Delta_{d\ell aj} E_{d\ell} \end{aligned}$$

where $\Delta_*^t(t)\Delta_*(t) \neq I$. It is desired to design n_s resilient decentralized controllers of the form

$$\begin{aligned} \dot{\xi}_j(t) &= [A_r + \Delta A_{rj}]\xi_j(t) + [A_{dr} + \Delta A_{drj}]x_j(t - \tau_j(t)) + Bu_j(t) \\ &+ \ [K_o + \Delta K_{oj}][y_j(t) - C_{rj}\xi_j(t)], \\ u_j(t) &= [K_c + \Delta K_{cj}][\xi_j(t) \ + \int_{t-\tau_j(t)}^{t} A_{dr}\xi_j(s)\, ds] \end{aligned}$$

$$\begin{aligned} \Delta A_{rj}(t) &= D_r \Delta_{rj} E_r, \ \ \Delta A_{drj}(t) = D_{dr} \Delta_{drj} E_{dr}, \\ \Delta K_{oj}(t) &= D_o \Delta_{oj} E_o, \ \ \Delta K_{rj}(t) = D_c \Delta_{cj} E_c \end{aligned}$$

where $\Delta_*^t(t)\Delta_*(t) \neq I$. Develop a procedure to determine the observer matrices $A_r,\ A_{dr}$ and unknown gain matrices $K_o,\ K_c$ such that the observer-based controlled system is asymptotically stable.

8.7 Notes and References

We have considered the resilient decentralized feedback stabilization problem of a class of nonlinear interconnected discrete-time systems. This class of systems has uncertain nonlinear perturbations satisfying quadratic constraints that are functions of the overall state vector. Decentralized state and output feedback schemes have been developed and analyzed such that the overall closed-loop system guarantees global stability condition, derived in terms of local subsystem variables. Incorporating feedback gain perturbations, new resilient decentralized feedback schemes have been subsequently developed. The proposed approach has been formulated within the framework of convex optimization over LMIs. Simulation results have illustrated the effectiveness of the proposed decentralized output feedback controllers.

Decentralized delay-dependent stability and stabilization methods have been developed for a class of linear interconnections of time-delay plants subjected to convex-bounded parametric uncertainties and coupled with time-delay interconnections. We have constructed an appropriate decentralized Lyapunov-Krasovskii functional to exhibit the delay-dependent dynamics at the subsystem level. Decentralized LMIs-based delay-dependent stability conditions have been derived such that every local subsystem of the linear interconnected delay system is robustly asymptotically stable with a $\gamma-$level $\mathcal{L}_2-gain$. A decentralized state feedback stabilization scheme has been designed such that the family of closed-loop feedback subsystems enjoys the delay-dependent asymptotic stability with a prescribed $\gamma-$level $\mathcal{L}_2$ gain for each subsystem. The decentralized feedback gains are determined by convex optimization over LMIs and all the developed results are tested on a representative example.

The chapter contributes to the solution of the overlapping $\mathcal{H}_\infty$ resilient state feedback control design for a class of nonlinear continuous-time uncertain state-delayed nominally linear systems. Time-varying unknown norm-bounded parameter uncertainties are considered in the system and the controller. Conditions preserving closed-loop systems expansion-contraction relations guaranteeing the $\mathcal{H}_\infty$ disturbance attenuation bounds have been proved. They are derived in terms of conditions on complementary matrices. An LMI delay-independent procedure has been supplied as a control design tool. The results have been specialized into an overlapping decentralized control setting. It means that the presented method leads to the control design of robust $\mathcal{H}_\infty$ resilient state tridiagonal feedback controllers.

In [258], extensive discussions were carried on incorporating gain perturbations. The decentralized versions of some of these results need further investigation.

Chapter 9

Decentralized Sliding-Mode Control

In this chapter, we continue further into the decentralized control techniques for interconnected systems, where we focus herein on methods for designing classes of decentralized controllers based on sliding-mode control (SMC) theory. Our effort is divided into two sections: in the first section we deal with a class of nonlinear interconnected systems with nonlinear uncertain subsystems considered where matched and mismatched uncertainties are both dealt with. In the second section, global decentralized stabilization of a class of interconnected delay systems is considered where both known and uncertain interconnections involve time delay. Matched and mismatched interconnections are treated separately to reduce the conservatism. In both sections, the approach followed allows a more general structure for the interconnections and uncertainty bounds than other literature in this area. The conservatism in the results is reduced by fully using system output information and the uncertainty bounds.

9.1 Robust Sliding-Mode Control

Large-scale systems are often modeled as dynamic equations composed of interconnections of lower-dimensional subsystems. One of the main features of these systems is that they are often spatially distributed, and thus the information transfer among subsystems may incur high cost or even encounter practical limitations. Moreover, system state variables are often not fully available for practical systems. Some state variables may be difficult/costly to measure and sometimes have no physical meaning and thus cannot be measured at all. It may be possible to use an observer to estimate unknown states, but this approach faces the twin problems of requiring more hardware resources and greatly increasing the dimension of the system. In turn, this brings about further difficulties especially for large-scale systems. It is therefore pertinent to study decentralized control for large-scale interconnected systems using output feedback.

In this regard, many of the decentralized output feedback control methods [302, 377, 378] where Lyapunov approaches are used to form the control scheme and strict structural conditions are imposed on the nominal subsystems together with some strong limitations on the admissible interconnections. On the other hand, adaptive control techniques are employed in [140, 141] with parametric uncertainty. The corresponding results can only be applied to certain systems with special structure. Sliding mode control has been used successfully by many researchers [66, 67, 365, 395], mostly focusing on centralized control which is hard to implement in large-scale interconnected systems. Sliding mode control schemes for large-scale systems have also been reported in [109, 172].

9.1.1 Introduction and preliminaries

In this section, the focus is on a decentralized output feedback control strategy based on sliding-mode techniques. A class of nonlinear large-scale interconnected systems with matched and mismatched uncertainties is considered. No statistical information about the uncertainties is imposed. Furthermore, the bounding functions on the uncertainties and interconnections take a more general structure than that imposed by other authors. Nonlinear interconnections are considered along with nonlinear nominal subsystems.

To set the scene for subsequent analysis, we consider a linear system

$$\dot{x} = Ax + Bu \tag{9.1}$$
$$y = Cx \tag{9.2}$$

where $x \in \Re^n, u \in \Re^m, y \in \Re^p$ are the states, inputs, and outputs, respectively, and assume $m \leq p < n$. The triple (A, B, C) comprises constant matrices of appropriate dimensions with B and C both being of full rank.

For system (9.1)-(9.2), it is assumed that $\text{rank}(CB) = m$. Then it can be shown [65] that a coordinate transformation $\tilde{x} = \tilde{T}x$ exists such that the system triple (A, B, C) with respect to the new coordinates $\tilde{x}$ has the following structure:

$$\tilde{A} = \begin{bmatrix} \tilde{A}_{11} & \tilde{A}_{12} \\ \tilde{A}_{21} & \tilde{A}_{22} \end{bmatrix}, \quad \tilde{B} = \begin{bmatrix} 0 \\ B_2 \end{bmatrix}, \quad \tilde{C} = \begin{bmatrix} 0 & \breve{T} \end{bmatrix} \tag{9.3}$$

where $\tilde{A} \in \Re^{(n-m)\times(n-m)}$, $B_2 \in \Re^{m\times m}$ is nonsingular and $\breve{T} \in \Re^{p\times p}$ is orthogonal. Further, it is assumed that system $(\tilde{A}_{11}, \tilde{A}_{12}, \tilde{C}_1)$ with C_1 defined by

$$\tilde{C}_1 = [0_{(p-m)\times(n-p)} \quad I_{p-m}] \tag{9.4}$$

is output feedback stabilizable, that is, there exists a matrix $K \in \Re^{m\times(p-m)}$ such that $\tilde{A}_{11} - \tilde{A}_{12}K\tilde{C}_1$ is stable. A necessary condition for $(\tilde{A}_{11}, \tilde{A}_{12}, \tilde{C}_l)$ to be stabilizable is that [65] the invariant zeros of (A, B, C) lie in the open left half-plane and a sliding surface

$$FCx = 0 \tag{9.5}$$

is proposed, where $F = F_2[K \quad I_m]\breve{T}^t$ and $F_2 \in \Re^{m\times m}$ is any nonsingular matrix. A coordinate change is then introduced based on the nonsingular transformation $z = \hat{T}\tilde{x}$ with $\hat{T}$ defined by

$$\hat{T} = \begin{bmatrix} I_{n-m} & 0 \\ K\tilde{C}_1 & I_m \end{bmatrix}$$

Then in the new coordinates z, system (9.1)-(9.2) has the following form:

$$\hat{A} = \begin{bmatrix} A_{11} & A_{12} \\ A_{21} & A_{22} \end{bmatrix}, \quad \hat{B} = \begin{bmatrix} 0 \\ B_2 \end{bmatrix}, \quad \hat{C} = \begin{bmatrix} 0 & C_2 \end{bmatrix} \tag{9.6}$$

where $A_{l1} = \tilde{A}_{11} - \tilde{A}_{12}K\tilde{C}_1$ is stable, $C_2 \in \Re^{p\times p}$ is nonsingular, and C satisfies $F\hat{C} = [0 \quad F_2]$ with F_2 nonsingular. The canonical form in (9.6) can be obtained from a systematic algorithm together with any output feedback pole placement algorithm of choice; see [65, 67] for more details.

9.1.2 System description

We consider a nonlinear large-scale system formed by n_s coupled subsystems as follows:

$$\begin{aligned} \dot{x}_j &= A_j x_j + f_j(x_j) + B_j(u_j + \Delta g_j(x_j)) + H_j(x) && (9.7) \\ y_j &= C_j x_j, \quad j = 1, 2, ..., n_s && (9.8) \end{aligned}$$

where $x = \text{col}(x_1, x_2, ..., x_N)$, $x_j \in \Re^{n_j}$, $u_j \in \Re^{m_j}$, $y_j \in \Re^{p_j}$ are the states, inputs, and outputs of the jth subsystem, respectively, and $m_j \leq p_j < n_j$. The triple (A_j, B_j, C_j) represents constant matrices of appropriate dimensions with B_j and C_j being of full rank. The function $f_j(x_j)$ represents known nonlinearities in the ith subsystem. The matched uncertainty of the jth isolated subsystem is denoted by $\Delta g_j(x_j)$ and $H_j(x)$ represents system interconnections including all mismatched uncertainties. The functions are all assumed to be continuous in their arguments.

Assumption 9.1.1 *rank*$(C_j,\ B_j) = m_j$ *for* $j = 1, 2, ..., n_s$.

It follows that there exists a coordinate transformation $\tilde{x}_j = \tilde{T}_j x_j$ such that the triple (A_j, B_j, C_j) with respect to the new coordinates has the structure

$$\begin{bmatrix} \tilde{A}_{i11} & \tilde{A}_{i12} \\ \tilde{A}_{i21} & \tilde{A}_{i22} \end{bmatrix}, \quad \begin{bmatrix} 0 \\ \tilde{B}_{i2} \end{bmatrix}, \quad \begin{bmatrix} 0 & \tilde{C}_{i2} \end{bmatrix} \tag{9.9}$$

where $\tilde{A}_{i11} \in \Re^{(n_j-m_j)\times(n_j-m_j)}$, $\tilde{B}_{i2} \in \Re^{m_j\times m_j}$, and $\tilde{C}_{i2} \in \Re^{p_j\times p_j}$ $for j = 1, 2, ..., n_s$.

Assumption 9.1.2 *For the triples* $(\tilde{A}_{j11}, \tilde{A}_{j12}, \bar{C}_{j2})$ *with*

$$\bar{C}_{j2} = [0_{(p_j-m_j)\times(n_j-p_j)} \quad I_{(p_j-m_j)}]$$

there exist matrices K_j *such that* $\tilde{A}_{j11} - \tilde{A}_{j12} K_j \bar{C}_{j2}$ *are stable for* $j = 1, ..., n_s$.

Remark 9.1.3 *It must be emphasized that* ***Assumptions 9.1.1*** *and* ***9.1.2*** *are based on the linear part of the individual nominal subsystems to guarantee the existence of the output sliding surface. Observe that* ***Assumption 9.1.2*** *demands* $(\tilde{A}_{j11}, \tilde{A}_{j12}, \bar{C}_{j2})$ *instead of* (A_j, B_j, C_j) *to be output feedback stabilizable. As shown in [65], a necessary condition for this is that the triple* (A_j, B_j, C_j) *is minimum phase.*

Assumption 9.1.4 *Suppose that* $f_j(x_j)$ *has the decomposition* $f_j(x_j) = \Gamma_j(y_j)x_j$*, where* $\Gamma_j \in \Re^{n_j\times n_j}$ *is a continuous function matrix for* $j = l, 2, ..., n_s$.

Assumption 9.1.5 *There exist known continuous functions* $\rho_j(\cdot)$ *and* $\eta_j(\cdot)$ *such that for* $i = l, 2, ..., n_s$

$$|\Delta g_j(x_j)| \leq \rho_j(y_j),$$
$$|H_j(x)| \leq \eta_j(x)$$

where $\eta_j(\cdot)$ *satisfies* $\eta_j(x) \leq \beta_j(x)|x|$ *for some continuous function* β_j.

Remark 9.1.6

1. *If* $f_j(0) = 0$ *and* f_j *is sufficiently smooth, then the decomposition* $f_j(x_j) = \Gamma_j(y_j)x_j$ *is guaranteed. The restriction on* f_j *imposed here is the requirement that* $\Gamma_j(x_j) = \Gamma_j(y_j)$.

2. **Assumption 9.1.5** *ensures that all uncertainties in (9.7)-(9.8) are bounded by known functions, and the matched uncertainty is bounded by a function of the output. In this way, a more general bound for* $\|\Delta g_j(x_j)\|$ *than* $\rho_j(y_j)$ *is allowed which is only used to show that the effect of the matched uncertainty can be eliminated completely if its bound can be described by a function of the system output. Additionally, the uncertainties and interconnections in system (9.7)-(9.8) have more general forms than the parametric structure in [140, 141], and no special requirement for the structure of the uncertainty and interconnection terms is imposed.*

9.1.3 Stability analysis

It is observed under **Assumptions 9.1.1** and **9.1.2** that there exist matrices F_j such that for $j = 1, 2, ..., n_s$ the system

$$\dot{x}_j = A_j x_j + B_j u_j$$

when restricted to the surface $F_j C_j x_j = 0$, is stable. The composite sliding surface for the interconnected system (9.7)-(9.8) is chosen as

$$\sigma(x) = 0 \tag{9.10}$$

with $\sigma(x) \equiv: \mathrm{col}(\sigma_1(x_1), \sigma_2(x_2), ..., \sigma_N(x_{n_s}))$ and

$$\sigma_j(x_j) \;=\; F_j C_j x_j \;=\; F_j y_j \tag{9.11}$$

where the F_j are obtained from the algorithm given in [65]. Subsequently, under **Assumptions 9.1.1-9.1.4** there exist nonsingular coordinate transformations $z_j = T_j x_j$ such that in the new coordinates $z = \mathrm{col}(z_1, z_2, ..., z_{n_s})$ system (9.7)-(9.8) has the following form:

$$\begin{aligned} \dot{z}_j &= \begin{bmatrix} A_{j11} & A_{j12} \\ A_{j21} & A_{j22} \end{bmatrix} z_j + \begin{bmatrix} \Gamma_{j1}(y_j) & * \\ * & * \end{bmatrix} z_j \\ & \begin{bmatrix} 0 \\ B_{j2} \end{bmatrix} (u_j + \Delta g_j(T_j^{-1} z_j)) + T_j H_j(T^{-1} z), \qquad (9.12) \\ y_j &= \begin{bmatrix} 0 & C_{j2} \end{bmatrix} z_j, \quad j = 1, 2, ..., n_s \qquad (9.13) \end{aligned}$$

where $A_{j11} = \tilde{A}_{j11} - \tilde{A}_{j12} K_j \bar{C}_{j2}$ is stable, $\Gamma_{j1}(y_j) \in \Re^{(n_j - m_j)\times(n_j - m_j)}$ and the $*$ are subblocks of $T_j \Gamma_j(\cdot) T_j^{-1}$ that play no part in the subsequent analysis, $T^{-1} \equiv: \mathrm{diag} T_1^{-1}, T_2^{-1}, ..., T_{n_s}^{-1}$ and the square submatrices $B_{j2} \in \Re^{m_j \times m_j}$ and $C_{j2} \in \Re^{p_j \times p_j}$ are nonsingular. Furthermore, $F_j[0 \quad C_{j2}] = [0 \quad F_{j2}]$ where

$F_{j2} \in \Re^{m_j \times m_j}$ is nonsingular. Since A_{j11} is stable for $j = 1, ..., n_s$, for any $Q_j > 0$, the following Lyapunov equation has a unique solution $P_j > 0$ such that

$$A_{j11}^t P_j + P_j A_{j11} = -Q_j, \quad j = 1, 2, ..., n_s \tag{9.14}$$

Proceeding further and in order to fully use the available structure characteristics, partition $z_j = \text{col}(z_{j1}, z_{j2})$ with $z_{j1} \in \Re^{n_j - m_j}$ and $z_{j2} \in \Re^{m_j}$. It follows that in the new coordinate z, the switching function $F_j C_j x_j$ can be described by $F_j[0 \quad C_{j2}]z_j = F_{j2} z_{j2}$, and from the nonsingularity of F_{j2}, it follows that the sliding surface (9.11) becomes $z_{j2} = 0$ with $j = 1, 2, ..., n_s$. Further, partition C_{j2} and z_{j1} as $C_{j2} = [C_{j21} \quad C_{j22}]$ and $z_{j1} = \text{col}(z_{j11}, z_{j12})$, respectively, where $C_{j21} \in \Re^{p_j \times (p_j - m_j)}$, $z_{j11} \in \Re^{n_j - p_j}$, and $z_{j12} \in \Re^{p_j - m_j}$. It is observed that $y_j = C_{j21} z_{j12} + C_{j22} z_{j2}$. By restricting system (9.12)-(9.13) to move on the sliding surface $z_{j2} = 0$, the sliding mode has the following form:

$$\dot{z}_{j1} = A_{j11} z_{i1} + \Gamma_{j1}((C_{j21} z_{j12}) z_{j1} + W_j(z_{11}, ..., z_{n_s 1})) \tag{9.15}$$

where $W_j(z_{11}, ..., z_{n_s 1})$ is the first $(n_j - m_j)$ component of

$$T_j H_j(T^{-1} z)|_{(z_{12}, ..., z_{n_s 2}) = 0}$$

From **Assumption 9.1.5**, it is easy to find a function $\gamma_j(z_{11}, z_{21}, ..., z_{n_s 1})$ depending $\eta_j(T^{-1} x)$ and T_j such that

$$\|W_j(z_{11}, ..., z_{n_s 1})\| \leq \gamma_j(z_{11}, ..., z_{n_s 1}) \sum_{j=1}^{n_s} \|z_{j1}\| \tag{9.16}$$

It is readily evident that (9.15) is a reduced-order interconnected system composed of n_s subsystems with dimension $n_j - m_j$ for which the following stability result is established.

Theorem 9.1.7 *Consider the sliding mode dynamics (9.15). Under* ***Assumptions 9.1.1-9.1.5****, the sliding motion is asymptotically stable if there exists a domain*

$$\Omega = \{(z_{11}, z_{21}, ..., z_{n_s 1}) | \| z_j 1 \| \leq d_j, i = 1, 2, ..., N\}$$

for some constants $d_j > 0$ such that $M^t + M > 0$ in $\Omega \backslash \{0\}$ where $M = (m_{ij})_{n_s \times n_s}$ and for $j, k = 1, 2, ..., n_s$

$$m_{jk} = \begin{cases} \lambda(Q_j) - \|R_j(C_{j21} z_{j12}\| - 2\|P_j\| \gamma_j(\cdot), & j = k \\ -2\|P_j\| \gamma_j(\cdot), & j \neq k \end{cases} \tag{9.17}$$

with P_j and Q_j satisfying (9.14), $R_j(\cdot) :\equiv P_j \Gamma_{j1}(\cdot) + \Gamma_{j1}^t(\cdot) P_j$ with $\Gamma_{j1}(\cdot)$ given by (9.12), and $\gamma_j(\cdot)$ determined by (9.16).

Proof 9.1.8 *It suffices to prove that system (9.15) is asymptotically stable. Consider the Lyapunov function candidate*

$$V(z_{11}, z_{21}, ..., z_{Nn_s1}) = \sum_{i=1}^{n_s} z_{j1}^t P_j z_{j1} \tag{9.18}$$

Evaluating the time derivative of $V(z_{11}, z_{21}, ..., z_{n_s1})$ *along the trajectories of system (9.15) using (9.14) gives*

$$\begin{aligned} \dot{V} = &\sum_{i=1}^{N} \{-z_{i1}^t Q_j z_{i1} + z_{i1}^t (P_j \Gamma_{i1}(\cdot) P_j) z_{i1} \\ &+ 2z_{i1}^t P_j W_j(\cdot)\} \end{aligned} \tag{9.19}$$

By (9.16), it follows that

$$\begin{aligned} \dot{V} \leq\; & \sum_{i=1}^{N} \{-\underline{\lambda}(Q_j)\|z_{i1}\|^2 + \|P_j \Gamma_{i1}(\cdot) + \Gamma_{i1}^t(\cdot) P_j\| \|z_{i1}\|^2 \\ & +2\|z_{i1}\| \|P_j\| \|W_j(\cdot)\|\} \\ \leq\; & \sum_{i=1}^{N} \{-\underline{\lambda}(Q_j)\|z_{i1}\|^2 + \|R_j(\cdot)\| \|z_{i1}\|^2\} \\ & +2\sum_{i=1}^{N} \{\|z_{i1}\| \|P_j\| \gamma_j(\cdot) \sum_{j=1}^{N} \|z_{j1}\|\} \\ =\; & -\sum_{i=1}^{N} \{\underline{\lambda}(Q_j) - \|R_j(\cdot)\| - 2\|P_j\|\gamma_j(\cdot)\} \|z_{i1}\|^2 \\ & +2\sum_{i=1}^{N} \sum_{j=1, j\neq i}^{N} \|P_j\|\gamma_j(\cdot)\|z_{i1}\| \|z_{j1} \\ =\; & -\frac{1}{2} Y^t (M^t + M) Y \end{aligned} \tag{9.20}$$

where

$$Y \equiv: \mathrm{col}(\|z_{ii}\|, ..., \|z_{n_s1}\|)$$

Since by assumption $M^t + M > 0$ *in* $\Omega\{0\}$*, the proof is concluded.*

Remark 9.1.9 *It must be observed that the function matrix* M *in* **Theorem 9.1.7** *only depends on the sliding mode variables* z_{j1} *instead of* z_j*, with* $j = 1, 2, ..., n_s$*. When* $\Gamma_{j1}(\cdot)$ *and the bounds for all mismatched parts can* *be expressed as functions of z_{j1} with $j = 1, 2, ..., n_s$, a global stability result is obtained. In [377, 378] the corresponding matrix M depends on all the state variables. Moreover, since Γ_{j1} only depends on z_{j12}, $R_j(\cdot)$ is reduced to a null matrix if $p_j = m_j$. This greatly relaxes $\Gamma_j(\cdot)$. In addition, the matrices F_j are completely determined by choice of K_j which is required to ensure that $\tilde{A}_{j11} - \tilde{A}_{j12}K_j\bar{C}_{j2}$ is stable. From the foregoing analysis, it is observed that the matrix M depends on R_j, γ_j, P_j, and Q_j, where the quantities R_j and γ_j are determined by the given system, whereas P_j depends on Q_j, which can be designed freely. From (9.17) a reasonable selection is the one that minimizes $\lambda_M(P_j)/\lambda_m(Q_j)$ which assists in making $M^2 + M$ diagonally dominant and having all positive diagonal elements.*

9.1.4 Decentralized sliding-mode control

Our goal now is to design a decentralized output feedback sliding-mode control such that the system state is driven to the sliding surface (9.10). In terms of the interconnected system (9.7)-(9.8), the corresponding condition is described by [109]

$$\sum_{i=1}^{N} \frac{\sigma_j^t(x_j)\dot{\sigma}_j(x_j)}{\|\sigma_j(x_j)\|} < 0 \tag{9.21}$$

where $\sigma_j(x_j)$ is defined by (9.11). The result of comparing system (9.7)-(9.8) with (9.12)-(9.13) yields $C_j = [0 \quad C_{j2}]T_j = C_{j2}[0 \quad I_p]T_j$ where C_{j2} is given by (9.13). Then

$$x_j = T_j^{-1}T_jx_j = T_j^{-1}\begin{bmatrix}(T_jx_j)_1 \\ C_{i2}^{-1}y_j\end{bmatrix} \tag{9.22}$$

where $(T_jx_j)_1$ is the first $n_j - p_j$ components of T_jx_j. Now, consider system (9.7)-(9.8) in the domain

$$\Theta \equiv: \Theta_1 \times \Theta_2 \times ... \times \Theta_{n_s}, \quad \Theta_j \in \Re^{n_j}$$

and explicitly

$$\Theta_j \equiv: \{x_j | x_j \in \Re^{n_j}, \quad \|(T_jx_j)_1\| \leq \mu_j\} \tag{9.23}$$

for some positive constant μ_j with $j = 1, 2, ..., n_s$. We now let $i = 1, 2, ..., n_s$:

$$F_jC_j(A_j+\Gamma_j(y_j))T_j^{-1} := [\Upsilon_{j1}(y_j) \quad \Upsilon_{i2}(y_j)], \quad \Upsilon_{j1} \in \Re^{m_j\times(n_j-p_j)} \tag{9.24}$$

In view of

$$F_jC_jB_j = F_jC_jT_j^{-1}T_jB_j = [0 \quad F_{j2}]\begin{bmatrix}0 \\ B_{i2}\end{bmatrix} = F_{j2}B_{j2}$$

it is readily seen that $F_jC_jB_j$ is nonsingular. Then, the following control law is proposed:

$$\begin{aligned} u_i = &- (F_jC_jB_j)^{-1}\frac{F_jy_j}{\|F_jy_j\|}\{\|\Upsilon_{i2}(\cdot)C_{i2}^{-1}y_j\| + \frac{\varepsilon_j}{2}\|\Upsilon_{i1}(\cdot)\|^2 \\ &+ \frac{\mu_j^2}{2\varepsilon_j} + \|F_jC_jB_j\|\rho_j(y_j) + k_j(y_j)\} \end{aligned} \tag{9.25}$$

for $j = 1, 2, ..., n_s$ where $\varepsilon_j > O$ is an adjustable constant; F_j and ρ_j are defined by (9.11) and **Assumption 9.1.5**, respectively; and $k_j(y_j)$ is the control gain to be designed later. The following result stands out.

Theorem 9.1.10 *Consider the nonlinear interconnected system (9.7)-(9.8). Under* ***Assumptions 9.1.1-9.1.5****, the decentralized control (9.25) drives the system (9.7)-(9.8) to the composite sliding surface (9.10) and maintains a sliding motion for the domain* $\Theta\{0\}$, $k_j(y_j)$ *satisfies*

$$\sum_{i=1}^{N} k_j(y_j) - \sum_{i=1}^{N} \eta_j(x)\|F_jC_j\| > 0 \tag{9.26}$$

where F_j *and* η_j *are determined by (9.11) and* ***Assumption 9.1.5****, respectively, and* Θ *is defined by (9.23).*

Proof: We need to be prove that the composite reachability condition (9.21) is satisfied. From (9.11) and **Assumption 9.1.4**, we have

$$\begin{aligned} \dot{\sigma}_j(x_j) =& F_jC_j(A_j + \Gamma_j(y_j))x_j + F_jC_jB_j(u_j + \Delta g_j(x_j)) \\ &+ F_jC_jH_j(x) \end{aligned} \tag{9.27}$$

for $i = 1, 2, ..., N$. Substituting (9.25) into (9.27), we obtain

$$\begin{aligned} \frac{\sigma_j^t\dot{\sigma}_j}{\|\sigma_j\|} =& \frac{(F_jy_j)^t}{\|F_jy_j\|}\{F_jC_j(A_j + \Gamma_j(\cdot))x_j + F_jC_jH_j(x) \\ &+ F_jC_jB_j\Delta g_j(x_j)\} - \|\Upsilon_{i2}(y_j)C_{j2}^{-1}y_j\| \\ &- \frac{\varepsilon_j}{2}\|\Upsilon_{i1}(\cdot)\|^2 - \frac{\mu_j^2}{2\varepsilon_j} - \|F_jC_jB_j\|\rho_j(\cdot) - k_j(\cdot). \end{aligned} \tag{9.28}$$

From (9.22) and the algebraic inequality $ab \le (\varepsilon/2)a^2 + (b^2/2\varepsilon)$ for $\varepsilon > 0$, see the Appendix, for $j = 1, 2, ..., n_s$.

$$\begin{aligned}
&\|F_j C_j (A_j + \Gamma_j(\cdot)\| \\
&= \|F_j C_j (A_j + \Gamma_j(\cdot)) T_j^{-1} \begin{bmatrix} (T_j x_j)_1 \\ C_{i2}^{-1} y_j \end{bmatrix} \\
&= \|\Upsilon_{j1}(y_j)(T_j x_j)_1 + \Upsilon_{i2}(\cdot) C_{j2}^{-1} y_j \\
&\le \|\Upsilon_{j2}(y_j) C_{j2}^{-1} y_j\| + \frac{\varepsilon_j}{2} \|\Upsilon_{j1}(y_j)\|^2 + \frac{\|(T_j x_j)_1}{2\varepsilon_j}
\end{aligned} \tag{9.29}$$

Also

$$\|F_j C_j B_j \Delta g_j(x_j)\| \;\le\; \|F_j C_j B_j\| \rho_j(y_j) \tag{9.30}$$
$$\|F_j C_j H_j(x)\| \;\le\; \|F_j C_j\| \eta_j(x) \tag{9.31}$$

Substituting (9.29)-(9.31) into (9.28), we have

$$\begin{aligned}
\frac{\sigma_j^t(x_j)\dot{\sigma}_j(x_j)}{\|\sigma_j(x_j)\|} &\le \frac{\|(T_j x_j)_1\|^2}{2\varepsilon_j} - \frac{\mu_j^2}{2\varepsilon_j} - k_j(\cdot) + \eta_j(\dot{)}\|F_j C_j\| \\
&\le -k_j(\cdot) + \eta_j(\cdot)\|F_j C_j\|
\end{aligned}$$

in the domain Θ_j. Then, if $k_j(y_j)$ is selected to satisfy (9.26), it follows that in the domain Θ, the reachability condition (9.21) is satisfied. Hence, the result follows.

Remark 9.1.11 *We observe that the decompositions (9.22) and (9.24) attempt to use as much as possible the known quantities associated with $f_j(x_j)$, that is, $\Gamma_j(y_j)$, in the control law, to reduce conservatism. The effect of $\Upsilon_{j2}(y_j)C_{j2}^{-1}$ from $\gamma_j(y_j)$ is canceled out completely by the nonlinear control (9.25) and from which it can be readily seen that ε_j can be used to adjust the control amplitude to some extent by considering the size of Θ_j and the value of $\Upsilon_{j2}(y_j)$ if a global result is not available. It can also be concluded that the reaching condition (9.26) is satisfied theoretically in any compact domain of the origin if high gain control is allowed. Generally speaking, the larger the domain that is required, the higher the required control gain. This is in contrast with the results of [377, 378], where the conclusion can only be satisfied in a small domain about the origin.*

The following result follows.

Corollary 9.1.12 *For system (9.7)-(9.8), suppose* ***Assumptions 9.1.1-9.1.5*** *are satisfied, and* $M^t + M > 0$ *by* ***Theorem 9.1.7*** *with* $M \in \Re^{\sum_{i=1}^{N}(n_j - m_j)}\{0\}$. *Then*

(i) The sliding mode dynamics (15) are globally asymptotically stable.

(ii) The closed-loop system composed of (9.7)-(9.8) and the control law

$$\begin{aligned} u_j(y_j) = &- (F_j C_j B_j)^{-1} \frac{F_j y_j}{\|F_j y_j\|}\{\|\Upsilon_{i2}(y_j) C_{j2}^{-1} y_j\| \\ &+ \|F_j C_j B_j\| \rho_j(y_j) + k_j(y_j)\} \end{aligned} \tag{9.32}$$

is globally asymptotically stable if $\Upsilon_{j1}(y_j) = 0$ *and* $\|H_j(x)\| \leq \sum_{j=1}^{n_s} \upsilon_{jk}(y_j)$ *for some continuous* υ_{jk} *with* $j = 1, 2, ..., n_s$.

Proof:

(i) From the structure of the Lyapunov function (9.18), the result is obtained directly from **Theorem 9.1.7**.

(ii) From the proof of **Theorem 9.1.10** and considering the control law (9.18), the expressions $(\varepsilon_j/2)\|\Upsilon_{i1}(y_j)\|^2 + (\mu_j^2/2\varepsilon_j)$ are introduced to counteract $\Upsilon_{j1}(y_j)(T_j x_j)_1$. In fact, this is unnecessary if $\Upsilon_{j1}(y_j) = 0$. In this case, (9.25) reduces to (9.1.4) and it suffices to choose $k_j(y_j) > \sum_{j=1}^{N} \|F_j C_j\| \upsilon_{jk}(y_j)$ for $j = 1, 2, ..., n_s$. By parallel development to **Theorem 9.1.10**, it can be shown that the corresponding reachability condition is satisfied. Therefore, the controlled trajectories of system (9.7)–(9.8) are steered toward the sliding surface (9.10) globally and remain on the surface thereafter. In combination with (i), the conclusion (ii) is obtained.

9.1.5 Illustrative example 9.1

Consider the interconnected system composed of two third-order subsystems

$$\begin{aligned} \dot{x}_1 \;&=\; \begin{bmatrix} -7.9 & 0 & 1 \\ 0 & -6.9 & 1 \\ 1 & 0 & 0 \end{bmatrix} x_1 + \begin{bmatrix} y_{12}^2 x_{12} \\ y_{11} x_{12} \\ x_{11} \sin y_{11} \end{bmatrix} \\ &+ \begin{bmatrix} 0 \\ 0 \\ 1 \end{bmatrix} (u_1 + \Delta g_1(x_1)) + H_1(x) \end{aligned} \tag{9.33}$$

$$\dot{x}_2 = \begin{bmatrix} -5.9 & 0 & 1 \\ 0 & -6.9 & 1 \\ 1 & 0 & 0 \end{bmatrix} x_2 + \begin{bmatrix} x_{12} \\ y_{21}x_{23} \\ 0 \end{bmatrix} + \begin{bmatrix} 0 \\ 0 \\ 1 \end{bmatrix} (u_2 + \Delta g_2(x_2)) + H_2(x) \tag{9.34}$$

$$y_j = \begin{bmatrix} 1 & 1 & 0 \\ 0 & 0 & 1 \end{bmatrix} x_j \quad j = 1, 2 \tag{9.35}$$

where $x_j = \text{col}(x_{j1}, x_{j2}, x_{j3})$ and $y_j = \text{col}(y_{jl}, y_{j2})$ are, respectively, the state variables and outputs of the ith subsystem for $j = 1, 2$. The uncertainties are assumed to satisfy

$$\begin{aligned} \|\Delta g_1(x_1)\| &\le \|y_1\| \sin^2 y_{12} := \rho_1(y_1), \\ \|\Delta g_2(x_2)\| &\le (y_{21} + y_{22})^2 := \rho_2(y_2) \end{aligned}$$

and the interconnections satisfy

$$\begin{aligned} \|H_1(x)\| &\le (x_{21} + x_{22})^2 \sin^2 x_{13} := \eta_1(x) \\ \|H_2(x)\| &\le \frac{1}{4}(|x_{13}| + |x_{23}| + |x_{11} + x_{12}|) := \eta_2(x) \end{aligned}$$

Clearly **Assumption 9.1.1** is satisfied since $C_j B_j = [0\ 1]^t$ and thus is full rank. Applying the algorithm of [65], the coordinate transformation $\tilde{x}_j = \tilde{T}_j x_j$ with

$$\tilde{T}_1 = \tilde{T}_2 = \begin{bmatrix} 1 & 0 & 0 \\ 1 & 1 & 0 \\ 0 & 0 & 1 \end{bmatrix}$$

yields the canonical form (9.9) as follows:

$$\begin{bmatrix} \tilde{A}_{111} & \tilde{A}_{112} \\ \tilde{A}_{121} & \tilde{A}_{122} \end{bmatrix} = \begin{bmatrix} -7.9 & 0 & \vdots 1 \\ -1 & -6.9 & \vdots 2 \\ \cdots & \cdots & \cdots \\ 1 & 0 & \vdots 0 \end{bmatrix} \tag{9.36}$$

$$\begin{bmatrix} \tilde{A}_{211} & \tilde{A}_{212} \\ \tilde{A}_{221} & \tilde{A}_{222} \end{bmatrix} = \begin{bmatrix} -5.9 & 0 & \vdots 1 \\ 1 & -5.9 & \vdots 2 \\ \cdots & \cdots & \cdots \\ 1 & 0 & \vdots 0 \end{bmatrix} \tag{9.37}$$

with

$$\tilde{B}_{12} = \tilde{B}_{22} \quad \text{and} \quad \tilde{C}_{12} = \tilde{C}_{22} = I_2 \tag{9.38}$$

It is easy to check that **Assumption 9.1.2** is satisfied with the choice $K_1 = K_2 = 0$ due to the stability of $\tilde{A}_{111}$ and $\tilde{A}_{211}$. Since (9.36)-(9.38) already has the canonical form (9.12), (9.13), it follows that

$$T_1 = \tilde{T}_j, \quad A_{j11} = \tilde{A}_{j11}, \quad B_{j2} = \tilde{B}_{j2}, \quad C_{j2} = \tilde{C}_{j2}$$

for $j : 1, 2$. In addition, the components of the nonlinearities

$$\Gamma_{11}(y_1) = \begin{bmatrix} -y_{12}^2 & y_{12}^2 \\ -y_{12}^2 & y_{12}^2 \end{bmatrix} \quad \text{and} \quad \Gamma_{21}(y_2) = \begin{bmatrix} 1 & 0 \\ 1 & 0 \end{bmatrix}$$

According to [65], we have

$$F_1 = F_2 = [0 \quad 1]$$

and the designed sliding surface from (9.10) is described by $\sigma = 0$ which implies $y_{j2} = 0$ for $j = 1, 2$. Choosing $Q_1 = Q_2 = 7.5 I_2$ and solving the Lyapunov equation (9.14) yield

$$P_1 = \begin{bmatrix} 0.5032 & -0.0377 \\ \bullet & 0.5698 \end{bmatrix}, \quad P_2 = \begin{bmatrix} 0.6735 & 0.0440 \\ \bullet & 0.5988 \end{bmatrix}$$

By considering the switching surface $\sigma = 0$, it is straightforward to show that $\gamma_l = 0$, and $\|R_1\| = 0$ since $\Gamma_{11}(y_1)|_{\sigma=0} = 0$. By direct computation, $\|R_{21}\| = 1.6598$. Furthermore, since $\|T_j\| = 1.618$ and $\|H_2(z)\|_{\sigma=0} \leq \frac{1}{4}\|z\|, \gamma_2 = 1.618/4 = 0.4045$. Then from (9.17), it follows that

$$M^t + M = \begin{bmatrix} 15.98 & -0.56 \\ \bullet & 11.68 \end{bmatrix} > 0$$

This implies, from **Theorem 9.1.7**, the designed sliding mode is globally asymptotically stable. It follows that

$$\begin{aligned} &\eta_1(x)\|F_1 C_1\| + \eta_2(x)\|F_2 C|2\| \\ &= y_{21}^2 \sin^2 y_{12} + \frac{1}{4}(|y_{12}| + |y_{22}| + |y_{11}|) \\ &\leq \frac{1}{4}(|y_{11}| + |y_{12}|) + \frac{1}{2}\sin^4 y_{12} + \frac{1}{2} y_{21}^4 + \frac{1}{4}|y_{22}| \end{aligned}$$

and to satisfy the reachability condition (9.26), the control gains can be chosen as

$$\begin{aligned} k_1(y_1) &= \frac{1}{4}(|y_{11}| + |y_{12}|) + \frac{1}{2}\sin^4 y_{12} + \alpha_1 \\ k_2(y_2) &= \frac{1}{2}(y_{21}^4 + \frac{1}{4}|y_{22}| + \alpha_2 \end{aligned}$$

where α_1 and α_2 are any positive constants. From (9.24), $\Upsilon_{11}(y_1) = 1 + \sin y_{11}$ and $\Upsilon_{21}(y_2) = 1$. Therefore, by **Corollary 9.1.12**, system (9.33)-(9.35) is globally stabilizable by the control

$$\begin{aligned} u_1 &= -\frac{y_{12}}{|y_{12}|}(|1+\sin y_{11}||y_{11}+y_{l2}| + \|y_1\| + \sin^2 y_{12} + k_1) \\ u_2 &= -\frac{y_{22}}{|y_{22}|}(|y_{21}+y_{22}| + (y_{21}+y_{22})^2 + k_2) \end{aligned}$$

It should be emphasized that the example above yields a global result and has the following characteristics:

(i) the nonlinearity $f_j(x_j)$ cannot be bounded by a linear function of $\|x_j\|$ globally;

(ii) the interconnection is nonlinear with unknown structure;

(iii) the interconnection and $f_j(x_j)$ are both mismatched;

(iv) the condition $FCA = XC$ employed in [395] is not satisfied for any X for the system (A_j, B_j, C_j) with $j = 1, 2$. Therefore, existing decentralized output feedback schemes [140, 172, 302] are not applicable here. Furthermore, the schemes of [65, 159, 395] cannot be used even if a centralized control scheme is allowed.

9.2 Delay-Dependent Sliding-Mode Control

In this section, we direct attention to the problem of global decentralized stabilization of a class of interconnected delay systems, where both known and uncertain interconnections involve time delay. Matched and mismatched interconnections are considered separately to reduce the conservatism. A composite sliding surface is designed and the stability of the associated sliding motion, which is governed by a time-delay interconnected system, is analyzed based on the Razumikhin Lyapunov approach. A decentralized static output feedback variable structure control which is delay-dependent synthesized to drive the interconnected system to the sliding surface globally.

Recently, various control approaches have been employed to deal with time-delay control systems and the reader is referred to Chapter 6 for a recent account. Sliding-mode control, as one of the discontinuous control approaches, is completely robust to so-called matched uncertainty, and can be used to deal with mismatched uncertainty [380]. Since sliding-mode dynamics are reduced-order, it is possible to reduce the conservatism in the stability analysis of the sliding motion. This has motivated the application of sliding-mode techniques to time-delay systems [90, 285]. However most of the published work focuses on centralized control systems. The results of applying sliding-mode techniques to time-delayed interconnected systems are very few. In the limited available literature [113, 320, 379], it is required that all states are available and the interconnections are linear and matched. The assumption that all the terms involving time-delay are matched means that the sliding mode is no longer a time-delay system.

In what follows, extending the material of Chapter 6, a class of nonlinear interconnected systems with time-varying delays is considered, where the time-delay appears not only in the isolated subsystems, but also in the interconnections. The interconnections are separated as matched and mismatched parts and are dealt with separately to reduce the conservatism. By using appropriate coordinate transformations, sliding-mode dynamics are derived which are reduced-order time-delayed interconnected systems. Sufficient conditions are developed using a Razumikhin-Lyapunov approach such that the sliding motion is uniformly globally asymptotically stable. A decentralized control scheme based on only output information is proposed to drive the system to the composite sliding surface. It is shown that the effect of the interconnections can be canceled by designing an appropriate decentralized controller under appropriate conditions.

9.2.1 System description

Consider a time-varying delayed interconnected system composed of n_s subsystems, where the jth subsystem has n_j-th order and is described by

$$\dot{x}_j = A_j x_j + B_j\Big(u_j + G_j(t, x_j, x_j(t - d_j(t)))\Big) + \sum_{k=1\, k\neq j}^{n} \Big(H_{jk} y_j(t - d_j(t)) + \Delta H_{ij}(t, x_j, x_j(t - d_j(t)))\Big) \quad (9.39)$$

$$y_j = C_j x_j, \quad j = 1, 2, ..., n_s \quad (9.40)$$

where $x_j \in \Re^{n_j}$, $u_j \in \Re^{m_j}$, and $y_j \in \Re^{p_j}$ with $m_j \leq p_j < n_j$ are the state variables, inputs, and outputs, of the ith subsystem, respectively. The triples $(A_j, B)i, C_j)$ and $H_{jk} \in \Re^{n_j \times p_j}$ with $i \neq j$ represent constant matrices of appropriate dimensions with B_j and C_j of full rank. The function $G_j(.)$ is a matched nonlinearity in the ith subsystem. $\sum_{k=1,k\neq j}^{n} H_{jk} y_k(t - d_k(t))$ and $\sum_{k=1,k\neq j}^{n} \Delta H_{jk}(t, x_k, x_k(t - d_k(t)))$ are, respectively, a known interconnection and an uncertain interconnection of the ith subsystem. The factor $d_j(t)$ is the time-varying delay which is known continuous, nonnegative, and bounded in $\Re^+$, and thus $d_j := \sup_{t\in\Re^+}\{d_j(t)\} < \infty$. The initial conditions are given by

$$x_j(t) = \phi_j(t), \quad t \in [-\bar{d}_j, 0]$$

for $j = 1, 2, ..., n$. All the nonlinear functions are assumed to be smooth enough such that the unforced system has a unique continuous solution.

For convenience, a variable/vector with subscript d_j is introduced to denote the time-delayed variable throughout this section, for example, $x_{jd_j}(t)$ denotes $x_j(t - d_j(t))$ and $y_{jd_j}(t)$ represents $y_j(t - d_j(t))$. It is known that the sliding motion is insensitive to matched uncertainty and hence it is useful to separate the treatment of matched interconnections from mismatched interconnections to reduce conservatism. Accordingly, we consider the decompositions of the interconnections:

$$H_{jk} = H^a_{jk} + H^b_{jk} \tag{9.41}$$

$$\Delta H_{jk}(t) = \Delta H^a_{jk}(t, x_k, x_{kd_k}) + \Delta H^b_{jk}(t, x_k, x_{kd_k}) \tag{9.42}$$

where

$$H^a_{jk} = B_j D_{jk} \tag{9.43}$$

$$\Delta H^a_{jk}(t, x_k, x_{kd}) = B_j \Delta\Theta_{jk}(t, x_k, x_{kd_k}) \tag{9.44}$$

where for some $\Delta\Theta_{jk}(.) \in \Re^{m_j}$ and $D_{jk} \in \Re^{m_j \times p_j}$ where $\Delta\Theta_{jk}(.)$ is uncertain for $i \neq j,\ i, j = 1, 2, ..., n$. It must be asserted that the decompositions in (9.41) and (9.42) that satisfy (9.43) and (9.44) can be readily obtained from basic matrix operations.

Assumption 9.2.1 *There exist known continuous functions $g_j(.), \alpha_{jk}(.)$, and $\beta_{jk}(.)$ such that*

$$\|G_j(t, x, x_{id_j})\| \leq g_j(t, y_j, y_{jd_j}) \tag{9.45}$$

$$\|\Delta\Theta_{jk}(t, x_j, x_{kd_k})\| \leq \alpha_{jk}(t, y_j, y_{jd_j})\|y_{jd_j}\| \tag{9.46}$$

$$\|\Delta H^b_{jk}(t, x_j, x_{kd_k})\| \leq \beta_{jk}(t, y_j, \|y_{jd_j}\|)\|y_{jd_j}\| \tag{9.47}$$

for $j \neq k,\ j, k = 1, 2, ..., n_s$ where $\beta_j k(., ., r)$ is nondecreasing with respect to the variable r in $\Re^+$.

It follows from (9.42), (9.44), (9.46), and (9.47) that there exist known continuous functions $\rho_{jk}(.)$ such that

$$\|\Delta H_{jk}(t, x_k, x_{kd_k})\| \leq \rho_{jk}(t, y_k, y_{kd_k})\|y_{kd_k}\| \ for \ j, k = 1, 2, ..., n_s, \ j \neq k \tag{9.48}$$

It must be observed that **Assumption 9.2.1** limits the type of uncertainties that can be tolerated by the interconnected system to be bounded by functions of the system outputs. Since the interconnections involve time-delays, the bounding functions are nonlinear and depend on the delayed information.

Assumption 9.2.2 *rank*$(C_j B_j) = m_j$ *for* $i = 1, 2, ..., n_s$.

Assumption 9.2.2, in light of [67], implies that there exists a nonsingular linear coordinate transformation

$$\tilde{x}_j = \tilde{T}_j x_j$$

such that the triple (A_j, B_j, C_j) with respect to the new coordinates has the structure

$$\tilde{A}_j = \begin{bmatrix} \tilde{A}_{j1} & \tilde{A}_{j2} \\ \tilde{A}_{j3} & \tilde{A}_{j4} \end{bmatrix}, \quad \tilde{B}_j = \begin{bmatrix} 0 \\ \tilde{B}_{j2} \end{bmatrix}, \quad \tilde{C}_j = [0 \quad \tilde{C}_{j2}] \tag{9.49}$$

where $\tilde{A}_{j1} \in \Re^{(n_j - m_j)\times(n_j - m_j)}$, $\tilde{B}_{j2} \in \Re^{m_j \times mi}$ is nonsingular, and $\tilde{C}_{j2} \in \Re^{p_j \times p_j}$ is orthogonal.

Assumption 9.2.3 *The triple* $(\tilde{A}_{j1}, \tilde{A}_{j2}, \Xi_j)$ *is output feedback stabilizable, where for* $j = 1, 2, ..., n_s$

$$\Xi_j := [0_{(p_j - m_j)\times(n_j - p_j)} \quad I_{p_j - m_j}]$$

We note that $\tilde{A}_{j1}$, ..., $\tilde{A}_{j4}$, $\tilde{B}_{j2}$, $\tilde{C}_{j2}$ are dependent on the coordinate transformation employed. **Assumptions 9.2.2** and **9.2.3** together describe inherent properties of the triplet (A_j, B_j, C_j).

More importantly, **Assumption 9.2.3** implies that there exists a matrix $\tilde{K}_j$ such that $A_{j1} = \tilde{A}_{j1} - \tilde{A}_{j2}\tilde{K}_j\Xi_j$ is stable. From [67], the coordinate transformation

$$\tilde{x}_j \mapsto z_j = \bar{T}_j \tilde{x}_j$$

where

$$\bar{T}_j = \begin{bmatrix} I & 0 \\ -\tilde{K}_j\Xi_j & I \end{bmatrix} \quad i = 1, 2, ..., n_s \tag{9.50}$$

will transform the triple (A_j, B_j, C_j) to the following form in the new coordinate system

$$\begin{bmatrix} A_{j1} & A_{j2} \\ A_{j3} & A_{j4} \end{bmatrix}, \quad \begin{bmatrix} 0 \\ B_{j2} \end{bmatrix}, \quad [0 \quad C_{j2}] \tag{9.51}$$

where $A_{i1} = \tilde{A}_{j1} - \tilde{A}_{j2}\tilde{K}_j\Xi_j$ is stable and both $B_{j2} \in \Re^{m_j \times m_j}$ and $C_{j2} \in \Re^{p_j \times p_j}$ are nonsingular. In the sequel, we assume that $m_j < p_j$. In case that $m_j = p_j$, then **Assumption 9.2.3** can then be replaced by the standard assumption that $\tilde{A}_{j1}$ is stable.

Remark 9.2.4 *Assumptions* ***9.2.2*** *and* ***9.2.3*** *are limitations on the triple* (A_j, B_j, C_j). *Effectively, they ensure the existence of the output sliding surface.* **Assumption 9.2.2** *requires* $(\tilde{A}_{i1}, \tilde{A}_{i2}, \Xi_j)$ *instead of* (A_j, B_j, C_j) *to be output feedback stabilizable. Note that the former is related to a system with order* $n_j - m_j$, *while the latter is an* n_j*-order system. Sometimes this reduced order problem is more amenable to solution, for example, if the matrix triple is related to a one-input and two-output system, the output feedback problem reduces to a classical root-locus investigation.*

In the sequel, we seek to design a variable structure control law of the form

$$u_j = u_j\left(t, y_j, y_{jd_j}\right) \tag{9.52}$$

based on sliding-mode techniques such that the associated closed-loop system formed by applying the control law (9.52) to the interconnected system (9.39) and (9.40) is globally uniformly asymptotically stable even in the presence of the uncertainties and time-delays. It is evident that the control u_j in (9.52) depends on the time t, the jth subsystem output y_j, and the delayed output y_{jd_j}, which is available since the delay $d(t) := (d_1(t),\ d_2(t),\ ..., d_{n_s}(t))$ is assumed to be known.

9.2.2 Stability analysis

In the following, we consider that **Assumptions 9.2.2** and **9.2.3** hold. Let

$$F_j := F_{j2}\begin{bmatrix} K_j & I_{m_j} \end{bmatrix} \tilde{C}_{j2}^t, \quad j = 1, 2, ..., n_s \tag{9.53}$$

where $F_{j2} \in \Re^{m_j \times m_j}$ is any nonsingular matrix, K_j is related to $\tilde{K}_j$, and $\tilde{C}_{j2}$ is given in (9.49). It has been shown in [67] that F_j defined in (9.53) satisfies

$$F_j \begin{bmatrix} 0 & \tilde{C}_{j2} \end{bmatrix} = \begin{bmatrix} 0 & F_{j2} \end{bmatrix} \tag{9.54}$$

where $F_{j2} \in \Re^{m_j \times m_j}$ is nonsingular. Then, for the interconnected system (9.39) and (9.40), consider the composite sliding surface defined by

$$\{\mathrm{col}(x_1, x_2, ..., x_{n_s}) \ \mid \ S_j(x_j) = 0, \quad j = 1, 2, ..., n_s\} \tag{9.55}$$

where

$$S_j(x_j) := F_j C_j x_j = F_j y_j \tag{9.56}$$

and the matrices F_j for $j = 1, 2, ..., n$ are given in (9.53). Under **Assumptions 9.2.2** and **9.2.3**, it follows from the foregoing analysis that there exists a coordinate transformation

$$x_j \mapsto z_j = T_j x_j,\ j = 1, 2, ..., n_s$$

such that in the new coordinates

$$z = \text{col}(z_1, z_2, ..., z_{n_s})$$

system (9.39) and (9.40) can be described by

$$\begin{aligned}
\dot{z}_j &= \begin{bmatrix} A_{j1} & A_{j2} \\ A_{j3} & A_{j4} \end{bmatrix} z_j + \begin{bmatrix} 0 \\ B_{j2} \end{bmatrix} (u_j + G_j(t, T_j^{-1} z_j, T_j^{-1} z_{jd_j}) \\
&+ \sum_{k=1,k\neq j}^{n_s} (D_{jk} y_{kd_k} + \Delta\Theta_{jk}(t, T_k^{-1} z_k, T_k^{-1} z_{kd_k}))) \\
&+ \sum_{k=1,k\neq j}^{n_s} T_j (H_{jk}^b y_{kd_k} + \Delta H_{jk}^b (t, T_k^{-1} z_k, T_k^{-1} z_{kd_k})) \qquad (9.57) \\
y_j &= \begin{bmatrix} 0 & C_{j2} \end{bmatrix} z_j, \quad j = 1, 2, ..., n_s \qquad (9.58)
\end{aligned}$$

where $A_{j1} = \tilde{A}_{j1} - \tilde{A}_{j2} K_j \Xi_j \in \Re^{n_j - m_j}$ is stable, both $B_{j2} \in \Re^{m_j \times m_j}$ and $C_{j2} \in \Re^{p_j \times p_j}$ are nonsingular, $H_{jk}^a(.)$ and $H_{jk}^b(.)$, $\Delta\Theta_{jk}(.)$ and $\Delta H_{jk}^b(.)$ are given in (9.41), (9.44), and (9.42), respectively. Since A_{j1} is stable, for any $Q_j > 0$, the Lyapunov equation

$$A_{j1}^t P_j + P_j A_{j1} = -Q_j \qquad (9.59)$$

has a unique solution $P_j > 0$ for $j = 1, 2, ..., n$. For convenience, we partition

$$T_j \equiv: \begin{bmatrix} T_{j1} \\ T_{j2} \end{bmatrix}, \quad T_j^{-1} \equiv: \begin{bmatrix} W_{j1} & W_{j2} \end{bmatrix} \qquad (9.60)$$

where $T_{j1} \in \Re^{(n_j - m_j) \times n_j}$ and $W_{j1} \in \Re^{n_j \times (n_j - m_j)}$. Combining (9.56) and (9.57), we get

$$\begin{aligned}
\dot{z}_{j1} &= A_{j1} z_{j1} + A_{j2} z_{j2} + \sum_{k=1,k\neq j}^{n} T_{j1} (H_{jk}^b y_{kd_k} \\
&+ \Delta H_{jk}^b (t, T_k^{-1} z_k, T_k^{-1} z_{kd_k})) \qquad (9.61) \\
\dot{z}_{j2} &= A_{j3} z_{j1} + A_{j4} z_{j2} + B_{j2} (u_j + G_j(.) \\
&+ \sum_{k=1,k\neq j}^{n} (D_{jk} y_{kd_k} + \Delta\Theta_{jk}(t, T_k^{-1} z_j, T_k^{-1} z_{kd_k}))) \\
&+ \sum_{k=1,k\neq j}^{n} T_{i1} \left(H_{jk}^b y_{kd_k} + \Delta H_{jk}^b (t, T_k^{-1} z_k, T_k^{-1} z_{kd_k}) \right) \qquad (9.62) \\
y_j &= \begin{bmatrix} 0 & C_{j2} \end{bmatrix} z_j, \quad i = 1, 2, ..., n_s \qquad (9.63)
\end{aligned}$$

where $z_j := \text{col}(z_{i1}, z_{i2})$ with $z_{i1} \in \Re^{n_j - m_j}$ and $z_{i2} \in \Re^{m_j}$. From (9.56), (9.63), and (9.53), we obtain

$$\begin{aligned} S_j(x_j) &= F_j y_j = F_j \begin{bmatrix} 0 & C_{j2} \end{bmatrix} z_j \\ &= \begin{bmatrix} 0 & F_{j2} \end{bmatrix} \begin{bmatrix} z_{j1} \\ z_{j2} \end{bmatrix} = F_{j2} z_{j2} \end{aligned}$$

where $F_{j2} \in \Re^{m_j \times m_j}$ is nonsingular. Therefore, in the new coordinates z, the sliding surface (9.55) can be described by

$$\{\text{col}(z_1, z_2, ..., z_{n_s}) \quad | \quad z_{j2} = 0, \quad i = 1, 2, ..., n_s\}$$

In a similar way, we partition the output distribution matrix in (9.63) as

$$\begin{bmatrix} 0 & C_{j2} \end{bmatrix} = [\underbrace{0 \quad C_{j21}}_{C_{js}} \quad | \quad C_{j22}] \tag{9.64}$$

where $C_{js} := \begin{bmatrix} 0 & C_{j21} \end{bmatrix} \in \Re^{p_j \times (n_j - m_j)}$ and $C_{j22} \in \Re^{p_j \times m_j}$. When constrained along the sliding surface, the jth subsystem output y_j is then described by

$$y_{js} := \begin{bmatrix} 0 & C_{j21} \end{bmatrix} z_{j1} = C_{js} z_{j1}, \quad j = 1, 2, ..., n_s \tag{9.65}$$

Now we take a look at the structure of system (9.61)-(9.62) in view of the partition for T_j^{-1} in (9.60). We observe that the sliding-mode dynamics of system (9.39)-(9.40) associated with the sliding surface (9.55)-(9.56) can be expressed by

$$\begin{aligned} \dot{z}_{j1} &= A_{j1} z_{j1} + \sum_{k=1, k\neq j}^{n} T_{j1}(H^b_{jk} C_{ks} z_{k1d_k} \\ &+ \Delta H^b_{jk}(t, W_{k1} z_{k1}, W_{k1} z_{k1d_k})) \end{aligned} \tag{9.66}$$

where W_{j1} is defined by the partition (9.60) for $j = 1, 2, ..., n_s$.

Theorem 9.2.5 *Assume that* **Assumptions 9.2.1 - 9.2.3** *hold. Then, for the sliding surface in (9.55), the sliding motion of the interconnected time-delay system (9.39)-(9.40) is governed by dynamical system (9.66). Moreover, the sliding motion is globally uniformly asymptotically stable if*

(i) The $n_j \times n_j$ symmetric matrix

$$N_j := Q_j - P_j T_{j1}\Big(\sum_{k=1, k\neq j}^{n_s} \frac{1}{\varepsilon_j} H^b_{jk} C_{ks} P_k^{-1} C^t_{ks} (H^b_{jk})^t \Big) T^t_{j1} P_j - \sum_{k=1, k\neq j}^{n_s} \varepsilon_k P_k > 0$$

for some $\varepsilon_j > 0$ *for* $j = 1, 2, ..., n_s$

(ii) The $n \times n$ *matrix function* $M + M^t > 0$ *where* $M = [m_{jk}(.)]_{n \times n}$ *is defined by*

$$m_{jk}(.) := \begin{cases} \lambda_{\min}(N_j), & j = k \\ -\gamma_j \beta_{jk}(t, C_{js} z_{j1} \gamma_k \|C_{js}\| \|z_{k1}\|) \|P_j T_{j1}\| \|C_{ks}\|, & j \neq k \end{cases}$$

for some $\gamma_j > 1$ *for* $j, k = 1, 2, ..., n_s$ *and*

$$\mu := \inf_{z_{11}, ..., z_{n_s 1}} \{\lambda_{\min}(M + M^t)\} > 0$$

Proof: It suffices to focus on the system described in (9.66) and show that it is globally uniformly asymptotically stable. Consider the Lyapunov function candidate

$$V(z_{11}, z_{21}, ..., z_{n_s 1}) = \sum_{j=1}^{n_s} z_{j1}^t P_j z_{j1}$$

where $P_j > 0$ satisfies (9.59) for $j = 1, 2, ..., n_s$. A little algebra shows that the time derivative of $V(.)$ along the trajectories of system (9.66) using (9.59) is given by

$$\begin{aligned} \dot{V} &= -\sum_{j=1}^{n_s} z_{j1}^t Q_j z_{j1} + 2 \sum_{j=1}^{n} \sum_{k=1, k \neq j}^{n_s} z_{j1}^t P_j T_{j1} H_{jk}^b C_{ks} z_{k1d_k} \\ &+ 2 \sum_{j=1}^{n_s} \sum_{k=1, k \neq j}^{n_s} z_{j1}^t P_j T_{j1} \Delta H_{jk}^b(.) \end{aligned} \qquad (9.67)$$

By **Inequality 2** of the Appendix, for any $\varepsilon_j > 0$, we have

$$\begin{aligned} 2 z_{j1}^t P_j T_{j1} H_{jk}^b C_{ks} z_{k1d_k} &\leq \frac{1}{\varepsilon_k} z_{j1}^t P_j T_{j1} H_{jk}^b C_{ks} P_k^{-1} (T_{j1} H_{jk}^b C_{ks})^t P_j z_{j1} \\ &+ \varepsilon_k z_{k1d_k}^t P_k z_{k1d_k} \end{aligned} \qquad (9.68)$$

Since the z_{j1} for $j = 1, 2, ..., n$ are independent of each other, it is clear that

$$V(z_{11d_1}, z_{21d_2}, ..., z_{n_s 1d_n}) \leq q \, V(z_{11}, z_{21}, ..., z_{n_s 1})$$

for $q > 1$, is equivalent to

$$z_{j1d_j}^t P_j z_{j1d_j} \leq q \, z_{j1}^t P_j z_{j1}, \quad j = 1, 2, ..., n_s \qquad (9.69)$$

which implies

$$\|z_{j1d_j}\| \leq \gamma_j \|z_{j1}\|, \quad j = 1, 2, ..., n_s \tag{9.70}$$

for some positive constants $\gamma_j > 1$ with $j = 1, 2, ..., n$. It follows from (9.65) and (9.70) that

$$\|y_{isd_j}\| = \|C_{js} z_{j1d_j}\| \leq \gamma_j \|C_{js}\| \|z_{ij}\| \tag{9.71}$$

Since $\beta_{jk}(t, y_{ks}, r)$ is nondecreasing with respect to the variable $r \in \Re^+$, it follows from (9.47), (9.71), and (9.65) that

$$P_j T_{j1} \Delta H^b_{jk}(.) \leq \gamma_k \beta_{jk}(t, C_{ks} z_{k1}, \gamma_k \|C_{ks}\| \|z_{k1}\|) \|P_j T_{j1}\| \|C_{ks}\| \|z_{k1}\| \\ (j \neq k) \tag{9.72}$$

Therefore, from (9.47), (9.68), and (9.72), it follows that when

$$V(z_{11d_1}, z_{21d_2}, ..., z_{n_s 1d_n}) \leq qV(z_{11}, z_{21}, ..., z_{n_s 1})$$

$$\begin{aligned}
\dot{V} \leq & -\sum_{j=1}^{n_s} z^t_{j1} Q_j z_{j1} \\
& + \sum_{j=1}^{n_s} \sum_{k=1,k\neq j}^{n_s} \left(\frac{1}{\varepsilon_k} z^t_{j1} P_j T_{j1} H^b_{jk} C_{ks} P_k^{-1} (T_{j1} H^b_{jk} C_{ks})^t P_j z_{j1} \right) \\
& + \sum_{j=1}^{n_s} \sum_{k=1,k\neq j}^{n_s} \varepsilon_k z^t_{k1} P_k T_{k1} \\
& + 2\sum_{j=1}^{n_s} \sum_{k=1,k\neq j}^{n_s} \gamma_k \beta_{jk}(t, C_{ks} z_{k1}, \gamma_k \|C_{ks}\| \|z_{k1}\|) \|P_j T_{i1}\| \|C_{ks}\| \|z_{k1}\| \|z_{j1}\| \\
\leq & -\sum_{j=1}^{n_s} z^t_{j1} N_j z_{j1} \\
& + 2\sum_{j=1}^{n_s} \sum_{k=1,k\neq j}^{n_s} \gamma_k \beta_{jk}(t, C_{ks} z_{k1}, \gamma_k \|C_{ks}\| \|z_{k1}\|) \|P_j T_{i1}\| \|C_{ks}\| \|z_{k1}\| \|z_{j1}\| \\
= & -\frac{1}{2} \mathbf{Z}(M + M^t) \mathbf{Z}^t \\
\leq & -\frac{1}{2} \inf_{z_{11},\ldots,z_{n1}} \{\lambda_{\min}(M + M^t)\} \sum_{j=1}^{n_s} \|z_{j1}\|^2 && (9.73) \\
= & -\frac{1}{2} \mu \|\mathbf{z}\|^2 && (9.74)
\end{aligned}$$

where

$$\mathbf{z} = \mathrm{col}(z_1, z_2, ..., z_n),\ Z = [||z_{11}|| \, ||z_{21}|| ... ||z_{n_s 1}||]$$

and the fact $z_{j1}^t N_j z_{j1} \geq \lambda_{\min}(N)||z_{j1}||^2$ has been used. Since $\mu > 0$, the conclusion follows from the basic Razumikhin theorem.

Remark 9.2.6 *If $m_{r0} = p_{r0}$ for the r_0-th $(1 \leq r_0 \leq n)$ subsystem, then $y_{r0} = z_{r0 2}$, and thus from (9.65), $y_{r0} s = 0$. In this case, from the proof of* ***Theorem 9.2.5****, condition (i) can be weakened by*

$$\begin{aligned} N_j = \; & Q_j - P_j T_{i1} (\textstyle\sum_{j=1, j\neq i, r0}^{n} \frac{1}{\varepsilon_j} H_{jk}^b C_{js} P_j^{-1} C_{js}^t (H_{jk}^b)^t) \\ & - \textstyle\sum_{j=1, j\neq i, r0}^{n} \varepsilon_j P_j > 0, \quad i = 1, 2, ..., n \end{aligned}$$

9.2.3 Reachability analysis

We now proceed to design a decentralized static output feedback sliding-mode control such that the system states are steered along the sliding surface (9.55) and (9.56). For interconnected system (9.39) and (9.40), a well-known reachability condition is described by [109]:

$$\sum_{j=1}^{n} \frac{S_j^t(x_j) S_j(x_j)}{||S_j(x_j)||} < 0 \tag{9.75}$$

where the switching function $S_j(.)$ is defined by (9.56). To proceed further, the following condition is imposed on system (9.39)-(9.40).

Assumption 9.2.7 *The matrix equation $\Gamma_j C_j = C_j A_j$ is solvable for Γ_j with $i = 1, 2, ..., n_s$.*

Remark 9.2.8 ***Assumption 9.2.7*** *implies that there exist matrices Γ_j with $i = 1, 2, ..., n_s$ such that*

$$C_j^t \Gamma_j^t \; = \; A_j^t C_j^t,\ j = 1, 2, ..., n_s$$

This ensures that the linear algebraic equations

$$C_j^t \Gamma_j^\ell \; = \; ({}_j^t C_j^t)^\ell,\ j = 1, 2, ..., n_s,\ \ell = 1, 2, ..., p_j$$

are solvable for Γ_j^ℓ.

The following result will be used in the subsequent analysis.

Lemma 9.2.9 *Assume* $H_{jk} \in \Re^{n_j \times p_j}$ *with* n_j *and* p_j *positive integer numbers, and* $x = \text{col}(x_1, x_2, ..., x_n)$ *where* $x_j \in \Re^{n_j}$ *for* $i = 1, 2, ..., n$. *Then*

$$(1) \quad ||x|| \leq x^t sgn(x)$$

$$(2) \quad \sum_{j=1}^{n_s} \sum_{k=1,k\neq j}^{n_s} H_{jk} x_k = \sum_{j=1}^{n_s} \left(\sum_{k=1,k\neq j}^{n_s} H_{kj} \right) x_j$$

where $sgn(.)$ *denotes the usual vector signum function.*

Proof: Condition (1) follows from [381]. To show condition (2) we proceed from the fact that $\sum_{j=1}^{n_s} \sum_{k=1}^{n_s} H_{jk} x_k = \sum_{k=1}^{n_s} \sum_{j=1}^{n_s} H_{jk} x_k$. It follows that

$$\sum_{j=1}^{n_s} \sum_{k=1,k\neq j}^{n_s} H_{jk} x_k =$$

$$\sum_{j=1}^{n_s} \sum_{k=1}^{n_s} H_{jk} x_k - H_{11} x_1 - H_{22} x_2 - ... - H_{n_s n_s} x_{n_s}$$

$$= \sum_{j=1}^{n_s} \sum_{k=1}^{n_s} H_{jk} x_k - \sum_{k=1}^{n} H_{kk} x_k$$

$$= \sum_{j=1}^{n_s} (\sum_{k=1}^{n_s} H_{jk} - H_{kk}) x_k$$

$$= \sum_{j=1}^{n_s} \sum_{k=1,k\neq j}^{n_s} H_{kj}) x_j$$

The main result is now provided.

Theorem 9.2.10 *Consider the interconnected time-delay system (9.39) and (9.40). Under* ***Assumptions 9.2.1-9.2.7****, there exists a global delay-dependent static output feedback decentralized control law that drives the system (9.39) and (9.40) to the composite sliding surface (9.55) and (9.56) and maintains a sliding motion on it thereafter.*

Proof: Since the triple in (9.51) is obtained from (A_j, B_j, C_j) using the transformation $z_j = T_j x_j$, it follows that

$$T_j B_j = \begin{bmatrix} 0 \\ B_{j2} \end{bmatrix}, \quad C_j T_j^{-1} = [0 \quad C_{j2}]$$

where both $B_{j2} \in \Re^{m_j \times m_j}$ and $C_{j2} \in \Re^{p_j \times p_j}$ are nonsingular. In view of (9.54), we have

$$F_j C_j B_j = F_j [0 \quad C_{j2}] \begin{bmatrix} 0 \\ B_{j2} \end{bmatrix} = F_{j2} B_{j2}$$

which shows that $F_j C_j B_j$ is nonsingular because F_{j2} and B_{j2} are nonsingular. Construct a variable structure control

$$\begin{aligned} u_j &= -(F_j C_j B_j)^{-1} F_j \Gamma_j y_j - (F_j C_j B_j)^{-1} \Big(\|F_j C_j B_j\| g_j(t, y_j, y_{id_j}) \\ &+ \eta_j \Big) \text{sgn}(F_j y_j) \\ &- (F_j C_j B_j)^{-1} \Big(\sum_{k=1, k \neq j}^{n_s} \|F_k C_k H_{jki}\|) \|y_{jd_j}\| \text{sgn}(F_j y_j) \Big) \\ &- (F_j C_j B_j)^{-1} \Big(\sum_{k=1, k \neq j}^{n_s} \|F_k C_k\| \rho_{kj}(.)) \|y_j d_j\| \text{sgn}(F_j y_j) \Big) \end{aligned} \tag{9.76}$$

where $g_j(.)$, β_{jk}, and $\rho_{jk}(.)(i \neq j)$ are defined by (9.45), (9.47), and (9.48), respectively; η_j can be chosen as any positive constant for $j, k = 1, 2, ..., n_s$ and the symbol sgn(.) denotes the usual signum vector function. From (9.39), (9.56), and (9.48), it follows that under **Assumption 9.2.7**

$$\begin{aligned} &\sum_{j=1}^{n_s} \frac{S_j^t(xi) \dot{S}_j(x_j)}{\|S_j(x_j)\|} \\ \leq\ & \sum_{j=1}^{n_s} \frac{(F_j y_j)^t}{\|F_j y_j\|} F_j \Gamma_j y_j \\ +\ & \sum_{j=1}^{n_s} \frac{(F_j y_j)^t}{\|F_j y_j\|} F_j C_j B_j (u_j + G_j(t, x_j, x_{jd_j})) \\ +\ & \sum_{j=1}^{n_s} \sum_{k=1, k \neq j}^{n_s} \|F_j C_j H_{jk}\| \|y_{kd_k}\| \\ +\ & \sum_{j=1}^{n_s} \sum_{k=1, k \neq j}^{n_s} \rho_{jk}(t, y_k, y_{kd_k}) \|F_j C_j\| \|y_{kd_k}\| \end{aligned} \tag{9.77}$$

Substituting the control of (9.74) into (9.77) and using **Lemma 9.2.9**, it follows from (9.45) that

$$
\begin{aligned}
& \sum_{j=1}^{n_s} \frac{S_j^t(x_j)\dot{S}_j(x_j)}{\|S_j(x_j)\|} \\
\leq \;& \sum_{j=1}^{n_s}\Big(-\frac{(F_jy_j)^t \mathrm{sgn}(F_jy_j)}{\|F_jy_j\|}\|F_jC_jB_j\|g_j(t,y_j,y_{id_j}) \\
+ \;& \|F_jC_jB_j\|\|G_j(t,x_j,x_{jd_j})\|\Big) - \sum_{j=1}^{n_s} \frac{(F_jy_j)^t \mathrm{sgn}(F_jy_j)}{\|F_jy_j\|}\eta_j \\
- \;& \sum_{j=1}^{n_s} \frac{(F_jy_j)^t \mathrm{sgn}(F_jy_j)}{\|F_jy_j\|}\Big(\sum_{k=1,k\neq j}^{n_s} \|F_kC_kH_{kj}\|\Big)\|y_{jd_j}\| \\
+ \;& \sum_{j=1}^{n_s}\left(\sum_{k=1,k\neq j}^{n_s} \|F_jC_jH_{kj}\|\right)\|y_{jd_j}\| \\
- \;& \sum_{j=1}^{n_s} \frac{(F_jy_j)^t \mathrm{sgn}(F_jy_j)}{\|F_jy_j\|}\sum_{k=1,k\neq j}^{n_s} \rho_{kj}(t,y_j,y_{jd_j})\|F_jC_j\|\|y_{kd_k}\| \\
+ \;& \sum_{j=1}^{n_s}\sum_{k=1,k\neq j}^{n_s} \rho_{jk}(t,y_k,y_{kd_k})\|F_kC_k\|\|y_kd_k\| < 0
\end{aligned}
$$

which concludes the proof.

From sliding mode control theory, it follows that **Theorems 9.2.5** and **9.2.10** together show that the developed decentralized control law (9.76) uniformly asymptotically stabilizes system (9.39)-(9.40) globally.

9.2.4 Illustrative example 9.2

Consider an interconnected time-delay system of the type (9.39)-(9.40) with the following data:

$$
\begin{aligned}
A_1 &= \begin{bmatrix} -8 & 0 & 1 \\ 0 & -8 & 1 \\ 1 & 1 & 0 \end{bmatrix}, \quad B_1 = \begin{bmatrix} 0 \\ 0 \\ 1 \end{bmatrix}, \\
G_1 &= (x_{11}+x_{12})^2 x_{13d_1} \sin x_{12d_1} \Delta g_1(.), \quad H_{12} = \begin{bmatrix} 0 & -1 \\ 1 & 0 \\ 4 & 1 \end{bmatrix},
\end{aligned}
$$

$$A_2 = \begin{bmatrix} -6 & 0 & 1 \\ 0 & -6 & 1 \\ 1 & 1 & 0 \end{bmatrix}, \quad B_2 = \begin{bmatrix} 0 \\ 0 \\ 1 \end{bmatrix},$$

$$G_2 = \Delta g_2(.), \quad H_{21}(.) = \begin{bmatrix} 4(x_{11d_1} + x_{12d_1})(\sin x_{13})^2 \Delta h_1(.) \\ 4(\sin x_{13})^2 x_{13d_1} (\Delta h_1(.))^2 \\ (x_{11} + x_{12}) x_{13d_1} (\sin x_{11d_1})^2 \Delta h_2(.) \end{bmatrix},$$

$$C_1 = \begin{bmatrix} 1 & 1 & 0 \\ 0 & 0 & 1 \end{bmatrix}, \quad C_1 = \begin{bmatrix} 1 & 1 & 0 \\ 0 & 0 & 1 \end{bmatrix}$$

where $x_1 = \mathrm{col}(x_{11}, x_{12}, x_{13}) \in \Re^3$ and $x_2 = \mathrm{col}(x_21, x_22, x_23) \in \Re^3$, $u_1 \in \mathcal{R}$ and $u_2 \in \mathcal{R}$, and $y_1 = \mathrm{col}(y_{11}, y_{12}) \in \Re^2$ and $y_2 = \mathrm{col}(y_{21}, y_{22}) \in \Re^2$ are, respectively, the state variables, inputs and outputs of the system. The uncertainties $\Delta g_1(.)$, $\Delta g_2(.)$, $\Delta h_1(.)$, and $\Delta h_2(.)$ are assumed to satisfy

$$\begin{bmatrix} |\Delta g_1(.)| \leq (x_{13d_1})^2 \sin^2(x_{11} + x_{12}) & |\Delta h_1(.)| \leq 1 \\ |\Delta g_2(.)| \leq |x_{21d_2} + x_{22d_2}| x_{23d_2}^2 \sin^2 x_{22} & |\Delta h_2(.)| \leq |y_{1d_1}| \end{bmatrix}$$

Note that$\Delta H_{12} = 0$ and $\Delta H_{21} = 0$ and the decomposition in the form of (9.41) and (9.42) is given by

$$H_{12}^a = \begin{bmatrix} 0 & 0 \\ 0 & 0 \\ 4 & 1 \end{bmatrix}, \quad H_{12}^b = \begin{bmatrix} 0 & -1 \\ 1 & 0 \\ 0 & 0 \end{bmatrix} \tag{9.78}$$

$$\Delta H_{12}^a = \Delta H_{12}^b = 0, \quad H_{21}^a = H_{21}^b = 0 \tag{9.79}$$

$$H_{21}^a = \begin{bmatrix} 0 \\ 0 \\ (x_{11} + x_{12}) x_{13d_1} (\sin x_1 1 d_1|)^2 \Delta h_2(.) \end{bmatrix} \tag{9.80}$$

$$H_{21}^b = \begin{bmatrix} 4(x_{11d_1} + x_{12d_1})(\sin x13)^2 \Delta h_1(.) \\ 4(\sin^2(x_{13})) x 13 d_1 (\Delta h_1(.))^2 \\ 0 \end{bmatrix} \tag{9.81}$$

It is easy to check that **Assumption 9.2.1** is satisfied with

$$\begin{aligned} g_1(.) &= y_{11}^2 |y_{12d_1}|^3 \sin^2(y_1), \quad g_2(.) = |y_{21d_2}| y_{22}^2, \\ \alpha_{12}(.) &= 0, \quad \alpha_{21}(.) = |y_{11} y_{12d_1}| \\ \beta_{12}(.) &= 0, \quad \beta_{21}(.) = 4 \sin^2(y_{12}), \\ \rho_{12}(.) &= 0, \quad |rho_2 1(.) = \sqrt{16 \sin^4(y_{12}) + y_{11}^2} \end{aligned}$$

and **Assumption 9.2.2** holds. Algebraic manipulation shows that **Assumption 9.2.3** is satisfied with $K_1 = \tilde{K}_2 = 0$, and the coordinate transformation $z_j = T_j x_j,\ j = 1, 2$ is given by

$$T_1 = \begin{bmatrix} 0.7071 & 0.7071 & 0 \\ -1 & -1 & 0 \\ 0 & 0 & -1 \end{bmatrix}, \quad T_2 = \begin{bmatrix} 0.7071 & 0.7071 & 0 \\ -1 & -1 & 0 \\ 0 & 0 & -1 \end{bmatrix}$$

The sliding surface matrices are $F_1 = \begin{bmatrix} 0 & -1 \end{bmatrix}$ and $F_2 = \begin{bmatrix} 0 & -1 \end{bmatrix}$.

Let $Q_1 = Q_2 = I_2$. The corresponding solutions to the Lyapunov equations (9.59) are

$$P_1 = \begin{bmatrix} 0.0625 & 0 \\ 0 & 0.0625 \end{bmatrix}, \quad P_2 = \begin{bmatrix} 0.0833 & 0 \\ 0 & 0.0833 \end{bmatrix}$$

Further, let $\gamma_1 = \gamma_2 = 1.1$. By direct computation, it is easy to show that

$$N_1 = \begin{bmatrix} 0.9108 & -0.0083 \\ -0.0083 & 0.9049 \end{bmatrix}, \quad N_2 = \begin{bmatrix} 0.9375 & 0 \\ 0 & 0.9375 \end{bmatrix}$$

and

$$M = \begin{bmatrix} 0.8991 & 0 \\ -0.5185 \sin^2(y_{12}) y_{12} & 0.9375 \end{bmatrix}$$

It is easy to verify that the conditions in **Theorem 9.2.5** are satisfied globally. Accordingly, the sliding motion associated with the sliding surface

$$\{(x_{11}, x_{12}, x_{13}, x_{21}, x_{22}, x_{23}) | x_{13} = 0, x_{23} = 0\}$$

is globally uniformly asymptotically stable. Moreover, letting

$$\Gamma_1 = \begin{bmatrix} -8 & 2 \\ 1 & 0 \end{bmatrix} \text{ and } \Gamma_2 = \begin{bmatrix} -8 & amp; 2 \\ 2 & 2 \end{bmatrix}$$

It is quite evident that **Assumption 9.2.7** holds. Therefore, the decentralized control law in (9.76) stabilizes the interconnected system under consideration globally.

9.3 Problem Set VII

Problem VII.1: Consider the following uncertain interconnected time-delay system

$$\begin{aligned} \dot{x}_j &= A_j x_j + A_{dj} x_j(t - \tau_j(t)) + B_j(u_j + \xi_j(t, x_j, u_j)), \\ y_j &= C_j x_j, \quad j = 1, 2, ..., n_s \end{aligned}$$

where $x_j \in \Re^{n_j},\ u_j \in \Re^{m_j},\ y_j \in \Re^{p_j}$ are the state, input, and output vectors, respectively. The time-varying delay is bounded $0 \leq \tau_j(t) \leq d_j$. The matrices $B_j,\ C_j$ are of full rank and the unknown functions $\xi_j(.,.,.)$ are assumed to satisfy

$$||\xi_j(t, x_j, u_j)|| \ \leq \ \alpha_j ||u_j|| \ + \ \beta_j(t, y_j), \ \ \alpha_j < 1 \quad j = 1, 2, ..., n_s$$

for some known function $\beta_j(t, y_j)$. Assume that $rank(C_j B_j) = m_j$. Apply a similar procedure of Section 9.1 to derive a SMC based on the local output information that guarantees asymptotic stability of the closed-loop system.

Problem VII.2: We consider a nonlinear discrete-time large-scale system formed by n_s coupled subsystems as follows:

$$\begin{aligned} x_j(k+1) &= A_j x_j + f_j(x_j) + B_j(u_j + \Delta g_j(x_j)) + H_j(x), \\ y_j &= C_j x_j, \quad j = 1, 2, ..., n_s \end{aligned}$$

where, by similarity to the continuous-time case, $x = \mathrm{col}(x_1, x_2, ..., x_N), x_j \in \Re^{n_j}, u_j \in \Re^{m_j}, y_j \in \Re^{p_j}$ are the states, inputs, and outputs of the jth subsystem, respectively, and $m_j \leq p_j < n_j$. The triple (A_j, B_j, C_j) represents constant matrices of appropriate dimensions with B_j, and C_j is of full rank such that $rank(C_j,\ B_j) = m_j \quad for\ j = 1, 2, ..., n_s$. Assume that $f_j(x_j) = \Gamma_j(y_j) x_j,\ \Gamma_j \in \Re^{n_j \times n_j}$. Considering that there exist known continuous functions $\rho_j(\cdot)$ and $\eta_j(\cdot)$ such that for $i = l, 2, ..., n_s$

$$|\Delta g_j(x_j)| \leq \rho_j(y_j), \ \ |H_j(x)| \leq \eta_j(x)$$

where $\eta_j(\cdot)$ satisfies $\eta_j(x) \leq \beta_j(x)|x|$ for some continuous function β_j. Develop a complete analysis and feedback control design to render the closed-loop system asymptotically stable. Illustrate the basic differences with regards to the continuous-time case.

Problem VII.3: Consider interconnected discrete-time-delay system composed of n_s subsystems where the jth subsystem is described by

$$\begin{aligned} & x_j(k+1) = A_j x_j + B_j\Big(u_j + G_j(k, x_j, x_j(k - \tau_j))\Big) + \\ & \sum_{m=1\, m\neq j}^{n} \Big(H_{jm} y_m(t - \tau_m) + \Delta H_{jm}(k, x_m, x_m(k - \tau_m))\Big), \\ & y_j = C_j x_j, \quad j = 1, 2, ..., n_s \end{aligned}$$

where $x_j \in \Re^{n_j}$, $u_j \in \Re^{m_j}$, and $y_j \in \Re^{p_j}$ with $m_j \leq p_j < n_j$ being the state variables, inputs, and outputs of the ith subsystem, respectively. The triples (A_j, B_j, C_j) and $H_{jm} \in \Re^{n_j \times p_m}$ with $m \neq j$ represent constant matrices of appropriate dimensions with B_j and C_j of full rank. The function $G_j(.)$ is a matched nonlinearity in the jth subsystem. The terms

$$\sum_{m=1, m\neq j}^{n} H_{jm} y_m(t - \tau_m), \qquad \sum_{m=1, m\neq j}^{n} \Delta H_{jm}(k, x_m, x_m(t - \tau_m))$$

are, respectively, a known interconnection and an uncertain interconnection of the ith subsystem. The factor τ_m is the bounded time-varying delay. In addition, consider the decompositions of the interconnections:

$$\begin{aligned}
H_{jm} &= H^a_{jm} + H^b_{jm}, \\
\Delta H_{jm}(t) &= \Delta H^a_{jm}(k, x_k, x_m(k - \tau_m)), \\
&+ \Delta H^b_{jm}(k, x_m, x_m(k - \tau_m)), \\
H^a_{jm} &= B_j D_{jm}, \\
\Delta H^a_{jm}(k, x_m, x_m(k - \tau_m)) &= B_j \Delta\Theta_{jm}(k, x_m, x_m(k - \tau_m))
\end{aligned}$$

and assume that there exist known continuous functions $g_j(.)$, $\alpha_{jk}(.)$, and $\beta_{jk}(.)$ such that

$$\begin{aligned}
\|G_j(k, x_m, x_m(k - \tau_m))\| &\leq g_j(k, y_m, y_m(k - \tau_m)), \\
\|\Delta\Theta_{jm}(k, x_m, x_m(k - \tau_m))\| &\leq \alpha_{jm}(k, y_m, y_m(k - \tau_m))\|y_m(k - \tau_m)\|, \\
\|\Delta H^b_{jk}(k, x_m, x_m(k - \tau_m))\| &\leq \beta_{jm}(k, y_m, \|y_m(k - \tau_m)\|)\|y_m(k - \tau_m)\|
\end{aligned}$$

for $j \neq m$, $j, m = 1, 2, ..., n_s$ where $\beta_{jm}(., ., r)$ is nondecreasing with respect to the variable r in $\Re^+$. Develop a complete analysis and feedback control design to render the closed-loop system asymptotically stable. Illustrate the basic differences with regards to the continuous-time case.

Problem VII.4: A class of uncertain interconnected time-delay systems is described by

$$\dot{x}_j = A_j x_j + \sum_{k=1, k\neq j}^{n_s} A_{jk}(t) x_k(t - \tau_k(t)) + B_j(u_j + g_j(t, x_j, u_j, \sigma_j))$$

where $x_j \in \Re^{n_j}$, $u_j \in \Re^{m_j}$, $\sigma_j \in \Re^{p_j}$ are the state, input, and uncertain parameter vectors, respectively. The time-varying delay is bounded $0 \leq \tau_j(t) \leq \tau_M$. The matched perturbations are summarized by $g_j(t, x_j, u_j, \sigma_j) \in \Re^{m_j}$, the matrix pairs $(A_j,\ B_j),\ j = 1, ..., n_s$ are completely controllable. Let the delayed uncertainties be matched and satisfy the unknown functions $\xi_j(., ., .)$ as assumed to satisfy

$$A_{jk} = B_j G_{jk}(t), \quad ||G_{jk}(t)|| \leq \alpha_{jk}, \ \alpha_{jk} \geq 0, \quad j, k = 1, 2, ..., n_s$$

Consider the case where the lumped perturbations $g_j(t, x_j, u_j, \sigma_j)$ are bounded in the form

$$||g_j(t, x_j, u_j, \sigma_j)|| \leq \rho_j ||x_j(t)|| + \phi_j ||u_j(t)|| + \psi_j$$

In terms of the local switching surface

$$\Omega_j(t) = B_j^{\dagger} x_j(t) - \int_0^t (B_j^{\dagger} A_j + K_j) x_j(s) ds$$

where $\Omega_j(t)\Re^{m_j},\ K_j \in \Re^{M_j \times n_j},\ B_j^{\dagger} \in \Re^{M_j \times n_j}$ with $B_j^{\dagger}$ being the generalized inverse of B_j and the gain matrix K_j satisfies $\lambda_j(A_j + B_j K_j) < 0$. Apply a similar procedure of Section 9.2 to derive an appropriate SMC based on the local output information that guarantees asymptotic stability of the closed-loop system. Illustrate your results on the data:

$$\begin{aligned}
A_1 &= \begin{bmatrix} 0 & 1 \\ -2 & 1 \end{bmatrix}, \ B_1 = \begin{bmatrix} 0 \\ 1 \end{bmatrix}, \\
A_2 &= \begin{bmatrix} 0 & 1 \\ -3 & 2 \end{bmatrix}, \ B_2 = \begin{bmatrix} 0 \\ 1 \end{bmatrix}, \\
A_{12} &= \begin{bmatrix} 0 & 0 \\ sin(t) & 0.3cos(2t) \end{bmatrix}, \\
A_{21} &= \begin{bmatrix} 0 & 0 \\ 1 + sin(t) & 1 + 0.3cos(2t) \end{bmatrix}, \\
||g_1(t, x_1, u_1, \sigma_1)|| &\leq (0.5 + 0.2sin(t))||x_1(t) + 0.2cos(t)||u_1(t)|| + 0.4, \\
||g_2(t, x_2, u_2, \sigma_2)|| &\leq (1 - 2sin(3t))||x_2(t) + 0.2exp^{cos(t)}||u_2(t)|| \\
&\quad + 0.2sin(t), \\
\tau_1(t) &= (|0.4 - 0.1sin(t))|, \ \tau_2(t) = (|0.3 - 0.2cos(2t))|, \\
\tau_M &= 0.5, \\
\phi_1 &= 0.205, \ \phi_2 = 0.544
\end{aligned}$$

9.4 Notes and References

This chapter has adopted a sliding-mode control strategy to stabilize a class of nonlinear interconnected systems with mismatched uncertainty that uses only static output feedback. A composite sliding surface is first constructed and subsequently, a decentralized control scheme is synthesized. A reachability condition is developed for the interconnected system. In certain situations global results can also be obtained and it is shown that the results are applicable to a wide class of interconnected systems.

The methods and results have been given for systems in continuous-time domain. We observed that it requires high-speed discontinuous action to steer the states of a system into a sliding surface and to maintain subsequent motion along this surface. Applying a discrete-time sliding-mode approach would have the benefit of reducing the chattering phenomenon that arises in computer-control applications. It is suggested to pursue further research work on applying delta operator within sliding-mode control for interconnected systems. Extending the work to time-interval-delay systems will support the method presented in this chapter. Also, incorporating the results of [381, 382, 403] will yield improved development of sliding-mode control techniques for interconnected systems with and without time-delays.

Chapter 10

Decentralized Filtering-I

In this chapter, we continue our evaluation of the development of large-scale systems theories and techniques, where we focus herein on decentralized filtering and estimation techniques with particular considerations to robustness and resilience issues.

10.1 Introduction

For dynamical systems, the topic of robust filtering arose in view of the desire to determine estimates of nonmeasurable state variables when the system undergoes parametric uncertainties. From this perspective, robust filtering can be viewed as an extension of the celebrated Kalman filter to uncertain dynamical systems. The past decade has witnessed major developments in robust filtering problems [24, 75, 83, 278, 313, 375]. There have been two basic categories of approaches: the first is based on the Riccati equation (RE) through the search of appropriate scaling parameters and the other relies on feasibility testing of linear matrix inequalities (LMIs) formulation. The later approach has gained widespread popularity in view of the development of the interior point algorithm for convex optimization. When dealing with large-scale or interconnected systems, the topic becomes more compounded as one has to be careful with the issues of dimensionality, nonlinearities, and/or robustness. For a constructive approach to decentralized control of large-scale systems, see the results reported in [324, 328]. It seems that the topic of decentralized filtering receives little attention.

In another research front and during the course of filter implementation based on different design algorithms, it turns out that the filters can be sensitive with respect to errors in the filter coefficients [60, 257, 264, 387]. The sources for this include, but are not limited to, imprecision in analogue-digital conversion, fixed word length, finite resolution instrumentation, and numerical roundoff errors. Similar concern was raised earlier when using observers [61].

The objective of this section is to address two problems for a class of interconnected discrete-time nonlinear systems. The first problem deals with robust decentralized filtering and the second problem studies resilient decentralized filtering. It is needless to stress that the phrase "resilient" implies robustness with respect to plant parametric uncertainties and against gain perturbations in filter matrices. Specifically, the robust ℓ_∞ filtering design problem is treated for a class of interconnected discrete-time systems with uncertain function of nonlinear perturbations by exploiting the decentralized information structure constraint. We develop LMIs-based conditions for designing the robust decentralized filters. Then by considering additive filter gain perturbations, we solve the resilient decentralized filtering and express the results as convex minimization over LMIs. Simulation examples are provided to illustrate the theoretical developments.

10.2 Problem Statement

A class of nonlinear interconnected discrete-time systems with state-delay Σ composed of N coupled subsystems $\Sigma_j,\ j = 1, .., n_s$ is represented by:

$$\Sigma: \ x(k+1) = Ax(k) + g(k, x(k)) + Bw(k) \tag{10.1}$$
$$y(k) = Cx(k) + Dw(k) \tag{10.2}$$
$$z(k) = Lx(k) \tag{10.3}$$

where $k \in Z_+ := \{0, 1, ...\}$, $x = (x_1^t, ..., x_{n_s}^t)^t \in \Re^n$, $n = \sum_{j=1}^{n_s} n_j$ is the interconnected system state, $y = (y_1^t, ..., y_{n_s}^t)^t \in \Re^m$, $m = \sum_{j=1}^{n_s} m_j$ is the measured output of the interconnected system, $z = (z_1^t, ..., z_{n_s}^t)^t \in \Re^p$, $p = \sum_{j=1}^{n_s} p_j$ is the vector of state combination to be estimated, $w = (w_1^t, ..., w_{n_s}^t)^t \in \Re^p$, $p = \sum_{j=1}^{n_s} p_j$ is a vector with zero-mean input noise and having unity power spectrum density matrix. The initial condition $x(0)$ is a zero-mean random variable uncorrelated with $w(k)$ for all $k \geq 0$. The associated matrices are real constants and modeled as $A = diag\{A_1, .., A_{n_s}\}$, $A_j \in \Re^{n_j \times n_j}$, $B = diag\{B_1, .., B_{n_s}\}$, $B_j \in \Re^{n_j \times p_j}$, $C = diag\{C_1, .., C_{n_s}\}$, $C_j \in \Re^{m_j \times n_j}$, $D = diag\{D_1, .., D_{n_s}\}$, $D_j \in \Re^{m_j \times p_j}$, $L = diag\{L_1, .., L_{n_s}\}$, $L_j \in \Re^{p_j \times n_j}$, which describe the nominal system. The function $g : Z_+ \times \Re^n \times \Re^n \rightarrow \Re^{n_j}$, $g(k, x(k)) = (g_1^t(k, x(k)), .., g_{n_s}^t(k, x(k)))^t$ is a vector function piecewise-continuous in its arguments. In the sequel, we assume that this function is uncertain and the available information is that, in the domains of continuity **G**, it satisfies the quadratic inequality

$$g^t(k,x(k))g(k,x(k)) \leq x^t(k)\widetilde{G}^t\widetilde{\Phi}^{-1}\widetilde{G}x(k) \tag{10.4}$$

where $\widetilde{G} = [\widetilde{G}_1^t, .., \widetilde{G}_N^t]^t$, $\widetilde{G}_j \in \Re^{r_j \times n}$ are constant matrices such that $g_j(k,0,0) = 0$ and $x = 0$ is an equilibrium of system (10.1).

Exploiting the structural form of system (10.1), the jth subsystem model Σ_j can be described by

$$\begin{aligned} \Sigma_j: \ x_j(k+1) &= A_j x_j(k) + B_j w_j(k) + g_j(k,x(k)) \\ y_j(k) &= C_j x_j(k) + D_j w(k) \\ z_j(k) &= L_j x_j(k) \end{aligned} \tag{10.5}$$

where $x_j(k) \in \Re^{n_j}$, $w_j(k) \in \Re^{p_j}$, $z_j(k) \in \Re^{p_j}$, and $y_j(k) \in \Re^{m_j}$ are the subsystem state, disturbance input, linear combination of states, and measured output, respectively. The function $g_j : \mathsf{Z}_+ \times \Re^n \times \Re^n \to \Re^{n_j}$ is a piecewise-continuous vector function in its arguments and in line of (10.4) it satisfies the quadratic inequality

$$g_j^t(k,x(k))\ g_j(k,x(k)) \leq \phi_j^2\ x^t(k)\widetilde{G}_j^t\widetilde{G}_j x(k) \tag{10.6}$$

where $\phi_j > 0$ are bounding parameters such that $\widetilde{\Phi} = diag\{\phi_1^{-2}I_{r_1}, .., \phi_{n_s}^{-2}I_{r_{n_s}}\}$ where I_{m_j} represents the $m_j \times m_j$ identity matrix. From (10.4) and (10.6), it is always possible to find matrices Φ such that

$$g^t(k,x(k))g(k,x(k)) \ \leq \ x^t(k)G^t\Phi^{-1}Gx(k) \tag{10.7}$$

where $G = diag\{G_1, .., G_{n_s}\}$, $\Phi = diag\{\delta_1 I_{r_1}, .., \delta_{n_s} I_{r_{n_s}}\}$, and $\delta_j = \phi_j^{-2}$.

Our objective in this chapter is to develop new tools for the linear filter design of a class of interconnected discrete-time systems with uncertain function of nonlinear perturbations (10.4)-(10.7) by exploiting the decentralized information structure constraint. We seek to establish complete LMI-based procedures for the linear decentralized filtering by basing all the computations at the subsystem level.

10.3 Robust Subsystem Filters

In the filtering problem considered hereafter, we consider the feasible set C_f as the set of all linear shift-invariant operators with state-space realization at the

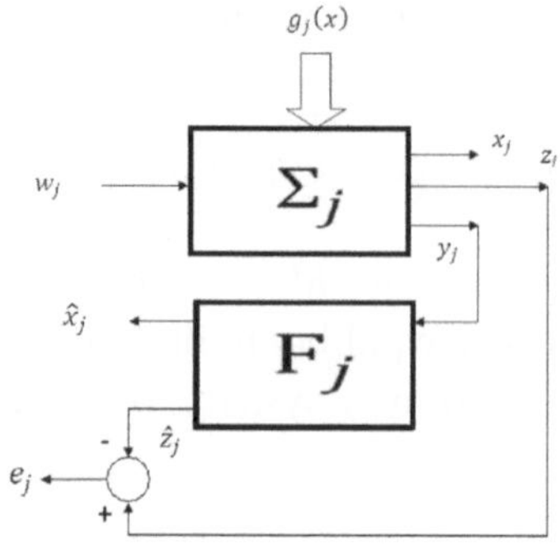

Figure 10.1: Filtered subsystem model

subsystem level of the form:

$$\begin{aligned} \hat{x}_j(k+1) &= E_j\hat{x}_j(k) + F_j\, y_j(k), \qquad \hat{x}(0) = 0 \\ \hat{z}_j(k) &= H_j\hat{x}_j(k) \end{aligned} \tag{10.8}$$

where the matrices $E_j \in \Re^{s_j \times s_j}$, $F_j \in \Re^{s_j \times m_j}$, and $H_j \in \Re^{p_j \times s_j}$ and the scalar $s_j > 0$ are the design parameters. Recall from (10.1)-(10.3) and (10.8) that all matrices are shift-invariant since we are focusing on the stationary case. Connecting the filter (10.8) to the subsystem (10.5), we obtain the filtering error subsystem

$$\begin{aligned} \zeta_j(k+1) &= \tilde{A}_j\zeta_j(k) + \tilde{B}_j w_j(k) + \tilde{g}_j(k, x(k)) \\ e_j(k) &= \tilde{L}_j\zeta_j(k) \end{aligned} \tag{10.9}$$

where the subsystem estimation error $e_j(k) := z_j(k) \; - \; \hat{z}_j(k)$ and

$$\begin{aligned} \tilde{g}_j(k, x(k)) \; &:= \; \begin{bmatrix} g_j(k, x(k)) \\ 0 \end{bmatrix}, \; \zeta_j := \begin{bmatrix} x_j \\ \hat{x}_j \end{bmatrix}, \; \tilde{L} := [L_j \quad - H_j] \\ \tilde{A}_j \; &:= \; \begin{bmatrix} A_j & 0 \\ F_jC_j & E_j \end{bmatrix}, \; \tilde{B} := \begin{bmatrix} B_j \\ F_jD_j \end{bmatrix} \end{aligned} \tag{10.10}$$

A schematic diagram of the filtered subsystem is displayed in Figure 10.1.

10.3.1 Robust ℓ_∞ filter

Considering the filtering error subsystem (10.9), our goal in this part is to establish tractable conditions for designing the subsystem filter to guarantee the global asymptotic stability of the origin ($x = 0$) for all $g(k, x(k)) \in \mathbf{G}$ and achieving induced ℓ_2 disturbance attenuation γ according to the following definition.

Definition 10.3.1 *Given $\gamma_j > 0$, the filtering error subsystem (10.10) is said to be robustly asymptotically stable with an induced ℓ_2 disturbance attenuation γ if it is asymptotically stable for all admissible uncertainties satisfying (10.7) and under zero initial conditions $||e_j(k)||_2 \;<\; \gamma\; ||w_j(k)||_2$ for all nonzero $w_j(k) \in \ell_2[0,\infty)$.*

The main result of subsystem filter design is established by the following theorem.

Theorem 10.3.2 *The filtering error subsystem (10.10) is robustly asymptotically stable for all nonlinear uncertainties satisfying (10.4) if there exist matrices $0 < \mathcal{Y}_{aj}^t = \mathcal{Y}_{aj} \in \Re^{n_j \times n_j}$, $0 < \mathcal{Y}_{oj}^t = \mathcal{Y}_{oj} \in \Re^{s_j \times s_j}$, $\mathcal{Y}_{cj} \in \Re^{s_j \times s_j}$, $\mathcal{Y}_{ej} \in \Re^{s_j \times n_j}$, $\mathcal{Y}_{sj} \in \Re^{s_j \times m_j}$, $H_j \in \Re^{p_j \times s_j}$ and scalars $\delta_j > 0$, $\gamma_j > 0$ such that the following convex optimization problem is feasible:*

$$\min_{\mathcal{Y}_{aj},\mathcal{Y}_{oj},\mathcal{Y}_{cj},\mathcal{Y}_{ej},\mathcal{Y}_{sj},H_j,\delta_j} \quad \sum_{j=1}^{N} \delta_j + \gamma_j$$

$$subject\ to: \ \{\mathcal{Y}_{aj} > 0,\ \mathcal{Y}_{oj} > 0\}_{j=1}^{N},$$

$$diag\Big(\Gamma_1, ..., \Gamma_N\Big) < 0,$$

$$\Gamma_j = \begin{bmatrix} -\Psi_{1j} & \Psi_{2j} & 0 & \Psi_{2j} & \Psi_{4j} & \Psi_{5j} \\ \bullet & -\Psi_{6j} & \Psi_{3j} & 0 & 0 & 0 \\ \bullet & \bullet & -\gamma_j^2 I_j & \Psi_{3j}^t & 0 & 0 \\ \bullet & \bullet & \bullet & -\Psi_{1j} & 0 & 0 \\ \bullet & \bullet & \bullet & \bullet & -\delta_j I_j & 0 \\ \bullet & \bullet & \bullet & \bullet & \bullet & -I_j \end{bmatrix} \qquad (10.11)$$

where

$$\Psi_{1j} = \begin{bmatrix} \mathcal{Y}_{aj} & 0 \\ \bullet & \mathcal{Y}_{oj} \end{bmatrix}, \ \Psi_{2j} = \begin{bmatrix} A_j^t \mathcal{Y}_{aj} & \mathcal{Y}_{ej}^t \\ 0 & \mathcal{Y}_{cj}^t \end{bmatrix}, \ \Psi_{3j} = \begin{bmatrix} \mathcal{Y}_{aj} B_j \\ \mathcal{Y}_{sj} D_j \end{bmatrix},$$

$$\Psi_{4j} = \begin{bmatrix} G_j^t \\ \mathcal{Y}_{sj} D_j \end{bmatrix}, \ \Psi_{5j} = \begin{bmatrix} L_j^t \\ -H_j^t \end{bmatrix},$$

$$\Psi_{6j} = \begin{bmatrix} I_j - \mathcal{Y}_{aj} & 0 \\ \bullet & I_j - \mathcal{Y}_{oj} \end{bmatrix} \qquad (10.12)$$

Moreover, the filter gains are given by $E_j = \mathcal{Y}_{oj}^{-1} \mathcal{Y}_{cj}$, $F_j = \mathcal{Y}_{oj}^{-1} \mathcal{Y}_{sj}$, H_j.

Proof: Introduce the Lyapunov functional (LF):

$$V(k) = \sum_{j=1}^{N} V_j(k) = \sum_{j=1}^{N} \left[\zeta_j^t(k) \mathcal{P}_j \zeta_j(k) \right] \tag{10.13}$$

where $0 < \mathcal{P}_j^t = \mathcal{P}_j \in \Re^{(n_j+s_j)\times(n_j+s_j)}$ are weighting matrices for $j = 1,..,N$. A straightforward computation gives the first-difference of $\Delta V(k) = V(k+1) - V(k)$ along the solutions of (10.9) as:

$$\begin{aligned}
\Delta V(k) &= \sum_{j=1}^{N} [\tilde{A}_j \zeta_j(k) + \tilde{B}_j w_j(k) + \tilde{g}_j(k, x(k))]^t \mathcal{P}_j \\
& \quad [\tilde{A}_j \zeta_j(k) + \tilde{B}_j w_j(k) + \tilde{g}_j(k, x(k))] \\
& - \sum_{j=1}^{N} \zeta_j^t(k) \mathcal{P}_j \zeta_j(k) \ := \ \xi^t(k)\, \Xi \, \xi(k) \qquad (10.14) \\
\Xi &= diag(\Xi_1, ..., \Xi_N), \ \Xi_j = \begin{bmatrix} \tilde{A}_j^t \mathcal{P}_j \tilde{A}_j - \mathcal{P}_j & \tilde{A}_j^t \mathcal{P}_j \tilde{B}_j & \tilde{A}_j^t \mathcal{P}_j \\ \bullet & \tilde{B}_j^t \mathcal{P}_j \tilde{B}_j & \tilde{B}_j^t \mathcal{P}_j \\ \bullet & \bullet & \mathcal{P}_j \end{bmatrix}, \\
\xi(k) &= [\xi_1^t(k), ..., \xi_N^t(k)]^t, \\
\xi_j^t(k) &= \left[\ \zeta_j^t(k) \quad w_j^t(k) \quad \tilde{g}_j^t(k, x(k)) \ \right] \qquad (10.15)
\end{aligned}$$

In terms of $\tilde{G}_j = [G_j \quad 0]$ and recalling that (10.7) can be conveniently written as

$$\xi^t \, diag\left(\begin{bmatrix} -\phi_1^2 \, \tilde{G}_1^t \tilde{G}_1 & 0 & 0 \\ \bullet & 0 & 0 \\ \bullet & \bullet & I_1 \end{bmatrix}, ..., \begin{bmatrix} -\phi_N^2 \, \tilde{G}_N^t \tilde{G}_N & 0 & 0 \\ \bullet & 0 & 0 \\ \bullet & \bullet & I_N \end{bmatrix} \right) \xi \leq 0 \tag{10.16}$$

then the sufficient condition of stability $\Delta V(k) < 0$ implies that $\Xi < 0$. By resorting to the **S**-procedure, inequalities (10.14) and (10.16) can be rewritten together as

$$\begin{aligned}
& \mathcal{P}_1 > 0, ..., \mathcal{P}_N > 0, \ \omega_1 \geq 0, ..., \omega_N \geq 0, \\
& diag\left(\Pi_1, ..., \Pi_N \right) < 0, \\
& \Pi_j = \begin{bmatrix} \tilde{A}_j^t \mathcal{P}_j \tilde{A}_j - \mathcal{P}_j + \omega_j \phi_j^2 \tilde{G}_j^t \tilde{G}_j & \tilde{A}_j^t \mathcal{P}_j \tilde{B}_j & \tilde{A}_j^t \mathcal{P}_j \\ \bullet & \tilde{B}_j^t \mathcal{P}_j \tilde{B}_j & \tilde{B}_j^t \mathcal{P}_j \\ \bullet & \bullet & \mathcal{P}_j - \omega_j I_j \end{bmatrix}
\end{aligned} \tag{10.17}$$

which describes nonstrict LMIs since $\omega_j \geq 0$. Recall from [29] that minimization under nonstrict LMIs corresponds to the same result as minimization under strict LMIs when both strict and nonstrict LMI constraints are feasible. Moreover, if there is a solution for (10.17) for $\omega_j = 0$, there will be also a solution for some $\omega_j > 0$ and sufficiently small ϕ_j. Therefore, we safely replace $\omega_j \geq 0$ by $\omega_j > 0$. Equivalently, with $\mathcal{Y}_j = \omega_j^{-1}\,\mathcal{P}_j$ and multiplying by ω_j^{-1}, we may further rewrite (10.17) in the form

$$\mathcal{Y}_1 > 0, ..., \mathcal{Y}_N > 0, \quad diag\left(\bar{\Pi}_1, ..., \bar{\Pi}_N\right) < 0,$$

$$\bar{\Pi}_j = \begin{bmatrix} \tilde{A}_j^t \mathcal{Y}_j \tilde{A}_j - \mathcal{Y}_j + \phi_j^2 \tilde{G}_j^t \tilde{G}_j & \tilde{A}_j^t \mathcal{Y}_j \tilde{B}_j & \tilde{A}_j^t \mathcal{Y}_j \\ \bullet & \tilde{B}_j^t \mathcal{Y}_j \tilde{B}_j & \tilde{B}_j^t \mathcal{Y}_j \\ \bullet & \bullet & \mathcal{Y}_j - I_j \end{bmatrix} \tag{10.18}$$

Consider the ℓ_2-performance measure

$$J = \sum_{j=1}^{N} J_j = \sum_{j=1}^{N} \sum_{m=0}^{\infty} \left(e_j^t(m) e_j(m) - \gamma^2 w_j^t(m) w_j(m) \right)$$

For any $w_k \in \ell_2(0, \infty) \neq 0$ and zero initial condition $x_o = 0$ (hence $V_j(0) = 0$), we have

$$\begin{aligned} J_j &= \sum_{m=0}^{\infty} \left(e_j^t(m) e_j(m) - \gamma^2 w_j^t(m) w_j(m) \ + \Delta V_j|_{(10.9)} \right) \\ &\quad - \sum_{j=0}^{\infty} \Delta V_j|_{(10.9)} \\ &= \sum_{j=0}^{\infty} \left(e_j^t(m) e_j(m) - \gamma^2 w_j^t(m) w_j(m) \ + \Delta V_j|_{(10.9)} \right) \ - \ V_j(\infty) \\ &\leq \sum_{j=0}^{\infty} \left(e_j^t(m) e_j(m) - \gamma^2 w_j^t(m) w_j(m) \ + \Delta V_j|_{(10.9)} \right) \end{aligned} \tag{10.19}$$

where $\Delta V_j|_{(10.9)}$ defines the Lyapunov difference along the solutions of system (10.9). By combining (10.9) and (10.19), it can easily be shown by algebraic manipulation that

$$\begin{aligned} & e_j^t(m)e_j(m) - \gamma^2 w_j^t(m)w_j(m) \; + \Delta V_j|_{(10.9)} = \\ & \zeta_j^t(k)\tilde{L}_j^t\tilde{L}_j\zeta_j(k) - \gamma^2 w_j^t w_j \; + \xi_j^t \,\bar{\Pi}_j\, \xi_j \; := \; \xi_k^t \,\widehat{\Pi}_j\, \xi_k^t \\ & \widehat{\Pi}_j = \\ & \begin{bmatrix} \tilde{A}_j^t \mathcal{Y}_j \tilde{A}_j - \mathcal{Y}_j + \phi_j^2 \tilde{G}_j^t \tilde{G}_j + \tilde{L}_j^t \tilde{L}_j & \tilde{A}_j^t \mathcal{Y}_j \tilde{B}_j & \tilde{A}_j^t \mathcal{Y}_j \\ \bullet & \tilde{B}_j^t \mathcal{Y}_j \tilde{B}_j - \gamma^2 I_j & \tilde{B}_j^t \mathcal{Y}_j \\ \bullet & \bullet & \mathcal{Y}_j - I_j \end{bmatrix} \end{aligned} \tag{10.20}$$

Focusing on $\widehat{\Pi}_j$, using the change of variable $\delta_j = \phi_j^{-2}$ and followed by Schur complement operations we arrive at

$$\overline{\Pi}_j = \begin{bmatrix} -\mathcal{Y}_j & \tilde{A}_j^t \mathcal{Y}_j & 0 & \tilde{A}_j^t \mathcal{Y}_j & \tilde{G}_j^t & \tilde{L}_j^t \\ \bullet & \mathcal{Y}_j - I_j & \mathcal{Y}_j \tilde{B}_j & 0 & 0 & 0 \\ \bullet & \bullet & -\gamma^2 I_j & \tilde{B}_j^t \mathcal{Y}_j & 0 & 0 \\ \bullet & \bullet & \bullet & -\mathcal{Y}_j & 0 & 0 \\ \bullet & \bullet & \bullet & \bullet & -\delta_j I_j & 0 \\ \bullet & \bullet & \bullet & \bullet & \bullet & -I_j \end{bmatrix} \tag{10.21}$$

Introducing the change of variables

$$\begin{aligned} \mathcal{Y}_j &= \begin{bmatrix} \mathcal{Y}_{aj} & 0 \\ \bullet & \mathcal{Y}_{oj} \end{bmatrix}, \; \mathcal{Y}_{cj} = \mathcal{Y}_{oj} E_j, \\ \mathcal{Y}_{ej} &= \mathcal{Y}_{oj} F_j C_j, \; \mathcal{Y}_{sj} = \mathcal{Y}_{oj} F_j \end{aligned} \tag{10.22}$$

to (10.18), we obtain LMI (10.11) which establishes the internal asymptotic stability with ℓ_2 disturbance attenuation, and eventually leads to $||e_j(k)||_2 < \gamma||w_j(k)||_2$ for all nonzero $w_j(k) \in \ell_2[0,\infty)$. Therefore, the proof is completed.

Remark 10.3.3 *It should be noted that the family of N linear filters in the form (10.8) guarantees the decentralized reproduction of the linear combination of the interconnected system's state with disturbance attenuation level* $\sum_{j=1}^{N} \delta_j$ *under the information structure constraint (10.7) with maximal* ϕ_j. *Observe also that the filter order is* $1 < s_j \leq n_j$ *and thus covers both the full- and reduced-order cases. The performance of these subsystem filters will be demonstrated in the simulation studies.*

10.3.2 Illustrative example 10.1

In order to illustrate the theoretical results thus far, we consider a model of the type (10.1)-(10.3) with the following data:

$$A = \begin{bmatrix} -0.8 & 0 & 0.1 & 0 & 0 & 0 \\ 0 & 0.7 & 0.1 & 0 & 0 & 0 \\ -0.1 & 0 & 0.1 & 0 & 0 & 0 \\ 0 & 0 & 0 & -0.6 & 0 & 0.2 \\ 0 & 0 & 0 & 0 & -0.5 & 0.1 \\ 0 & 0 & 0 & -0.3 & 0 & 0.2 \end{bmatrix}, \; C^t = \begin{bmatrix} 1 & 0 & 0 & 0 \\ 1 & 0 & 0 & 0 \\ 0 & 1 & 0 & 0 \\ 0 & 0 & 1 & 0 \\ 0 & 0 & 1 & 0 \\ 0 & 0 & 0 & 1 \end{bmatrix}$$

$$B = \begin{bmatrix} 0.15 & 0 \\ 0.25 & 0 \\ 0.1 & 0 \\ 0 & 0.3 \\ 0 & 0.1 \\ 0 & 0.2 \end{bmatrix}, \; D = \begin{bmatrix} 0.1 & 0 \\ 0.1 & 0 \\ 0 & 0.2 \\ 0 & 0.2 \end{bmatrix}, \; L^t = \begin{bmatrix} 0.4 & 0 \\ 0.3 & 0 \\ 0.2 & 0 \\ 0 & 0.1 \\ 0 & 0.2 \\ 0 & 0.3 \end{bmatrix}$$

$$g(k, x(k)) = \begin{bmatrix} 0 & 0 & -1 & 0 & 0 & 0 \\ 0 & 1 & 0 & 0 & 0 & 0 \\ 1 & 0 & 0 & 0 & 0 & -1 \\ 0 & 0 & 1 & 0 & 0 & 0 \\ 0 & -1 & 0 & 0 & 0 & 0 \\ -1 & 0 & 0 & 0 & 0 & 1 \end{bmatrix} e(k, x(k))x,$$

$$e(k, x(k)) = \Re^7 \to [0, 0.6]$$

In the numerical computations of the robust ℓ_∞ filter, we considered three cases and the ensuing results are given by

$$s_1 = 3, \; s_2 = 3 \; (Full(third) - order filter),$$

$$\delta_1 = 0.1559, \quad \gamma_1 = 1.9865, \quad \delta_2 = 0.1558, \quad \gamma_2 = .9865$$

$$E_1 = \begin{bmatrix} 0.0154 & 0.0171 & 0.0056 \\ 0.0254 & 0.2241 & -0.0460 \\ 0.0093 & -0.1142 & 0.3236 \end{bmatrix}, \; F_1 = \begin{bmatrix} 0.7355 & 0.1328 \\ 0.1444 & -0.5201 \\ 0.8498 & -0.4101 \end{bmatrix},$$

$$H_1 = \begin{bmatrix} 0.5048 & -0.0381 \\ 0.6740 & -0.0444 \\ -0.5714 & 0.1014 \end{bmatrix},$$

$$E_2 = \begin{bmatrix} 0.0154 & 0.0171 & 0.0056 \\ 0.0254 & 0.2241 & -0.0460 \\ 0.0093 & -0.1142 & 0.3236 \end{bmatrix}, \; F_2 = \begin{bmatrix} 0.7355 & 0.1328 \\ 0.1444 & -0.5201 \\ 0.8498 & -0.4101 \end{bmatrix},$$

$$
\begin{aligned}
H_2 &= \begin{bmatrix} 0.5048 & -0.0381 \\ 0.6740 & -0.0444 \\ -0.5714 & 0.1014 \end{bmatrix}, \\
s_1 &= 2, \;\; s_2 = 2 \;\; (Reduced(second) - order filter), \\
\delta_1 &= 0.4973, \;\; \gamma_1 = 2.6485, \;\; \delta_2 = 0.4973, \;\; \gamma_2 = 2.6485 \\
E_1 &= \begin{bmatrix} 0.1309 & 0.0321 \\ 0.0418 & 0.2378 \end{bmatrix}, \; F_1 = \begin{bmatrix} 0.6438 & 0.1277 \\ 0.4882 & -0.1305 \end{bmatrix}, \\
H_1 &= \begin{bmatrix} 0.8674 & -0.0472 \\ -0.7364 & 0.0887 \end{bmatrix},
\end{aligned}
$$

$$
\begin{aligned}
E_2 &= \begin{bmatrix} 0.1874 & 0.0574 \\ 0.1134 & -0.2630 \end{bmatrix}, \; F_2 = \begin{bmatrix} 0.8765 & 0.2008 \\ 0.5648 & -0.7101 \end{bmatrix}, \\
H_2 &= \begin{bmatrix} 0.9106 & -0.3811 \\ 0.8840 & -0.9614 \end{bmatrix}, \\
s_1 &= 1, \;\; s_2 = 1 \;\; (Reduced(first) - order filter), \\
\delta_1 &= 0.7384, \;\; \gamma_1 = 4.5724, \;\; \delta_2 = 0.7384, \;\; \gamma_2 = 4.5724 \\
E_1 &= [0.1352], \; F_1 = \begin{bmatrix} 0.6948 & -0.6603 \end{bmatrix}, \; H_1 = \begin{bmatrix} 0.4483 \\ -0.7455 \end{bmatrix}, \\
E_2 &= [0.2241], \; F_2 = \begin{bmatrix} 0.7782 & -0.5301 \end{bmatrix}, \; H_2 = \begin{bmatrix} 0.3877 \\ -0.8013 \end{bmatrix}
\end{aligned}
$$

From the ensuing results, we observe that the full-order filtered subsystem performs better than the reduced-order filtered subsystems in the sense of yielding smaller values of $\gamma-$ and $\delta-$levels. This is quite expected in view of the additional degrees of freedom the full-order filter has over the reduced ones.

10.4 Resilient Subsystem Filters

In this section, we are going to study the issue of filter implementation. It turns out that implementing the designed filter (10.8) for a single subsystem, much in line with the controller design counterpart, one might encounter inaccuracies or parameter perturbations that eventually lead to the resilient filtering problem [258]. Indeed, the problem is further compounded for the case of an interconnected system. In this regard, this problem has been overlooked in the literature on discrete-time systems and/or interconnected systems. For the continuous-time case, Kalman filtering with respect to estimator gain perturbations that was considered in [387], linear resilient filtering was treated in [257] and resilient

L_2/L_∞ filtering for state-delay polytopic systems was developed in [264]. In the sequel, we generalize the result of [257, 258, 387] in different ways by considering the resilient linear filtering of discrete-time interconnected systems with respect to perturbations in the estimator gains and cast the filter design conditions in computationally tractable LMI forms.

Incorporating the additive perturbations in the filter gains, we consider in the sequel the feasible set $\mathcal{C}_f$ as the set of all resilient shift-invariant operators with state-space realization at the subsystem level of the form:

$$\begin{aligned} \hat{x}_j(k+1) &= [E_j + \Delta E_j(k)]\hat{x}_j(k) + [F_j + \Delta F_j(k)]\; y_j(k), \\ &= E_{j\Delta}\hat{x}(k) + F_{j\Delta}\; y(t), \qquad \hat{x}_j(0) = 0 \qquad (10.23) \\ \hat{z}_j(k) &= [H_j + \Delta H_j(k)]\hat{x}_j(k) \;=\; H_{j\Delta}\hat{x}(k) \qquad (10.24) \end{aligned}$$

where the matrices $E_j,\ F_j,\ H_j$ are the design parameters as introduced before and the additive gain perturbations are represented by

$$\begin{aligned} &\begin{bmatrix} \Delta E_j(k) \\ \Delta H_j(k) \end{bmatrix} = \begin{bmatrix} M_{aj} \\ M_{cj} \end{bmatrix} \Delta_j(k)\; N_{aj},\;\; \Delta F_j(k) = M_{aj}\Delta_j(k)\; N_{cj}, \\ &\Delta_j^t(k)\; \Delta_j(k) \;\leq\; I_j, \;\; \forall\; k \qquad (10.25) \end{aligned}$$

with the matrices $M_{aj} \in \Re^{s_j \times \alpha_j},\ N_{aj} \in \Re^{\beta_j \times s_j},\ M_{cj} \in \Re^{p_j \times \alpha_j},\ \Delta_j \in \Re^{\alpha_j \times \beta_j}$, and $N_{cj} \in \Re^{\beta_j \times m_j}$ as constants.

Connecting the filter (10.23)-(10.24) to the system (10.8), we can express the filtering error subsystem as follows:

$$\begin{aligned} \zeta_j(k+1) &= \tilde{A}_{j\Delta}\zeta_j(k) + \tilde{B}_{j\Delta}w_j(k) + \tilde{g}_j(k, x(k)) \\ e_j(k) &= \tilde{L}_{j\Delta}\zeta_j(k) \qquad (10.26) \end{aligned}$$

where the subsystem estimation error $e_j(k) := z_j(k) \;-\; \hat{z}_j(k)$ and

$$\begin{aligned} \tilde{A}_{j\Delta} \;&:=\; \begin{bmatrix} A_j & 0 \\ F_{j\Delta}C_j & E_{j\Delta} \end{bmatrix}, \\ &= \begin{bmatrix} A_j & 0 \\ F_jC_j & E_j \end{bmatrix} + \begin{bmatrix} C_j^tN_{cj}^t \\ N_{aj}^t \end{bmatrix} \Delta_j^t(k) \begin{bmatrix} 0 & M_{aj}^t\mathcal{Y}_{oj} \end{bmatrix}, \\ &= \begin{bmatrix} A_j & 0 \\ F_jC_j & E_j \end{bmatrix} + \alpha_j\Delta_j^t(k)\beta_j^t, \\ \tilde{B}_{j\Delta} \;&:=\; \begin{bmatrix} B_j \\ F_{j\Delta}D_j \end{bmatrix} = \begin{bmatrix} B_j \\ F_jD_j \end{bmatrix} + \begin{bmatrix} 0 \\ M_{aj} \end{bmatrix} \Delta_j(k)N_{cj}D_j, \end{aligned}$$

$$
\begin{aligned}
\tilde{L}_{j\Delta} &= [L_j \quad - H_{j\Delta}] = [L_j \quad - H_j] + M_{cj}\Delta_j(k)[0 \quad - N_{aj}], \\
&= [L_j \quad - H_{j\Delta}] = [L_j \quad - H_j] + M_{cj}\Delta_j(k)\pi_j^t, \\
\mathcal{Y}_j\tilde{B}_{j\Delta} &= \begin{bmatrix} \mathcal{Y}_{aj}B_j \\ \mathcal{Y}_{oj}F_jD_j \end{bmatrix} + \begin{bmatrix} 0 \\ \mathcal{Y}_{oj}M_{aj} \end{bmatrix} \Delta_j(k)N_{cj}D_j, \\
&= \begin{bmatrix} \mathcal{Y}_{aj}B_j \\ \mathcal{Y}_{oj}F_jD_j \end{bmatrix} + \beta_j\Delta_j(k)N_{cj}D_j
\end{aligned} \tag{10.27}
$$

10.4.1 Resilient ℓ_∞ filter

Considering the filtering error subsystem (10.26), our immediate task is to establish tractable conditions for designing the subsystem filter to achieve induced ℓ_2 disturbance attenuation γ according to **Definition 10.3.1** for all possible gain perturbations (10.25). It follows from the foregoing analysis and in particular (10.19) that the filtering error subsystem (10.26) has ℓ_2 disturbance attenuation level γ_j provided that the following LMI,

$$
\begin{aligned}
\overline{\Pi}_{j\Delta} &= \begin{bmatrix} -\mathcal{Y}_j & \tilde{A}_{j\Delta}^t\mathcal{Y}_j & 0 & \tilde{A}_{j\Delta}^t\mathcal{Y}_j & \tilde{G}_j^t & \tilde{L}_{j\Delta}^t \\ \bullet & \mathcal{Y}_j - I_j & \mathcal{Y}_j\tilde{B}_{j\Delta} & 0 & 0 & 0 \\ \bullet & \bullet & -\gamma^2 I_j & \tilde{B}_{j\Delta}^t\mathcal{Y}_j & 0 & 0 \\ \bullet & \bullet & \bullet & -\mathcal{Y}_j & 0 & 0 \\ \bullet & \bullet & \bullet & \bullet & -\delta_j I_j & 0 \\ \bullet & \bullet & \bullet & \bullet & \bullet & -I_j \end{bmatrix} \\
&= \begin{bmatrix} -\mathcal{Y}_j & \tilde{A}_j^t\mathcal{Y}_j & 0 & \tilde{A}_j^t\mathcal{Y}_j & \tilde{G}_j^t & \tilde{L}_j^t \\ \bullet & \mathcal{Y}_j - I_j & \mathcal{Y}_j\tilde{B}_j & 0 & 0 & 0 \\ \bullet & \bullet & -\gamma^2 I_j & \tilde{B}_j^t\mathcal{Y}_j & 0 & 0 \\ \bullet & \bullet & \bullet & -\mathcal{Y}_j & 0 & 0 \\ \bullet & \bullet & \bullet & \bullet & -\delta_j I_j & 0 \\ \bullet & \bullet & \bullet & \bullet & \bullet & -I_j \end{bmatrix} \\
&+ \begin{bmatrix} 0 & \Delta\tilde{A}_j^t\mathcal{Y}_j & 0 & \Delta\tilde{A}_j^t\mathcal{Y}_j & 0 & \Delta\tilde{L}_j^t \\ \bullet & 0 & \mathcal{Y}_j\Delta\tilde{B}_j & 0 & 0 & 0 \\ \bullet & \bullet & 0 & \Delta\tilde{B}_j^t\mathcal{Y}_j & 0 & 0 \\ \bullet & \bullet & \bullet & 0 & 0 & 0 \\ \bullet & \bullet & \bullet & \bullet & 0 & 0 \\ \bullet & \bullet & \bullet & \bullet & \bullet & 0 \end{bmatrix} < 0
\end{aligned} \tag{10.28}
$$

has a feasible solution for all possible gain perturbations (10.25). Applying **Fact 1** and using (10.27), it can be shown that inequality (10.28) is majorized into the form

$$\overline{\Pi}_{j\Delta} \leq \begin{bmatrix} -\mathcal{Y}_j & \tilde{A}_j^t\mathcal{Y}_j & 0 & \tilde{A}_j^t\mathcal{Y}_j & \tilde{G}_j^t & \tilde{L}_j^t \\ \bullet & \mathcal{Y}_j - I_j & \mathcal{Y}_j\tilde{B}_j & 0 & 0 & 0 \\ \bullet & \bullet & -\gamma^2 I_j & \tilde{B}_j^t\mathcal{Y}_j & 0 & 0 \\ \bullet & \bullet & \bullet & -\mathcal{Y}_j & 0 & 0 \\ \bullet & \bullet & \bullet & \bullet & -\delta_j I_j & 0 \\ \bullet & \bullet & \bullet & \bullet & \bullet & -I_j \end{bmatrix}$$

$$+ \ \varepsilon_j \begin{bmatrix} \alpha_j \\ 0 \\ D_j^t N_{cj}^t \\ 0 \\ 0 \\ 0 \end{bmatrix} \begin{bmatrix} \alpha_j \\ 0 \\ D_j^t N_{cj}^t \\ 0 \\ 0 \\ 0 \end{bmatrix}^t + \varepsilon_j^{-1} \begin{bmatrix} 0 \\ \beta_j \\ 0 \\ \beta_j \\ 0 \\ 0 \end{bmatrix} \begin{bmatrix} 0 \\ \beta_j \\ 0 \\ \beta_j \\ 0 \\ 0 \end{bmatrix}^t \tag{10.29}$$

$$+ \ \sigma_j^{-1} \begin{bmatrix} \pi_j \\ 0 \\ 0 \\ 0 \\ 0 \\ 0 \end{bmatrix} \begin{bmatrix} \pi_j \\ 0 \\ 0 \\ 0 \\ 0 \\ 0 \end{bmatrix}^t + \sigma_j \begin{bmatrix} 0 \\ 0 \\ 0 \\ 0 \\ 0 \\ M_{cj} \end{bmatrix} \begin{bmatrix} \pi_j \\ 0 \\ 0 \\ 0 \\ 0 \\ M_{cj} \end{bmatrix}^t < 0 \tag{10.30}$$

for some $\varepsilon_j > 0, \ \sigma_j > 0$. Schur complement operations cast inequality (10.30) into the form

$$\begin{bmatrix} \Pi_1 & \Pi_2 \\ \bullet & \Pi_3 \end{bmatrix} < 0$$

$$\Pi_1 = \begin{bmatrix} -\mathcal{Y}_j & \tilde{A}_j^t\mathcal{Y}_j & 0 & \tilde{A}_j^t\mathcal{Y}_j & \tilde{G}_j^t \\ \bullet & \mathcal{Y}_j - I_j & \mathcal{Y}_j\tilde{B}_j & 0 & \\ \bullet & \bullet & -\gamma^2 I_j & \tilde{B}_j^t\mathcal{Y}_j & 0 \\ \bullet & \bullet & \bullet & -\mathcal{Y}_j & 0 \\ \bullet & \bullet & \bullet & \bullet & -\delta_j I_j \end{bmatrix},$$

$$\Pi_2 = \begin{bmatrix} \tilde{L}_j^t & \varepsilon_j \alpha_j & 0 & \pi_j & 0 \\ 0 & 0 & \beta_j & 0 & 0 \\ 0 & \varepsilon_j D_j^t N_{cj}^t & 0 & 0 & 0 \\ 0 & 0 & \beta_j & 0 & 0 \\ 0 & 0 & 0 & 0 & 0 \end{bmatrix},$$

$$\Pi_3 = \begin{bmatrix} -I_j & 0 & 0 & 0 & \sigma_j M_{cj} \\ \bullet & -\varepsilon_j I_j & 0 & 0 & 0 \\ \bullet & \bullet & -\varepsilon_j I_j & 0 & 0 \\ \bullet & \bullet & \bullet & -\sigma_j I_j & 0 \\ \bullet & \bullet & \bullet & \bullet & -\sigma_j I_j \end{bmatrix} \tag{10.31}$$

The main result of subsystem filter design is established by the following theorem.

Theorem 10.4.1 *The filtering error subsystem (10.9) is robustly asymptotically stable for all nonlinear uncertainties satisfying (10.7) if there exist matrices* $0 < \mathcal{Y}_{aj}^t = \mathcal{Y}_{aj} \in \Re^{n_j \times n_j}$, $0 < \mathcal{Y}_{oj}^t = \mathcal{Y}_{oj} \in \Re^{s_j \times s_j}$, $\mathcal{Y}_{cj} \in \Re^{s_j \times s_j}$, $\mathcal{Y}_{ej} \in \Re^{s_j \times n_j}$, $\mathcal{Y}_{sj} \in \Re^{s_j \times m_j}$, $H_j \in \Re^{p_j \times s_j}$ *and scalars* $\delta_j > 0$, $\gamma_j > 0$ *such that the following convex optimization problem is feasible:*

$$\min_{\mathcal{Y}_{aj},\mathcal{Y}_{oj},\mathcal{Y}_{cj},\mathcal{Y}_{ej},\mathcal{Y}_{sj},H_j,\delta_j} \quad \sum_{j=1}^{N} \delta_j + \gamma_j$$

$$subject\ to: \ \{\mathcal{Y}_{aj} > 0,\ \mathcal{Y}_{oj} > 0\}_{j=1}^{N},$$

$$diag\Big(\Upsilon_1, ..., \Upsilon_N\Big) < 0,$$

$$\Upsilon_j = \begin{bmatrix} \Upsilon_{j1} & \Upsilon_{j2} \\ \bullet & \Upsilon_{j3} \end{bmatrix}$$

$$\Upsilon_{j1} = \begin{bmatrix} -\Psi_{1j} & \Psi_{2j} & 0 & \Psi_{2j} & \Psi_{4j} \\ \bullet & -\Psi_{6j} & \Psi_{3j} & 0 & 0 \\ \bullet & \bullet & -\gamma^2 I_j & \Psi_{3j}^t & 0 \\ \bullet & \bullet & \bullet & -\Psi_{1j} & 0 \\ \bullet & \bullet & \bullet & \bullet & -\delta_j I_j \end{bmatrix}$$

$$\Upsilon_{j2} = \begin{bmatrix} \Psi_{5j} & \varepsilon_j \alpha_j & 0 & \pi_j & 0 \\ 0 & 0 & \beta_j & 0 & 0 \\ 0 & \varepsilon_j D_j^t N_{cj}^t & 0 & 0 & 0 \\ 0 & 0 & \beta_j & 0 & 0 \\ 0 & 0 & 0 & 0 & 0 \end{bmatrix}$$

$$\Upsilon_{j3} = \begin{bmatrix} -I_j & 0 & 0 & 0 & \sigma_j M_{cj} \\ \bullet & -\varepsilon_j I_j & 0 & 0 & 0 \\ \bullet & \bullet & -\varepsilon_j I_j & 0 & 0 \\ \bullet & \bullet & \bullet & -\sigma_j I_j & 0 \\ \bullet & \bullet & \bullet & \bullet & -\sigma_j I_j \end{bmatrix} \quad (10.32)$$

where $\Psi_{1j},\ \Psi_{2j},\ \Psi_{3j},\ \Psi_{4j},\ \Psi_{5j},\ \Psi_{6j}$ *are given by (10.12). Moreover, the filter gains are given by* $E_j = \mathcal{Y}_{oj}^{-1}\mathcal{Y}_{cj},\ F_j = \mathcal{Y}_{oj}^{-1}\mathcal{Y}_{sj},\ H_j$.

10.4.2 Illustrative example 10.2

We reconsider Example 1 with $s_j = 3$ (full-order filter) and

$$M_{a1} = \begin{bmatrix} 0.7 \\ 0.5 \\ 0.3 \end{bmatrix},\ M_{a2} = \begin{bmatrix} 0.3 \\ 0.6 \\ 0.4 \end{bmatrix},\ N_{a1}^t = \begin{bmatrix} 0.2 \\ 0.2 \\ 0.2 \end{bmatrix},\ N_{a2}^t = \begin{bmatrix} 0.3 \\ 0.3 \\ 0.3 \end{bmatrix},$$

$$M_{c1} = \begin{bmatrix} 0.4 \\ 0.4 \\ 0.4 \end{bmatrix},\ M_{c2} = \begin{bmatrix} 0.6 \\ 0.1 \\ 0.5 \end{bmatrix},\ N_{c1}^t = \begin{bmatrix} 0.4 \\ 0.1 \\ 0.1 \end{bmatrix},\ N_{c2}^t = \begin{bmatrix} 0.1 \\ 0.4 \\ 0.5 \end{bmatrix}$$

The feasible solution of the convex minimization problem (10.32) gives the full-order resilient filters as summarized by

$$\begin{aligned} s_1 &= 3,\ s_2 = 3\ (Full(third) - order filter), \\ \delta_1 &= 0.1341,\ \ \gamma_1 = 1.6885,\ \ \delta_2 = 0.1341,\ \ \gamma_1 = 1.6885 \\ E_1 &= \begin{bmatrix} 0.0149 & -0.0117 & 0.0105 \\ 0.0254 & -0.3301 & -0.0380 \\ -0.0078 & 0.1062 & 0.2424 \end{bmatrix},\ F_1 = \begin{bmatrix} 0.6665 & 0.1256 \\ -0.1336 & 0.4901 \\ -0.5377 & 0.2351 \end{bmatrix}, \\ H_1 &= \begin{bmatrix} -0.3777 & 0.1474 \\ -0.0650 & 0.3987 \\ 0.1875 & -0.4243 \end{bmatrix} \\ E_2 &= \begin{bmatrix} 0.0146 & -0.0132 & 0.0114 \\ 0.0307 & -0.3412 & -0.1055 \\ -0.0108 & 0.1055 & 0.2637 \end{bmatrix},\ F_2 = \begin{bmatrix} 0.7015 & 0.1225 \\ -0.1401 & 0.4871 \\ -0.5426 & 0.2440 \end{bmatrix}, \\ H_2 &= \begin{bmatrix} -0.3667 & 0.1884 \\ -0.0690 & 0.4112 \\ 0.1905 & -0.4008 \end{bmatrix} \end{aligned}$$

Had we considered the design of reduced-order resilient filters, we would have used the additive perturbation matrices as

$$\begin{aligned}
s_1 &= 2, \;\; s_2 = 2 \;\; (Reduced(second) - order filter), \\
M_{a1} &= \begin{bmatrix} 0.5 \\ 0.3 \end{bmatrix}, \; M_{a2} = \begin{bmatrix} 0.3 \\ 0.4 \end{bmatrix}, \; N^t_{a1} = \begin{bmatrix} 0.2 \\ 0.2 \end{bmatrix}, \; N^t_{a2} = \begin{bmatrix} 0.3 \\ 0.3 \end{bmatrix}, \\
M_{c1} &= \begin{bmatrix} 0.4 \\ 0.4 \end{bmatrix}, \; M_{c2} = \begin{bmatrix} 0.1 \\ 0.5 \end{bmatrix}, \; N^t_{c1} = \begin{bmatrix} 0.1 \\ 0.1 \end{bmatrix}, \; N^t_{c2} = \begin{bmatrix} 0.1 \\ 0.5 \end{bmatrix} \\
s_1 &= 1, \;\; s_2 = 1 \;\; (Reduced(first) - order filter), \\
M_{a1} &= [0.5], \; M_{a2} = [0.4], \; N^t_{a1} = [0.2], \; N_{a2} = [0.3], \; M_{c1} = [0.4], \\
M_{c2} &= [0.5], \\
N_{c1} &= [0.1], \; N_{c2} = [0.5]
\end{aligned}$$

In this case, the feasible solution of the convex minimization problem (10.32) is obtained as

$$\begin{aligned}
s_1 &= 2, \;\; s_2 = 2 \;\; (Reduced(second) - order filter), \\
\delta_1 &= 0.4327, \quad \gamma_1 = 2.3307, \quad \delta_2 = 0.4327, \quad \gamma_1 = 2.3307 \\
E_1 &= \begin{bmatrix} 0.1298 & 0.0445 \\ 0.0512 & 0.2446 \end{bmatrix}, \; F_1 = \begin{bmatrix} 0.7140 & 0.2083 \\ -0.1312 & 0.5120 \end{bmatrix}, \\
H_1 &= \begin{bmatrix} 0.7554 & -0.0733 \\ -0.8055 & -0.0717 \end{bmatrix} \\
E_2 &= \begin{bmatrix} -0.0748 & 0.1766 \\ -0.1409 & 0.2550 \end{bmatrix}, \; F_2 = \begin{bmatrix} 0.7126 & 0.2111 \\ -0.4568 & -0.7103 \end{bmatrix}, \\
H_2 &= \begin{bmatrix} 0.7807 & -0.4411 \\ 0.6642 & -0.8837 \end{bmatrix} \\
s_1 &= 1, \;\; s_2 = 1 \;\; (Reduced(first) - order filter), \\
\delta_1 &= 0.6498, \quad \gamma_1 = 4.0694, \quad \delta_2 = 0.6498, \quad \gamma_1 = 4.0694 \\
E_1 &= [0.1245], \; F_1 = \begin{bmatrix} 0.6776 & -0.7102 \end{bmatrix}, \; H_1 = \begin{bmatrix} 0.3925 \\ -0.6614 \end{bmatrix} \\
E_2 &= [0.2198], \; F_2 = \begin{bmatrix} 0.6962 & -0.4991 \end{bmatrix}, \; H_2 = \begin{bmatrix} 0.3695 \\ -0.7891 \end{bmatrix}
\end{aligned}$$

On comparing the ensuing results of the resilient and robust filters, it is found that we observe that the resilient filtered subsystem performs better than the robust filtered subsystem in the sense of yielding smaller values of $\gamma-$ and $\delta-$levels. In fact, in this numerical example, we attained 11-14% improvement and this is true for full- and reduced-order cases.

10.5 Problem Set VIII

Problem VIII.1: Consider a linear large-scale system represented by:

$$\begin{aligned} \Sigma_j : \; \dot{x}_j(t) &= A_j x_j(t) + B_j u_j(t) + \sum_{k=1,k\neq j}^{n_s} A_{jk} x_k(t) \\ y_j(t) &= C_j x_j(t) \end{aligned}$$

where $x_j(t) \in \Re^{n_j}$, $u_j(t) \in \Re^{m_j}$, and $y_j(t) \in \Re^{p_j}$ are the subsystem state, control input, and measured output, respectively. In addition, matrices $A_j,\ B_j,\ C_j,\ A_j, A_{jk}$ are all constant matrices of appropriate dimensions with $rank(C_j) = p_j,\ j = 1, ..., n_s$. It is desired to design q local observer to estimate the unmeasurable components of $x_j,\ j = 1, ..., n_s$ using only local inputs and local output measurements. One possible form would be

$$\begin{aligned} \dot{w}_j(t) &= F_j w_j(t) + L_j u_j(t) + E_j y_j(t) \\ \tilde{x}_j(t) &= M_j w_j(t) + N_j y_j(t) \end{aligned}$$

where $F_j,\ L_j,\ E_j,\ M_j$, and N_j are the design matrices to be determined. Using the following data

$$\begin{aligned} A_1 &= \begin{bmatrix} -1 & 0 & 0 \\ -1 & 0 & 1 \\ 1 & -2 & -1 \end{bmatrix},\ B_1 = \begin{bmatrix} 0 & 1 \\ 1 & 0 \\ 1 & 1 \end{bmatrix}, \\ A_2 &= \begin{bmatrix} -1 & 0 & 0 \\ -1 & -2 & 0 \\ 0 & 0 & -4 \end{bmatrix},\ B_2 = \begin{bmatrix} 0 & -1 \\ 1 & 0 \\ 0 & 1 \end{bmatrix}, \\ A_{12} &= \begin{bmatrix} 0 & 0 & 0 \\ 0 & 0 & 0 \\ -1 & 1 & 1 \end{bmatrix},\ A_{21} = \begin{bmatrix} 0 & 0 & 0 \\ -8 & 1 & -1 \\ 4 & -0.5 & 0.5 \end{bmatrix}, \\ C_1 &= \begin{bmatrix} 0 & 1 & 0 \\ 0 & 0 & 1 \end{bmatrix},\ C_2 = \begin{bmatrix} 0 & 1 & 0 \\ 0 & 0 & 1 \end{bmatrix} \end{aligned}$$

evaluate your results.

Problem VIII.2: Consider a collection of n_s coupled discrete-time systems of the form

$$\begin{aligned}\Sigma_j: \ x_j(k+1) &= A_j x_j(k) + B_j w_j(k) + \sum_{m=1, m\neq j}^{n_s} A_{jm} x_m(k) \\ y_j(k) &= C_j x_j(k) + D_j w(k) \\ z_j(k) &= L_j x_j(k), \ \ j = 1, ..., n_s\end{aligned}$$

where $x_j(k) \in \Re^{n_j}$, $w_j(k) \in \Re^{p_j}$, $z_j(k) \in \Re^{p_j}$, and $y_j(k) \in \Re^{m_j}$ are the subsystem state, disturbance input, linear combination of states, and measured output, respectively. It is required to develop a class of subsystem filters of the form

$$\begin{aligned}\hat{x}_j(k+1) &= E_j \hat{x}_j(k) + F_j \, y_j(k), \qquad \hat{x}(0) = 0 \\ \hat{z}_j(k) &= H_j \hat{x}_j(k)\end{aligned}$$

where the matrices $E_j \in \Re^{s_j \times s_j}$, $F_j \in \Re^{s_j \times m_j}$, and $H_j \in \Re^{p_j \times s_j}$ and the scalar $s_j > 0$ are the design parameters such that collection of the filtered subsystems are asymptotically stable with an $\ell_2 - \ell_\infty$ performance measure.

Problem VIII.3: Develop a decentralized resilient ℓ_∞ filter for the system addressed in **Problem VIII.2** with state-space realization at the subsystem level of the form

$$\begin{aligned}\hat{x}_j(k+1) &= E_j[I_j + \Delta E_j(k)]\hat{x}_j(k) + F_j[I_j + \Delta F_j(k)] \, y_j(k), \\ \hat{z}_j(k) &= H_j[I_j + \Delta H_j(k)]\hat{x}_j(k)\end{aligned}$$

where the matrices $E_j, \ F_j, \ H_j$ are the design parameters and the multiplicative gain perturbations are represented by

$$\begin{aligned}\begin{bmatrix} \Delta E_j(k) \\ \Delta H_j(k) \end{bmatrix} &= \begin{bmatrix} M_{aj} \\ M_{cj} \end{bmatrix} \Delta_j(k) \, N_{aj}, \ \ \Delta F_j(k) = M_{aj} \Delta_j(k) \, N_{cj}, \\ \Delta_j^t(k) \, \Delta_j(k) &\leq I_j, \ \ \forall \ k\end{aligned}$$

where the matrices are constant matrices with appropriate dimensions.

10.6 Notes and References

In this chapter, we have developed new tools for robust and resilient filter design of a class of interconnected discrete-time systems with uncertain function of nonlinear perturbations. These tools essentially exploit the decentralized information structure constraint. Specifically, we have established complete solution procedures for linear robust filtering with ℓ_∞ performance and linear resilient filtering in the face of additive gain variations by basing all the computations at the subsystem level. By a suitable convex analysis, it has been shown that both design problems can be converted into convex minimization problems over linear matrix inequalities (LMIs). Several extensions are possible as outlined in the set of problems. While this chapter focused on decentralized filtering within a deterministic setting, the next chapter will discuss decentralized filtering within a stochastic setting.

Chapter 11

Decentralized Filtering-II

In this chapter, we approach the end of our guided tour into techniques and methods of large-scale systems, where we focus on decentralized filtering and fault detection based on overlapping decomposition. The material covered is divided into two sections. In the first section, the optimal state estimation in a large-scale linear interconnected dynamical system is considered and a decentralized computational structure is developed. The decentralized filter uses hierarchical structure to perform successive orthogonalization on the measurement subspace of each sub-system in order to provide the optimal estimate. This ensures substantial savings in computation time. In addition, since only low-order subsystem machine equations are manipulated at each stage, numerical inaccuracies are reduced and the filter remains stable for even high-order system. In the second section, the problem of fault detection in linear, stochastic, interconnected dynamic systems is examined. The objective addresses a design approach based on employing a set of decentralized filters at the subsystem level resulting from overlapping decomposition. The malfunctioning sensors can be detected and isolated by comparing the estimated values of a single state from different Kalman filters.

11.1 Decentralized Kalman Filtering

Throughout the book we have tackled the problems of optimization and control of large-scale systems comprising interconnected dynamical subsystems from a deterministic point of view [271]-[331]. The objective has been to trace the development of decentralized control methods. On the other hand, the work reported in [8, 40, 98, 307] has emphasized the importance of the stochastic control problem. Filter structure for large systems has previously been suggested in [8, 98, 332, 335]. However, the filter structures proposed in [8, 333] provide estimates that are suboptimal while the optimal decentralized filter [98] is not readily extendable to a system comprising more than two subsystems. It is quite natural to seek the design of a decentralized structure that provides optimal estimation and that is applicable to systems comprising n_s subsystems.

11.1.1 Basis of filtering method

It is well known that the most appealing property of the global Kalman filter from a practical viewpoint is its recursive nature. Essentially, this recursive property of the filter arises from the fact that if an estimate exists based on measurements up to that instant, then upon receiving another set of measurements, one could subtract from these measurements that part that could be anticipated from the results of the first measurements, that is, the updating is based on the part of the new data that is orthogonal to the old data. In searching for an appropriate decentralized filter for a system comprising lower order interconnected subsystems, one endeavors to perform this orthogonalization on a subsystem-by-subsystem basis, that is, the optimal estimates of the state of subsystem one is obtained by successively orthogonalizing the error based on a new measurement for subsystems $\{1, 2, 3, \ldots, n_s\}$ with respect to the Hilbert space formed by all measurements of all the subsystems up to that instant. Much computational saving would be expected to result by using this successive orthogonalization procedure since, at each stage, only low-order subspaces are manipulated. The actual orthogonalization procedure that is performed in the Kalman filter is based on the following theorem [181].

Theorem 11.1.1 *Let β be a member of space H of random variables which is a closed subspace of L_2 and let $\hat{\beta}$ denote its orthogonal projection on a closed subspace y_1 of H (thus, $\hat{\beta}_1$ is the best estimate of β in $\mathcal{Y}_1$). Let y_2 be an m vector of random variables generating a subspace $\mathcal{Y}_2$ of H and let $\hat{y}_2$ denote the m-dimensional vector of the projections of the components of y_2 on to $\mathcal{Y}_1$ (thus $\hat{y}_2$ is the vector of best estimates of y_2 in $\mathcal{Y}_1$). Let $\tilde{y}_2 = y_2 - \hat{y}_2$.*

Then the projection of β onto the subspace $\mathcal{Y}_1 \oplus \mathcal{Y}_2$, denoted $\hat{\beta}$, is

$$\hat{\beta} = \hat{\beta}_1, + \boldsymbol{E}\{\beta\tilde{y}_2^T\}[\boldsymbol{E}\{\tilde{y}_2\tilde{y}_2^T\}]^{-1}\tilde{y}_2 \tag{11.1}$$

*where **E** is the expected value.*

Equation (11.1) can be interpreted as $\hat{\beta}$ *is* $\hat{\beta}_1$ *plus the best estimate of* β *in the subspace* $\tilde{y}_2$ *generated by* $\tilde{y}_2$.

Next consider the system comprising n_s interconnected linear dynamical subsystems defined by

$$x_j(k+1) = \phi_{jj}x_j(k) + \sum_{k=1;k\neq j}^{n_s} \phi_{jk}x_j(k) + w_j(k) \quad j = 1,2,\ldots,n_s \quad (11.2)$$

with the outputs given by

$$y_j(k+1) = H_j x_j(k+1) + v_j(k+1) \quad j = 1,\ldots,n_s \quad (11.3)$$

where $w_j,\ v_j$ are uncorrelated zero mean Gaussian white noise sequences with covariances $Q_j,\ R_j$, respectively. Consider the Hilbert space $\mathcal{Y}$ formed by the measurements of the overall systems. At the instant $k+1$, this space is denoted by $\mathcal{Y}(k+1)$. The optimal minimum variance estimate $\hat{x}(k+1|k+1)$ is given by

$$\begin{aligned}\hat{x}(k+1|k+1) &= \mathbf{E}\{x(k+1)|\mathcal{Y}(k+1)\} \\ &= \mathbf{E}\{x(k+1)|\mathcal{Y}(k)\} + E\{x(k+1)|\tilde{y}(k+1|k)\} \quad (11.4)\end{aligned}$$

This equation spells out, in algebraic terms, the geometrical result of **Theorem 11.1.1**. This motivates the idea of generating a filter by decomposing the second term, that is, $\mathbf{E}\{x(k+l)|\tilde{y}(k+1|k)\}$, such that the optimal estimate $\tilde{x}(k+1|k+1)$ is given using the two terms by considering the estimate as the orthogonal projection of $x_j(k+1)$ taken on the Hilbert space generated by

$$\begin{aligned}&\mathcal{Y}(k) \oplus \tilde{\mathcal{Y}}_1(k+1|k) \oplus \tilde{\mathcal{Y}}_2^1(k+1|k+1) \\ &\oplus \tilde{\mathcal{Y}}_3^2(k+1|k+1) \oplus \cdots \oplus \tilde{\mathcal{Y}}_{n_s}^{n_s-1}(k+1|k+1)\end{aligned}$$

where $\tilde{\mathcal{Y}}_j^{j-1}(k+1|k+1)$ is the subspace generated by the subspace of measurements $\mathcal{Y}_j(k+1)$ and the projection of it on the subspaces generated by $\mathcal{Y}(k) + \mathcal{Y}_1(k+1) + \mathcal{Y}_2(k+1) + \ldots + \mathcal{Y}_{j-1}(k+1)$, which leads to the following theorem.

Theorem 11.1.2 *The optimal estimate $\hat{x}_j(k+1|k+1)$ of the ith subsystem is given by the projection of $x_j(k+1)$ on the space generated by all measurements up to k $(\mathcal{Y}(k))$ and projection of $x_j(k+1)$ on the subspace generated by*

$$\tilde{\mathcal{Y}}_1(k+1|k) \oplus \tilde{\mathcal{Y}}_2^1(k+1|k+1) + \ldots \oplus \tilde{\mathcal{Y}}_{n_s}^{n_s-1}(k+1|k+1)$$

Proof: Rewrite (11.4) as

$$\hat{x}_j(k+1|k+1) = \mathbf{E}\{x_j(k+1)|\mathcal{Y}(k), y_1(k+1), y_2(k+1), \ldots$$
$$y_j(k+1), y_{j+1}(k+1), \ldots y_{n_s}(k+1)\}$$
$$= \mathbf{E}\{x_j(k+1)|\mathcal{Y}(k), y_1(k+1), y_2(k+1) + \ldots$$
$$+ y_j(k+1), y_{i+1}(k+1), \ldots y_{N-1}(k+1)\}$$
$$+ \mathbf{E}\{x_j(k+1)|\tilde{y}_N^{N-1}(k+1|k+1)\}$$

where

$$\tilde{y}_N^{N-1}(k+1|k+1) = y_N(k+1)$$
$$-E\{y_N(k+1)|\mathcal{Y}(k), y_1(k+1), \ldots y_{N-1}(k+1)\}$$

Equivalently stated,

$$\begin{aligned}
\hat{x}_j(k+1|k+1) &= \mathbf{E}\{x_j(k+1)|\mathcal{Y}(k)\} \\
&+ \mathbf{E}\{x_j(k+1)|\tilde{y}_1(k+1|k)\} \\
&+ \sum_{r=2}^{n_s} \mathbf{E}\{x_j(k+1)|\tilde{y}_r^{r-1}(k+1|k+1)\}
\end{aligned}$$

which establishes the desired result.

Using the idea of successive orthogonalization of the spaces defined above, the algebraic structure of the decentralized filter is described in the next section.

11.1.2 The algebraic structure

In order to develop the filter equations for the overall systems comprising n_s interconnected subsystems, write the equations for the overall systems as

$$\begin{aligned}
x(k+1) &= \phi(k+1,k)x(k) + w(k) \\
y(k+1) &= H(k+1)x(k+1) + v(k+1)
\end{aligned}$$

To display the subsystems structure more clearly, we decompose the above equations as

$$\begin{aligned}
x_j(k+1) &= \sum_{k=1}^{n_s} \phi_{jk}(k+1,k)x_k(k) + w_j(k) \\
y_j(k+1) &= H_j(k+1)x_j(k+1) + v_j(k+1)
\end{aligned}$$

Then the optimal state prediction for the *jth* subsystem is given by

$$\hat{x}_j(k+1|k) = \sum_{k=1}^{n_s} \phi_{jk}(k+1,k)\hat{x}_j(k|k) \tag{11.5}$$

Now, by definition of the prediction errors, $\tilde{x}_j(k+1|k) = x_j(k+1) - \hat{x}_j(k+1|k)$. A recursive expression for the covariance of the prediction error can be written as

$$\begin{aligned} P_{jj}(k+1|k) &= \sum_{k=1}^{n_s}\sum_{r=1}^{n_s} \phi_{jk}(k+1,k) \\ &\quad \cdot P_{\tilde{x}_k\tilde{x}_r}\phi_{ir}^t(k+1,k) + Q_j(k) \\ &= \sum_{k=1}^{n_s} \phi_{jk}(k+1,k) \\ &\quad \cdot \left\{ \sum_{r=1}^{n_s} P_{\tilde{x}_k\tilde{x}_r}(k|k)\phi_{jr}^t(k+1,k) \right\} + Q_j(k) \end{aligned} \tag{11.6}$$

Also,

$$\begin{aligned} P_{jk}(k+1|k) &= \sum_{r=1}^{n_s}\sum_{\ell=1}^{n_s} \phi_{jr}(k+1,k)P_{r\ell}(k|k)\phi_{k\ell}^t(k+1,k) \\ &= \sum_{r=1}^{n_s} \phi_{jr}(k+1,k)\left\{ \sum_{\ell=1}^{n_s} P_{r\ell}(k|k)\phi_{k\ell}^t(k+1,k) \right\} \end{aligned} \tag{11.7}$$

It is easy to show, using the proof of **Theorem 11.1.2**, that

$$\begin{aligned} \hat{x}_j(k+1|k+1) &= \hat{x}_j(k+1|k+1)_\ell \\ &+ \sum_{r=\ell+1}^{n_s} P_{x_j\tilde{y}_r^{r-1}}(k+1|k+1) \\ &\quad \cdot P^{-1}_{\tilde{y}_r^{r-l}\tilde{y}_r^{r-1}}(k+1|k+1)\tilde{y}_r^{r-1}(k+1|k+1) \end{aligned} \tag{11.8}$$

where

$$\begin{aligned} \hat{x}_j(k+1|k+1)_\ell &= \hat{x}_j(k+1|k+1)_{\ell-1} \\ &\quad \cdot k_{j\ell}^{\ell-1}(k+1)\tilde{y}_\ell^{\ell-1}(k+1|k+1) \end{aligned} \tag{11.9}$$

and

$$\begin{aligned} P_{jj}(k+1|k+1)_l &= P_{jj}(k+1|k+1)_{\ell-1} \\ &- K_{j\ell}^{\ell-1}(k+1)P_{\tilde{y}_\ell^{\ell-1}\tilde{x}_{j\ell-1}}(k+1|k+1) \end{aligned} \tag{11.10}$$

where

$$K_{ji_\ell}^{\ell-1}(k+1) = P_{\tilde{x}_{j_{\ell-1}}\tilde{y}_\ell^{\ell-1}}(k+1|k+1) \cdot P^{-1}_{\tilde{y}_\ell^{\ell-l}\tilde{y}_\ell^{\ell-1}}(k+1|k+1) \tag{11.11}$$

$$\tilde{y}_\ell^{\ell-1}(k+1|k+1) = \tilde{y}_\ell^{\ell-2}(k+1|k+1) - K_{\ell-1}^{\ell-2}(k+1)\tilde{y}_{\ell-1}^{\ell-2}(k+1|k+1) \tag{11.12}$$

$$K_{\ell_{\ell-1}}^{\ell-2} = P_{\tilde{y}_\ell^{\ell-2}\tilde{y}_\ell^{\ell-2}}(k+1|k+1)P^{-1}_{\tilde{y}_{\ell-1}^{\ell-2}\tilde{y}_{\ell-1}^{\ell-2}}(k+1|k+1) \tag{11.13}$$

$$P_{\tilde{y}_\ell^{\ell-1}\tilde{y}_\ell^{\ell-1}}(k+1|k+1) = P_{\tilde{y}_\ell^{\ell-2}\tilde{y}_\ell^{\ell-2}}(k+1|k+1) - K_{\ell_{\ell-1}}^{\ell-2}(k+1)P_{\tilde{y}_{\ell-1}^{\ell-2}\tilde{y}_{\ell-1}^{\ell-2}}(k+1|k+1) \tag{11.14}$$

$$\begin{aligned} &P_{\tilde{y}_\ell^{\ell-1}\tilde{y}_\ell^{\ell-1}}(k+1|k+1) = H_\ell P_{\tilde{x}_\ell^{\ell-1}\tilde{x}_\ell^{\ell-1}} \\ &\cdot(k+1|k+1)H_\ell^t + R_\ell(k+1) \\ &P_{\tilde{x}_{j_{\ell-1}}\tilde{y}_\ell^{\ell-1}}(k+1|k+1) = P_{\tilde{x}_\ell^{\ell-1}\tilde{x}_\ell^{\ell-1}}(k+1|k+1) \\ &\cdot H_\ell^t(k+1) \end{aligned} \tag{11.15}$$

$$P_{jk}(k+1|k+1)_l = P_{jk}(k+1|k+1)_{\ell-1} - K_{j_\ell}^{\ell-1}(k+1)P_{\tilde{y}_\ell^{\ell-1}\tilde{x}_k^{\ell-1}}(k+1|k+1) \tag{11.16}$$

In effect, (11.5)-(11.7) and (11.9)-(11.16) provide the algebraic equations of the new filter Implementation of the algorithm for one step of filter consists of the following:

1. From (11.5), (11.6), and (11.7) we compute the prediction estimate as well as its covariance matrix.

2. Put $\ell = 1$ [note that $\hat{x}_j(k+1|k+1)_0 = \hat{x}_j(k+1|k)$ and $P_{jk}(k+1|k+1)_0 = P_{jk}(k+1|k) j = 1,\dots,n_s\ k = 1,\dots,n_s$]. From (11.9)-(11.16), we compute the filtered estimate $x_j(k+1|k+1)_\ell$, and the corresponding covariance matrix.

3. If $\ell = N$ the resulting estimate is the optimal Kalman estimate and the covariance matrix is the minimum error covariance matrix; otherwise go to Step 2.

We note that although the foregoing algorithm and the global Kalman filter are algebraically equivalent, the numerical properties of the decomposed filter are significantly better. In particular, the decentralized algorithm is amenable to parallel processing [99]. Additionally, a good measure of the computation time requirements of the global Kalman filter and the decentralized filter is given by the number of elementary multiplication operations involved. To provide a comparison, we assume that $x \in \Re^n$, $y \in \Re^m$ **x**, then the number of multiplications required under the assumption that ***H*** is block diagonal and each subsystem has the same number of states and outputs is

$$1.5n^2 + 1.5n^3 \quad + \quad nm\Big(\frac{1}{n_s} + \frac{m+1}{2n_s} + m + 1 + \frac{n+1}{2}\Big) + \frac{m^2(3m+1)}{2}$$

where n_s is the number of subsystem.

On the other hand, assume that all subsystems have an equal number of state variables n/n_s and an equal number of measurements m/n_s. Then the number of multiplications required is [99]:

$$\begin{aligned} 1.5n^2 + 1.5n^3 \quad &+ \quad n_s + \left\{ \frac{mn}{{n_s}^2} + \frac{mn(m+N)}{2{n_s}^3} + \frac{m^2(\frac{3m}{n_s}+1)}{2{n_s}^2} \right. \\ &+ \quad N\left[\frac{n^2 m}{{n_s}^3} + \frac{nm^2}{{n_s}^3} + \frac{nm}{{n_s}^2} + \frac{nm(n+n_s)}{2{n_s}^3} \right] \\ &+ \quad \left. \frac{n_s(n_s-1)}{2} . \frac{n^2 m}{{n_s}^3} \right\} \end{aligned}$$

For high-order systems, substantial savings in computation time is quite evident.

11.1.3 Application to multimachine system

The multimachine power system under consideration consists of 11 coupled machines. The model of an n-machine system consists of a set of nonlinear equations that can be written for the *ith* machine as [99]:

$$M_j \ddot{\delta}_j = P_j - \sum_{\substack{j=1 \\ j \neq i}}^{n} b_{ij} sin\delta_{ij} \quad i = 1, \ldots, n_s$$

where *M* is the inertia, *P* is the power injected, b_{ij} is the interconnection variable, and δ is the angle.

For the system to be completely controllable and completely observable the *nth* machine is taken as reference. Then, by subtracting the *nth* equation from the equation of each machine, a model can be constructed for the *n* machine system.

For small perturbations, the nonlinear model can be linearized about the equilibrium point so that the linear equations that result can be written in discrete form:

$$x(k+l) = Ax(k) + \zeta$$

For the 11-machine system, *A* is the 20 × 20 matrix given below and ζ is a zero mean Gaussian white noise vector. For the *ith* machine, the observation equation is given by

$$y_j = \begin{bmatrix} 0 & 1 \end{bmatrix} \begin{bmatrix} x_{1i} \\ x_{2i} \end{bmatrix} + \begin{bmatrix} u_j \end{bmatrix}$$

where y_j is the speed of the *ith* machine and u_j is also a zero mean Gaussian white noise vector sequence.

$$A = \begin{bmatrix} A_1 & A_2 \\ & \\ A_3 & A_4 \end{bmatrix}$$

with

$$A_1 = \begin{bmatrix} A_{11} & \cdots & A_{15} \\ \vdots & \vdots & \vdots \\ A_{51} & \cdots & A_{55} \end{bmatrix}, \; A_2 = \begin{bmatrix} A_{16} & \cdots & A_{110} \\ \vdots & \vdots & \vdots \\ A_{56} & \cdots & A_{510} \end{bmatrix},$$

$$A_3 = \begin{bmatrix} A_{61} & \cdots & A_{65} \\ \vdots & \vdots & \vdots \\ A_{101} & \cdots & A_{105} \end{bmatrix}, \; A_4 = \begin{bmatrix} A_{66} & \cdots & A_{610} \\ \vdots & \vdots & \vdots \\ A_{106} & \cdots & A_{1010} \end{bmatrix}$$

along with the numerical values

$$A_{11} = \begin{bmatrix} 1 & 0.006 \\ -0.587 & 1 \end{bmatrix}, \; A_{12} = \begin{bmatrix} 0 & 0 \\ 0.055 & 0 \end{bmatrix}, \; A_{13} = \begin{bmatrix} 0 & 0 \\ 0.034 & 0 \end{bmatrix}$$

$$A_{14} = \begin{bmatrix} 0 & 0 \\ 0.016 & 0 \end{bmatrix}, \; A_{15} = \begin{bmatrix} 0 & 0 \\ 0.025 & 0 \end{bmatrix}, \; A_{16} = \begin{bmatrix} 0 & 0 \\ -0.010 & 0 \end{bmatrix}$$

$$A_{17} = \begin{bmatrix} 0 & 0 \\ -0.017 & 0 \end{bmatrix}, \; A_{18} = \begin{bmatrix} 0 & 0 \\ -0.034 & 0 \end{bmatrix}, \; A_{19} = \begin{bmatrix} 1.0 & 0.006 \\ -0.070 & 0 \end{bmatrix}$$

$$A_{110} = \begin{bmatrix} 0 & 0 \\ -0.079 & 0 \end{bmatrix}, \; A_{21} = \begin{bmatrix} 0 & 0 \\ 0.099 & 0 \end{bmatrix}, \; A_{22} = \begin{bmatrix} 1 & 0.006 \\ -0.8772 & 1 \end{bmatrix}$$

$$
\begin{aligned}
A_{23} &= \begin{bmatrix} 0 & 0 \\ 0.06 & 0 \end{bmatrix}, A_{24} = \begin{bmatrix} 0 & 0 \\ 0.055 & 0 \end{bmatrix}, A_{25} = \begin{bmatrix} 0 & 0 \\ 0.042 & 0 \end{bmatrix} \\
A_{26} &= \begin{bmatrix} 0 & 0 \\ 0.017 & 0 \end{bmatrix}, A_{27} = \begin{bmatrix} 0 & 0 \\ 0.009 & 0 \end{bmatrix}, A_{28} = \begin{bmatrix} 0 & 0 \\ -0.013 & 0 \end{bmatrix} \\
A_{29} &= \begin{bmatrix} 0 & 0 \\ -0.044 & 0 \end{bmatrix}, A_{210} = \begin{bmatrix} -0.059 & 0 \\ 0 & 0 \end{bmatrix}, A_{31} = \begin{bmatrix} 0 & 0 \\ 0.066 & 0 \end{bmatrix} \\
A_{32} &= \begin{bmatrix} 0 & 0 \\ 0.064 & 0 \end{bmatrix}, A_{33} = \begin{bmatrix} 1.0 & 0.006 \\ -0.725 & 1.0 \end{bmatrix}, A_{34} = \begin{bmatrix} 0 & 0 \\ 0.039 & 0 \end{bmatrix} \\
A_{35} &= \begin{bmatrix} 0 & 0 \\ 0.025 & 0 \end{bmatrix}, A_{36} = \begin{bmatrix} 0 & 0 \\ 0.003 & 0 \end{bmatrix}, A_{37} = \begin{bmatrix} 0 & 0 \\ -0.007 & 0 \end{bmatrix} \\
A_{38} &= \begin{bmatrix} 0 & 0 \\ -0.029 & 0 \end{bmatrix}, A_{39} = \begin{bmatrix} 0 & 0 \\ -0.073 & 0 \end{bmatrix}, A_{310} = \begin{bmatrix} -0.072 & 0 \\ 0 & 0 \end{bmatrix} \\
A_{41} &= \begin{bmatrix} 0 & 0 \\ 0.053 & 0 \end{bmatrix}, A_{42} = \begin{bmatrix} 0 & 0 \\ 0.113 & 0 \end{bmatrix}, A_{43} = \begin{bmatrix} 0 & 0 \\ 0.048 & 0 \end{bmatrix} \\
A_{44} &= \begin{bmatrix} 1 & 0.006 \\ -0.825 & 1 \end{bmatrix}, A_{45} = \begin{bmatrix} 0 & 0 \\ 0.025 & 0 \end{bmatrix}, A_{46} = \begin{bmatrix} 0 & 0 \\ 0.017 & 0 \end{bmatrix} \\
A_{47} &= \begin{bmatrix} 0 & 0 \\ 0.007 & 0 \end{bmatrix}, A_{48} = \begin{bmatrix} 0 & 0 \\ -0.019 & 0 \end{bmatrix}, A_{49} = \begin{bmatrix} 0 & 0 \\ -0.048 & 0 \end{bmatrix} \\
A_{410} &= \begin{bmatrix} 0 & 0 \\ -0.064 & 0 \end{bmatrix}, A_{51} = \begin{bmatrix} 0 & 0 \\ 0.071 & 0 \end{bmatrix}, A_{52} = \begin{bmatrix} 0 & 0 \\ 0.054 & 0 \end{bmatrix} \\
A_{53} &= \begin{bmatrix} 0 & 0 \\ 0.037 & 0 \end{bmatrix}, A_{54} = \begin{bmatrix} 0 & 0 \\ 0.027 & 0 \end{bmatrix}, A_{55} = \begin{bmatrix} 1 & 0.006 \\ -0.779 & 1 \end{bmatrix} \\
A_{56} &= \begin{bmatrix} 0 & 0 \\ 0.0162 & 0 \end{bmatrix}, A_{57} = \begin{bmatrix} 0 & 0 \\ -0.004 & 0 \end{bmatrix}, A_{58} = \begin{bmatrix} 0 & 0 \\ -0.00083 & 0 \end{bmatrix} \\
A_{59} &= \begin{bmatrix} 0 & 0 \\ -0.059 & 0 \end{bmatrix}, A_{510} = \begin{bmatrix} 0 & 0 \\ -0.056 & 0 \end{bmatrix}, A_{61} = \begin{bmatrix} 0 & 0 \\ 0.012 & 0 \end{bmatrix} \\
A_{62} &= \begin{bmatrix} 0 & 0 \\ 0.012 & 0 \end{bmatrix}, A_{63} = \begin{bmatrix} 0 & 0 \\ 0.011 & 0 \end{bmatrix}, A_{64} = \begin{bmatrix} 0 & 0 \\ 0.008 & 0 \end{bmatrix} \\
A_{65} &= \begin{bmatrix} 0 & 0 \\ 0.011 & 0 \end{bmatrix}, A_{66} = \begin{bmatrix} 1 & 0.006 \\ -0.623 & 1 \end{bmatrix}, A_{67} = \begin{bmatrix} 0 & 0 \\ 0.012 & 0 \end{bmatrix} \\
A_{68} &= \begin{bmatrix} 0 & 0 \\ -0.006 & 0 \end{bmatrix}, A_{69} = \begin{bmatrix} 0 & 0 \\ -0.057 & 0 \end{bmatrix}, A_{610} = \begin{bmatrix} 0 & 0 \\ -0.062 & 0 \end{bmatrix} \\
A_{71} &= \begin{bmatrix} 0 & 0 \\ 0.044 & 0 \end{bmatrix}, A_{72} = \begin{bmatrix} 0 & 0 \\ 0.064 & 0 \end{bmatrix}, A_{73} = \begin{bmatrix} 0 & 0 \\ 0.034 & 0 \end{bmatrix}
\end{aligned}
$$

$$
\begin{aligned}
A_{74} &= \begin{bmatrix} 0 & 0 \\ 0.047 & 0 \end{bmatrix}, \; A_{75} = \begin{bmatrix} 0 & 0 \\ 0.034 & 0 \end{bmatrix}, \; A_{76} = \begin{bmatrix} 0 & 0 \\ 0.074 & 0 \end{bmatrix} \\
A_{77} &= \begin{bmatrix} 1 & 0.006 \\ -1.042 & 1 \end{bmatrix}, \; A_{78} = \begin{bmatrix} 0 & 0 \\ 0.035 & 0 \end{bmatrix}, \; A_{79} = \begin{bmatrix} 0 & 0 \\ 0.045 & 0 \end{bmatrix} \\
A_{710} &= \begin{bmatrix} 0 & 0 \\ -0.2750 & 0 \end{bmatrix}, \; A_{81} = \begin{bmatrix} 0 & 0 \\ 0.0012 & 0 \end{bmatrix}, \; A_{82} = \begin{bmatrix} 0 & 0 \\ -0.0019 & 0 \end{bmatrix} \\
A_{83} &= \begin{bmatrix} 0 & 0 \\ 0.004 & 1 \end{bmatrix}, \; A_{84} = \begin{bmatrix} 0 & 0 \\ -0.002 & 0 \end{bmatrix}, \; A_{85} = \begin{bmatrix} 0 & 0 \\ 0.003 & 0 \end{bmatrix} \\
A_{86} &= \begin{bmatrix} 0 & 0 \\ 0.008 & 0 \end{bmatrix}, \; A_{87} = \begin{bmatrix} 0 & 0 \\ -0.004 & 0 \end{bmatrix}, \; A_{88} = \begin{bmatrix} 1 & 0.006 \\ -0.558 & 1 \end{bmatrix} \\
A_{89} &= \begin{bmatrix} 0 & 0 \\ -0.0420 & 0 \end{bmatrix}, \; A_{810} = \begin{bmatrix} 0 & 0 \\ 0 & 0 \end{bmatrix}, \; A_{91} = \begin{bmatrix} 0 & 0 \\ 0.0317 & 0 \end{bmatrix} \\
A_{92} &= \begin{bmatrix} 0 & 0 \\ 0.047 & 0 \end{bmatrix}, \; A_{93} = \begin{bmatrix} 0 & 0 \\ 0.025 & 0 \end{bmatrix}, \; A_{94} = \begin{bmatrix} 0 & 0 \\ 0.035 & 0 \end{bmatrix} \\
A_{95} &= \begin{bmatrix} 0 & 0 \\ 0.025 & 0 \end{bmatrix}, \; A_{96} = \begin{bmatrix} 0 & 0 \\ 0.038 & 0 \end{bmatrix}, \; A_{97} = \begin{bmatrix} 0 & 0 \\ 0.046 & 0 \end{bmatrix} \\
A_{98} &= \begin{bmatrix} 0 & 0 \\ 0.050 & 0 \end{bmatrix}, \; A_{99} = \begin{bmatrix} 1 & 0.006 \\ -0.587 & 1 \end{bmatrix}, \; A_{910} = \begin{bmatrix} 0 & 0 \\ -0.003 & 0 \end{bmatrix} \\
A_{101} &= \begin{bmatrix} 0 & 0 \\ 0.010 & 0 \end{bmatrix}, \; A_{102} = \begin{bmatrix} 0 & 0 \\ 0.068 & 0 \end{bmatrix}, \; A_{103} = \begin{bmatrix} 0 & 0 \\ 0.006 & 0 \end{bmatrix} \\
A_{104} &= \begin{bmatrix} 0 & 0 \\ 0.005 & 0 \end{bmatrix}, \; A_{105} = \begin{bmatrix} 0 & 0 \\ 0.011 & 0 \end{bmatrix}, \; A_{106} = \begin{bmatrix} 0 & 0 \\ 0.010 & 0 \end{bmatrix} \\
A_{107} &= \begin{bmatrix} 0 & 0 \\ 0.003 & 0 \end{bmatrix}, \; A_{108} = \begin{bmatrix} 0 & 0 \\ 0.010 & 0 \end{bmatrix} \\
A_{109} &= \begin{bmatrix} 0 & 0 \\ -0.006 & 0 \end{bmatrix}, \; A_{1010} = \begin{bmatrix} 1 & 0.006 \\ -0.587 & 1 \end{bmatrix}
\end{aligned}
$$

The covariances of ζ and u are given by $Q_j = 5 \times I_2$ and $R_j = I_2$, respectively; $P_0 = 25 \times I$ is the initial covariance matrix. The initial estimate was taken to be zero while the initial states were all taken to be 10.

The global and decentralized filters are then simulated to illustrate the computational behavior. The ensuing results for the global filter are plotted in Figures 11.1 - 11.3 for the first three states and the corresponding estimates. The corresponding states and estimates using the decentralized filter are depicted in Figures 11.4 - 11.6. It should be observed that the global Kalman filter is numerically unstable while the developed decentralized solution is stable. This is essentially due to buildup of numerical errors when implementing a $20th$-order

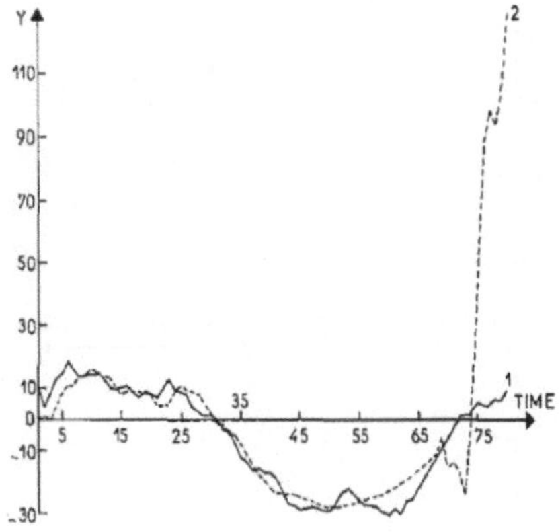

Figure 11.1: The global Kalman filter: first state and estimate

Kalman filter. These numerical inaccuracies are avoided in the case of the coordinated hierarchical solution as it implements only second-order subsystems at each stage, thereby yielding stable filter.

In what follows, we direct attention to an approach for fault detection in linear, stochastic, interconnected dynamic systems, based on designing a set of decentralized filters for the subsystems resulting from overlapping decomposition of the overall large-scale system. The malfunctioning sensors are detected and isolated by comparing the estimated values of a single state from different Kalman filters.

11.2 Decentralized Fault Detection

An important problem facing engineers in industrial complexes, which has attracted scientists and researchers in the field of systems science, is the problem of failure detection in running systems. A fault detection technique for large-scale deterministic systems has been developed [338], using an overlapping decomposition method. A clearly mentioned before, this concept has been utilized [124] for constructing decentralized control strategies for interconnected systems, such that each local subsystem has a range of information that may partially overlap with that of other subsystems [129]. The central idea behind the technique was the design of a set of Luenberger observers [182] for the overlapping subsystems. Failures can be detected and localized by comparing the estimates of the same states via the different observers.

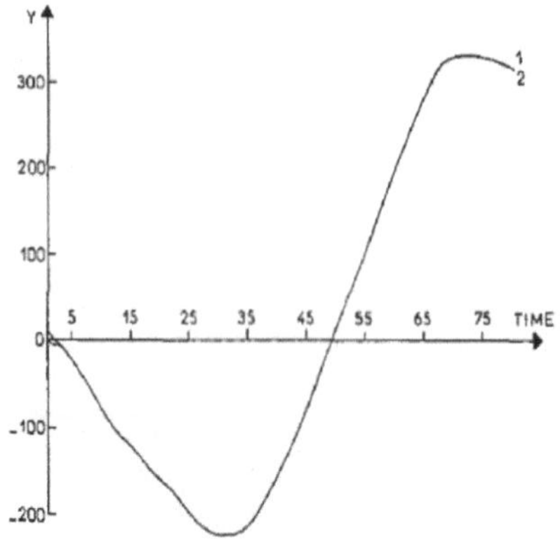

Figure 11.2: The global Kalman filter: second state and estimate

In this section, we extend the above technique to the important case of fault detection in stochastic large-scale systems. We will benefit from the appealing result of the foregoing section and proceed to design a set of decentralized Kalman filters based on the overlapping subsystems that constitute the overall system considered. Discrepancies between the estimated values of the same state from different Kalman filters point to the malfunctioning sensors.

11.2.1 System model

Consider a large-scale linear interconnected system **S**, which is described by the following state and output equations:

$$x(k+1) = Ax(k) + Bw(k) \tag{11.17}$$
$$y(k+1) = Hx(k+1) + v(k+1) \tag{11.18}$$

where $x(k) \in \Re^n$ is the state vector, $w(k) \in \Re^m$ is the input noise vector, $y(k+1) \in \Re^q$ is the output measurement vector, and $u(k+1) \in \Re^q$ is the noise disturbing the output. **A**, **B**, and **H** are the system matrices of appropriate dimensions, in which **B** and **H** are assumed to be block diagonal matrices with n_s blocks corresponding to the n_s subsystems.

For the above system given by (11.17) and (11.18), we made the following assumptions:

1. $w(k)$ and $v(k)$ *are Gaussian random vectors with zero mean and covariances given by* $\mathbf{E}\{w(k)w^t(j)\} = Q\delta_{kj}, \mathbf{E}\{v(k)v^t(j)\} = R\delta_{kj}$.

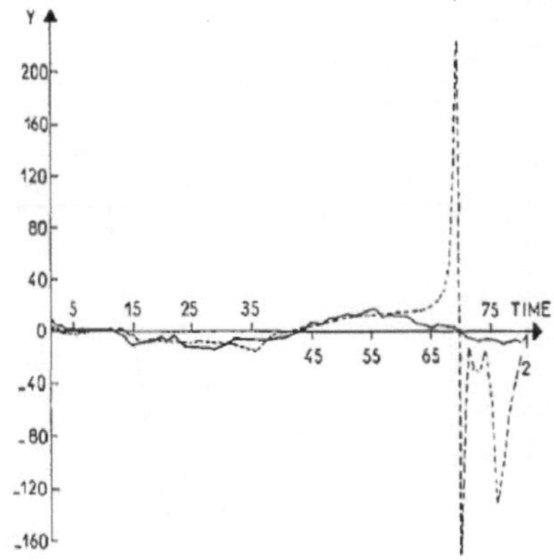

Figure 11.3: The global Kalman filter: third state and estimate

2. *The disturbance vectors are uncorrelated, that is,* $\mathbf{E}\{v(k)w^t(j)\} = 0 \; \forall k, j$.

3. *The initial state vector* $x(0)$ *is a Gaussian random vector with mean* $\boldsymbol{E}\{x(O)\} = \mu_o$ *and covariance* $\mathbf{E}\{[x(O) - \mu_0][x(0) - \mu_0]^t\} = p(0)$.

4. $x(0)$ *and the noise vectors* $u(k)$ *and* $w(k)$ *are uncorrelated, that is,* $\mathbf{E}\{x(0)v^t(k)\} = 0, \mathbf{E}\{x(O)w^t(k)\} = 0 \forall k$.

11.2.2 Overlapping decomposition

Now, we proceed to identify the subsystem models. On considering the state and output measurement models (11.17) and (11.18), we express the *ith* subsystem $\mathbf{S}_j$ in the form

$$x_j(k+1) = A_j x_j(k) + B_j w_j(k) + \sum_{k=1}^{n_s} A_{jk} x_k(k) \qquad (11.19)$$

$$y_j(k+1) = H_j x_j(k+1) + v_j(k+1) \qquad (11.20)$$

where

$$x_j \in \Re^{n_j}, \; w_j \in \Re^{m_j}, \; y_j \in \Re^{q_j}, \; v_j \in \Re^{q_j}$$

and

$$\sum_{j=1}^{n_s} n_j = n, \sum_{j=1}^{n_s} m_j = m, \sum_{j=1}^{n_s} q_j = q$$

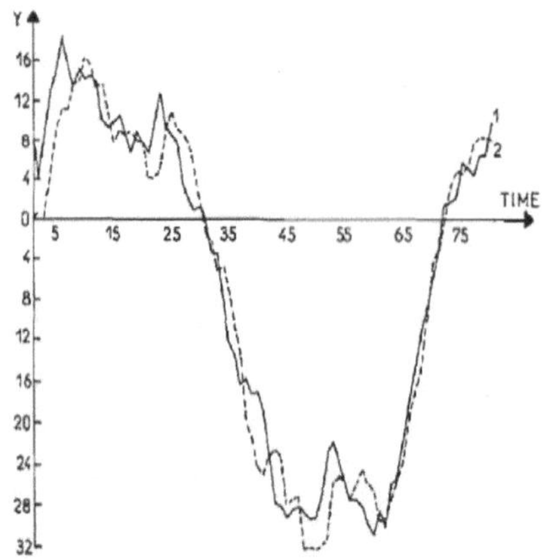

Figure 11.4: The decentralized Kalman filter: first state and estimate

The state vector of the jth subsystem $x_j(k+1)$ is partitioned into two parts:

$$x_j(k+1) = \begin{bmatrix} x_{j1}(k+1) \\ \cdots\cdots\cdots\cdots \\ x_{j2}(k+1) \end{bmatrix}, \quad j = 1, 2, \ldots, n_s \tag{11.21}$$

where

$$x_{i1} \in \Re^{n_{j,1}},\ x_{j2} \in \Re^{n_{j,2}} \text{ and } \quad n_j = n_{j,1} + n_{j,2}$$

To comply with the principles of overlapping decomposition, each subsystem, except the first one, is expanded to include the second state vector of the preceding subsystem. The result is an overlapping decomposition scheme for the overall system for which the state vector of the *ith* expanded subsystem is given by

$$\tilde{x}_j(k+1) = \begin{bmatrix} x_{j-1,2}(k+1) \\ x_{j,1}(k+1) \\ x_{j,2}(k+1) \end{bmatrix}; \quad j = 2, 3, \ldots, n_s \tag{11.22}$$

Consequently, the state and output equations of the expanded overall system $\tilde{\mathrm{S}}$ are expressed by:

$$\tilde{x}(k+1) = \tilde{A}\tilde{x}(k) + \tilde{B}w(k) \tag{11.23}$$
$$\tilde{y}(k+1) = \tilde{H}\tilde{x}(k+1) + v(k+1) \tag{11.24}$$

where $\tilde{x}(k+1) \in \Re^n$ is the overall expanded state vector. Following the results of Chapter 5, we have the fundamental relations

$$\tilde{A} = TAT^t + M,\ \tilde{B} = TB + N,\ \tilde{H} = HT^t + L,\ \tilde{x} = Tx$$

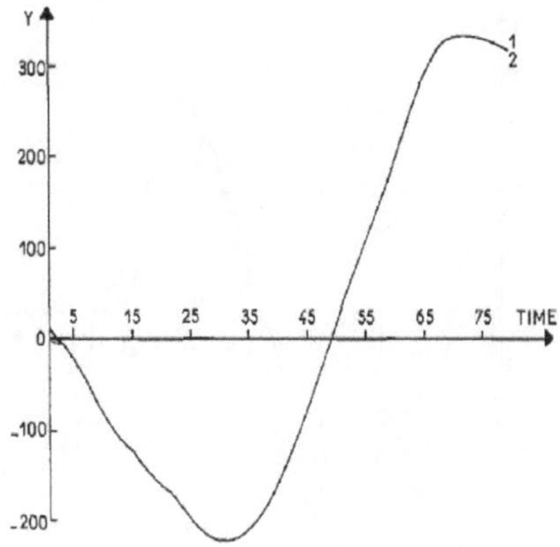

Figure 11.5: The decentralized Kalman filter: second state and estimate

M, N, and L are complementary matrices of appropriate dimensions and T^t is the generalized inverse of $T \in \Re^{\tilde{n} \times n}$ which is a transformation matrix given by

$$T = \begin{bmatrix} I_{j,1} & & & & & \\ & I_{j,2} & & & & \\ & I_{j,2} & & 0 & & \\ & & I_{2,1} & & & \\ & 0 & & \ddots & & \\ & & & & I_{n_s,1} & \\ & & & & & I_{n_s,2} \end{bmatrix} \tag{11.25}$$

and

$$T' = (T^t T)^{-1} \tag{11.26}$$

where $I_{j,1}$ is an identity matrix with dimension $(n_j - n_{j,2}) \times (n_j - n_{j,2})$ and $I_{j,2}$ is an identity matrix with dimension $n_{j,2} \times n_{j,2}$, $j = 1, 2, 3, \ldots, n_s$.

11.2.3 Fault detection technique

In what follows, we consider the problem of detecting the malfunctioning sensors of the augmented system $\tilde{S}$, which comprises n_s overlapping subsystems. This will be carried out through the design of decentralized Kalman filters for the subsystems and by comparing the estimated states, which are obtained by two successive filters for each subsystem.

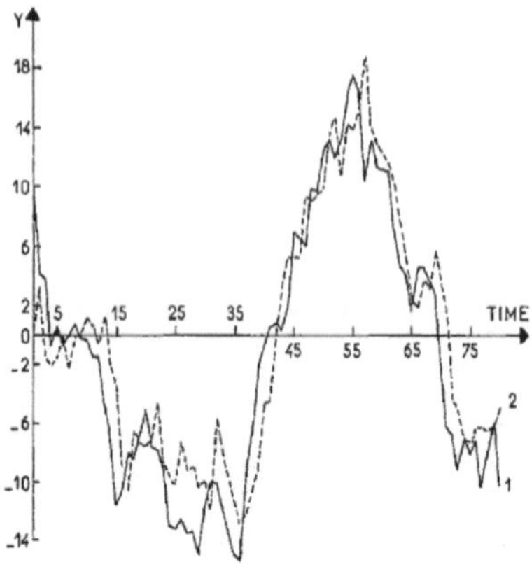

Figure 11.6: The decentralized Kalman filter: third state and estimate

Neglecting the interactions with other subsystems for the time being, the equations of the *ith* expanded subsystem $\tilde{S}_j$ are given by

$$\begin{aligned} \tilde{x}_j(k+1) &= \tilde{A}_j\tilde{x}_j(k) + \tilde{B}_j w_j(k) &(11.27)\\ \tilde{y}_j(k+1) &= \tilde{H}_j\tilde{x}_j(k) + v_j(k+1) &(11.28)\end{aligned}$$

It is interesting to emphasize that the results obtained for suboptimal decentralized controllers can be applied by duality to the problem of decentralized estimators [224]. Thus, in order to estimate the state $\tilde{x}_j(k)$, we construct a set of decentralized Kalman filters as follows:

$$\begin{aligned} \hat{x}_j(k|k) &= \tilde{A}_j\hat{x}_j(k-1|k-1) + K_j\bar{y}_j(k) &(11.29)\\ \bar{y}_j(k) &= y_j(k) - \tilde{H}_j\tilde{A}_j\hat{x}_j(k-1|k-1) &(11.30)\end{aligned}$$

where K_j is the gain matrix computed by

$$k_j = P_j(k|k-1) - \tilde{H}^t[\tilde{H}_jP_j(k|k-1)\tilde{H}_j^t + R_j]^{-1} \quad (11.31)$$

and $P_j(k|k-1)$ is the covariance matrix of the predicted error, such that

$$P_j(k/k-1) = \tilde{A}_jP_j(k/k)\tilde{A}_j^t + \tilde{B}_jQ_j\tilde{B}_j^t \quad (11.32)$$

Also, $P_j(k/k)$ is the filtered error covariance matrix given by

$$P_j(k|k) = P_j(k|k-1) - K_j\tilde{H}_jP_j(k|k-1) \quad (11.33)$$

Owing to overlapping decomposition, the state vectors $\tilde{x}_j$ and $\tilde{x}_{i-1}$ share the part $\tilde{x}_{i-1,2}$, that is,

$$\tilde{x}_{j-1} = \begin{bmatrix} \tilde{x}_{j-2,2} \\ \tilde{x}_{j-1,1} \\ \tilde{x}_{j-1,2} \end{bmatrix} \quad \text{and} \quad \tilde{x}_j = \begin{bmatrix} \tilde{x}_{j-1,2} \\ \tilde{x}_{j,1} \\ \tilde{x}_{j,2} \end{bmatrix}$$

Let $K_1, K_2, \ldots, K_{n_s}$ denote the N gain matrices of the local Kalman filters, and $[\hat{x}_{j-1,2}(k/k)]_{s.s.(j-1)}, [\hat{x}_{j-1,2}(k/k)]_{s.s.(j)}$ represent the estimated values of the state vector $\tilde{x}_{j-1,2}$ from the filters of subsystems $j-1$ and j, respectively. During normal operation of the overall system we have

$$\| [\hat{x}_{j-1,2}(k|k)]_{s.s.(j-1)} - [\hat{x}_{j-1,2}(k|k)]_{s.s.(j)} \| < \epsilon \tag{11.34}$$

where ϵ is a sufficiently small positive number and $j = 2, 3, \ldots, n_s$.

Interestingly enough, if one of the n_s subsystem sensors is malfunctioning, the above condition will then be violated. Thus, by inspecting the validity of (11.34), we could not only detect the sensor failure among the n_s subsystems, but also know which one has failed.

It is important to emphasize that overlapping decomposition can be used for the design of state feedback decentralized controllers which provide redundancy in the overall system, hence yielding improved system reliability. In this case, the estimated states from the set of decentralized Kalman filters can be used to obtain the overlapping controllers. Therefore, the design procedure can be used for both control and failure detection purposes. Figure 11.7 represents the structure of this failure detection scheme, in which the comparison of the state variables takes place at the higher level in a hierarchical structure of the system.

11.2.4 Simulation results

To illustrate the developed fault-detection technique, we consider a simulation example of a ladder network that may represent a high-voltage transmission network or an LC filtering circuit. The state dynamics and output equations are given in the form

$$\dot{x}(t) = \mathbf{G}x(t) + \mathbf{F}w(t) + c \tag{11.35}$$

$$y(t) = \mathbf{H}x(t) + v(t) \tag{11.36}$$

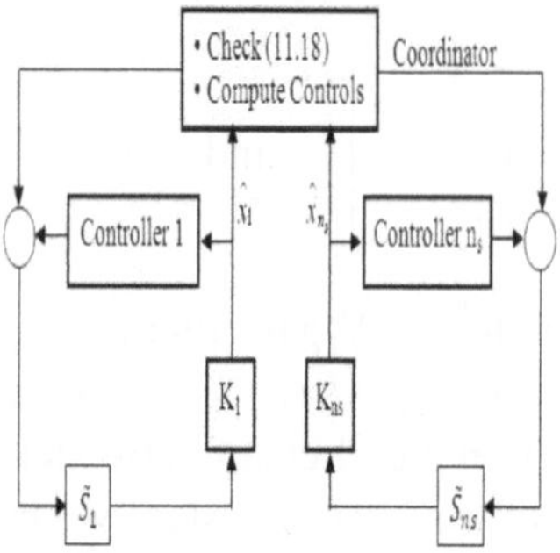

Figure 11.7: Structure of failure detection scheme

where $x, w \in \Re^6$; $y, v \in \Re^3$ and $\mathbf{G}$, $\mathbf{F}$, $\mathbf{H}$, and c are given by

$$\mathbf{G} = \begin{bmatrix} -1.5 & 1.2 & 0.0 & 0.0 & 0.0 & 0.0 \\ 0.8 & -2.4 & 1.6 & 0.0 & 0.0 & 0.0 \\ 0.0 & 1.6 & -2.4 & 0.8 & 0.0 & 0.0 \\ 0.0 & 0.0 & 0.6 & -1.2 & 0.6 & 0.0 \\ 0.0 & 0.0 & 0.0 & 0.8 & -2.4 & 1.6 \\ 0.0 & 0.0 & 0.0 & 0.0 & 4.8 & -4.8 \end{bmatrix}$$

$\mathbf{F}$ is a 6×6 identity matrix.

$$\mathbf{H} = \begin{bmatrix} 0.0 & 1.0 & 0.0 & 0.0 & 0.0 & 0.0 \\ 0.0 & 0.0 & 0.0 & 1.0 & 0.0 & 0.0 \\ 0.0 & 0.0 & 0.0 & 0.0 & 0.0 & 1.0 \end{bmatrix}$$

$$c = [0.291\ 0.008\ -0.136\ 0.0\ 0.024\ 0.0]^t$$

The above model is then discretized, assuming that $w(t)$ and $V(t)$ change only at equally spaced sampling instants. Thus, the discrete model for a sampling period $T = 0.001$ s is given by

$$x(k+1) = \mathbf{A}x(k) + w(k) + \mathbf{c} \quad (11.37)$$

$$y(k+1) = \mathbf{H}x(k+1) + v(k+1) \quad (11.38)$$

where

$$\mathbf{A} = \begin{bmatrix} 0.985 & 0.012 & 0.000 & 0.000 & 0.000 & 0.000 \\ 0.008 & 0.976 & 0.016 & 0.000 & 0.000 & 0.000 \\ 0.000 & 0.016 & 0.976 & 0.008 & 0.000 & 0.000 \\ 0.000 & 0.000 & 0.006 & 0.988 & 0.006 & 0.000 \\ 0.000 & 0.000 & 0.000 & 0.008 & 0.976 & 0.016 \\ 0.000 & 0.000 & 0.000 & 0.000 & 0.048 & 0.952 \end{bmatrix}$$

$$\mathbf{c} = \begin{bmatrix} 0.00291 & 0.00008 & -0.00136 & 0.0 & 0.00024 & 0.0 \end{bmatrix}^t$$

$w(k)$ is assumed to be a zero mean Gaussian random vector with covariance

$$\mathbf{Q} = \text{diag} \begin{bmatrix} 0.001 & 0.001 & 0.001 & 0.001 & 0.001 & 0.001 \end{bmatrix}$$

$v(k)$ is assumed to be a Gaussian random vector with zero mean and covariance

$$\mathbf{R} = \text{diag} \begin{bmatrix} 0.001 & 0.001 & 0.001 \end{bmatrix}$$

The initial state $x(0)$ is a Gaussian random vector with zero mean and covariance matrix $P(O)$, given by

$$P(0) = \text{diag} \begin{bmatrix} 0.526 & 0.526 & 0.526 & 0.526 & 0.526 & 0.526 \end{bmatrix}$$

For the purpose of failure detection, the overlapping decomposition is applied to the system (11.37) and (11.38), resulting in the following expanded system:

$$\tilde{x}(k+1) = \tilde{A}\tilde{x}(k) + \tilde{w}(k) + \tilde{c} \quad (11.39)$$

$$y(k+1) = \tilde{H}\tilde{x}(k+1) + v(k+1) \quad (11.40)$$

where

$$\tilde{c} = \begin{bmatrix} 0.0029 & 0.0008 & 0.0008 & -0.00136 & 0.0 & 0.0 & 0.024 & 0.0 \end{bmatrix}^t$$

$$\tilde{\mathbf{A}} = \begin{bmatrix} 0.985 & 0.012 & 0.000 & 0.000 & 0.000 & 0.000 & 0.000 & 0.000 \\ 0.008 & 0.976 & 0.000 & 0.016 & 0.000 & 0.000 & 0.000 & 0.000 \\ \hdashline 0.008 & 0.000 & 0.976 & 0.016 & 0.000 & 0.000 & 0.000 & 0.000 \\ 0.000 & 0.000 & 0.016 & 0.976 & 0.008 & 0.000 & 0.000 & 0.000 \\ 0.000 & 0.000 & 0.000 & 0.006 & 0.988 & 0.000 & 0.006 & 0.000 \\ \hdashline 0.000 & 0.000 & 0.000 & 0.000 & 0.000 & 0.988 & 0.006 & 0.000 \\ 0.000 & 0.000 & 0.000 & 0.000 & 0.000 & 0.008 & 0.976 & 0.016 \\ 0.000 & 0.000 & 0.000 & 0.000 & 0.000 & 0.000 & 0.048 & 0.952 \end{bmatrix}$$

$$\tilde{\mathbf{H}} = \begin{bmatrix} 0.0 & 1.0 & 0.0 & 0.0 & 0.0 & 0.0 & 0.0 & 0.0 \\ \cdots & \cdots & \cdots & \cdots & \cdots & \cdots & \cdots & \cdots \\ 0.0 & 0.0 & 0.0 & 0.0 & 1.0 & 0.0 & 0.0 & 0.0 \\ \cdots & \cdots & \cdots & \cdots & \cdots & \cdots & \cdots & \cdots \\ 0.0 & 0.0 & 0.0 & 0.0 & 0.0 & 0.0 & 0.0 & 1.0 \end{bmatrix}$$

$$\tilde{x} = \begin{bmatrix} x_1 & x_2 & x_2 & x_3 & x_4 & x_4 & x_5 & x_6 \end{bmatrix}^t$$

$$y = \begin{bmatrix} y_1 & y_2 & y_3 \end{bmatrix}^t$$

The overall system is decomposed into three interconnected subsystems, as shown by the dotted lines in $\tilde{A}$ and $\tilde{x}$. The decentralized Kalman filters are computed for the subsystems. The steady-state gain matrices of the filters are given below:

$$\text{Subsystem 1} : \mathbf{k_1} = \begin{bmatrix} 0.15722 \\ 0.61377 \end{bmatrix}$$

$$\text{Subsystem 2} : \mathbf{k_2} = \begin{bmatrix} 0.08593 \\ 0.13266 \\ 0.61600 \end{bmatrix}$$

$$\text{Subsystem 3} : \mathbf{k_3} = \begin{bmatrix} 0.17980 \\ 0.40430 \\ 0.62008 \end{bmatrix}$$

The computed gain matrices were used to estimate the system states in the following cases:

1. Normal operation. For this case, we have

$$|[\tilde{x}_2(k/k)]_{s.s.1} - [\tilde{x}_2(k/k)]_{s.s.2}| < \epsilon_1$$

and

$$|[\hat{x}_4(k/k)]_{s.s.1} - [\hat{x}_2(k/k)]_{s.s.3}| < \epsilon_2$$

The results obtained are shown in Figures 11.8 and 11.9, and ϵ_1 and ϵ_2 are constants which are usually determined by the experience of the designer. In this application, ϵ_1 and ϵ_2 were taken to be equal to 0.1.

In the following cases, failures of the sensors of subsystems (11.17), (11.18), and (11.19) are assumed to occur one at a time. Sufficient time is given for the estimators to produce satisfactory values of the estimated states before the injection of sensor failures. To simulate these failures, the output of the respective sensor falls to zero after 1.2 s.

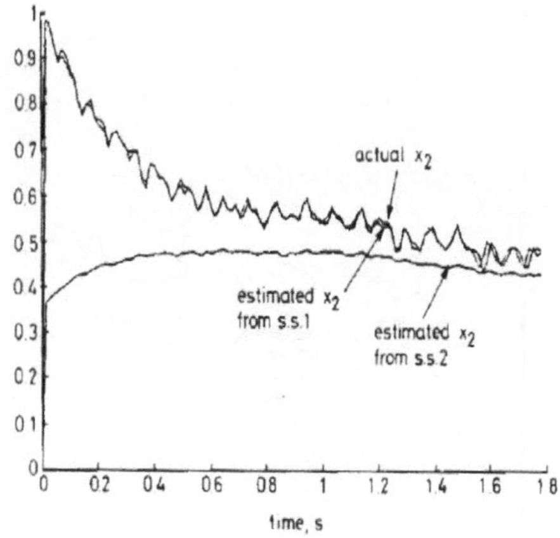

Figure 11.8: Case I: Actual x_2 and estimated x_2

2. The sensors of subsystem 1 have failed $(y_1 = 0)$ after 1.2 s (1200 recursions). As a result, we have

$$|[\tilde{x}_2(k/k)]_{s.s.1} - [\tilde{x}_2(k/k)]_{s.s.2}| > \epsilon_1$$

and

$$|[\hat{x}_4(k/k)]_{s.s.2} - [\hat{x}_4(k/k)]_{s.s.3}| < \epsilon_2$$

The estimate of x_2 from subsystem 1 is highly affected, whereas that from subsystem 2 is not. Consequently, the difference between these two estimates exceeds ϵ_1. The estimates of x_4 from subsystems 2 and 3 are not affected by this failure. The results are shown in Figures 11.10 and 11.11, respectively.

3. The sensors of subsystem 2 have failed $(y_2 = 0)$ after 1.2 s (1200 recursions), thus:

$$|[\tilde{x}_2(k/k)]_{s.s.1} - [\tilde{x}_2(k/k)]_{s.s.2}| > \epsilon_1$$

and

$$|[\hat{x}_4(k/k)]_{s.s.2} - [\hat{x}_4(k/k)]_{s.s.3}| > \epsilon_2$$

In this case, as shown by Figures 11.12 and 11.13, the difference between the estimates of x2 obtained by subsystems 1 and 2, as well as that between the estimates of x_4 from subsystems 2 and 3, exceeds the prespecified values of ϵ_l and ϵ_2.

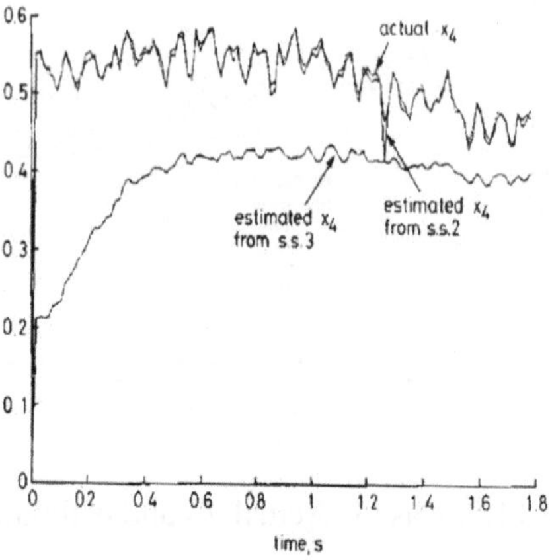

Figure 11.9: Case I:Actual x_4 and estimated x_4

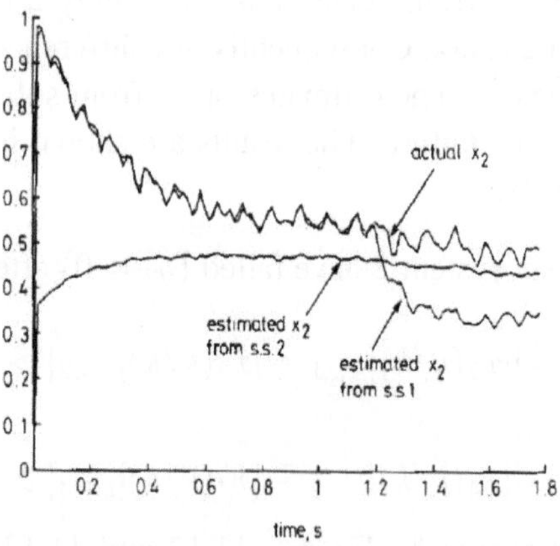

Figure 11.10: Case II: Actual x_2 and estimated x_2

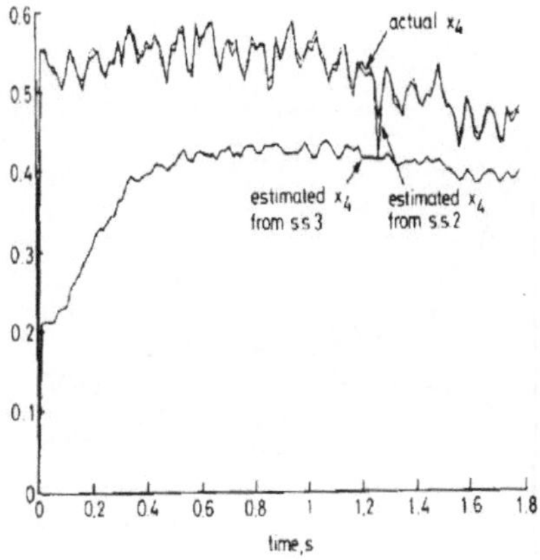

Figure 11.11: Case II: Actual x_4 and estimated x_4

4. The sensors of subsystem 3 have failed $(y_3 = 0)$ after 1.2 s (1200 recursions), yielding

$$|[\hat{x}_2(k/k)]_{s.s.1} - [\hat{x}_2(k/k)]_{s.s.2}| < \epsilon_1$$

and

$$|[\hat{x}_4(k/k)]_{s.s.2} - [\hat{x}_4(k/k)]_{s.s.3}| > \epsilon_2$$

As shown in Figures 11.14 and 11.15, the estimates of x_2 are not affected, whereas the difference between the estimates of x_4 exceeds ϵ_2. Figures 11.16 - 11.21 show the estimated states in the different cases of affected sensors. The obtained results, demonstrated by the previous figures, indicate that the proposed fault detection scheme not only detects failure, but can also localize the malfunctioning sensors in the system. This is achieved by overlapping decomposition of the system, and the design of a set of decentralized Kalman filters for the subsystems. It is shown that, by comparing the estimates of the states shared by the overlapped subsystems, it is possible to detect and isolate the malfunctioning sensors in the system. The results of the simulation example emphasize the efficiency of the proposed technique.

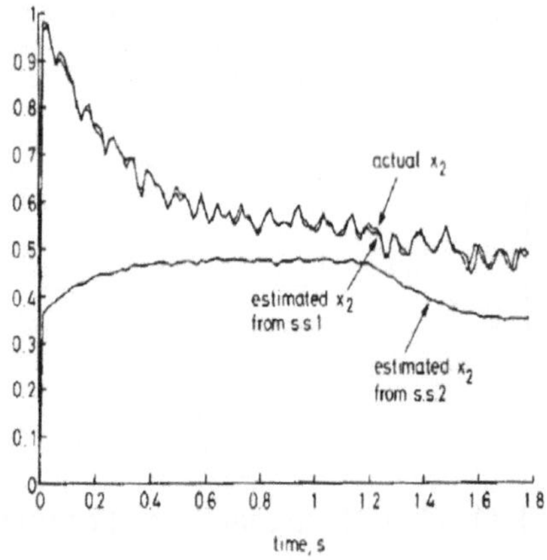

Figure 11.12: Case III: Actual x_2 and estimated x_2

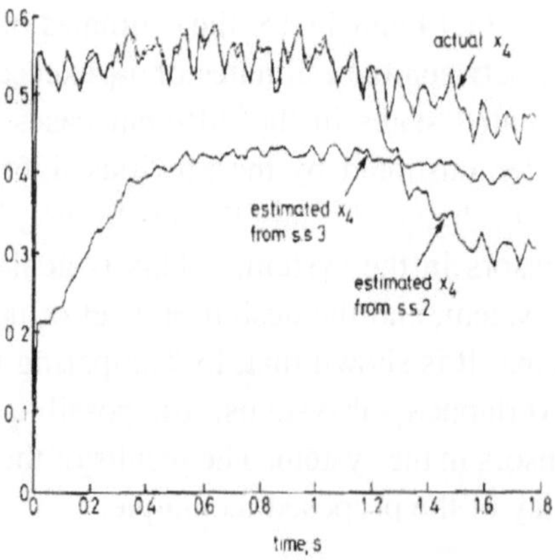

Figure 11.13: Case III: Actual x_4 and estimated x_4

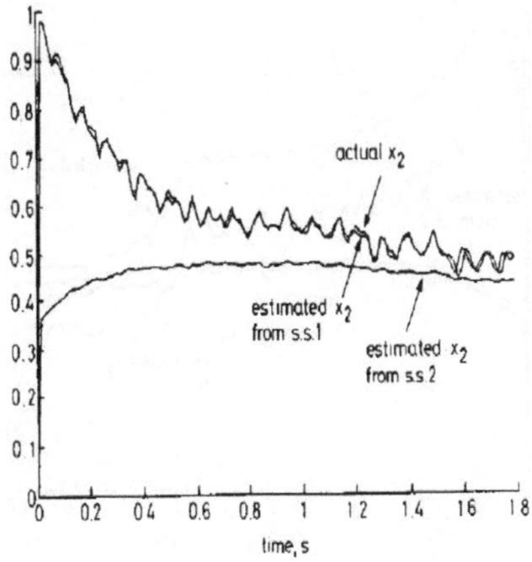

Figure 11.14: Case IV: Actual x_4 and estimated x_4

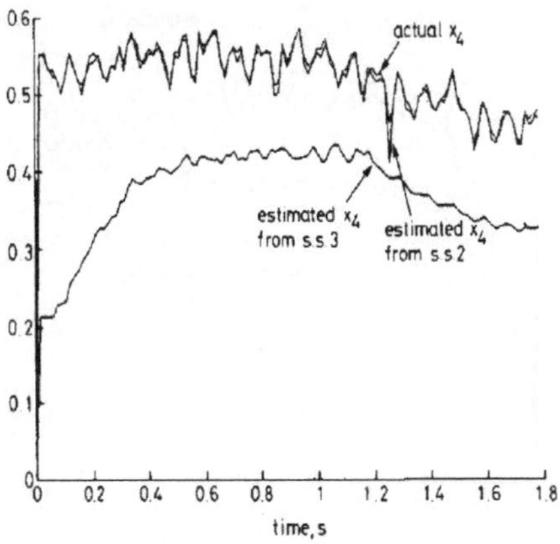

Figure 11.15: Case IV: Actual x_4 and estimated x_4

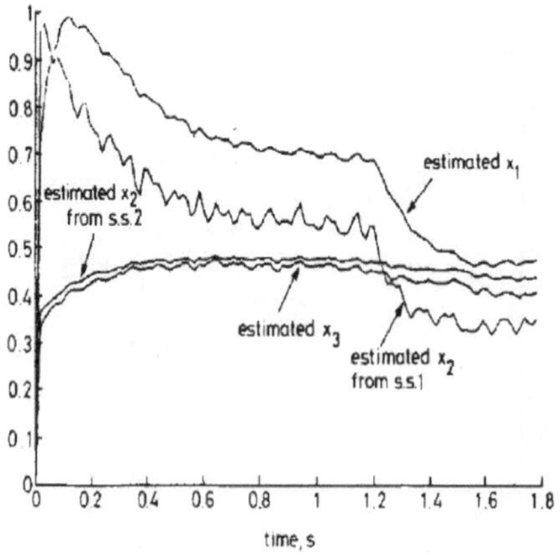

Figure 11.16: Case II: Estimated x_1, x_2, and x_3

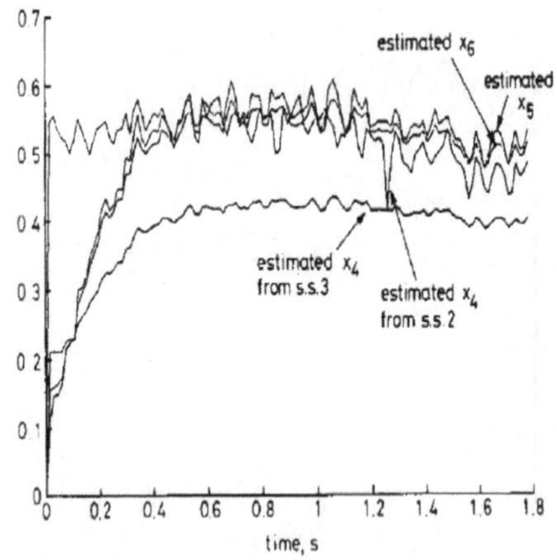

Figure 11.17: Case II: Estimated x_4, x_5, and x_6

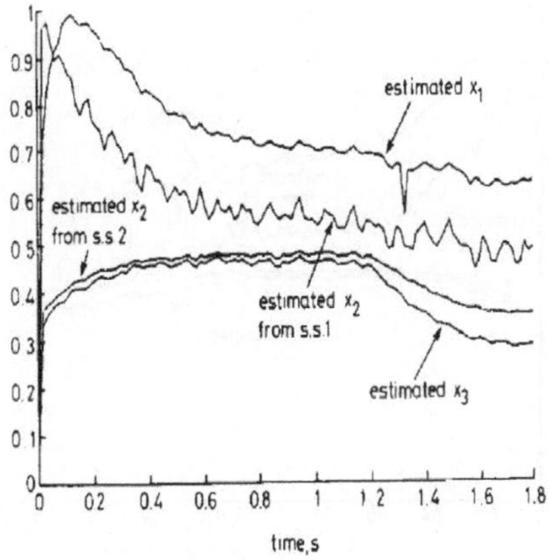

Figure 11.18: Case III: Estimated x_1, x_2, and x_3

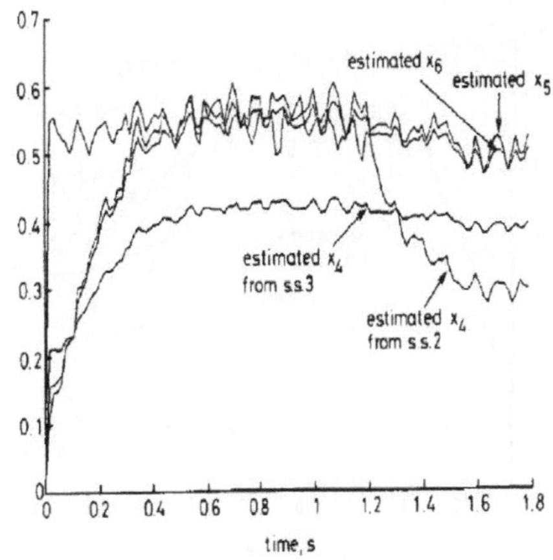

Figure 11.19: Case III: Estimated x_4, x_5, and x_6

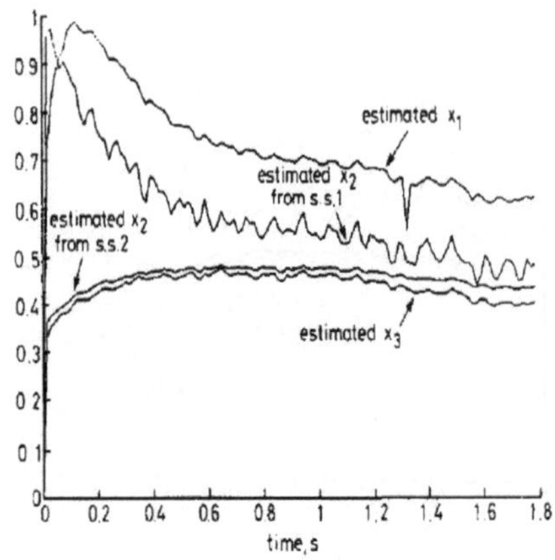

Figure 11.20: Case IV: Estimated x_1, x_2, and x_3

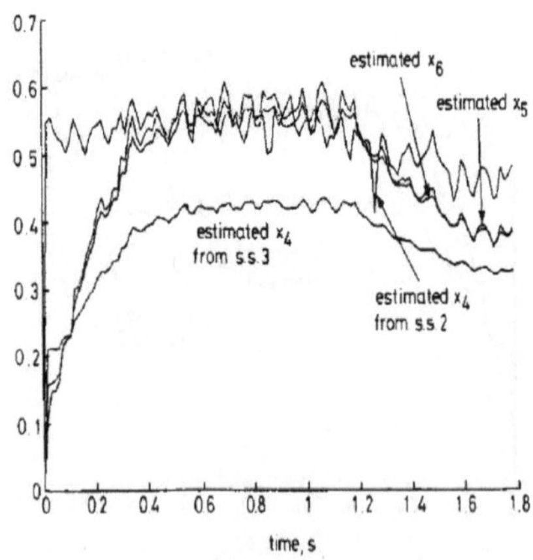

Figure 11.21: Case IV: Estimated x_4, x_5, and x_6

11.3 Problem Set IX

Problem IX.1: Consider a collection of n_s linear coupled static models of the form

$$z_j = H_{jj}\theta_j + \sum_{m=1,m\neq j}^{n_s} H_{jm}\theta_m + \psi_j, \; j = 1, ..., n_s$$

where $z_j \in \Re^{n_j}$, $\theta_j \in \Re^{m_j}$, and $\psi_j \in \Re^{m_j}$ are the subsystem measurement vector, the constant parameter vector, and the vector of measurement errors, respectively. Also, $H_{jj} \in \Re^{n_j \times m_j}$ is the modulation matrix with known elements. It is assumed that θ_j, ψ_j are independent Gaussian random vectors with the parameters

$$\mathbf{E}\theta_j = \mu_j, \; var\theta_j = P_j, \; \mathbf{E}\psi_j = 0, \; var\psi_j = Q_j$$

The objective is to derive a nonrecursive two-level algorithm that computes the optimum minimum variance estimate. You may extend the ideas of multiple projection approach. Provide an assessment of the computational requirements.

Problem IX.2: Consider the model of **Problem IX.1**. Develop a two-level recursive algorithm for the minimum variance estimator and evaluate the structure of the different covariance matrices. Show the performance of the algorithm on the following single-input single-output discrete transfer function with unknown coefficients

$$\frac{y(z)}{u(z)} = \frac{b_o + b_1 z^{-1} + \cdots + b_n z^{-n}}{1 + a_1 z^{-1} + \cdots + a_n z^{-n}}$$

Consider two specific model parameters:

$$\begin{aligned} Model\ 1 &= b_o = 0, \; b_1 = 1, \; b_2 = 0.5, \; a_1 = -1.5, \; a_2 = 0.7 \\ Model\ 2 &= b_o = 1, \; b_1 = 0.1, \; b_2 = -0.8, \; b_3 = 0, \; b_4 = -0.45, \\ & a_1 = -2.3, \; a_2 = 1.93, \; a_3 = -0.713, \; a_4 = 0.11, \\ & a_5 = -0.006 \end{aligned}$$

and use appropriate values for P_j and Q_j.

Problem IX.3: A decentralized control system in standard feedback configuration with the actuator and decentralized controller are in cascade with the process in the forward path and the sensor in the feedback path is considered. Let the state-space realization be given by

$$\begin{aligned} \dot{x}(t) &= A\,x(t) + B\,u_a(t), \\ y(t) &= C\,x(t) \end{aligned}$$

where $x \in \Re^n$ is the state vector, $u_a \in \Re^m$ is the actuator output vector, and $y \in \Re^m$ is the process output vector. The outputs from the controller and the sensor are $u \in \Re^m$ and $y_s \in \Re^m$, respectively. The model of actuator and sensor failures can be represented by

$$\begin{aligned} u_a(t) &= E_a\,x(t) + f_a, \\ y_s &= E_s\,y(t) + f_s \end{aligned}$$

where $E_a,\ E_s \in \Re^{n \times n}$ are the diagonal actuator and sensor fault matrices that influence the closed-loop stability. Design the decentralized output-feedback controller

$$u_a(t) = F\,y(t) + r, \quad F \in \Im \subset \Re^{m \times m}$$

such that the resulting closed-loop system is asymptotically stable, where $\Im \subset \Re^{m \times m}$ denotes the set of matrices that have prespecified structures. Discuss the features of several controller forms in detecting several fault patterns.

Problem IX.4: Consider a linear discrete model that includes possible loss of control effectiveness of the form

$$\begin{aligned} x_{k+1} &= A_k\,x_k + B_k\,u_k + E_k \gamma_k + w_k, \\ y_{k+1} &= C_k\,x_{k+1} + v_{k+1} \end{aligned}$$

where $x_k \in \Re^n$, $u_k \in \Re^m$, and $y_{k+1} \in \Re^p$ are the state, control input, and output variables, respectively, and w_k and v_{k+1} denote the white noise sequences of uncorrelated Gaussian random vectors with zero means and covariance matrices Q_k and R_{k+1}, respectively. The initial state x_o is specified as a random Gaussian vector with means ϕ and covariance P_o. The matrices $E_k,\ \gamma_k$ are specified by

$$\begin{aligned} E_k &= B_k\,U_k,\ U_k = Blockdiagonal[u_k^1,\ u_k^2,\ ...,\ u_k^m], \\ \gamma_k &= \mathbf{col}[\gamma_k^1,\ \gamma_k^2,\ ...,\ \gamma_k^m] \end{aligned}$$

The objective is to examine the loss in control effectiveness by introducing the control effectiveness factors such that

$$\gamma_{k+1} = \gamma_k + w_k, \quad -1 \le \gamma_{jk} \le 0,\ j = 1, ..., m$$

with the mean and covariance of γ_k being 0 and S, respectively. Develop the minimum variance solution by a direct application of a two-stage Kalman filter algorithm. Then examine the application of controller

$$u_k = -F_k \hat{x}_{k|k} - H_k \hat{\gamma}_{k|k}$$

by invoking the separation theorem and study the behavior of the closed-loop system.

11.4 Notes and References

In the first part of this chapter, an improved decentralized algorithm has been developed for optimal state estimation in large-scale linear interconnected dynamical systems. The algorithm essentially embodies a series of successive orthogonalization on the measurement subspaces for each subsystem within a hierarchical computational structure in order to provide the optimal estimate. In turn, this ensures substantial savings in computation time. Additionally, since only low-order subsystem equations are manipulated at each stage, numerical inaccuracies are drastically reduced and the filter remains stable for even high-order systems. The hierarchical structure is ideally suited for implementation using a parallel processing system and the effectiveness of the algorithm has been demonstrated on multiple machine power systems. Ample extensions demand further investigation to improve the efficiency and widen the applicability of decentralized Kalman filtering.

In the second part of this chapter, we have developed a decentralized approach for failure detection in large-scale interconnected stochastic dynamic systems. The approach exploits the advantages of overlapping decomposition technique and the flexible design of a set of decentralized Kalman filters for the subsystems. By comparing the estimates of the states shared by the overlapped subsystems, it is shown in principle to achieve the detection and isolation of the malfunctioning sensors in the system.

Chapter 12

Epilogue

In this closing chapter, it is our hope that an agreement must have be reached that *decentralized systems* would be the most convenient designation for large-scale systems, complex systems, or interconnected systems since the chief goal in these systems is to deploy decentralization in the analysis, control, filtering, and processing tasks. Equivalently stated, the effort of any task is *distributed* among various units that are cooperating to achieve the desired objective. In what follows, a critical evaluation of the achieved versus remaining results is made with reference to the major ideas including decomposition, coordination, decentralization, and overlapping. In particular, our purpose hereafter is to shed more light on some prevailing aspects and potential remarks in decentralized systems.

12.1 Processing Aspects

Initially, we note that a comparison between distributed information processing (DIP) and centralized information processing (CIP) schemes reveals the following features:

1. *The computations within DIP schemes are much easier because less information is needed and manipulated.*

2. *Parallel computation of DIP is possible without the need for complete synchronization.*

3. *The computation is reliable of DIP as far as failures in the computing elements are concerned.*

This eventually leads us to the first potential remark that decentralized schemes can be applied by means of low-cost computing facilities. However, the underlying off-line effort behind DIP is quite tremendous. The problem of decomposing the overall analytical and design problems adequately is crucial in order to guarantee that the resulting subproblems are relatively autonomous (or have some degrees of freedom) if strong coordination is to be avoided. This in turn poses a fundamental compromise between off-line efforts and on-line processing in the sense that to make sacrifice at the beginning would be to gain advantages at the end.

Therefore the degree of success made in developing large-scale systems theories has to be contrasted with regard to its contribution to the solution of this fundamental problem.

12.2 Modeling and Analysis Aspects

A predominant characteristic of large-scale systems (LSS) is the underlying structure of different interacting subsystems. This is reflected by examining the models explained in Chapter 3. It is quite natural to seek clarification of the role in which the performance of LSS depends on the properties of the subsystems and the interaction relations. It must be emphasized that the features, merits, and demerits of the coupling pattern are of primary concern; particularly, one has to investigate whether the subsystems are weakly or strongly coupled and whether the couplings turn out to be cooperative or competitive in nature. By and large, our guided tour throughout the book has shown that interconnected systems can be classified into three main groups as follows.

12.2.1 Serially coupled systems

In typical systems appearing in water resources and petrochemical industries, the subsystems have a unidirectional pattern of interaction arising from the mass or energy flow. In those cases, the model structures might dictate a technique for the underlying analysis and design problems including coordinated control. In this way, the analysis and design subproblems are directly associated with the subsystems in addition to *a coordinator* that supervises the overall job.

12.2.2 Weakly coupled subsystems

As demonstrated throughout the book, we mean here that LSS is comprised of interacting subsystems but the interactions influence the subsystem behavior weakly. By and large, interactions are weak if the links have low reinforcements or if the time-scales of the subsystems are quite different. There are numerous measures for weak-coupling; see Chapter 3. Analytical and design problems of LSS with weakly coupled subsystems can be decomposed according to the subsystem structure of the overall system, but the resulting subproblems are still dependent upon each other. Iterative procedures are invoked to reduce this interdependence.

12.2.3 Strongly coupled subsystems

This is evidently the general case. Due to the strong interdependence among subsystems, analytical and design problems cannot be effectively divided along the boundaries of the subsystems. Rather, each subsystem has to be considered within its own surrounding, or some approximation of it. In this regard, diagonal dominance (Chapter 3), overlapping decomposition (Chapter 5), or particular methods utilizing the symmetric structure of the system (Chapter 7) can be used to derive subproblems with weak interdependence.

With regards to the available decomposition methods, some prominant features are important to mention for the purpose of their evaluation and future extension:

1. *Decomposition methods rely heavily on structural properties of the overall system. Looked at in this light, graph theoretic methods seem to provide a promising direction for LSS with high dimension and uncertainties. Such a direction has not been fully exploited.*

2. *The boundaries defining individual subsystems appear to be a matter of engineering intuition. The appropriate criteria would be to select those parts of the overall system that identify subsystems with weak interactions, that belong to the slow or fast part of the system, or that have to be used as the overlapping parts. Strictly speaking, the question of where to place the boundaries of the subsystems in a practical example has not been satisfactorily answered and demands further research activities.*

3. *The degree of success of a given decomposition method hinges upon the appropriate design of the associated control stations to form a decentralized feedback system in which the couplings between the subsystems turn out to be weak. This is a sort of a posteriori judgment and deserves further investigations of reasonable decompositions that start from the closed-loop system.*

4. *The literature is nearly void of techniques to establish modeling and identification schemes of LSS that avoid setting up a model of the overall system but rather end up directly with separate models of the subsystems including their specific surroundings. Indeed, this is a challenging approach. Opposite to the prevailing trend of deriving the models of the subsystems from the overall system description, the ultimate objective would be new methods that have to be characterized to select that part or those properties of LSS that have to be referred to in the analysis of a certain subsystem or the design of a certain local control station.*

12.2.4 Multivariable design methods

We have learned from Chapter 4 that analytical and design methods based on the multivariable control theory have been satisfactorily extended to decentralized systems. It turns out that fixed modes occur as direct consequence of the on-line information structure constraints of decentralized control and therefore occur more for decentralized control than for centralized control. The design principles of pole assignment and optimal control have been extended to decentralized controllers (Chapters 3 and 4). It was shown clearly how the overall system behaves under the control of several independent control agents.

The extent of applicability of the multivariable (centralized) design principles to LSS poses major difficulties. The main reasons for this are:

1. *A complete model of the LSS has to be known,*

2. *Most of the manipulations have to be carried out with this large-order model,* and

3. *The design goals are formulated for the overall system.*

Strictly speaking, these design methods are, therefore, applicable only for relatively *small* systems, where decentralized controllers are to be used and where the dimensionality and uncertainty of the plant does not pose serious difficulties in the analytical and design problems. Exception to this is the treatment of decentralized systems with hierarchical structures; see Chapter 4.

12.2.5 Design of decentralized systems

Experience has shown that new problems arise whenever the information structure used in the off-line procedures of analysis and design is somehow constrained. Decentralized control is realized in the form of several control agents (independent with no communication) although, in reality, the subproblems resulting from a decomposition procedure of the overall problem are usually coupled (weakly or strongly). This approach could be phrased as *multicontroller structure* approach.

A major concern arises here: How can the discrepancy conflicts between the solutions of the subproblems be resolved?

Given the goals of the subsystem and the subsystem interactions, the subproblems can generally be cooperative or competitive. Theoretical investigations showed that in the former case, conflict resolution represents a minor problem, so that no or only a simple coordination is necessary. On the other hand in the latter case, major conflicts occur and pose the main difficulties for the solution.

Conflict resolution within a decentralized information structure has to be made possible by appropriately organizing the analytical or design process. Regarding the construction of analytical and design schemes with decentralized information structure can incorporate some sequencing order among control agents, in which case each design step has a centralized information structure. Since only one control station is considered at a time, the subsystem controller can be designed as a centralized feedback. The specification of the sequencing order is crucial to guarantee that the designed decentralized controller will eventually satisfy the design specifications stated for the overall system. Much more work is needed to derive design specifications for the kth design step from the given overall system specifications.

Alternatively, with the overall system being split into weakly coupled subsystems the design problem can then be decomposed accordingly and solved while ignoring the subsystem interactions. The control stations obtained from independent design problems are used together to constitute a first-level of a coordinated control. On the second level, there is another supervisory controller to accommodate the interaction pattern; see Chapter 3.

12.2.6 Application aspects

As repeatedly emphasized throughout the book, high dimensionality, information structure constraints, model uncertainty, gain perturbations, and time-delays characterize the complexity of large-scale systems. Depending on the practical application under consideration, one aspect or another dominates. All the methods described in this volume take account of these aspects to a different extent and thus have their specific fields of application.

A major facet pertaining to the large body of analytical and design methods is that the practically prevailing question of *which method should be applied to the practical problem at hand* is quite open. Engineering intuition and human experience are certainly needed in order to assess modeling and measurement information concerning its importance for the solution of a given problem and to structure the plant and the analytical and design processes accordingly.

In conclusion, the theory of feedback control of decentralized systems contributes to the application of two distinct but related areas: the control of large (interconnected), spatially distributed dynamical processes, and distributed computing for implementing control algorithms. It essentially provides the control engineer with the technical background and design algorithms that facilitate an efficient solution of complex control problems by using modern computing facilities.

Chapter 13

Appendix

In this appendix, we collect some useful mathematical inequalities and lemmas that have been extensively used throughout the book.

13.1 Stability Notions

In this section, we present some definitions and results pertaining to stability of dynamical systems. A detailed account of this topic can be found in [9, 176].

Definition 13.1.1 *A function of x and t is a* **carathedory function** *if, for all $t \in \Re$, it is continuous in x and for all $x \in \Re^n$, it is Lebesgue measurable in t.*

13.1.1 Practical stabilizability

Given the uncertain dynamical system

$$\begin{aligned} \dot{x}(t) &= [A + \Delta A(r) + M]x(t) + [B + \Delta B(s)]u(t) \\ &+ Cv(t) + H(t, x, r), \quad x(0) = x_o \end{aligned} \tag{13.1}$$

$$y(t) = x(t) + w(t) \tag{13.2}$$

where $x \in \Re^n$, $u \in \Re^m$, $y =\in \Re^n$, $v \in \Re^s$, $w \in \Re^n$ are the state, control, measured state, disturbance, and measurement error of the system, respectively, and $r \in \Re^p$, $s \in \Re^q$ are the uncertainty vectors. System (13.1)-(13.2) is said to be *practically stabilizable* if, given $\mathbf{d} > 0$, there is a control law $g(.,.) : \Re^m \times \Re \rightarrow \Re^m$, for which, given any admissible uncertainties r, s, disturbances $w \in \Re^n$, $v \in \Re^s$, any initial time $t_o \in \Re$, and any initial state $x_o \in \Re^n$, the following conditions hold:

1. *The closed-loop system*

$$\begin{aligned}\dot{x}(t) &= [A + \Delta A(r) + M]x(t) + [B + \Delta B(s)]g(y,t) \\ &+ Cv(t) + H(t,x,r) \end{aligned} \qquad (13.3)$$

   possesses a solution $x(.) : [t_o, t_1] \rightarrow \Re^n,\ x(t_o) = x_o$.

2. *Given any* $\nu > 0$ *and any solution* $x(.) : [t_o, t_1] \rightarrow \Re^n,\ x(t_o) = x_o$ *of system (13.3) with* $||x_o|| \leq \nu,$ *there is a constant* $d(\nu) > 0$ *such that* $||x(t)|| \leq d(\nu),\ \forall t \in [t_o, t_1]$.

3. *Every solution* $x(.) : [t_o, t_1] \rightarrow \Re^n$ *can be continued over* $[t_o,\ \infty)$.

4. *Given any* $\bar{d} \geq \mathbf{d},$ *any* $\nu > 0$ *and solution* $x(.) : [t_o, t_1] \rightarrow \Re^n,\ x(t_o) = x_o$ *of system (13.3) with* $||x_o|| \leq \nu,$ *there exists a finite time* $T(\bar{d}, \nu) < \infty,$ *possibly dependent on* ν *but not on* t_o*, such that* $||x(t)|| \leq \bar{d},\ \forall t \geq t_o + T(\bar{d}, \nu)$.

5. *Given any* $d \geq \mathbf{d}$ *and any solution* $x(.) : [t_o, t_1] \rightarrow \Re^n,\ x(t_o) = x_o$ *of system (13.3) there is a constant* $\delta(d) > 0$ *such that* $||x(t_o)|| \leq \delta d$ *implies* $||x(t)|| \leq \bar{d},\ \forall t \geq t_o$.

13.1.2 Razumikhin stability

A continuous function $\alpha : [0, a) \longmapsto [0, \infty)$ is said to belong to class $\mathcal{K}$ if it is strictly increasing and $\alpha(0) = 0$. Further, it is said to belong to class $\mathcal{K}_\infty$ if $a = \infty$ and $\lim_{r \rightarrow \infty}\ \alpha(r) = \infty$.

Consider a time-delay system

$$\dot{x}(t) = f(t, x(t - d(t)) \qquad (13.4)$$

with an initial condition

$$x(t) = (t), \quad t \in [-\bar{d}, 0]$$

where the function vector $f : \Re^+ \times \mathcal{C}_{[-\bar{d},0]} \mapsto \Re^n$ takes $\mathcal{R}\times$ (bounded sets of $\mathcal{C}_{[-\bar{d},0]}$) into bounded sets in $\Re^n$; $d(t)$ is the time-varying delay and $d := \sup_{t \in \Re^+}\{d(t)\} < \infty$. The symbol $\mathcal{C}_{[a,b]}$ represents the set of $\Re^n$-valued continuous function on $[a, b]$.

Lemma 13.1.2 *If there exist class* $\mathcal{K}_\infty$ *functions* $\zeta_1(.)$ *and* $\zeta_2(.),$ *a class* $\mathcal{K}$ *function* $\zeta_3(.)$ *and a function* $V_1(.) : [-\bar{d}, \infty] \times \Re^n \mapsto \Re^+$ *satisfying*

$$\zeta_1(||x||)\ \leq\ V_1(t,x)\ \leq\ \zeta_2(||x||), \quad t \in \Re^+, \quad x \in \Re^n$$

such that the time derivative of V_1 *along the solution of the system (13.4) satisfies*

$$\dot{V}_1(t,x) \leq -\zeta_3(||x||) \text{ if } \quad V_1(t+d, x(t+d)) \leq V_1(t, x(t)) \qquad (13.5)$$

for any $d \in [-\bar{d}, 0]$, *then system (13.4) is uniformly stable. If, in addition,*

$$\zeta_3(\tau) > 0, \ \tau > 0$$

and there exists a continuous nondecreasing function $\xi(\tau) > 0, \ \tau > 0$ *such that (13.5) is strengthened to*

$$\dot{V}_1(t,x) \leq -\zeta_3(||x||) \text{ if } \quad V_1(t+d, x(t+d)) \leq \xi(V_1(t, x(t))) \qquad (13.6)$$

for any $d \in [-\bar{d}, 0]$, *then system (13.4) is uniformly asymptotically stable. Further, if, in addition,* $\lim_{\tau\to\infty} \ \zeta_1(\tau) = \infty$, *then system (13.4) is globally uniformly asymptotically stable.*

The proof of this lemma can be found in [93].

Lemma 13.1.3 *Consider system (13.4). If there exists a function*

$$V_o(x) = x^t P x, \ P > 0$$

such that for $d \in [-\bar{d}, 0]$ *the time derivative of* V_o *along the solution of system (13.4) satisfies*

$$\dot{V}_o(t,x) \leq -q_1 \|x\|^2 \text{ if } \quad V_o(x(t+d)) \leq q_2 V_o(x(t)) \qquad (13.7)$$

for some constants $q_1 > 0$ *and* $q_2 > 1$, *then the system (13.4) is globally uniformly asymptotically stable.*

Proof: Since $P > 0$, it is clear that

$$\lambda_{\min}(P)\|x\|^2 \leq V_o(x) \leq \lambda_{\max}(P)\|x\|^2$$

Let $\zeta_1(\tau) = \lambda_{\min}(P)\tau^2$ and $\zeta_2(\tau) = \lambda_{\max}(P)\tau^2$. It is easy to see that both $\zeta_1(.)$ and $zeta_2(.)$ are class $\mathcal{K}_\infty$ functions and

$$\zeta_1(\|x\|) \leq V_0(x) \leq \zeta_2(\|x\|), \quad x\Re^n$$

Further, let $zeta_3() = -q_1 \ \tau^2$ and $\xi(\tau) = q_2 \ \tau$. It is evident from $q_1 > 0$ and $q_2 > 1$ that for $\tau > 0$,

$$\xi(\tau) > \text{and } \zeta_3(\tau) > 0$$

Hence, the conclusion follows from (13.7).

13.2 Basic Inequalities

All mathematical inequalities are proved for completeness. They are termed facts due to their high frequency of usage in the analytical developments.

13.2.1 Inequality 1

For any real matrices Σ_1 , Σ_2, and Σ_3 with appropriate dimensions and $\Sigma_3^t \Sigma_3 \le I$, it follows that

$$\Sigma_1\Sigma_3\Sigma_2 + \Sigma_2^t\Sigma_3^t\Sigma_1^t \le \alpha\, \Sigma_1\Sigma_1^t + \alpha^{-1}\, \Sigma_2^t\Sigma_2, \quad \forall \alpha > 0$$

Proof: This inequality can be proved as follows. Since $\Phi^t\Phi \ge 0$ holds for any matrix Φ, then take Φ as

$$\Phi = [\alpha^{1/2}\, \Sigma_1 - \alpha^{-1/2}\, \Sigma_2]$$

Expansion of $\Phi^t\Phi \ge 0$ gives $\forall \alpha > 0$

$$\alpha\, \Sigma_1\Sigma_1^t + \alpha^{-1}\, \Sigma_2^t\Sigma_2 - \Sigma_1^t\Sigma_2 - \Sigma_2^t\Sigma_1 \ge 0$$

which by simple arrangement yields the desired result.

13.2.2 Inequality 2

Let Σ_1, Σ_2, Σ_3 and $0 < R = R^t$ be real constant matrices of compatible dimensions and $H(t)$ be a real matrix function satisfying $H^t(t)H(t) \le I$. Then for any $\rho > 0$ satisfying $\rho\Sigma_2^t\Sigma_2 < R$, the following matrix inequality holds:

$$(\Sigma_3 + \Sigma_1 H(t)\Sigma_2)R^{-1}(\Sigma_3^t + \Sigma_2^t H^t(t)\Sigma_1^t) \le \rho^{-1}\Sigma_1\Sigma_1^t + \Sigma_3\Big(R - \rho\Sigma_2^t\Sigma_2\Big)^{-1}\Sigma_3^t$$

Proof: The proof of this inequality proceeds like the previous one by considering that

$$\Phi = [(\rho^{-1}\, \Sigma_2\Sigma_2^t)^{-1/2}\Sigma_2 R^{-1}\Sigma_3^t - (\rho^{-1}\, \Sigma_2\Sigma_2^t)^{-1/2} H^t(t)\Sigma_1^t]$$

Recall the following results:

$$\rho\Sigma_2^t\Sigma_2 < R,$$

$$[R - \rho\Sigma_2^t\Sigma_2]^{-1} = [R^{-1} + R^{-1}\Sigma_2^t[\rho^{-1}I - \Sigma_2 R^{-1}\Sigma_2^t]^{-1}\Sigma_2 R^{-1}\Sigma_2$$

and

$$H^t(t)H(t) \leq I \Longrightarrow H(t)H^t(t) \leq I$$

Expansion of $\Phi^t\Phi \geq 0$ under the condition $\rho\Sigma_2^t\Sigma_2 < R$ with standard matrix manipulations gives

$$\begin{aligned}
&\Sigma_3 R^{-1}\Sigma_2^t H^t(t)\Sigma_1^t + \Sigma_1 H(t)\Sigma_2 R^{-1}\Sigma_3^t + \Sigma_1 H(t)\Sigma_2\Sigma_2^t H^t(t)\Sigma_1^t \leq \\
&\rho^{-1}\Sigma_1 H(t)H^t(t)\Sigma_1^t + \Sigma_3^t R^{-1}\Sigma_2[\rho^{-1}I\ \Sigma_2\Sigma_2^t]^{-1}\Sigma_2 R^{-1}\Sigma_3^t \Longrightarrow \\
&(\Sigma_3 + \Sigma_1 H(t)\Sigma_2)R^{-1}(\Sigma_3^t + \Sigma_2^t H^t(t)\Sigma_1^t) - \Sigma_3 R^{-1}\Sigma_3^t \leq \\
&\rho^{-1}\Sigma_1 H(t)H^t(t)\Sigma_1^t + \Sigma_3^t R^{-1}\Sigma_2[\rho^{-1}I\ -\ \Sigma_2\Sigma_2^t]^{-1}\Sigma_2 R^{-1}\Sigma_3^t \Longrightarrow \\
&(\Sigma_3 + \Sigma_1 H(t)\Sigma_2)R^{-1}(\Sigma_3^t + \Sigma_2^t H^t(t)\Sigma_1^t) \leq \\
&\Sigma_3[R^{-1} + \Sigma_2[\rho^{-1}I\ -\ \Sigma_2\Sigma_2^t]^{-1}\Sigma_2 R^{-1}]\Sigma_3^t + \\
&\rho^{-1}\Sigma_1 H(t)H^t(t)\Sigma_1^t = \\
&\rho^{-1}\Sigma_1 H(t)H^t(t)\Sigma_1^t + \Sigma_3\Big(R - \rho\Sigma_2^t\Sigma_2\Big)^{-1}\Sigma_3^t
\end{aligned}$$

which completes the proof. $\nabla\nabla\nabla$

13.2.3 Inequality 3

For any real vectors β, ρ and any matrix $Q^t = Q > 0$ with appropriate dimensions, it follows that

$$-2\rho^t\beta \ \leq\ \rho^t\, Q\, \rho + \beta^t\, Q^{-1}\, \beta$$

Proof: Starting from the fact that

$$[\rho + Q^{-1}\beta]^t\, Q\, [\rho + Q^{-1}\beta]\ \geq 0\ ,\ \ Q > 0$$

which when expanded and arranged yields the desired result. $\nabla\nabla\nabla$

13.2.4 Inequality 4 (Schur complements)

Given a matrix Ω composed of constant matrices $\Omega_1, \Omega_2, \Omega_3$ where $\Omega_1 = \Omega_1^t$ and $0 < \Omega_2 = \Omega_2^t$ as follows:

$$\Omega\ =\ \begin{bmatrix} \Omega_1 & \Omega_3 \\ \Omega_3^t & \Omega_2 \end{bmatrix}$$

We have the following results:

(A) $\Omega \geq 0$ if and only if either

$$\left\{ \begin{array}{rcl} \Omega_2 & \geq & 0 \\ \Pi & = & \Upsilon \Omega_2 \\ \Omega_1 - \Upsilon \, \Omega_2 \, \Upsilon^t & \geq & 0 \end{array} \right. \tag{13.8}$$

or

$$\left\{ \begin{array}{rcl} \Omega_1 & \geq & 0 \\ \Pi & = & \Omega_1 \Lambda \\ \Omega_2 - \Lambda^t \, \Omega_1 \, \Lambda & \geq & 0 \end{array} \right. \tag{13.9}$$

hold where Λ, Υ are some matrices of compatible dimensions.

(B) $\Omega > 0$ if and only if either

$$\left\{ \begin{array}{rcl} \Omega_2 & > & 0 \\ \Omega_1 - \Omega_3 \, \Omega_2^{-1} \, \Omega_3^t & > & 0 \end{array} \right.$$

or

$$\left\{ \begin{array}{rcl} \Omega_1 & \geq & 0 \\ \Omega_2 - \Omega_3^t \, \Omega_1^{-1} \, \Omega_3 & > & 0 \end{array} \right.$$

hold where Λ, Υ are some matrices of compatible dimensions.

In this regard, matrix $\Omega_3 \, \Omega_2^{-1} \, \Omega_3^t$ is often called the Schur complement $\Omega_1(\Omega_2)$ in Ω.

Proof: **(A)** To prove (13.8), we first note that $\Omega_2 \geq 0$ is necessary. Let $z^t = [z_1^t \;\; z_2^t]$ be a vector partitioned in accordance with Ω. Thus we have

$$z^t \, \Omega \, z = z_1^t \Omega_1 z_1 + 2 z_1^t \Omega_3 z_2 + z_2^t \Omega_2 z_2 \tag{13.10}$$

Select z_2 such that $\Omega_2 z_2 = 0$. If $\Omega_3 z_2 \neq 0$, let $z_1 = -\pi \Omega_3 z_2$, $\pi > 0$. Then it follows that

$$z^t \, \Omega \, z = \pi^2 \, z_2^t \Omega_3^t \Omega_1 \Omega_3 z_2 - 2\pi \, z_2^t \Omega_3^t \Omega_3 z_2$$

which is negative for a sufficiently small $\pi > 0$. We thus conclude $\Omega_1 \, z_2 = 0$ which then leads to $\Omega_3 z_2 = 0$, $\forall \, z_2$ and consequently

$$\Omega_3 = \Upsilon \, \Omega_2 \tag{13.11}$$

for some Υ.

Since $\Omega \geq 0$, the quadratic term $z^t \, \Omega \, z$ possesses a minimum over z_2 for any z_1. By differentiating $z^t \, \Omega \, z$ from (13.10) wrt z_2^t, we get

$$\frac{\partial (z^t \, \Omega \, z)}{\partial z_2^t} = 2\Omega_3^t \, z_1 + 2\Omega_2 \, z_2 = 2\Omega_2 \, \Upsilon^t \, z_1 + 2\Omega_2 \, z_2$$

Setting the derivative to zero yields

$$\Omega_2 \, \Upsilon \, z_1 \; = \; -\Omega_2 \, z_2 \tag{13.12}$$

Using (13.11) and (13.12) in (13.10), it follows that the minimum of $z^t \, \Omega \, z$ over z_2 for any z_1 is given by

$$\min_{z_2} \; z^t \, \Omega \, z = z_1^t [\Omega_1 \; - \; \Upsilon \, \Omega_2 \, \Upsilon^t] z_1$$

which proves the necessity of $\Omega_1 \; - \; \Upsilon \, \Omega_2 \, \Upsilon^t \geq 0$.

On the other hand, we note that the conditions (13.8) are necessary for $\Omega \geq 0$ and since together they imply that the minimum of $z^t \, \Omega \, z$ over z_2 for any z_1 is nonnegative, they are also sufficient.

Using a similar argument, conditions (13.9) can be derived as those of (13.8) by starting with Ω_1.

The proof of **(B)** follows as a direct corollary of **(A)**.

13.2.5 Inequality 5

For any quantities u and v of equal dimensions and for all $\eta_t \; = \; i \; \in \; \mathcal{S}$, it follows that the following inequality holds:

$$||u \, + \, v||^2 \; \leq \; [1 + \beta^{-1}] \; ||u||^2 \; + \; [1 + \beta]||v||^2 \tag{13.13}$$

for any scalar $\beta > 0, \quad i \in \mathcal{S}$

Proof: Since

$$\begin{aligned} &[u \, + \, v]^t \; [u \, + \, v] = \\ &u^t \; u \, + \, v^t \; v \, + \, 2 \; u^t \; v \end{aligned} \tag{13.14}$$

it follows by taking norm of both sides of (13.14) for all $i \in \mathcal{S}$ that

$$||u \, + \, v||^2 \; \leq \; ||u||^2 \; + \; ||v||^2 \; + \; 2 \; ||u^t \; v|| \tag{13.15}$$

We know from the triangle inequality that

$$2 \; ||u^t \; v|| \; \leq \; \beta^{-1} \; ||u||^2 \; + \; \beta \; ||v||^2 \tag{13.16}$$

On substituting (13.16) into (13.15), it yields (13.13).

13.2.6 Inequality 6

Given matrices $0 < \mathcal{Q}^t = \mathcal{Q},\ \mathcal{P} = \mathcal{P}^t,$ then it follows that

$$- \mathcal{P}\mathcal{Q}^{-1}\mathcal{P} \leq - 2\,\mathcal{P} + \mathcal{Q} \tag{13.17}$$

This can be easily established by considering the algebraic inequality

$$(\mathcal{P} - \mathcal{Q})^t \mathcal{Q}^{-1}(\mathcal{P} - \mathcal{Q}) \geq 0$$

and expanding to get

$$\mathcal{P}\mathcal{Q}^{-1}\mathcal{P} - 2\mathcal{P} + \mathcal{Q} \geq 0 \tag{13.18}$$

which when manipulating, yields (13.17). An important special case is obtained when $\mathcal{P} \equiv I$, that is,

$$- \mathcal{Q}^{-1} \leq - 2\,I + \mathcal{Q} \tag{13.19}$$

This inequality proves useful when using Schur complements to eliminate the quantity $\mathcal{Q}^{-1}$ from the diagonal of an LMI without alleviating additional math operations.

13.3 Lemmas

The basic tools and standard results that are utilized in robustness analysis and resilience design in the different chapters are collected hereafter.

Lemma 13.3.1 *The matrix inequality*

$$- \Lambda + S\,\Omega^{-1}\,S^t < 0 \tag{13.20}$$

holds for some $0 < \Omega = \Omega^t \in \Re^{n\times n},$ *if and only if*

$$\begin{bmatrix} -\Lambda & S\mathcal{X} \\ \bullet & -\mathcal{X} - \mathcal{X}^t + \mathcal{Z} \end{bmatrix} < 0 \tag{13.21}$$

holds for some matrices $\mathcal{X} \in \Re^{n\times n}$ *and* $\mathcal{Z} \in \Re^{n\times n}$.

Proof: ($\Longrightarrow$) By Schur complements, inequality (13.20) is equivalent to

$$\begin{bmatrix} -\Lambda & S\Omega^{-1} \\ \bullet & -\Omega^{-1} \end{bmatrix} < 0 \tag{13.22}$$

Setting $\mathcal{X} = \mathcal{X}^t = \mathcal{Z} = \Omega^{-1}$, we readily obtain inequality (13.21).
($\Longleftarrow$) Since the matrix $[I \quad S]$ is of full rank, we obtain

$$\begin{bmatrix} I \\ S^t \end{bmatrix}^t \begin{bmatrix} -\Lambda & S\mathcal{X} \\ \bullet & -\mathcal{X} - \mathcal{X}^t + \mathcal{Z} \end{bmatrix} \begin{bmatrix} I \\ S^t \end{bmatrix} < 0 \Longleftrightarrow$$
$$-\Lambda + S\,\mathcal{Z}\,S^t < 0 \Longleftrightarrow -\Lambda + S\,\Omega^{-1}\,S^t < 0, \quad \mathcal{Z} = \Omega^{-1} \qquad (13.23)$$

which completes the proof.

Lemma 13.3.2 *The matrix inequality*

$$A\mathcal{P} + \mathcal{P}A^t + D^t\,\mathcal{R}^{-1}\,D + \mathcal{M} < 0 \qquad (13.24)$$

holds for some $0 < \mathcal{P} = \mathcal{P}^t \in \Re^{n\times n}$, *if and only if*

$$\begin{bmatrix} A\mathcal{V} + \mathcal{V}^tA^t + \mathcal{M} & \mathcal{P} + A\mathcal{W} - \mathcal{V} & D^t\mathcal{R} \\ \bullet & -\mathcal{W} - \mathcal{W}^t & 0 \\ \bullet & \bullet & -\mathcal{R} \end{bmatrix} < 0 \qquad (13.25)$$

holds for some $\mathcal{V} \in \Re^{n\times n}$ *and* $\mathcal{W} \in \Re^{n\times n}$.
Proof: ($\Longrightarrow$) By Schur complements, inequality (13.24) is equivalent to

$$\begin{bmatrix} A\mathcal{P} + \mathcal{P}A^t + \mathcal{M} & D^t\mathcal{R} \\ \bullet & -\mathcal{R} \end{bmatrix} < 0 \qquad (13.26)$$

Setting $\mathcal{V} = \mathcal{V}^t = \mathcal{P},\ \mathcal{W} = \mathcal{W}^t = \mathcal{R}$, it follows from **Lemma (13.3.1)** with Schur complements that there exists $\mathcal{P} > 0,\ \mathcal{V},\ \mathcal{W}$ such that inequality (13.25) holds.
($\Longleftarrow$) In a similar way, Schur complements to inequality (13.25) imply that:

$$\begin{bmatrix} A\mathcal{V} + \mathcal{V}^tA^t + \mathcal{M} & \mathcal{P} + A\mathcal{W} - \mathcal{V} & D^t\mathcal{R} \\ \bullet & -\mathcal{W} - \mathcal{W}^t & 0 \\ \bullet & \bullet & -\mathcal{R} \end{bmatrix} < 0$$
$$\Longleftrightarrow \begin{bmatrix} I \\ A \end{bmatrix} \begin{bmatrix} A\mathcal{V} + \mathcal{V}^tA^t + \mathcal{M} + D^t\mathcal{P}^{-1}D & \mathcal{P} + A\mathcal{W} - \mathcal{V} \\ \bullet & -\mathcal{W} - \mathcal{W}^t \end{bmatrix}$$
$$\begin{bmatrix} I \\ A \end{bmatrix}^t < 0 \Longleftrightarrow A\mathcal{P} + \mathcal{P}A^t + D^t\,\mathcal{P}^{-1}\,D + \mathcal{M} < 0\ ,\ \mathcal{V} = \mathcal{V}^t \qquad (13.27)$$

which completes the proof.

The following lemmas are found in [294].

Lemma 13.3.3 *Given any* $x \in \Re^n$*:*

$$\max \{[x^t \, RH\Delta G \, x]^2 : \Delta \in \Re\} \; = \; x^t \, RHH^tR \, x \;\; x^t \, G^tG \, x$$

Lemma 13.3.4 *Given matrices* $0 \leq X = X^t \in \Re^{p\times p}$, $Y = Y^t < 0 \in \Re^{p\times p}$, $0 \leq Z = Z^t \in \Re^{p\times p}$, *such that*

$$[\xi^t \, Y \, \xi]^2 \; - \; 4 \, [\xi^t \, X \, \xi \;\; \xi^t \, Z \, \xi]^2 \; > \; 0$$

for all $0 \neq \xi \in \Re^p$ *is satisfied. Then there exists a constant* $\alpha > 0$ *such that*

$$\alpha^2 \, X \, + \, \alpha \, Y \, + \, Z \, < \, 0$$

Lemma 13.3.5 *For a given two vectors* $\alpha \in \Re^n$, $\beta \in \Re^m$ *and matrix* $\mathcal{N} \in \Re^{n\times m}$ *defined over a prescribed interval* Ω*, it follows for any matrices* $X \in \Re^{n\times n}$, $Y \in \Re^{n\times m}$, *and* $Z \in \Re^{m\times m}$ *the following inequality holds:*

$$-\,2 \int_\Omega \alpha^t(s) \, \mathcal{N} \, \beta(s) \, ds \; \leq \; \int_\Omega \left[\begin{array}{c} \alpha(s) \\ \beta(s) \end{array}\right]^t \left[\begin{array}{cc} X & Y - \mathcal{N} \\ Y^t - \mathcal{N}^t & Z \end{array}\right] \left[\begin{array}{c} \alpha(s) \\ \beta(s) \end{array}\right] ds$$

where

$$\left[\begin{array}{cc} X & Y \\ Y^t & Z \end{array}\right] \geq 0$$

An algebraic version of **Lemma 13.3.5** is stated below.

Lemma 13.3.6 *For a given two vectors* $\alpha \in \Re^n$, $\beta \in \Re^m$ *and matrix* $\mathcal{N} \in \Re^{n\times m}$ *defined over a prescribed interval* Ω*, it follows for any matrices* $X \in \Re^{n\times n}$, $Y \in \Re^{n\times m}$, *and* $Z \in \Re^{m\times m}$ *the following inequality holds:*

$$\begin{aligned} -\,2 \, \alpha^t \, \mathcal{N} \, \beta \;\; & \leq \;\; \left[\begin{array}{c} \alpha \\ \beta \end{array}\right]^t \left[\begin{array}{cc} X & Y - \mathcal{N} \\ Y^t - \mathcal{N}^t & Z \end{array}\right] \left[\begin{array}{c} \alpha \\ \beta \end{array}\right] \\ & = \;\; \alpha^t X \alpha + \beta^t (Y^t - \mathcal{N}^t) \alpha + \alpha^t (Y - \mathcal{N}) \beta + \beta^t Z \beta \end{aligned}$$

subject to

$$\left[\begin{array}{cc} X & Y \\ Y^t & Z \end{array}\right] \geq 0$$

Lemma 13.3.7 *Let* $0 < Y = Y^t$ *and* $M,\ N$ *be given matrices with appropriate dimensions. Then it follows that*

$$Y + M\,\Delta\,N + N^t\,\Delta^t\,M^t \;<\; 0, \;\forall\; \Delta^t\,\Delta \leq I$$

holds if and only if there exists a scalar $\varepsilon > 0$ *such that*

$$Y + \varepsilon\,M\,M^t + \varepsilon^{-1}\,N^t\,N \;<\; 0$$

In the following lemma, we let $X(z) \in \Re^{n\times p}$ be a matrix function of the variable z. A matrix $X_*(z)$ is called the orthogonal complement of $X(z)$ if $X^t(z)\,X_*(z) = 0$ and $X(z)\,X_*(z)$ is nonsingular (of maximum rank).

Lemma 13.3.8 *Let* $0 < L = L^t$ *and* $X,\ Y$ *be given matrices with appropriate dimensions. Then it follows that the inequality*

$$L(z) + X(z)\,P\,Y(z) + Y^t(z)\,P^t\,X^t(z) \;>\; 0 \qquad (13.28)$$

holds for some P *and* $z = z_o$ *if and only if the following inequalities*

$$X_*^t(z)\,L(z)\,X_*(z) \;>\; 0, \quad Y_*^t(z)\,L(z)\,Y_*(z) \;>\; 0 \qquad (13.29)$$

hold with $z = z_o$.

It is significant to observe that feasibility of matrix inequality (13.28) with variables P and z is equivalent to the feasibility of (13.29) with variable z and thus the matrix variable P has been eliminated from (13.28) to form (13.29). Using Finsler's lemma, we can express (13.29) in the form

$$L(z) - \beta\,X(z)\,X^t(z) \;>\; 0, \quad L(z) - \beta\,Y(z)\,Y^t(z) \;>\; 0 \quad (13.30)$$

for some $\beta \in \Re$.

The following is a statement of the reciprocal projection lemma.

Lemma 13.3.9 *Let* $P > 0$ *be a given matrix. The following statements are equivalent:*
i) $\mathcal{M} + Z + Z^t \;<\; 0$
ii) the LMI problem

$$\begin{bmatrix} \mathcal{M} + P - (V + V^t) & V^t + Z^t \\ V + Z & -P \end{bmatrix} < \; 0$$

is feasible with respect to the general matrix V.

A useful lemma that is frequently used in overbounding given inequalities is presented.

Lemma 13.3.10 *For matrices X and Y and $K^t = K \geq 0$ of appropriate dimensions with $K^\dagger$ being the Moore-Penrose generalized inverse of matrix K, the following lemma is proved:*
$X^t K K^\dagger Y + Y^t K K^\dagger X \leq X^t K X + Y^t K^\dagger Y$
In particular, if x and y are vectors and $K^t = K > 0$, then
$x^t y \leq (1/2) x^t K x + (1/2) y^t K^{-1} y$
Proof: Let the real Schur decomposition of K be $K = U^t V U$ where $U = U^{-t}$ is orthogonal and $V = diag(\lambda_1, ..., \lambda_n)$ is the diagonal matrix of eigenvalues. The lemma then follows from

$$\begin{aligned} 0 &\leq [\sqrt{V} U X - \sqrt{V^\dagger} U^{-t} Y]^t [\sqrt{V} U X - \sqrt{V^\dagger} U^{-t} Y] \\ &\leq X^t U^t V U X + Y^t U^{-1} V^\dagger U Y - Y^t U^{-1} \sqrt{V^\dagger} \sqrt{V} U X \\ &- X^t U^t \sqrt{V} \sqrt{V^\dagger} U^{-t} X \end{aligned}$$

as $U^{-1}\sqrt{V^\dagger}\sqrt{V} U = U^t \sqrt{V}\sqrt{V^\dagger} U^{-t} = K K^\dagger$ and $K^\dagger = U^{-1} V^\dagger U$.

13.4 Linear Matrix Inequalities

It has been shown that a wide variety of problems arising in system and control theory can conveniently reduce to a few standard convex or quasi convex optimization problems involving linear matrix inequalities (LMIs). The resulting optimization problems can then be solved numerically very efficiently using commercially available interior-point methods.

13.4.1 Basics

One of the earliest LMIs arises in Lyapunov theory. It is well known that the differential equation

$$\dot{x}(t) = A\, x(t) \tag{13.31}$$

has all of its trajectories converge to zero (stable) if and only if there exists a matrix $P > 0$ such that

$$A^t P + A P < 0 \tag{13.32}$$

This leads to the LMI formulation of stability, that is, *a linear time-invariant system is asymptotically stable if and only if there exists a matrix $0 < P = P^t$*

satisfying the LMIs

$$A^t\,P + A\,P \; < \; 0, \quad P > 0$$

Given a vector variable $x \in \Re^n$ and a set of matrices $0 < G_j = G_j^t \in \Re^{n\times n}$, $j = 0, ..., p$, then a basic compact formulation of a linear matrix inequality is

$$G(x) \; := \; G_0 + \sum_{j=1}^{p} x_j\, G_j \;\; > \; 0 \tag{13.33}$$

Notice that (13.33) implies that $v^t G(x) v > 0 \;\forall 0 \neq v \in \Re^n$. More importantly, the set $\{x \;|G(x) \; > \; 0$ is convex. Nonlinear (convex) inequalities are converted to LMI form using Schur complements in the sense that

$$\begin{bmatrix} Q(x) & S(x) \\ \bullet & R(x) \end{bmatrix} > 0 \tag{13.34}$$

where $Q(x) = Q^t(x),\; R(x) = R^t(x),\; S(x)$ depend affinely on x, is equivalent to

$$R(x) > 0, \quad Q(x) - S(x) R^{-1}(x) S^t(x) \; > \; 0 \tag{13.35}$$

More generally, the constraint

$$Tr[S^t(x)\; P^{-1}(x)\; S(x)] \; < \; 1, \quad P(x) \; > \; 0$$

where $P(x) = P^t(x) \in \Re^{n\times n},\; S(x) \in \Re^{n\times p}$ depend affinely on x, is handled by introducing a new (slack) matrix variable $Y(x) = Y^t(x) \in\in \Re^{p\times p}$ and the LMI (in x and Y):

$$TrY \; < \; 1, \quad \begin{bmatrix} Y & S(x) \\ \bullet & P(x) \end{bmatrix} \; > \; 0 \tag{13.36}$$

Most of the time, our LMI variables are matrices. It should be clear from the foregoing discussions that a quadratic matrix inequality (QMI) in the variable P can be readily expressed as a linear matrix inequality (LMI) in the same variable.

13.4.2 Some standard problems

Here we provide some common convex problems that we encountered throughout the monograph. Given an LMI $G(x) \; > \; 0$, the corresponding LMI problem (LMIP) is to

find a feasible $x \equiv x^f$ *such that* $G(x^f) > 0$, or
determine that the LMI is infeasible.

It is obvious that this is a convex feasibility problem.

The generalized eigenvalue problem (GEVP) is to minimize the maximum generalized eigenvalue of a pair of matrices that depend affinely on a variable, subject to an LMI constraint. GEVP has the general form

$$\begin{aligned} &minimize \ \lambda \\ &subject\ to \ \ \lambda B(x) - A(x) > 0 \ , \ \ B(x) > 0, \\ &\qquad\qquad C(x) > 0 \end{aligned} \tag{13.37}$$

where A, B, C are symmetric matrices that are affine functions of x. Equivalently stated,

$$\begin{aligned} &minimize \ \lambda_M[A(x), B(x)] \\ &subject\ to \ \ B(x) > 0 \ , \ \ C(x) > 0 \end{aligned} \tag{13.38}$$

where $\lambda_M[X, Y]$ denotes the largest generalized eigenvalue of the pencil $\lambda Y - X$ with $Y > 0$. This is a quasi-convex optimization problem since the constraint is convex and the objective, $\lambda_M[A(x), B(x)]$, is quasi-convex.

The eigenvalue problem (EVP) is to minimize the maximum eigenvalue of a matrix that depends affinely on a variable, subject to an LMI constraint. EVP has the general form

$$\begin{aligned} &minimize \ \lambda \\ &subject\ to \ \ \lambda I - A(x) > 0, \ \ B(x) > 0 \end{aligned} \tag{13.39}$$

where A, B are symmetric matrices that are affine functions of the optimization variable x. This is a convex optimization problem.
EVPs can appear in the equivalent form of minimizing a linear function subject to an LMI, that is,

$$\begin{aligned} &minimize \ c^t x \\ &subject\ to \ \ G(x) > 0 \end{aligned} \tag{13.40}$$

where $G(x)$ is an affine function of x. Examples of $G(x)$ include

$$PA + A^tP + C^tC + \gamma^{-1}PBB^tP < 0, \quad P > 0$$

It should be stressed that the standard problems (LMIPs, GEVPs, EVPs) are tractable, from both theoretical and practical viewpoints:
They can be solved in polynomial time.
They can be solved in practice very efficiently using commercial softwares.

13.4.3 The S-procedure

In some design applications, we faced the constraint that some quadratic function be negative whenever some other quadratic function is negative. In such cases, this constraint can be expressed as an LMI in the data variables defining the quadratic functions.

Let $G_o, ..., G_p$ be quadratic functions of the variable $\xi \in \Re^n$:

$$G_j(\xi) := \xi^t R_j \xi \ + \ 2u_j^t \xi \ + \ v_j \ , \quad j = 0, ..., p, \quad R_j = R_j^t$$

We consider the following condition on $G_o, ..., G_p$:

$$G_o(\xi) \ \leq \ 0 \ \forall \xi \quad such \ that \quad G_j(\xi) \ \geq \ 0, \quad j = 0, ..., p \qquad (13.41)$$

It is readily evident that if there exist scalars $\omega_1 \geq 0, ..., \omega_p \geq 0$ such that

$$\forall \xi, \quad G_o(\xi) \ - \ \sum_{j=1}^{p} \omega_j \ G_j(\xi) \ \geq \ 0 \qquad (13.42)$$

then inequality (13.41) holds. Observe that if the functions $G_o, ..., G_p$ are affine, then Farkas lemma states that (13.41) and (13.42) are equivalent. Interestingly enough, inequality (13.42) can be written as

$$\begin{bmatrix} R_o & u_o \\ \bullet & v_o \end{bmatrix} - \sum_{j=1}^{p} \omega_j \begin{bmatrix} R_j & u_j \\ \bullet & v_j \end{bmatrix} \ \geq \ 0 \qquad (13.43)$$

The foregoing discussions were stated for nonstrict inequalities. In the case of strict inequality, we let $R_o, ..., R_p \in \Re^{n \times n}$ be symmetric matrices with the following qualifications:

$$\xi^t R_o \xi \ > \ 0 \ \forall \xi \quad such \ that \quad \xi^t G_j \xi \ \geq \ 0, \quad j = 0, ..., p \qquad (13.44)$$

Once again, it is obvious that if there exist scalars $\omega_1 \geq 0, ..., \omega_p \geq 0$ such that

$$\forall \xi, \quad G_o(\xi) \ - \ \sum_{j=1}^{p} \omega_j \ G_j(\xi) \ > \ 0 \qquad (13.45)$$

then inequality (13.44) holds. Observe that (13.45) is an LMI in the variables $R_o, \omega_1, ..., \omega_p$.

It should be remarked that the S-procedure dealing with nonstrict inequalities allows the inclusion of constant and linear terms. In the strict version, only quadratic functions can be used.

13.5 Some Continuous Lyapunov-Krasovskii Functionals

In this section, we provide some Lyapunov-Krasovskii functionals and their time-derivatives which are of common use in stability studies throughout the text.

$$V_1(x) = x^t P x + \int_{-\tau}^{0} x^t(t+\theta) Q x(t+\theta)\, d\theta \tag{13.46}$$

$$V_2(x) = \int_{-\tau}^{0} \left[\int_{t+\theta}^{t} x^t(\alpha) R x(\alpha)\, d\alpha \right] d\theta \tag{13.47}$$

$$V_3(x) = \int_{-\tau}^{0} \left[\int_{t+\theta}^{t} \dot{x}^t(\alpha) W \dot{x}(\alpha)\, d\alpha \right] d\theta \tag{13.48}$$

where x is the state vector, τ is a constant delay factor, and the matrices $0 < P^t = P,\ 0 < Q^t = Q,\ 0 < R^t = R,\ 0 < W^t = W$ are appropriate weighting factors.

Standard matrix manipulations lead to

$$\dot{V}_1(x) = \dot{x}^t P x + x^t P \dot{x} + x^t(t) Q x(t) - x^t(t-\tau) Q x(t-\tau) \tag{13.49}$$

$$\begin{aligned} \dot{V}_2(x) &= \int_{-\tau}^{0} \left[x^t(t) R x(t) - x^t(t+\alpha) R x(t+\alpha) \right] d\,\theta \\ &= \tau\, x^t(t) R x(t) - \int_{-\tau}^{0} x^t(t+\theta) R x(t+\theta) \Big] d\,\theta \end{aligned} \tag{13.50}$$

$$\dot{V}_3(x) = \tau\, \dot{x}^t(t) W x(t) - \int_{t-\tau}^{t} \dot{x}^t(\alpha) W \dot{x}(\alpha)\, d\alpha \tag{13.51}$$

13.6 Some Formulas on Matrix Inverses

This section presents some useful formulas for inverting matrix expressions in terms of the inverses of their constituents.

13.6.1 Inverse of block matrices

Let A be a square matrix of appropriate dimension and partitioned in the form

$$A = \begin{bmatrix} A_1 & A_2 \\ A_3 & A_4 \end{bmatrix} \tag{13.52}$$

where both A_1 and A_4 are square matrices. If A_1 is invertible, then

$$\Delta_1 = A_4 - A_3 A_1^{-1} A_2$$

is called the Schur complement of A_1. Alternatively, if A_4 is invertible, then

$$\Delta_4 = A_1 - A_2 A_4^{-1} A_3$$

is called the Schur complement of A_4.

It is well known that matrix A is invertible if and only if either

$$A_1 \quad and \quad \Delta_1 \quad are \ invertible$$

or

$$A_4 \quad and \quad \Delta_4 \quad are \ invertible$$

Specifically, we have the following equivalent expressions:

$$\begin{bmatrix} A_1 & A_2 \\ A_3 & A_4 \end{bmatrix}^{-1} = \begin{bmatrix} \Upsilon_1 & -A_1^{-1}A_2\Delta_1^{-1} \\ -\Delta_1^{-1}A_3A_1^{-1} & \Delta_1^{-1} \end{bmatrix} \tag{13.53}$$

or

$$\begin{bmatrix} A_1 & A_2 \\ A_3 & A_4 \end{bmatrix}^{-1} = \begin{bmatrix} \Delta_4^{-1} & -\Delta_4^{-1}A_2A_4^{-1} \\ -A_4^{-1}A_3\Delta_4^{-1} & \Upsilon_4 \end{bmatrix} \tag{13.54}$$

where

$$\begin{aligned} \Upsilon_1 &= A_1^{-1} + A_1^{-1}A_2\Delta_1^{-1}A_3A_1^{-1} \\ \Upsilon_4 &= A_4^{-1} + A_4^{-1}A_3\Delta_4^{-1}A_2A_4^{-1} \end{aligned} \tag{13.55}$$

Important special cases are

$$\begin{bmatrix} A_1 & 0 \\ A_3 & A_4 \end{bmatrix}^{-1} = \begin{bmatrix} A_1^{-1} & 0 \\ -A_4^{-1}A_3A_1^{-1} & A_4^{-1} \end{bmatrix} \tag{13.56}$$

and

$$\begin{bmatrix} A_1 & A_2 \\ 0 & A_4 \end{bmatrix}^{-1} = \begin{bmatrix} A_1^{-1} & -A_1^{-1}A_2A_4^{-1} \\ 0 & A_4^{-1} \end{bmatrix} \tag{13.57}$$

13.6.2 Matrix inversion lemma

Let $A \in \Re^{n\times n}$ and $C \in \Re^{m\times m}$ be nonsingular matrices. By using the definition of matrix inverse, it can be easily verified that

$$[A + B\,C\,D]^{-1} = A^{-1} - A^{-1}\,B\,[D\,A^{-1}B + C^{-1}]^{-1}\,DA^{-1} \tag{13.58}$$

13.7 Some Discrete Lyapunov-Krasovskii Functionals

In this section, we provide a general form of discrete Lyapunov-Krasovskii functionals and their first-difference which can be used in stability studies of discrete-time throughout the text.

$$\begin{aligned}
V(k) &= V_o(k) + V_a(k) + V_c(k) + V_m(k) + V_n(k) \\
V_o(k) &= x^t(k)\mathcal{P}_\sigma x(k),\; V_a(k) = \sum_{j=k-d(k)}^{k-1} x^t(j)\mathcal{Q}_\sigma x(j) \\
V_c(k) &= \sum_{j=k-d_m}^{k-1} x^t(j)\mathcal{Z}_\sigma x(j) + \sum_{j=k-d_M}^{k-1} x^t(j)\mathcal{S}_\sigma x(j) \\
V_m(k) &= \sum_{j=-d_M+1}^{-d_m} \sum_{m=k+j}^{k-1} x^t(m)\mathcal{Q}_\sigma x(m) \\
V_n(k) &= \sum_{j=-d_M}^{-d_m-1} \sum_{m=k+j}^{k-1} \delta x^t(m)\mathcal{R}_{a\sigma}\delta x(m) \\
&+ \sum_{j=-d_M}^{-1} \sum_{m=k+j}^{k-1} \delta x^t(m)\mathcal{R}_{c\sigma}\delta x(m)
\end{aligned} \tag{13.59}$$

where

$$\begin{aligned}
0 < \mathcal{P}_\sigma &= \sum_{j=1}^{N}\lambda_j\mathcal{P}_j,\; 0 < \mathcal{Q}_\sigma = \sum_{j=1}^{N}\lambda_j\mathcal{Q}_j,\; 0 < \mathcal{S}_\sigma = \sum_{j=1}^{N}\lambda_j\mathcal{S}_j, \\
0 < \mathcal{Z}_\sigma &= \sum_{j=1}^{N}\lambda_j\mathcal{Z}_j,\; 0 < \mathcal{R}_{a\sigma} = \sum_{j=1}^{N}\lambda_j\mathcal{R}_{aj},\; 0 < \mathcal{R}_{c\sigma} = \sum_{j=1}^{N}\lambda_j\mathcal{R}_{cj}
\end{aligned} \tag{13.60}$$

are weighting matrices of appropriate dimensions. Consider now a class of discrete-time systems with interval-like time-delays that can be described by

$$\begin{aligned}
x(k+1) &= A_\sigma x(k) + D_\sigma x(k-d_k) + \Gamma_\sigma \omega(k) \\
z(k) &= C_\sigma x(k) + G_\sigma x(k-d_k) + \Sigma_\sigma \omega(k)
\end{aligned} \tag{13.61}$$

where $x(k) \in \Re^n$ is the state, $z(k) \in \Re^q$ is the controlled output, and $\omega(k) \in \Re^p$ is the external disturbance that is assumed to belong to $\ell_2[0,\infty)$. In the sequel, it is assumed that d_k is time-varying and satisfying

$$d_m \le d_k \le d_M \tag{13.62}$$

where the bounds $d_m > 0$ and $d_M > 0$ are constant scalars. The system matrices contain uncertainties that belong to a real convex bounded polytopic model of the type

$$[A_\sigma,\ D_\sigma, ..., \Sigma_\sigma] \in \widehat{\Xi}_\lambda := \left\{ [A_\lambda,\quad D_\lambda, ..., \Sigma_\lambda] \right.$$
$$\left. = \sum_{j=1}^{N} \lambda_j [A_j,\ D_j, ..., \Sigma_j], \lambda \in \Lambda \right\} \tag{13.63}$$

where Λ is the unit simplex

$$\Lambda := \left\{ (\lambda_1, \cdots, \lambda_N) : \sum_{j=1}^{N} \lambda_j = 1\ ,\ \lambda_j \geq 0 \right\} \tag{13.64}$$

Define the vertex set $\mathcal{N} = \{1, ..., N\}$. We use $\{A, ..., \Sigma\}$ to imply generic system matrices and $\{A_j, ..., \Sigma_j,\ j \in \mathcal{N}\}$ to represent the respective values at the vertices. In what follows, we provide a definition of exponential stability of system (13.61):

A straightforward computation gives the first-difference of $\Delta V(k) = V(k+1) - V(k)$ along the solutions of (13.61) with $\omega(k) \equiv 0$ as:

$$\begin{aligned}
\Delta V_o(k) &= x^t(k+1)\mathcal{P}_\sigma x(k+1) - x^t(k)\mathcal{P}_\sigma x(k) \\
&= [A_\sigma x(k) + D_\sigma x(k-d_k)]^t \mathcal{P}_\sigma [A_\sigma x(k) + D_\sigma x(k-d_k)] \\
&- x^t(k)\mathcal{P}_\sigma x(k) \\
\Delta V_a(k) &\leq x^t(k)\mathcal{Q}x(k) - x^t(k-d(k))\mathcal{Q}x(k-d(k)) \\
&+ \sum_{j=k-d_M+1}^{k-d_m} x^t(j)\mathcal{Q}x(j) \\
\Delta V_c(k) &= x^t(k)\mathcal{Z}x(k) - x^t(k-d_m)\mathcal{Z}x(k-d_m) + x^t(k)\mathcal{S}x(k) \\
&- x^t(k-d_M)\mathcal{S}x(k-d_M) \\
\Delta V_m(k) &= (d_M - d_m)x^t(k)\mathcal{Q}x(k) - \sum_{j=k-d_M+1}^{k-d_m} x^t(k)\mathcal{Q}x(k) \\
\Delta V_n(k) &= (d_M - d_m)\delta x^t(k)\mathcal{R}_a \delta x(k) + d_M \delta x^t(k)\mathcal{R}_c \delta x(k) \\
&- \sum_{j=k-d_M}^{k-d_m-1} \delta x^t(j)\mathcal{R}_a \delta x(j) - \sum_{j=k-d_M}^{k-1} \delta x^t(j)\mathcal{R}_c \delta x(j)
\end{aligned} \tag{13.65}$$

13.8 Additional Inequalities

A basic inequality that has been frequently used in the stability analysis of time-delay systems is called *Jensen's Inequality* or *the Integral Inequality*, a detailed account of which is available in [93].

Lemma 13.8.1 *For any constant matrix $0 < \Sigma \in \Re^{n\times n}$, scalar $\tau_* < \tau(t) < \tau^+$, and vector function $\dot{x} : [-\tau^+, -\tau_*] \rightarrow \Re^n$ such that the following integration is well defined, then it holds that*

$$-(\tau^+ - \tau_*)\int_{t-\tau^+}^{t-\tau_*} \dot{x}^t(s)\Sigma\dot{x}(s)ds \le \begin{bmatrix} x(t-\tau_*) \\ x(t-\tau^+) \end{bmatrix}^t \begin{bmatrix} -\Sigma & \Sigma \\ \bullet & -\Sigma \end{bmatrix} \begin{bmatrix} x(t-\tau_*) \\ x(t-\tau^+) \end{bmatrix}$$

Building on **Lemma 13.8.1**, the following lemma specifies a particular inequality for quadratic terms.

Lemma 13.8.2 *For any constant matrix $0 < \Sigma \in \Re^{n\times n}$, scalar $\tau_* < \tau(t) < \tau^+$, and vector function $\dot{x} : [-\tau^+, -\tau_*] \rightarrow \Re^n$ such that the following integration is well defined, then it holds that*

$$-\ (\tau^+ - \tau_*)\int_{t-\tau^+}^{t-\tau_*} \dot{x}^t(s)\ \Sigma\ \dot{x}(s)\ ds \ \le\ \xi^t(t)\ \Upsilon\ \xi(t)$$

$$\xi(t) = \begin{bmatrix} x(t-\tau_*) \\ x(t-\tau(t)) \\ x(t-\tau^+) \end{bmatrix}^t, \quad \Upsilon = \begin{bmatrix} -\Sigma & \Sigma & 0 \\ \bullet & -2\Sigma & \Sigma \\ \bullet & \bullet & -\Sigma \end{bmatrix}$$

Proof: Considering the case $\tau_* < \tau(t) < \tau^+$ and applying the Leibniz-Newton formula, it follows that

$$\begin{aligned}
&-\ (\tau^+ - \tau_*)\int_{t-\tau^+}^{t-\tau_*} \dot{x}^t(s)\ \Sigma\ \dot{x}(s)\ ds\ -\ (\tau^+ - \tau_*)\int_{t-\tau(t)}^{t-\tau_*} \\
&\times\ \left[\dot{x}^t(s)\ \Sigma\ \dot{x}(s)\ ds + \int_{t-\tau^+}^{t-\tau(t)} \dot{x}^t(s)\ \Sigma\ \dot{x}(s)\ ds\right] \\
&\le\ -\ (\tau(t) - \tau_*)\int_{t-\tau(t)}^{t-\tau_*} \left[\dot{x}^t(s)\ \Sigma\ \dot{x}(s)\ ds\right. \\
&-\ \left.(\tau^+ - \tau(t))\int_{t-\tau^+}^{t-\tau(t)} \dot{x}^t(s)\ \Sigma\ \dot{x}(s)\ ds\right] \\
&\le\ -\int_{t-\tau(t)}^{t-\tau_*} \dot{x}^t(s)\ ds\ \Sigma \int_{t-\tau(t)}^{t-\tau_*} \dot{x}^t(s)\ ds
\end{aligned}$$

$$
\begin{aligned}
- & \int_{t-\tau_+}^{t-\tau(t)} \dot{x}^t(s)\, ds\, \Sigma \int_{t-\tau_+}^{t-\tau(t)} \dot{x}^t(s)\, ds \\
= & \ [x(t-\tau_*) \ - \ x(t-\tau(t))]^t\, \Sigma\, [x(t-\tau_*) \ - \ x(t-\tau(t))] \\
- & \ [x(t-\tau(t)) \ - \ x(t-\tau^+)]^t\, \Sigma\, [x(t-\tau(t)) \ - \ x(t-\tau^+)]
\end{aligned}
$$

which completes the proof.

Lemma 13.8.3 *For any constant matrix $0 < \Sigma \in \Re^{n\times n}$, scalar η, any $t \in [0,\infty)$, and vector function $g : [t-\eta, t] \rightarrow \Re^n$ such that the following integration is well defined, then it holds that*

$$
\Big(\int_{t-\eta}^{t} g(s)\, ds \Big)^t \Sigma \int_{t-\eta}^{t} g(s)\, ds \leq \eta \int_{t-\eta}^{t} g^t(s)\, \Sigma\, g(s)\, ds \qquad (13.66)
$$

Proof: It is simple to show that for any $s \in [t-\eta, t]$, $t \in [0,\infty)$, and Schur complements

$$
\begin{bmatrix} g^t(s)\Sigma g(s) & g^t(s) \\ \bullet & \Sigma^{-1} \end{bmatrix} \geq 0
$$

Upon integration, we have

$$
\begin{bmatrix} \int_{t-\eta}^{t} g^t(s)\Sigma g(s) ds & \int_{t-\eta}^{t} g^t(s) ds \\ \bullet & \eta\Sigma \end{bmatrix} \geq 0
$$

By Schur complements, we obtain inequality (13.66).

The following lemmas show how to produce equivalent LMIs by an elimination procedure.

Lemma 13.8.4 *There exists $\mathcal{X}$ such that*

$$
\begin{bmatrix} \mathcal{P} & \mathcal{Q} & \mathcal{X} \\ \bullet & \mathcal{R} & \mathcal{Z} \\ \bullet & \bullet & \mathcal{S} \end{bmatrix} > 0 \qquad (13.67)
$$

if and only if

$$
\begin{bmatrix} \mathcal{P} & \mathcal{Q} \\ \bullet & \mathcal{R} \end{bmatrix} > 0, \ \begin{bmatrix} \mathcal{R} & \mathcal{Z} \\ \bullet & \mathcal{S} \end{bmatrix} > 0 \qquad (13.68)
$$

Proof: Since LMIs (13.68) form subblocks on the principal diagonal of LMI (13.67), necessity is established. To show sufficiency, apply the congruence transformation

$$\begin{bmatrix} I & 0 & 0 \\ \bullet & I & 0 \\ 0 & -V^t R^{-1} & I \end{bmatrix}$$

to LMI (13.67); it is evident that (13.67) is equivalent to

$$\begin{bmatrix} \mathcal{P} & \mathcal{Q} & \mathcal{X} - \mathcal{Q}\mathcal{R}^{-1}\mathcal{Z} \\ \bullet & \mathcal{R} & 0 \\ \bullet & \bullet & \mathcal{S} - \mathcal{Z}^t\mathcal{R}^{-1}\mathcal{Z} \end{bmatrix} > 0 \tag{13.69}$$

Clearly (13.68) is satisfied for $\mathcal{X} = \mathcal{Q}\mathcal{R}^{-1}\mathcal{Z}$ if (13.68) is satisfied in view of Schur complements.

Lemma 13.8.5 *There exists $\mathcal{X}$ such that*

$$\begin{bmatrix} \mathcal{P} & \mathcal{Q} + \mathcal{X}\mathcal{G} & \mathcal{X} \\ \bullet & \mathcal{R} & \mathcal{Z} \\ \bullet & \bullet & \mathcal{S} \end{bmatrix} > 0 \tag{13.70}$$

if and only if

$$\begin{bmatrix} \mathcal{P} & \mathcal{Q} \\ \bullet & \mathcal{R} - \mathcal{V}\mathcal{G} - \mathcal{G}^t\mathcal{V}^t + \mathcal{G}^t\mathcal{Z}\mathcal{G} \end{bmatrix} > 0,$$
$$\begin{bmatrix} \mathcal{R} - \mathcal{V}\mathcal{G} - \mathcal{G}^t\mathcal{V}^t + \mathcal{G}^t\mathcal{Z}\mathcal{G} & \mathcal{V} - \mathcal{G}^t\mathcal{Z} \\ \bullet & \mathcal{Z} \end{bmatrix} > 0 \tag{13.71}$$

Proof: Applying the congruence transformation

$$\begin{bmatrix} I & 0 & 0 \\ 0 & I & 0 \\ 0 & -\mathcal{G} & I \end{bmatrix}$$

to LMI (13.70) and using **Lemma 13.8.4**, we readily obtain the results.

Lemma 13.8.6 *There exists $0 < \mathcal{X}^t = \mathcal{X}$ such that*

$$\begin{bmatrix} \mathcal{P}_a + \mathcal{X} & \mathcal{Q}_a \\ \bullet & \mathcal{R}_a \end{bmatrix} > 0,$$
$$\begin{bmatrix} \mathcal{P}_c - \mathcal{X} & \mathcal{Q}_c \\ \bullet & \mathcal{R}_c \end{bmatrix} > 0 \tag{13.72}$$

if and only if

$$\begin{bmatrix} \mathcal{P}_a + \mathcal{P}_c & \mathcal{Q}_a & \mathcal{Q}_c \\ \bullet & \mathcal{R}_a & 0 \\ \bullet & \bullet & \mathcal{R}_c \end{bmatrix} > 0 \tag{13.73}$$

Proof: It is obvious from Schur complements that LMI (13.73) is equivalent to

$$\begin{aligned} &\mathcal{R}_a > 0, \quad \mathcal{R}_c > 0 \\ &\Xi = \mathcal{P}_a + \mathcal{P}_c - \mathcal{Q}_a \mathcal{R}_a^{-1} \mathcal{Q}_a^t - \mathcal{Q}_c \mathcal{R}_c^{-1} \mathcal{Q}_c^t > 0 \end{aligned} \tag{13.74}$$

On the other hand, LMI (13.72) is equivalent to

$$\begin{aligned} &\mathcal{R}_a > 0, \quad \mathcal{R}_c > 0 \\ &\Xi_a = \mathcal{P}_a + \mathcal{X} - \mathcal{Q}_a \mathcal{R}_a^{-1} \mathcal{Q}_a^t > 0, \\ &\Xi_c = \mathcal{P}_c - \mathcal{X} - \mathcal{Q}_c \mathcal{R}_c^{-1} \mathcal{Q}_c^t > 0 \end{aligned} \tag{13.75}$$

It is readily evident from (13.74) and (13.75) that $\Xi = \Xi_a + \Xi_c$ and hence the existence of $\mathcal{X}$ satisfying (13.75) implies (13.74). By the same token, if (13.74) is satisfied, $\mathcal{X} = \mathcal{Q}_a \mathcal{R}_a^{-1} \mathcal{Q}_a^t - \mathcal{P}_a - \frac{1}{2}\Xi$ yields $\Xi_a = \Xi_c = \Xi_a = \frac{1}{2}\Xi$ and (13.75) is satisfied.

Bibliography

[1] Al-Fuhaid, A. S., M. S. Mahmoud and F. A. Saleh, "Stabilization of Power Systems by Decentralized Systems and Control Theory", *Electric Machines and Power Systems*, vol. 21, no. 3, 1993, pp. 293–318.

[2] Aly, G. S. and H. Ottertum, "Dynamic Benhaviour of Mixer-Settlers, Part III: Testing Mathematical Models", *J. Appl. Chem. Biotechnol.*, vol. 23, 1973, pp. 643–651.

[3] Anderson, B. D. O. and J. B. Moore, *Linear Optimal Control*, Prentice-Hall, Inc., Englewood Cliffs, NJ, 1971.

[4] Anderson, B. D. O. and J. B. Moore, "Time-Varying Feedback Laws for Decentralized Control", *IEEE Trans. Automat. Control*, vol. AC-26, 1981, pp. 1133–1138.

[5] Anderson, B. D. O. and D. J. Clements, "Algebraic Characterization of Fixed Modes in Decentralized Control", *Automatica*, vol. 17, 1981, pp. 703–712.

[6] Anderson, B. D. O., "Transfer Function Matrix Description of Decentralized Fixed Modes", *IEEE Trans. Automat. Control*, vol. AC-27, 1982, pp. 1176–1182.

[7] Anderson, B. D. O. and A. Dehghani, "Challenges of Adaptive Control-Past, Permanent and Future", *Annual Reviews in Control*, vol. 32, 2008, pp. 123–135.

[8] Arafeh, S. and K. P. Sage, "Multi-Level Discrete Time System Identification in Large Scale Systems," *Int. J. Syst. Sci*, vol. 5, no. 8, 1974, pp. 753–791.

[9] Bahnasawi, A. A. and M. S. Mahmoud, *Control of Partially-Known Dynamical Systems*, Springer-Verlag, Berlin, 1989.

[10] Bahnasawi, A. A., A. S. Al-Fuhaid and M. S. Mahmoud, "Decentralized and Hierarchical Control of Interconnected Uncertain Systems", *Proc. IEE Part D*, vol. 137, 1990, pp. 311–321.

[11] Bahnasawi, A. A., A. S. Al-Fuhaid and M. S. Mahmoud, "A New Hierarchical Control Structure for a Class of Uncertain Discrete Systems", *Control Theory and Advanced Technology*, vol. 6, 1990, pp. 1–21.

[12] Bakule, L. and J. Lunze, "Decentralized Design of Feedback Control for Large Scale Systems", *Kybernetika*, vol. 24, no. 3–6, 1988, pp. 1–100.

[13] Bakule, L. and J. Rodellar, "Decentralized Control and Overlapping Decomposition of Mechanical Systems. Part 1: System Decomposition. Part 2: Decentralized Stabilization", *Int. J. Control*, vol. 61, 1995, pp. 559–587.

[14] Bakule, L. and J. Rodellar, "Decentralized Control Design of Uncertain Nominally Symmetric Composite Systems", *IEE Proc. Control Theory and Applications*, vol. 143, no. 6, 1996, pp. 530–535.

[15] Bakule, L., J. Rodellar and J. M. Rossell, "Structure of Expansion-Contraction Matrices in the Inclusion Principle for Dynamic Systems," *SIAM J. Matrix Anal. Appl.*, vol. 21, no. 4, 2000, pp. 1136–1155.

[16] Bakule, L., J. Rodellar and J. M. Rossell, "Generalized Selection of Complementary Matrices in the Inclusion Principle," *IEEE Tran. Automat. Contro.*, vol. AC-45, no. 6, 2000, pp. 1237–1243.

[17] Bakule, L., J. Rodellar and J. M. Rossell, "Overlapping Quadratic Optimal Control of Linear Time-Varying Commutative Systems," *SIAM J. Control and Optimization*, vol. 40, no. 5, 2002, pp. 1611–1627.

[18] Bakule, L., "Reduced-Order Control Design of Time-Delayed Uncertain Symmetric Composite Systems", *Proc. 10th IFAC/IFORS/IMACS/IFIP Symposium on Large Scale Systems: Theory and Applications*, Osaka, Japan, 2004, pp. 130–135.

[19] Bakule, L., F. P. Crainiceanu, J. Rodellar and J. M. Rossell, "Overlapping Reliable Control for a Cable-Stayed Bridge Benchmark", *IEEE Trans. Control Sys. Tech.*, vol. 13, no. 4, 2005, pp. 663–669.

[20] Bakule, L., "Complexity-Reduced Guaranteed Cost Control Design for Delayed Uncertain Symmetrically Connected Systems", *Proc. American Control Conference*, 2005, pp. 2590–2595.

[21] Bakule, L., J. Rodellar and J. M. Rossell, "Robust Overlapping Guaranteed Cost Control of Uncertain State-Delay Discrete-Time Systems", *IEEE Trans. Automat. Contro.*, vol. 51, no. 12, 2006, pp. 1943–1950.

[22] Bakule, L., "Decentralized Control: An Overview", *Annual Rev. Control*, vol. 32, 2008, pp. 87–98.

[23] Benitez-Read, J. S. and M. Jamshidi, "Adaptive Input-Output Linearizing Control of Nuclear Reactor", *C-TAT Journal*, vol. 8, no. 3, 1992, pp. 525–545.

[24] Bernstein, D. S. and W. M. Haddad, "Steady-State Kalman Filtering with an $\mathcal{H}_\infty$ Error Bound", *Systems and Control Letters*, vol. 16, 1991, pp. 309–317.

[25] Bernstein, D., "Some Open Problems in Matrix Theory Arising in Linear Systems and Control", *Linear Algebra Applicat.*, vol. 164, 1992, pp. 409–432.

[26] Bertsekas, D. P. and J. N. Tsitsiklis, *Parallel and Distributed Computation*, Prentice-Hall, Upper Saddle River, NJ, 1989.

[27] Blondel, V., M. Gevers and A. Lindquist, "Survey on the State of Systems and Control", *European J. Control*, vol. 1, 1995, pp. 5–23.

[28] Boukas, E. K. and M. S. Mahmoud, "A Practical Approach to Control of Nonlinear Discrete-Time State-Delay Systems", *Optimal Control Appl. and Methods*, vol. 28, no. 5, 2007, pp. 397–417.

[29] Boyd, S., L. El Ghaoui, E. Feron and V. Balakrishnan, *Linear Matrix Inequalities in Systems and Control Theory*, SIAM Studies in Applied Mathematics, Philadelphia, PA, 1994.

[30] Brasch, F. M. and J. B. Pearson, "Pole Placement Using Dynamic Compensators", *IEEE Trans. Automatic Control*, vol. 15, no. 1, 1970, pp. 34–43.

[31] Cantoni, M., E. Weyer, Y. Li, S. K. Oai, I. Mareels and M. Ryan, "Control of Large-Scale Irrigation Networks", *Proc. IEEE*, vol. 59, no. 1, 2007, pp. 75–91.

[32] Chandra, R. S., C. Langbort and R. D'Andrea, "Distributed Control Design with Robustness to Small Time Delays", *Systems & Control Letters*, vol. 58, no. 4, 2009, pp. 296–303.

[33] Chapman, J. W., M. D. Ilic, C. A. King, L. Eng and H. Kaufman, "Stabilizing a Multimachine Power System via Decentralized Feedback Linearizing Excitation Control", *IEEE Trans. Power Systems*, vol. 8, no. 3, 1993, pp. 830–839.

[34] Chen, H. B., J. K. Shiau and J. H. Chow, "Simultaneous $\mathcal{H}_\infty$-Suboptimal Control Design", *Proc. American Control Conference*, Seattle, WA, 1995, pp. 1946–1950.

[35] Chen, C. T., *Introduction to Linear System Theory*, Prentice-Hall, New York, 1996.

[36] Chen, X. B. and S. S. Stankovic, "Decomposition and Decentralized Control of Systems with Multi-Overlapping Structure", *Automatica*, vol. 41, 2005, pp. 1765–1772.

[37] Chen, J. D., "Stability Criteria for Large-Scale Time-Delay Systems: The LMI Approach and the Genetic Algorithms", *Control and Cybernetics*, vol. 35, no. 2, 2006, pp. 291–301.

[38] Chen, J. D., "LMI-Based Stability Criteria for Large-Scale Time-Delay Systems", *JSME Int. Journal*, series C, vol. 49, no. 1, 2006, pp. 225–229.

[39] Cheng, C. and Q. Zhao, "Reliable Control of Uncertain Delayed Systems with Integral Quadratic Constraints", *IEE Proc. Control Theory Appl.*, vol. 151, no. 6, 2004, pp. 790–796.

[40] Chong, C. Y. and M. Athans, "On the periodic coordination of linear stochastic systems", *Proc. 6th IFAC World Congress*, Boston, 1975.

[41] Chu, D. and D. D. Siljak, "A Canonical Form for the Inclusion Principle of Dynamic Systems", *SIAM Journal on Control and Optimization*, vol. 44, no. 3, 2005, pp. 969–990.

[42] Cohen, G. and A. Benveniste, "Decentralization without Coordination," *Ricarche Ricarche Di Automatica*, vol. 5, no. 1, 1974, pp. 16–29.

[43] Corfmat, J. P. and A. S. Morse, "Decentralized Control of Linear Multivariable Systems", *Automatica*, vol. 12, 1976, pp. 479–496.

[44] Davis, M., "Markov Models and Optimization", Chapman & Hall, London, 1992.

[45] Davison, E. J. and S. H. Wang, "Properties of Linear Time-Invariant Multivariable Systems Subject to Arbitrary Output and State Feedback", *IEEE Trans. Automat. Contro.*, vol. AC-18, 1973, pp. 24–32.

[46] Davison, E. J., "A Generalization of the Output Control of Linear Multivariable Systems with Unmeasurable Arbitrary Disturbances", *IEEE Trans. Automat. Contro.*, vol. AC-20, 1975, pp. 788–792.

[47] Davison, E. J. and A. Goldenberg, "The Robust Decentralized Control of a General Servomechanism Problem: The Servo-Compensator", *Automatica*, vol. 11, 1975, pp. 461–471.

[48] Davison, E. J., "The Robust Decentralized Control of a General Servomechanism Problem", *IEEE Trans. Automat. Contro.*, vol. AC-21, 1976, pp. 114–124.

[49] Davison, E. J., "Decentralized Stabilization and Regulation in Large Multivariable Systems", in Ho, Y. C. and S. K. Mitter (eds.) *Directions in Large Scale Systems*, Plenum Press, New York, 1976, pp. 303–323.

[50] Davison, E. J., "The Robust Decentralized Servomechanism Problem with Extra Stabilizing Control Agents", *IEEE Trans. Automat. Contro.*, vol. AC-22, 1977, pp. 256–259.

[51] Davison, E. J., "Recent Results on Decentralized Control of Large Scale Multivariable Systems", *IFAC Symposium on Multivariable Technological Systems*, Fredericton, Canada, 1977, pp. 1–10.

[52] Davison, E. J., "The Robust Decentralized Control of a Servomechanism Problem for Composite Systems with Input-Output Interconnections", *EEE Trans. Automat. Contro.*, vol. AC-23, 1979, pp. 325–327.

[53] Davison, E. J. and W. Gesing, "Sequential Stability and Optimization of Large Scale Decentralized Systems", *Automatica*, vol. 15, 1979, pp. 307–324.

[54] Davison, E. J. and U. Ozguner, "Synthesis of Decentralized Robust Servomechanism Problem Using Local Models Fixed Modes", *IEEE Trans. Automat. Contro.*, vol. AC-27, 1982, pp. 583–600.

[55] Davison, E. J. and U. Ozguner, "Characterizations of Decentralized Fixed Modes for Interconnected Systems", *Automatica*, vol. 19, no. 2, 1982, pp. 169–182.

[56] Davison, E. J. and T. N. Chang, "Decentralized Stabilization and Pole Assignment for General Proper Systems", *IEEE Trans. Automat. Contro.*, vol. AC-35, 1990, pp. 652–664.

[57] D'Andrea, R. and G. E. Dullerud, "Distributed Control Design for Spatially Interconnected Systems", *IEEE Trans. Automat. Contro.*, vol. 48, no. 9, 2003, pp. 1478–1495.

[58] de Souza, C. E. and M. D. Fragoso, "H_∞ Control for Linear Systems with Markovian Jumping Parameters," *Control-Theory and Advanced Technology*, vol. 9, no. 2, 1993, pp. 457–466.

[59] Dorato, P. and R. K. Yedavalli, *Recent Advances in Robust Control*, IEEE Press, New York, 1990.

[60] Dorato, P. "Non-Fragile Controllers Design: An Overview", *Proceedings the American Control Conference*, Philadelphia, PA, 1998, pp. 2829–2831.

[61] Doyle, J. C. and G. Stein, "Robustness with Observers", *IEEE Trans. Automat. Contro.*, vol. 24, 1979, pp. 607–611.

[62] Doyle, J. C., K. Glover, P. P. Khargonekar and B. A. Francis, "State-Space Solutions to Standard $\mathcal{H}_2$ and $\mathcal{H}_\infty$ Control Problems", *IEEE Trans. Automat. Contro.*, vol. 34, 1989, pp. 831–847.

[63] Elliott, R. J. and D. D. Sworder, "Control of a Hybrid Conditionally Linear Gaussian Processes", *J. Optimization Theory and Applications*, vol. 74, 1992, pp. 75–85.

[64] El Ghaoui, L. and S. Niculescu, Eds., *Advances in Linear Matrix Inequalities Methods in Control.* SIAM, Philadelphia, PA, 2000.

[65] Edwards, C. and Spurgeon, S. K., "Sliding Mode Stabilization of Uncertain Systems Using Only Output Information", *Int. J. Control*, vol. 62, no. 5, 1995, pp. 1129–1144.

[66] Edwards, C. and Spurgeon, S. K., "Sliding-Mode Output Feedback Controller Design Using Linear Matrix Inequalities", *IEEE Trans. Auto. Control*, vol. 46, no. 1, 2001, pp. 115–119.

[67] Edwards, C. and S. K. Spurgeon, *Sliding Mode Control: Theory and Applications*, Taylor & Francis, London, 1998.

[68] Edwards, C., X. G. Yan and S. K. Spurgeon, "On the Solvability of the Constrained Lyapunov Problem", *IEEE Trans. on Automat. Control*, vol. 52, no. 10, 2007, pp.1982–1987.

[69] Even, S. *Graph Algorithms*, Computer Science Press, London, 1979.

[70] Famularo, D., P. Dorato, C. T. Abdallah, W. M. Haddad and A. Jadbabie, "Robust Non-Fragile LQ Controllers: The Static State Feedback Case", *Int. J. Control*, vol. 73, no. 2, 2000, pp. 159–165.

[71] Fax, J. A. and R. M. Murray, "Information Flow and Cooperative Control of Vehicle Formations", *IEEE Trans. Automat. Contro.*, vol. 49, no. 9, 2004, pp. 1465–1476.

[72] Fleming, W., S. Sethi and M. Soner, "An Optimal Stochastic Production Planning Problem with Randomly Fluctuating Demand", SIAM J. Control and Optimization, vol. 25, 1987, pp. 1494–1502.

[73] Feng, X., K. A. Loparo, Y. Ji and H. J. Chizeck, "Stochastic Stability Properties of Jump Linear Systems", IEEE Trans. Automatic Control, vol. 37, 1992, pp. 38–53.

[74] Fridman, E. and U. Shaked, "Delay-Dependent Stability and $\mathcal{H}_\infty$ Control: Constant and Time-Varying Delays", *Int. J. Control*, vol. 76, 2003, pp. 48–60.

[75] Fu, M., C. E. de Souza and L. Xie, "H_∞-Estimation for Uncertain Systems", *Int. J. Robust and Nonlinear Control*, vol. 2, 1992, pp. 87–105.

[76] Gahinet, P., A. Nemirovski, A. L. Laub and M. Chilali, *LMI Control Toolbox*, The Math Works, Inc., Boston, MA, 1995.

[77] Gahinet, P. and P. Apkarian, "A Linear Matrix Inequality Approach to $\mathcal{H}_\infty$ Control", *Int. J. Robust and Nonlinear Control*, vol. 4, 1994, pp. 421–448.

[78] Gao, H., J. Lam, C. Wang and Y. Wang, "Delay-Dependent Output-Feedback Stabilization of Discrete-Time Systems with Time-Varying State Delay", *IEE Proc. Control Theory and Applic.*, vol. 151, no. 6, 2004, pp. 691–698.

[79] Gao, H. and T. Chen, "New Results on Stability of Discrete-Time Systems with Time-Varying State-Delay", *IEEE Trans. Automat. Contro.*, vol. 52, no. 2, 2007, pp. 328–334.

[80] Gavel, D. T. and D. D. Siljak, "Decentralized Adaptive Control: Structural Conditions for Stability", *IEEE Trans. Automatic Control*, vol. 34, no. 4, 1989, pp. 413–426.

[81] Grateloup, G. and A. Titli, "Two-Level Dynamic Optimization Methods", *J. Optimization Theory and Applications*, vol. 15, no. 5, 1975, pp. 533–547.

[82] Geromel, J. C., J. Bernussou and P. Peres, "Decentralized Control through Parameter Space Optimization", *Automatica*, vol. 30, 1994, pp. 1565–1578.

[83] Geromel, J. C., J. Bernussou and M. C. de Oliveira, "$\mathcal{H}_\infty$Norm Optimization with Constrained Dynamic Output Feedback Controllers: Decentralized and Reliable Control", *IEEE Trans. Automat. Contro.*, vol. 44, no. 5, 1999, pp. 1449–1454.

[84] Geromel, J. C., P. L. D. Peres and S. R. de Souza "A Convex Approach to the Mixed $\mathcal{H}_2/\mathcal{H}_\infty$ Control Problem for Discrete-Time Uncertain Systems", *SIAM J. Control Optim.*, vol. 33, 1995, pp. 1816–1833.

[85] Ghosh, S., S. K. Das and G. Ray, "Decentralized Stabilization of Uncertain Systems with Interconnection and Feedback Delay: An LMI Approach", *IEEE Trans. Automat. Contro.*, vol. 54, no. 4, 2009, pp. 905–912.

[86] Godbole, D. N. and J. Lygeros, "Longitudinal Control of the Lead Car of a Platoon", *IEEE Trans. Vehicular Tech.*, vol. 43, no. 4, 1994, pp. 1125–1135.

[87] Golub, G. H. and C. F. Van Loan, *Matrix Computations*, John Hopkins Studies in the Math Sciences, Baltimore, 1996.

[88] Gong, Z., "Decentralized Robust Control of Uncertain Interconnected Systems with Prescribed Degree of Exponential Convergence", *IEEE Trans. Automat. Contro.*, vol. 40, no. 4, 1995, pp. 704–707.

[89] Gong, Z. and M. Aldeen, "Stabilization of Decentralized Control Systems", *J. Math Systems, Estimation and Control*, vol. 7, 1997, pp. 1–16.

[90] Gouaisbaut, F., Y. Blanco and J. P. Richard, "Robust Sliding Mode Control of Non-Linear Systems with Delay: A Design via Polytopic Formulation", *Int. J. Control*, vol. 77, no. 2, 2004, pp. 206–215.

[91] Grigoriadis, K. and R. Skelton, "Low-Order Design for LMI Problems Using Alternating Projection Methods", *Automatica*, vol. 32, 1997, pp. 837–852.

[92] Gu, G., "Stabilizability Conditions of Multivariable Uncertain Systems via Output Feedback", *IEEE Trans. Automat. Contro.*, vol. 35, 1990, pp. 925–927.

[93] Gu, K., V. L. Kharitonov and J. Chen, *Stability of Time-Delay Systems*, Birkhauser, Boston, 2003.

[94] Gu, K. Q., "Discretized Lyapunov Functional for Uncertain Systems with Multiple Time-Delay", *Int. J. Control*, vol. 72, no. 16, 1999, pp.1436–1445.

[95] Guo, Y., D. J. Hill and Y. Wang, "Nonlinear Decentralized Control of Large-Scale Power Systems", *Automatica*, vol. 36, 2000, pp. 1275–1289.

[96] Haddad, W. M. and J. R. Corrado, "Robust Resilient Dynamic Controllers for Systems with Parametric Uncertainty and Controller Gain Variations", *Proceedings the American Control Conference*, Philadelphia, PA, 1998, pp. 2837–2841.

[97] Haimes, Y. Y., *Hierarchical Analyses of Water Resources Systems*, McGraw-Hill, New York, 1977.

[98] Hassan, M. F. "Optimal Kalman Filter for Large Scale Systems Using the Prediction Approach", *IEEE Trans. Syst. Man Cybern.*, vol. SMC-6, 1976, pp. .

[99] Hassan, M. F., G. Salut, M. G. Singh and A. Titli, "A Decentralized Computational Algorithm for the Global Kalman Filter", *IEEE Trans. Automat. Contro.*, vol. AC-23, 1978, pp. 262–268.

[100] Hassan, M. F., M. S. Mahmoud and M. I. Younis, "A Dynamic Leontief Modeling Approach to Management for Optimal Utilization in Water Resources Systems", *IEEE Trans. Syst. Man Cybern.*, vol. SMC-11, 1981, pp. 552–558.

[101] Hassan, M. F., M. S. Mahmoud, M. G. Singh and M. P. Spathopolous, "A Two-Level Parameter Estimation Algorithm Using the Multiple Projection Approach", *Automatica*, vol. 18, 1982, pp. 621–630.

[102] Hayakawa, T., H. Ishii and K. Tsumura, "Adaptive Quantized Control for Linear Uncertain Discrete-Time Systems", *Automatica*, vol. 45, 2009, pp. 692–700.

[103] He, Y., Q. G. Wang, L. H. Xie and C. Lin, "Further Improvement of Free-Weighting Matrices Technique for Systems with Time-Varying Delay", *IEEE Trans. Automat. Contro.*, vol. 52, no. 2, 2007, pp. 293–299.

[104] Himmelblau, D. M., Ed., *Decomposition of Large-Scale Problems*, Elsevier, New York, 1973.

[105] Ho, D. W. C. and G. Lu, "Robust Stabilization for a Class of Discrete-Time Non-Linear Systems via Output Feedback: The Unified LMI Approach", *Int. J. Control*, vol. 76, no. 1, 2003, pp. 105–115.

[106] Ho, Y. C. and S. K. Mitter, Eds., *Directions in Large Scale Systems*, Plenum Press, New York, 1976.

[107] Hovd, M. and S. Skogestad, "Control of Symmetrically Interconnected Plants", *Automatica*, vol. 30, no. 6, 1994, pp. 957–973.

[108] Hsieh, C. S., "Reliable Control Design Using a Two-Stage Linear Quadratic Reliable Control", *IEE Proc. Control Theory Appl.*, vol. 150, no. 1, 2003, pp. 77–82.

[109] Hsu, K. C., "Decentralized Variable-Structure Control Design for Uncertain Large-Scale Systems with Series Nonlinearities", *Int. J. Control*, vol. 68, no. 6, 1997, pp.1231–1240.

[110] Hu, Z., "Decentralized Stabilization of Large-Scale Interconnected Systems with Delays", *IEEE Trans. Automat. Contro.*, vol. 39, no. 5, 1994, pp. 180–182.

[111] Hua, C., X. Guan and P. Shi, "Robust Decentralized Adaptive Control for Interconnected Systems with Time Delays", *J. Dyn. Systems, Measurements and Control*, vol. 127, 2005, pp. 656–662.

[112] Hua, C., X. Guan and P. Shi, "Decentralized Robust Model Reference Adaptive Control for Interconnected Time-Delay Systems", *J. Computer and Applied Math.*, vol. 193, 2006, pp. 383–396.

[113] Hua, C. and X. Guan, "Output Feedback Stabilization for Time-Delay Non-Linear Interconnected Systems Using Neural Networks", *IEEE Trans. Neural Networks*, vol. 19, no. 4, 2008, pp. 673–688.

[114] Hua, C. C., Q. G. Wang and X. P. Guan, "Exponential Stabilization Controller Design for Interconnected Time Delay Systems", *Automatica*, vol. 44, 2008, pp. 2600–2606.

[115] Huang, P. and Sundareshan, M. K., "A New Approach to the Design of Reliable Decentralized Control Schemes for Large-Scale Systems", *Proc. IEEE Int. Conf. Circuits and Systems*, Houston, TX, 1980, pp. 678–680.

[116] Huang, S., J. Lam and G. H. Yang, "Reliable Linear-Quadratic Control for Symmetric Composite Systems", *Int. J. Systems Science*, vol. 32, no. 1, 2001, pp. 73–82.

[117] Huang, S., J. Lam, G. H. Yang and S. Zhang, "Fault Tolerant Decentralized $\mathcal{H}_\infty$ Control for Symmetric Composite Systems", *IEEE Trans. Automat. Contro.*, vol. 44, no. 11, 1999, pp. 2108–2114.

[118] Ibrir, S., W. F. Xie and C. Y. Su, "Observer-Based Control of Discrete-Time Lipschitzian Nonlinear Systems: Application to One-Link Joint Robot", *Int. Journal of Control*, vol. 78, no. 6, 2005, pp. 385–395.

[119] Iftar, A. and Ozguner, U, "An Optimal Control Approach to the Decentralized Robust Servomechanism Problem", *IEEE Trans. Automat. Contro.*, vol. 34, 1989, pp. 1268–1271.

[120] Iftar, A. and U. Ozguner, "Contractible Controller Design and Optimal Control with State and Input Inclusion", *Automatica*, vol. 26, 1990, pp. 593–597.

[121] Iftar, A., "Decentralized Estimation and Control with Overlapping Input, State and Output Decomposition", *Automatica*, vol. 29, 1993, pp. 511–516.

[122] Iftar, A. and U. Ozguner, "Overlapping Decompositions, Expansions, Contractions, and Stability of Hybrid Systems", *IEEE Trans. Automatic Control*, vol. 43, no. 8, 1998, pp. 1040–1055.

[123] Iftar, A. and E. J. Davison, "Decentralized Control Strategies for Dynamic Routing", *Optimal Control Applications and Methods*, vol. 23, 2002, pp. 329–335.

[124] Ikeda, M. and D. D. Šiljak, "Overlapping Decompositions, Expansions and Contractions of Dynamic Systems", *Large Scale Syst.*, vol. 1, 1980, pp. 29–38.

[125] Ikeda, M. and D. D. Šiljak, "Decentralized Stabilization of Linear Time-Varying Systems", *IEEE Trans. Automat. Contro.*, vol. AC-25, 1980, pp. 106–107.

[126] Ikeda, M. and D. D. Šiljak, "On Decentrally Stabilizable Large-Scale Systems", *Automatica*, vol. 16, 1980, pp. 331–334.

[127] Ikeda, M. and D. D. Šiljak, "Overlapping Decompositions, Expansions and Contractions of Dynamic Systems", *Large Scale Systems*, vol. 1, 1980, pp. 29–38.

[128] Ikeda, M. and D. D. Šiljak, "Generalized Decompositions of Dynamic Systems and Vector Lyapunov Functions", *IEEE Trans. Automat. Contro.*, vol. 26, 1981, pp. 1118–1125.

[129] Ikeda, M., D. D. Šiljak and D. E. White, "Decentralized Control with Overlapping Informations Sets", *J. Optimization Theory and Applic.*, vol. 34, no. 2, 1981, pp. 279–309.

[130] Ikeda, M., D. D. Šiljak and D. E. White, "An Inclusion Principle for Dynamic Systems", *IEEE Trans. Automat. Contro.*, vol. 29, 1984, pp. 244–249.

[131] Ikeda, M. and D. D. Šiljak, "Overlapping Decentralized Control with Input, State and Output Inclusion", *Control Theory Adv. Tech.*, vol. 2, 1986, pp. 155–172.

[132] Ioannou, P. A., "Decentralized Adaptive Control of Interconnected Systems", *IEEE Trans. Automat. Contro.*, vol. AC-31, no. 4, 1986, pp. 291–298.

[133] Istepanian, R. and J. Whidborne, Eds., *Digital Controller Implementation and Fragility*, Springer-Verlag, New York, 2001.

[134] Iwasaki, T. and R. Skelton, "Parametrization of all Stabilizing Controllers via Quadratic Lyapunov Functions", *J. Optim. Theory Applicat.*, vol. 77, 1995, pp. 291–307.

[135] Iwasaki, T., R. Skelton and J. Geromel, "Linear Quadratic Suboptimal Control with Static Output Feedback", *Syst. Control Lett.*, vol. 23, 1994, pp. 421–430.

[136] Jamshidi, M., *Large-Scale Systems: Modeling, Control and Fuzzy Logic*, Prentice-Hall, New York, 1997.

[137] Javdan, M. R., "A Unified Theory of Optimal Multilevel Control," *Int. J. Control*, vol. 22, no. 4, 1975, pp. 517–527.

[138] Javdan, M. R., "Extension of Dual Coordination to a Class of Nonlinear Systems", *Int. J. Control*, vol. 24, no. 4, 1976, pp. 551–571.

[139] Ji, Y. and H. J. Chizeck, "Controllability, Stabilizability and Continuous-Time Markovian Jump Linear-Quadratic Control", *IEEE Trans. Automatic Control*, vol. 35, 1990, pp. 777–788.

[140] Jain, S. and F. Khorrami, "Decentralized Adaptive Output Feedback Design for Large-Scale Nonlinear Systems", *IEEE Trans. Automat. Contro.*, vol. 42, no. 5, 1997, pp. 729–735.

[141] Jiang, Z. P., "Decentralized and Adaptive Nonlinear Tracking of Large-Scale Systems via Output Feedback", *IEEE Trans. Automat. Contro.*, vol. 45, no. ll, 2000, pp. 2122–2128.

[142] Jiang, Z. P., D. W. Repperger and D. J. Hill, "Decentralized Nonlinear Output-Feedback Stabilization with Disturbance Attenuation", *IEEE Trans. Automatic Control*, vol. 46, no. 10, 2001, pp. 1623–1629.

[143] Jiang, Z. P., "Decentralized Disturbance Attenuating Output-Feedback Trackers for Large-Scale Nonlinear Systems", *Automatica*, vol. 38, 2002, pp. 1407–1415.

[144] Jiang, X. and Q. L. Han, "$\mathcal{H}_\infty$ Control for Linear Systems with Interval Time-Varying Delay", *Automatica*, vol. 41, no. 12, 2005, pp. 2099–2106.

[145] Jiao, X. and T. Shen, "Adaptive Feedback Control of Nonlinear Time-Delay Systems: The LaSalle-Razumikhin-Based Approach", *IEEE Trans. Automat. Contro.*, vol. 50, no. 11, 2005, pp. 1909–1913.

[146] Jing, X. J., D. L. Tan and Y. C. Wang, "An LMI Approach to Stability of Systems with Severe Time-Delay", *IEEE Trans. Automat. Contro.*, vol. 49, no. 7, 2004, pp. 1192–1195.

[147] Kailath, T., *Linear Systems*, Prentice-Hall, Englewood Cliffs, NJ, 1980.

[148] Kalsi, K., J. Lian and S. H. Zak, "On Decentralized Control of Nonlinear Interconnected Systems", *Int. J. Control*, vol. 82, no. 3, 2009, pp. 541–554.

[149] Karafyllis, I., "Robust Global Stabilization by Means of Discrete-Delay Output Feedback", *Systems & Control Letters*, vol. 57, 2008, pp. 987–995.

[150] Kawka, P. A. and A. G. Alleyne, "Stability and Feedback Control of Wireless Networked Systems", *Proc. American Contr. Conf.*, Portland, OR, 2005, pp. 2953–2959.

[151] Keel, L. H. and S. P. Bhattacharyya, "Robust, Fragile, or Optimal ?", *IEEE Trans. Automat. Control*, vol. 42, 1997, pp. 1098–1105.

[152] Khalil, H. K., *Nonlinear Systems*, Second Edition, Prentice-Hall, Englewood Cliffs, NJ, 2002.

[153] Khargonekar, P. P., M. A. Rotea and E. Baynes, "Mixed $\mathcal{H}_2/\mathcal{H}_\infty$ Filtering", *Int. J. Robust and Nonlinear Control*, vol. 6, 1996, pp. 313–330.

[154] Kobayashi, Y., M. Ikeda and Y. Fujisaki, "Stability of Large Space Structures Preserved under Failures and Local Controllers", *IEEE Trans. Automat. Control*, vol. 52, no. 2, 2007, pp. 318–322.

[155] Kokotovic, P. A., "Applications of Singular Perturbation Techniques to Control Problems", *SIAM Review*, vol. 26, no. 4, 1984, pp. 501–550.

[156] Krishnamurthy, P. and F. Khorrami, "Decentralized Control and Disturbance Attenuation for Large-Scale Nonlinear Systems in Generalized Output-Feedback Canonical Form", *Automatica*, vol. 9, 2003, pp. 1923–1933.

[157] Kushner, H., *Stochastic Stability and Control*, Academic Press, New York, 1967.

[158] Kwakernaak, H. W. and R. Sivan, *Linear Optimal Control Systems*, Wiley, New York, 1972.

[159] Kwan, C. M. "Further Results on Variable Output Feedback Controllers", *IEEE Trans. Auto. Control*, vol. 46, no. 9, 2001, pp. 1505–1508.

[160] Lam, J. and G. H. Yang, "Balanced Model Reduction of Symmetric Composite Systems", *Int. J. Control*, vol. 65, no. 6, 1996, pp. 1031–1043.

[161] Lam, J., G.-H. Yang, S.-Y. Zhang and J. Wang, "Reliable Control Using Redundant Controllers", *IEEE Trans. Automat. Contro.*, vol. 43, 1998, pp. 1588–1593.

[162] Langenhop, C. E., "On Generalized Inverses of Matrices", *SIAM L Appl. Math*, vol. 15, 1967, pp. 1239–1246.

[163] Langbort, C. L., R. S. Chandra and R. D'Andrea, "Decentralized Stabilization of Linear Continuous or Discrete-Time Systems with Delays in the Interconnection", *IEEE Trans. Automat. Contro.*, vol. 49, no. 9, 2004, pp. 1502–1519.

[164] Lasdon, L. S., *Optimization Theory for Large Systems*, Macmillan, New York, 1970.

[165] Lavaei, J. and A. G. Aghdam, "A Necessary and Sufficient Condition for the Existence of a LTI Stabilizing Decentralized Overlapping Controller", *Proc. the 45th IEEE Conf. on Decision and Control*, San Diego, CA, 2006, pp. 6179–6186.

[166] Lavaei, J. and A. G. Aghdam, "Elimination of Fixed Modes by Means of High-Performance Constrained Periodic Control", *Proc. the 45th IEEE Conf. on Decision and Control*, San Diego, CA, 2006, pp. 4441–4447.

[167] Lavaei, J. and A. G. Aghdam, "Characterization of Decentralized and Quotient Fixed Modes via Graph Theory", *Proc. American Control Conference*, New York, 2007, pp. 790–795.

[168] Lavaei, J. and A. G. Aghdam, "Simultaneous LQ Control of a Set of LTI Systems Using Constrained Generalized Sampled-Data Hold Functions", *Automatica*, vol. 43, 2007, pp. 274–280.

[169] Lavaei, J. and A. G. Aghdam, "High-performance Decentralized Control Design for General Interconnected Systems with Applications in Cooperative Control", *Int. J. Control*, vol. 80, 2007, pp. 935–951.

[170] Lavaei, J., A. Momeni and A. G. Aghdam, "A Model Predictive Decentralized Control Scheme with Reduced Communication Requirement for Spacecraft Formation", *IEEE Trans. Control Systems Technology*, vol. 16, 3008, pp. 268–278.

[171] Lavaei, J., A. Momeni and A. G. Aghdam, "LQ Suboptimal Decentralized Controllers with Disturbance Rejection Property for Hierarchical Systems", *Int. J. Control*, vol. 81, 2008, pp. 1720–1732.

[172] Lee, J. L., "On the Decentralized Stabilization of Interconnected Variable Structure Systems Using Output Feedback", *J. Franklin Institute*, vol. 332, no. 5, 1995, pp. 595–605.

[173] Lee, K. H., "Robust Decentralized Stabilization of a Class of Linear Discrete-Time Systems with Nonlinear Interactions", *Int. J. Control*, vol. 80, 2007, pp. 1544–1551.

[174] Lee, T. N. and U. L. Radovic, "Decentralized Stabilization of Linear Continuous or Discrete-Time Systems with Delays in the Interconnection", *IEEE Trans. Automat. Contro.*, vol. 33, no. 5, 1989, pp. 757–760.

[175] Lee, S. R. and M. Jamshidi, "Stability Analysis for Large-Scale Time Delay Systems via the Matrix Lyapunov Function", *Kybernetika*, vol. 28, no. 4, 1992, pp. 271–283.

[176] Leitmann, G., "Guaranteed Ultimate Boundedness for a Class of Uncertain Linear Dynamical Systems", *IEEE Trans. Automat. Contro.*, vol. 23, no. 9, 1978, pp. 1109–1110.

[177] Leitmann, G., "Guaranteed Asymptotic Stability for Some Linear Systems with Bounded Uncertainties", *ASME J. Dyn. Syst. Meas. & Control*, vol. 101, no. 3, 1979, pp. 212–219.

[178] Leibfritz, F. and E. M. Mostafa, "Trust Region Methods for Solving the Optimal Output Feedback Design Problem", *Int. J. Control*, vol. 76, 2003, pp. 501–519.

[179] Leros, A. P. and P. P. Groumpos, "Time-Invariant BAS-Decentralized Large-Scale Linear Regulator Problem", *Int. J. Control*, vol. 46, no. 1, 1987, pp. 129–152.

[180] Liu, X., "Output Regulation of Strongly Coupled Symmetric Composite Systems", *Automatica*, vol. 28, no. 5, 1992, pp. 1037–1041.

[181] Luenberger, D. G., *Optimization by Vector Space Method*, New York, Wiley, 1969.

[182] Luenberger, D. G., "An Introduction to Observers", *IEEE Trans. Automat. Control.*, vol. AC-16, no. 3, 1971, pp. 596–602.

[183] Levine, W. S. and M. Athans, "On the Optimal Error Regulation of a String of Moving Vehicles," *IEEE Trans. Automat. Contro.*, vol. AC-11, 1966, pp. 355–361.

[184] Levine, W. S. and M. Athans, "On the Determination of the Optimal Constant Output Feedback Gains for Linear Multivariable Systems," *IEEE Trans. Automat. Contr.*, vol. AC-15, 1970, pp. 44–48.

[185] Li, R. H. and M. G. Singh, "Information Structures in Deterministic Decentralized Control Problems", *IEEE Trans. Syst., Man and Cybern.*, vol. SMC-13, 1983, pp. 162–166.

[186] Lin, Q., G. Wang and T. H. Lee, "A Less Conservative Robust Stability Test for Linear Uncertain Time-Delay Systems", *IEEE Trans. Automat. Contr.*, vol. 51, 2006, pp. 87–91.

[187] Litkouhi, B. and H. K. Khalil, "Infinite-Time Regulators for Singularly Perturbed Difference Equations", *Int. J. Control*, vol. 39, 1984, pp. 567–598.

[188] Litkouhi, B. and H. K. Khalil, "Multirate and Composite Control of Two-Time-Scale Discrete-Time Systems", *IEEE Trans. Automat. Control*, vol. AC-30, 1985, pp. 645–651.

[189] Liu, F., P. Jiang, H. Su and J. Chu, "Robust $\mathcal{H}_\infty$ Control for Time-Delay Systems with Additive Controller Uncertainty", *Proc. the 4th World Congress on Intelligent Control and Automation*, Shanghai, P. R. China, 2002, pp. 1718–1722.

[190] Lu, Q., Y. Sun, Z. Xu and T. Mochizuki, "Decentralized Nonlinear Optimal Excitation Control", *IEEE Trans. Power Systems*, vol. 11, no. 4, 1996, pp. 1957–1962.

[191] Lunze, J., "Dynamics of Strongly Coupled Symmetric Composite Systems", *Int. J. Control*, vol. 44, no. 6, 1986, pp. 1617–1640.

[192] Lunze, J., *Feedback Control of Large-Scale Systems*, Prentice-Hall, London, 1992.

[193] Luyben, W. L., *Chemical Reactor Design and Control*, John Wiley & Sons, New Jersey, 2007.

[194] Mahmoud, M. S., "Multilevel Systems Control and Applications: A Survey", *IEEE Trans. Systems, Man and Cybernetics*, vol. SMC-7, no. 3, 1977, pp. 125–143.

[195] Mahmoud, M. S., "Closed-Loop Multilevel Control for Large Nonlinear Systems via Invariant Embedding Techniques", *J. Computers and Electrical Eng.*, vol. 4, 1977, pp. 525–543.

[196] Mahmoud, M. S., W. G. Vogt and M. H. Mickle, "Multilevel Control and Optimization Using Generalized Gradients Techniques", *Int. J. Control*, vol. 25, 1977, pp. 525–543.

[197] Mahmoud, M. S., W. G. Vogt and M. H. Mickle, "Feedback Multilevel Control for Continuous Dynamic Systems", *J. Franklin Institute*, vol. 303, no. 5, 1977, pp. 453–471.

[198] Mahmoud, M. S., "A Class of Optimization Techniques for Linear State Regulators", *Int. J. Systems Science*, vol. 8, no. 5, 1977, pp. 513–537.

[199] Mahmoud, M. S., "Dynamic Feedback Methodology for Interconnected Control Systems", *Int. J. Control*, vol. 29, no. 5, May 1979, pp. 881–898.

[200] Mahmoud, M. S., "A Quantitative Comparison between Two Decentralized Control Approaches", *Int. J. Control*, vol. 28, no. 2, 1978, pp. 261–275.

[201] Mahmoud, M. S., W. G. Vogt and M. H. Mickle, "Decomposition and Coordination Methods for Constrained Optimization", *J. Optimization Theory and Applications*, vol. 28, 1979, pp. 549–594.

[202] Mahmoud, M. S., "Dynamic Multilevel Methodology for Interconnected Control Systems", *Int. J. Control*, vol. 29, no. 5, 1979, pp. 881–898.

[203] Mahmoud, M. S., "Dynamic Multilevel Optimization for a Class of Nonlinear Systems", *Int. J. Control*, vol. 30, no. 6, 1979, pp. 927–948.

[204] Mahmoud, M. S., M. F. Hassan and M. G. Singh, "A New Hierarchical Approach to the Joint Problem of Systems Identification and Optimization", *Large Scale Systems Theory and Applications*, vol. 1, 1980, pp. 159–166.

[205] Mahmoud, M. S., "Multilevel Systems: Information Flow in Large Linear Problems", in *Handbook of Large Scale Systems Engineering Applications*, Madan G. Singh and Andre Title (eds.), North Holland, Amsterdam, 1979, pp. 96–109.

[206] Mahmoud, M. S. and M. G. Singh, *Large-Scale Systems Modelling*, Pergamon Press, London, 1981.

[207] Mahmoud, M. S. and M. G. Singh, "Decentralized State Reconstruction of Interconnected Discrete Systems", *Large Scale Systems*, vol. 2, 1981, pp. 151–158.

[208] Mahmoud, M. S., "Design of Observer-Based Controllers for a Class of Discrete Systems", *Automatica*, vol. 18, no. 3, 1982, pp. 323–329.

[209] Mahmoud, M. S., "Order Reduction and Control of Discrete Systems", *Proc. IEE Part D*, vol. 129, no. 4, 1982, pp. 129–135.

[210] Mahmoud, M. S., "Structural Properties of Discrete Systems with Slow and Fast Modes", *Large Scale Systems Theory and Applications*, vol. 3, no. 4, 1982, pp. 227–236.

[211] Xinogalas, T. C., M. S. Mahmoud and M. G. Singh, "Hierarchical Computation of Decentralized Gains for Interconnected Systems", *Automatica*, vol. 18, 1982, pp. 474–478.

[212] Mahmoud, M. S., "Multi-Time Scale Analysis in Discrete Systems", *J. Engineering and Applied Sciences*, vol. 2, no. 4, 1983, pp. 301–315.

[213] Mahmoud, M. S., "Comments on Singular-Perturbation Method for Discrete Models of Continuous Systems in Optimal Control", *Proc. IEE Part D*, vol. 130, no. 3, 1983, pp. 136.

[214] Mahmoud, M. S. and Y. Chen, "Design of Feedback Controllers by Two-Stage Methods", *Applied Math Modeling*, vol. 7, no. 3, 1983, pp. 163–168.

[215] Hassan, M. F., M. S. Mahmoud and M. I. Younis, "Linear Multilevel Estimators for Stabilization of Discrete Systems", *Int. J. Systems Science*, vol. 14, no. 7, 1983, pp. 731–743.

[216] Fawzy, A. S., M. S. Mahmoud and O. R. Hinton, "Hierarchical Techniques for On Line Microprocessor Control of an Industrial Fermentation Process", *Int. J. Systems Science*, vol. 14, 1983, pp. 19–29.

[217] Mahmoud, M. S. and M. G. Singh, *Discrete Systems: Analysis, Optimization and Control*, Springer-Verlag, Berlin, 1984.

[218] Mahmoud, M. S. and A. S. Fawzy, "On the Hierarchical Algorithms for Nonlinear Systems with Bounded Control", *Optimal Control Applications and Methods*, vol. 5, 1984, pp. 275–288.

[219] Mahmoud, M. S., N. M. Khraishi and H. A. Othman, "Bounds on Suboptimality in Discrete Aggregated Models", *Large Scale Systems Theory and Applications*, vol. 8, 1985, pp. 19–32.

[220] Mahmoud, M. S., Y. Chen and M.G. Singh, "On the Eigenvalue Assignment in Discrete Systems with Slow and Fast Modes", *Int. J. Systems Science*, vol. 16, no. 1, 1985, pp. 168–187.

[221] Mahmoud, M. S., N. M. Khraishi and H. A. Othman, "Bounds on Sub-optimality in Discrete Aggregated Models", *Large Scale Systems Theory and Applications*, vol. 8, no. 1, 1985, pp. 19–32.

[222] Mahmoud, M. S. and M. G. Singh, "On the Use of Reduced-Order Models in Output Feedback Design of Discrete Systems", *Automatica*, vol. 21, no. 4, 1985, pp. 485–489.

[223] Othman, H. A., N. M. Khraishi and M. S. Mahmoud, "Discrete Regulators with Time-Scale Separation", *IEEE Trans. Automat. Control*, vol. AC-30, no. 6, 1985, pp. 293–297.

[224] Mahmoud, M. S., M. F. Hassan and M. G. Darwish, *Large Scale Control Systems: Theories and Techniques*, Marcel Dekker Inc., New York, 1985.

[225] Mahmoud, M. S., M. F. Hassan and S. J. Saleh, "Decentralized Structures for Stream Water Quality Control Problems", *Optimal Control Applications and Methods*, vol. 6, 1985, pp. 167–186.

[226] Mahmoud, M. S. and S. J. Saleh, "Regulation of Water Quality Standards in Streams by Decentralized Control", *Int. J. Control*, vol. 41, 1985, pp. 525–540.

[227] Mahmoud, M. S. and M. G. Singh, "On the Use of Reduced-Order Models in Output Feedback Design of Discrete Systems", *Automatica*, vol. 21, 1985, pp. 485–489.

[228] Mahmoud, M. S., "Stabilization of Discrete Systems with Multiple Time Scales", *IEEE Trans. Automat. Control*, vol. AC-31, no. 2, 1986, pp. 159–162.

[229] Mahmoud, M. S., Y. Chen and M. G. Singh, "Discrete Two-Time Scale Systems", *Int. J. System Science*, vol. 17, no. 8, 1986, pp. 1187–1207.

[230] Mahmoud, M. S., Y. Chen and M. G. Singh, "A Two-Stage Output Feedback Design", *Proc. IEE Part D*, vol. 133, no. 6, 1986, pp. 279–284.

[231] Mahmoud, M. S. and M. F. Hassan, "A Decentralized Water Quality Control Scheme", *IEEE Trans. Systems, Man and Cybernetics*, vol. SMC-16, 1986, pp. 694–702.

[232] Mahmoud, M. S. and J. A. Assiri, "Performance Analysis of Two-Level Structures on Finite Precision Machines", *Automatica*, vol. 22, 1986, pp. 371–375.

[233] Nassar, A. M. and M. S. Mahmoud, "Implementation of Two-Level Algorithms Using Fixed-Point Arithmetic", *Int. J. Systems Science*, vol. 17, 1986, pp. 1279–1292.

[234] Mahmoud, M. S. and M. F. Hassan, "A Decentralized Water Quality Control Scheme", *IEEE Trans. Systems, Man and Cybernetics*, vol. SMC-16, 1986, pp. 694–702.

[235] Mahmoud, M. S., Y. Chen and M. G. Singh, "A Two-Stage Output Feedback Design", *Proc. IEE Part D*, vol. 133, 1986, pp. 279–284.

[236] Mahmoud, M. S., M. F. Hassan and S. J. Saleh, "Decentralized Structures for Stream Water Quality Control Problems", *Optimal Control Applications and Methods*, vol. 6, 1987, pp. 167–186.

[237] Mahmoud, M. S., "Costate Coordination Structure", in *Encyclopedia of Systems and Control*, vol. 2, Pergamon Press, Oxford, 1987, pp. 859–862.

[238] Mahmoud, M. S., "Singular Perturbations: Discrete Version", in *Encyclopedia of Systems and Control*, vol. 7, Pergamon Press, Oxford, 1987, pp. 4429–4434.

[239] Mahmoud, M. S. and S. Z. Eid, "Optimization of Freeway Traffic Control Problems" *Optimal Control Applications and Methods*, vol. 9, 1988, pp. 37–49.

[240] Mahmoud, M. S., "Discrete Systems with Multiple-Time Scales", in *Control and Dynamic Systems* Series, vol. 27, C. T. Leondes (ed.), Academic Press, New York, 1988, pp. 307–367.

[241] Mahmoud, M. S., "Algorithms for Decentralized Hierarchical Systems with Application to Stream Water Quality", in *Control and Dynamic Systems* Series, vol. 29, C. T. Leondes (ed.), Academic Press, New York, 1989, pp. 283–302.

[242] Mahmoud, M. S., "Algorithms for Decentralized Hierarchical Systems", in *Control and Dynamic Systems* Series, vol. 30, C. T. Leondes (ed.), Academic Press, New York, 1989, pp. 217–245.

[243] Mahmoud, M. S., "Singular Perturbations: Discrete Version", in *Concise Encyclopedia of Modeling and Simulation*, Pergamon Press, Oxford, 1989.

[244] Mahmoud, M. S., "Dynamic Decentralized Stabilization for a Class of Multi-Stage Processes", *Automatica*, vol. 24, no. 3, 1989, pp. 421–425.

[245] Mahmoud, M. S., *Computer-Operated Systems Control*, Marcel Dekker Inc., New York, 1991.

[246] Mahmoud, M. S., "Stabilizing Control for a Class of Uncertain Interconnected Systems", *IEEE Trans. Automat. Contro.*, vol. 39, no. 12, 1994, pp. 2484–2488.

[247] Mahmoud, M. S. and A. A. Bahnasawi, "Control Design of Uncertain Systems with Slow and Fast Modes", *J. Systems Analysis and Modeling Simulation*, vol. 17, no. 1, 1994, pp. 67–83.

[248] Mahmoud, M. S., "Guaranteed Stabilization of Interconnected Discrete-Time Systems", *Int. J. Systems Science*, vol. 26, no. 1, 1995, pp. 337–358.

[249] Mahmoud, M. S., "Adaptive Stabilization of a Class of Interconnected Systems", *Int. J. Computers and Electrical Engineering*, vol. 23, 1997, pp. 225–238.

[250] Mahmoud, M. S. and S. Kotob, "Adaptive Decentralized Model-Following Control", *Systems Analysis and Modelling Analysis*, vol. 27, 1997, pp. 169–210.

[251] Mahmoud, M. S. and S. Bingulac, "Robust Design of Stabilizing Controllers for Interconnected Time-Delay Systems", *Automatica*, vol. 34, 1998, pp. 795–800.

[252] Mahmoud, M. S. and M. Zribi, "Robust and $\mathcal{H}_\infty$ Stabilization of Interconnected Systems with Delays", *IEE Proc. Control Theory Appl.*, vol. 145, 1998, pp. 558–567.

[253] Mahmoud, M. S., "Robust Stability and Stabilization of a Class of Nonlinear Systems with Delays", *J. Mathematical Problems in Engineering*, vol. 4, no. 2, 1998, pp. 165–185.

[254] Mahmoud, M. S. and M. Zribi, "$\mathcal{H}_\infty$ Controllers for Time-Delay Systems Using Linear Matrix Inequalities", *J. Optimization Theory and Applications*, vol. 100, 1999, pp. 89–123.

[255] Mahmoud, M. S., *Robust Control and Filtering for Time-Delay Systems*, Marcel-Dekker, New York, 2000.

[256] Mahmoud, M. S. and A. Ismail, "Control of Electric Power Systems", *Int. J. Systems Analysis and Modeling Simulation*, vol. 43. no. 12, 2003, pp. 1639–1673.

[257] Mahmoud, M. S., "Robust Linear Filtering of Uncertain Systems", *Automatica*, vol. 40, 2004, pp. 1797–1802.

[258] Mahmoud, M. S., *Resilient Control of Uncertain Dynamical Systems*, Springer, Berlin, 2004.

[259] Mahmoud, M. S. and A. Ismail, "Robust Control Redesign of an Inverted Wedge", *Proc. 2nd IEEE GCC Conf.*, Bahrain, December 2–5, 2004, pp. 37–46.

[260] Mahmoud, M. S. and A. Ismail, "Robust Control Redesign of an Inverted Wedge", *Proc. 2nd IEEE GCC Conf.*, Bahrain, December 2–5, 2004, pp. 37–46.

[261] Mahmoud, M. S., "New Results on Robust Control Design of Discrete-Time Uncertain Systems", *IEE Proc. Control Theory and Appli.*, vol. 152, no. 4, 2005, pp. 453–459.

[262] Mahmoud, M. S., "Resilient Adaptive Control of Polytopic Delay Systems", *IMA J. Math Control and Info*, vol. 22, no. 4, 2005, pp. 200–225.

[263] Mahmoud, M. S. and A. Ismail, "Passivity and Passivication of Interconnected Time-Delay Models of Reheat Power Systems", *J. Math Problems in Engineering*, vol. 13, no. 4, 2006, pp. 1–21.

[264] Mahmoud, M. S., "Resilient $\mathcal{L}_2/\mathcal{L}_\infty$ Filtering of Polytopic Systems with State-Delays", *IET Control Theory and Appli.*, vol. 1, no. 2007, pp. 141–154.

[265] Mahmoud, M. S., "Improved Stability and Stabilization Approach to Linear Interconnected Time-Delay Systems", *Optimal Control Applications and Methods*, vol. 31, no. 3, 2010, pp. 81–92.

[266] Mahmoud, M. S. and N. B. Almutairi, "Decentralized Stabilization of Interconnected Systems with Time-Varying Delays", *European J. Control*, vol. 15, no. 6, 2009, pp. 624–633.

[267] Makila, P. M. and H. T. Toivonen, "Computational Methods for Parametric LQ Problems: A Survey", *IEEE Trans. Automat. Contro.*, vol. 32, 1987, pp. 658–671.

[268] Massioni, P. and M. Verhaegen, "Distributed Control for Identical Dynamically Coupled Systems: A Decomposition Approach", *IEEE Trans. Automat. Contro.*, vol. 54, no. 1, 2009, pp. 124–135.

[269] Mao, C. and J. H. Yang, "Decentralized Output Tracking for Linear Uncertain Interconnected Systems", *Automatica*, vol. 31, 1995, pp. 151–154.

[270] McFarlane, A. C. J., *Complex Variable Methods for Linear Multivariable Feedback Systems*, Taylor & Francis, London, 1980.

[271] Mesarovic, M. D., D. Macko and Y. Takahara, *Theory of Hierarchical Multilevel Systems*, Academic Press, New York, 1970.

[272] Mesbahi, M., "A Semi-Definite Programming Solution of the Least Order Dynamic Output Feedback Synthesis Problem", *Proc. 38th IEEE Conf. Decision and Control*, Phoenix, AZ, 1999, pp. 1851–1856.

[273] Michel, A. N. and R. K. Miller, *Qualitative Analysis of Large-Scale Dynamical Systems*, Academic Press, New York, 1977.

[274] Mirkin, B. M. and P. O. Gutman, "Adaptive Coordinated Decentralized Control of State Delayed Systems with Actuator Failures", *Asian Journal of Control*, vol. 8, no. 4, 2006, pp. 441–448.

[275] Moheimani, O. and I. Petersen, "Optimal Quadratic Guaranteed Cost Control of a Class of Uncertain Time-Delay Systems", *IEE Proc. Control Theory Appl.*, vol. 144, no. 2, 1997, pp. 183–188.

[276] Monk, J. and J. Comfort, "Mathematical Model of an Internal Combustion Engine and Dynamometer Test Rig", *Measurement Control*, vol. 3, 1970, 3, pp. T93–T100.

[277] Munro, N. and S. N. Hirbod, "Multivariable Control of an Engine and Dynamometer Test Rig", *Proc. Seventh IFAC Congress*, Helsinki, 1978, pp. 369–376.

[278] Nagpal, K. M. and P. P. Khargonekar, "Filtering and Smoothing in $\mathcal{H}_\infty$ Setting", *IEEE Trans. Automat. Contro.*, vol. 36, 1991, pp. 152–166.

[279] Naidu, D. S. and A. K. Rao, *Singular Perturbation Analysis of Discrete Control Systems*, Springer-Verlag, New York, 1985.

[280] Naidu, D. S., "Singular Perturbations and Time-Scales in Control Theory and Applications: An Overview", *Dynamics of Continuous, Discrete and Impulsive Systems Series B: Applications & Algorithms*, vol. 9, 2002, pp. 233–278.

[281] Narendra, K. S., N. Oleng and S. Mukhopadhyay, "Decentralized Adaptive Control with Partial Communication", *IEE Proc. Control Theory Appl.*, vol. 153, no. 5, 2006, pp. 546–555.

[282] Nguang, S. and P. Shi, "Delay-Dependent $\mathcal{H}_\infty$ Filtering for Uncertain Time Delay Nonlinear Systems: An LMI Approach", *Proc. American Contr. Conf.*, Minneapolis, MN, 2006, pp. 5043–5048.

[283] Nian, X. and R. Li, "Robust Stability of Uncertain Large-Scale Systems with Time-Delay", In *J. Systems Sciences*, vol. 32, 2001, pp. 541–544.

[284] Nicholson, N., "Dynamic Optimization of a Boiler", *Proc. IEE*, vol. 3, 1964, pp. 1479–1486.

[285] Niu, Y., D. W. C. Ho, and J. Lam, "Robust Integral Sliding Mode Control for Uncertain Stochastic Systems with Time-Varying Delay", *Automatica*, vol. 41, no. 5, 2005, pp. 873–880.

[286] Ohta, Y. and D. D. Siljak, "An Inclusion Principle for Hereditary Systems", *J. Math. Anal. Appl.*, vol. 98, 1984, pp. 581–598.

[287] Oucheriah, S., "Decentralized Stabilization of Large-Scale Systems with Multiple Delays in the Interconnection", *Int. J. Control*, vol. 73, 2000, pp. 1213–1223.

[288] Pagilla, P. R., "Robust Decentralized Control of Large-Scale Interconnected Systems: General Interconnections", *Proc. American Control Conference*, San Diego, 1999, pp. 4527–4531.

[289] Pagilla, P. R. and Y. Zhu, "A Decentralized Output Feedback Controller for a Class of Large-Scale Interconnected Nonlinear Systems", *J. Dynamic Systems, Measurement, and Control*, vol. 127, 2005, pp. 167–172.

[290] Park, P. G. and J. W. Ko, "Stability and Robust Stability for Systems with a Time-Varying Delay", *Automatica*, vol. 43, 2007, pp. 1855–1858.

[291] Patel, R. V. and M. Toda, "Quantitative Measure of Robustness for Multivariable Systems", *Proc. Joint Auto. Control Conf.*, San Francisco, CA, TPS-A.

[292] Pearson, J. D., "Dynamic Decomposition Techniques", in *Optimization Methods for Large-Scale Systems*, Wismer, D. A. (ed.), McGraw-Hill, New York, 1971.

[293] Peaucelle, D., D. Arzelier and C. Farges, "LMI Results for Resilient State-Feedback with $\mathcal{H}_\infty$ Performance," *Proc. the 43rd IEEE Conference on Decision and Control*, Paradise Island, Bahamas, December 2004, pp. 400–405.

[294] Petersen, I. R., "A Stabilization Algorithm for a Class of Uncertain Linear Systems", *Systems & Control Letters*, vol. 8, 1987, pp. 351–357.

[295] Phohomsiri, P., F. E. Udwadia and H. Von Bremmen, "Time-Delayed Positive Velocity Feedback Control Design for Active Control of Structures," *J. Engineering Mechanics*, vol. 132, no. 6, 2006, pp. 690–703.

[296] Pujol, G., J. Rodellar, J. M. Rossell and F. Fozo, "Decentralized Reliable Guaranteed Cost Control of Uncertain Systems: An LMI Design", *IET Control Theory Appl.*, vol. 1, no. 3, 2007, pp. 779–785.

[297] Quazza, G., "Large-Scale Control Problems in Electric Power Systems", *Automatica* 13, 1977, pp. 579–593.

[298] Rautert, T. and E. W. Sachs, "Computational Design of Optimal Output Feedback Controllers", *SIAM J. Optim.*, vol. 7, 1996, pp. 1117–1125.

[299] Richard, J. P., "Time-Delay Systems: An Overview of Some Recent Advances and Open Problems", *Automatica*, vol. 39, no. 10, 2003, pp. 1667–1694.

[300] Rotkowitz, M. and S. Lall, "Decentralized Control Information Structures Preserved under Feedback", *Proc. 41st IEEE Conf. Dec. and Control*, Las Vegas, NV, 2002, pp. 569–575.

[301] Rotkowitz, M. and S. Lall, "A Characterization of Convex Problems in Decentralized Control", *IEEE Trans. Automatic Control*, vol. 51, 2006, pp. 274–286.

[302] Saberi, A. and Khalil, H., "Decentralized Stabilization of Interconnected Systems Using Output Feedback", *Int. J. Control*, vol. 4l, no. 6, 1985, pp. 1461–1475.

[303] Saeks, R., Ed., *Large Scale Dynamic Systems*, Point Lobos Press, No. Hollywood, CA, 1976.

[304] Saeks, R., "On the Decentralized Control of Interconnected Dynamic Systems", *IEEE Trans. Automat. Contro.*, vol. AC-24, 1979, pp. 269–271.

[305] Sage, A. P., *Optimum Systems Control*, Prentice-Hall, Englewood Cliffs, NJ, 1968.

[306] Sage, A. P., *Methodology for Large-Scale Systems*, McGraw-Hill, New York, 1977.

[307] Sandell, N. P. , P. Varaiya, M. Athans and M. G. Safonov, "A Survey of Decentralized Control Methods for Large-Scale Systems", *IEEE Trans. Automat. Contro.*, vol. AC-23, 1978, pp. 108–128.

[308] Sastry, S. S., *Nonlinear Systems: Analysis, Stability and Control*, Springer, New York, 1999.

[309] Savastuk, S. K. and D. D. Siljak, "Optimal Decentralized Control", *Proc. American Control Conference*, Baltimore, MD, 1994, pp. 3369–3373.

[310] Sezer, M. E. and D. D. Siljak, "On Structural Decomposition and Stabilization of Large-Scale Control Systems", *IEEE Trans. Automat. Contro.*, vol. AC-26, 1981, pp. 439–444.

[311] Sezer, M. E. and D. D. Siljak, "Nested Epsilon-Decomposition of Clustering Complex Systems" , *Automatica*, vol. 22, 1986, pp. 321–331.

[312] Scorletti, G. and G. Duc, "An LMI Approach to Decentralized $\mathcal{H}_\infty$ Control", *Int. J. Control*, vol. 74, no. 3, 2001, pp. 211–224.

[313] Shaked, U. and C. E. de Souza, "Robust Minimum Variance Filtering", *IEEE Trans. Signal Processing*, vol. 43, 1995, pp. 2474–2483.

[314] Scherer, C., "Mixed $\mathcal{H}_2/\mathcal{H}_\infty$ Control", in *Trends in Control: A European Perspective*, A. Isidori (ed.), Springer-Verlag, Berlin, pp. 171–216.

[315] Scherer, C., P. Gahinet and M. Chilali, "Multiobjective Output-Feedback Control via LMI Optimization", *IEEE Trans. Automat. Contro.*, vol. 42, 1997, pp. 896–911.

[316] Shi, L. S., C. K. Ko, Z. Jin, D. Gayme, V. Gupta, S. Waydo and R. M. Murray, "Decentralized Control across Bit-Limited Communication Channels: An Example", *Proc. American Contr. Conf.*, Portland, ME, 2005, pp. 3348–3353.

[317] Shi, P. and E. K. Boukas, "H_∞ Control for Markovian Jumping Linear Systems with Parametric Uncertainty", *J. Optimization Theory and Applications*, vol. 95, no. 1, 1997, pp. 75–99.

[318] Shi, P., E. K. Boukas and R. K. Agarwal, "Control of MMarkovian Jump Discrete-Time Systems with Norm Bounded Uncertainty and Unknown Delays", *IEEE Trans. Automat. Contro.*, vol. 44, no. 11, 1999, pp. 2139–2144.

[319] Shi, P., E. K. Boukas and R. K. Agarwal, "Kalman Filtering for Continuous-Time Uncertain Systems with Markovian Jumping Parameters", *IEEE Trans. Automat. Contro.*, vol. 44, no. 8, 1999, pp. 1592–1597.

[320] Shyu, K., W. Liu and Hsu K. "Design of Large-Scale Time-Delayed Systems with Dead-Zone Input via Variable Structure Control", *Automatica*, vol. 41, no. 7, 2005, pp. 1239–1246.

[321] Šiljak, D. D., "On Stability of Large-Scale Systems under Structural Perturbations", *IEEE Trans. Systems, Man and Cybernetics*, vol. SMC-3, no. 4, 1973, pp. 415–417.

[322] Šiljak, D. D. and M. K. Sundareshan, "A Multilevel Optimization of Large-Scale Dynamic Systems", *IEEE Trans. Automat. Contro.*, vol. 21, no. 2, 1976, pp. 79–84.

[323] Šiljak, D. D., *Large-Scale Dynamic Systems: Stability and Structure*, North-Holland, Amsterdam, 1978.

[324] Šiljak, D. D., *Decentralized Control of Complex Systems*, Academic, Cambridge, 1991.

[325] Šiljak, D. D., "Decentralized Control and Computations: Status and Prospects", *Annual Reviews in Control*, vol. 20, 1996, pp. 131–141.

[326] Šiljak, D. D. and D. M. Stipanovic, "Robust Stabilization of Nonlinear Systems: The LMI Approach", *Math. Prob. Eng.*, vol. 6, 2000, pp. 461–493.

[327] Šiljak, D. D., D. M. Stipanovic and A. I. Zecevic, "Robust Decentralized Turbine/Governor Control Using Linear Matrix Inequalities", *IEEE Trans. Power Syst.*, vol. 17, 2002, pp. 715–722.

[328] Šiljak, D. D., D. M. Stipanovic and A. I. Zecevic, "Robust Decentralized Turbine/Governor Using Linear Matrix Inequalities", *IEEE Trans. Power Syst.*, vol. 17, 2002, pp. 715–722.

[329] Šiljak, D. D. and A. I. Zecevic, "Control of Large-Scale Systems: Beyond Decentralized Feedback", *Preprints of the 10th IFAC/IFORS/IMACS/-IFIP Symposium on Large Scale Systems: Theory and Applications*, vol. 1, Osaka, Japan, 2004, pp. 1–10.

[330] Šiljak, D. D. and A. I. Zecevic, "Control of Large-Scale Systems: Beyond Decentralized Feedback", *Annual Reviews in Control*, vol. 29, 2005, pp. 169–179.

[331] Singh, M. G., S. A. W. Drew and J. F. Coales, "Comparisons of Practical Hierarchical Control Methods for Interconnected Dynamical Systems", *Automatica*, vol. 11, no. 4, 1973, pp. 331–350.

[332] Singh, M. G., "A New Algorithm for the On-Line Multilevel Control of Large Interconnected Systems with Fast Dynamics," *Int. J. Control*, vol. 21, no. 4, 1975, pp. 587–597.

[333] Singh, M. G., "Multi-Level State Estimation," *Int. J. Syst. Sci.*, vol. 6, no. 6, 1975, pp. 535–555.

[334] Singh, M. G., M. F. Hassan and A. Titli, "Multilevel Feedback Control for Interconnected Dynamical Systems Using the Prediction Principle", *IEEE Trans. Systems, Man and Cybernetics*, vol. SMC-6, no. 4, 1976, pp. 233–239.

[335] Singh, M. D., *Dynamical Hierarchical Control*, North-Holland, Amsterdam, 1977.

[336] Singh, M. G. and A. Titli, *Systems: Decomposition, Optimization and Control*, Pergamon Press, Oxford, 1978.

[337] Singh, M. D. and A. Title, Eds., *Large Scale Systems: Theory and Applications*, North-Holland, Amsterdam, 1979.

[338] Singh, M. G., M. F. Hassan, Y. L. Chen and O. R. Pan, "New Approach to Failure Detection in Large-Scale Systems", *IEE Proc. D, Control Theory Appl.*, vol. 130, 1983, pp. 117–121.

[339] Srichander, R. and B. K. Walker, "Stochastic Analysis for Continuous-Time Fault-Tolerant Control Systems", *Int. J. Control*, vol. 57, no. 2, 1989, pp. 433–452.

[340] Smith, N. J. and A. P. Sage, "An Introduction to Hierarchical Systems Theory," *Comput. and Elec. Engrg.*, vol. 1, 1973, pp. 55–71.

[341] Smith, R. S. and F. Y. Hadaegh, "Closed-Loop Dynamics of Cooperative Vehicle Formations with Parallel Estimators and Communication", *IEEE Trans. Automat. Contro.*, vol. 52, no. 8, 2007, pp. 1404–1405.

[342] Sojoudi, S., J. Lavaei and A. G. Aghdam, "Optimal Information Flow Structure for Control of Interconnected Systems", *Proc. American Control Conference*, New York, 2007, pp. 1508–1513.

[343] Sourias, D. D. and V. Manousiouthakis, "Best Achievable Decentralized Performance", *IEEE Trans. Automatic Control*, vol. 40, 1995, pp. 1858–1871.

[344] Spooner, M. D. and K. Passino, *Stable Adaptive Systems*, North-Holland, Amsterdam, 2002.

[345] Stankovic, S. S. and D. D. Siljak, "Contractibility of Overlapping Decentralized Control", *Systems & Control Letters*, vol. 44, 2001, pp. 189–200.

[346] Stankovic, S. S., D. M. Stipanovic and D. D. Siljak, "Decentralized Dynamic Output Feedback for Robust Stabilization of a Class of Nonlinear Interconnected Systems", *Automatica*, vol. 43, 2007, pp. 861–867.

[347] Stankovic, S. S. and D. D. Siljak, "Robust Stabilization of Nonlinear Interconnected Systems by Decentralized Dynamic Output Feedback", *Systems & Control Letters*, vol. 58, no. 4, 2009, pp. 271–275.

[348] Stanojevic, M. J. and D. D. Siljak, "Robust Stability and Stabilization of Discrete-Time Non-Linear Systems: The LMI Approach", *Int. J. Control*, vol. 74, 2001, pp. 873–879.

[349] Stanojevic, M. J. and D. D. Siljak, "Contractibility of Overlapping Decentralized Control", *Systems and Control Letters*, vol. 44, 2001, pp. 189–199.

[350] Stanković, S. S., M. J. Stipanović and D. D. Šiljak, "Decentralized Overlapping Control of a Platoon of Vehicles", *IEEE Trans. Control System Technology*, vol. 8, 2000, pp. 816–832.

[351] Stipanovic, D. M. and D. D. Siljak, "Connective Stability of Discontinuous Dynamic Systems", *J. Optimization Theory and Applic.*, vol. 115, no. 3, 2002, pp. 711–726.

[352] Stipanovic, D. M., G. Inalhan, R. Teo and C. J. Tomlin, "Decentralized Overlapping Control of a Formation of Unmanned Aerial Vehicles", *Automatica*, vol. 40, 2004, pp. 1285–1296.

[353] Sundareshan, M. K., "Exponential Stabilization of Large-Scale Systems: Decentralized and Multilevel Schemes", *IEEE Trans. Syst. Man. Cyber.*, vol. SMC-7, 1977, pp. 478–483.

[354] Sundareshan, M. K., "Generation of Multilevel Control and Estimation Schemes for Large-Scale Systems: A Perturbation Approach", *IEEE Trans. Syst. Man. Cyber.*, vol. SMC-7, 1977, pp. 144–152.

[355] Sundareshan, M. K., "Decentralized Observation in Large-Scale Systems", *IEEE Trans. Syst. Man. Cyber.*, vol. SMC-7, 1977, pp. 863–864.

[356] Sundareshan, M. K. and R. M. Elbanna, "Qualitative Analysis and Decentralized Controllers Synthesis for a Class of Large-Scale Systems with Symmetrically Interconnected Subsystems", *Automatica*, vol. 27, no. 2, 1991, pp. 383–388.

[357] Swarnakar, A., H. J. Marquez and T. Chen, "A Design Framework for Overlapping Controllers and Its Application", *Proc. 46th IEEE Conf. Dec. and Control*, New Orleans, LA, 2007, pp. 2809–2814.

[358] Syrmos, V. L., C. T. Abdallah, P. Dorato and K. Grigoriadis, "Static Output Feedback: A Survey", *Automatica*, vol. 33, 1997, pp. 125–137.

[359] Tarokh, M. and Jamshidi, M., "Elimination of Decentralized Fixed Modes with Minimum Number of Interconnection Gains", *Large Scale Systems*, vol. 2, 1987, pp. 207–215.

[360] Trofino-Neto, A. and V. Kucera, "Stabilization via Static Output Feedback", *IEEE Trans. Automat. Contro.*, vol. 38, 1993, pp. 764–765.

[361] Udwadia, F. E. and R. Kumar, "Time-Delayed Control of Classically Damped Structural Systems", *Int. J. Control*, vol. 60, 1994, pp. 687–713.

[362] Udwadia F. E., H. von Bremen, R. Kumar and M. Hosseini, "Time Delayed Control of Structural Systems", *Int. J. Earthquake Engineering and Structural Dynamics*, vol. 32, 2003, pp. 495–535.

[363] Udwadia, F. E. and P. Phohomsiri, "Active Control of Structures Using Time Delayed Positive Feedback Proportional Control Designs", *Structural Control and Health Monitoring*, vol. 13, no. 1, 2006, pp. 536–552.

[364] Udwadia, F. E., H. von Bremen and P. Phohomsiri, "Time-Delayed Control Design for Active Control of Structures: Principles and Applications", *Struct. Control Health Monit.*, vol. 14, no. 1, 2007, pp. 27–61.

[365] Utkin, V. I., *Sliding Modes and Their Applications to Variable Structure Systems*, MIR Publication House, Moscow, 1978.

[366] Veillette, R. J., J. V. Medanic and W. R. Perkins, "Design of Reliable Control Systems", *IEEE Trans. Automat. Contro.*, vol. 37, no. 3, 1992, pp. 290–304.

[367] Veillette, R. J., "Reliable Linear Quadratic State Feedback Control", *Automatica*, vol. 31, 1995, pp. 137–143.

[368] Vidyasagar, M. and N. Viswanadham, "Algebraic Design Techniques for Reliable Stabilization", *IEEE Trans. Automat. Contro.*, vol. AC-17, 1982, pp. 1085–1095.

[369] Voulgaris, P. G., "A Convex Characterization of Classes of Problems in Control with Specific Interaction and Communication Structures", *Proc. American Contr. Conf.*, Arlington, VA, 2001, pp. 3128–3133.

[370] Wang, S. H. and E. J. Davison, "On the Stabilization of Decentralized Control Systems", *IEEE Trans. Automat. Contro.*, vol. 18, no. 2, 1973, pp. 473–478.

[371] Wang, Y., C. de-Souze and L. H. Xie, "Decentralized Output Feedback Control of Interconnected Uncertain Delay Systems", *Proc. 12th IFAC Congress*, Australia, vol. 6, 1993, pp. 38–42.

[372] Watanabe, K., E. Nobuyama and A. Kojima, "Recent Advances in Control of Time Delay Systems—A Tutorial Review", *Proc. 35th IEEE Conf. Dec. and Contr.*, Kobe, Japan, 1996, pp. 2083–2089.

[373] Wismer, D. A., Ed., *Optimization Methods for Large-Scale Systems with Applications*, McGraw-Hill, New York, 1971.

[374] Wu, H., "Decentralized Adaptive Robust Control for a Class of Large-Scale Systems Including Delayed State Perturbations in the Interconnections", *IEEE Trans. Automat. Contro.*, vol. 47, no. 10, 2002, pp. 1745–1751.

[375] Xie, L. H., C. E. de Souza and M. Fu, "$\mathcal{H}_\infty$ Estimation for Discrete-Time Linear Uncertain Systems", *Int. J. Robust and Nonlinear Control*, vol. 1, 1991, pp. 11–23.

[376] Xu, S. and J. Lam, "A Survey of Linear Matrix Inequality Techniques in Stability of Delay Systems", *Int. J. Systems Science*, vol. 39, 2008, pp. 1095–1113.

[377] Yan, X. G., J. J. Wang, X. Y. Li and S. Y. Zhang, "Decentralized Output Feedback Robust Stabilization for a Class of Nonlinear Interconnected Systems with Similarity", *IEEE Trans. Automat. Contro.*, vol. 43, no. 2, 1998, pp. 294–299.

[378] Yan, X. G. and L. Xie, "Reduced-Order Control for a Class of Nonlinear Similar Interconnected Systems with Mismatched Uncertainty", *Automatica*, vol. 39, no. l, 2003, pp. 91–99.

[379] Yan, J. J., "Memoryless Adaptive Decentralized Sliding Mode Control for Uncertain Large-Scale Systems with Time-Varying Delays", *ASME Journal Dynamics Systems, Measurement and Control*, vol. 125, no. 2, 2003, pp. 172–176.

[380] Yan, X. G., C. Edwards and S. K. Spurgeon, "Decentralized Robust Sliding-Mode Control for a Class of Nonlinear Interconnected Systems by Static Output Feedback", *Automatica*, vol. 40, no. 4, 2004, pp. 613–620.

[381] Yan, X. G. and C. Edwards, "Adaptive Sliding-Mode Observer-Based Fault Reconstruction for Nonlinear Systems with Parametric Uncertainties", *IEEE Trans. on Industrial Electronics*, vol. 55, no. 11, 2008, pp. 4029–4036.

[382] Yan, X. G. and C. Edwards, "Robust Decentralized Actuator Fault Detection and Estimation for Large-Scale Systems Using Sliding-Mode Observer", *Int. J. Control*, vol. 81, no. 4, 2008, pp. 591–606.

[383] Yang, G. H., J. Wang, C. B. Soh and J. Lam, "Design of Reliable Controllers for Symmetric Composite Systems: Primary Contingency Case", *Proc. the 35th IEEE Conference on Decision and Control*, Kobe, Japan, 1996, pp. 3612–3617.

[384] Yang, G. H. and S. Zhang, "Stabilizing Controllers for Uncertain Symmetric Composite Systems", *Automatica*, vol. 31, no. 2, 1995, pp. 337–340.

[385] Yang, G. H., S. Zhang, J. Lam and J. Wang, "Reliable Control Using Redundant Controllers, *IEEE Trans. Automat. Contro.*, vol. 43, no. 11, 1998, pp. 1588–1593.

[386] Yang, T. C., "An Algorithm to Find the Smallest Possible Values in a Matrix to Satisfy the M-Matrix Condition", *IEEE Trans. Automat. Contro.*, vol. 42, no. 12, 1997, pp. 1738–1740.

[387] Yang, G. H. and J. L. Wang, "Robust Resilient Kalman Filtering for Uncertain Linear Systems with Estimator Gain Uncertainty", *IEEE Trans. Automat. Contro.*, vol. 46, 2001, pp. 343–348.

[388] Yang, G.-H., J. L. Wang and Y. C. Soh, "Reliable $\mathcal{H}_\infty$ Controller Design for Linear Systems", *Automatica*, vol. 37, 2001, pp. 717–725.

[389] Yang, G. H., J. L. Wang and C. Lin, "$\mathcal{H}_\infty$ Control for Linear Systems with Additive Controller Gain Variations", *Int. J. Control*, vol. 73, 2000, pp. 1500–1506.

[390] Yang, G.-H. and J. Wang, "Non-Fragile $\mathcal{H}_\infty$ Control for Linear Systems with Multiplicative Controller Gain Variations", *Automatica*, vol. 37, 2001, pp. 727–737.

[391] Yee, J. -S., G.-H. Wang and J. Wang, "Non-Fragile Guaranteed Cost Control for Discrete-Time Uncertain Linear Systems", *Int. J. Systems Science*, vol. 32, no. 7, 2001, pp. 845–853.

[392] Yüksel, S., H. Hindi and L. Crawford, "Optimal Tracking with Feedback-Feedforward Control Separation over a Network", *Proc. American Contr. Conf.*, Minneapolis, MN, 2006, pp. 3500–3506.

[393] Yüksel, S. and T. Basar, "On the Absence of Rate Loss in Decentralized Sensor and Controller Structure for Asymptotic Stability", *Proc. American Contr. Conf.*, Minneapolis, MN, 2006, pp. 5562–5567.

[394] Yüksel, S. and T. Basar, "Communication Constraints for Decentralized Stabilizability with Time-Invariant Policies", *IEEE Trans. Automatic Control*, vol. 52, no. 6, 2007, pp. 1060–1066.

[395] Zak, S. H. and S. Hui, "On Variable Structure Output Feedback Controllers for Uncertain Dynamic Systems", *IEEE Trans. Automat. Contro.*, vol. 38, no. 10, 1993, pp. 1509–1512.

[396] Zecevic, A. I. and D. D. Siljak, "Stabilization of Nonlinear Systems with Moving Equilibria", *IEEE Trans. Automat. Contro.*, vol. 48, 2003, pp. 1036–1040.

[397] Zecevic, A. I. and D. D. Siljak, "Design of Robust Static Output Feedback for Large-Scale Systems", *IEEE Trans. Automatic Control*, vol. 49, no. 11, 2004, pp. 2040–2044.

[398] Zecevic, A. I., G. Neskovic and D. D. Siljak, "Robust Decentralized Exciter Control with Linear Feedback", *IEEE Trans. Power Syst.*, vol. 19, 2004, pp. 1096–1103.

[399] Zecevic, A. I. and D. D. Siljak, "Global Low-Rank Enhancement of Decentralized Control for Large-Scale Systems", *IEEE Trans. Automatic Control*, vol. 50, no. 5, 2005, pp. 740–744.

[400] Zhai, G., M. Ikeda and Y. Fujisaki, "Decentralized $\mathcal{H}_\infty$ Controller Design: A Matrix Inequality Approach Using a Homotopy Method", *Automatica*, vol. 37, 2001, pp. 565–572.

[401] Zhao, Q. and J. Jiang, "Reliable State Feedback Control System Design with Actuator Failures", *Automatica*, vol. 34, 1998, pp. 1267–1272.

[402] Zhong, W. S., G. M. Dimirovski and J. Zhao, "Decentralized Synchronization of an Uncertain Complex Dynamical Network", *Proc. American Contr. Conf.*, New York, NY, 2007, pp. 1437–1442.

[403] Zhou, J. "Decentralized Adaptive Control for Large-Scale Time-Delay Systems with Dead-Zone Input", *Automatica*, vol. 44, no. 7, 2008, pp. 1790–1799.

[404] Zhou, K. and J. C. Doyle, *Essentials of Robust Control*, Prentice-Hall, Englewood Cliffs, NJ, 1998.

[405] Zhou, B. and G. R. Duan, "Global Stabilization of Linear Systems via Bounded Controls", *Systems & Control Letters*, vol. 58, 2009, pp. 54–61.

[406] Zhu, Y. and P. R. Pagilla, "Decentralized Output Feedback Control of a Class of Large-Scale Interconnected Systems", *IMA J. Math Control and Info.*, vol. 24, 2007, pp. 57–69.

[407] Zribi, M., M. S. Mahmoud, M. Karkoub and T. Li "$\mathcal{H}_\infty$-Controllers for Linearized Time-Delay Power Systems", *IEE Proc. Gene. Trans. and Distribution*, vol. 147, no. 6, 2000, pp. 401–408.

Index